Fundamentals of Linear Circuits

Fundamentals of Linear Circuits

THOMAS L. FLOYD

MERRILL, AN IMPRINT OF
MACMILLAN PUBLISHING COMPANY
NEW YORK

MAXWELL MACMILLAN CANADA
TORONTO

MAXWELL MACMILLAN INTERNATIONAL
NEW YORK ■ OXFORD ■ SINGAPORE ■ SYDNEY

Cover art: Tech-Graphics
Editor: David Garza
Developmental Editor: Carol Hinklin Robison
Production Editor: Rex Davidson
Art Coordinator: Peter A. Robison
Text Designer: Anne Flanagan
Cover Designer: Russ Maselli
Production Buyer: Patricia A. Tonneman
Insert art: Tech-Graphics

This book was set in Times Roman by The Clarinda Company and was printed and bound by Arcata Graphics/Halliday. The cover was printed by Lehigh Press, Inc.

Copyright © 1992 by Macmillan Publishing Company, a division of Macmillan, Inc. Merrill is an imprint of Macmillan Publishing Company.

Printed in the United States of America

All rights reserved. No part of this book may be reproduced or transmitted in any form or by any means, electronic or mechanical, including photocopy, recording, or any information storage and retrieval system, without permission in writing from the Publisher.

Macmillan Publishing Company
866 Third Avenue
New York, NY 10022

Macmillan Publishing Company is part of the
Maxwell Communication Group of Companies.

Maxwell Macmillan Canada, Inc.
1200 Eglinton Avenue East, Suite 200
Don Mills, Ontario M3C 3N1

Library of Congress Cataloging-in-Publication Data
Floyd, Thomas L.
    Fundamentals of linear circuits / Thomas L. Floyd.
      p.   cm.
    Includes index.
    ISBN 0-02-338481-6
    1. Electronic circuits. 2. Solid state electronics. I. Title.
TK7867.F58 1992
621.381′5—dc20   91-39639
                                          CIP

Printing: 1 2 3 4 5 6 7 8 9   Year: 2 3 4 5

MERRILL'S INTERNATIONAL SERIES IN ENGINEERING TECHNOLOGY

Adamson	*Applied Pascal for Technology,* 0-675-20771-1
	The Electronics Dictionary for Technicians, 0-02-300820-2
	Microcomputer Repair, 0-02-300825-3
	Structured BASIC Applied to Technology, 0-675-20772-X
	Structured C for Technology, 0-675-20993-5
	Structured C for Technology (w/disks), 0-675-21289-8
Antonakos	*The 68000 Microprocessor: Hardware and Software Principles and Applications,* 0-675-21043-7
Asser/Stigliano/ Bahrenburg	*Microcomputer Servicing: Practical Systems and Troubleshooting,* 0-675-20907-2
	Microcomputer Theory and Servicing, 0-675-20659-6
	Lab Manual to accompany Microcomputer Theory and Servicing, 0-675-21109-3
Aston	*Principles of Biomedical Instrumentation and Measurement,* 0-675-20943-9
Bateson	*Introduction to Control System Technology, Third Edition,* 0-675-21010-0
Beach/Justice	*DC/AC Circuit Essentials,* 0-675-20193-4
Berlin	*Experiments in Electronic Devices to accompany Floyd's Electronic Devices and Electronic Devices: Electron Flow Version, Third Edition,* 0-02-308422-7
	The Illustrated Electronics Dictionary, 0-675-20451-8
Berlin/Getz	*Experiments in Instrumentation and Measurement,* 0-675-20450-X
	Fundamentals of Operational Amplifiers and Linear Integrated Circuits, 0-675-21002-X
	Principles of Electronic Instrumentation and Measurement, 0-675-20449-6
Berube	*Electronic Devices and Circuits Using MICRO-CAP II,* 0-02-309160-6
Bogart	*Electronic Devices and Circuits, Second Edition,* 0-675-21150-6
Bogart/Brown	*Experiments in Electronic Devices and Circuits, Second Edition,* 0-675-21151-4
Boylestad	*DC/AC: The Basics,* 0-675-20918-8
	Introductory Circuit Analysis, Sixth Edition, 0-675-21181-6
Boylestad/Kousourou	*Experiments in Circuit Analysis, Sixth Edition,* 0-675-21182-4
	Experiments in DC/AC Basics, 0-675-21131-X
Brey	*Microprocessors and Peripherals: Hardware, Software, Interfacing, and Applications, Second Edition,* 0-675-20884-X
	The Intel Microprocessors—8086/8088, 80186, 80286, 80386, and 80486—Architecture, Programming, and Interfacing, Second Edition, 0-675-21309-6
Broberg	*Lab Manual to accompany Electronic Communication Techniques, Second Edition,* 0-675-21257-X
Buchla	*Digital Experiments: Emphasizing Systems and Design, Second Edition,* 0-675-21180-8
	Experiments in Electric Circuits Fundamentals, Second Edition, 0-675-21409-2
	Experiments in Electronics Fundamentals: Circuits, Devices, and Applications, Second Edition, 0-675-21407-6

Buchla/McLachlan	*Applied Electronic Instrumentation and Measurement*, 0-675-21162-X
Ciccarelli	*Circuit Modeling: Exercises and Software, Second Edition*, 0-675-21152-2
Cooper	*Introduction to VersaCAD*, 0-675-21164-6
Cox	*Digital Experiments: Emphasizing Troubleshooting, Second Edition*, 0-675-21196-4
Croft	*Getting a Job: Resume Writing, Job Application Letters, and Interview Strategies*, 0-675-20917-X
Davis	*Technical Mathematics*, 0-675-20338-4
	Technical Mathematics with Calculus, 0-675-20965-X
	Study Guide to accompany Technical Mathematics, 0-675-20966-8
	Study Guide to accompany Technical Mathematics with Calculus, 0-675-20964-1
Delker	*Experiments in 8085 Microprocessor Programming and Interfacing*, 0-675-20663-4
Floyd	*Digital Fundamentals, Fourth Edition*, 0-675-21217-0
	Electric Circuits Fundamentals, Second Edition, 0-675-21408-4
	Electronic Devices, Third Edition, 0-675-22170-6
	Electronic Devices, Electron Flow Version, 0-02-338540-5
	Electronics Fundamentals: Circuits, Devices, and Applications, Second Edition, 0-675-21310-X
	Fundamentals of Linear Circuits, 0-02-338481-6
	Principles of Electric Circuits, Electron Flow Version, Second Edition, 0-675-21292-8
	Principles of Electric Circuits, Third Edition, 0-675-21062-3
Fuller	*Robotics: Introduction, Programming, and Projects*, 0-675-21078-X
Gaonkar	*Microprocessor Architecture, Programming, and Applications with the 8085/8080A, Second Edition*, 0-675-20675-8
	The Z80 Microprocessor: Architecture, Interfacing, Programming, and Design, 0-675-20540-9
Gillies	*Instrumentation and Measurements for Electronic Technicians*, 0-675-20432-1
Goetsch	*Industrial Supervision: In the Age of High Technology*, 0-675-22137-4
Goetsch/Rickman	*Computer-Aided Drafting with AutoCAD*, 0-675-20915-3
Goody	*Programming and Interfacing the 8086/8088 Microprocessor*, 0-675-21312-6
Hubert	*Electric Machines: Theory, Operation, Applications, Adjustment, and Control*, 0-675-21136-0
Humphries	*Motors and Controls*, 0-675-20235-3
Hutchins	*Introduction to Quality: Management, Assurance and Control*, 0-675-20896-3
Keown	*PSpice and Circuit Analysis*, 0-675-22135-8
Keyser	*Materials Science in Engineering, Fourth Edition*, 0-675-20401-1
Kirkpatrick	*The AutoCAD Book: Drawing, Modeling and Applications, Second Edition*, 0-675-22288-5
	Industrial Blueprint Reading and Sketching, 0-675-20617-0
Kraut	*Fluid Mechanics for Technicians*, 0-675-21330-4
Kulathinal	*Transform Analysis and Electronic Networks with Applications*, 0-675-20765-7
Lamit/Lloyd	*Drafting for Electronics*, 0-675-20200-0
Lamit/Wahler/Higgins	*Workbook in Drafting for Electronics*, 0-675-20417-8

Lamit/Paige	*Computer-Aided Design and Drafting,* 0-675-20475-5
Laviana	*Basic Computer Numerical Control Programming, Second Edition,* 0-675-21298-7
MacKenzie	*The 8051 Microcontroller,* 0-02-373650-X
Maruggi	*Technical Graphics: Electronics Worktext, Second Edition,* 0-675-21378-9
	The Technology of Drafting, 0-675-20762-2
	Workbook for the Technology of Drafting, 0-675-21234-0
McCalla	*Digital Logic and Computer Design,* 0-675-21170-0
McIntyre	*Study Guide to accompany Electronic Devices, Third Edition and Electronic Devices: Electron Flow Version,* 0-02-379296-5
	Study Guide to accompany Electronics Fundamentals, Second Edition, 0-675-21406-8
Miller	*The 68000 Microprocessor Family: Architecture, Programming, and Applications, Second Edition,* 0-02-381560-4
Monaco	*Essential Mathematics for Electronics Technicians,* 0-675-21172-7
	Introduction to Microwave Technology, 0-675-21030-5
	Laboratory Activities in Microwave Technology, 0-675-21031-3
	Preparing for the FCC General Radiotelephone Operator's License Examination, 0-675-21313-4
	Student Resource Manual to accompany Essential Mathematics for Electronics Technicians, 0-675-21173-5
Monssen	*PSpice with Circuit Analysis,* 0-675-21376-2
Mott	*Applied Fluid Mechanics, Third Edition,* 0-675-21026-7
	Machine Elements in Mechanical Design, Second Edition, 0-675-22289-3
Nashelsky/Boylestad	*BASIC Applied to Circuit Analysis,* 0-675-20161-6
Panares	*A Handbook of English for Technical Students,* 0-675-20650-2
Pfeiffer	*Proposal Writing: The Art of Friendly Persuasion,* 0-675-20988-9
	Technical Writing: A Practical Approach, 0-675-21221-9
Pond	*Introduction to Engineering Technology,* 0-675-21003-8
Quinn	*The 6800 Microprocessor,* 0-675-20515-8
Reis	*Digital Electronics Through Project Analysis,* 0-675-21141-7
	Electronic Project Design and Fabrication, Second Edition, 0-02-399230-1
	Laboratory Manual for Digital Electronics Through Project Analysis, 0-675-21254-5
Rolle	*Thermodynamics and Heat Power, Third Edition,* 0-675-21016-X
Rosenblatt/Friedman	*Direct and Alternating Current Machinery, Second Edition,* 0-675-20160-8
Roze	*Technical Communication: The Practical Craft,* 0-675-20641-3
Schoenbeck	*Electronic Communications: Modulation and Transmission, Second Edition,* 0-675-21311-8
Schwartz	*Survey of Electronics, Third Edition,* 0-675-20162-4
Sell	*Basic Technical Drawing,* 0-675-21001-1
Smith	*Statistical Process Control and Quality Improvement,* 0-675-21160-3
Sorak	*Linear Integrated Circuits: Laboratory Experiments,* 0-675-20661-8
Spiegel/Limbrunner	*Applied Statics and Strength of Materials,* 0-675-21123-9

Stanley, B.H.	*Experiments in Electric Circuits, Third Edition,* 0-675-21088-7
Stanley, W.D.	*Operational Amplifiers with Linear Integrated Circuits, Second Edition,* 0-675-20660-X
Subbarao	*16/32-Bit Microprocessors: 68000/68010/68020 Software, Hardware, and Design Applications,* 0-675-21119-0
Tocci	*Electronic Devices: Conventional Flow Version, Third Edition,* 0-675-20063-6
	Fundamentals of Pulse and Digital Circuits, Third Edition, 0-675-20033-4
	Introduction to Electric Circuit Analysis, Second Edition, 0-675-20002-4
Tocci/Oliver	*Fundamentals of Electronic Devices, Fourth Edition,* 0-675-21259-6
Webb	*Programmable Logic Controllers: Principles and Applications, Second Edition,* 0-02-424970-X
Webb/Greshock	*Industrial Control Electronics,* 0-675-20897-1
Weisman	*Basic Technical Writing, Sixth Edition,* 0-675-21256-1
Wolansky/Akers	*Modern Hydraulics: The Basics at Work,* 0-675-20987-0
Wolf	*Statics and Strength of Materials: A Parallel Approach,* 0-675-20622-7

PREFACE

Fundamentals of Linear Circuits provides thorough, comprehensive, and practical coverage of electronic devices, circuits, and applications. The extensive troubleshooting coverage and innovative system application sections serve as very important and necessary links between theory and the real world.

This book is divided into two basic parts. Chapters 1 through 4 cover discrete devices and circuits, while Chapters 5 through 14 deal with linear integrated circuits, with considerable emphasis on the operational amplifier.

A BRIEF OVERVIEW

Basic semiconductor theory and the concept of the pn junction diode are introduced in Chapter 1. Various types of diodes and their applications are covered in Chapter 2. Bipolar junction transistors, field-effect transistors, and thyristors are covered in Chapter 3. The discrete devices and circuits coverage is completed with Chapter 4, which covers amplifiers and oscillators.

The coverage of linear integrated circuits begins with an introduction to operational amplifiers (op-amps) in Chapter 5. Op-amp response is covered in Chapter 6, and basic op-amp circuits including comparators, summing amplifiers, integrators, and differentiators is the topic of Chapter 7. Chapter 8 is a coverage of active filters using the op-amp, and Chapter 9 deals with signal generators (oscillators) and timers. Power supply circuits are covered in Chapter 10. Special amplifiers including instrumentation amplifiers, isolation amplifiers, OTAs, log and antilog amplifiers are introduced in Chapter 11. Communications circuits are covered in Chapter 12, which includes basic AM and FM receiver principles, linear multipliers, mixers, and phase-locked loops. Chapter 13 is an introduction to data conversion circuits such as the analog switch, sample-and-hold circuits, D/A converters, A/D converters, V/F converters, and F/V converters. Finally, Chapter 14 focuses on various types of transducers, associated measurement circuits, and the zero-voltage switch.

FEATURES

- An innovative system application section in each chapter (except Chapter 1)
- A functional full-color insert keyed to selected system applications
- System-related chapter openers
- Significant troubleshooting coverage
- Functional use of second color
- Standard resistance values used throughout

- An introductory message at the beginning of each section that sets the tone for that section
- A practice exercise for each example
- End-of-chapter problems
- Margin logos indicating troubleshooting problems and color insert references
- Multiple-choice self-tests
- Performance-based chapter objectives
- Data sheets available in Appendix A

The ancillary package includes the following:

- Transparency and transparency master package
- Instructor's Resource Manual with system applications worksheet masters

ILLUSTRATION OF FEATURES WITHIN EACH CHAPTER

CHAPTER OPENER As shown in Figure P-1, each chapter begins with a two-page opener. The left page contains a listing of the sections within the chapter, the chapter objectives, and a brief introduction. The right page presents a preview of the system application that will be the focus of the last section of the chapter and provides several specialized objectives oriented to this feature.

Preface

List of performance-based objectives.

System block diagram with highlighted circuit board.

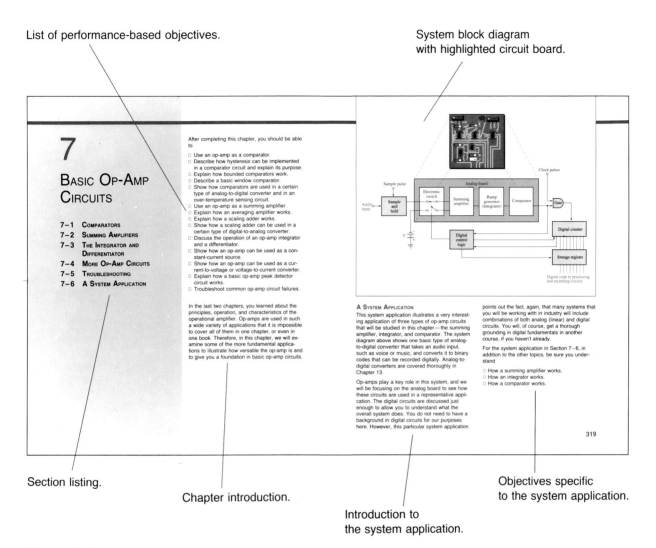

Section listing.

Chapter introduction.

Introduction to the system application.

Objectives specific to the system application.

Figure P–1
Chapter opener

PREFACE

SECTION OPENER AND SECTION REVIEW Each section within a chapter begins with a brief introduction that highlights the material to be covered or provides a general overview. Each section ends with a set of review questions that focus on the key concepts presented in the section. Answers to these review questions are given at the end of the chapter. Figure P-2 illustrates these two features.

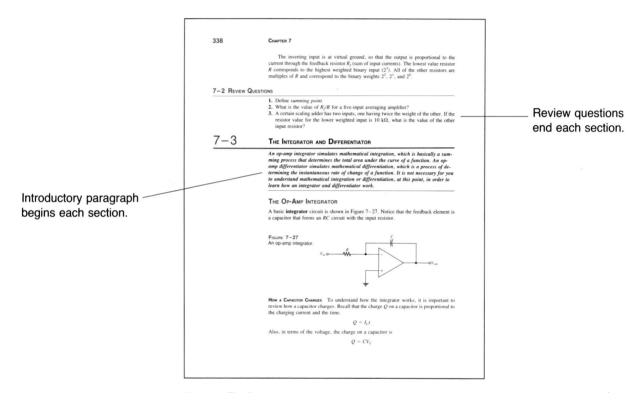

FIGURE P-2
Section opener and section review

EXAMPLES AND PRACTICE EXERCISES Frequent examples help to illustrate and clarify basic concepts. At the end of each example is a practice exercise, which is intended to help reinforce or expand on the example in some way. The nature of the practice exercises varies. Some require the student to repeat the procedure demonstrated in the example but with a different set of values or conditions. Others focus on a more limited part of the example or ask questions that encourage further thought beyond the procedure contained in the example. Answers to all practice exercises are given at the end of the chapter. A typical example and practice exercise are shown in Figure P-3.

Preface

xiii

Each example begins with a colored horizontal rule and box.

Each example contains a practice exercise related to the example.

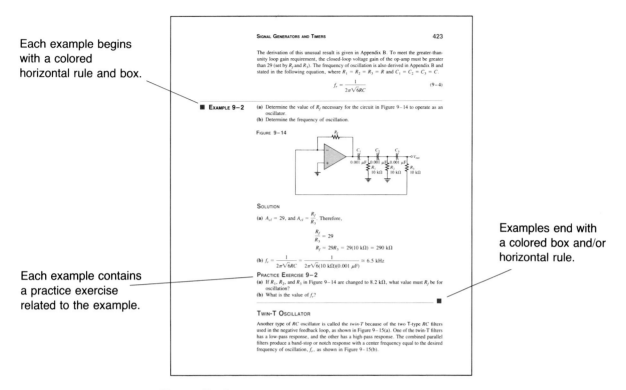

Examples end with a colored box and/or horizontal rule.

FIGURE P-3
An example and practice exercise

SYSTEM APPLICATION The last section of each chapter (except Chapter 1) is a system application in which a certain circuit board in a "real-world" system is the focus of several on-the-job type activities. Certain activities require the student to troubleshoot the circuit board for specified faults, including interpretation of instrument readings in the color insert. Generally, the circuit board relates directly to some or all of the material covered in the chapter. In some cases, however, the student is required to "stretch" a bit and, as a result, to learn something new.

Results and answers for the activities in the system application sections are provided only in the Instructor's Resource Manual, where a set of worksheet masters for appropriate activities is also provided. These can be photocopied for student hand-outs.

Opener includes a list of objectives.

"On the Test Bench" provides a detailed look at the pc board as if it were removed from the system for testing and analysis.

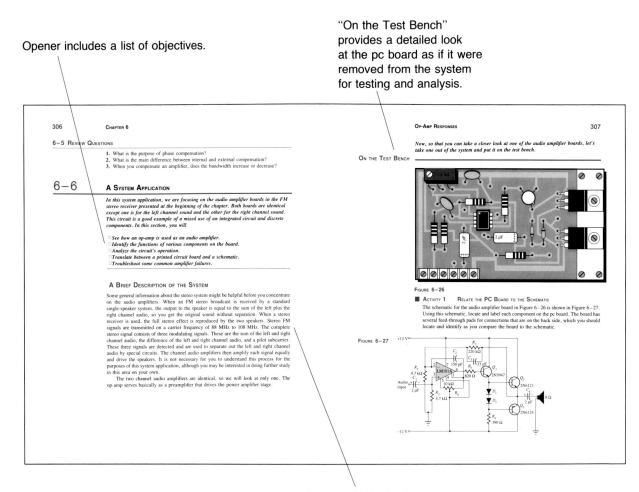

An overall introduction to the system application is provided before a particular pc board circuit is focused on.

FIGURE P–4
A system application section

PREFACE

The overall objectives of the system application are

- To provide a transition between theoretical concepts and real-world circuitry.
- To help provide a "physical" sense of the devices and circuits studied in the chapter.
- To increase student skills with on-the-job activities.
- To help answer the question, "Why do I need to know this?"

A typical system application section is shown in Figure P–4.

A series of activities involves the student in working with pc boards and schematics, circuit analysis, report writing, troubleshooting, and test setups.

A logo marks the troubleshooting activity as well as the troubleshooting problems at the end of the chapter.

A logo marks those special assignment activities that are related to the color insert section.

FIGURE P–4
Continued

FULL-COLOR INSERT Three selected system applications are related to the full-color insert using a special assignment activity marked by a color insert logo. The color insert consists of circuit board test set-ups that either require the student to troubleshoot the board based on instrument readings or to determine instrument settings for testing the board for proper operation.

CHAPTER END MATTER At the end of each chapter is a summary, glossary, formula list, multiple-choice self-test, and sectionalized problem set, as well as answers to section review questions and to practice exercises. Terms that appear boldface in the text are defined in the glossary.

SUGGESTIONS FOR USE

Fundamentals of Linear Circuits can be used to accommodate different scheduling and program needs. Some suggestions are as follows.

Option 1 For those programs that cover discrete devices and circuits in a separate course, the first four chapters of this book can be omitted or used for review and reference. Chapters 5 through 14 provide for a one term linear integrated circuits course with considerable emphasis on the op-amp.

Option 2 For those programs requiring an emphasis on linear integrated circuits with a minimum but thorough coverage of discrete devices and circuits, the entire fourteen chapters of the book provide a complete course.

SYSTEM APPLICATION The system application is an extremely versatile tool for providing both motivation and real-world experiences in the classroom. The variety of systems is intended to give the student an appreciation for the wide range of applications for electronic devices.

Although these system applications can be treated as optional, it is highly recommended that they be included in your course. System applications can be used as

- ☐ An integral part of the chapter for the purpose of relating devices to a realistic system and for establishing a useful purpose for the device(s). All or selected activities can be assigned and discussed in class or turned in for a grade.
- ☐ A separate out-of-class assignment to be turned in for extra credit.
- ☐ An in-class activity to promote and stimulate discussion and interaction among students and between students and the instructor.
- ☐ A case in point to help answer the question on the mind of most students: "Why do I need to know this?"

A NOTE TO THE STUDENT

The material in this preface is intended to help both you and your instructor make the most effective use of this textbook as a teaching and learning tool. Although you should certainly read everything in this preface, this part is especially for you, the student.

I am sure that you realize that knowledge and skills are not obtained easily or without effort. Much hard work is required to properly prepare yourself for any career, and electronics is no exception. You should use this book as more than just a reference. You must really dig in by reading, thinking, and doing. Don't expect every concept or procedure to become immediately clear. Some topics may take several readings, working many problems, and much help from your instructor before you really understand them.

Work through each example step-by-step and then do the associated practice exercise. Answer the review questions at the end of each section. If you don't understand an example or if you can't answer a question, go back into the section until you can. Check your answers at the end of the chapter. The multiple-choice self-tests at the end of each chapter are a good way to check your overall comprehension and retention of the subjects covered. You should do the self-test before you start the problems. Check your answers at the end of the book.

The problem sets at the end of each chapter (except Chapter 1) provide exercises with varying degrees of difficulty. In any technical field, it is very important that you work lots of problems. Working through a problem gives you a level of insight and understanding that reading or classroom lectures alone do not provide. Never think that you fully understand a concept or procedure by simply watching or listening to someone else. In the final analysis, you must do it yourself and you must do it to the best of your ability.

A LOOK BACK

Now, before you begin your study of electronic devices and circuits, let's briefly look back at the beginnings of electronics and some of the important developments that have led to the electronics technology that we have today. It is always good to have a sense of the history of your career field. The names of many of the early pioneers in electricity and electromagnetics still live on in terms of familiar units and quantities. Names such as Ohm, Ampere, Volta, Farad, Henry, Coulomb, Oersted, and Hertz are some of the better known examples. More widely known names such as Franklin and Edison are also very significant in the history of electricity and electronics because of their tremendous contributions.

THE BEGINNING OF ELECTRONICS

The early experiments in electronics involved electric currents in glass vacuum tubes. One of the first to conduct such experiments was a German named Heinrich Geissler (1814–1879). Geissler removed most of the air from a glass tube and found that the tube glowed when there was an electric current through it. Around 1878, British scientist Sir William Crookes (1832–1919) experimented with tubes similar to those of Geissler. In his experiments, Crookes found that the current in the vacuum tubes seemed to consist of particles.

Thomas Edison (1847–1931), experimenting with the carbon-filament light bulb he had invented, made another important finding. He inserted a small metal plate in the bulb. When the plate was positively charged, there was a current from the filament to the plate. This device was the first thermionic diode. Edison patented it but never used it.

The electron was discovered in the 1890s. The French physicist Jean Baptiste Perrin

(1870–1942) demonstrated that the current in a vacuum tube consists of the movement of negatively charged particles in a given direction. Some of the properties of these particles were measured by Sir Joseph Thomson (1856–1940), a British physicist, in experiments he performed between 1895 and 1897. These negatively charged particles later became known as electrons. The charge on the electron was accurately measured by an American physicist, Robert A. Millikan (1868–1953), in 1909. As a result of these discoveries, electrons could be controlled, and the electronic age was ushered in.

PUTTING THE ELECTRON TO WORK A vacuum tube that allowed electrical current in only one direction was constructed in 1904 by British scientist John A. Fleming. The tube was used to detect electromagnetic waves. Called the Fleming valve, it was the forerunner of the more recent vacuum diode tubes. Major progress in electronics, however, awaited the development of a device that could boost, or amplify, a weak electromagnetic wave or radio signal. This device was the audion, patented in 1907 by Lee deForest, an American. It was a triode vacuum tube capable of amplifying small electrical ac signals.

Two other Americans, Harold Arnold and Irving Langmuir, made great improvements in the triode vacuum tube between 1912 and 1914. About the same time, deForest and Edwin Armstrong, an electrical engineer, used the triode tube in an oscillator circuit. In 1914, the triode was incorporated in the telephone system and made the transcontinental telephone network possible. The tetrode tube was invented in 1916 by Walter Schottky, a German. The tetrode, along with the pentode (invented in 1926 by Dutch engineer Tellegen), greatly improved the triode. The first television picture tube, called the kinescope, was developed in the 1920s by Vladimir Sworykin, an American researcher.

During World War II, several types of microwave tubes were developed that made possible modern microwave radar and other communications systems. In 1939, the magnetron was invented in Britain by Henry Boot and John Randall. In the same year, the klystron microwave tube was developed by two Americans, Russell Varian and his brother Sigurd Varian. The traveling-wave tube (TWT) was invented in 1943 by Rudolf Komphner, an Austrian-American.

SOLID-STATE ELECTRONICS The crystal detectors used in early radios were the forerunners of modern solid-state devices. However, the era of solid-state electronics began with the invention of the transistor in 1947 at Bell Labs. The inventors were Walter Brattain, John Bardeen, and William Shockley. Figure P–5 shows these three men.

In the early 1960s, the integrated circuit (IC) was developed. It incorporated many transistors and other components on a single small chip of semiconductor material. Integrated circuit technology has been continuously developed and improved, allowing increasingly more complex circuits to be built on smaller chips.

Around 1965, the first integrated general-purpose operational amplifier was introduced. This low-cost, highly versatile device incorporated nine transistors and twelve resistors in a small package. It proved to have many advantages over comparable discrete component circuits in terms of reliability and performance. Since this introduction, the IC operational amplifier has become a basic building block for a wide variety of linear systems.

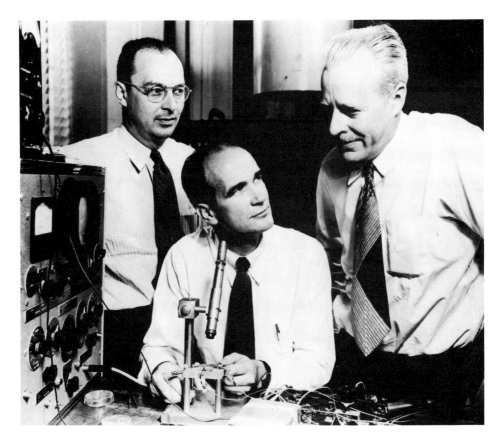

FIGURE P-5
Nobel Prize winners Drs. John Bardeen, William Shockley, and Walter Brattain, shown left to right, with apparatus used in their first investigations that led to the invention of the transistor. The trio received the 1956 Nobel Physics award for their invention of the transistor, which was announced by Bell Laboratories in 1948. (Courtesy of Bell Laboratories)

ACKNOWLEDGMENTS

Fundamentals of Linear Circuits was made possible by the efforts and talents of many people. I want to express my appreciation to Carol Robison, Dave Garza, Steve Helba, Rex Davidson, Pete Robison, Anne Flanagan, and Russ Maselli of Merrill, an imprint of Macmillan Publishing Company. Also, thanks to Lois Porter for a great job in editing the manuscript.

I am grateful to the many contributors: John Berry, DeVry Institute of Technology—Dallas; Gary Bryan, DeVry Institute of Technology—Phoenix; Vincent Ceppaluni, University of Maryland; Phillip Chiarelli, Electronics Institute; Jim Davis, Muskingum

Area Technical College; Mel Duvall, Sacramento City College; Gary Pfeiffer, DeVry Institute of Technology—Columbus; Elvin Stepps, University of Cincinnati; Les Taylor, DeVry Institute of Technology—Woodbridge; and Jim Thompson, American River College.

Also thanks to Gary Snyder of Bently Nevada for his assistance in checking the accuracy of the manuscript.

Tom Floyd

Contents

1
Introduction to Semiconductors — 1

- 1–1 Atoms 2
- 1–2 The Atomic Structure of Semiconductors 3
- 1–3 Covalent Bonds 6
- 1–4 Conduction in Semiconductor Materials 7
- 1–5 N-Type and P-Type Semiconductors 11
- 1–6 PN Junctions 13
- 1–7 Biasing the Diode 16
- 1–8 Diode Characteristics 21

2
Diodes and Applications — 32

- 2–1 Half-Wave Rectifiers 34
- 2–2 Full-Wave Rectifiers 37
- 2–3 Rectifier Filters 43
- 2–4 Diode Clipping and Clamping Circuits 47
- 2–5 Zener Diodes 54
- 2–6 Varactor Diodes 62
- 2–7 LEDs and Photodiodes 65
- 2–8 The Diode Data Sheet 68
- 2–9 Troubleshooting 72
- 2–10 A System Application 76

3
Transistors and Thyristors — 96

- 3–1 Bipolar Junction Transistors (BJTs) 98
- 3–2 Voltage-Divider Bias 103

xxi

3–3 The Bipolar Transistor As an Amplifier 105
3–4 The Bipolar Transistor As a Switch 112
3–5 BJT Parameters and Ratings 114
3–6 The Junction Field-Effect Transistor (JFET) 117
3–7 JFET Characteristics 119
3–8 The Metal Oxide Semiconductor FET (MOSFET) 124
3–9 FET Biasing 127
3–10 Unijunction Transistors (UJTs) 131
3–11 Thyristors 133
3–12 Transistor Packages and Terminal Identification 135
3–13 Troubleshooting 137
3–14 A System Application 143

4
AMPLIFIERS AND OSCILLATORS   162

4–1 Common-Emitter Amplifiers 164
4–2 Common-Collector Amplifiers 170
4–3 Common-Base Amplifiers 174
4–4 FET Amplifiers 176
4–5 Multistage Amplifiers 184
4–6 Class A Amplifier Operation 187
4–7 Class B Push-Pull Amplifier Operation 191
4–8 Class C Amplifier Operation 197
4–9 Oscillators 200
4–10 Troubleshooting 205
4–11 A System Application 207

5
OPERATIONAL AMPLIFIERS   222

5–1 Introduction to Operational Amplifiers 224
5–2 The Differential Amplifier 226
5–3 Op-Amp Data Sheet Parameters 235
5–4 Negative Feedback 242
5–5 Op-Amp Configurations with Negative Feedback 244
5–6 Effects of Negative Feedback on Op-Amp Impedances 249
5–7 Bias Current and Offset Voltage Compensation 254

5–8   Troubleshooting   258
5–9   A System Application   261

6
Op-Amp Responses   280

6–1   Basic Concepts   282
6–2   Op-Amp Open-Loop Response   287
6–3   Op-Amp Closed-Loop Response   290
6–4   Positive Feedback and Stability   292
6–5   Op-Amp Compensation   298
6–6   A System Application   306

7
Basic Op-Amp Circuits   318

7–1   Comparators   320
7–2   Summing Amplifiers   332
7–3   The Integrator and Differentiator   338
7–4   More Op-Amp Circuits   344
7–5   Troubleshooting   347
7–6   A System Application   351

8
Active Filters   370

8–1   Basic Filter Responses   372
8–2   Filter Response Characteristics   376
8–3   Active Low-Pass Filters   379
8–4   Active High-Pass Filters   383
8–5   Active Band-Pass Filters   386
8–6   Active Band-Stop Filters   393
8–7   Filter Response Measurements   395
8–8   A System Application   397

9
Signal Generators and Timers   412

9–1   Definition of an Oscillator   414
9–2   Oscillator Principles   415

- 9–3 Sine Wave Oscillators 417
- 9–4 Nonsinusoidal Oscillators 425
- 9–5 The 555 Timer As an Oscillator 431
- 9–6 The 555 Timer As a One-Shot 438
- 9–7 A System Application 442

10
Power Supply Circuits — 454

- 10–1 Voltage Regulation 456
- 10–2 Basic Series Regulators 458
- 10–3 Basic Shunt Regulators 464
- 10–4 Basic Switching Regulators 468
- 10–5 Integrated Circuit Voltage Regulators 476
- 10–6 Applications of IC Voltage Regulators 481
- 10–7 A System Application 488

11
Special Amplifiers — 500

- 11–1 Instrumentation Amplifiers 502
- 11–2 Isolation Amplifiers 508
- 11–3 Operational Transconductance Amplifiers (OTAs) 514
- 11–4 Log and Antilog Amplifiers 522
- 11–5 Analog Multipliers and Dividers 529
- 11–6 A System Application 535

12
Communications Circuits — 552

- 12–1 Basic Receivers 554
- 12–2 The Linear Multiplier 559
- 12–3 Amplitude Modulation 566
- 12–4 The Mixer 573
- 12–5 AM Demodulation 576
- 12–6 IF and Audio Amplifiers 577
- 12–7 Frequency Modulation 581
- 12–8 The Phase-Locked Loop (PLL) 586
- 12–9 A System Application 596

13
DATA CONVERSION CIRCUITS — 616

- 13–1 Analog Switches 618
- 13–2 Sample-and-Hold Amplifiers 623
- 13–3 Interfacing the Analog and Digital Worlds 627
- 13–4 Digital-to-Analog (D/A) Conversion 632
- 13–5 Basic Concepts of Analog-to-Digital (A/D) Conversion 638
- 13–6 Analog-to-Digital (A/D) Conversion Methods 642
- 13–7 Voltage-to-Frequency (V/F) and Frequency-to-Voltage (F/V) Converters 655
- 13–8 Troubleshooting 665
- 13–9 A System Application 671

14
MEASUREMENT AND CONTROL CIRCUITS — 692

- 14–1 RMS-to-DC Converters 694
- 14–2 Angle-Measuring Circuits 699
- 14–3 Temperature-Measuring Circuits 707
- 14–4 Strain-Measuring and Pressure-Measuring Circuits 723
- 14–5 Power-Control Circuits 730
- 14–6 A System Application 733

Appendix A
Data Sheets — 747

Appendix B
Derivations of Selected Equations — 800

Answers to Self-Tests — 805

Answers to Selected Odd-Numbered Problems — 807

Glossary — 817

Index — 825

Color Insert: Circuit Boards and Instrumentation for Selected System Applications

This special 8-page full-color insert (which follows page 452) provides realistic printed circuit boards and test instruments for special Test Bench assignments related to certain System Applications.

This book is dedicated to electronics students, in the hope that it will help to properly prepare them for their future as useful and productive citizens.

1

INTRODUCTION TO SEMICONDUCTORS

1–1 ATOMS
1–2 THE ATOMIC STRUCTURE OF SEMICONDUCTORS
1–3 COVALENT BONDS
1–4 CONDUCTION IN SEMICONDUCTOR MATERIALS
1–5 N-TYPE AND P-TYPE SEMICONDUCTORS
1–6 PN JUNCTIONS
1–7 BIASING THE DIODE
1–8 DIODE CHARACTERISTICS

After completing this chapter, you should be able to

☐ Discuss the significance of electron shells.
☐ Describe ionization.
☐ Explain the basic structure of silicon and germanium.
☐ Discuss how atoms bond together to form crystals.
☐ Explain how current in a material is related to the energy levels within an atom's structure.
☐ Define p-type and n-type semiconductors.
☐ Describe how a pn junction is formed and how it is used.
☐ Explain what a diode is.
☐ Describe forward and reverse bias of a diode.
☐ Interpret a diode characteristic curve.
☐ Define *anode* and *cathode*.
☐ Identify the terminals of a diode.

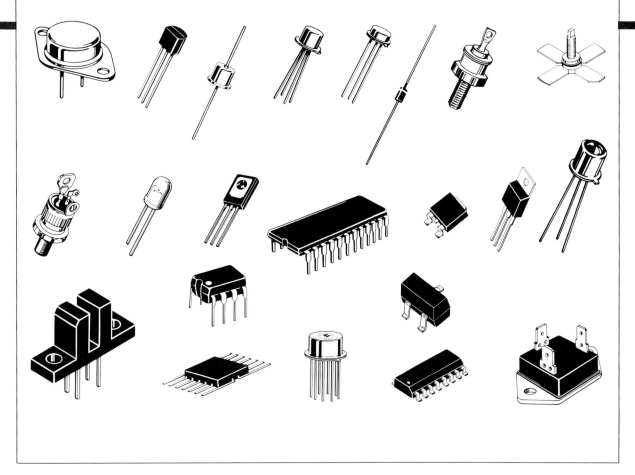

To acquire a basic understanding of semiconductors, you must have some knowledge of atomic theory and the structure of semiconductor materials. In this chapter, you will learn about atomic theory. We will discuss the basic materials used in manufacturing both **discrete devices**, such as diodes and transistors, and integrated circuits. We will introduce pn junctions, an important concept essential for the understanding of diode and transistor operation. Also, other diode characteristics are introduced, and you will learn how to properly use a diode in a circuit.

Most electronic devices are made of semiconductor material. The devices are shown above in a variety of common packaging configurations and include diodes, transistors, and various types of integrated circuits. In electronic systems, individual devices such as these are interconnected to form circuits that are designed to function in a specific way by properly utilizing the characteristics of each device. In the following chapters, you will learn how the various devices are used in certain system applications.

Art copyright of Motorola, Inc. Used by permission.

1–1 ATOMS

All matter is made up of atoms; and all atoms are made up of electrons, protons, and neutrons. In order to better understand how semiconductors work, you need to know something about the atom. In this section and the next one, you will learn about the structure of the atom, electron orbits and shells, valence electrons, ions, and the two major semiconductor materials—silicon and germanium. This material is important because the configuration of certain electrons in an atom is the key factor in determining how a given material conducts electrical current.

An **atom** is the smallest particle of an element that retains the characteristics of that element. Each known element has atoms that are different from the atoms of all other elements. This gives each element a unique atomic structure. According to the classical Bohr model, atoms have a planetary type of structure that consists of a central nucleus surrounded by orbiting electrons as illustrated in Figure 1–1. The nucleus consists of positively charged particles called *protons* and uncharged particles called *neutrons*. **Electrons** are the basic particles of negative charge.

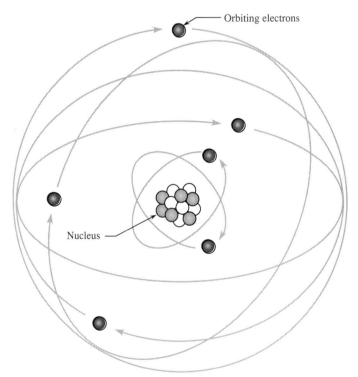

FIGURE 1–1
The Bohr model of an atom.

Each type of atom has a certain number of electrons and protons that distinguishes it from the atoms of all other elements. For example, the simplest atom is that of hydrogen. It has one proton and one electron, as shown in Figure 1–2(a). The helium atom, shown

Introduction to Semiconductors

in Figure 1–2(b), has two protons and two neutrons in the nucleus orbited by two electrons.

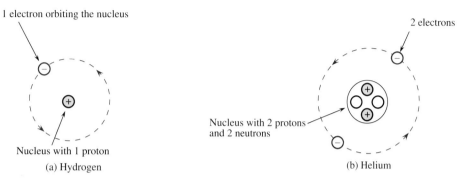

FIGURE 1–2
Hydrogen and helium atoms.

Atomic Number and Weight

All elements are arranged in the periodic table of the elements in order according to their atomic number, which equals the number of electrons in an electrically balanced (neutral) atom. The elements can also be arranged by their atomic weight, which is approximately the number of protons and neutrons in the nucleus. For example, hydrogen has an atomic number of one and an atomic weight of one. The atomic number of helium is two, and its atomic weight is four. In their normal, or neutral, state, all atoms of a given element have the same number of electrons as protons; the positive charges cancel the negative charges, and the atom has a net charge of zero.

1–1 Review Questions

1. Describe an atom.
2. What is an electron?

1–2 The Atomic Structure of Semiconductors

The basic concept of the structure of an atom was introduced in the last section. In the rest of this chapter, atomic theory is extended to semiconductor materials that are used in electronic devices such as diodes and transistors. This coverage lays the foundation for a good understanding of how semiconductor devices function.

Electron Shells and Orbits

Electrons orbit the nucleus of an atom at certain distances from the nucleus. Electrons near the nucleus have less **energy** than those in more distant orbits. Only discrete (separate and distinct) values of electron energies exist within atomic structures. Therefore, electrons must orbit only at discrete distances from the nucleus.

Each discrete distance (orbit) from the nucleus corresponds to a certain energy level. In an atom, orbits are grouped into energy bands known as *shells*. A given atom has a fixed number of shells. Each shell has a fixed maximum number of electrons at permissible energy levels (orbits). The differences in energy levels within a shell are much smaller than the difference in energy between shells. The shells are designated *K, L, M, N*, and so on, with *K* being closest to the nucleus. This concept is illustrated in Figure 1–3.

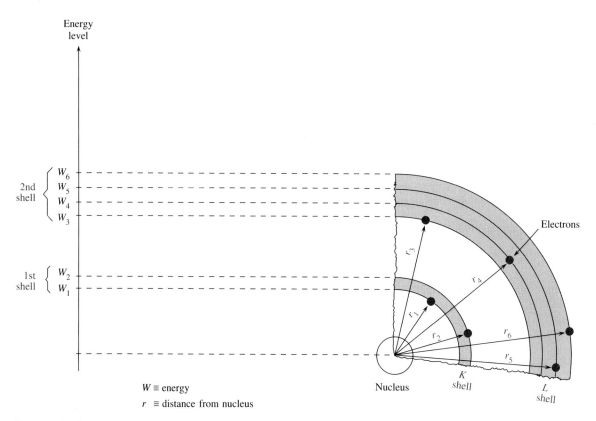

FIGURE 1–3
Energy levels increase as distance from the nucleus of the atom increases.

VALENCE ELECTRONS

Electrons in orbits farther from the nucleus are less tightly bound to the atom than those closer to the nucleus. This is because the force of attraction between the positively charged nucleus and the negatively charged electron decreases with increasing distance.

Electrons with the highest energy levels exist in the outermost shell of an atom and are relatively loosely bound to the atom. These valence electrons contribute to chemical

Introduction to Semiconductors

reactions and bonding within the structure of a material. The **valence** of an atom is the number of electrons in its outermost shell.

Ionization

When an atom absorbs energy from a heat source or from light, for example, the energy levels of the electrons are raised. When an electron gains energy, it moves to an orbit farther from the nucleus. Since the valence electrons possess more energy and are more loosely bound to the atom than inner electrons, they can jump to higher orbits more easily when external energy is absorbed.

If a valence electron acquires a sufficient amount of energy, it can be completely removed from the outer shell and the atom's influence. The departure of a valence electron leaves a previously neutral atom with an excess of positive charge (more protons than electrons). The process of losing a valence electron is known as **ionization** and the resulting positively charged atom is called a positive **ion.** For example, the chemical symbol for hydrogen is H. When it loses its valence electron and becomes a positive ion, it is designated H^+. The escaped valence electron is called a **free electron.**

When a free electron falls into the outer shell of a neutral hydrogen atom, the atom becomes negatively charged (more electrons than protons) and is called a negative ion, designated H^-.

Silicon and Germanium Atoms

Two types of widely used semiconductor materials are **silicon** and **germanium.** Both the silicon and the germanium atoms have four valence electrons. They differ in that silicon has 14 protons in its nucleus and germanium has 32. Figure 1–4 shows a representation of the atomic structure for both materials.

Figure 1–4
Silicon and germanium atoms.

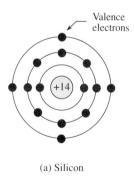
(a) Silicon

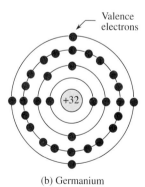
(b) Germanium

1–2 Review Questions

1. What are the components of an atom?
2. What is a valence electron?
3. What is a free electron?

1-3 COVALENT BONDS

*When silicon atoms combine to form a solid material, they arrange themselves in a fixed pattern called a **crystal**. The atoms within the crystal structure are held together by covalent bonds, which are created by the interaction of the valence electrons of each atom. A solid chunk of silicon is a crystalline material.*

Figure 1–5 shows how each silicon atom positions itself with four adjacent atoms. A silicon atom with its four valence electrons shares an electron with each of its four neighbors. This effectively creates eight valence electrons for each atom and produces a state of chemical stability. Also, this sharing of valence electrons produces the **covalent** bonds that hold the atoms together; each shared electron is attracted equally by the two adjacent atoms that share it. Covalent bonding of a pure (intrinsic) silicon crystal is shown in Figure 1–6. Bonding for germanium is similar because it also has four valence electrons.

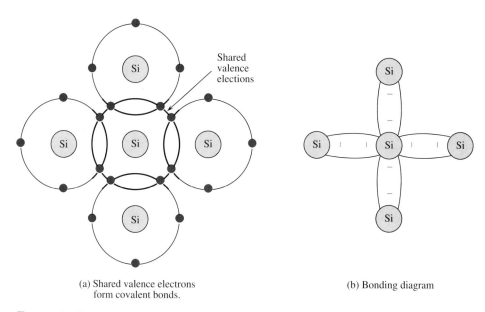

(a) Shared valence electrons form covalent bonds.

(b) Bonding diagram

FIGURE 1–5
Covalent bonds in silicon.

1-3 REVIEW QUESTIONS

1. What are two semiconductor materials?
2. What is a covalent bond?

Introduction to Semiconductors

Figure 1–6
Covalent bonds in a pure (intrinsic) silicon crystal.

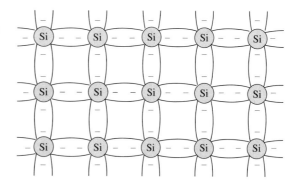

1–4

Conduction in Semiconductor Materials

How a material conducts electrical current is very important in understanding how electronic devices operate. You can't really understand the operation of a device such as a diode or transistor without knowing something about the basic current phenomenon. In this section, you will see how conduction occurs and why some materials are better conductors than others.

As you have seen, the electrons of an atom can exist only within prescribed energy bands. Each shell around the nucleus corresponds to a certain energy band and is separated from adjacent shells by energy gaps, in which no electrons can exist. This condition, shown in Figure 1–7 for an unexcited (no external energy) silicon crystal, occurs only at absolute zero temperature.

Figure 1–7
Energy band diagram for unexcited silicon crystal. There are no electrons in the conduction band.

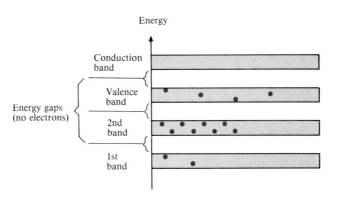

Conduction Electrons and Holes

A pure silicon crystal at room temperature derives heat (thermal) energy from the surrounding air, causing some valence electrons to gain sufficient energy to jump the gap from the valence band into the conduction band, becoming free electrons. This situation is illustrated in the energy diagram of Figure 1–8(a) and in the bonding diagram of Figure 1–8(b).

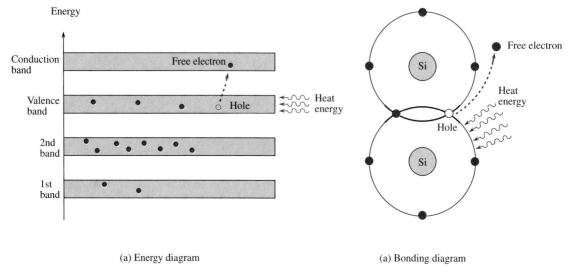

(a) Energy diagram   (a) Bonding diagram

FIGURE 1–8
Creation of an electron-hole pair in an excited silicon atom. An electron in the conduction band is a free electron.

When an electron jumps to the conduction band, a vacancy is left in the valence band. This vacancy is called a **hole.** For every electron raised to the conduction band by thermal or light energy, there is one hole left in the valence band, creating what is called an *electron-hole pair*. **Recombination** occurs when a conduction-band electron loses energy and falls back into a hole in the valence band.

In summary, a piece of pure silicon at room temperature has, at any instant, a number of conduction-band (free) electrons that are unattached to any atom and are essentially drifting randomly throughout the material. Also, an equal number of holes are created in the valence band when these electrons jump into the conduction band, as illustrated in Figure 1–9.

GERMANIUM VERSUS SILICON

The situation in a germanium crystal is similar to that in silicon except that, because of its atomic structure, pure germanium has more free electrons than silicon and therefore a higher conductivity. Silicon, however, is the favored **semiconductor** material and is used far more widely than germanium. One reason for its wide usage is that silicon can be used at a much higher temperature than germanium.

ELECTRON AND HOLE CURRENT

When a voltage is applied across a piece of intrinsic silicon, as shown in Figure 1–10, the thermally generated free electrons in the conduction band are easily attracted toward the positive end. This movement of free electrons is one type of current in a semiconductor material and is called *electron current*.

Another type of current occurs at the valence level, where the holes created by the free electrons exist. Electrons remaining in the valence band are still attached to their

FIGURE 1–9
Electron-hole pairs in a silicon crystal. Free electrons are being continuously generated while some recombine with holes.

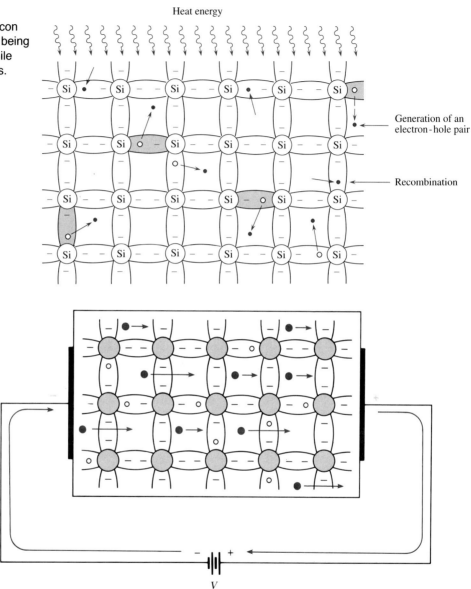

FIGURE 1–10
Electron current in intrinsic silicon is produced by movement of thermally generated free electrons.

atoms and are not free to move randomly in the crystal structure. However, a valence electron can move into a nearby hole, with little change in its energy level, thus leaving another hole where it came from. Effectively, the hole has moved from one place to another in the crystal structure, as illustrated in Figure 1–11. This current is called *hole current*.

FIGURE 1–11

Hole current in intrinsic silicon. In this illustration the hole is effectively moving from right to left.

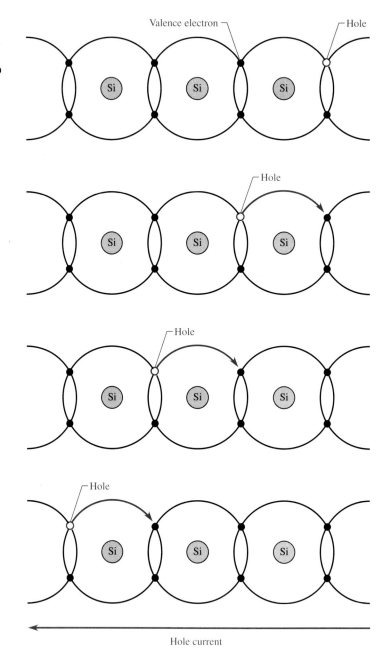

SEMICONDUCTORS, CONDUCTORS, AND INSULATORS

In an **intrinsic** (pure) **semiconductor,** there is a relatively small number of free electrons, so neither silicon nor germanium is very useful in its intrinsic state. Pure semiconductors are neither good insulators nor good conductors because current in a material depends directly on the number of free electrons.

INTRODUCTION TO SEMICONDUCTORS

A comparison of the energy bands in Figure 1–12 for the three types of materials shows the essential differences among them regarding conduction. The energy gap for an insulator is so wide that very few electrons acquire enough energy to jump into the conduction band. The valence band and the conduction band in a conductor (such as copper) overlap so that there are always many conduction electrons; even without the application of external energy, many valence electrons already have enough energy to jump into the conduction band. A semicondutor, as Figure 1–12(b) shows, has an energy gap that is much narrower than that in an insulator.

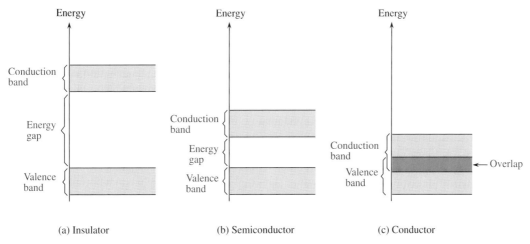

FIGURE 1–12

Energy diagrams for three categories of materials.

1–4 REVIEW QUESTIONS

1. In the atomic structure of a semiconductor, within which energy band do free electrons exist? Within which energy band do valence electrons exist?
2. How are holes created in an intrinsic semiconductor?
3. Why is current established more easily in a semiconductor than in an insulator?

1–5 N-TYPE AND P-TYPE SEMICONDUCTORS

Semiconductor materials do not conduct current well and are of very little value in their intrinsic state because of the limited number of free electrons in the conduction band and holes in the valence band. Pure silicon (or germanium) must be modified by increasing the free electrons and holes to increase its conductivity and make it useful in electronic devices. This is done by adding impurities to the intrinsic material as you will see in this section. The two types of semiconductor materials, n-type and p-type, are the key building blocks for all types of electronic devices.

Doping

The conductivity of silicon and germanium can be drastically increased by the controlled addition of impurities to the pure semiconductor material. This process, called **doping,** increases the number of current carriers (electrons or holes), thus increasing the conductivity and decreasing the resistivity. The two categories of impurities are *n-type* and *p-type*.

N-Type Semiconductor

To increase the number of conduction-band electrons in pure silicon, pentavalent impurity atoms are added. These are atoms with five valence electrons, such as arsenic, phosphorus, and antimony.

As illustrated in Figure 1–13, each pentavalent atom (antimony, in this case) forms covalent bonds with four adjacent silicon atoms. Four of the antimony atom's valence electrons are used to form the covalent bonds, leaving one extra electron. This extra electron becomes a conduction electron because it is not attached to any atom. The number of conduction electrons can be controlled by the amount of impurity atoms added to the silicon. A conduction electron created by this doping process does not leave a hole in the valence band.

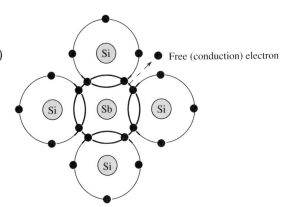

Figure 1–13
Pentavalent impurity atom in a silicon crystal. An antimony (Sb) impurity atom is shown in the center. The extra electron becomes a free electron.

Since most of the current carriers are electrons, silicon (or germanium) doped in this way is an n-type semiconductor material where the *n* stands for the negative charge on an electron. The electrons are called the *majority carriers* in n-type material. Although the great majority of current carriers in n-type material are electrons, there are a few holes that are created when electron-hole pairs are thermally generated. These holes are not produced by the addition of the pentavalent impurity atoms. Holes in n-type material are called *minority carriers*.

P-Type Semiconductor

To increase the number of holes in pure silicon, trivalent impurity atoms are added. These are atoms with three valence electrons, such as aluminum, boron, and gallium.

As illustrated in Figure 1–14, each trivalent atom (boron, in this case) forms covalent bonds with four adjacent silicon atoms. All three of the boron atom's valence electrons are

INTRODUCTION TO SEMICONDUCTORS

used in the covalent bonds; and, since four electrons are required, a hole is formed with each trivalent atom. The number of holes can be controlled by the amount of trivalent impurity atoms added to the silicon. A hole created by this doping process is not accompanied by a conduction (free) electron.

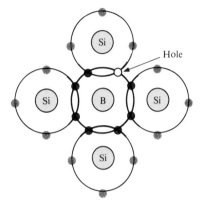

FIGURE 1–14
Trivalent impurity atom in a silicon crystal. A boron (B) impurity atom is shown in the center.

Since most of the current carriers are holes, silicon (or germanium) doped in this way is a p-type semiconductor material because holes can be thought of as positive charges. The holes are the majority carriers in p-type material. Although the majority of current carriers in p-type material are holes, there are a few electrons that are created when electron-hole pairs are thermally generated. These free electrons are not produced by the addition of the trivalent impurity atoms. Electrons in p-type material are called *minority carriers*.

1–5 REVIEW QUESTIONS

1. How is an n-type semiconductor formed?
2. How is a p-type semiconductor formed?
3. What is a majority carrier?

1–6 PN JUNCTIONS

If we take a block of silicon and dope half of it with a pentavalent impurity and the other half with a trivalent impurity, a boundary called the pn junction is formed between the resulting n-type and the p-type portions. Amazingly, this pn junction is the thing that allows diodes and transistors to work. This section and the next one will give you some insight into the pn junction and provide a springboard into the next chapter where you will learn about the diode, a very important electronic device.

When a piece of silicon is doped so that half is n-type and the other half is p-type, a pn junction is formed between the two regions, as shown in Figure 1–15(a). The n region has many free electrons (majority carriers) and only a few thermally generated holes (minority carriers). The p region has many holes (majority carriers) and only a few thermally generated free electrons (minority carriers). This is shown in Figure 1–15(b). The **pn junction** is fundamental to the operation of diodes, transistors, and other solid-state devices.

FIGURE 1–15
Basic pn structure in a diode at the instant of junction formation. Both majority and minority carriers are shown.

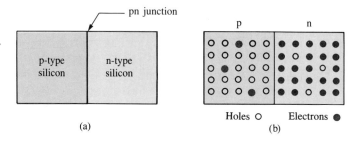

THE DEPLETION LAYER

With no external voltage, the conduction electrons in the n region are aimlessly drifting in all directions. At the instant of junction formation, some of the conduction electrons near the junction drift across into the p region and recombine with holes near the junction.

For each electron that crosses the junction and recombines with a hole, a pentavalent atom is left with a net positive charge in the n region near the junction, making it a positive ion. Also, when the electron recombines with a hole in the p region, a trivalent atom acquires a net negative charge, making it a negative ion.

As a result of this recombination process, a large number of positive and negative ions builds up near the pn junction. As this buildup occurs, the conduction electrons in the n region must overcome both the attraction of the positive ions and the repulsion of the negative ions in order to migrate into the p region. Thus, as the ion layers build up, the area on both sides of the junction becomes essentially depleted of any conduction electrons or holes and is known as the *depletion layer*. This condition is illustrated in Figure 1–16. When an equilibrium condition is reached, the depletion layer has widened to a point where no more electrons can cross the pn junction.

FIGURE 1–16
PN junction equilibrium condition. The few electrons in the p region (solid dots) and the few holes (open circles) in the n region are the minority carriers created by thermally produced electron-hole pairs.

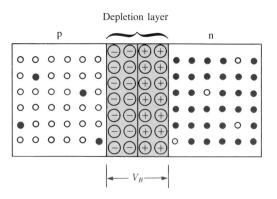

The existence of the positive and negative ions on opposite sides of the junction creates a **barrier potential** (V_B) across the depletion layer, as indicated in Figure 1–16. At 25°C, the barrier potential is approximately 0.7 V for silicon and 0.3 V for germanium.

Introduction to Semiconductors

As the junction temperature increases, the barrier potential decreases, and vice versa. The barrier potential is significant because it determines the amount of external voltage that must be applied across the pn junction to produce current. You will see this more clearly when we discuss biasing of the pn junction.

Energy Diagram of the PN Junction

Now, we will look at the operation of the pn junction in terms of its energy level. First, consider the pn junction at the instant of its formation. The energy bands of the trivalent impurity atoms in the p-type material are at a slightly higher level than those of the pentavalent impurity atom in the n-type material, as shown in Figure 1–17. They are higher because the core attraction for the valence electrons (+3) in the trivalent atom is less than the core attraction for the valence electrons (+5) in the pentavalent atom. Thus, the trivalent valence electrons are in a slightly higher orbit and, thus, at a higher energy level.

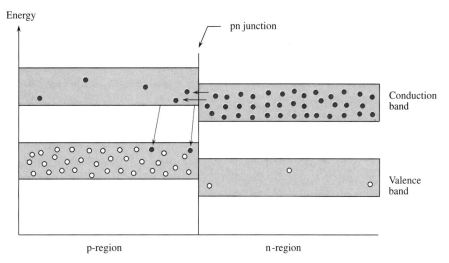

Figure 1–17
PN junction energy diagram as diffusion across the junction begins.

Notice in Figure 1–17 that there is some overlap of the conduction bands in the p and n regions and also some overlap of the valence bands in the p and n regions. This overlap permits the electrons of higher energy near the top of the n-region conduction band to begin diffusing across the junction into the lower levels of the p-region conduction band. As soon as an electron diffuses across the junction, it recombines with a hole in the valence band. As diffusion continues, the depletion layer begins to form. Also, the energy bands in the n region "shift" down as the electrons with higher energy are lost to diffusion. When the top of the n-region conduction band reaches the same level as the bottom of the p-region conduction band, diffusion ceases and the equilibrium condition is reached. This condition is shown in terms of energy levels in Figure 1–18. There is an energy gradient across the depletion layer rather than an abrupt change in energy level.

16 CHAPTER 1

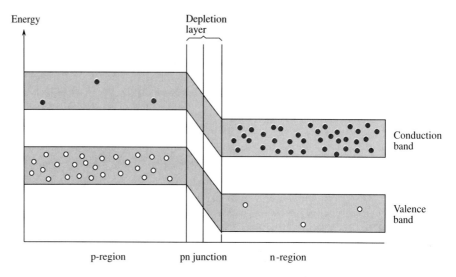

FIGURE 1–18
Energy diagram levels at equilibrium. The n-region bands are shifted down from their original levels because the higher-energy electrons are lost to diffusion.

1–6 REVIEW QUESTIONS

1. What is a pn junction?
2. When p and n regions are joined, a depletion layer forms. Describe the characteristics of the depletion layer.
3. Are there any majority carriers in the depletion layer?
4. What is the value of the barrier potential for silicon?

1–7 BIASING THE DIODE

As you have learned, there is no current across a pn junction at equilibrium. The primary usefulness of the pn junction diode is its ability to allow current in only one direction and to prevent current in the other direction as determined by the bias. There are two bias conditions for a pn junction—forward and reverse. Either of these conditions is created by connecting an external dc voltage in the proper direction across the diode.

FORWARD BIAS

The term **bias** in electronics refers to a fixed dc voltage that sets the operating conditions for a semiconductor device. *Forward bias* is the condition that permits current across a pn junction. Figure 1–19 shows a dc voltage connected in a direction to forward-bias the diode. Notice that the negative **terminal** of the battery is connected to the n region (called the **cathode**), and the positive terminal is connected to the p region (called the **anode**).

Introduction to Semiconductors

Figure 1–19
Forward-bias connection. The purpose of the resistor is to limit the current in order to prevent damage to the diode.

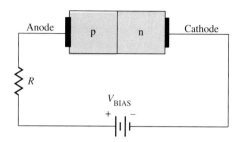

This is how forward bias works: The negative terminal of the battery pushes the conduction electrons in the n region toward the junction, while the positive terminal pushes the holes in the p region also toward the junction. Recall that like charges repel each other. When it overcomes the barrier potential, the external bias voltage source provides the n-region electrons with enough energy to penetrate the depletion layer and cross the junction, where they combine with the p-region holes. As electrons leave the n region, more flow in from the negative terminal of the source. Thus, current through the n region is formed by the movement of conduction electrons (majority carriers) toward the junction. When the conduction electrons enter the p region and combine with holes, they become valence electrons. Then they move as valence electrons from hole to hole toward the positive anode connection. The movement of these valence electrons essentially creates a movement of holes in the opposite direction as you learned earlier. Thus, current in the p region is formed by the movement of holes (majority carriers) toward the junction. Figure 1–20 illustrates current in a forward-biased diode.

Figure 1–20
Electron flow in a forward-biased pn junction diode.

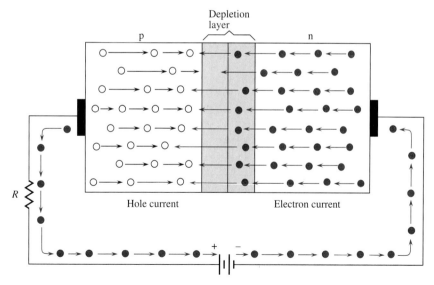

EFFECT OF BARRIER POTENTIAL ON FORWARD BIAS The barrier potential of the depletion layer can be thought of acting as a small battery that opposes bias, as illustrated in Figure 1–21.

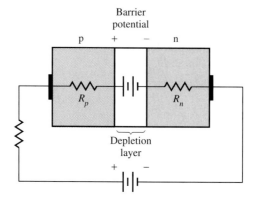

FIGURE 1–21
Barrier potential and bulk resistance equivalent for a pn junction diode.

The resistances R_p and R_n represent the bulk resistance of the p and n materials. Keep in mind that the barrier potential is not a voltage source and cannot be measured with a voltmeter; it only has the effect of a battery when forward bias is applied.

The external bias voltage must overcome the barrier potential before the diode conducts, as illustrated in Figure 1–22. Conduction occurs at approximately 0.7 V for silicon and 0.3 V for germanium. Once the diode is conducting in the forward direction, the voltage drop across it remains at approximately the barrier potential and changes very little with changes in forward current (I_F) except for bulk resistance effects. The bulk resistances are usually only a few ohms and result in only a small voltage drop when the diode conducts. Often this drop can be neglected.

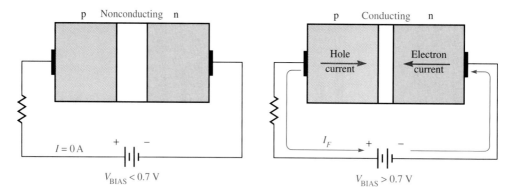

FIGURE 1–22
The bias voltage must overcome the barrier potential for the diode to conduct current.

ENERGY DIAGRAM FOR FORWARD BIAS Forward bias raises the energy levels of the conduction electrons in the n region, allowing them to move into the p region and combine with holes in the valence band. This condition is shown in Figure 1–23.

Introduction to Semiconductors

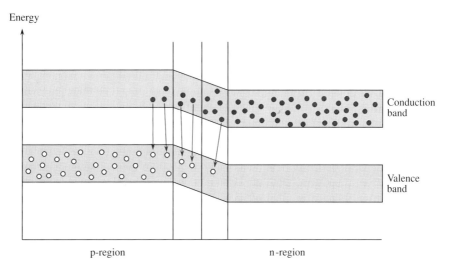

Figure 1–23
Energy diagram for forward bias, showing recombination in the depletion layer and in the p region as conduction electrons move across the junction.

Reverse Bias

Reverse bias is the condition that prevents current across the pn junction. Figure 1–24 shows a dc voltage source connected to reverse-bias the diode. Notice that the negative terminal of the battery is connected to the p region, and the positive terminal to the n region.

Figure 1–24
Reverse-bias connection.

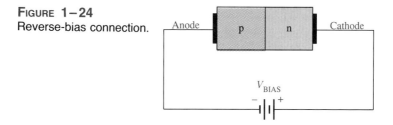

This is how reverse bias works: The negative terminal of the battery attracts holes in the p region away from the pn junction, while the positive terminal also attracts electrons away from the junction. As electrons and holes move away from the junction, the depletion layer begins to widen; more positive ions are created in the n region, and more negative ions are created in the p region, as shown in Figure 1–25(a). The depletion layer widens until the potential difference across it equals the external bias voltage. At this point, the holes and electrons stop moving away from the junction, and majority current ceases, as indicated in Figure 1–25(b). The initial flow of majority carriers away from the junction is called *transient current* and lasts only for a very short time upon application of reverse bias. When the diode is reverse-biased, the depletion layer effectively acts as an insulator between the layers of oppositely charged ions, forming an effective capacitance, as illustrated in Figure 1–25(c). Since the depletion layer widens with increased reverse-biased voltage, the capacitance decreases, and vice versa. This internal capacitance is called the *depletion-layer capacitance*.

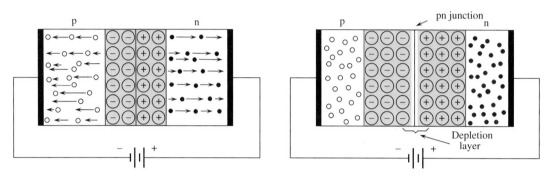

(a) Transient current at initial application of reverse bias

(b) Current ceases when barrier potential equals bias voltage

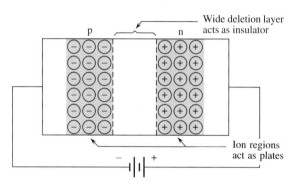

(c) Depletion layer widens as reverse bias increases

FIGURE 1–25
Reverse bias.

REVERSE CURRENT As you have learned, majority current very quickly becomes zero when reverse bias is applied. There is, however, a very small current produced by minority carriers during reverse bias. Germanium, as a rule, has a greater reverse current than silicon. This current is typically in the μA or nA range. A relatively small number of thermally produced electron-hole pairs exist in the depletion layer. Under the influence of the external voltage, some electrons manage to diffuse across the pn junction before recombination. This process establishes a small minority carrier current throughout the material.

The reverse current is dependent primarily on the junction temperature and not on the amount of reverse-biased voltage. A temperature increase causes an increase in reverse current.

Introduction to Semiconductors

Reverse Breakdown If the external reverse-bias voltage is increased to a large enough value, avalanche breakdown occurs. Here is what happens: Assume that one minority conduction-band electron acquires enough energy from the external source to accelerate it toward the positive end of the diode. During its travel, it collides with an atom and imparts enough energy to knock a valence electron into the conduction band. There are now two conduction-band electrons. Each will collide with an atom, knocking two more valence electrons into the conduction band. There are now four conduction-band electrons which, in turn, knock four more into the conduction band. This rapid multiplication of conduction-band electrons, known as an *avalanche effect,* results in a rapid buildup of reverse current.

Most diodes normally are not operated in reverse breakdown and can be damaged if they are. However, a particular type of diode, a zener diode, is specially designed for reverse-breakdown operation (see Chapter 2).

1-7 Review Questions

1. What are the two bias conditions?
2. Which bias condition produces majority carrier current?
3. Which bias condition produces a widening of the depletion region?
4. Which carriers produce the current during avalanche breakdown?

1-8 Diode Characteristics

In this section, you will learn that the characteristic curve graphically shows how the diode works and you will see the diode symbol that is used in circuit schematics. Three diode approximations are discussed. Each approximation represents the diode at a different level of accuracy so that you can use the one most appropriate for a given situation. In some cases, the lowest level of accuracy is all that is needed and additional details only complicate the situation. In other cases, you need the highest level of accuracy so that all factors can be taken into account.

Diode Characteristic Curve

As you learned in the last section, a **diode** conducts current when it is forward-biased if the bias voltage exceeds the barrier potential, and the diode prevents current when it is reverse-biased at less than the breakdown voltage.

Figure 1-26 is a graph of diode current versus voltage. The upper right quadrant of the graph represents the forward-biased condition. As you can see, there is essentially no forward current (I_F) for forward voltages (V_F) below the barrier potential. As the forward voltage approaches the value of the barrier potential (typically 0.7 V for silicon and 0.3 V for germanium), the current begins to increase. Once the forward voltage reaches the barrier potential, the current increases drastically and must be limited by a series resistor. The voltage across the forward-biased diode remains approximately equal to the barrier potential, but increases slightly with forward current.

FIGURE 1–26
Diode characteristic curve (silicon).

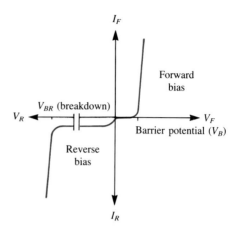

The lower left quadrant of the graph represents the reverse-biased condition. As the reverse voltage increases to the left, the current remains near zero until the breakdown voltage is reached. When breakdown occurs, there is a large reverse current which, if not limited, can destroy the diode. Typically, the breakdown voltage is greater than 50 V for most rectifier diodes. Most diodes should not be operated in reverse breakdown.

DIODE SYMBOL

Figure 1–27(a) shows the standard schematic symbol for a general-purpose diode. The arrow points in the direction opposite the electron flow. The two terminals of the diode are the anode and cathode. When the anode is positive with respect to the cathode, the diode is forward-biased and current is from cathode to anode, as shown in Figure 1–27(b). Remember that when the diode is forward-biased, the barrier potential, V_B, always appears between the anode and cathode, as indicated in the figure. When the anode is negative with respect to the cathode, the diode is reverse-biased, as shown in Figure 1–27(c). The bias battery is designated V_{BB} and is not the same as the barrier potential.

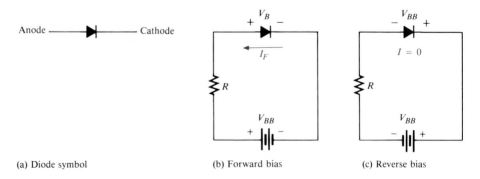

(a) Diode symbol   (b) Forward bias   (c) Reverse bias

FIGURE 1–27
Diode schematic symbol and bias circuits. V_{BB} is the bias battery voltage, and V_B is the barrier potential. The resistor limits the forward current to a safe value.

INTRODUCTION TO SEMICONDUCTORS

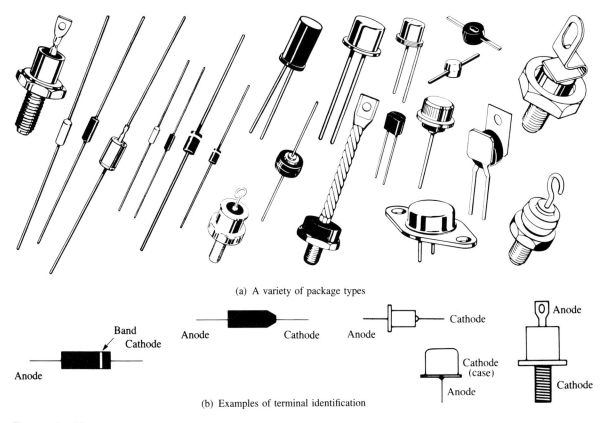

FIGURE 1–28
Typical diodes. Part (a) copyright of Motorola, Inc. Used by permission.

Some typical diodes are shown in Figure 1–28(a) to illustrate the variety of physical structures. Figure 1–28(b) illustrates terminal identification.

TESTING THE DIODE WITH A MULTIMETER

The internal battery in analog ohmmeters will forward-bias or reverse-bias a diode, permitting a quick and simple check for proper functioning. Many digital multimeters have a diode test position.

To check the diode in the forward direction, connect the positive meter lead to the anode and the negative lead to the cathode, as shown in Figure 1–29(a). When a diode is forward-biased, its internal resistance is low (typically less than 100 Ω).

When the meter leads are reversed, as shown in Figure 1–29(b), the internal ohmmeter battery reverse-biases the diode, and a very large resistance value (ideally, infinite) is indicated. The pn junction is shorted if a low resistance is indicated in both bias conditions; it is open if a very high resistance is read for both checks. In the diode test position on many digital multimeters, the forward diode voltage is displayed when the diode is good, as shown in Figure 1–29(c).

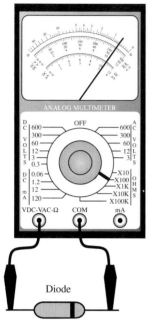
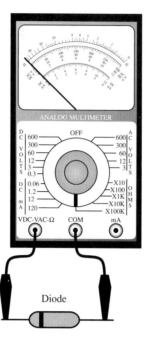
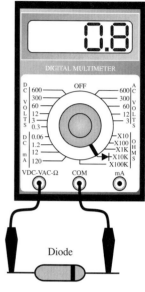

(a) Forward check with analog meter gives very low resistance.

(b) Reverse check with analog meter gives very high resistance.

(c) Diode check position on digital meter gives forward voltage (barrier potential plus drop across forward resistance).

FIGURE 1–29
Checking a semiconductor diode with a multimeter.

DIODE APPROXIMATIONS

THE IDEAL MODEL The simplest way to visualize diode operation is to think of it as a switch. When forward-biased, the diode ideally acts as a closed (on) switch, and when reverse-biased, it acts as an open (off) switch, as shown in Figure 1–30. The characteristic curve for this approximation is also shown. Note that the forward voltage and the reverse current are always zero in the ideal case.

This ideal model, of course, neglects the effect of the barrier potential, the internal resistances, and other parameters. However, in many cases it is accurate enough, particularly when the bias voltage is at least ten times greater than the barrier potential.

THE BARRIER POTENTIAL MODEL The next higher level of accuracy is the barrier potential model. In this approximation, the forward-biased diode is represented as a closed switch in series with a small "battery" equal to the barrier potential V_B (0.7 V for Si and 0.3 V for Ge), as shown in Figure 1–31(a). The positive end of the equivalent battery is toward the anode. Keep in mind that the barrier potential is not a voltage source and cannot be measured with a voltmeter; rather it only has the effect of a battery when forward bias is applied because the forward-bias voltage, V_{BB}, must overcome the barrier potential before the diode begins to conduct current. The reverse-biased diode is represented by an open switch, as in the ideal case, because the barrier potential does not affect reverse bias, as shown in Figure 1–31(b). The characteristic curve for this ideal model is shown in Figure

INTRODUCTION TO SEMICONDUCTORS

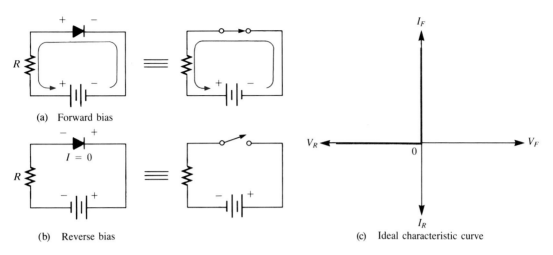

FIGURE 1–30
Ideal approximation of the diode as a switch.

FIGURE 1–31
Diode approximation including the barrier potential.

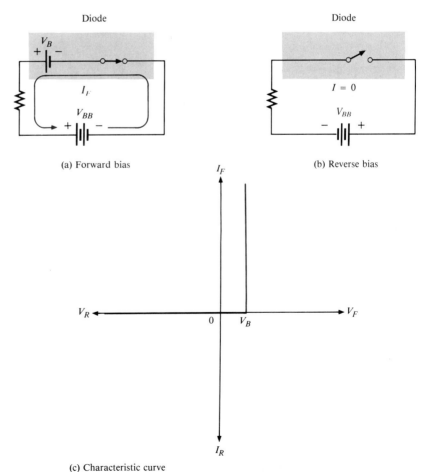

1–31(c). In this textbook, the barrier potential approximation is included in analysis unless otherwise stated.

THE COMPLETE MODEL One more level of accuracy will be considered at this point. Figure 1–32(a) shows the forward-biased diode model with both the barrier potential and the low forward (bulk) resistance. Figure 1–32(b) shows how the high reverse resistance affects the reverse-biased model. The characteristic curve is shown in Figure 1–32(c).

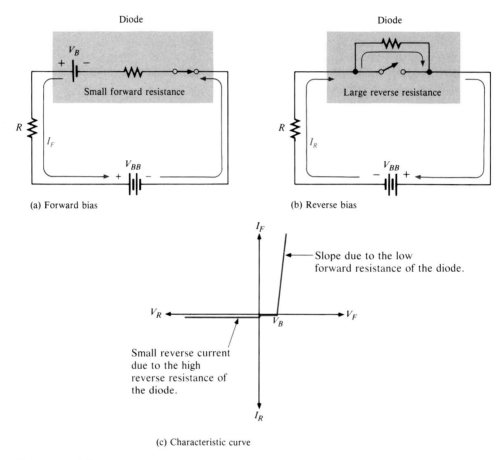

FIGURE 1–32
Diode approximation including barrier potential, forward resistance, and reverse resistance.

Other parameters such as junction capacitance and breakdown voltage become important only under certain operating conditions and will be considered only where appropriate.

Introduction to Semiconductors

1-8 Review Questions

1. What are the two conditions under which the diode is operated?
2. Under what condition is the diode never intentionally operated?
3. What is the simplest way to visualize a diode?
4. To accurately represent a diode, what factors must be included?

Summary

- An atom is described as consisting of a nucleus containing protons and neutrons orbited by electrons.
- Protons are positive, neutrons are neutral, and electrons are negative.
- Atomic shells are energy bands.
- The outermost shell containing electrons is the valence shell.
- Silicon and germanium are semiconductor materials. Silicon is the most predominant.
- Atoms within a crystal structure are held together with covalent bonds.
- Electron-hole pairs are thermally produced.
- The process of adding impurities to an intrinsic (pure) semiconductor to increase and control conductivity is called *doping*.
- A p-type semiconductor is doped with trivalent impurity atoms.
- An n-type semiconductor is doped with pentavalent impurity atoms.
- The depletion layer is a region adjacent to the pn junction containing no majority carriers.
- Forward bias permits majority carrier current through the pn junction.
- Reverse bias prevents majority carrier current.
- A pn structure is called a diode.
- Reverse current is due to thermally produced electron-hole pairs.
- Reverse breakdown occurs when the reverse-biased voltage exceeds a specified value.

Glossary

Anode The more positive terminal of a diode or other electronic device.

Atom The smallest particle of an element possessing the unique characteristics of that element.

Barrier potential The inherent voltage across the depletion layer of a pn junction.

Bias The application of dc voltage to a diode or other electronic device to produce a desired mode of operation.

Cathode The more negative terminal of a diode or other electronic device.

Covalent Related to the bonding of two or more atoms by the interaction of their valence electrons.

Crystal The pattern or arrangement of atoms forming a solid material.

Diode An electronic device that permits current in only one direction.

Discrete device An individual electrical or electronic component that must be used in combination with other components to form a complete functional circuit.

Doping The process of imparting impurities to an intrinsic semiconductor material in order to control its conduction characteristics.

Electron The basic particle of negative electrical charge in matter.

Energy The ability to do work.

Free electron A valence electron that has broken away from its parent atom and is free to move from atom to atom within the atomic structure of a material.

Germanium A semiconductor material.

Hole The absence of an electron.

Intrinsic semiconductor A pure material with relatively few free electrons.

Ion An atom that has gained or lost a valence electron resulting in a net positive or negative charge.

Ionization The removal or addition of an electron from or to an atom so that the resulting atom (ion) has a net positive or negative charge.

PN junction The boundary between n-type and p-type materials.

Recombination The process of a free electron falling into a hole in the valence band of an atom.

Semiconductor A material that has a conductance value between that of a conductor and that of an insulator. Silicon and germanium are examples.

Silicon A semiconductor material used in diodes and transistors.

Terminal An external contact point on an electronic device.

Valence Related to the outer shell or orbit of an atom.

SELF-TEST

1. The charge on electrons is
   (a) positive   (b) negative   (c) neutral   (d) variable
2. The number of valence electrons in both silicon and germanium is
   (a) two   (b) eight   (c) four   (d) eighteen
3. When an atom loses or gains a valence electron, the atom becomes
   (a) covalent   (b) neutral   (c) a crystal   (d) an ion

4. Atoms within a crystal are held together by
   - (a) atomic glue
   - (b) subatomic particles
   - (c) covalent bonds
   - (d) the valence band
5. Free electrons exist in the
   - (a) valence band
   - (b) conduction band
   - (c) lowest band
   - (d) recombination band
6. A hole is
   - (a) a vacancy in the valence band left by an electron
   - (b) a vacancy in the conduction band
   - (c) a positive electron
   - (d) a conduction-band electron
7. The widest energy gap between the valence band and the conduction band occurs in
   - (a) semiconductors
   - (b) insulators
   - (c) conductors
   - (d) a vacuum
8. The process of adding impurity atoms to a pure semiconductor material is called
   - (a) recombination
   - (b) crystalization
   - (c) bonding
   - (d) doping
9. The two types of current in a semiconductor material are called
   - (a) positive current and negative current
   - (b) electron current and conventional current
   - (c) electron current and hole current
   - (d) forward current and reverse current
10. The majority carriers in an n-type semiconductor are
    - (a) electrons
    - (b) holes
    - (c) positive ions
    - (d) negative ions
11. The pn junction is found in
    - (a) semiconductor diodes
    - (b) transistors
    - (c) all semiconductor materials
    - (d) a and b
12. In a semiconductor diode, the region near the pn junction consisting of positive and negative ions is called the
    - (a) neutral zone
    - (b) recombination region
    - (c) depletion layer
    - (d) diffusion area
13. A fixed dc voltage that sets the operating condition of a semiconductor device is called the
    - (a) bias
    - (b) depletion voltage
    - (c) battery
    - (d) barrier potential
14. In a semiconductor diode, the two bias conditions are
    - (a) positive and negative
    - (b) blocking and nonblocking
    - (c) open and closed
    - (d) forward and reverse
15. When a diode is forward-biased, it is
    - (a) blocking current
    - (b) conducting current

(c) similar to an open switch (d) similar to a closed switch
(e) a and c (f) b and d

16. The voltage across a forward-biased silicon diode is approximately
   (a) 0.7 V (b) 0.3 V
   (c) 0 V (d) dependent on the bias voltage

17. In Figure 1–33, identify the forward-biased diode(s).
   (a) D_1 (b) D_2 (c) D_3
   (d) D_1 and D_3 (e) D_2 and D_3

FIGURE 1–33

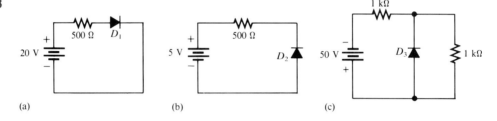

(a) (b) (c)

18. When the positive lead of an ohmmeter is connected to the cathode of a diode and the negative lead is connected to the anode, the meter reads
   (a) a very low resistance
   (b) an infinitely high resistance
   (c) a high resistance initially, decreasing to about 100 Ω
   (d) a gradually increasing resistance

ANSWERS TO REVIEW QUESTIONS

SECTION 1–1
1. The smallest particle that retains the characteristics of its element
2. The smallest particle of negative charge

SECTION 1–2
1. Electrons, protons, neutrons
2. An electron in the outermost shell (valance band)
3. An electron that has broken free from the valence band

SECTION 1–3
1. Silicon and germanium
2. The sharing of electrons with neighboring atoms

Introduction to Semiconductors

Section 1–4
1. Conduction band; valence band
2. An electron is thermally raised to the conduction band, leaving a hole in the valence band.
3. The energy gap between the valence band and the conduction band is narrower for a semiconductor.

Section 1–5
1. By the addition of pentavalent atoms to the intrinsic semiconductor material
2. By the addition of trivalent atoms to the intrinsic semiconductor material
3. The particle in greatest abundance: electrons in n-type material and holes in p-type material

Section 1–6
1. The boundary between n-type and p-type materials
2. Devoid of majority carriers, contains only positive and negative ions
3. No
4. 0.7 V

Section 1–7
1. Forward, reverse
2. Forward
3. Reverse
4. Minority

Section 1–8
1. Forward, reverse
2. Reverse breakdown
3. As an ideal switch
4. Barrier potential and bulk resistances

2
DIODES AND APPLICATIONS

2–1 HALF-WAVE RECTIFIERS
2–2 FULL-WAVE RECTIFIERS
2–3 RECTIFIER FILTERS
2–4 DIODE CLIPPING AND CLAMPING CIRCUITS
2–5 ZENER DIODES
2–6 VARACTOR DIODES
2–7 LEDs AND PHOTODIODES
2–8 THE DIODE DATA SHEET
2–9 TROUBLESHOOTING
2–10 A SYSTEM APPLICATION

After completing this chapter, you should be able to

☐ Recognize a half-wave rectifier circuit and explain how it works.
☐ Determine the average value of a half-wave rectified signal.
☐ Recognize a center-tapped full-wave rectifier circuit and explain how it works.
☐ Recognize a full-wave bridge rectifier and explain how it works.
☐ Determine the average value of a full-wave rectified signal.
☐ Determine the PIV (peak inverse voltage) across the diodes in full-wave rectifiers.
☐ Use a capacitor-input filter to smooth out a rectified voltage.
☐ Define *ripple voltage* and explain what causes it.
☐ Select a filter capacitor to minimize ripple voltage.
☐ Recognize other types of power supply filters.
☐ Use the zener diode as a voltage regulator.
☐ Explain line regulation and load regulation.
☐ Explain how a varactor diode is used as a variable capacitor.
☐ Discuss the basic principles of light-emitting diodes (LEDs) and photodiodes.

In this chapter, we discuss the applications of diodes in converting ac to dc by the process known as *rectification*. Half-wave and full-wave rectification are introduced, and you will study the basic circuits. We examine the limitations of diodes used in rectifier applications, and you will learn about diode clipping circuits and dc restoring (clamping) circuits.

In addition to rectifier diodes, we introduce zener diodes and their applications in voltage regulation. Varactor diodes, light-emitting diodes, and photodiodes and their applications also are discussed.

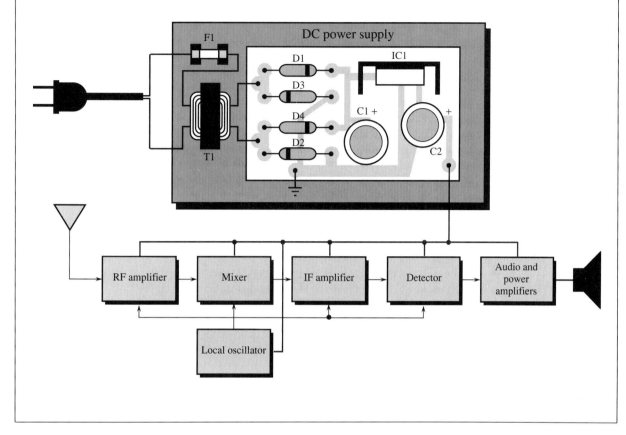

A System Application

The **power supply** is an important part of any electronic system because it supplies the dc voltage and current necessary for all the other circuits in the system to operate. You will learn how a typical power supply in an electronic system works. The diagram on this page shows a dc power supply as part of a radio receiver system. As you can see, the output of the power supply goes to all parts of the system and provides the necessary bias voltage and operating power for the diodes, transistors, and other devices in the amplifiers and other circuitry to function properly.

For the system application in Section 2–10, in addition to the other topics, be sure you understand

☐ How a diode works.
☐ The parameters and ratings of diodes.
☐ How to read a diode data sheet.
☐ What rectification of ac voltage means.
☐ How rectifiers work.
☐ What a filter is.
☐ Why filters are important in rectifiers.

2–1 HALF-WAVE RECTIFIERS

Because of their ability to conduct current in one direction and block current in the other direction, diodes are used in circuits called **rectifiers** *that convert ac voltage into dc voltage. Rectifiers are found in all dc power supplies that operate from an ac voltage source. Power supplies are an essential part of all electronic systems from the simplest to the most complex. In this section, you will study the most basic type of rectifier, the* **half-wave rectifier.**

Figure 2–1 illustrates the process called *half-wave rectification*. In part (a), an ac source is connected to a load resistor through an ideal diode. Let's examine what happens during one cycle of the input voltage. When the sine wave input voltage goes positive, the diode is forward-biased and conducts current to the load resistor, as shown in part (b). The current produces a voltage across the load, which has the same shape as the positive

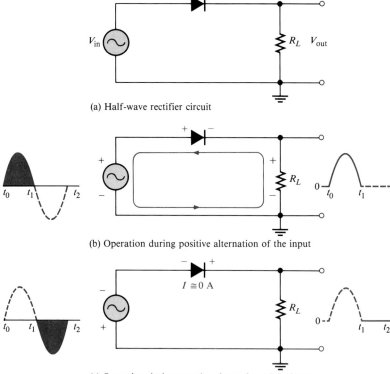

FIGURE 2–1
Operation of half-wave rectifier. The diode is considered ideal.

(a) Half-wave rectifier circuit

(b) Operation during positive alternation of the input

(c) Operation during negative alternation of the input

(d) Half-wave output voltage for three input cycles

half-cycle of the input voltage. When the input voltage goes negative during the second half of its cycle, the diode is reverse-biased. There is no current, so the voltage across the load resistor is zero, as shown in part (c). The net result is that only the positive half-cycles of the ac input voltage appear across the load, making the output a pulsating dc voltage, as shown in part (d).

AVERAGE VALUE OF THE HALF-WAVE OUTPUT VOLTAGE

The average (dc) value of the half-wave output voltage is the value that would be indicated by a dc voltmeter. It can be calculated with the following equation where V_p is the peak value. The derivation is given in Appendix B.

$$V_{avg} = \frac{V_p}{\pi} \tag{2-1}$$

Figure 2–2 shows the half-wave voltage with its average value indicated by the colored dashed line.

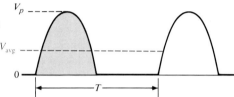

FIGURE 2–2
Average value of half-wave rectified signal.

EXAMPLE 2–1

What is the average (dc) value of the half-wave rectified voltage waveform in Figure 2–3?

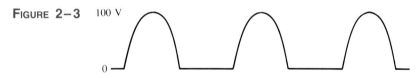

FIGURE 2–3   100 V

SOLUTION

$$V_{avg} = \frac{V_p}{\pi} = \frac{100 \text{ V}}{\pi} = 31.83 \text{ V}$$

PRACTICE EXERCISE 2–1
Determine the average value of the half-wave voltage if its peak amplitude is 12 V.

EFFECT OF THE BARRIER POTENTIAL ON THE HALF-WAVE RECTIFIER OUTPUT

In the previous discussion, the diode was considered ideal. When the diode barrier potential is taken into account, here is what happens: During the positive half-cycle, the input voltage must overcome the barrier potential before the diode becomes forward-biased. For a silicon diode, this results in a half-wave output with a peak value that is 0.7 V less than the peak value of the input (0.3 V less for a germanium diode), as shown in Figure 2–4. The expression for peak output voltage is

$$V_{p(out)} = V_{p(in)} - 0.7 \text{ V} \qquad (2\text{–}2)$$

In working with diode circuits, you will find that it is sometimes practical to neglect the effect of barrier potential when the peak value of the applied voltage is much greater than the barrier potential (at least ten times. Some people use 100 times). As mentioned before, we will always use silicon diodes and take the barrier potential into account unless stated otherwise.

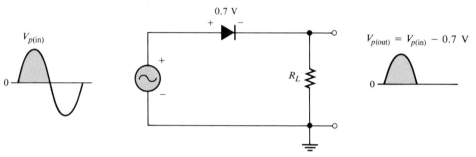

FIGURE 2–4
Effect of barrier potential on half-wave rectified output voltage (silicon diode shown).

EXAMPLE 2–2

Determine the peak output voltage of the rectifier circuit in Figure 2–5 for the indicated input voltage. The diode is silicon, and the barrier potential should be included.

FIGURE 2–5

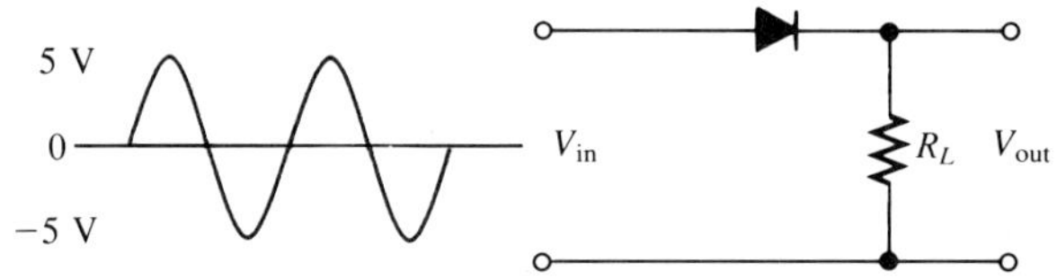

SOLUTION
The peak half-wave output voltage is

$$V_p = 5 \text{ V} - 0.7 \text{ V} = 4.3 \text{ V}$$

PRACTICE EXERCISE 2–2
Determine the peak output voltage for the rectifier in Figure 2–5 if the peak input is 3 V.

PEAK INVERSE VOLTAGE (PIV)

The maximum value of reverse voltage, sometimes designated as *peak inverse voltage* (PIV), occurs at the peak of the negative alternation of the input cycle when the diode is reverse-biased. This condition is illustrated in Figure 2–6. The PIV equals the peak value of the input voltage, and the diode must be capable of withstanding this amount of repetitive reverse voltage.

DIODES AND APPLICATIONS

FIGURE 2–6
The PIV occurs at the peak of the half-cycle when the diode is reverse-biased. In this circuit, the PIV occurs at the peak of the negative half-cycle.

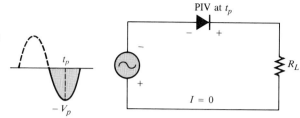

2–1 REVIEW QUESTIONS

1. At what point on the input cycle does the PIV occur in a half-wave rectifier?
2. For a half-wave rectifier, there is current through the load for approximately what percentage of the input cycle?
3. What is the average value of the voltage shown in Figure 2–7?

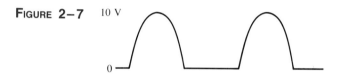

FIGURE 2–7

2–2 FULL-WAVE RECTIFIERS

Although half-wave rectifiers have some applications, the full-wave rectifier is the most commonly used type in dc power supplies. In this section, you will use what you learned about half-wave rectification and expand it to full-wave rectifiers. You will learn about two types of full-wave rectifiers, center-tapped and bridge.

The difference between full-wave and half-wave rectification is that a **full-wave rectifier** allows unidirectional current to the load during the entire input cycle, and the half-wave rectifier allows this only during one half of the cycle. The result of full-wave rectification is a dc output voltage that pulsates every half-cycle of the input, as shown in Figure 2–8.

FIGURE 2–8
Full-wave rectification.

Since the number of positive alternations that make up a full-wave rectified voltage is twice that of the half-wave voltage, the average value for a full-wave rectified sine-wave voltage is twice that of the half-wave, expressed as follows:

$$V_{avg} = \frac{2V_p}{\pi} \qquad (2\text{–}3)$$

EXAMPLE 2–3

Find the average value of the full-wave rectified voltage in Figure 2–9.

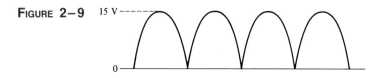

FIGURE 2–9

SOLUTION

$$V_{avg} = \frac{2V_p}{\pi} = \frac{2(15\ V)}{\pi} = 9.55\ V$$

PRACTICE EXERCISE 2–3

Find the average value of the full-wave rectified voltage if its peak is 155 V. On what function setting would a multimeter measure this value?

THE CENTER-TAPPED FULL-WAVE RECTIFIER

The center-tapped (CT) full-wave rectifier circuit uses two diodes connected to the secondary of a center-tapped transformer, as shown in Figure 2–10. The input signal is coupled through the transformer to the secondary. Half of the total secondary voltage appears between the **center tap** and each end of the secondary winding as shown.

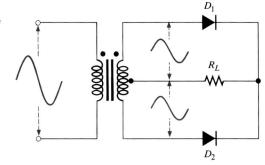

FIGURE 2–10
A center-tapped (CT) full-wave rectifier.

For a positive half-cycle of the input voltage, the polarities of the secondary voltages are as shown in Figure 2–11(a). This condition forward-biases the upper diode D_1 and reverse-biases the lower diode D_2. The current path is through D_1 and the load resistor, as indicated.

For a negative half-cycle of the input voltage, the voltage polarities on the secondary are as shown in Figure 2–11(b). This condition reverse-biases D_1 and forward-biases D_2. The current path is through D_2 and the load resistor, as indicated.

Because the current during both the positive and the negative portions of the input cycle is in the same direction through the load, the output voltage developed across the load is a full-wave rectified dc voltage.

DIODES AND APPLICATIONS

FIGURE 2–11
Basic operation of a center-tapped full-wave rectifier. Note that the current through the load resistor is in the same direction during the entire input cycle.

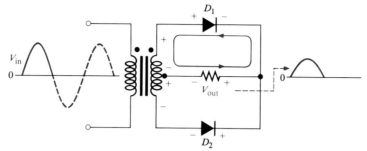

(a) During positive half-cycles, D_1 is forward-biased and D_2 is reverse-biased.

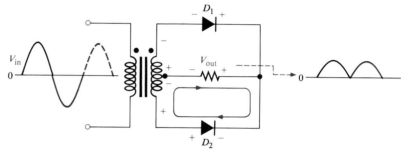

(b) During negative half-cycles, D_2 is forward-biased and D_1 is reverse-biased.

EFFECT OF THE TURNS RATIO ON THE FULL-WAVE OUTPUT VOLTAGE If the turns ratio of a transformer is 1, the peak value of the rectified output voltage equals half the peak value of the primary input voltage less the barrier potential (we will sometimes refer to the barrier potential as the diode drop). This value occurs because half of the input voltage appears across each half of the secondary winding.

In order to obtain an output voltage equal to the input (less the barrier potential), you must use a step-up transformer with a turns ratio of 2 (1:2). In this case, the total secondary voltage is twice the primary voltage, so the voltage across each half of the secondary is equal to the input.

PEAK INVERSE VOLTAGE (PIV) Each diode in the full-wave rectifier is alternately forward-biased and then reverse-biased. The maximum reverse voltage that each diode must withstand is the peak value of the total secondary voltage (V_s), as illustrated in Figure 2–12. When the total secondary voltage V_s has the polarity shown, the anode of D_1 is $+V_s/2$ and the anode of D_2 is $-V_s/2$. Since D_1 is forward-biased, its cathode is at the same voltage as its anode ($+V_s/2$, neglecting the barrier potential); this is also the voltage on the cathode of D_2. The total reverse voltage across D_2, therefore, is

$$\frac{+V_s}{2} - \left(\frac{-V_s}{2}\right) = \frac{V_s}{2} + \frac{V_s}{2} = V_s$$

Since $V_s = 2V_{out}$, the peak inverse voltage across either diode in the center-tapped full-wave rectifier is

$$PIV = 2V_{p(out)} \tag{2–4}$$

FIGURE 2–12
PIV across a diode is twice the peak value of the output voltage.

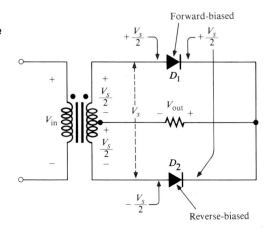

EXAMPLE 2–4

(a) Show the voltage waveforms across the secondary winding and across R_L when a 25 V peak sine wave is applied to the primary winding in Figure 2–13. This circuit is the same as the one in Figure 2–10, except the center tap and one end of R_L are connected to ground.
(b) What minimum PIV rating must the diodes have?

FIGURE 2–13

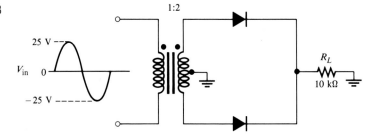

SOLUTION
(a) The waveforms are shown in Figure 2–14.
(b) The total peak secondary voltage is

$$V_{p(s)} = \left(\frac{N_2}{N_1}\right)V_{p(in)} = (2)25 \text{ V} = 50 \text{ V}$$

FIGURE 2–14

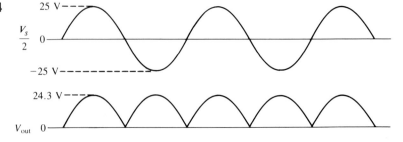

There is a 25 V peak across each half of the secondary. The output load voltage has a peak value of 25 V, less the 0.7 V drop across the diode. Each diode must have a minimum PIV rating of 50 V − 0.7 V = 49.3 V.

Practice Exercise 2-4
What diode PIV rating is required to handle a peak input of 160 V in Figure 2–13?

The Full-Wave Bridge Rectifier

The full-wave bridge rectifier uses four diodes, as shown in Figure 2–15. When the input cycle is positive as in part (a), diodes D_1 and D_2 are forward-biased and conduct current in the direction shown. A voltage is developed across R_L which looks like the positive half of the input cycle. During this time, diodes D_3 and D_4 are reverse-biased.

When the input cycle is negative, as in Figure 2–15(b), diodes D_3 and D_4 are forward-biased and conduct current in the same direction through R_L as during the positive half-cycle. During the negative half-cycle, D_1 and D_2 are reverse-biased. A full-wave rectified output voltage appears across R_L as a result of this action.

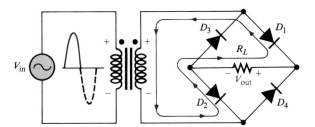

(a) During positive half-cycle of the input, D_1 and D_2 are forward-biased and conduct current. D_3 and D_4 are reverse-biased.

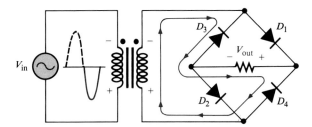

(b) During negative half-cycle, D_3 and D_4 are forward-biased and conduct current. D_1 and D_2 are reverse-biased.

Figure 2–15
Operation of full-wave bridge rectifier.

Bridge Output Voltage During the positive half-cycle of the secondary voltage, diodes D_1 and D_2 are forward-biased. Neglecting the diode drops, the total secondary voltage, V_s, appears across the load resistor. The same is true when D_3 and D_4 are forward-biased during the negative half-cycle. Thus,

$$V_{out} = V_s \qquad (2-5)$$

As you can see in Figure 2–15, two diodes are always in series with the load resistor during both the positive and the negative half-cycles. If these diode drops are taken into account, the output voltage (with silicon diodes) is

$$V_{out} = V_s - 1.4 \text{ V} \qquad (2-6)$$

Peak Inverse Voltage (PIV) Assuming that D_1 and D_2 are forward-biased, let us examine the reverse voltage across D_3 and D_4. Visualizing D_1 and D_2 as shorts (ideally), as in Figure 2–16, you can see that D_3 and D_4 have a peak inverse voltage equal to the peak secondary voltage. Since the output voltage (neglecting diode drops) is equal to the secondary voltage, we have

$$\text{PIV} \cong V_{p(out)} \qquad (2\text{–}7)$$

The PIV rating of the bridge diodes is half that required for the center-tapped configuration.

FIGURE 2–16
PIV in a bridge rectifier during the positive half-cycle of the input voltage.

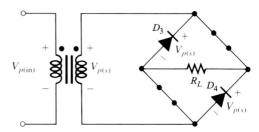

EXAMPLE 2–5

(a) Determine the peak output voltage for the bridge rectifier in Figure 2–17.
(b) What minimum PIV rating is required for the silicon diodes?

FIGURE 2–17

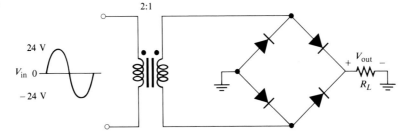

SOLUTION
(a) The peak output voltage is (taking into account the two diode drops)

$$V_{p(out)} = V_s - 1.4\ \text{V} = \left(\frac{N_2}{N_1}\right)V_{p(in)} - 1.4\ \text{V} = \left(\frac{1}{2}\right)24\ \text{V} - 1.4\ \text{V}$$
$$= 12\ \text{V} - 1.4\ \text{V} = 10.6\ \text{V}$$

(b) The PIV for each diode is

$$\text{PIV} \cong V_{p(out)} = 10.6\ \text{V}$$

PRACTICE EXERCISE 2–5
Determine the peak output voltage for the bridge rectifier in Figure 2–17 if the peak primary voltage is 160 V. What is the PIV rating for the diodes?

2-2 REVIEW QUESTIONS

1. What is the average value of a full-wave rectified voltage with a peak value of 60 V?
2. Which type of full-wave rectifier has the greater output voltage for the same input voltage and transformer turns ratio?
3. For a given output voltage, is the PIV for bridge rectifier diodes less than or greater than the PIV for center-tapped rectifier diodes?

2-3 RECTIFIER FILTERS

A power supply filter greatly reduces the fluctuations in the output voltage of a half-wave or full-wave rectifier and produces a nearly constant-level dc voltage. The reason for filtering is that electronic circuits require a constant source of dc voltage and current to provide power and biasing for proper operation. Filtering is done using capacitors, inductors, or combinations of both, as you will see in this section.

In most power supply applications, the standard 60 Hz ac power line voltage must be converted to a sufficiently constant dc voltage. The 60 Hz pulsating dc output of a half-wave rectifier or the 120 Hz pulsating output of a full-wave rectifier must be filtered to reduce the large voltage variations. Figure 2-18 illustrates the filtering concept showing a smooth dc output voltage. The full-wave rectifier voltage is applied to the input of the **filter,** and, ideally, a constant dc level appears on the output.

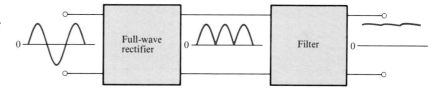

FIGURE 2-18
Basic block diagram of a power supply that converts 60 Hz ac to dc.

CAPACITOR-INPUT FILTER

A half-wave rectifier with a capacitor-input filter is shown in Figure 2-19. We will use the half-wave rectifier to illustrate the filtering principle; then, we will expand the concept to the full-wave rectifier.

During the positive first quarter-cycle of the input, the diode is forward-biased allowing the capacitor to charge to within a diode drop of the input peak, as illustrated in Figure 2-19(a). When the input begins to decrease below its peak, as shown in Figure 2-19(b), the capacitor retains its charge and the diode becomes reverse-biased. During the remaining part of the cycle, the capacitor can discharge only through the load resistance at a rate determined by the $R_L C$ time constant. The larger the time constant, the less the capacitor will discharge.

During the first quarter of the next cycle, the diode again will become forward-biased when the input voltage exceeds the capacitor voltage by approximately a diode drop, as illustrated in Figure 2-19(c).

FIGURE 2–19

Operation of a half-wave rectifier with a capacitor-input filter.

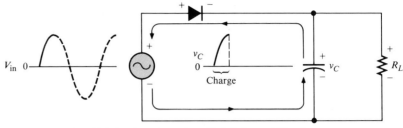

(a) Initial charging of capacitor (diode is forward-biased).

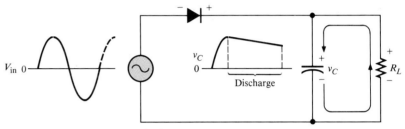

(b) Capacitor discharges through R_L after peak of the positive alternation (diode is reverse-biased).

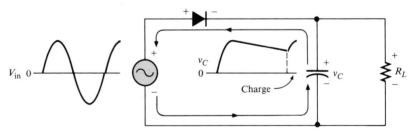

(c) Capacitor charges back to the peak (diode is forward-biased).

RIPPLE VOLTAGE As you have seen, the capacitor quickly charges at the beginning of a cycle and slowly discharges after the positive peak (when the diode is reverse-biased). The variation in the output voltage due to the charging and discharging is called the **ripple voltage**. The smaller the ripple voltage, the better the filtering action, as illustrated in Figure 2–20.

For a given input frequency, the output frequency of a full-wave rectifier is twice that of a half-wave rectifier. As a result, a full-wave rectifier is easier to filter. When filtered, the full-wave rectified voltage has less ripple voltage than does a half-wave voltage for the same load resistance and capacitor values. Less ripple voltage occurs because the capacitor discharges less during the shorter interval between full-wave pulses, as shown in Figure 2–21. A good rule of thumb for effective filtering is to make $R_L C \geq 10T$, where T is the period of the rectified voltage.

RIPPLE FACTOR The *ripple factor*, r, is an indication of the effectiveness of the filter and is defined as the ratio of the ripple voltage, V_r, to the dc (average) value of the filter output voltage, V_{dc}.

FIGURE 2–20
Half-wave ripple voltage (colored waveform).

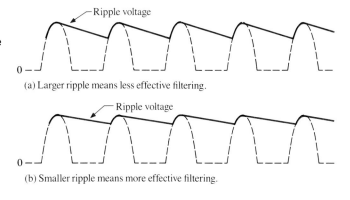

(a) Larger ripple means less effective filtering.

(b) Smaller ripple means more effective filtering.

FIGURE 2–21
Comparison of ripple voltages for half-wave and full-wave signals with the same filter capacitor and derived from the same sine-wave input.

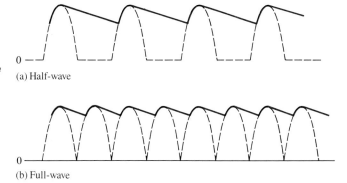

(a) Half-wave

(b) Full-wave

These parameters are illustrated in Figure 2–22. The ripple factor expressed as a percentage is

$$r = \left(\frac{V_r}{V_{dc}}\right)100\% \qquad (2\text{–}8)$$

The lower the ripple factor, the better the filter. The ripple factor can be decreased by increasing the value of the filter capacitor.

FIGURE 2–22
V_r and V_{dc} determine the ripple factor.

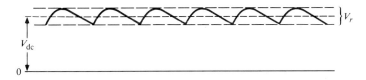

SURGE CURRENT IN THE CAPACITOR-INPUT FILTER Before the switch in Figure 2–23(a) is closed, the filter capacitor is uncharged. At the instant the switch is closed, voltage is connected to the bridge and the capacitor appears as a short, as shown. An initial surge of current is produced through the two forward-biased diodes. The worst-case situation occurs when the switch is closed at a peak of the secondary voltage and a maximum surge current ($I_{S(max)}$) is produced, as illustrated in part (a).

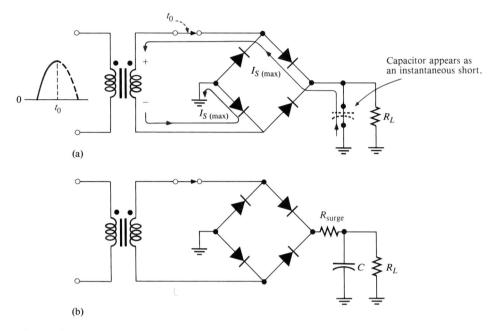

FIGURE 2–23
Surge current in a capacitor-input filter.

It is possible that the surge current could destroy the diodes, and for this reason a surge-limiting resistor is sometimes connected, as shown in Figure 2–23(b). The value of this resistor must be small compared to R_L. Also, the diodes must have a forward current rating such that they can withstand the momentary surge of current.

THE LC FILTER

When a choke is added to the filter input, as in Figure 2–24, an additional reduction in the ripple voltage is achieved. The choke has a high reactance at the ripple frequency, and the capacitive reactance is low compared to both X_L and R_L. The two reactances form an ac voltage divider that tends to significantly reduce the ripple from that of a straight capacitor-input filter.

FIGURE 2–24
Rectifier with an LC filter.

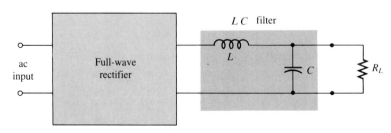

Diodes and Applications

It should be noted at this time that an *LC* filter produces an output with a dc value approximately equal to the average value of the rectified input. The capacitor-input filter, however, produces an output with a dc value approximately equal to the peak value of the input.

Another point of comparison is that the amount of ripple voltage in the capacitor-input filter varies inversely with the load resistance. Ripple voltage in the *LC* filter is essentially independent of the load resistance and depends only on X_L and X_C, as long as X_C is sufficiently less than R_L.

π-Type and T-Type Filters

A one-section π-type filter is shown in Figure 2–25(a). It can be thought of as a capacitor filter followed by an *LC* filter. The T-type filter in Figure 2–25(b) is basically an *LC* filter followed by an inductor filter. They combine the peak filtering action of the single capacitor or inductor filter with the reduced ripple and load independence of the *LC* filter.

FIGURE 2–25
π-type and T-type *LC* filters.

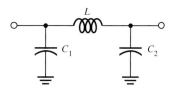

(a) π-type

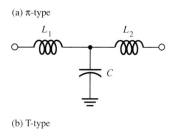

(b) T-type

2–3 Review Questions

1. A 60 Hz sine wave is applied to a half-wave rectifier. What is the output frequency? What is the output frequency for a full-wave rectifier?
2. What causes the ripple voltage on the output of a capacitor-input filter?
3. The load resistance of a capacitor-filtered full-wave rectifier is reduced. What effect does this reduction have on the ripple voltage?
4. What is one advantage of an *LC* filter over a capacitor filter? What is one disadvantage?

2–4 Diode Clipping and Clamping Circuits

Diode circuits are sometimes used to clip off portions of signal voltages above or below certain levels; these circuits are called **clippers** *or* **limiters**. *Another type of diode circuit is used to restore a dc level to an electrical signal; these are called* **clampers**. *Both diode circuits will be examined in this section.*

Diode Clippers

Figure 2–26(a) shows a diode circuit that clips off the positive part of the input signal. As the input signal goes positive, the diode becomes forward-biased. Since the cathode is at ground potential (0 V), the anode cannot exceed 0.7 V (assuming silicon). Thus, point A is clipped at +0.7 V when the input exceeds this value.

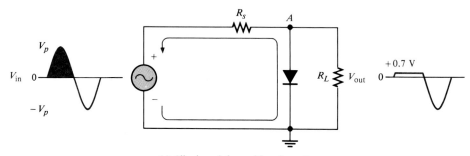

(a) Clipping of the positive alternation

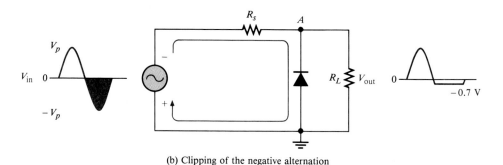

(b) Clipping of the negative alternation

FIGURE 2–26
Diode clipping (limiting) circuits.

When the input goes back below 0.7 V, the diode reverse-biases and appears as an open. The output voltage looks like the negative part of the input, but with a magnitude determined by the R_s and R_L voltage divider as follows:

$$V_{out} = \left(\frac{R_L}{R_s + R_L}\right)V_{in}$$

If R_s is small compared to R_L, then $V_{out} \cong V_{in}$.

Turn the diode around, as in Figure 2–26(b), and the negative part of the input is clipped off. When the diode is forward-biased during the negative part of the input, point A is held at −0.7 V by the diode drop. When the input goes above −0.7 V, the diode is no longer forward-biased and a voltage appears across R_L proportional to the input.

■ **EXAMPLE 2–6**   What would you expect to see displayed on an oscilloscope connected as shown in Figure 2–27? The time base on the scope is set to show one and one-half cycles.

FIGURE 2-27

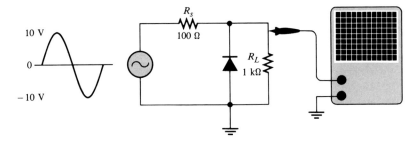

SOLUTION

The diode conducts when the input voltage goes below -0.7 V. Thus, we have a negative clipper with a peak output voltage determined by the following equation.

$$V_{p(out)} = \left(\frac{R_L}{R_s + R_L}\right)V_{p(in)} = \left(\frac{1\ k\Omega}{1.1\ k\Omega}\right)10\ V$$
$$= 9.09\ V$$

The scope will display an output waveform as shown in Figure 2–28.

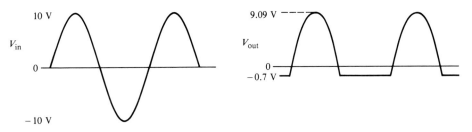

FIGURE 2-28
Waveforms for Figure 2–27.

PRACTICE EXERCISE 2-6

Describe the output waveform for Figure 2–27 if the diode is germanium and R_L is changed to 680 Ω.

ADJUSTMENT OF THE CLIPPING LEVEL

To adjust the level at which a signal voltage is clipped, add a bias voltage in series with the diode, as shown in Figure 2–29. The voltage at point A must equal $V_{BB} + 0.7$ V before the diode will conduct. Once the diode begins to conduct, the voltage at point A is limited to $V_{BB} + 0.7$ V so that all input voltage above this level is clipped off, as shown in the figure.

If the bias voltage is varied up or down, the clipping level changes correspondingly, as shown in Figure 2–30. If the polarity of the bias voltage is reversed, as in Figure 2–31,

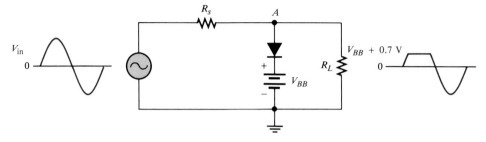

FIGURE 2–29
Positively biased clipper.

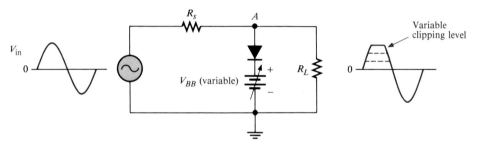

FIGURE 2–30
Positive clipper with variable bias.

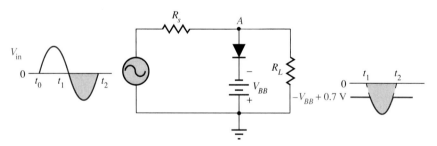

FIGURE 2–31

voltages above $-V_{BB} + 0.7$ V are clipped, resulting in an output waveform as shown. The diode is reverse-biased only when the voltage at point A goes below $-V_{BB} + 0.7$ V.

If it is necessary to clip off voltage below a specified negative level, then the diode and bias battery must be connected as in Figure 2–32. In this case, the voltage at point A must go below $-V_{BB} - 0.7$ V to forward-bias the diode and initiate clipping action, as shown.

Diodes and Applications

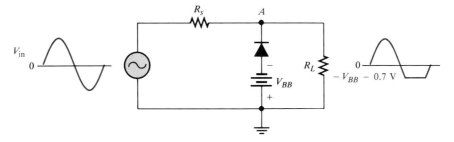

Figure 2–32

Example 2–7 Figure 2–33 shows a circuit combining a positive-biased clipper with a negative-biased clipper. Determine the output waveform. The diodes are silicon.

Figure 2–33

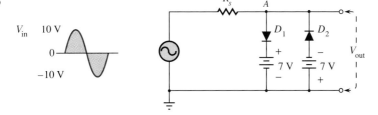

Solution
When the voltage at point A reaches +7.7 V, diode D_1 conducts and clips the waveform at +7.7 V. Diode D_2 does not conduct until the voltage reaches −7.7 V. Therefore, positive voltages above +7.7 V and negative voltages below −7.7 V are clipped off. The resulting output waveform is shown in Figure 2–34.

Figure 2–34
Output waveform for Figure 2–33.

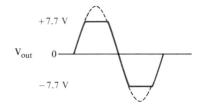

Practice Exercise 2–7
Determine the output waveform in Figure 2–33 if both dc sources are 10 V and the input has a peak value of 20 V.

Diode Clampers

The purpose of a clamper is to add a dc level to an ac signal. Clampers are sometimes known as *dc restorers*. Figure 2–35 shows a diode clamper that inserts a positive dc level. To understand the operation of this circuit, consider the first negative half-cycle of the input voltage. When the input initially goes negative, the diode is forward-biased, allowing the capacitor to charge to near the peak of the input ($V_{p(in)} - 0.7$ V), as shown in Figure 2–35(a). Just past the negative peak, the diode becomes reverse-biased because the cathode is held near $V_{p(in)}$ by the charge on the capacitor.

The capacitor can discharge only through the high resistance of R_L. Thus, from the peak of one negative half-cycle to the next, the capacitor discharges very little. The amount that is discharged, of course, depends on the value of R_L. For good clamping action, the RC time constant should be at least ten times the period of the input frequency.

The net effect of the clamping action is that the capacitor retains a charge approximately equal to the peak value of the input less the diode drop. The capacitor voltage acts essentially as a battery in series with the input signal, as shown in Figure 2–35(b). The dc voltage of the capacitor adds to the input voltage by superposition, as shown in Figure 2–35(c).

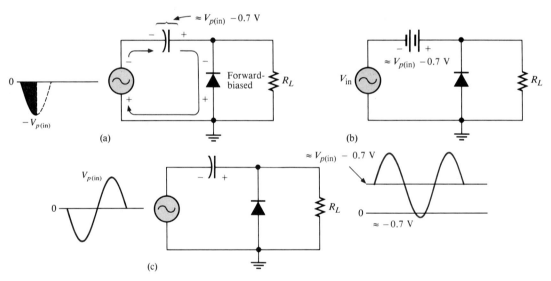

Figure 2–35
Positive clamping operation.

If the diode is turned around, a negative dc voltage is added to the input signal, as shown in Figure 2–36. If necessary, the diode can be biased to adjust the clamping level.

A Clamper Application A clamping circuit is often used in television receivers as a dc restorer. The incoming composite video signal is normally processed through capacitively coupled amplifiers that eliminate the dc component, thus losing the black and white reference levels and the blanking level. Before being applied to the picture tube, these reference levels must be restored. Figure 2–37 illustrates this process in a general way.

DIODES AND APPLICATIONS

FIGURE 2–36
Negative clamping.

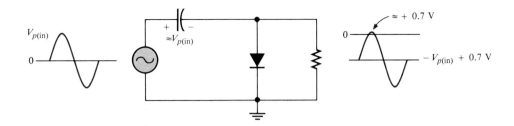

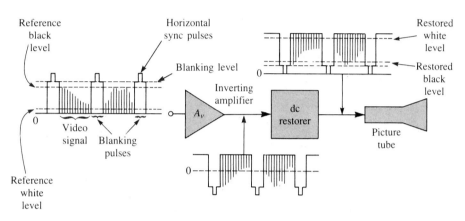

FIGURE 2–37
Clamping circuit (dc restorer) in a TV receiver.

EXAMPLE 2–8

What is the output voltage that you would expect to observe across R_L in the clamping circuit of Figure 2–38? Assume that RC is large enough to prevent significant capacitor discharge.

FIGURE 2–38

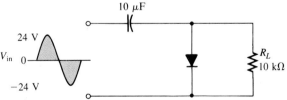

SOLUTION
Ideally, a negative dc value equal to the input peak less the diode drop is inserted by the clamping circuit.

$$V_{dc} \cong -(V_{p(in)} - 0.7 \text{ V}) = -(24 \text{ V} - 0.7 \text{ V}) = -23.3 \text{ V}$$

Actually, the capacitor will discharge slightly between peaks, and, as a result, the output voltage will have an average value of slightly less than that calculated above.

The output waveform goes to approximately 0.7 V above ground, as shown in Figure 2–39.

FIGURE 2–39
Output waveform for Figure 2–38.

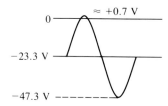

PRACTICE EXERCISE 2–8
What is the output voltage that you would observe across R_L in Figure 2–38 if $C = 22\ \mu F$ and $R_L = 18\ k\Omega$?

2–4 REVIEW QUESTIONS

1. Discuss how diode clippers and diode clampers differ in terms of their function.
2. What is the difference between a positive clipper and a negative clipper?
3. What is the maximum voltage across an unbiased positive silicon diode clipper during the positive alternation of the input voltage?
4. To limit the output of a positive clipper to 5 V when a 10 V peak input is applied, what value must the bias voltage be? Silicon diodes are used in the circuit.
5. What component in a clamper circuit effectively acts as a battery?

2–5 ZENER DIODES

A major application for zener diodes is voltage regulation in dc power supplies. In this section, you will see how the zener diode maintains a nearly constant dc voltage under the proper operating conditions. You will learn the conditions and limitations for properly using the zener diode and the factors that affect its performance.

FIGURE 2–40
Zener diode symbol.

Figure 2–40 shows the schematic symbol for a zener diode. The **zener diode** is a silicon pn junction device that differs from the rectifier diode in that it is designed for operation in the reverse breakdown region. The breakdown voltage of a zener diode is set by carefully controlling the doping level during manufacture. From the discussion of the diode characteristic curve in Chapter 1, recall that when a diode reaches reverse breakdown, its voltage remains almost constant even though the current may change drastically. This volt-ampere characteristic is shown again in Figure 2–41.

ZENER BREAKDOWN

There are two types of reverse breakdown in a zener diode, avalanche and zener. Avalanche breakdown occurs in rectifier diodes at a sufficiently high reverse voltage. Zener breakdown occurs in a zener diode at low reverse voltages. A zener diode is heavily doped to reduce the breakdown voltage, causing a very narrow depletion layer. As a

Diodes and Applications

Figure 2-41
General diode characteristic.

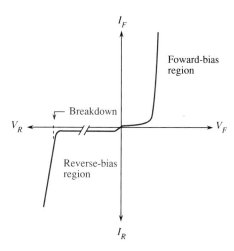

result, an intense electric field exists within the depletion layer. Near the breakdown voltage (V_Z), the field is intense enough to pull electrons from their valence bands and create current.

Zener diodes with breakdown voltages of less than approximately 5 V operate predominantly in zener breakdown. Those with breakdown voltages greater than approximately 5 V operate predominantly in avalanche breakdown. Both types, however, are called *zener diodes*. Zeners with breakdown voltages of 1.8 V to 200 V are commercially available.

BREAKDOWN CHARACTERISTICS Figure 2-42 shows the reverse portion of the characteristic curve of a zener diode. Notice that as the reverse voltage (V_R) is increased, the reverse current (I_R) remains extremely small up to the "knee" of the curve. At this point, the breakdown effect begins; the internal zener resistance (R_Z) begins to decrease as the reverse current increases rapidly. From the bottom of the knee, the zener breakdown voltage (V_Z) remains essentially constant although it increases slightly as I_Z increases.

FIGURE 2-42
Reverse characteristic of a zener diode. V_Z is usually specified at the zener test current, I_{ZT}, and is designated V_{ZT}.

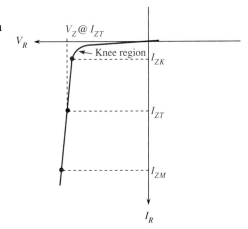

This regulating ability is the key feature of the zener diode. *A zener diode maintains a nearly constant voltage across its terminals over a specified range of reverse current values.*

A minimum value of reverse current, I_{ZK}, must be maintained in order to keep the diode in regulation. You can see on the curve that when the reverse current is reduced below the knee of the curve, the voltage changes drastically and regulation is lost. Also, there is a maximum current, I_{ZM}, above which the diode may be damaged.

Thus, basically, the zener diode maintains a nearly constant voltage across its terminals for values of reverse current ranging from I_{ZK} to I_{ZM}. A nominal zener voltage, V_{ZT}, is usually specified on a data sheet at a value of reverse current called the *zener test current, I_{ZT}.*

ZENER EQUIVALENT CIRCUIT

Figure 2–43(a) shows the ideal approximation of a zener diode in reverse breakdown. It acts simply as a battery having a value equal to the nominal zener voltage. Figure 2–43(b) represents the practical equivalent of a zener, where the zener resistance (R_Z) is included. Since the voltage curve is not ideally vertical, a change in zener current (ΔI_Z) produces a small change in zener voltage (ΔV_Z), as illustrated in Figure 2–43(c).

FIGURE 2–43
Zener diode equivalent circuits.

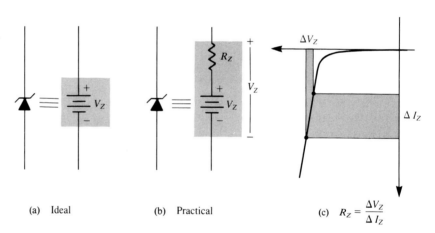

(a) Ideal   (b) Practical   (c) $R_Z = \dfrac{\Delta V_Z}{\Delta I_Z}$

The ratio of ΔV_Z to ΔI_Z is the resistance, as expressed in the following equation:

$$R_Z = \frac{\Delta V_Z}{\Delta I_Z} \qquad (2\text{–}9)$$

Normally, R_Z is specified at I_{ZT}, the zener test current. In most cases, we will assume that R_Z is constant over the full linear range of zener-current values.

EXAMPLE 2–9

A certain zener diode exhibits a 50 mV change in V_Z for a 2 mA change in I_Z on the linear portion of the characteristic curve between I_{ZK} and I_{ZM}. What is the zener resistance?

SOLUTION

$$R_Z = \frac{\Delta V_Z}{\Delta I_Z} = \frac{50 \text{ mV}}{2 \text{ mA}} = 25 \ \Omega$$

Practice Exercise 2–9
Calculate the zener resistance if the zener voltage changes 100 mV for a 20 mA change in zener current.

Example 2–10
A certain zener diode has a resistance of 5 Ω. The data sheet gives $V_{ZT} = 6.8$ V at $I_{ZT} = 20$ mA, $I_{ZK} = 1$ mA, and $I_{ZM} = 50$ mA. What is the voltage across its terminals when the current is 30 mA? What is the voltage when $I = 10$ mA?

Solution
Figure 2–44 represents the diode. The 30 mA current is a 10 mA increase above $I_{ZT} = 20$ mA.

$$\Delta I_Z = +10 \text{ mA}$$
$$\Delta V_Z = \Delta I_Z R_Z = (10 \text{ mA})(5 \text{ }\Omega) = +50 \text{ mV}$$

The change in voltage due to the increase in current above the I_{ZT} value causes the zener terminal voltage to increase. The zener voltage for $I_Z = 30$ mA is

$$V_Z = 6.8 \text{ V} + \Delta V_Z = 6.8 \text{ V} + 50 \text{ mV} = 6.85 \text{ V}$$

The 10 mA current is a 10 mA decrease below $I_{ZT} = 20$ mA.

$$\Delta I_Z = -10 \text{ mA}$$
$$\Delta V_Z = \Delta I_Z R_Z = (-10 \text{ mA})(5 \text{ }\Omega) = -50 \text{ mV}$$

The change in voltage due to the decrease in current below I_{ZT} causes the zener terminal voltage to decrease. The zener voltage for $I_Z = 10$ mA is

$$V_Z = 6.8 \text{ V} - \Delta V_Z = 6.8 \text{ V} - 50 \text{ mV} = 6.75 \text{ V}$$

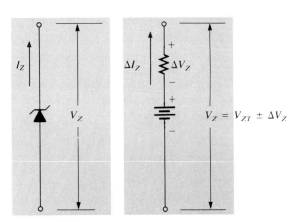

Figure 2–44

Practice Exercise 2–10
Repeat the analysis for a diode with $V_{ZT} = 12$ V at $I_{ZT} = 50$ mA, $I_{ZK} = 0.5$ mA, $I_{ZM} = 100$ mA, and $R_Z = 20$ Ω for a current of 20 mA and for a current of 80 mA.

ZENER VOLTAGE REGULATION

Zener diodes are widely used for **voltage regulation.** Figure 2–45 illustrates how a zener diode can be used to regulate a varying dc voltage to keep it at a constant level. This process is called *input* or **line regulation.**

As the input voltage varies (within limits), the zener diode maintains a nearly constant voltage across the output terminals. However, as V_{IN} changes, I_Z will change proportionally, and therefore the limitations on the input variation are set by the minimum and maximum current values (I_{ZK} and I_{ZM}) with which the zener can operate and on the condition that $V_{IN} > V_Z$. R is the series current-limiting resistor. The bar graph on the DMM symbols indicates the relative values and trends. Many modern DMMs provide analog bar graph displays in addition to the digital readout.

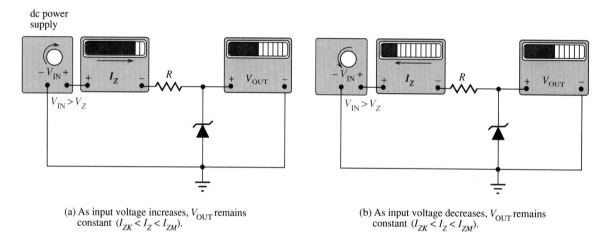

(a) As input voltage increases, V_{OUT} remains constant ($I_{ZK} < I_Z < I_{ZM}$).

(b) As input voltage decreases, V_{OUT} remains constant ($I_{ZK} < I_Z < I_{ZM}$).

FIGURE 2–45
Zener regulation of a varying input voltage.

For example, suppose that the zener diode in Figure 2–46 can maintain regulation over a range of current values from 4 mA to 40 mA. For the minimum current, the voltage across the 1 kΩ resistor is

$$V_R = (4 \text{ mA})(1 \text{ k}\Omega) = 4 \text{ V}$$

Since

$$V_R = V_{IN} - V_Z$$

then

$$V_{IN} = V_R + V_Z = 4 \text{ V} + 10 \text{ V} = 14 \text{ V}$$

For the maximum current, the voltage across the 1 kΩ resistor is

$$V_R = (40 \text{ mA})(1 \text{ k}\Omega) = 40 \text{ V}$$

Diodes and Applications

Therefore,

$$V_{IN} = 40 \text{ V} + 10 \text{ V} = 50 \text{ V}$$

As you can see, this zener diode can regulate an input voltage from 14 V to 50 V and maintain approximately a 10 V output. (The output will vary slightly because of the zener resistance.)

Figure 2-46

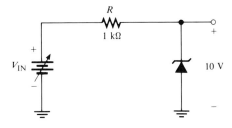

Example 2-11   Determine the minimum and the maximum input voltages that can be regulated by the zener diode in Figure 2-47. Assume that $I_{ZK} = 1$ mA, $I_{ZM} = 15$ mA, $V_{ZT} = 5.1$ V at $I_{ZT} = 8$ mA, and $R_Z = 10 \; \Omega$.

Figure 2-47

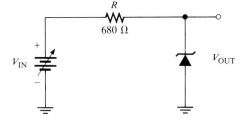

Solution
The equivalent circuit is shown in Figure 2-48.

Figure 2-48
Equivalent of the circuit in Figure 2-47.

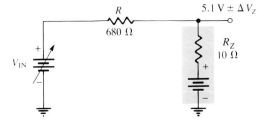

At $I_{ZK} = 1$ mA, the output voltage is

$$\begin{aligned} V_{OUT} &= 5.1 \text{ V} - \Delta V_Z = 5.1 \text{ V} - \Delta I_Z R_Z \\ &= 5.1 \text{ V} - (7 \text{ mA})(10 \; \Omega) \\ &= 5.1 \text{ V} - 0.070 \text{ V} \\ &= 5.03 \text{ V} \end{aligned}$$

Therefore,

$$V_{IN(min)} = I_{ZK}R + V_{OUT}$$
$$= (1 \text{ mA})(680 \text{ }\Omega) + 5.03 \text{ V}$$
$$= 5.71 \text{ V}$$

At $I_{ZM} = 15$ mA, the output voltage is

$$V_{OUT} = 5.1 \text{ V} + \Delta V_Z$$
$$= 5.1 \text{ V} + (7 \text{ mA})(10 \text{ }\Omega)$$
$$= 5.1 \text{ V} + 0.07 \text{ V}$$
$$= 5.17 \text{ V}$$

Therefore,

$$V_{IN(max)} = I_{ZM}R + V_{OUT}$$
$$= (15 \text{ mA})(680 \text{ }\Omega) + 5.17 \text{ V}$$
$$= 15.37 \text{ V}$$

PRACTICE EXERCISE 2–11
Determine the minimum and maximum input voltages that can be regulated by the zener diode in Figure 2–47. $I_{ZK} = 0.5$ mA, $I_{ZM} = 60$ mA, $V_{ZT} = 6.8$ V at 30 mA, and $R_Z = 18$ Ω. ■

REGULATION WITH A VARYING LOAD Figure 2–49 shows a zener regulator with a variable load resistor across the terminals. The zener diode maintains a nearly constant voltage across R_L as long as the zener current is greater than I_{ZK} and less than I_{ZM}. This process is called **load regulation**.

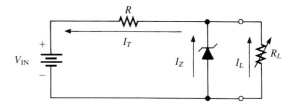

FIGURE 2–49
Zener regulation with a variable load.

NO LOAD TO FULL LOAD When the output terminals are open ($R_L = \infty$), the load current is zero and all of the current is through the zener. When a load resistor is connected, part of the total current is through the zener and part through R_L.

As R_L is decreased, the load current (I_L) increases and I_Z decreases. The zener diode continues to regulate until I_Z reaches its minimum value, I_{ZK}. At this point, the load current is maximum. The following example illustrates.

■ **EXAMPLE 2–12** Determine the minimum and the maximum load curents for which the zener diode in Figure 2–50 will maintain regulation. What is the minimum R_L that can be used? $V_Z = 12$ V, $I_{ZK} = 3$ mA, and $I_{ZM} = 90$ mA. Assume that $R_Z = 0$ Ω and V_Z remains a constant 12 V over the range of current values, for simplicity.

DIODES AND APPLICATIONS

FIGURE 2–50

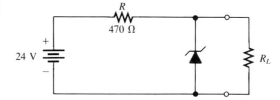

SOLUTION
When $I_L = 0$ A, I_Z is maximum and equal to the total circuit current, I_T.

$$I_{Z(max)} = I_T = \frac{V_{IN} - V_Z}{R} = \frac{24 \text{ V} - 12 \text{ V}}{470 \text{ }\Omega} = 25.53 \text{ mA}$$

Since this is much less than I_{ZM}, 0 A is an acceptable minimum value for I_L. That is,

$$I_{L(min)} = 0 \text{ A}$$

The maximum value of I_L occurs when I_Z is minimum ($I_Z = I_{ZK}$), so we can solve for $I_{L(max)}$ as follows:

$$I_{L(max)} = I_T - I_{ZK} = 25.53 \text{ mA} - 3 \text{ mA} = 22.53 \text{ mA}$$

The minimum value of R_L is

$$R_{L(min)} = \frac{V_Z}{I_{L(max)}} = \frac{12 \text{ V}}{22.53 \text{ mA}} = 533 \text{ }\Omega$$

PRACTICE EXERCISE 2–12
Find the minimum and maximum load currents for which the circuit in Figure 2–50 will maintain regulation. Determine the minimum R_L that can be used. $V_Z = 3.3$ V (constant), $I_{ZK} = 2$ mA, $I_{ZM} = 75$ mA, and $R_Z = 0$ Ω.

PERCENT REGULATION The percent regulation is a figure of merit used to specify the performance of a voltage regulator. It can be in terms of input (line) regulation or load regulation.

The *percent line regulation* specifies how much change occurs in the output voltage for a given change in input voltage. It is usually expressed as a percent change in V_{OUT} for a 1 V change in V_{IN} (%/V).

$$\text{Percent line regulation} = \frac{\Delta V_{OUT}}{\Delta V_{IN}} \times 100\% \quad (2\text{–}10)$$

The *percent load regulation* specifies how much change occurs in the output voltage over a certain range of load current values, usually from minimum current (no load, NL) to maximum current (full load, FL). It is normally expressed as a percentage and can be calculated with the following formula:

$$\text{Percent load regulation} = \frac{V_{NL} - V_{FL}}{V_{FL}} \times 100\% \quad (2\text{–}11)$$

where V_{NL} is the output voltage with no load, and V_{FL} is the output voltage with full (maximum) load.

EXAMPLE 2–13

A certain regulator has a no-load output voltage of 6 V and a full-load output of 5.82 V. What is the percent load regulation?

SOLUTION

$$\text{Percent load regulation} = \frac{V_{NL} - V_{FL}}{V_{FL}} \times 100\%$$

$$= \frac{6 \text{ V} - 5.82 \text{ V}}{5.82 \text{ V}} \times 100\%$$

$$= 3.09\%$$

PRACTICE EXERCISE 2–13
(a) If the no-load output voltage of a regulator is 24.8 V and the full-load output is 23.9 V, what is the percent load regulation?
(b) If the output voltage changes 0.035 V for a 2 V change in the input voltage, what is the line regulation expressed as %/V?

2–5 REVIEW QUESTIONS

1. How are zener diodes normally operated?
2. A certain 10 V zener diode has a resistance of 8 Ω at 30 mA. What is the terminal voltage?
3. What is the difference between input (line) regulation and load regulation?
4. In a zener diode regulator, for what value of load resistance is the zener current a maximum?
5. A zener regulator has an output voltage of 12 V with no load and 11.9 V with full load. What is the percent load regulation?

2–6 VARACTOR DIODES

Varactor diodes are also known as variable-capacitance diodes because the junction capacitance varies with the amount of reverse-bias voltage. Varactors are specifically designed to take advantage of this variable-capacitance characteristic. The capacitance can be changed by changing the reverse voltage. These devices are commonly used in electronic tuning circuits used in communications systems.

A **varactor** is basically a reverse-biased pn junction that utilizes the inherent capacitance of the depletion layer. The depletion layer, created by the reverse bias, acts as a capacitor dielectric because of its nonconductive characteristic. The p and n regions are conductive and act as the capacitor plates, as illustrated in Figure 2–51.

When the reverse-bias voltage increases, the depletion layer widens, effectively increasing the dielectric thickness and thus decreasing the capacitance. When the reverse-bias voltage decreases, the depletion layer narrows, thus increasing the capacitance. This action is shown in Figure 2–52(a) and (b). A general curve of capacitance versus voltage is shown in Figure 2–52(c).

DIODES AND APPLICATIONS

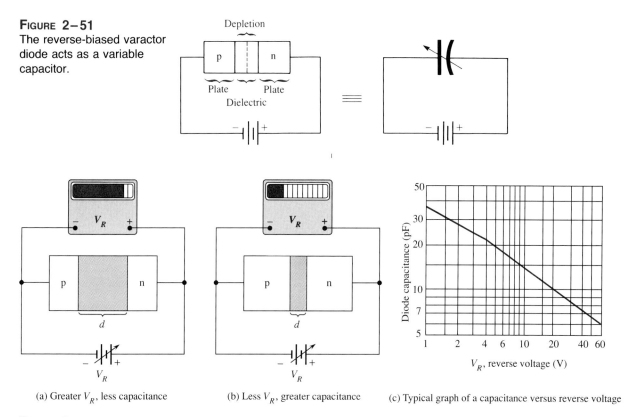

FIGURE 2–51
The reverse-biased varactor diode acts as a variable capacitor.

(a) Greater V_R, less capacitance (b) Less V_R, greater capacitance (c) Typical graph of a capacitance versus reverse voltage

FIGURE 2–52
Varactor diode capacitance varies with reverse voltage.

Recall that capacitance is determined by the plate area (A), dielectric constant (ϵ), and dielectric thickness (d), as expressed in the following formula:

$$C = \frac{A\epsilon}{d} \qquad (2-12)$$

In a varactor diode, the capacitance parameters are controlled by the method of doping in the depletion layer and the size and geometry of the diode's construction. Varactor capacitances typically range from a few picofarads to a few hundred picofarads.

Figure 2–53(a) shows a common symbol for a varactor, and Figure 2–53(b) shows a simplified equivalent circuit. R_S is the reverse series resistance, and C_V is the variable capacitance.

FIGURE 2–53
Varactor diode.

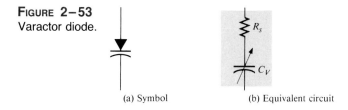

(a) Symbol (b) Equivalent circuit

APPLICATIONS

A major application of varactors is in tuning circuits. For example, electronic tuners in TV and other commercial receivers utilize varactors as one of their elements.

When used in a resonant circuit, the varactor acts as a variable capacitor, thus allowing the resonant frequency to be adjusted by a variable voltage level, as illustrated in Figure 2–54 where two varactor diodes provide the total variable capacitance in a parallel resonant (tank) circuit. V_C is a variable dc voltage that controls the reverse bias and therefore the capacitance of the diodes.

FIGURE 2–54
Varactors in a resonant circuit.

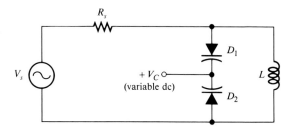

Recall that the resonant frequency of the tank circuit is

$$f_r \cong \frac{1}{2\pi\sqrt{LC}} \qquad (2\text{–}13)$$

This approximation is valid for $Q \geq 10$.

■ **EXAMPLE 2–14**

The capacitance of a certain varactor can be varied from 5 pF to 50 pF. The diode is used in a tuned circuit similar to that shown in Figure 2–54. Determine the tuning range for the circuit if $L = 10$ mH.

SOLUTION
The equivalent circuit is shown in Figure 2–55. Notice that the varactor capacitances are in series.

FIGURE 2–55

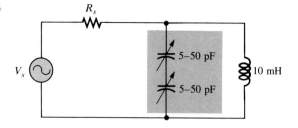

The minimum total capacitance is

$$C_{T(\text{min})} = \frac{C_{1(\text{min})}C_{2(\text{min})}}{C_{1(\text{min})} + C_{2(\text{min})}} = \frac{(5 \text{ pF})(5 \text{ pF})}{10 \text{ pF}} = 2.5 \text{ pF}$$

Diodes and Applications

The maximum resonant frequency, therefore, is

$$f_{r(max)} = \frac{1}{2\pi\sqrt{LC_{T(min)}}} = \frac{1}{2\pi\sqrt{(10 \text{ mH})(2.5 \text{ pF})}} \cong 1 \text{ MHz}$$

The maximum total capacitance is

$$C_{T(max)} = \frac{C_{1(max)}C_{2(max)}}{C_{1(max)} + C_{2(max)}} = \frac{(50 \text{ pF})(50 \text{ pF})}{100 \text{ pF}} = 25 \text{ pF}$$

The minimum resonant frequency, therefore, is

$$f_{r(min)} = \frac{1}{2\pi\sqrt{LC_{T(max)}}} = \frac{1}{2\pi\sqrt{(10 \text{ mH})(25 \text{ pF})}} \cong 318 \text{ kHz}$$

Practice Exercise 2–14
Determine the tuning range for Figure 2–55 if $L = 100$ mH.

2–6 Review Questions

1. What is the purpose of a varactor diode?
2. Based on the general curve in Figure 2–52(c), what happens to the diode capacitance when the reverse voltage is increased?

2–7 LEDs and Photodiodes

In this section, two types of optoelectronic devices—the light-emitting diode (LED) and the photodiode—are introduced. As the name implies, the LED is a light emitter. The photodiode, on the other hand, is a light detector. We will examine the characteristics of both devices, and you will see an example of their use in a system application at the end of the chapter.

The Light-Emitting Diode (LED)

The basic operation of the **light-emitting diode (LED)** is as follows: When the device is forward-biased, electrons cross the pn junction from the n-type material and recombine with holes in the p-type material. Recall that these free electrons are in the conduction band and at a higher energy level than the holes in the valence band. When recombination takes place, the recombining electrons release energy in the form of heat and light. A large exposed surface area on one layer of the semiconductor material permits the photons to be emitted as visible light. Figure 2–56 illustrates this process which is called *electroluminescence*.

The semiconductor materials used in LEDs are gallium arsenide (GaAs), gallium arsenide phosphide (GaAsP), and gallium phosphide (GaP). Silicon and germanium are not used because they are essentially heat-producing materials and are very poor at producing light. GaAs LEDs emit infrared (IR) radiation, which is nonvisible. GaAsP produces either red or yellow visible light, and GaP emits red or green visible light.

The symbol for an LED is shown in Figure 2–57.

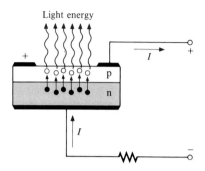

FIGURE 2–56
Electroluminescence in an LED.

FIGURE 2–57
Symbol for an LED.

The LED emits light in response to a sufficient forward current (I_F), as shown in Figure 2–58(a). The amount of power output translated into light is directly proportional to the forward current, as indicated in Figure 2–58(b). Typical LEDs are shown in Figure 2–58(c).

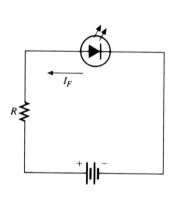

(a) Forward-biased operation

(b) Typical light output versus forward current

(c) Copyright of Motorola, Inc. Used by permission.

FIGURE 2–58
Operation of an LED.

APPLICATIONS LEDs are commonly used for indicator lamps and readout displays on a wide variety of instruments, ranging from consumer appliances to scientific apparatus. A common type of display device using LEDs is the seven-segment display. Combinations of the segments form the ten decimal digits. Also, IR-emitting diodes are employed in optical coupling applications, often in conjunction with fiber optics.

THE PHOTODIODE

The **photodiode** is a pn junction device that operates in reverse bias, as shown in Figure 2–59(a), where I_λ is the reverse current. Note the schematic symbol for the photodiode. The photodiode has a small transparent window that allows light to strike the pn junction. An alternate photodiode symbol is shown in Figure 2–59(b).

Recall that when reverse-biased, a rectifier diode has a very small reverse leakage current. The same is true for the photodiode. The reverse-biased current is produced by thermally generated electron hole pairs in the depletion layer, which are swept across the

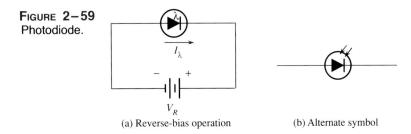

FIGURE 2–59
Photodiode.

(a) Reverse-bias operation (b) Alternate symbol

junction by the electric field created by the reverse voltage. In a rectifier diode, the reverse current increases with temperature due to an increase in the number of electron hole pairs.

In a photodiode, the reverse current increases with the light intensity at the pn junction. When there is no incident light, the reverse current (I_λ) is almost negligible and is called the *dark current*. An increase in the amount of light energy (measured in lumens per square meter, lm/m^2) produces an increase in the reverse current, as shown by the graph in Figure 2–60(a). For a given value of reverse-bias voltage, Figure 2–60(b) shows a set of characteristic curves for a typical photodiode.

FIGURE 2–60
Typical photodiode characteristics.

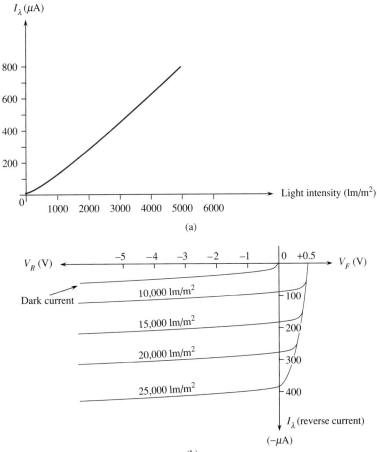

From the characteristic curve in Figure 2–60(b), the dark current for this particular device is approximately 25 μA at a reverse-bias voltage of 3 V. Therefore, the reverse resistance of the device with no incident light is

$$R_R = \frac{V_R}{I_\lambda} = \frac{3 \text{ V}}{25 \text{ μA}} = 120 \text{ k}\Omega$$

At 25,000 lm/m², the current is approximately 375 μA at −3 V. The resistance under this condition is

$$R_R = \frac{V_R}{I_\lambda} = \frac{3 \text{ V}}{375 \text{ μA}} = 8 \text{ k}\Omega$$

These calculations show that the photodiode can be used as a variable-resistance device controlled by light intensity.

Figure 2–61 illustrates that the photodiode allows essentially no reverse current (except for a very small dark current) when there is no incident light. When a light beam strikes the photodiode, it conducts an amount of reverse current that is proportional to the light intensity.

FIGURE 2–61
Operation of a photodiode.

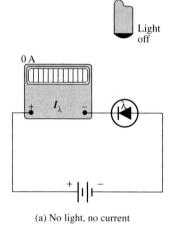
(a) No light, no current

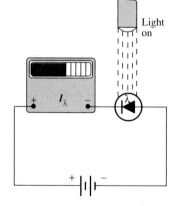
(b) When there is incident light, resistance decreases and reverse current flows.

2–7 REVIEW QUESTIONS

1. How does an LED differ from a photodiode?
2. List the semiconductor materials used in LEDs.
3. There is a very small reverse current in a photodiode under no-light conditions. What is this current called?

2–8 THE DIODE DATA SHEET

A manufacturer's data sheet gives detailed information on a device so that it can be used properly in a given application. A typical data sheet provides maximum ratings, electrical characteristics, mechanical data, and graphs of various parameters. In this section, we use a specific example to illustrate a typical data sheet.

Table 2–1 shows the maximum ratings for a certain series of rectifier diodes (1N4001 through 1N4007). These are the absolute maximum values under which the diode can be operated without damage to the device. For greatest reliability and longer life, the diode should always be operated well under these maximums. Generally, the maximum ratings are specified at 25°C and must be adjusted downward for greater temperatures.

TABLE 2–1

Rating	Symbol	1N4001	1N4002	1N4003	1N4004	1N4005	1N4006	1N4007	Unit
Peak repetitive reverse voltage Working peak reverse voltage DC blocking voltage	V_{RRM} V_{RWM} V_R	50	100	200	400	600	800	1000	V
Nonrepetitive peak reverse voltage	V_{RSM}	60	120	240	480	720	1000	1200	V
rms reverse voltage	$V_{R(rms)}$	35	70	140	280	420	560	700	V
Average rectified forward current (single-phase, resistive load, 60 Hz, $T_A = 75°C$)	I_0	1.0							A
Nonrepetitive peak surge current (surge applied at rated load conditions)	I_{FSM}	30 (for 1 cycle)							A
Operating and storage junction temperature range	T_j, T_{stg}	−65 to +175							°C

An explanation of some of the parameters from Table 2–1 follows.

V_{RRM} The maximum reverse peak voltage that can be applied repetitively across the diode. Notice that in this case, it is 50 V for the 1N4001 and 1 kV for the 1N4007. This is the same as PIV.

V_R The maximum reverse dc voltage that can be applied across the diode.

V_{RSM} The maximum reverse peak value of nonrepetitive (one-cycle) voltage that can be applied across the diode.

I_0 The maximum average value of a 60 Hz full-wave rectified forward current.

I_{FSM} The maximum peak value of nonrepetitive (one-cycle) forward current. The graph in Figure 2–62 expands on this parameter to show values for more than one cycle at temperatures of 25°C and 175°C. The dashed lines represent values where typical failures occur. Notice what happens on the lower solid line when ten cycles of I_{FSM} are applied. The limit is 15 A rather than the one-cycle value of 30 A.

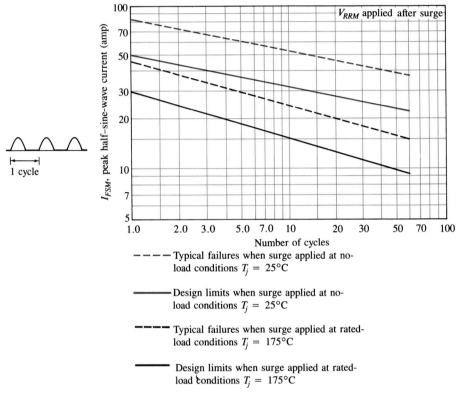

FIGURE 2-62
Nonrepetitive surge capability.

Table 2-2 lists typical and maximum values of certain electrical characteristics. These items differ from the maximum ratings in that they are not selected by design but are the result of operating the diode under specified conditions. A brief explanation of these parameters follows.

TABLE 2-2
Electrical characteristics

Characteristic and Conditions	Symbol	Typical	Maximum	Unit
Maximum instantaneous forward voltage drop ($I_F = 1$ A, $T_j = 25°C$)	v_F	0.93	1.1	V
Maximum full-cycle average forward voltage drop ($I_0 = 1$ A, $T_L = 75°C$, 1-in. leads)	$V_{F(avg)}$	—	0.8	V
Maximum reverse current (rated dc voltage) $T_j = 25°C$ $T_j = 100°C$	I_R	0.05 1.0	10.0 50.0	μA
Maximum full-cycle average reverse current ($I_0 = 1$ A, $T_L = 75°C$, 1-in. leads)	$I_{R(avg)}$	—	30.0	μA

DIODES AND APPLICATIONS

v_F The instantaneous voltage across the forward-biased diode when the forward current is 1 A at 25°C. Figure 2–63 shows how the forward voltages vary with forward current.

$V_{F(avg)}$ The maximum forward voltage drop averaged over a full cycle.

I_R The maximum current when the diode is reverse-biased with a dc voltage.

$I_{R(avg)}$ The maximum reverse current averaged over one cycle (when reverse-biased with an ac voltage).

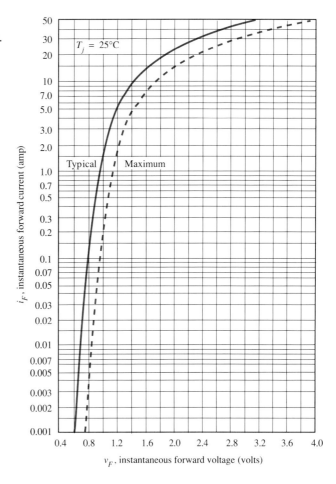

FIGURE 2–63 Forward voltage.

The mechanical data for these particular diodes as they appear on a typical data sheet are shown in Figure 2–64 on page 72.

2–8 REVIEW QUESTIONS

1. List the three diode ratings categories.
2. Identify each of the following parameters:
   (a) V_F (b) I_R (c) I_0

FIGURE 2-64
Mechanical data.

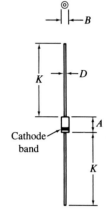

	Millimeters		Inches	
DIM	Min	Max	Min	Max
A	5.97	6.60	0.235	0.260
B	2.79	3.05	0.110	0.120
D	0.76	0.86	0.030	0.034
K	27.94	–	1.100	–

Mechanical characteristics

Case: Transfer Molded Plastic
Maximum lead temperature for soldering purposes: 350°C, ⅜″ from case for 10 seconds at 5 lbs. tension
Finish: All external surfaces are corrosion-resistant, leads are readily solderable
Polarity: Cathode indicated by color band
Weight: 0.40 Grams (approximately)

2-9 TROUBLESHOOTING

Several types of failures can occur in power supply rectifiers. In this section, we will examine some possible failures and the effects they would have on a circuit's operation.

OPEN DIODE

A half-wave rectifier with a diode that has opened (a common failure mode) is shown in Figure 2–65. In this case, you would measure 0 V dc across the load resistor, as depicted.

FIGURE 2-65
Test for an open diode in a half-wave rectifier with capacitor-input filter shows 0 V at the output.

Now consider the full-wave, center-tapped rectifier in Figure 2–66. Assume that diode D_1 has failed open. With an oscilloscope connected to the output, as shown in part (a), you would observe the following: You would see a larger-than-normal ripple voltage at a frequency of 60 Hz rather than 120 Hz. Disconnecting the filter capacitor, you would

FIGURE 2–66
Symptoms of an open diode in a full-wave, center-tapped rectifier.

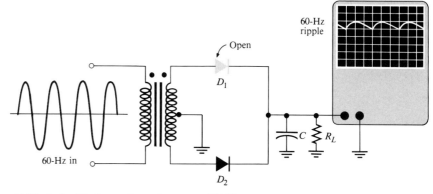

(a) Ripple should be less and have a frequency of 120 Hz. Instead it is greater in amplitude with a frequency of 60 Hz.

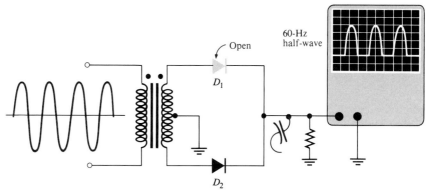

(b) With C removed, output should be a full-wave 120-Hz signal. Instead it is a 60-Hz, half-wave voltage.

observe a half-wave rectified voltage, as in part (b). Now let's examine the reason for these observations. If diode D_1 is open, there will be current through R_L only during the negative half-cycle of the input signal. During the positive half-cycle, an open path prevents current through R_L. The result is a half-wave voltage, as illustrated.

With the filter capacitor in the circuit, the half-wave signal will allow it to discharge more than it would with a normal full-wave signal, resulting in a larger ripple voltage. Basically, the same observations would be made for an open failure of diode D_2.

An open diode in a bridge rectifier would create symptoms identical to those just discussed for the center-tapped rectifier. As illustrated in Figure 2–67, the open diode would prevent current through R_L during half of the input cycle (in this case, the negative half). As a result, there would be a half-wave output and an increased ripple voltage at 60 Hz, as discussed before.

SHORTED DIODE

A shorted diode is one that has failed such that it has a very low resistance in both directions. If a diode suddenly became shorted in a bridge rectifier, it is likely that a sufficiently high current would exist during one half of the input cycle such that the

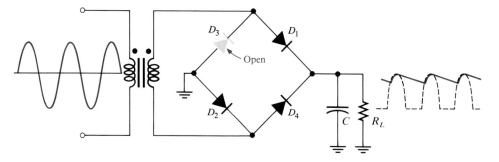

FIGURE 2-67
Effect of an open diode in a bridge rectifier is a half-wave operation that results in an increased ripple voltage at 60 Hz.

shorted diode itself would burn open or the other diode in series with it would open. The transformer could also be damaged, as illustrated in Figure 2-68 with D_1 shorted.

In part (a) of Figure 2-68, current is supplied to the load through the shorted diode during the first positive half-cycle, just as though it were forward-biased. During the

FIGURE 2-68
Effects of a shorted diode in a bridge rectifier.

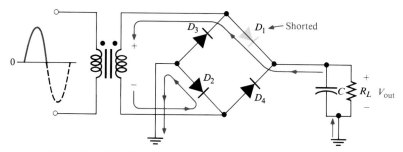

(a) Positive half-cycle: The shorted diode acts as a forward-biased diode, so the load current is normal.

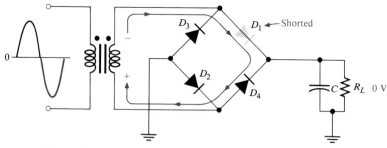

(b) Negative half-cycle: The shorted diode produces a short circuit across the source. As a result D_1, D_4, or the transformer secondary will probably burn open.

Diodes and Applications

negative half-cycle, the current is shorted through D_1 and D_4, as shown in part (b). Again, damage to the transformer is possible unless properly fused. It is likely that this excessive current would burn either or both of the diodes open. If only one of the diodes opened, you would still observe a half-wave voltage on the output. If both diodes opened (D_1 and D_4 in this case), there would be no voltage developed across the load. These conditions are illustrated in Figure 2–69.

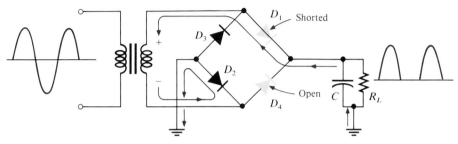

(a) One open diode produces a half-wave output.

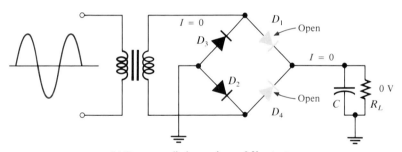

(b) Two open diodes produce a 0-V output.

FIGURE 2–69
Effects of open diodes.

SHORTED OR LEAKY FILTER CAPACITOR

A shorted capacitor would most likely cause some or all of the diodes in a full-wave rectifier to open due to excessive current or the fuse would blow. In any event, there would be no dc voltage on the output.

A leaky capacitor can be represented by a leakage resistance in parallel with the capacitor, as shown in Figure 2–70(a). The effect of the leakage resistance is to reduce the discharging time constant, causing an increase in ripple voltage on the output, as shown in Figure 2–70(b).

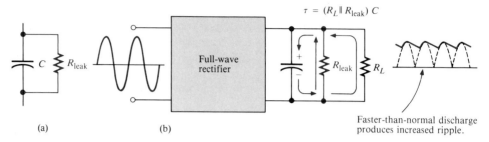

FIGURE 2–70
Effects of a leaky filter capacitor on the output of a full-wave rectifier.

2–9 REVIEW QUESTIONS

1. What effect would an open D_2 produce in the rectifier of Figure 2–66?
2. You are checking a 60 Hz full-wave bridge rectifier and observe that the output has a 60 Hz ripple. What failure(s) do you suspect?
3. You observe that the output ripple of a full-wave rectifier is much greater than normal but its frequency is still 120 Hz. What component do you suspect?

2–10 A SYSTEM APPLICATION

The purpose of the dc power supply in any electronic system is to provide to all parts of the system a constant dc voltage and sufficient dc current from which all the circuits operate. In other words, the dc power supply energizes the system. In this section, you will be dealing with the dc power supply in the radio receiver system shown at the opening of the chapter and you will

☐ *See how a diode rectifier is used in a system application.*
☐ *See how the filter operates in a system application.*
☐ *See how the 110 V ac voltage from a standard outlet is converted to the dc supply voltage.*
☐ *Translate between a printed circuit board and a schematic.*
☐ *Troubleshoot some common power supply failures.*

At the opening of this chapter, you saw the dc power supply printed circuit board with the components assembled on it. This part of the system represents a principal application of the diode and the full-wave bridge rectifier as well as the capacitor filter. An additional component in the power supply is the voltage **regulator.** This device is covered in detail in a later chapter, so we won't spend any time on it here except to state its purpose in the power supply.

DIODES AND APPLICATIONS

Now, so that you can take a closer look at the power supply, let's take it out of the system and put it on the test bench.

ON THE TEST BENCH

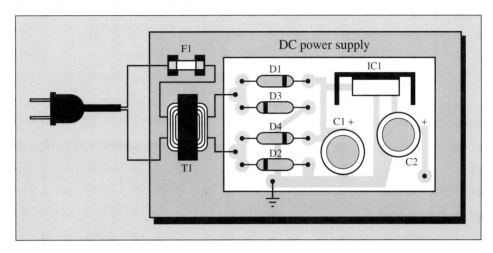

FIGURE 2–71

IDENTIFYING THE COMPONENTS

On the pc board shown in Figure 2–71, the diodes are labeled D1, D2, D3, and D4. The banded end of the diode is the cathode, as indicated in Figure 2–72(a). The filter capacitors are labeled C1 and C2. These are cylindrical electrolytic capacitors as shown in Figure 2–72(b). The 3-terminal voltage regulator is an **integrated circuit** device labeled IC1. The thick bracket-shaped object is a heat sink for conducting heat away from the device. A pictorial view is shown in Figure 2–72(c) to give you an idea of its shape. Again, we don't get into IC voltage regulators until later in the book. Also mounted in the power supply are the transformer (T1) and the fuse (F1).

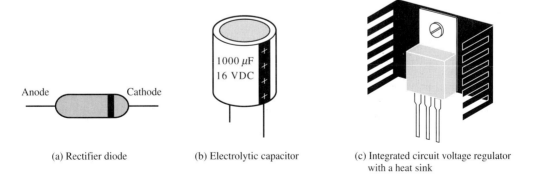

(a) Rectifier diode   (b) Electrolytic capacitor   (c) Integrated circuit voltage regulator with a heat sink

FIGURE 2–72

CHAPTER 2

■ **ACTIVITY 1** **RELATE THE PC BOARD TO THE SCHEMATIC**

Carefully follow the conductive traces on the pc board to see how the components are interconnected. Compare each point (A through J) on the pc board with the corresponding point on the schematic in Figure 2–73. This exercise will develop your skill in going from a pc board to a schematic or vice versa—*a very important skill for a technician*. Notice that some of the components on the pc board have been made to appear "transparent." This lets you see connections and traces under the components for the purposes of this exercise. Of course, on an actual pc board, you can't see through the components. For each point on the pc board, place the letter of the corresponding point or points on the schematic.

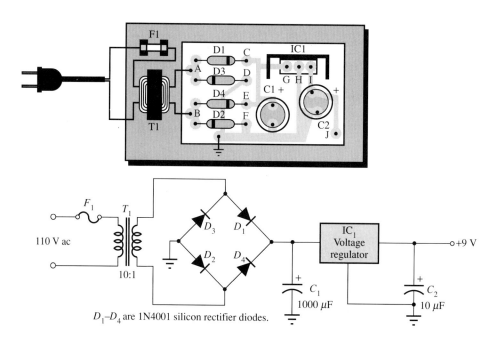

FIGURE 2–73

■ **ACTIVITY 2** **ANALYZE THE POWER SUPPLY CIRCUIT**

With 110 V rms applied to the input, determine what the voltages should be at each point indicated (1, 2, 3, and 4) in Figure 2–74 using the oscilloscope or digital multimeter as indicated.

■ **ACTIVITY 3** **WRITE A TECHNICAL REPORT**

Discuss the detailed operation from input to output including voltage values. Tell what each component does and why it is in the circuit. Since we have not studied voltage regulators yet, all you need to know at this point is that the voltage regulator (IC1) takes the filtered output voltage from the rectifier and produces a constant +9 V output under varying load conditions.

DIODES AND APPLICATIONS 79

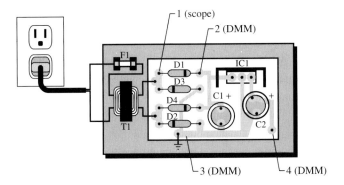

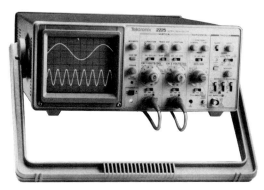

FIGURE 2-74
Photos courtesy of Tektronix, Inc.

 ■ **ACTIVITY 4** TROUBLESHOOT THE POWER SUPPLY FOR EACH OF THE FOLLOWING PROBLEMS BY STATING THE PROBABLE CAUSE OR CAUSES (REFER TO FIGURE 2-74)

1. A full-wave 120 Hz voltage with a peak value of about 14 V at point 2.
2. A large 120 Hz ripple voltage with a peak of about 14 V at point 2.
3. No voltage at point 1.
4. No voltage at point 2.
5. A half-wave 60 Hz voltage with a peak of about 14 V at point 2.
6. No voltage at point 4.

 ■ **ACTIVITY 5** TEST BENCH SPECIAL ASSIGNMENT

Go to Test Bench 1 in the color insert section (which follows page 452) and carry out the assignment stated there.

2-10 REVIEW QUESTIONS

1. List the major parts that make up a basic dc power supply.
2. What is the purpose of a power supply in a system application?

3. What is the input voltage to a typical power supply?
4. From the data sheet for the 1N4001 diode in Appendix A, determine its PIV.
5. Are the diodes used in this power supply operating close to their PIV rating?
6. What is the purpose of capacitor C_1 in the power supply?
7. What is the purpose of capacitor C_2 in the power supply?
8. Why is the voltage regulator connected to a heat sink?
9. What do you think would happen if C_2 were removed?

Summary

- The single diode in a half-wave rectifier conducts for half of the input cycle.
- The output frequency of a half-wave rectifier equals the input frequency.
- The average (dc) value of a half-wave rectified signal is 0.318 ($1/\pi$) times its peak value.
- The PIV (peak inverse voltage) is the maximum voltage appearing across the diode in reverse bias.
- Each diode in a full-wave rectifier conducts for half of the input cycle.
- The output frequency of a full-wave rectifier is twice the input frequency.
- The basic types of full-wave rectifier are center-tapped and bridge.
- The output voltage of a center-tapped, full-wave rectifier is approximately one-half of the total secondary voltage.
- The PIV for each diode in a center-tapped, full-wave rectifier is twice the output voltage.
- The output voltage of a bridge rectifier equals the total secondary voltage.
- The PIV for each diode in a bridge rectifier is half that required for the center-tapped configuration and is approximately equal to the peak output voltage.
- A capacitor-input filter provides a dc output approximately equal to the peak of the input.
- Ripple voltage is caused by the charging and discharging of the filter capacitor.
- The smaller the ripple, the better the filter.
- An LC filter provides improved ripple reduction over the capacitor-input filter.
- An LC filter produces a dc output voltage approximately equal to the average value of the rectified input.
- Diode clippers cut off voltage above or below specified levels.
- Diode clampers add a dc level to an ac signal.
- The zener diode operates in reverse breakdown.
- There are two breakdown mechanisms in a zener diode, avalanche breakdown and zener breakdown.
- When $V_Z < 5$ V, zener breakdown is predominant.
- When $V_Z > 5$ V, avalanche breakdown is predominant.

DIODES AND APPLICATIONS

- A zener diode maintains an essentially constant voltage across its terminals over a specified range of zener currents.
- Zener diodes are used as shunt voltage regulators.
- Regulation of output voltage over a range of input voltages is called *input* or *line regulation*.
- Regulation of output voltage over a range of load currents is called *load regulation*.
- The smaller the percent regulation, the better.
- A varactor diode acts as a variable capacitor under reverse-biased conditions.
- The capacitance of a varactor varies inversely with reverse-biased voltage.
- Diode symbols are shown in Figure 2–75.

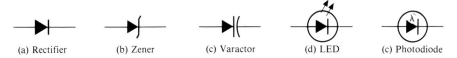

(a) Rectifier   (b) Zener   (c) Varactor   (d) LED   (e) Photodiode

FIGURE 2–75
Diode symbols.

GLOSSARY

Center tap A connection at the midpoint of the secondary of a transformer.

Clamper A circuit that adds a dc level to an ac signal. A dc restorer.

Clipper A circuit that removes part of a waveform above or below a specified level. A limiter.

Filter A type of electrical circuit that passes certain frequencies and rejects all others.

Full-wave rectifier A circuit that converts an alternating sine wave into a pulsating dc consisting of both halves of a sine wave for each input cycle.

Half-wave rectifier A circuit that converts an alternating sine wave into a pulsating dc consisting of one-half of a sine wave for each input cycle.

Integrated circuit (IC) A type of circuit in which all the components are constructed on a single tiny chip of silicon.

Light-emitting diode (LED) A type of diode that emits light when there is forward current.

Line regulation The percent change in output voltage for a given change in line (input) voltage.

Load regulation The percent change in output voltage for a given change in load current.

Photodiode A diode whose reverse resistance changes with incident light.

Power supply An electronic instrument that produces voltage, current, and power from the ac power line or batteries in a form suitable for use in various applications to power electronic equipment.

Rectifier An electronic circuit that converts ac into pulsating dc.

Regulator An electronic circuit that maintains an essentially constant output voltage with a changing input voltage or load.

Ripple voltage The variation in the dc voltage on the output of a filtered rectifier caused by the slight charging and discharging action of the filter capacitor.

Varactor A diode that is used as a voltage-variable capacitor.

Voltage regulation The process of maintaining an essentially constant output voltage over variations in input voltage or load.

Zener diode A type of diode that operates in reverse breakdown (called zener breakdown) to provide voltage regulation.

FORMULAS

(2–1) $\quad V_{avg} = \dfrac{V_p}{\pi} \quad$ Half-wave average value

(2–2) $\quad V_{p(out)} = V_{p(in)} - 0.7 \text{ V} \quad$ Peak half-wave rectifier ouput

(2–3) $\quad V_{avg} = \dfrac{2V_p}{\pi} \quad$ Full-wave average value

(2–4) $\quad \text{PIV} = 2V_{p(out)} \quad$ Diode peak inverse voltage, center-tapped rectifier (neglecting V_B)

(2–5) $\quad V_{out} = V_s \quad$ Bridge full-wave output (neglecting diode drops)

(2–6) $\quad V_{out} = V_s - 1.4 \text{ V} \quad$ Bridge full-wave output (including diode drops)

(2–7) $\quad \text{PIV} \cong V_{p(out)} \quad$ Diode peak inverse voltage, bridge rectifier

(2–8) $\quad r = \left(\dfrac{V_r}{V_{dc}}\right)100\% \quad$ Ripple factor

(2–9) $\quad R_Z = \dfrac{\Delta V_Z}{\Delta I_Z} \quad$ Zener resistance

(2–10) $\quad \text{Percent line regulation} = \dfrac{\Delta V_{OUT}}{\Delta V_{IN}} \times 100\% \quad$ Line regulation

(2–11) $\quad \text{Percent load regulation} = \dfrac{V_{NL} - V_{FL}}{V_{FL}} \times 100\% \quad$ Load regulation

DIODES AND APPLICATIONS

(2–12) $\quad C = \dfrac{A\epsilon}{d}$ $\qquad\qquad\qquad$ Capacitance

(2–13) $\quad f_r = \dfrac{1}{2\pi\sqrt{LC}}$ $\qquad\qquad$ Resonant frequency

SELF-TEST

1. The process of converting ac to pulsating dc is called
   (a) clipping   (b) charging   (c) rectification   (d) filtering
2. The output frequency of a half-wave rectifier with a 60 Hz sinusoidal input is
   (a) 30 Hz   (b) 60 Hz   (c) 120 Hz   (d) 0 Hz
3. The number of diodes used in a half-wave rectifier is
   (a) one   (b) two   (c) three   (d) four
4. If a 75 V peak sine wave is applied to a half-wave rectifier, the peak inverse voltage across the diode is
   (a) 75 V   (b) 150 V   (c) 37.5 V   (d) 0 V
5. The output frequency of a full-wave rectifier with a 60 Hz sinusoidal input is
   (a) 30 Hz   (b) 60 Hz   (c) 120 Hz   (d) 0 Hz
6. Two types of full-wave rectifier are
   (a) single diode and dual diode   (b) primary and secondary
   (c) forward-biased and reverse-biased   (d) center-tapped and bridge
7. When a diode in a center-tapped rectifier opens, the output is
   (a) 0 V   (b) half-wave rectified
   (c) reduced in amplitude   (d) unaffected
8. During the positive half-cycle of the input voltage in a bridge rectifier,
   (a) one diode is forward-biased   (b) all diodes are forward-biased
   (c) all diodes are reverse-biased   (d) two diodes are forward-biased
9. The process of changing a half-wave or a full-wave rectified voltage to a constant dc voltage is called
   (a) filtering   (b) ac to dc conversion
   (c) damping   (d) ripple suppression
10. The small variation in the output voltage of a dc power supply is called
    (a) average voltage   (b) surge voltage
    (c) residual voltage   (d) ripple voltage
11. One measure of the effectiveness of a power supply filter is the
    (a) surge current   (b) ripple factor
    (c) peak value of the output   (d) average value of the output

12. A diode clipping circuit
    (a) removes part of a waveform
    (b) inserts a dc level
    (c) produces an output equal to the average value of the input
    (d) increases the peak value of the input
13. A clamping circuit is also known as a(an)
    (a) averaging circuit   (b) inverter
    (c) dc restorer   (d) ac restorer
14. The zener diode is designed for operation in
    (a) zener breakdown   (b) forward bias
    (c) reverse bias   (d) avalanche breakdown
15. Zener diodes are widely used as
    (a) current limiters   (b) power distributors
    (c) voltage regulators   (d) variable resistors
16. Varactor diodes are used as
    (a) variable resistors   (b) variable current sources
    (c) variable inductors   (d) variable capacitors
17. LEDs are based on the principle of
    (a) forward bias   (b) electroluminescence
    (c) photon sensitivity   (d) electron-hole recombination
18. In a photodiode, light produces
    (a) reverse current   (b) forward current
    (c) electroluminescence   (d) dark current

Problems

Section 2–1 Half-Wave Rectifiers

1. Calculate the average value of a half-wave rectified voltage with a peak value of 200 V.
2. Sketch the waveforms for the load current and voltage for Figure 2–76. Show the peak values. (The diode is silicon.)
3. Can a diode with a PIV rating of 5 V be used in the circuit of Figure 2–76?

FIGURE 2–76

DIODES AND APPLICATIONS

4. Determine the peak power delivered to R_L in Figure 2–77.

FIGURE 2–77

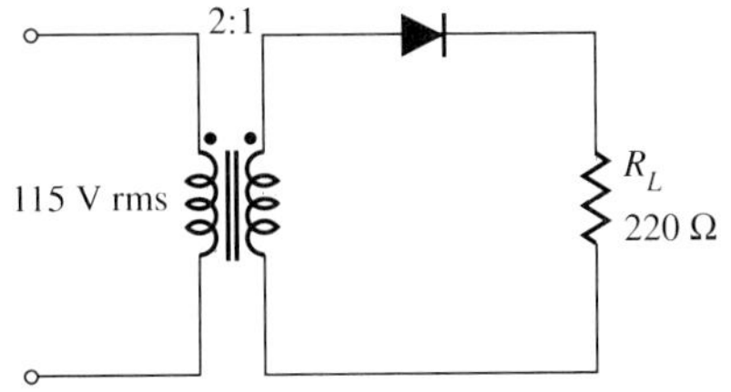

115 V rms, 2:1 transformer, R_L 220 Ω

SECTION 2–2 FULL-WAVE RECTIFIERS

5. Calculate the average value of a full-wave rectified voltage with a peak value of 75 V.
6. Consider the circuit in Figure 2–78.
   (a) What type of circuit is this?
   (b) What is the total peak secondary voltage?
   (c) Find the peak voltage across each half of the secondary.
   (d) Sketch the voltage waveform across R_L.
   (e) What is the peak current through each diode?
   (f) What is the PIV for each diode?

FIGURE 2–78

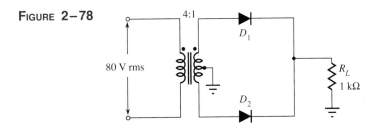

80 V rms, 4:1, D_1, D_2, R_L 1 kΩ

7. Calculate the peak voltage rating of each half of a center-tapped transformer used in a full-wave rectifier that has an average output voltage of 110 V.
8. Show how to connect the diodes in a center-tapped rectifier in order to produce a negative-going full-wave voltage across the load resistor.
9. What PIV rating is required for the diodes in a bridge rectifier that produces an average output voltage of 50 V?

SECTION 2–3 RECTIFIER FILTERS

10. The ideal dc output voltage of a capacitor-input filter is the (peak, average) value of the rectified input.
11. Refer to Figure 2–79 and sketch the following voltage waveforms in relationship to the input waveform: V_{AB}, V_{AD}, V_{BD}, and V_{CD}.

FIGURE 2-79

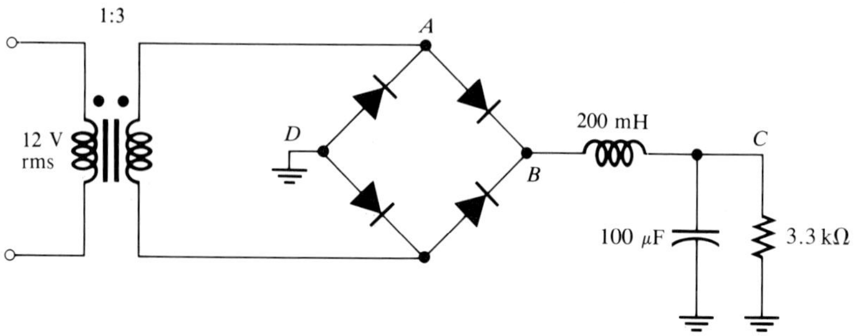

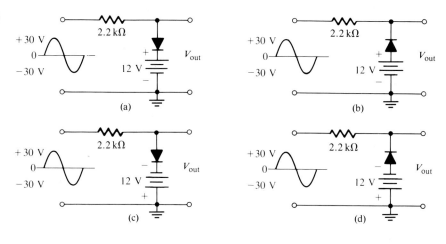

SECTION 2-4 DIODE CLIPPING AND CLAMPING CIRCUITS

12. Sketch the output waveforms for each circuit in Figure 2-80.

13. Describe the output waveform of each circuit in Figure 2-81. Assume that the *RC* time constant is much greater than the period of the input.

FIGURE 2-80

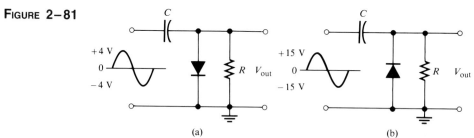

FIGURE 2-81

SECTION 2-5 ZENER DIODES

14. A certain zener diode has a $V_Z = 7.5$ V and an $R_Z = 5\ \Omega$ at a certain current. Sketch the equivalent circuit.

15. Determine the minimum input voltage required for regulation to be established in Figure 2-82. Assume an ideal zener diode with $I_{ZK} = 1.5$ mA and $V_Z = 14$ V.

FIGURE 2–82

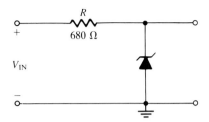

16. To what value must R be adjusted in Figure 2–83 to make $I_Z = 40$ mA? Assume that $V_Z = 12$ V at 30 mA and $R_Z = 30\ \Omega$.
17. A loaded zener regulator is shown in Figure 2–84. $V_Z = 5.1$ V at 35 mA, $I_{ZK} = 1$ mA, $I_{ZM} = 70$ mA, and $R_Z = 12\ \Omega$. Determine the minimum and maximum permissible load currents.
18. Find the percent load regulation in Problem 17.

FIGURE 2–83

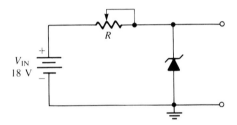

FIGURE 2–84

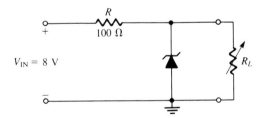

19. For the circuit of Problem 17, assume that the input voltage is varied from 6 V to 12 V. Determine the approximate percent line regulation with no load and with maximum load.
20. The no-load output voltage of a certain zener regulator is 8.23 V, and the full-load output is 7.98 V. Calculate the percent load regulation.

SECTION 2–6 VARACTOR DIODES

21. Figure 2–85 is a curve of reverse voltage versus capacitance for a certain varactor. Determine the change in capacitance if V_R varies from 5 V to 20 V.
22. Refer to Figure 2–85 and determine the value of V_R that produces 25 pF.

FIGURE 2–85

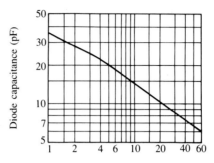

23. What capacitance value is required for each of the varactors in Figure 2–86 to produce a resonant frequency of 1 MHz?
24. At what value must the control voltage be set in Problem 23 if the varactors have the characteristic curve in Figure 2–85?

FIGURE 2–86

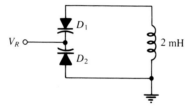

SECTION 2–7 LEDs AND PHOTODIODES

25. When the switch in Figure 2–87 is closed, will the microammeter reading increase or decrease? Assume that D_1 and D_2 are optically coupled.
26. With no incident light, a certain amount of reverse current flows in a photodiode. What is this current called?

FIGURE 2–87

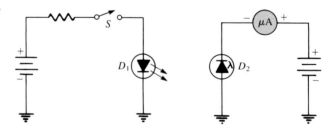

SECTION 2–8 THE DIODE DATA SHEET

27. From the data sheet in Appendix A, determine how much peak inverse voltage that a 1N1183A diode can withstand.
28. Repeat Problem 27 for a 1N1188A.

DIODES AND APPLICATIONS

29. If the peak output voltage of a full-wave bridge rectifier is 50 V, determine the minimum value of the surge limiting resistor required when 1N1183A diodes are used.

SECTION 2–9 TROUBLESHOOTING

30. From the meter readings in Figure 2–88, determine if the rectifier circuit is functioning properly. If it is not, determine the most likely failure(s).

31. Each part of Figure 2–89 shows oscilloscope displays of rectifier output voltages. In each case, determine whether or not the rectifier is functioning properly and, if it is not, what is (are) the most likely failure(s).

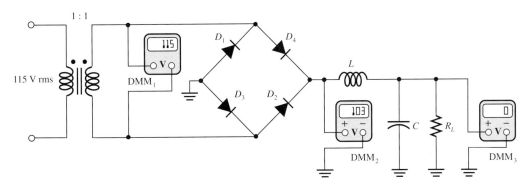

FIGURE 2–88

FIGURE 2–89

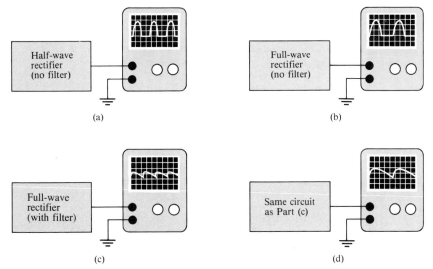

32. For each set of measured voltages at the points (1, 2, and 3) indicated in Figure 2–90, determine if they are correct and if not identify the most likely fault(s). State what you would do to correct the problem once it is isolated.

  (a) $V_1 = 110$ V rms, $V_2 \cong 30$ V dc, $V_3 \cong 12$ V dc

(b) $V_1 = 110$ V rms, $V_2 \cong 30$ V dc, $V_3 \cong 30$ V dc
(c) $V_1 = 0$ V, $V_2 = 0$ V, $V_3 = 0$ V
(d) $V_1 = 110$ V rms, $V_2 \cong 30$ V peak full-wave 120 Hz voltage, $V_3 \cong 12$ V, 120 Hz pulsating voltage
(e) $V_1 = 110$ V rms, $V_2 = 0$ V, $V_3 = 0$ V

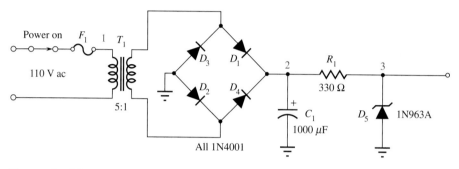

FIGURE 2–90

 33. Determine the most likely failure in the circuit board of Figure 2–91 for each of the following symptoms. State the corrective action you would take in each case. The transformer has a turns ratio of 1.

(a) No voltage at point 1.
(b) No voltage at point 2. 110 V rms at point 1.
(c) No voltage at point 3. 110 V rms at point 1.

FIGURE 2–91

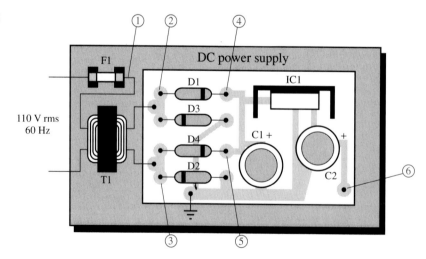

(d) 150 V rms at point 2. Input is correct at 110 V rms.

(e) 68 V rms at point 3. Input is correct at 110 V rms.

(f) A pulsating full-wave rectified voltage with a peak of 155.5 V at point 4.

(g) Excessive 120 Hz ripple voltage at point 5.

(h) Ripple voltage has a frequency of 60 Hz at point 4.

(i) No voltage at point 6.

34. In testing the power supply in Figure 2–92, you found the voltage at the positive side of C_1 to have a 60 Hz ripple with a greater than normal amplitude. Just to be sure, you replaced all of the diodes with known good ones. You check the point again to verify proper operation and it still has the 60 Hz ripple. What now?

FIGURE 2–92

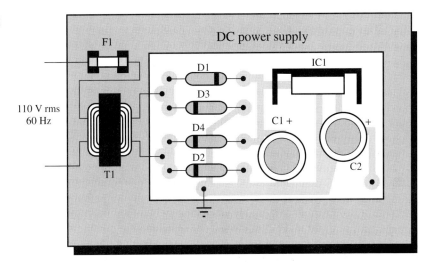

35. Draw the schematic for the circuit board in Figure 2–93 and determine what the correct output voltages should be.

36. The ac input of the circuit board in Figure 2–93 is connected to the secondary of a transformer with a turns ratio of 1 that is operating from the 110 V ac source. When you measure the output voltages, both are zero. What do you think has failed, and what is the reason for this failure?

37. Select a transformer turns ratio that will provide the greatest possible secondary voltage compatible with the circuit board in Figure 2–93.

92 ♦ CHAPTER 2

38. In Figure 2–93, V_{OUT2} is zero and V_{OUT1} is correct. What is (are) possible reasons for this?

FIGURE 2–93

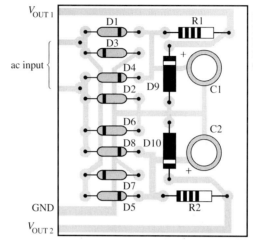

ANSWERS TO REVIEW QUESTIONS

SECTION 2–1
1. Peak of negative alternation
2. 50%
3. 3.18 V

SECTION 2–2
1. 38.2 V
2. Bridge
3. Less

SECTION 2–3
1. 60 Hz, 120 Hz
2. Capacitor charging and discharging slightly
3. It increases ripple.
4. Reduced ripple; less dc voltage output

SECTION 2–4
1. Clippers clip off or remove portions of a waveform. Clampers insert a dc level.
2. A positive clipper clips off positive voltages. A negative clipper clips off negative voltages.
3. 0.7 V appears across the diode.
4. The bias voltage must be 5 V − 0.7 V = 4.3 V.
5. The capacitor acts as a battery.

DIODES AND APPLICATIONS

SECTION 2–5
1. In breakdown
2. 10.24 V
3. *Line regulation:* Constant output voltage for varying input voltage. *Load regulation:* Constant output voltage for varying load current.
4. Maximum R_L
5. 0.833%

SECTION 2–6
1. A varactor diode is a variable capacitor.
2. The diode capacitance decreases.

SECTION 2–7
1. LEDs give off light when forward-biased; photodiodes respond to light when reverse-biased.
2. Gallium arsenide, gallium arsenide phosphide, gallium phosphide
3. Dark current

SECTION 2–8
1. Forward current and voltage; reverse current and voltage; thermal
2. (a) Maximum dc forward voltage drop   (b) Maximum reverse direct current
   (c) Average maximum forward current

SECTION 2–9
1. Half-wave output
2. Open diode
3. Leaky filter capacitor

SECTION 2–10
1. Transformer, rectifier, filter, and regulator
2. The power supply provides all parts of the system with dc voltage and current to operate the circuits.
3. 110 V rms (Portable power supplies use a battery.)
4. 50 V, designated V_{RRM}
5. No, they experience approximately 15 V PIV and they are rated at 50 V.
6. C_1 filters (smooths) the full-wave output of the rectifier.
7. C_2 provides additional filtering on the regulator output.
8. To dissipate the heat that it produces
9. The dc output voltage would not be affected. C_2 removes any noise or transient voltages that may get on the line.

Answers to Practice Exercises

2–1 3.82 V
2–2 2.3 V
2–3 98.68 V, DC VOLTS function
2–4 19.3 V including diode drop
2–5 78.6 V, ≈ 78.6 V
2–6 A positive peak of 8.72 V and clipped at −0.3 V
2–7 Clipped at +10.7 V and −10.7 V
2–8 Same as Figure 2–39
2–9 5 Ω
2–10 $V_Z = 12.6$ V at 80 mA, $V_Z = 11.4$ V at 20 mA
2–11 $V_{IN(min)} = 6.16$ V, $V_{IN(max)} = 48.14$ V
2–12 $I_{L(min)} = 0$ A, $I_{L(max)} = 42$ mA, $R_{L(min)} = 79$ Ω
2–13 **(a)** 3.77% **(b)** 1.75%/V
2–14 100.66 kHz to 318.31 kHz

3

TRANSISTORS AND THYRISTORS

3–1 BIPOLAR JUNCTION TRANSISTORS (BJTs)
3–2 VOLTAGE-DIVIDER BIAS
3–3 THE BIPOLAR TRANSISTOR AS AN AMPLIFIER
3–4 THE BIPOLAR TRANSISTOR AS A SWITCH
3–5 BJT PARAMETERS AND RATINGS
3–6 THE JUNCTION FIELD-EFFECT TRANSISTOR (JFET)
3–7 JFET CHARACTERISTICS
3–8 THE METAL OXIDE SEMICONDUCTOR FET (MOSFET)
3–9 FET BIASING
3–10 UNIJUNCTION TRANSISTORS (UJTs)
3–11 THYRISTORS
3–12 TRANSISTOR PACKAGES AND TERMINAL IDENTIFICATION
3–13 TROUBLESHOOTING
3–14 A SYSTEM APPLICATION

After completing this chapter, you should be able to

☐ Describe the basic construction of bipolar junction transistors (BJTs).
☐ Distinguish between npn and pnp transistors.
☐ Bias a BJT with dc voltages so that it can be operated as an amplifier.
☐ Define the transistor currents and explain how they are related.
☐ Explain how voltage-divider bias works in a transistor circuit.
☐ Interpret the characteristic curves of a transistor.
☐ Define the terms *cutoff* and *saturation*.
☐ Explain how a transistor produces voltage gain.
☐ Explain how a transistor can be used as a switch.
☐ Identify the significance of certain transistor parameters.
☐ Describe the basic construction and operation of junction field-effect transistors (JFETs).
☐ Describe the basic construction and operation of metal oxide semiconductor field-effect transistors.
☐ Explain how a MOSFET functions in the depletion and the enhancement modes.
☐ Properly bias an FET.
☐ Describe how a unijunction transistor (UJT) works and how it is used in a basic oscillator.
☐ Explain the basic operation of SCRs, triacs, and diacs and how they are used as control devices.
☐ Recognize various types of transistor packages and identify the transistor leads.
☐ Troubleshoot transistor bias circuits.

In this chapter, we discuss the two basic types of transistors—the bipolar junction transistor (BJT) and the field-effect transistor (FET). You will be introduced to the two major application areas of amplification and switching. Also, the unijunction transistor (UJT) will be introduced and common types of thyristors will be presented.

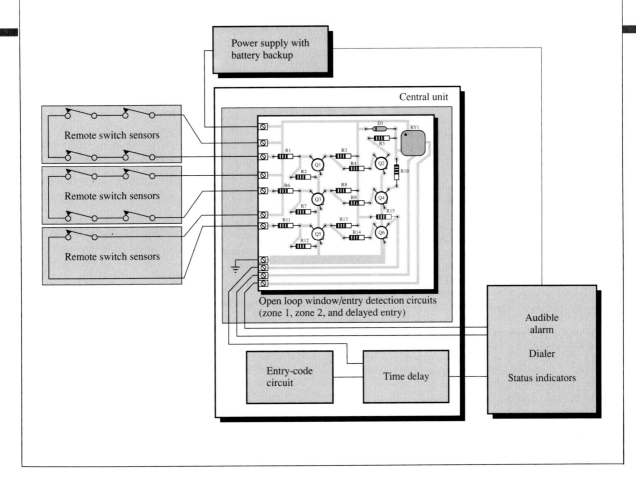

A System Application

Transistors are used in one form or another in just about every electronic system imaginable. A simple electronic security system is the focus of this chapter's system application section. You will see how transistors, operating in their switching modes, are used to perform a very basic function in the system shown in the diagram above. Also, you will see how the transistor is used to operate nonelectronic devices such as, in this case, the electromechanical relay.

For the system application in Section 3–14, in addition to the other topics, be sure you understand

- [ ] How a bipolar junction transistor works.
- [ ] How the currents in a transistor are calculated.
- [ ] The meaning of *saturation* and *cutoff*.
- [ ] The principles of biasing a transistor.
- [ ] How to read a data sheet.

3-1 BIPOLAR JUNCTION TRANSISTORS (BJTs)

The basic structure of the bipolar junction transistor, BJT, determines its operating characteristics. In this section, you will see how semiconductor materials are joined to form a transistor, and you will learn the standard transistor symbols. Also, you will see how important dc bias is to the operation of transistors in terms of setting up proper currents and voltages in a transistor circuit. Finally, two important parameters, α_{dc} and β_{dc}, are introduced.

The **bipolar junction transistor (BJT)** is constructed with three doped semiconductor regions separated by two pn junctions. The three regions are called **emitter, base,** and **collector.** The two types of bipolar transistors are shown in Figure 3–1. One type consists of two n regions separated by a p region (npn), and the other consists of two p regions separated by an n region (pnp).

FIGURE 3–1
Construction of bipolar junction transistors.

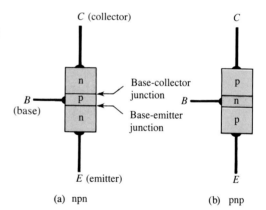

(a) npn (b) pnp

The pn junction joining the base region and the emitter region is called the *base-emitter junction*. The junction joining the base region and the collector region is called the *base-collector junction,* as indicated in Figure 3–1(a). A wire lead connects to each of the three regions, as shown. These leads are labeled E, B, and C for emitter, base, and collector, respectively. The base region is lightly doped and very narrow compared to the heavily doped emitter and collector materials. The reason for this is discussed in the next section.

Figure 3–2 shows the schematic symbols for the npn and pnp bipolar transistors. The term **bipolar** refers to the use of both holes and electrons as carriers in the transistor structure.

TRANSISTORS AND THYRISTORS

FIGURE 3–2
Standard bipolar junction transistor symbols.

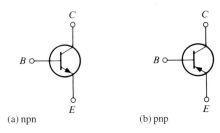

(a) npn   (b) pnp

TRANSISTOR BIASING

In order for the transistor to operate properly as an amplifier, the two pn junctions must be correctly biased with external voltages. We will use the npn transistor to illustrate transistor biasing. The operation of the pnp is the same as for the npn except that the roles of the electrons and holes, the bias voltage polarities, and the current directions are all reversed. Figure 3–3 shows the proper bias arrangement for both npn and pnp transistors. Notice that in both cases the base-emitter (*BE*) junction, is forward-biased and the base-collector (*BC*) junction is reverse-biased. This is called *forward-reverse bias*.

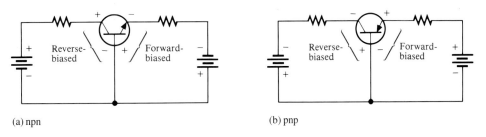

(a) npn   (b) pnp

FIGURE 3–3
Forward-reverse bias of a bipolar junction transistor.

Now, let's examine what happens inside the transistor when it is forward-reverse biased. The forward bias from base to emitter narrows the *BE* depletion layer, and the reverse bias from base to collector widens the *BC* depletion layer, as depicted in Figure 3–4(a). The n-type emitter region is teeming with conduction-band (free) electrons that easily diffuse across the *BE* junction into the p-type base region, just as in a forward-biased diode.

The base region is lightly doped and very narrow so that it has a very limited number of holes. Thus, only a small percentage of all the electrons flowing across the *BE* junction combine with the available holes. These relatively few recombined electrons flow out of the base lead as valence electrons, forming the small base current, I_B, as shown in Figure 3–4(b).

Most of the electrons flowing from the emitter into the narrow base region do not recombine and diffuse into the *BC* depletion layer. Once in this layer, they are pulled across the reverse-biased *BC* junction by the depletion layer field set up by the force of attraction between the positive and negative ions. Actually, you can think of the electrons

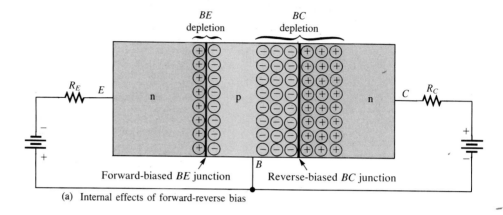
(a) Internal effects of forward-reverse bias

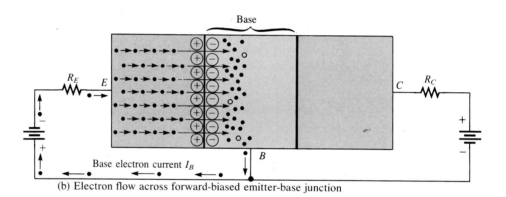
(b) Electron flow across forward-biased emitter-base junction

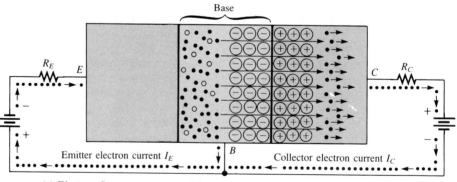
(c) Electron flow across reverse-biased base-collector junction

FIGURE 3–4
Illustration of BJT action. The base region is very narrow, but it is shown wider here for clarity.

as being pulled across the reverse-biased *BC* junction by the attraction of the positive ions on the other side, as illustrated in Figure 3–4(c). The electrons now move through the collector region, out through the collector lead, and into the positive terminal of the external dc source, thereby forming the collector current, I_C, as shown. The amount of collector current depends directly on the amount of base current and is essentially independent of the dc collector voltage.

TRANSISTOR CURRENTS

The directions of current in an npn and a pnp transistor are as shown in Figures 3–5(a) and 3–5(b), respectively. An examination of these diagrams shows that the emitter current is the sum of the collector and base currents, expressed as follows:

$$I_E = I_C + I_B \qquad (3-1)$$

As mentioned before, I_B is very small compared to I_E or I_C. The capital-letter subscripts indicate dc values.

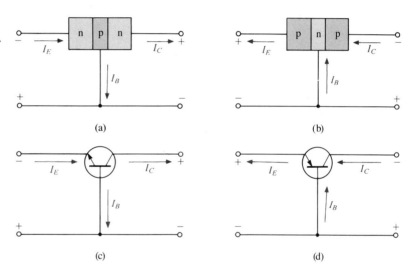

FIGURE 3–5
Current directions in a transistor.

DC ALPHA (α_{dc}) AND DC BETA (β_{dc})

Emitter, collector, and base currents are also related by two parameters: the dc **alpha** (α_{dc}), which is the ratio I_C/I_E and the dc **beta** (β_{dc}), which is the ratio I_C/I_B. β_{dc} is the direct current gain and is usually designated as h_{FE} on the transistor data sheets.

The collector current is equal to α_{dc} times the emitter current.

$$I_C = \alpha_{dc} I_E \qquad (3-2)$$

where α_{dc} typically has a value between 0.95 and 0.99.

The collector current is related to the base current by β_{dc}.

$$I_C = \beta_{dc} I_B \qquad (3-3)$$

where β_{dc} typically has a value between 20 and 200.

Transistor Voltages

The three dc voltages for the biased transistor in Figure 3–6 are the emitter voltage (V_E), the collector voltage (V_C), and the base voltage (V_B). These single-subscript voltages are with respect to ground. The collector voltage is equal to the dc supply voltage, V_{CC}, less the drop across R_C.

$$V_C = V_{CC} - I_C R_C \qquad (3\text{–}4)$$

The base voltage is equal to the emitter voltage plus the base-emitter junction barrier potential (V_{BE}), which is about 0.7 V for a silicon transistor.

$$V_B = V_E + V_{BE} \qquad (3\text{–}5)$$

In the configuration of Figure 3–6, the emitter is the common terminal, so $V_E = 0$ V.

FIGURE 3–6
Bias voltages.

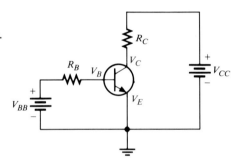

■ **EXAMPLE 3–1**   Determine I_B, I_C, I_E, V_B, and V_C in Figure 3–7, where β_{dc} is 50.

FIGURE 3–7

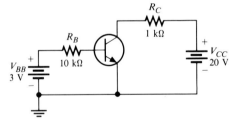

SOLUTION
Since V_E is ground, $V_B = 0.7$ V. The drop across R_B is $V_{BB} - V_B$, so I_B is calculated as follows:

$$I_B = \frac{V_{BB} - V_B}{R_B} = \frac{3\text{ V} - 0.7\text{ V}}{10\text{ k}\Omega} = 0.23\text{ mA}$$

Now we can find I_C, I_E, and V_C.

$$I_C = \beta_{dc} I_B = 50(0.23\text{ mA}) = 11.5\text{ mA}$$
$$I_E = I_C + I_B = 11.5\text{ mA} + 0.23\text{ mA} = 11.73\text{ mA}$$
$$V_C = V_{CC} - I_C R_C = 20\text{ V} - (11.5\text{ mA})(1\text{ k}\Omega) = 8.5\text{ V}$$

PRACTICE EXERCISE 3–1
Determine I_B, I_C, I_E, V_{CE}, and V_{CB} in Figure 3–7 for the following values: $R_B = 22$ kΩ, $R_C = 220$ Ω, $V_{BB} = 6$ V, $V_{CC} = 9$ V, and $\beta_{dc} = 90$.

3–1 REVIEW QUESTIONS

1. What are the three transistor terminals?
2. Define *forward-reverse bias*.
3. What is β_{dc}?
4. If I_B is 10 μA and β_{dc} is 100, what is the collector current?

3–2 VOLTAGE-DIVIDER BIAS

Next, we will study a method of biasing a transistor for linear operation using a resistive voltage divider. This is the most widely used biasing method, for reasons that you will discover in this section.

The voltage-divider bias configuration uses only a single dc source to provide forward-reverse bias to the transistor, as shown in Figure 3–8. Resistors R_1 and R_2 form a voltage divider that provides the base bias voltage. Resistor R_E allows the emitter to rise above ground potential.

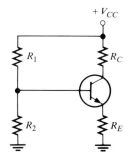

FIGURE 3–8
Voltage-divider bias.

The voltage divider is loaded by the resistance as viewed from the base of the transistor. In some cases, this loading effect is significant in determining the base bias voltage. We will now examine this arrangement in more detail.

INPUT RESISTANCE AT THE BASE

The approximate input resistance of the transistor, viewed from the base of the transistor in Figure 3–9, is derived as follows.

$$R_{IN} = \frac{V_B}{I_B}$$

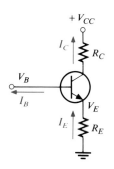

FIGURE 3–9
Circuit for deriving input resistance. The dc input resistance is V_B/I_B.

Neglecting the V_{BE} of 0.7 V, we have

$$V_B \cong V_E = I_E R_E$$

Since $I_E \cong I_C$ when $\alpha_{dc} \cong 1$, then
$$I_E \cong \beta_{dc} I_B$$
Substituting yields
$$V_B \cong \beta_{dc} I_B R_E$$
$$R_{IN} \cong \frac{\beta_{dc} I_B R_E}{I_B}$$
The I_B terms cancel, leaving
$$R_{IN} \cong \beta_{dc} R_E \qquad (3-6)$$

BASE VOLTAGE

Now, using the voltage-divider formula, the following equation gives the base voltage for the circuit in Figure 3–8:
$$V_B = \left(\frac{R_2 \| R_{IN}}{R_1 + R_2 \| R_{IN}}\right) V_{CC} \qquad (3-7)$$

Generally, if R_{IN} is at least ten times greater than R_2, then Equation (3–7) can be simplified as follows:
$$V_B \cong \left(\frac{R_2}{R_1 + R_2}\right) V_{CC} \qquad (3-8)$$

Once you have determined the base voltage, you can determine the emitter voltage V_E (for an npn transistor) as follows:
$$V_E = V_B - 0.7 \text{ V}$$

■ **EXAMPLE 3–2** Determine V_B, V_E, V_C, V_{CE}, I_B, I_E, and I_C in Figure 3–10.

FIGURE 3–10

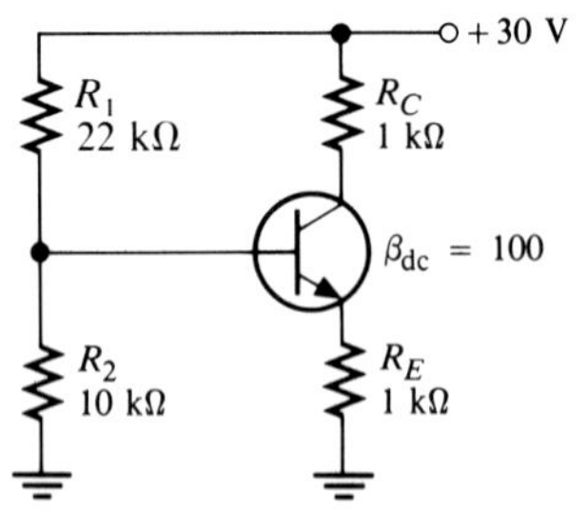

SOLUTION
The input resistance at the base is
$$R_{IN} \cong \beta_{dc} R_E = 100(1 \text{ k}\Omega) = 100 \text{ k}\Omega$$

Since R_{IN} is ten times greater than R_2, the base voltage is approximately

$$V_B \cong \left(\frac{R_2}{R_1 + R_2}\right) V_{CC} = \left(\frac{10 \text{ k}\Omega}{32 \text{ k}\Omega}\right) 30 \text{ V} = 9.375 \text{ V}$$

Therefore, $V_E = V_B - 0.7 \text{ V} \cong 8.675 \text{ V}$.

Now that we know V_E, we can find I_E by Ohm's law.

$$I_E = \frac{V_E}{R_E} = \frac{8.675 \text{ V}}{1 \text{ k}\Omega} = 0.008675 \text{ A} = 8.675 \text{ mA}$$

Since α_{dc} is so close to 1 for most transistors, it is a good approximation to assume that $I_C \cong I_E$. Thus,

$$I_C \cong 8.675 \text{ mA}$$

Using $I_C = \beta_{dc} I_B$ and solving for I_B, we get

$$I_B = \frac{I_C}{\beta_{dc}} \cong \frac{8.675 \text{ mA}}{100} = 0.08675 \text{ mA} = 86.75 \text{ }\mu\text{A}$$

Since we know I_C, we can find V_C.

$$V_C = V_{CC} - I_C R_C = 30 \text{ V} - (8.675 \text{ mA})(1 \text{ k}\Omega)$$
$$= 30 \text{ V} - 8.675 \text{ V} = 21.325 \text{ V}$$

Since V_{CE} is the collector-to-emitter voltage, it is the difference of V_C and V_E.

$$V_{CE} = V_C - V_E = 21.325 \text{ V} - 8.675 \text{ V}$$
$$= 12.65 \text{ V}$$

PRACTICE EXERCISE 3–2

Determine V_B, V_E, V_C, V_{CE}, I_B, I_E, and I_C taking into account R_{IN} at the base.

3–2 REVIEW QUESTIONS

1. How many dc voltage sources are required for voltage-divider bias?
2. In Figure 3–10, if R_2 is 4.7 kΩ, what is the value of V_B?

3–3

THE BIPOLAR TRANSISTOR AS AN AMPLIFIER

*The purpose of dc bias is to allow a transistor to operate as an **amplifier**. Thus, a transistor can be used to produce a larger signal using a smaller signal as a "pattern." In this section, we will discuss how a transistor acts as an amplifier. The subject of amplifiers is covered in more detail in the next chapter.*

First, we will examine the parameters and dc operating conditions that are important in the operation of a transistor amplifier circuit.

Collector Characteristic Curves

With a circuit such as that in Figure 3–11(a), a set of curves can be generated to show how I_C varies with V_{CE} for various values of I_B. These curves are the *collector characteristic curves*.

Notice that both V_{BB} and V_{CC} are adjustable. If V_{BB} is set to produce a specific value of I_B and V_{CC} is zero, then $I_C = 0$ and $V_{CE} = 0$. Now, as V_{CC} is gradually increased, V_{CE} will increase and so will I_C, as indicated on the color-shaded portion of the curve between points A and B in Figure 3–11(b).

When V_{CE} reaches approximately 0.7 V, the base-collector junction becomes reverse-biased and I_C reaches its full value determined by the relationship $I_C = \beta_{dc} I_B$. At this point, I_C levels off to an almost constant value as V_{CE} continues to increase. This action appears to the right of point B on the curve. Actually, I_C increases slightly as V_{CE} increases due to the widening of the base-collector depletion layer which results in fewer holes for recombination in the base region.

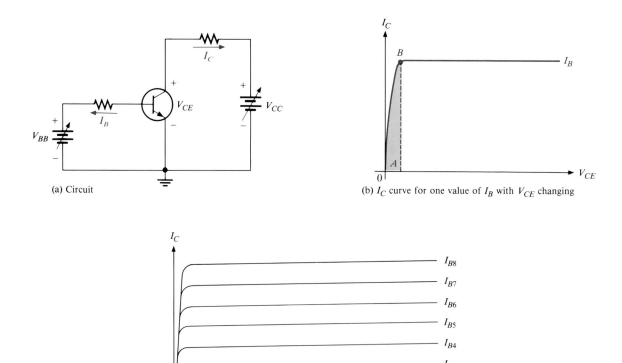

FIGURE 3–11
Collector characteristic curves.

By using other values of I_B, we can produce additional I_C versus V_{CE} curves, as shown in Figure 3–11(c). These curves constitute a *family* of collector curves for a given transistor.

EXAMPLE 3–3

Sketch the family of collector curves for the circuit in Figure 3–12 for $I_B = 5$ μA to 25 μA in 5-μA increments. Assume that $\beta_{dc} = 100$.

FIGURE 3–12

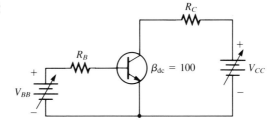

SOLUTION
Using the relationship $I_C = \beta_{dc} I_B$, values of I_C are calculated and tabulated in Table 3–1. The resulting curves are plotted in Figure 3–13.

TABLE 3–1

I_B	I_C
5 μA	0.5 mA
10 μA	1.0 mA
15 μA	1.5 mA
20 μA	2.0 mA
25 μA	2.5 mA

FIGURE 3–13

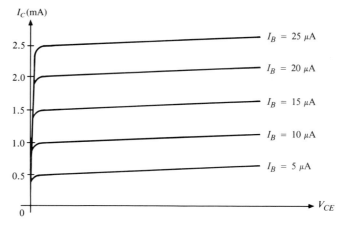

PRACTICE EXERCISE 3–3
Where would the curve for $I_B = 0$ appear on the graph?

CUTOFF AND SATURATION

When $I_B = 0$, the transistor is in **cutoff**. Under this condition, there is a very small amount of collector leakage current, I_{CEO}, due mainly to thermally produced carriers. In cutoff, both the base-emitter and the base-collector junctions are reverse-biased.

Now let's consider the condition known as **saturation.** When the base current is increased, the collector current also increases and V_{CE} decreases as a result of more drop across R_C. When V_{CE} reaches its saturation value, $V_{CE(sat)}$, the base-collector junction becomes forward-biased and I_C can increase no further even with a continued increase in I_B. At the point of saturation, the relation $I_C = \beta_{dc}I_B$ is no longer valid. For a transistor, $V_{CE(sat)}$ occurs somewhere below the knee of the collector curves; it is usually only a few tenths of a volt for silicon transistors and is often assumed to be zero for analysis purposes.

DC LOAD LINE OPERATION

A straight line drawn on the collector curves between the cutoff and saturation points of the transistor is called the *load line*. Once set up, the transistor always operates along this line. Thus, any value of I_C and the corresponding V_{CE} will fall on this line.

Now we will set up a load line for the circuit in Figure 3–14, so that you can learn what it tells us about the transistor operation. First the cutoff point on the load line is determined as follows: When the transistor is cut off, there is essentially no collector current. Thus, the collector-emitter voltage, V_{CE}, is equal to V_{CC}. In this case, $V_{CE} = 30$ V.

FIGURE 3–14

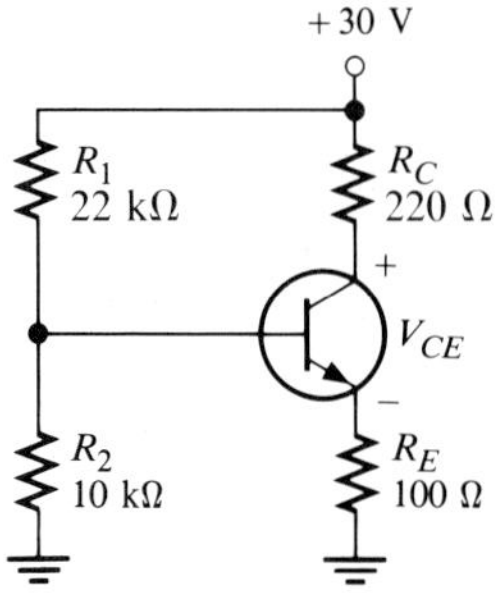

Next the saturation point on the load line is determined. When the transistor is saturated, V_{CE} is approximately zero. (Actually, it is usually a few tenths of a volt, but zero is a good approximation.) Therefore, all the V_{CC} voltage is dropped across $R_C + R_E$. From this we can determine the saturation value of collector current, $I_{C(sat)}$. This value is the maximum value for I_C. We cannot possibly increase it further without changing V_{CC}, R_C, or R_E. In Figure 3–14, the value of $I_{C(sat)}$ is $V_{CC}/(R_C + R_E)$, which is 93.75 mA.

Next the cutoff and saturation points are plotted on the assumed curves in Figure 3–15, and a straight line, which is the load line, is drawn between them.

Q POINT

The base current, I_B, is established by the base bias. The point at which the base current curve intersects the load line is the *quiescent* or ***Q* point** for the circuit. The coordinates of the Q point are the values for I_C and V_{CE} at that point, as illustrated in Figure 3–15.

Now we have completely described the dc operating conditions for the amplifier circuit. In the following paragraphs we discuss ac or signal conditions of an amplifier.

FIGURE 3–15
Load line for the circuit in Figure 3–14.

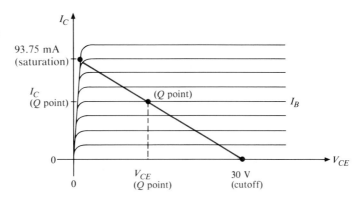

SIGNAL (ac) OPERATION OF AN AMPLIFIER

The purpose of the circuit in Figure 3–16 is to produce an output signal with the same waveform as the input signal but with a greater amplitude. This increase is called **amplification.** The figure shows an input signal, V_{in}, capacitively coupled to the base. The collector voltage is the output signal, as indicated. The input signal voltage causes the base current to vary at the same frequency above and below its dc value. This variation in base current produces a corresponding variation in collector current. However, the variation in collector current is much larger than the variation in base current because of the current **gain** through the transistor. The ratio of the collector current, I_c, to the base current, I_b, is designated β_{ac} (the ac beta) or h_{fe}.

$$\beta_{ac} = \frac{I_c}{I_b} \tag{3-9}$$

The value of β_{ac} normally differs slightly from that of β_{dc} for a given transistor. Remember that lowercase subscripts indicate ac currents and voltages and uppercase subscripts indicate dc currents and voltages.

FIGURE 3–16
An amplifier with voltage-divider bias with capacitively coupled input signal. V_{in} and V_{out} are with respect to ground.

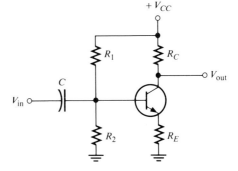

SIGNAL VOLTAGE GAIN OF AN AMPLIFIER

Now let us take the amplifier in Figure 3–16 and examine its voltage gain with a signal input. The output voltage is the collector voltage. The variation in collector current produces a variation in the voltage across R_C and a resulting variation in the collector voltage, as shown in Figure 3–17.

FIGURE 3–17
Voltage amplification.

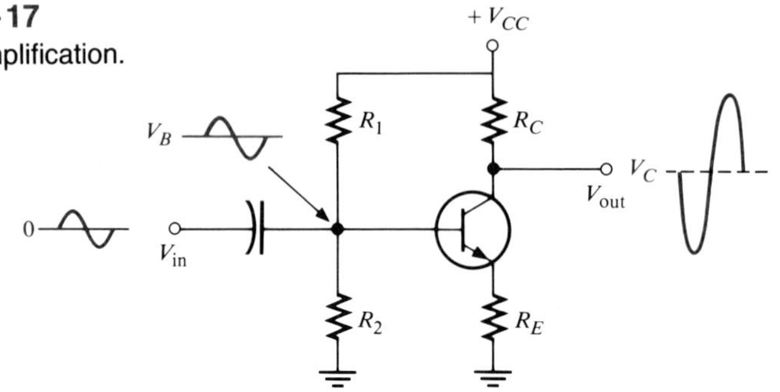

As the collector current increases, the $I_C R_C$ drop increases. This increase produces a decrease in collector voltage because $V_c = V_{CC} - I_C R_C$. Likewise, as the collector current decreases, the $I_C R_C$ drop decreases and produces an increase in collector voltage. Therefore, there is a 180° phase difference between the collector current and the collector voltage. The base voltage and collector voltage are also 180° out-of-phase, as indicated in Figure 3–17. This 180° phase difference between input and output is called an *inversion*.

The voltage gain A_v of the amplifier is V_{out}/V_{in}, where V_{out} is the signal voltage at the collector and V_{in} is the signal voltage at the base. Because the base-emitter junction is forward-biased, the signal voltage at the emitter is approximately equal to the signal voltage at the base. Thus, since $V_b \cong V_e$, the gain is approximately V_c/V_e or $I_c R_C / I_e R_E$. I_c and I_e are very close to the same value because the α_{ac} is close to 1. Therefore, they cancel, giving the voltage gain formula:

$$A_v \cong \frac{R_C}{R_E} \qquad (3\text{–}10)$$

A negative sign on A_v is often used to indicate inversion.

EXAMPLE 3–4

In Figure 3–18, a signal voltage of 50 mV rms is applied to the base.
(a) Determine the output voltage for the amplifier.
(b) Find the dc collector voltage on which the output signal voltage is riding.
(c) Sketch the output waveform.

FIGURE 3–18

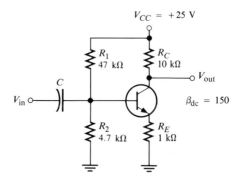

SOLUTION

(a) The signal voltage gain is

$$A_v \cong \frac{R_C}{R_E} = \frac{10 \text{ k}\Omega}{1 \text{ k}\Omega} = 10$$

The output signal voltage is the input signal voltage times the voltage gain.

$$V_{\text{out}} = A_v V_{\text{in}} = (10)(50 \text{ mV}) = 0.5 \text{ V rms}$$

(b) Next, we find the dc collector voltage.

$$R_{\text{IN}} \cong \beta_{\text{dc}} R_E = (150)(1 \text{ k}\Omega) = 150 \text{ k}\Omega$$

R_{IN} can be neglected because it is more than ten times R_2. Thus,

$$V_B \cong \left(\frac{R_2}{R_1 + R_2}\right) V_{CC} = \left(\frac{4.7 \text{ k}\Omega}{51.7 \text{ k}\Omega}\right) 25 \text{ V} = 2.27 \text{ V}$$

$$I_C \cong I_E = \frac{V_E}{R_E} = \frac{V_B - 0.7 \text{ V}}{1 \text{ k}\Omega} = 1.57 \text{ mA}$$

$$V_C = V_{CC} - I_C R_C = 25 \text{ V} - (1.57 \text{ mA})(10 \text{ k}\Omega) = 9.3 \text{ V}$$

This value is the dc level of the output. The peak value of the sine-wave output signal is

$$V_p = 1.414(0.5 \text{ V}) = 0.707 \text{ V}$$

(c) Figure 3–19 shows the waveform. The sine wave is "riding" on the 9.3 V dc level.

FIGURE 3–19

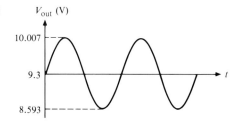

PRACTICE EXERCISE 3–4

If R_C is changed to 12 kΩ, what is the output waveform in Figure 3–18?

SIGNAL OPERATION ON THE LOAD LINE

We can obtain a graphical picture of an amplifier's operation by showing an example of signal variations on a set of collector curves with a load line, as shown in Figure 3–20. Let us assume that the dc Q-point values are as follows: $I_B = 40 \text{ } \mu\text{A}$, $I_C = 4 \text{ mA}$, and $V_{CE} = 8 \text{ V}$. The input signal varies the base current from a maximum of 50 μA to a minimum of 30 μA. The resulting variations in collector current and collector-to-emitter voltage are shown on the graph. The operation is linear as long as the variations do not reach cutoff at the lower end of the load line or saturation at the upper end.

FIGURE 3–20
Signal operation on the load line.

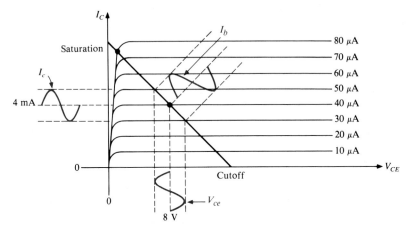

3–3 REVIEW QUESTIONS

1. What does the term *amplification* mean?
2. What is the Q point?
3. A certain transistor circuit has $R_C = 47$ kΩ and $R_E = 2.2$ kΩ. What is the approximate voltage gain?

3–4 THE BIPOLAR TRANSISTOR AS A SWITCH

In the previous section, we discussed the transistor as a linear amplifier. The second major application area is switching applications. When used as an electronic switch, a transistor is normally operated alternately in cutoff and saturation.

Figure 3–21 illustrates the basic operation of a transistor as a **switch.** In part (a), the transistor is cut off because the base-emitter junction is not forward-biased. In this condition, there is, ideally, an open between collector and emitter, as indicated by the switch equivalent. In part (b), the transistor is saturated because the base-emitter junction is forward-biased and the base current is large enough to cause the collector current to reach its saturated value. In this condition there is, ideally, a short between collector and emitter, as indicated by the switch equivalent. Actually, a voltage drop of a few tenths of a volt normally occurs, which is the saturation voltage.

FIGURE 3–21
Ideal switching action of a transistor.

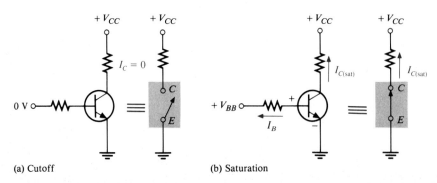

(a) Cutoff   (b) Saturation

CONDITIONS IN CUTOFF

As mentioned before, a transistor is in cutoff when the base-emitter junction is *not* forward-biased. Neglecting leakage current, all of the currents are zero, and V_{CE} is equal to V_{CC}.

$$V_{CE(\text{cutoff})} = V_{CC} \quad (3-11)$$

CONDITIONS IN SATURATION

When the emitter junction is forward-biased and there is enough base current to produce a maximum collector current, the transistor is saturated. Since V_{CE} is very small at saturation, an approximation for the collector current is

$$I_{C(\text{sat})} \cong \frac{V_{CC}}{R_C} \quad (3-12)$$

The minimum value of base current needed to produce saturation is

$$I_{B(\text{min})} = \frac{I_{C(\text{sat})}}{\beta_{\text{dc}}} \quad (3-13)$$

I_B should be significantly greater than $I_{B(\text{min})}$ to keep the transistor well into saturation.

EXAMPLE 3–5

(a) For the transistor switching circuit in Figure 3–22, what is V_{CE} when $V_{IN} = 0$ V?
(b) What minimum value of I_B is required to saturate this transistor if β_{dc} is 200? Assume $V_{CE(\text{sat})} = 0$ V.
(c) Calculate the maximum value of R_B when $V_{IN} = 5$ V.

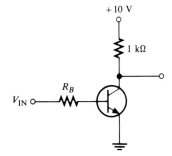

FIGURE 3–22

SOLUTION

(a) When $V_{IN} = 0$ V, the transistor is *off* and $V_{CE} = V_{CC} = 10$ V.
(b) Since $V_{CE(\text{sat})} = 0$ V,

$$I_{C(\text{sat})} \cong \frac{V_{CC}}{R_C} = \frac{10 \text{ V}}{1 \text{ k}\Omega} = 10 \text{ mA}$$

$$I_{B(\text{min})} = \frac{I_{C(\text{sat})}}{\beta_{\text{dc}}} = \frac{10 \text{ mA}}{200} = 0.05 \text{ mA}$$

This is the value of I_B necessary to drive the transistor to the point of saturation. Any further increase in I_B will drive the transistor deeper into saturation but will not increase I_C.

(c) When the transistor is saturated, $V_{BE} = 0.7$ V. The voltage across R_B is

$$V_{IN} - 0.7 \text{ V} = 4.3 \text{ V}$$

The maximum value of R_B needed to allow a minimum I_B of 0.05 mA is calculated by Ohm's law as follows:

$$R_B = \frac{V_{IN} - 0.7 \text{ V}}{I_B} = \frac{4.3 \text{ V}}{0.05 \text{ mA}} = 86 \text{ k}\Omega$$

PRACTICE EXERCISE 3-5
Determine the minimum value of I_B required to saturate the transistor in Figure 3-22 if β_{dc} is 125 and $V_{CE(sat)}$ is 0.2 V.

3-4 REVIEW QUESTIONS

1. When a transistor is used as a switching device, in what two states is it operated?
2. When does the collector current reach its maximum value?
3. When is the collector current approximately zero?
4. What are the two conditions that produce saturation?
5. When is V_{CE} equal to V_{CC}?

3-5 BJT PARAMETERS AND RATINGS

Two parameters, β_{dc} and α_{dc}, were introduced in Section 3-1. Although these parameters are related, β_{dc} is the most useful, and we will discuss it further in this section. Maximum transistor ratings are also discussed because they are important in establishing the limits within which a transistor must operate.

MORE ABOUT β_{dc}

Because the β_{dc} is a very important bipolar transistor parameter, we need to examine it further. β_{dc} is not really a constant but varies with both collector current and temperature. Keeping the junction temperature constant and increasing I_C causes β_{dc} to increase to a maximum. A further increase in I_C beyond this maximum point causes β_{dc} to decrease. If I_C is held constant and the temperature is varied, β_{dc} changes directly with the temperature. If the temperature goes up, β_{dc} goes up, and vice versa. Figure 3-23 shows the variation of β_{dc} with I_C and junction temperature (T_J) for a typical transistor.

A transistor data sheet usually specifies β_{dc} (h_{FE}) at specific I_C values. Even at fixed values of I_C and temperature, β_{dc} varies from device to device for a given transistor. The β_{dc} specified at a certain value of I_C is usually the minimum value, $\beta_{dc(min)}$, although the

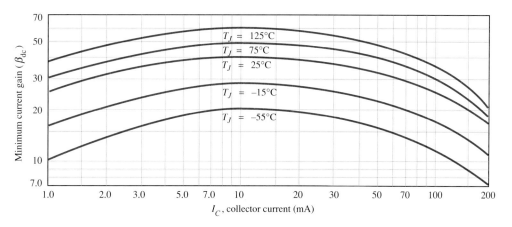

FIGURE 3–23
Variation of β_{dc} with I_C for several temperatures.

maximum and typical values are also sometimes specified. A typical transistor data sheet is shown in Appendix A.

MAXIMUM RATINGS

Like any other electronic device, the transistor has limitations on its operation. These limitations are stated in the form of maximum ratings and are normally specified on the manufacturer's data sheet. Typically, maximum ratings are given for collector-to-base voltage (V_{CB}), collector-to-emitter voltage (V_{CE}), emitter-to-base voltage (V_{EB}), collector current (I_C), and power dissipation (P_D).

The product of V_{CE} and I_C must not exceed the maximum power dissipation. Both V_{CE} and I_C cannot be maximum at the same time. If V_{CE} is maximum, I_C can be calculated as

$$I_C = \frac{P_{D(\text{max})}}{V_{CE}} \qquad (3\text{--}14)$$

If I_C is maximum, we can calculate V_{CE} by rearranging Equation (3–14) as follows:

$$V_{CE} = \frac{P_{D(\text{max})}}{I_C} \qquad (3\text{--}15)$$

For a given transistor, a maximum power dissipation curve can be plotted on the collector characteristic curves, as shown in Figure 3–24(a). These values are tabulated in Figure 3–24(b). For this transistor, $P_{D(\text{max})}$ is 0.5 W, $V_{CE(\text{max})}$ is 20 V, and $I_{C(\text{max})}$ is 50 mA. The curve shows that this particular transistor cannot be operated in the shaded portion of the graph. $I_{C(\text{max})}$ is the limiting rating between points A and B, $P_{D(\text{max})}$ is the limiting rating between points B and C, and $V_{CE(\text{max})}$ is the limiting rating between points C and D.

Figure 3–24
Maximum power dissipation curve.

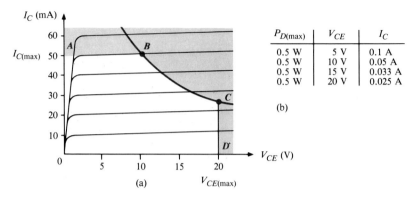

Example 3–6

The silicon transistor in Figure 3–25 has the following maximum ratings: $P_{D(max)} = 0.8$ W, $V_{CE(max)} = 15$ V, $I_{C(max)} = 100$ mA, $V_{CB(max)} = 20$ V, and $V_{EB(max)} = 10$ V. Determine the maximum value to which V_{CC} can be adjusted without exceeding a rating. Which rating would be exceeded first?

Figure 3–25

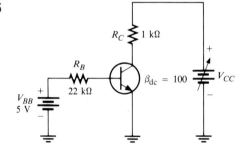

Solution
First find I_B so that I_C can be determined.

$$I_B = \frac{V_{BB} - V_{BE}}{R_B} = \frac{5\text{ V} - 0.7\text{ V}}{22\text{ k}\Omega} = 195.5\ \mu\text{A}$$

$$I_C = \beta_{dc} I_B = (100)(195.5\ \mu\text{A}) = 19.55\text{ mA}$$

I_C is much less than $I_{C(max)}$ and will not change with V_{CC}. It is determined only by I_B and β_{dc}.

The voltage drop across R_C is

$$V_{R_C} = I_C R_C = (19.55\text{ mA})(1\text{ k}\Omega) = 19.55\text{ V}$$

Now we can determine the maximum value of V_{CC} when $V_{CE} = V_{CE(max)} = 15$ V.

$$V_{R_C} = V_{CC} - V_{CE}$$

Thus,

$$V_{CC(\text{max})} = V_{CE(\text{max})} + V_{R_C} = 15 \text{ V} + 19.55 \text{ V} = 34.55 \text{ V}$$

V_{CC} can be increased to 34.55 V, under the existing conditions, before $V_{CE(\text{max})}$ is exceeded. However, we do not know whether or not $P_{D(\text{max})}$ has been exceeded at this point. Let's find out.

$$P_D = V_{CE(\text{max})} I_C = (15 \text{ V})(19.55 \text{ mA}) = 0.293 \text{ W}$$

Since $P_{D(\text{max})}$ is 0.8 W, it is *not* exceeded when $V_{CE} = 34.55$ V. Thus, $V_{CE(\text{max})}$ is the limiting rating in this case. If the base current is removed causing the transistor to turn off, $V_{CE(\text{max})}$ will be exceeded because the entire supply voltage, V_{CC}, will be dropped across the transistor.

PRACTICE EXERCISE 3–6
The transistor in Figure 3–25 has the following maximum ratings: $P_{D(\text{max})} = 0.5$ W, $V_{CE(\text{max})} = 25$ V, $I_{C(\text{max})} = 200$ mA, $V_{CB(\text{max})} = 30$ V, $V_{EB(\text{max})} = 15$ V. Determine the maximum value to which V_{CC} can be adjusted without exceeding a rating. Which rating would be exceeded first? ■

3–5 REVIEW QUESTIONS

1. Does the β_{dc} of a transistor increase or decrease with temperature?
2. Generally, what effect does an increase in I_C have on the β_{dc}?
3. What is the allowable collector current in a transistor with $P_{D(\text{max})} = 0.32$ W when $V_{CE} = 8$ V?

3–6 THE JUNCTION FIELD-EFFECT TRANSISTOR (JFET)

Recall that the bipolar junction transistor (BJT) is a current-controlled device; that is, the base current controls the amount of collector current. The **field-effect transistor (FET)** *is different; it is a voltage-controlled device in which the voltage at the gate terminal controls the amount of current through the device. Also, compared to the BJT, the FET has a very high input resistance, which makes it superior in certain applications.*

The **junction field-effect transistor (JFET)** is a type of FET that operates with a reverse-biased junction to control current in a channel. Depending on their structure, JFETs fall into either of two categories, n-channel or p-channel. Figure 3–26(a) shows the basic structure of an n-channel JFET. Wire leads are connected to each end of the n-channel; the **drain** is at the upper end and the **source** is at the lower end. Two p-type regions are diffused in the n-type material to form a channel, and both p-type regions are connected to the **gate** lead. In the remaining structure diagrams, the interconnection of both p-type regions is omitted for simplicity, with a connection to only one shown. A p-channel JFET is shown in Figure 3–26(b).

FIGURE 3–26
Basic structure of the two types of JFET.

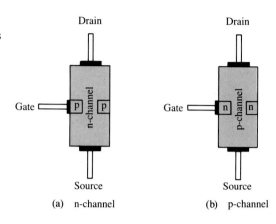

(a) n-channel

(b) p-channel

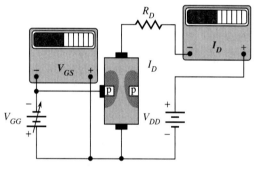

(a) JFET biased for conduction

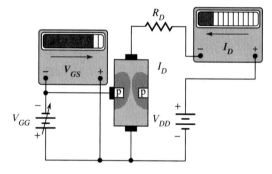

(b) Greater V_{GS} narrows the channel, thus decreasing I_D

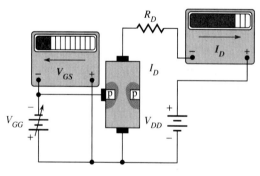

(c) Less V_{GS} widens channel and increases I_D

FIGURE 3–27
Effects of V_{GG} on channel width and drain current ($V_{GG} = V_{GS}$).

Basic Operation

To illustrate the operation of a JFET, Figure 3–27(a) shows bias voltages applied to an n-channel device. V_{DD} provides a drain-to-source voltage and supplies current from drain to source. V_{GG} sets the reverse-biased voltage between the gate and the source, as shown.

The JFET is *always* operated with the gate-to-source pn junction reverse-biased. Reverse-biasing of the gate-source junction with a negative gate voltage produces a depletion region in the n channel and thus increases its resistance. The channel width can be controlled by varying the gate voltage, and thereby the amount of drain current, I_D, can also be controlled. This concept is illustrated in Figure 3–27(b) and (c). The color shaded areas represent the depletion region created by the reverse bias. It is wider toward the drain end of the channel because the reverse-biased voltage between the gate and the drain is greater than that between the gate and the source.

JFET Symbols

The schematic symbols for both n-channel and p-channel JFETs are shown in Figure 3–28. Notice that the arrow on the gate points "in" for n-channel and "out" for p-channel.

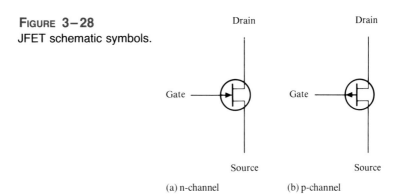

Figure 3–28
JFET schematic symbols.

(a) n-channel   (b) p-channel

3–6 Review Questions

1. What are the three terminals of a JFET?
2. Does an n-channel JFET require a positive, negative, or zero value for V_{GS}?
3. How is the drain current controlled in a JFET?

3–7 JFET Characteristics

In this section, you will see how the JFET operates as a voltage-controlled, constant-current device. You will also learn about cutoff and pinch-off as well as JFET input resistance and capacitance.

First let's consider the case where the gate-to-source voltage is zero ($V_{GS} = 0$ V). This voltage is produced by shorting the gate to the source, as in Figure 3–29(a) where both are grounded. As V_{DD} (and thus V_{DS}) is increased from 0 V, I_D will increase proportionally, as shown in the graph of Figure 3–29(b) between points A and B. In this region, the channel resistance is essentially constant because the depletion region is not large enough to have significant effect. This region is called the *ohmic region* because V_{DS} and I_D are related by Ohm's law.

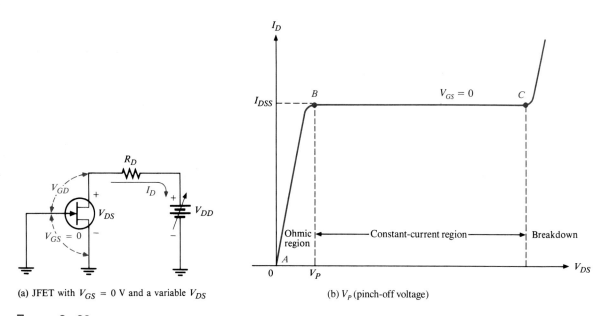

(a) JFET with $V_{GS} = 0$ V and a variable V_{DS}

(b) V_P (pinch-off voltage)

FIGURE 3–29
The drain characteristic curve of a JFET for $V_{GS} = 0$ V showing pinch-off.

At point B in Figure 3–29(b), the curve levels off and I_D becomes essentially constant. As V_{DS} increases from point B to point C, the reverse-bias voltage from gate to drain (V_{GD}) produces a depletion region large enough to offset the increase in V_{DS}, thus keeping I_D relatively constant.

PINCH-OFF

For $V_{GS} = 0$ V, the value of V_{DS} at which I_D becomes essentially constant (point B on the curve in Figure 3–29(b)) is the **pinch-off voltage, V_P**. For a given JFET, V_P has a fixed value. As you can see, a continued increase in V_{DS} above the pinch-off voltage produces an almost constant drain current. This value of drain current is I_{DSS} (Drain to Source current with gate Shorted) and is always specified on JFET data sheets. I_{DSS} is the *maximum* drain current that a specific JFET can produce regardless of the external circuit, and it is always specified for the condition, $V_{GS} = 0$ V.

Continuing along the graph in Figure 3–29(b), breakdown occurs at point C when I_D begins to increase very rapidly with any further increase in V_{DS}. Breakdown can result in irreversible damage to the device, so JFETs are always operated below breakdown and within the constant-current region (between points B and C on the graph).

V_{GS} CONTROLS I_D

Let's connect a bias voltage, V_{GG}, from gate to source as shown in Figure 3–30(a). As V_{GS} is set to increasingly more negative values by adjusting V_{GG}, a family of drain characteristic curves is produced as shown in Figure 3–30(b). Notice that I_D decreases as the magnitude of V_{GS} is increased to larger negative values. Also notice that, for each increase in V_{GS}, the JFET reaches pinch-off (where constant current begins) at values of V_{DS} less than V_P. So, the amount of drain current is controlled by V_{GS}.

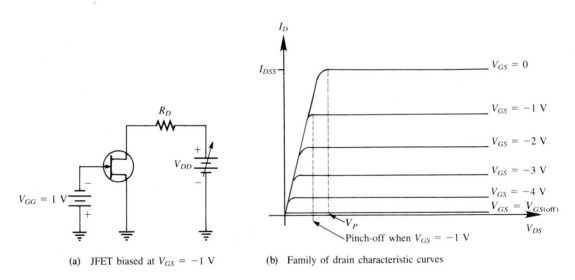

(a) JFET biased at $V_{GS} = -1$ V

(b) Family of drain characteristic curves

FIGURE 3–30
Pinch-off occurs at a lower V_{DS} as V_{GS} is increased to more negative values.

CUTOFF

The value of V_{GS} that makes I_D approximately zero is the cutoff value, $V_{GS(off)}$. The JFET must be operated between $V_{GS} = 0$ V and $V_{GS(off)}$. For this range of gate-to-source voltages, I_D will vary from a maximum of I_{DSS} to a minimum of almost zero.

As you have seen, for an n-channel JFET, the more negative V_{GS} is, the smaller I_D becomes in the constant-current region. When V_{GS} has a sufficiently large negative value, I_D is reduced to zero. This cutoff effect is caused by the widening of the depletion region to a point where it completely closes the channel as shown in Figure 3–31.

The basic operation of a p-channel JFET is the same as for an n-channel device except that it requires a negative V_{DD} and a positive V_{GS}.

FIGURE 3–31
JFET at cutoff.

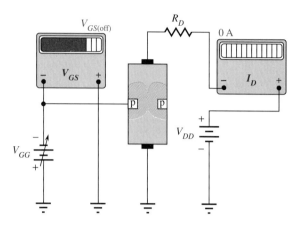

COMPARISON OF PINCH-OFF AND CUTOFF

As you have seen, there is definitely a difference between pinch-off and cutoff. There is also a connection. V_P is the value of V_{DS} at which the drain current becomes constant and is always measured at $V_{GS} = 0$ V. However, pinch-off occurs for V_{DS} values less than V_P when V_{GS} is nonzero. So, although V_P is a constant, the minimum value of V_{DS} at which I_D becomes constant varies with V_{GS}.

$V_{GS(off)}$ and V_P are always equal in magnitude but opposite in sign. A data sheet usually will give either $V_{GS(off)}$ or V_P, but not both. However, when you know one, you have the other. For example, if $V_{GS(off)} = -5$ V, then $V_P = +5$ V.

■ **EXAMPLE 3–7**

For the JFET in Figure 3–32, $V_{GS(off)} = -4$ V and $I_{DSS} = 12$ mA. Determine the *minimum* value of V_{DD} required to put the device in the constant-current region of operation.

FIGURE 3–32

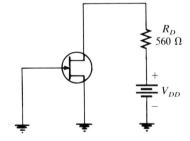

SOLUTION
Since $V_{GS(off)} = -4$, $V_P = 4$ V. The minimum value of V_{DS} for the JFET to be in its constant-current region is

$$V_{DS} = V_P = 4 \text{ V}$$

In the constant-current region with $V_{GS} = 0$ V,

$$I_D = I_{DSS} = 12 \text{ mA}$$

The drop across the drain resistor is

$$V_{RD} = (12 \text{ mA})(560 \text{ }\Omega) = 6.72 \text{ V}$$

Applying Kirchhoff's law around the drain circuit gives

$$V_{DD} = V_{DS} + V_{RD} = 4 \text{ V} + 6.72 \text{ V} = 10.72 \text{ V}$$

This is the value of V_{DD} to make $V_{DS} = V_P$ and put the device in the constant-current region.

PRACTICE EXERCISE 3–7

If V_{DD} is increased to 15 V, what is the drain current?

JFET INPUT RESISTANCE AND CAPACITANCE

A JFET operates with its gate-source junction reverse-biased. Therefore, the input resistance at the gate is very high. This high input resistance is one advantage of the JFET over the bipolar transistor. (Recall that a bipolar transistor operates with a forward-biased base-emitter junction.)

JFET data sheets often specify the input resistance by giving a value for the gate reverse current, I_{GSS}, at a certain gate-to-source voltage. The input resistance can then be determined using the following equation. The vertical lines indicate an absolute value (no sign).

$$R_{IN} = \left| \frac{V_{GS}}{I_{GSS}} \right| \qquad (3-16)$$

For example, the 2N3970 data sheet lists a maximum I_{GSS} of 250 pA for $V_{GS} = -20$ V at 25°C. I_{GSS} increases with temperature, so the input resistance decreases.

The input capacitance C_{iss} of a JFET is considerably greater than that of a bipolar transistor because the JFET operates with a reverse-biased pn junction. Recall that a reverse-biased pn junction acts as a capacitor whose capacitance depends on the amount of reverse voltage. For example, the 2N3970 has a maximum C_{iss} of 25 pF for $V_{GS} = 0$ V.

EXAMPLE 3–8 A certain JFET has an I_{GSS} of 1 nA for $V_{GS} = -20$ V. Determine the input resistance.

SOLUTION

$$R_{IN} = \left| \frac{V_{GS}}{I_{GSS}} \right| = \frac{20 \text{ V}}{1 \text{ nA}} = 20{,}000 \text{ M}\Omega$$

PRACTICE EXERCISE 3–8

Determine the minimum input resistance for the 2N3970 JFET.

3–7 REVIEW QUESTIONS

1. The drain-to-source voltage at the pinch-off point of a particular JFET is 7 V. If the gate-to-source voltage is zero, what is V_P?

2. The V_{GS} of a certain n-channel JFET is increased negatively. Does the drain current increase or decrease?
3. What value must V_{GS} have to produce cutoff in a p-channel JFET with a $V_P = -3$ V?

3-8 THE METAL OXIDE SEMICONDUCTOR FET (MOSFET)

The metal oxide semiconductor field-effect transistor (MOSFET) is the second category of field-effect transistor. The MOSFET differs from the JFET in that it has no pn junction structure; instead, the gate of the MOSFET is insulated from the channel by a silicon dioxide (SiO_2) layer. The two basic types of MOSFETs are depletion (D) and enhancement (E).

DEPLETION MOSFET (D-MOSFET)

Figure 3–33 illustrates the basic structure of D-MOSFETs. The drain and source are diffused into the substrate material and then connected by a narrow channel adjacent to the insulated gate. Both n-channel and p-channel devices are shown in the figure. We will use the n-channel device to describe the basic operation. The p-channel operation is the same, except the voltage polarities are opposite those of the n-channel.

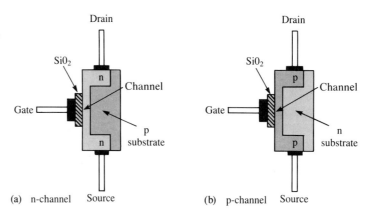

FIGURE 3–33
Basic structure of D-MOSFETs.

The D-MOSFET can be operated in either of two modes—the *depletion mode* or the *enhancement mode*—and is sometimes called a *depletion-enhancement MOSFET*. Since the gate is insulated from the channel, either a positive or a negative gate voltage can be applied. The D-MOSFET operates in the depletion mode when a negative gate-to-source voltage is applied and in the enhancement mode when a positive gate-to-source voltage is applied. These devices are generally operated in the depletion mode.

DEPLETION MODE Visualize the gate as one plate of a parallel-plate capacitor and the channel as the other plate. The silicon dioxide insulating layer is the dielectric. With a negative gate voltage, the negative charges on the gate repel conduction electrons from the channel, leaving positive ions in their place. Thereby, the n-channel is depleted of some of its electrons, thus decreasing the channel conductivity. The greater the negative voltage on the gate, the greater the depletion of n-channel electrons. At a sufficiently negative gate-to-source voltage, $V_{GS(off)}$, the channel is totally depleted and the drain current is zero. This depletion mode is illustrated in Figure 3–34(a). Like the n-channel JFET, the n-channel D-MOSFET conducts drain current for gate-to-source voltages between $V_{GS(off)}$ and 0 V. In addition, the D-MOSFET conducts for values of V_{GS} above 0 V.

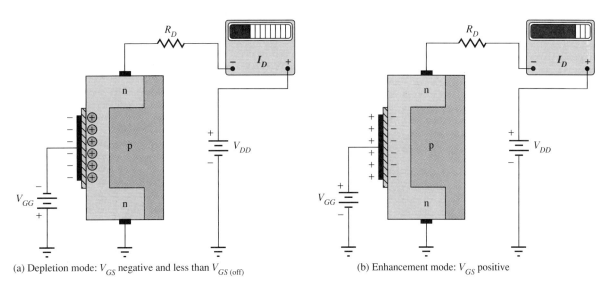

(a) Depletion mode: V_{GS} negative and less than $V_{GS\,(off)}$

(b) Enhancement mode: V_{GS} positive

FIGURE 3–34
Operation of n-channel D-MOSFET.

ENHANCEMENT MODE With a positive gate voltage, more conduction electrons are attracted into the channel, thus increasing (enhancing) the channel conductivity, as illustrated in Figure 3–34(b).

D-MOSFET SYMBOLS The schematic symbols for both the n-channel and the p-channel depletion MOSFETs are shown in Figure 3–35. The substrate, indicated by the arrow, is normally (but not always) connected internally to the source. An inward substrate arrow is for n-channel, and an outward arrow is for p-channel.

FIGURE 3–35
D-MOSFET schematic symbols.

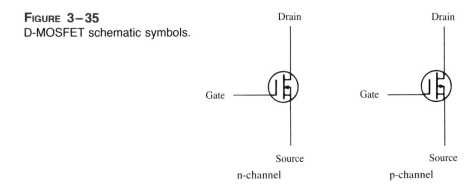

Enhancement (E-MOSFET)

This type of MOSFET operates *only* in the enhancement mode and has no depletion mode. It differs in construction from the D-MOSFET in that it has no physical channel. Notice in Figure 3–36(a) that the substrate extends completely to the SiO_2 layer.

For an n-channel device, a positive gate voltage above a threshold value, $V_{GS(th)}$, induces a channel by creating a thin layer of negative charges in the substrate region adjacent to the SiO_2 layer, as shown in Figure 3–36(b). The conductivity of the channel is enhanced by increasing the gate-to-source voltage, thus pulling more electrons into the channel. For any gate voltage below the threshold value, there is no channel.

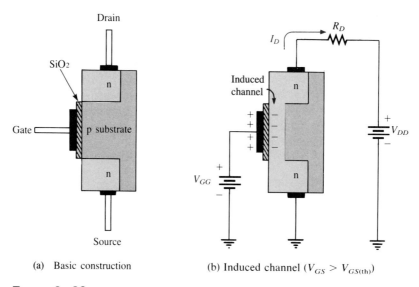

(a) Basic construction   (b) Induced channel ($V_{GS} > V_{GS(th)}$)

Figure 3–36
E-MOSFET construction and operation (n-channel).

The schematic symbols for the n-channel and p-channel E-MOSFETs are shown in Figure 3–37. The broken lines symbolize the absence of a physical channel.

Figure 3–37
E-MOSFET schematic symbols.

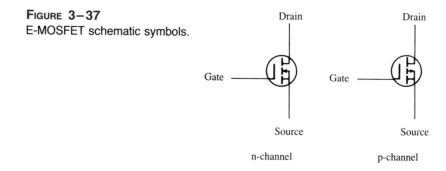

Handling Precautions

Because the gate of a MOSFET is insulated from the channel, the input resistance is extremely high (ideally infinite). The gate leakage current, I_{GSS}, for a typical MOSFET is the pA range, whereas the gate reverse current for a typical JFET is in the nA range.

The input capacitance, of course, results from the insulated gate structure. Excess static charge can accumulate because the input capacitance combines with the very high input resistance and can result in damage to the device as a result of electrostatic discharge (ESD). To avoid ESD and possible damage, the following precautions should be taken:

1. Metal oxide semiconductor (MOS) devices should be shipped and stored in conductive foam.
2. All instruments and metal benches used in assembly or testing should be connected to earth ground (round prong of wall outlets).
3. The assembler's or handler's wrist should be connected to earth ground with a length of wire and a high-value series resistor.
4. Never remove a MOS device (or any other device, for that matter) from the circuit while the power is on.
5. Do not apply signals to a MOS device while the dc power supply is off.

3–8 Review Questions

1. Name two types of MOSFETs, and describe the major difference in construction.
2. If the gate-to-source voltage in a depletion MOSFET is zero, is there current from drain to source?
3. If the gate-to-source voltage in an E-MOSFET is zero, is there current from drain to source?

3–9   FET Biasing

Using some of the FET characteristics discussed in the previous sections, we will now see how to dc bias FETs. The purpose of biasing is to select a proper dc gate-to-source voltage to establish a desired value of drain current.

Self-Biasing a JFET

Recall that a JFET must be operated such that the gate-source junction is always reverse-biased. This condition requires a negative V_{GS} for an n-channel JFET and a positive V_{GS} for a p-channel JFET. This can be achieved using the self-bias arrangements shown in Figure 3–38. Notice that the gate is biased at approximately 0 V by resistor R_G connected to ground. The reverse leakage current I_{GSS} does produce a very small voltage across R_G as indicated, but this can be neglected in most cases; it can be assumed that R_G has no voltage drop across it as we will do.

For the n-channel JFET in Figure 3–38(a), I_D produces a voltage drop across R_S and makes the source positive with respect to ground. Since $V_G = 0$ V and $V_S = I_D R_S$, the gate-to-source voltage is

$$V_{GS} = V_G - V_S = 0 - I_D R_S$$

FIGURE 3–38
Self-biased JFET ($I_S = I_D$ in all FETs).

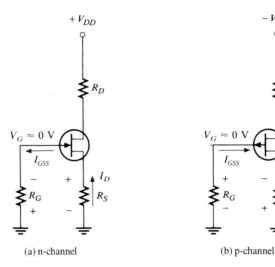

(a) n-channel (b) p-channel

Thus,

$$V_{GS} = -I_D R_S \qquad (3\text{–}17)$$

For the p-channel JFET shown in Figure 3–38(b), the current through R_S produces a negative voltage at the source, and therefore

$$V_{GS} = +I_D R_S \qquad (3\text{–}18)$$

In the following analysis, the n-channel JFET is used for illustration. Keep in mind that analysis of the p-channel JFET is the same except for opposite polarity voltages. The drain voltage with respect to ground is determined as follows:

$$V_D = V_{DD} - I_D R_D \qquad (3\text{–}19)$$

Since $V_S = I_D R_S$, the drain-to-source voltage is

$$V_{DS} = V_D - V_S$$
$$V_{DS} = V_{DD} - I_D(R_D + R_S) \qquad (3\text{–}20)$$

EXAMPLE 3–9 Find V_{DS} and V_{GS} in Figure 3–39, given that $I_D = 5$ mA.

SOLUTION
$$V_S = I_D R_S = (5 \text{ mA})(470 \text{ }\Omega) = 2.35 \text{ V}$$
$$V_D = V_{DD} - I_D R_D = 10 \text{ V} - (5 \text{ mA})(1 \text{ k}\Omega)$$
$$= 10 \text{ V} - 5 \text{ V} = 5 \text{ V}$$

Therefore,
$$V_{DS} = V_D - V_S = 5 \text{ V} - 2.35 \text{ V} = 2.65 \text{ V}$$

Since $V_G = 0$ V,
$$V_{GS} = V_G - V_S = 0 \text{ V} - 2.35 \text{ V} = -2.35 \text{ V}$$

TRANSISTORS AND THYRISTORS

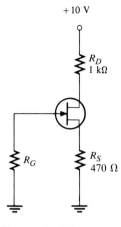

FIGURE 3–39

PRACTICE EXERCISE 3–9
Determine V_{DS} and V_{GS} in Figure 3–39 when $I_D = 8$ mA. Assume that $R_D = 860\ \Omega$, $R_S = 390\ \Omega$, and $V_{DD} = 12$ V.

D-MOSFET BIAS

Recall that D-MOSFETs can be operated with either positive or negative values of V_{GS}. A simple bias method is to set $V_{GS} = 0$ V so that an ac signal at the gate varies the gate-to-source voltage above and below this bias point. A MOSFET with 0 bias is shown in Figure 3–40. Since $V_{GS} = 0$, $I_D = I_{DSS}$ as indicated. The drain-to-source voltage is expressed as follows:

$$V_{DS} = V_{DD} - I_{DSS}R_D \qquad (3-21)$$

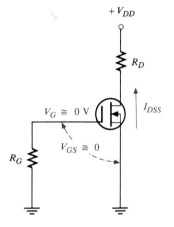

FIGURE 3–40
A zero-biased D-MOSFET.

■ **EXAMPLE 3–10** Determine the drain current in the circuit of Figure 3–41. The MOSFET data sheet gives $V_{GS(\text{off})} = -8$ V and $I_{DSS} = 12$ mA.

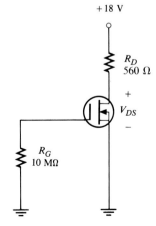

FIGURE 3–41

Solution
Since $I_D = I_{DSS} = 12$ mA, the drain-to-source voltage is calculated as follows:
$$V_{DS} = V_{DD} - I_{DSS}R_D = 18\text{ V} - (12\text{ mA})(560\ \Omega) = 11.28\text{ V}$$

Practice Exercise 3–10
Find V_{DS} in Figure 3–41 when $V_{GS(\text{off})} = -10$ V and $I_{DSS} = 20$ mA.

E-MOSFET Bias

Recall that E-MOSFETs must have a V_{GS} greater than the threshold value, $V_{GS(\text{th})}$. Figure 3–42 shows two ways to bias an E-MOSFET. (D-MOSFETs can also be biased using these methods.) An n-channel device is used for illustration. In either the drain-feedback or the voltage-divider bias arrangement, the purpose is to make the gate voltage more positive than the source by an amount exceeding $V_{GS(\text{th})}$.

Figure 3–42
E-MOSFET biasing arrangements.

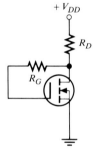
(a) Drain-feedback bias

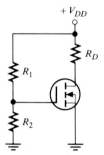
(b) Voltage-divider bias

In the drain-feedback bias circuit in Figure 3–42(a), there is negligible gate current and, therefore, no voltage drop across R_G. As a result, $V_{GS} = V_{DS}$.

Equations for the voltage-divider bias in Figure 3–42(b) are as follows:
$$V_{GS} = \left(\frac{R_2}{R_1 + R_2}\right)V_{DD}$$
$$V_{DS} = V_{DD} - I_D R_D$$

■ Example 3–11
Determine the amount of drain current in Figure 3–43. The MOSFET has a $V_{GS(\text{th})}$ of 3 V.

Solution
The meter indicates that $V_{GS} = 8.5$ V. Since this is a drain-feedback configuration, $V_{DS} = V_{GS} = 8.5$ V.
$$I_D = \frac{V_{DD} - V_{DS}}{R_D} = \frac{15\text{ V} - 8.5\text{ V}}{4.7\text{ k}\Omega} = 1.38\text{ mA}$$

FIGURE 3–43

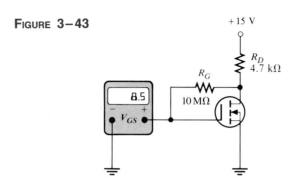

PRACTICE EXERCISE 3–11
Determine I_D if the meter in Figure 3–43 reads 5 V.

3–9 REVIEW QUESTIONS

1. Should a p-channel JFET have a positive or a negative V_{GS}?
2. In a certain self-biased n-channel JFET circuit, $I_D = 8$ mA and $R_S = 1$ kΩ. What is V_{GS}?
3. For a D-MOSFET biased at $V_{GS} = 0$ V, is the drain current equal to 0, I_{GSS}, or I_{DSS}?
4. For an n-channel E-MOSFET with $V_{GS(th)} = 2$ V, V_{GS} must be in excess of what value in order to conduct?

3–10 UNIJUNCTION TRANSISTORS (UJTs)

The unijunction transistor (UJT) *is a single pn junction device consisting of an emitter and two bases. It has no collector. The UJT is used mainly in switching and timing applications.*

Figure 3–44(a) shows the construction of a unijunction transistor. The base contacts are made to the n-type bar. The emitter lead is connected to the p-region. The UJT schematic symbol is shown in Figure 3–44(b).

FIGURE 3–44
Unijunction transistor (UJT).

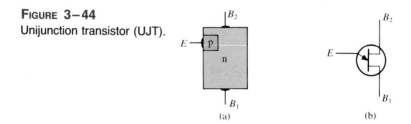

UJT OPERATION

In normal UJT operation, base 2 (B_2) and the emitter are biased positive with respect to base 1 (B_1). Figure 3–45 will illustrate the operation. This equivalent circuit represents

FIGURE 3–45
Equivalent circuit for a UJT.

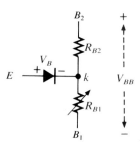

the internal UJT characteristics. The total resistance between the two bases, R_{BB}, is the resistance of the n-type material. R_{B1} is the resistance between point k and base 1. R_{B2} is the resistance between point k and base 2. The sum of these two resistances makes up the total resistance, R_{BB}. The diode represents the pn junction between the emitter and the n-type material.

The ratio R_{B1}/R_{BB} is designated η (the Greek letter eta) and is defined as the *intrinsic standoff ratio*. It takes an emitter voltage of $V_B + \eta V_{BB}$ to turn the UJT on. This voltage is called the *peak voltage*. Once the device is on, resistance R_{B1} drops in value. Thus, as emitter current increases, emitter voltage decreases because of the decrease in R_{B1}. This characteristic is the negative resistance characteristic of the UJT.

As the emitter voltage decreases, it reaches a value called the *valley voltage*. At this point, the pn junction is no longer forward-biased, and the UJT turns off.

AN APPLICATION

UJTs are commonly used in oscillator circuits. Figure 3–46 shows a typical circuit. Its operation is as follows: Initially the capacitor is uncharged and the UJT is off. When power is applied, the capacitor charges up exponentially. When it reaches the peak voltage, the UJT turns on and the capacitor begins to discharge as indicated. When the emitter reaches the valley voltage, the UJT turns off and the capacitor begins to charge again. The cycle repeats. Waveforms of the capacitor voltage and the R_1 voltage are shown in the diagram.

FIGURE 3–46
UJT oscillator circuit.

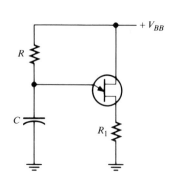
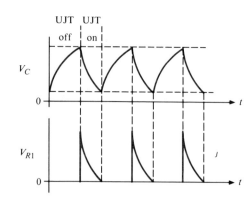

3-10 Review Questions

1. What does "UJT" stand for?
2. What are the terminals of a UJT?

3-11 Thyristors

Thyristors are devices constructed of four layers of semiconductor material. The types of thyristors covered in this section are the silicon-controlled rectifier (SCR), the diac, and the triac. These thyristor devices share certain characteristics. They act as open circuits capable of withstanding a certain rated voltage until triggered. When turned on (triggered), they become low-resistance current paths and remain so, even after the trigger is removed, until the current is reduced to a certain level or until they are turned off, depending on the type of device. Thyristors can be used to control the amount of ac power to a load and are used in lamp-dimming circuits, motor speed control, ignition systems, and charging circuits, to name a few. Unijunction transistors (UJTs) are often used as the trigger devices for thyristors.

Silicon-Controlled Rectifiers (SCRs)

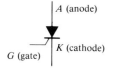

FIGURE 3-47
SCR symbol.

The **silicon-controlled rectifier (SCR)** has three terminals, as shown in the symbol in Figure 3-47. Like a rectifier diode, it is a unidirectional device, but the conduction of current is controlled by the gate, G. Current is from the cathode K to the anode A when the SCR is on. A positive voltage on the gate will turn the SCR on. The SCR will remain on as long as the current from the cathode to the anode is equal to or greater than a specified value called the *holding current*. When the current drops below the holding value, the SCR will turn off. It can be turned on again by a positive voltage on the gate.

SCR Construction The SCR is a four-layer device. It has two p- and two n-regions, as shown in Figure 3-48(a). This construction can be thought of as an npn and a pnp transistor interconnected as in part (b) of the figure. The upper n region is commonly shared by the base and the collector of the transistors.

FIGURE 3-48
Construction of an SCR and its transistor analogy.

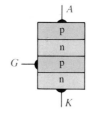

(a)

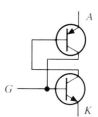

(b)

AN APPLICATION Figure 3–49 shows an SCR used to rectify and control the average power delivered to a load. The SCR rectifies the 60 Hz ac just as a conventional rectifier diode. It can conduct only during the positive half-cycle. The purpose of the phase controller is to produce a trigger pulse at the SCR gate in order to turn it on during any portion of the positive half-cycle of the input. If the SCR is turned on earlier in the half-cycle, more average power is delivered to the load. If the SCR is fired later in the half-cycle, less average power is delivered.

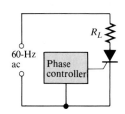
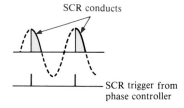

FIGURE 3–49
SCR controls average power to load.

TRIACS

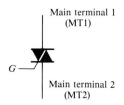

FIGURE 3–50
Triac symbol.

The **triac** is also known as a *bidirectional triode thyristor*. It is equivalent to two SCRs connected to allow current in either direction. The SCRs share a common gate. The symbol for a triac is shown in Figure 3–50.

APPLICATIONS Like the SCR, triacs are also used to control average power to a load by the method of phase control. The triac can be triggered such that the ac power is supplied to the load for a controlled portion of each half-cycle. During each positive half-cycle of the ac, the triac is off for a certain interval, called the *delay angle* (measured in degrees). Then it is triggered on and conducts current through the load for the remaining portion of the positive half-cycle, called the *conduction angle*. Similar action occurs on the negative half-cycle except that, of course, current is conducted in the opposite direction through the load. Figure 3–51 illustrates this action.

FIGURE 3–51
Basic triac phase control.

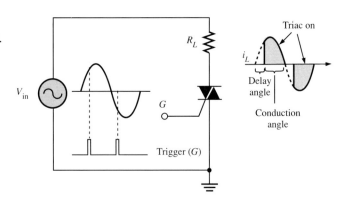

FIGURE 3–52
Diac symbol.

DIACS

The **diac** is a bidirectional device that does not have a gate. It conducts current in either direction when a sufficient voltage, called the *breakover potential,* is reached across the two terminals. The symbol for a diac is shown in Figure 3–52.

3–11 REVIEW QUESTIONS

1. What does "SCR" stand for?
2. How does an SCR basically work?
3. Name a typical application of a triac.
4. A diac is gate-controlled (T or F).

3–12 TRANSISTOR PACKAGES AND TERMINAL IDENTIFICATION

Transistors are available in a wide range of package types for various applications. Those with mounting studs or heat sinks are power transistors. Low- and medium-power transistors are usually found in smaller metal or plastic cases. Still another package classification is for high-frequency devices. As a technician, you often must check a transistor, either in the circuit or out of the circuit, for possible failure. First, of course, you must be able to recognize the transistor by its physical appearance. Also, you must be able to identify the base, collector, and emitter leads (or gate, drain, and source).

Transistors are found in a range of shapes and sizes. Figure 3–53 shows a few of the most typical case styles. Pin identification and construction diagrams are shown in Figure 3–54 for several types of transistor packages. Generally, the heat sink mounting or stud on power transistors is the collector terminal. On the metal "top hat" cases, the tab is closest to the emitter (or source) lead, and the collector (or drain) is often connected to the case.

FIGURE 3–53
Some common package configurations (copyright of Motorola, Inc. Used by permission).

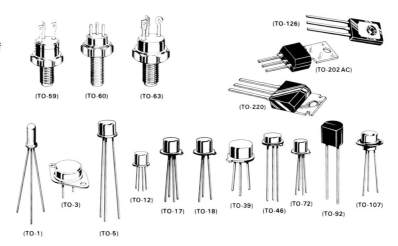

Chapter 3

FIGURE 3-54
Typical transistor packages, construction views, and examples of pin arrangements. For FETs, replace base with gate, collector with drain, and emitter with source (copyright of Motorola, Inc. Used by permission).

3-12 REVIEW QUESTIONS

1. Identify the leads on the bipolar transistors in Figure 3-55.

FIGURE 3-55

(a)

(b)

(c)

3-13 TROUBLESHOOTING

As you already know, a critical skill in electronics is the ability to identify a circuit malfunction and to isolate the failure to a single component if possible. In this section, the basics of troubleshooting transistor bias circuits and testing individual transistors are covered.

TROUBLESHOOTING A TRANSISTOR BIAS CIRCUIT

Several things can cause a failure in a transistor bias circuit. The most common faults are open base or collector resistor, loss of base or collector bias voltage, open transistor junctions, and open contacts or interconnections on the printed circuit board. Figure 3–56 illustrates a basic four-point check that will help isolate one of these faults in the circuit. The term *floating* refers to a point that is effectively disconnected from a "solid" voltage or ground. A very small, erratically fluctuating voltage is sometimes observed.

FIGURE 3–56
Test point troubleshooting guide for a basic transistor bias circuit.

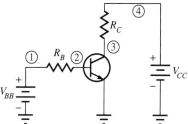

Symptom	Test point voltages			
	1	2	3	4
A. No V_{BB} Source shorted Source open	0 V Floating	0 V Floating	V_{CC} V_{CC}	V_{CC} V_{CC}
B. No V_{CC} Source shorted Source open	V_{BB} V_{BB}	0.7 V 0.7 V	0 V Floating	0 V Floating
C. R_{BN} open	V_{BB}	Floating	V_{CC}	V_{CC}
D. R_C open	V_{BB}	0.7 V	Floating	V_{CC}
E. Base-to-emitter junction open	V_{BB}	V_{BB}	V_{CC}	V_{CC}
F. Base-to-collector junction open	V_{BB}	0.7 V	V_{CC}	V_{CC}
G. Low β	V_{BB}	0.7 V	Higher than normal	V_{CC}

EXAMPLE 3–12

From the voltage measurements in Figure 3–57, what is the most likely problem?

FIGURE 3–57

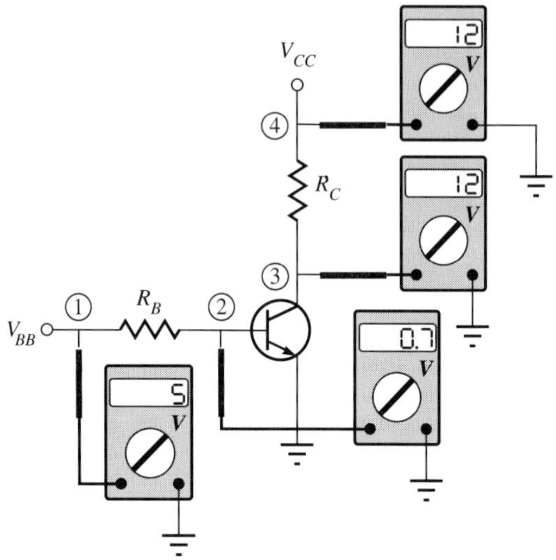

SOLUTION
Since $V_{CE} = 12$ V, the transistor is obviously in cutoff. The V_{BB} measurement appears OK and the voltage at the base is correct. Therefore, the collector-base junction of the transistor must be open. Replace the transistor and check the voltages again.

PRACTICE EXERCISE 3–12
What is the voltage at point 2 if the *BE* junction opens in Figure 3–57?

TRANSISTOR TESTING

An individual transistor can be tested either in-circuit or out-of-circuit. For example, let's say that an amplifier on a particular printed circuit (pc) board has malfunctioned. You perform a check and find that the symptoms are the same as Item F in the table of Figure 3–56, indicating an open from base to collector. This can mean either a bad transistor or an open in an external circuit connection. Your next step is to find out which one.

Good troubleshooting practice dictates that you do not remove a component from a circuit board unless you are reasonably sure that it is bad or you simply cannot isolate the problem down to a single component. When components are removed, there is a risk of damage to the pc board contacts and traces.

The next step is to do an in-circuit check of the transistor using a transistor tester similar to the ones shown in Figure 3–58. The three clip-leads are connected to the transistor terminals and the tester gives a GOOD/BAD indication.

CASE 1
If the transistor tests BAD, it should be carefully removed and replaced by a known good one. An out-of circuit check of the replacement device is usually a good idea, just to make

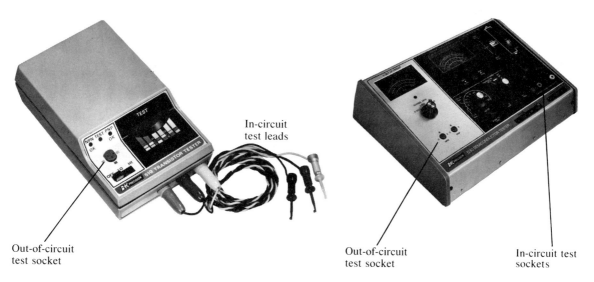

FIGURE 3–58
Transistor testers (courtesy of B & K Precision).

sure it is OK. The transistor is plugged into the socket on the transistor tester for out-of-circuit tests.

CASE 2

If the transistor tests GOOD in-circuit, examine the circuit board for a poor connection at the collector pad or for a break in the connecting trace. A poor solder joint often results in an open or a highly resistive contact. The physical point at which you actually measure the voltage is very important in this case. If you measure on the collector side of an external open, you will get the indication listed in Item D of the table in Figure 3–56. If you measure on the side of the external open closest to R_C, you will get the indication in Item F of the table. This situation is illustrated in Figure 3–59.

FIGURE 3–59
The indication of an open, when it is in the external circuit, depends on *where* you measure.

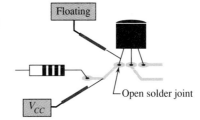

IMPORTANCE OF POINT-OF-MEASUREMENT IN TROUBLESHOOTING

It is interesting to look back and notice that if you had taken the initial measurement on the transistor lead itself and the open were external to the transistor, you would have measured V_{CC}. This would have indicated a bad transistor even before the tester was used.

CHAPTER 3

This simple concept emphasizes the importance of point-of-measurement in certain troubleshooting situations as further illustrated in Figure 3–60 for the case just discussed. This is also valid for opens at the base and emitter of a transistor.

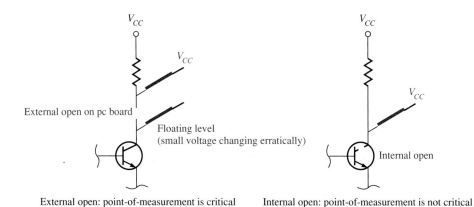

FIGURE 3–60
Importance of the point-of-measurement for certain open failures in transistor circuits.

EXAMPLE 3–13

What fault do the measurements in Figure 3–61 indicate?

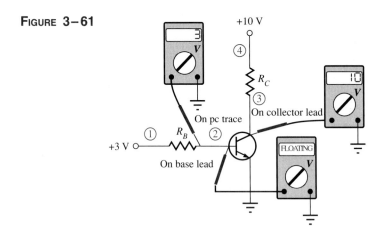

FIGURE 3–61

SOLUTION

The transistor is in cutoff, as indicated by the 10 V measurement on the collector lead. The base bias voltage of 3 V appears on the pc board contacts but not on the transistor lead. This indicates that there is an open external to the transistor between the two measured points. Check the solder joint at the base contact on the pc board. If the open were internal, there would be 3 V on the base lead.

Practice Exercise 3-13
If the meter in Figure 3-61 that now reads 3 V indicates a floating point, what is the most likely fault?

Out-of-Circuit Multimeter Test

If a transistor tester is not available, an analog multimeter can be used to perform a basic diode check on the transistor junctions for opens or shorts. For this test, a transistor can be viewed as two back-to-back diodes as shown in Figure 3-62. The *BC* junction is one diode and the *BE* junction is the other. Also many DMMs have a transistor test feature as Figure 3-63 indicates.

Figure 3-62
The transistor is viewed as two diodes for a junction check with a multimeter.

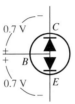
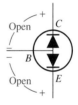

(a) Both junctions should read approximately 0.7 V when forward-biased.

(b) Both junctions should read open (ideally) when reverse-biased.

Figure 3-63
A DMM with a transistor test feature (courtesy of B & K Precision).

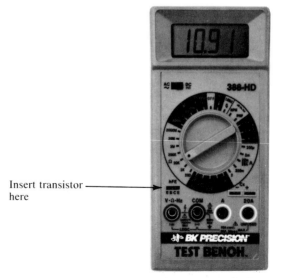

Insert transistor here

LEAKAGE MEASUREMENT

Very small leakage currents exist in all transistors and in most cases are small enough to neglect (usually nA). When a transistor is connected as shown in Figure 3–64(a) with the base open ($I_B = 0$), it is in cutoff. Ideally, $I_C = 0$; but actually there is a small current from collector to emitter, as mentioned earlier, called I_{CEO} (collector-to-emitter current with base open). This leakage current is usually in the nA range for silicon. A faulty transistor will often have excessive leakage current and can be checked in a transistor tester, which connects an ammeter as shown in part (a). Another leakage current in transistors is the reverse collector-to-base current, I_{CBO}. This is measured with the emitter open, as shown in Figure 3–64(b). If it is excessive, a shorted collector-base junction is likely.

FIGURE 3–64
Leakage current test circuits.

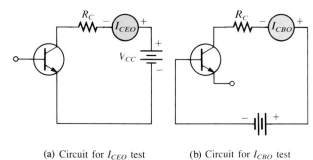

(a) Circuit for I_{CEO} test   (b) Circuit for I_{CBO} test

GAIN MEASUREMENT

In addition to leakage tests, the typical transistor tester also checks the β_{dc}. A known value of I_B is applied and the resulting I_C is measured. The reading will indicate the value of the I_C/I_B ratio, although in some units only a relative indication is given. Most testers provide for an in-circuit β_{dc} check, so that a suspected device does not have to be removed from the circuit for testing.

CURVE TRACERS

The *curve tracer* is an oscilloscope type of instrument that can display transistor characteristics such as a family of collector curves. In addition to the measurement and display of various transistor characteristics, diode curves can also be displayed, as well as the β_{dc}. A typical curve tracer is shown in Figure 3–65.

3–13 REVIEW QUESTIONS

1. If a transistor on a circuit board is suspected of being faulty, what should you do?
2. In a transistor bias circuit, such as the one in Figure 3–56, what happens if R_B opens?

FIGURE 3-65
Curve tracer (courtesy of Tektronix, Inc.).

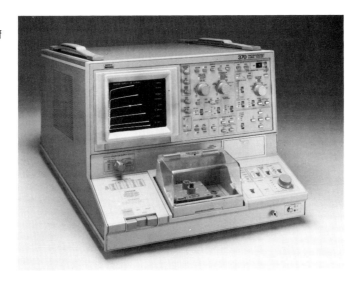

3. In a circuit such as the one in Figure 3–56, what are the base and collector voltages if there is an open between the emitter and ground?

3-14 A SYSTEM APPLICATION

At the opening of this chapter, you saw a block diagram for a basic electronic security system. The system has several parts, but in this section we are concerned with the transistor circuits that detect an open in the loops containing remote sensors. In this particular application, the transistors are used as switching devices, and you will

☐ *See how transistors are used in a switching application.*
☐ *See how a transistor is used to activate a relay.*
☐ *See how one transistor is used to drive another transistor.*
☐ *Translate between a printed circuit board and a schematic.*
☐ *Troubleshoot some common transistor circuit failures.*

A BRIEF DESCRIPTION OF THE SYSTEM

The circuit board shown in the system diagram in Figure 3–66 contains the transistor circuits for detecting when one of the remote switch sensors is open. There are three remote loops in this particular system. One loop protects all windows/doors in one area of the structure (Zone 1), and another loop protects all windows/doors in a second area (Zone 2). The third loop protects the main entry door. When an intrusion occurs, a switch sensor at the point of intrusion breaks contact and opens the loop. This causes the input to the detection circuit for that loop to go to a 0 V level and activate the circuit which, in turn, energizes the relay causing it to set off an alarm and/or initiate an automatic telephone dialing sequence. Either of the detection circuits for Zone 1 or Zone 2 can energize the common relay. The output of the detection circuit for the main entry goes to a time delay circuit to allow time for keying in an entry code that will disarm the system. If, after a sufficient time, no code or an improper code has been entered, the delay circuit will initiate an alarm.

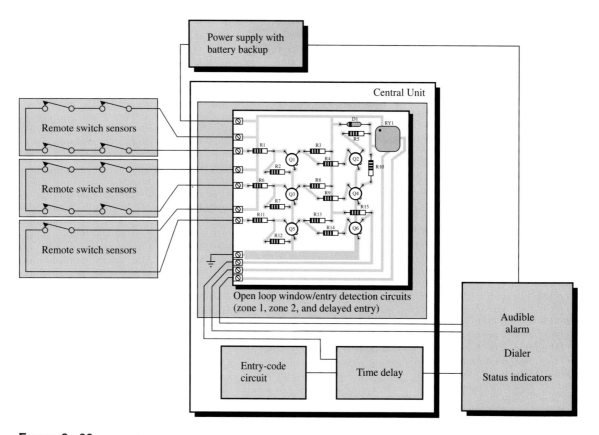

FIGURE 3–66

Although there are many aspects to a security system and a system can range from very simple to very complex, in this application we are concentrating only on the detection circuits board.

Now, so you can take a closer look at the detection circuits, let's take the board out of the system and put it on the test bench.

On the Test Bench

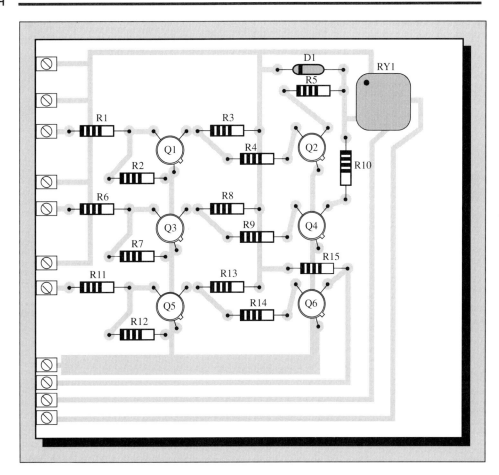

Figure 3–67

Identifying the Components

There are three separate detection circuits on the board in Figure 3–67, each consisting of two transistors and their associated resistors. The transistors are labeled Q1, Q2, and so forth. There is one relay labeled RY1 and a diode D1 for suppression of negative transients across the relay coil. As shown in Figure 3–68(a), the transistors are housed in TO-18 (TO-206AA) metal cans (remember, the emitter is closest to the tab). A detail of the relay is shown in Figure 3–68(b). You are already familiar with resistors and diodes.

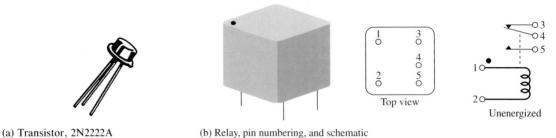

(a) Transistor, 2N2222A   (b) Relay, pin numbering, and schematic

FIGURE 3–68
Part (a) copyright of Motorola, Inc. Used by permission.

■ ACTIVITY 1   RELATE THE PC BOARD TO THE SCHEMATIC

Carefully follow the traces on the pc board to see how the components are interconnected. Compare each point (*A* through *X*) on the circuit board with the corresponding point on the schematic in Figure 3–69. For each point on the pc board, place the letter of the corresponding point or points on the schematic.

■ ACTIVITY 2   ANALYZE THE CIRCUITS

Calculate the amount of base current and collector current for each transistor when it is saturated. Refer to the β_{dc} (h_{FE}) information on the data sheet for the 2N2222 in Appendix A to make sure there is sufficient base current to keep the transistor well in saturation. Assume that the relay has a coil resistance of 15 Ω and requires a minimum of 400 mA to operate.

■ ACTIVITY 3   WRITE A TECHNICAL REPORT

Discuss the detailed operation of the transistor detection circuits on the pc board. Also, describe the inputs and outputs indicating what other areas of the system they connect to and describe the purpose served by each one.

■ ACTIVITY 4   DEVELOP A COMPLETE TEST PROCEDURE

Someone with no knowledge of how this circuitry works should be able to use your test procedure and verify that all of the circuits operate properly. The circuit board test is to be done on the test bench and not in the functional system, so you must simulate operational conditions.

STEP 1   Begin with the most basic and necessary thing—power.

STEP 2   Develop a detailed step-by-step procedure for simulating the external remote loops.

STEP 3   For each loop detector circuit, specify the points at which to measure voltage, the value of the voltage, and under what conditions it is to be measured. Detail each step required to fully check out all the circuits.

STEP 4   Specify in detail any other tests in addition to voltage measurements.

FIGURE 3-69

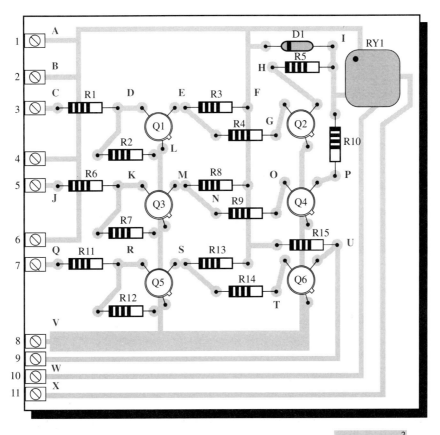

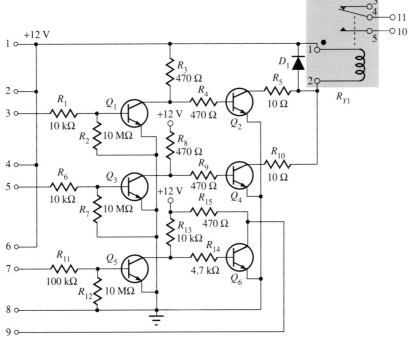

147

ACTIVITY 5 TROUBLESHOOT THE CIRCUIT BOARD FOR EACH OF THE FOLLOWING PROBLEMS BY STATING THE PROBABLE CAUSE OR CAUSES IN EACH CASE

1. Approximately 5.65 V at the collector of Q1 with the remote loop switch closed.
2. Relay is energized (contact from 10 to 11) continuously and independent of remote switches.
3. Collector lead of Q4 floating.
4. Pin 9 is always at 12 V. Q6 collector lead is also at 12 V.
5. When 12 V is applied to pin 5 with the closed loop switch, the base of Q3 is 0 V.
6. Approximately 3.6 V at the collector of Q5 when 12 V is applied to pin 7.

3–14 REVIEW QUESTIONS

1. State the basic purpose of the detector circuit board in the security system.
2. How many of the circuits on the board are identical?
3. When is there a contact closure between pins 10 and 11 on the pc board?
4. Based on your knowledge of coils and diodes, why do you think D1 is in the circuit?

SUMMARY

- A bipolar junction transistor (BJT) consists of three regions: emitter, base, and collector.
- The three regions of a BJT are separated by two pn junctions.
- The two types of bipolar transistor are the npn and the pnp.
- The term *bipolar* refers to two types of current: electron current and hole current.
- A field-effect transistor (FET) has three terminals: source, drain, and gate.
- A junction field-effect transistor (JFET) operates with a reverse-biased gate-to-source pn junction.
- JFETs have very high input resistance due to the reverse-biased gate-source junction.
- JFET current flows between the drain and the source through a channel whose width is controlled by the amount of reverse bias on the gate-source junction.
- The two types of JFETs are n-channel and p-channel.
- Metal oxide semiconductor field-effect transistors (MOSFETs) differ from JFETs in that the gate of a MOSFET is insulated from the channel.
- A depletion MOSFET (D-MOSFET) can operate with a positive, negative, or zero gate-to-source voltage.
- The D-MOSFET has a physical channel between the drain and the source.
- An enhancement MOSFET (E-MOSFET) can operate only when the gate-to-source voltage exceeds a threshold value.
- The enhancement MOSFET has no physical channel.
- Transistors are used as either amplifying devices or switching devices.
- Transistor and thyristor symbols are shown in Figure 3–70.

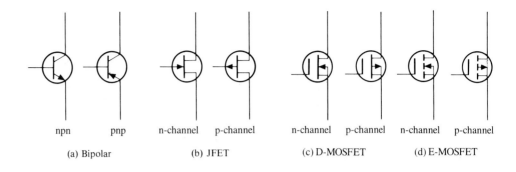

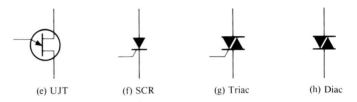

FIGURE 3–70
Transistor and thyristor symbols.

Glossary

Alpha (α)   The ratio of collector current to emitter current in a bipolar junction transistor.

Amplification   The process of producing a larger voltage, current, or power using a smaller input signal as a "pattern."

Amplifier   An electronic circuit having the capability of amplification and designed specifically for that purpose.

Base   One of the semiconductor regions in a bipolar junction transistor.

Beta (β)   The ratio of collector current to base current in a bipolar junction transistor.

Bipolar   Characterized by two pn junctions.

Bipolar junction transistor (BJT)   A transistor constructed with three doped semiconductor regions separated by two pn junctions.

Collector   One of the semiconductor regions in a BJT.

Cutoff   The nonconducting state of a transistor.

Diac   A semiconductor device that can conduct current in either of two directions when properly activated.

Drain   One of the three terminals of a field-effect transistor.

Emitter   One of the three semiconductor regions in a BJT.

Field-effect transistor (FET) A voltage-controlled device in which the voltage at the gate terminal controls the amount of current through the device.

Gain The amount of amplification.

Gate One of the three terminals of an FET.

Junction field-effect transistor (JFET) A type of FET that operates with a reverse-biased junction to control current in a channel.

MOSFET Metal-oxide semiconductor field-effect transistor; one of two major types of FET that operates without a pn junction.

Pinch-off voltage The value of the drain-to-source voltage of an FET at which the drain current becomes constant when the gate-to-source voltage is zero.

Q point The dc operating (bias) point of an amplifier.

Saturation The state of a BJT in which the collector has reached a maximum and is independent of the base current.

Silicon-controlled rectifier (SCR) A device that can be triggered on to conduct current in one direction.

Source One of the three terminals of a field-effect transistor.

Switch An electrical or electronic device for opening and closing a current path.

Thyristor A class of four-layer semiconductor devices.

Transistor A semiconductor device used for amplification and switching applications in electronic circuits.

Triac A bidirectional thyristor that can be triggered into conduction.

Unijunction transistor (UJT) A type of transistor consisting of an emitter and two bases.

FORMULAS

(3–1)	$I_E = I_C + I_B$	Bipolar transistor currents
(3–2)	$I_C = \alpha_{dc} I_E$	Relationship of collector and emitter currents
(3–3)	$I_C = \beta_{dc} I_B$	Relationship of collector and base currents
(3–4)	$V_C = V_{CC} - I_C R_C$	Collector voltage
(3–5)	$V_B = V_E + V_{BE}$	Base voltage
(3–6)	$R_{IN} \cong \beta_{dc} R_E$	Input resistance at base
(3–7)	$V_B = \left(\dfrac{R_2 \| R_{IN}}{R_1 + R_2 \| R_{IN}}\right) V_{CC}$	Base voltage with voltage-divider bias
(3–8)	$V_B \cong \left(\dfrac{R_2}{R_1 + R_2}\right) V_{CC}$	Approximate base voltage ($R_{IN} \gg R_2$)
(3–9)	$\beta_{ac} = \dfrac{I_c}{I_b}$	AC beta

TRANSISTORS AND THYRISTORS

(3–10) $\quad A_v \cong \dfrac{R_C}{R_E}$ $\qquad$ Voltage gain

(3–11) $\quad V_{CE(\text{cutoff})} = V_{CC}$ $\qquad$ V_{CE} at cutoff

(3–12) $\quad I_{C(\text{sat})} \cong \dfrac{V_{CC}}{R_C}$ $\qquad$ Collector saturation current

(3–13) $\quad I_{B(\text{min})} = \dfrac{I_{C(\text{sat})}}{\beta_{\text{dc}}}$ $\qquad$ Minimum base current for saturation

(3–14) $\quad I_C = \dfrac{P_{D(\text{max})}}{V_{CE}}$ $\qquad$ I_C for maximum V_{CE}

(3–15) $\quad V_{CE} = \dfrac{P_{D(\text{max})}}{I_C}$ $\qquad$ V_{CE} for maximum I_C

(3–16) $\quad R_{IN} = \left|\dfrac{V_{GS}}{I_{GSS}}\right|$ $\qquad$ JFET input resistance

(3–17) $\quad V_{GS} = -I_D R_S$ $\qquad$ Self-bias voltage for n-channel JFET

(3–18) $\quad V_{GS} = +I_D R_S$ $\qquad$ Self-bias voltage for p-channel JFET

(3–19) $\quad V_D = V_{DD} - I_D R_D$ $\qquad$ Drain voltage

(3–20) $\quad V_{DS} = V_{DD} - I_D(R_D + R_S)$ $\qquad$ Drain-to-source voltage

(3–21) $\quad V_{DS} = V_{DD} - I_{DSS} R_D$ $\qquad$ Drain-to-source voltage D-MOSFET

SELF-TEST

1. The n-type regions in an npn bipolar junction transistor are
   (a) collector and base
   (b) collector and emitter
   (c) base and emitter
   (d) collector, base, and emitter

2. The n-region in a pnp transistor is the
   (a) base   (b) collector   (c) emitter   (d) case

3. For normal operation of an npn transistor, the base must be
   (a) disconnected
   (b) negative with respect to the emitter
   (c) positive with respect to the emitter
   (d) positive with respect to the collector

4. The three currents in a BJT are
   (a) forward, reverse, neutral
   (b) drain, source, gate
   (c) alpha, beta, and sigma
   (d) base, emitter, and collector

5. Beta (β) is the ratio of
   (a) collector current to emitter current
   (b) collector current to base current
   (c) emitter current to base current
   (d) output voltage to input voltage

6. Alpha (α) is the ratio of
   - (a) collector current to emitter current
   - (b) collector current to base current
   - (c) emitter current to base current
   - (d) output voltage to input voltage
7. If the beta of a certain transistor is 30 and the base current is 1 mA, the collector current is
   - (a) 0.33 mA   (b) 1 mA   (c) 30 mA   (d) unknown
8. If the base current is increased,
   - (a) the collector current increases and the emitter current decreases
   - (b) the collector current decreases and the emitter current decreases
   - (c) the collector current increases and the emitter current does not change
   - (d) the collector current increases and the emitter current increases
9. When an n-channel JFET is biased for conduction, the gate is
   - (a) positive with respect to the source
   - (b) negative with respect to the source
   - (c) at the same voltage as the source
   - (d) at the same voltage as the drain
10. When the gate-to-source voltage of an n-channel JFET is made more positive, the drain current
    - (a) decreases   (b) increases   (c) stays constant   (d) becomes zero
11. When a negative gate-to-source voltage is applied to an n-channel MOSFET, it operates in the
    - (a) cutoff state
    - (b) saturated state
    - (c) enhancement mode
    - (d) depletion mode
12. A UJT consists of
    - (a) two emitters
    - (b) a collector and two emitters
    - (c) an emitter and two bases
    - (d) an emitter, base, and collector
13. An SCR is like a rectifier diode except that
    - (a) it can be triggered into conduction by a voltage at the gate
    - (b) it can conduct current in both directions
    - (c) it can handle more power
    - (d) it has four terminals
14. A triac is a device that
    - (a) can be triggered into conduction
    - (b) can conduct current in both directions
    - (c) (a) and (b)
    - (d) none of the above

PROBLEMS

SECTION 3–1 BIPOLAR JUNCTION TRANSISTORS (BJTs)

1. What is the exact value of I_C for $I_E = 5.34$ mA and $I_B = 475$ μA?
2. What is the α_{dc} when $I_C = 8.23$ mA and $I_E = 8.69$ mA?
3. A certain transistor has an $I_C = 25$ mA and an $I_B = 200$ μA. Determine the β_{dc}.
4. In a certain transistor circuit, the base current is 2% of the 30 mA emitter current. Determine the approximate collector current.
5. Find I_B, I_E, and I_C in Figure 3–71 given that $\alpha_{dc} = 0.98$ and $\beta_{dc} = 49$.

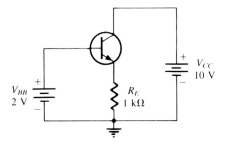

FIGURE 3–71

6. Determine the terminal voltages of each transistor with respect to ground for each circuit in Figure 3–72. Also determine V_{CE}, V_{BE}, and V_{BC}. $\beta_{dc} = 25$.

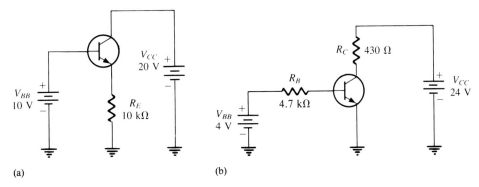

FIGURE 3–72

SECTION 3–2 VOLTAGE-DIVIDER BIAS

7. Determine I_B, I_C, and V_C in Figure 3–73.

FIGURE 3–73

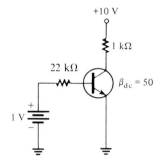

8. For the circuit in Figure 3–74, find V_B, V_E, I_E, I_C, and V_C.
9. In Figure 3–74, what is V_{CE}? What are the Q-point coordinates?

FIGURE 3–74

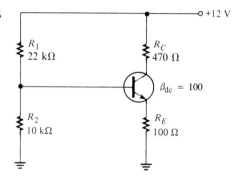

SECTION 3–3 THE BIPOLAR TRANSISTOR AS AN AMPLIFIER

10. A transistor amplifier has a voltage gain of 50. What is the output voltage when the input voltage is 100 mV?
11. To achieve an output of 10 V with an input of 300 mV, what voltage gain is required?
12. A 50 mV signal is applied to the base of a properly biased transistor with $R_E = 100\ \Omega$ and $R_C = 500\ \Omega$. Determine the signal voltage at the collector.

SECTION 3–4 THE BIPOLAR TRANSISTOR AS A SWITCH

13. Determine $I_{C(sat)}$ for the transistor in Figure 3–75. What is the value of I_B necessary to produce saturation? What minimum value of V_{IN} is necessary for saturation?
14. The transistor in Figure 3–76 has a β_{dc} of 50. Determine the value of R_B required to ensure saturation when V_{IN} is 5 V. What must V_{IN} be to cut off the transistor?

SECTION 3–5 BJT PARAMETERS AND RATINGS

15. Assume that β_{dc} in Figure 3–72(a) is 100. If β_{dc} changes from 100 to 150 due to a temperature increase, what is the change in collector current?

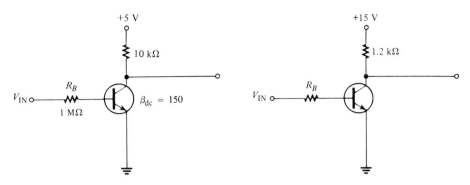

FIGURE 3–75   FIGURE 3–76

16. A certain transistor is to be operated at a collector current of 50 mA. How high can V_{CE} go without exceeding a $P_{D(max)}$ of 1.2 W?

SECTION 3–6 THE JUNCTION FIELD-EFFECT TRANSISTOR (JFET)

17. The V_{GS} of a p-channel JFET is increased from 1 V to 3 V.

(a) Does the depletion region narrow or widen?

(b) Does the resistance of the channel increase or decrease?

18. Why must the gate-to-source voltage of an n-channel JFET always be either zero or negative?

SECTION 3–7 JFET CHARACTERISTICS

19. A JFET has a specified pinch-off voltage of −5 V. When $V_{GS} = 0$, what is V_{DS} at the point where I_D becomes constant?

20. A certain n-channel JFET is biased such that $V_{GS} = -2$ V. What is the value of $V_{GS(off)}$ if V_P is specified to be 6 V?

21. A certain JFET data sheet gives $V_{GS(off)} = -8$ V and $I_{DSS} = 10$ mA. When $V_{GS} = 0$, what is I_D for values of V_{DS} above pinch-off? $V_{DD} = 15$ V.

22. A certain p-channel JFET has a $V_{GS(off)} = 6$ V. What is I_D when $V_{GS} = 8$ V?

23. The JFET in Figure 3–77 has a $V_{GS(off)} = -4$ V. Assume that you increase the supply voltage, V_{DD}, beginning at 0 until the ammeter reaches a steady value. What does the voltmeter read at this point?

FIGURE 3–77

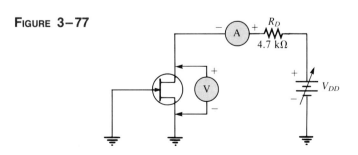

SECTION 3–8 THE METAL OXIDE SEMICONDUCTOR FET (MOSFET)

24. Sketch the schematic symbols for n-channel and p-channel D-MOSFETs and E-MOSFETs. Label the terminals.

25. Explain why both types of MOSFETs have an extremely high input resistance at the gate.

26. An n-channel D-MOSFET with a positive V_{GS} is operating in the _____ mode.

27. A certain E-MOSFET has a $V_{GS(th)} = 3$ V. What is the minimum V_{GS} for the device to turn on?

SECTION 3–9 FET BIASING

28. For each circuit in Figure 3–78, determine V_{DS} and V_{GS}.

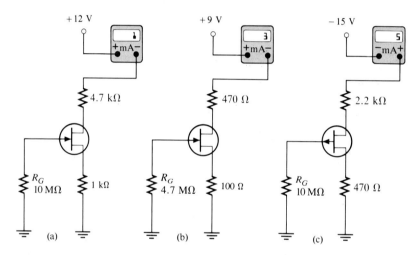

FIGURE 3–78

29. Determine in which mode (depletion or enhancement) each D-MOSFET in Figure 3–79 is biased.

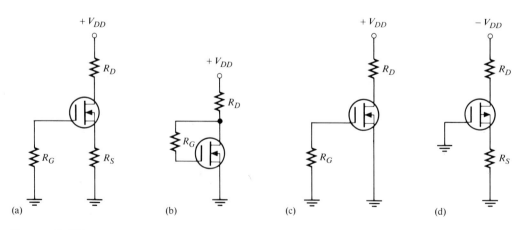

FIGURE 3–79

30. Each E-MOSFET in Figure 3–80 has a $V_{GS(th)}$ of +5 V or −5 V, depending on whether it is an n-channel or a p-channel device. Determine whether each MOSFET is *on* or *off*.

FIGURE 3–80

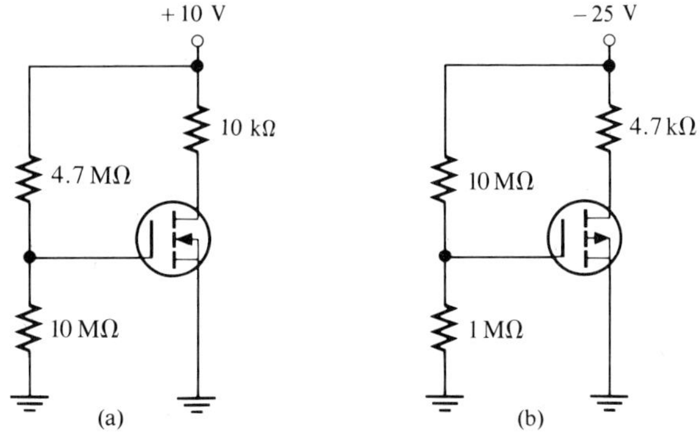

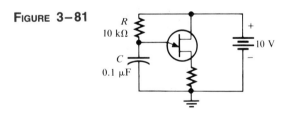

SECTION 3–10 UNIJUNCTION TRANSISTORS (UJTs)

31. In a UJT, if R_{BB} is 5 kΩ and R_{B1} is 3 kΩ, what is η?
32. What is the charging time constant in the oscillator circuit of Figure 3–81?
33. How long will the capacitor in Figure 3–81 initially charge if the peak voltage of the UJT is 6.3 V?
34. If the valley voltage of the UJT in Figure 3–81 is 2.23 V, what is the peak-to-peak value of the voltage across the capacitor, if the peak voltage is the same as in Problem 33? Sketch its general shape.

FIGURE 3–81

SECTION 3–11 THYRISTORS

35. Assume that the holding current for a particular SCR is 10 mA. What is the maximum value of R in Figure 3–82 necessary to keep the SCR in conduction once it is turned on? Neglect the drop across the SCR.
36. Repeat Problem 35 for a holding current of 500 μA.

FIGURE 3–82

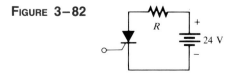

Section 3–13 Troubleshooting

37. In an out-of-circuit test of a good npn transistor, what should an analog ohmmeter indicate when its positive probe is touching the emitter and the negative probe is touching the base? When its positive probe is touching the base and the negative probe is touching the emitter?

38. What is the most likely problem, if any, in each circuit of Figure 3–83? Assume a β_{dc} of 75.

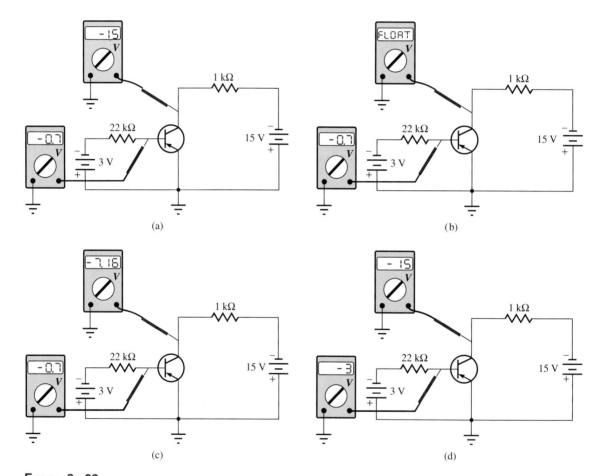

Figure 3–83

39. What is the value of the dc beta of each transistor in Figure 3–84?

40. This problem relates to the circuit board and schematic in Figure 3–69. A remote switch loop is connected between pins 2 and 3. When the remote switches are closed, the relay (RY_1) contacts between pin 10 and pin 11 are normally open. When a remote switch is opened, the relay contacts do not close. Determine the possible causes of this malfunction.

TRANSISTORS AND THYRISTORS

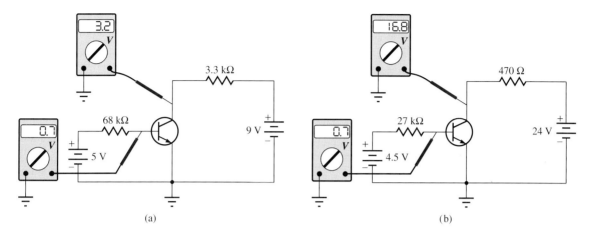

FIGURE 3–84

 41. This problem relates to Figure 3–69. The relay contacts remain closed between pins 10 and 11 no matter what any of the inputs are. This means that the relay is energized continuously. What are the possible faults?

 42. This problem relates to Figure 3–69. Pin 9 stays at approximately 0.1 V, regardless of the input at pin 7. What do you think is wrong? What would you check first?

Answers to Review Questions

Section 3–1
1. Emitter, base, collector
2. The base-emitter junction is forward-biased and the base-collector junction is reverse-biased.
3. Direct current gain
4. 1 mA

Section 3–2
1. One
2. 5.08 V

Section 3–3
1. The process of producing a larger amplitude signal using a smaller signal as a "pattern"
2. The dc operating point
3. 21.36

Section 3-4
1. Saturation and cutoff
2. At saturation
3. At cutoff
4. The base-emitter is forward-biased, and there is sufficient base current.
5. At cutoff

Section 3-5
1. Increase
2. β_{dc} increases with I_C to a certain value and then decreases.
3. 40 mA

Section 3-6
1. Drain, source, and gate
2. Negative
3. By V_{GS}

Section 3-7
1. 7 V
2. Decreases
3. +3 V

Section 3-8
1. Depletion MOSFET and enhancement only MOSFET. The D-MOSFET has a physical channel; the E-MOSFET does not.
2. Yes
3. No

Section 3-9
1. Positive
2. −8 V
3. I_{DSS}
4. 2 V

Section 3-10
1. Unijunction transistor
2. Emitter, base 1, and base 2

Section 3-11
1. Silicon-controlled rectifier
2. A positive voltage on the gate turns the SCR on, and the SCR conducts current from anode to cathode. If the current drops below the holding value, the SCR turns off.
3. Lamp dimmer, motor speed control, or ignition systems
4. False

Section 3–12
1. See Figure 3–85.

FIGURE 3–85

Section 3–13
1. First, test it in-circuit.
2. If R_B opens, the transistor is in cutoff.
3. The base voltage is V_{BB} and the collector voltage is V_{CC}.

Section 3–14
1. To detect when there is an open in one of the remote loops
2. Two, Zone 1 and Zone 2 are identical.
3. When the relay is activated as a result of an open switch in Zone 1 or Zone 2 loops
4. To clip off any negative voltage induced in the coil to prevent possible damage to a transistor

Answers to Practice Exercises

3–1 $I_B = 241\ \mu A$, $I_C = 21.7$ mA, $I_E = 21.94$ mA, $V_{CE} = 4.23$ V, $V_{CB} = 3.53$ V

3–2 $V_B = 8.77$ V, $V_E = 8.07$ V, $V_C = 21.93$ V
$V_{CE} = 13.86$ V, $I_B = 80.7\ \mu A$, $I_E = 8.07$ mA, $I_C \cong 8.07$ mA

3–3 Along the horizontal axis

3–4 A 0.85 V peak sine wave riding on a 6.16 V dc level

3–5 78.4 μA

3–6 $V_{CC(max)} = 44.55$ V, $V_{CE(max)}$ is exceeded first.

3–7 I_D remains at approximately 12 mA.

3–8 80,000 MΩ

3–9 $V_{DS} = 2$ V, $V_{GS} = -3.12$ V

3–10 6.8 V

3–11 2.13 mA

3–12 5 V

3–13 R_B open

4

Amplifiers and Oscillators

4–1 Common-Emitter Amplifiers
4–2 Common-Collector Amplifiers
4–3 Common-Base Amplifiers
4–4 FET Amplifiers
4–5 Multistage Amplifiers
4–6 Class A Amplifier Operation
4–7 Class B Push-Pull Amplifier Operation
4–8 Class C Amplifier Operation
4–9 Oscillators
4–10 Troubleshooting
4–11 A System Application

After completing this chapter, you should be able to

☐ Compare common-emitter (CE), common-collector (CC), and common-base (CB) amplifiers in terms of their configurations and operating characteristics.
☐ Determine how various circuit parameters affect the voltage gain of CE, CC, and CB amplifiers.
☐ Analyze BJT amplifiers for voltage gain, current gain, power gain, and input resistance.
☐ Explain the meaning of the term *emitter-follower* in relation to a common-collector amplifier.
☐ Describe what a Darlington pair is and explain its advantage.
☐ List the important characteristics of FET amplifiers.
☐ Show how multistage amplifiers are used to increase gain.
☐ Describe the differences that distinguish class A, class B, and class C amplifiers.
☐ Explain basic oscillator operation.
☐ Describe how Colpitts, Hartley, Clapp, crystal, and *RC* oscillators differ.
☐ Troubleshoot amplifiers.

As you learned in Chapter 3, the biasing of a transistor is purely a dc operation. The purpose of biasing, however, is to establish an operating point (Q point) about which variations in current and voltage can occur in response to an ac input signal.

When very small signal voltages must be amplified, such as from an antenna in a receiver, variations about the Q point of an amplifier are relatively small. Amplifiers designed to handle these small ac signals are called small-signal amplifiers. When large swings or variations in voltage and current about the Q point are required for power amplification, large-signal amplifiers are used. An example of a large-signal amplifier is the power amplifier that drives the speakers in a stereo system.

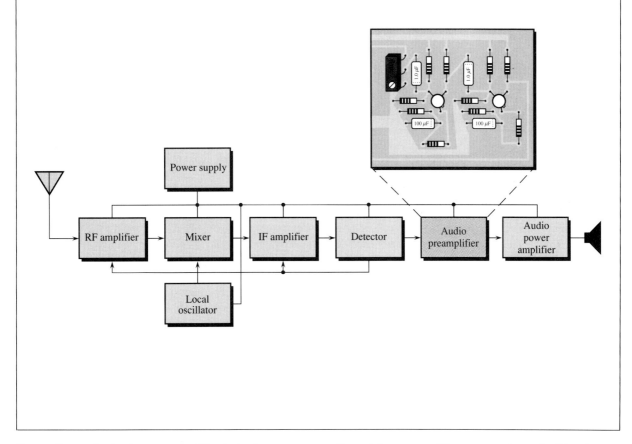

Regardless of whether an amplifier is in the small- or large-signal category, it will be in one of three configurations of amplifier circuits: common-emitter, common-collector, and common-base. For FETs the amplifier configurations are common-source, common-drain, and common-gate. Also, for any of the configurations, three basic modes of operation are possible—class A, class B, and class C—which are covered in this chapter.

Oscillators are also introduced in this chapter. An oscillator is a circuit that produces a sustained sine wave output without an input signal. It operates on the principle of positive feedback.

A System Application

The radio receiver is a system that most of us use every day. The block diagram for an AM superheterodyne receiver is shown above with the audio preamplifier as our focus in this chapter's system application. You do not need to understand the complete system in order to apply what you will learn in this chapter to the audio amplifier. Parts of this system use circuits that have not been covered yet. We will focus on some of these at appropriate points throughout the book.

The receiver selects amplitude-modulated (AM) frequencies between 535 kHz and 1605 kHz in the broadcast band and extracts the audio signal. The audio preamplifier amplifies the audio signal coming from the detector before it goes to the power amplifier. The power amplifier boosts the audio power sufficiently to drive the speaker.

For the system application in Section 4–11, in addition to the other topics, be sure you understand

- The principles of common-emitter amplifier operation.
- How loading affects an amplifier.
- The concept of multistage amplifiers.
- The use of capacitors for coupling and bypass functions.

4–1 COMMON-EMITTER AMPLIFIERS

The common-emitter (CE) is a type of bipolar amplifier configuration in which the emitter is at ac ground. The other two types of amplifier configuration, the common-collector and common-base, are covered in the following sections.

Figure 4–1 shows a typical common-emitter (CE) amplifier. The one shown has voltage-divider bias, although other types of bias methods are possible. C_1 and C_2 are coupling capacitors used to pass the signal into and out of the amplifier such that the source or load will not affect the dc bias voltages. C_3 is a bypass capacitor that shorts the emitter signal voltage (ac) to ground without disturbing the dc emitter voltage. Because of the bypass capacitor, the emitter is at signal ground (but not dc ground), thus making the circuit a **common-emitter** amplifier. The bypass capacitor is used to increase the signal voltage **gain.** (The reason why this increase occurs will be discussed next.) Notice that the input signal is applied to the base, and the output signal is taken off the collector. All capacitors are assumed to have a reactance of approximately zero at the signal frequency.

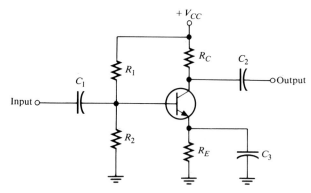

FIGURE 4–1
Typical common-emitter (CE) amplifier.

HOW A BYPASS CAPACITOR INCREASES VOLTAGE GAIN

The bypass capacitor shorts the signal around the emitter resistor, R_E, in order to increase the **voltage gain.** To understand why, let us consider the amplifier without the bypass capacitor and see what the voltage gain is. The CE amplifier with the bypass capacitor removed is shown in Figure 4–2.

As before, lowercase subscripts indicate signal (ac) voltages and signal (alternating) currents. The voltage gain of the amplifier is V_{out}/V_{in}. The output signal voltage is

$$V_{out} = I_c R_C$$

FIGURE 4–2
CE amplifier with bypass capacitor removed.

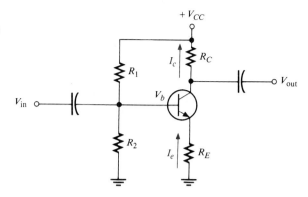

The signal voltage at the base is approximately equal to

$$V_b \cong V_{in} \cong I_e(r_e + R_E)$$

where r_e is the internal emitter resistance of the transistor. The voltage gain, A_v, can now be expressed as

$$A_v = \frac{V_{out}}{V_{in}} = \frac{I_c R_C}{I_e(r_e + R_E)}$$

Since $I_c \cong I_e$, the currents cancel and the gain is the ratio of the resistances.

$$A_v = \frac{R_C}{r_e + R_E} \qquad (4\text{–}1)$$

Keep in mind that this formula is for the CE without the bypass capacitor. If R_E is much greater than r_e, then $A_v \cong R_C/R_E$. This gain formula is similar to that presented in the last chapter, where r_e was neglected [Equation (3–10)].

If the bypass capacitor is connected across R_E, it effectively shorts the signal to ground, leaving only r_e in the emitter. Thus, the voltage gain of the CE amplifier with the bypass capacitor shorting R_E is

$$A_v = \frac{R_C}{r_e} \qquad (4\text{–}2)$$

Now, r_e is a very important transistor parameter because it determines the voltage gain of a CE amplifier in conjunction with R_C. A formula for estimating r_e is given in the following equation:

$$r_e \cong \frac{25 \text{ mV}}{I_E} \qquad (4\text{–}3)$$

Although the formula is simple, its derivation is not and is therefore reserved for Appendix B for those who are interested.

EXAMPLE 4–1

Determine the voltage gain of the amplifier in Figure 4–3 both with and without a bypass capacitor. $\beta_{dc} = \beta_{ac} = 150$.

FIGURE 4–3

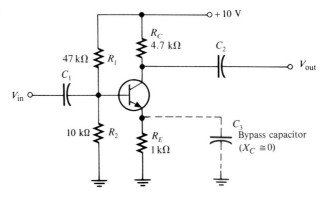

SOLUTION
First, we determine r_e. To do so, we need to find I_E. Thus,

$$V_B \cong \left(\frac{R_2}{R_1 + R_2}\right) V_{CC} = \left(\frac{10 \text{ k}\Omega}{47 \text{ k}\Omega + 10 \text{ k}\Omega}\right) 10 \text{ V} = 1.75 \text{ V}$$

$$V_E = V_B - 0.7 \text{ V} = 1.05 \text{ V}$$

$$I_E = \frac{V_E}{R_E} = \frac{1.05 \text{ V}}{1 \text{ k}\Omega} = 1.05 \text{ mA}$$

$$r_e \cong \frac{25 \text{ mV}}{I_E} \cong \frac{25 \text{ mV}}{1.05 \text{ mA}} = 23.81 \text{ }\Omega$$

The voltage gain without a bypass capacitor is

$$A_v = \frac{R_C}{r_e + R_E} = \frac{4.7 \text{ k}\Omega}{1023.81 \text{ }\Omega} = 4.59$$

The voltage gain with the bypass capacitor installed is

$$A_v = \frac{R_C}{r_e} = \frac{4.7 \text{ k}\Omega}{23.81 \text{ }\Omega} = 197$$

As you can see, the voltage gain is greatly increased by the addition of the bypass capacitor. In terms of decibels (dB), the voltage gain is

$$A_v = 20 \log(197) = 45.89 \text{ dB}$$

PRACTICE EXERCISE 4–1
What is the voltage gain with the bypass capacitor if $R_C = 5.6 \text{ k}\Omega$?

AMPLIFIERS AND OSCILLATORS

PHASE INVERSION

As we discussed in the last chapter, the output voltage at the collector is 180° out of phase with the input voltage at the base. Therefore, the CE amplifier is characterized by a phase inversion between the input and the output. This inversion is sometimes indicated by a negative voltage gain.

AC INPUT RESISTANCE

The dc input resistance R_{IN}, viewed from the base of the transistor, was developed in Section 3–2. The input resistance "seen" by the signal at the base is derived in a similar manner when the emitter resistor is bypassed to ground.

$$R_{in} = \frac{V_b}{I_b}$$

$$V_b = I_e r_e$$

$$I_e \cong \beta_{ac} I_b$$

$$R_{in} \cong \frac{\beta_{ac} I_b r_e}{I_b}$$

The I_b terms cancel, leaving

$$R_{in} \cong \beta_{ac} r_e \qquad (4-4)$$

TOTAL INPUT RESISTANCE TO A CE AMPLIFIER

Viewed from the base, R_{in} is the ac resistance. The actual resistance seen by the source includes that of bias resistors. We will now develop an expression for the total input resistance. The concept of ac ground was mentioned earlier. At this point it needs some additional explanation because it is important in the development of the formula for total input resistance, $R_{in(T)}$.

You have already seen that the bypass capacitor effectively makes the emitter appear as ground to the ac signal, because the X_C of the capacitor is nearly zero at the signal frequency. Of course, to a dc signal the capacitor looks like an open and thus does not affect the dc emitter voltage.

In addition to seeing ground through the bypass capacitor, the signal also sees ground through the dc supply voltage source, V_{CC}. It does so because there is zero signal voltage at the V_{CC} terminal. Thus, the $+V_{CC}$ terminal effectively acts as ac ground. As a result, the two bias resistors R_1 and R_2 appear in parallel to the ac input; one end of R_2 goes to actual ground and one end of R_1 goes to ac ground (V_{CC} terminal). Also, R_{in} at the base appears in parallel with $R_1 \| R_2$. This situation is illustrated in Figure 4–4.

The expression for the total input resistance to the CE amplifier as seen by the ac source is as follows:

$$R_{in(T)} = R_1 \| R_2 \| R_{in} \qquad (4-5)$$

R_C has no effect because of the reverse-biased, base-collector junction.

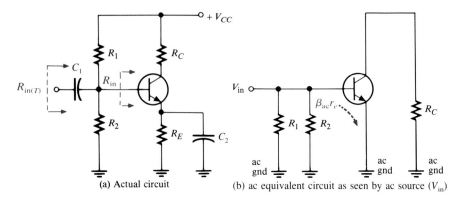

(a) Actual circuit

(b) ac equivalent circuit as seen by ac source (V_{in})

FIGURE 4–4
Total input resistance.

■ **EXAMPLE 4–2** Determine the total input resistance seen by the signal source in the CE amplifier in Figure 4–5. $\beta_{ac} = 150$.

FIGURE 4–5

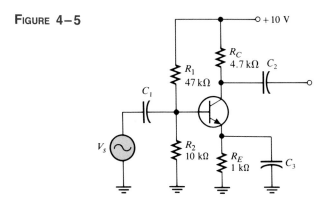

SOLUTION
In Example 4–1, we found r_e for the same circuit. Thus,

$$R_{in} = \beta_{ac} r_e = 150(23.81 \ \Omega) = 3571.5 \ \Omega$$
$$R_{in(T)} = R_1 \| R_2 \| R_{in} = 47 \ k\Omega \| 10 \ k\Omega \| 3571.5 \ \Omega = 2.49 \ k\Omega$$

PRACTICE EXERCISE 4–2
What is the total input resistance seen by the signal source in Figure 4–5 if C_3 is removed?

CURRENT GAIN

The signal **current gain** of a CE amplifier is

$$A_i = \frac{I_c}{I_s} \tag{4–6}$$

where I_s is the source current into the amplifier and is calculated by $V_s/R_{in(T)}$.

POWER GAIN

The **power gain** of a CE amplifier is the product of the voltage gain and the current gain.

$$A_p = A_v A_i \qquad (4\text{--}7)$$

■ **EXAMPLE 4–3** Determine the voltage gain, current gain, and power gain for the CE amplifier in Figure 4–6. $\beta_{ac} = 100$.

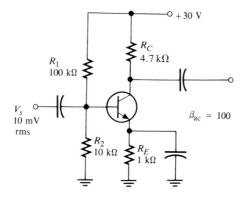

FIGURE 4–6

SOLUTION
First, we must find r_e. To do so, we must find I_E. We begin by calculating V_B. Since R_{IN} is ten times greater than R_2, it can be neglected. Thus,

$$V_B \cong \left(\frac{R_2}{R_1 + R_2}\right) V_{CC} = \left(\frac{10\ \text{k}\Omega}{110\ \text{k}\Omega}\right) 30\ \text{V} = 2.73\ \text{V}$$

$$I_E = \frac{V_E}{R_E} = \frac{V_B - 0.7\ \text{V}}{R_E} = \frac{2.03\ \text{V}}{1\ \text{k}\Omega} = 2.03\ \text{mA}$$

$$r_e = \frac{25\ \text{mV}}{I_E} = \frac{25\ \text{mV}}{2.03\ \text{mA}} = 12.32\ \Omega$$

The ac voltage gain is

$$A_v = \frac{R_C}{r_e} = \frac{4.7\ \text{k}\Omega}{12.32\ \Omega} = 381.5$$

We determine the signal current gain by first finding $R_{in(T)}$ to get I_s.

$$R_{in(T)} = R_1 \| R_2 \| \beta_{ac} r_e = 100\ \text{k}\Omega \| 10\ \text{k}\Omega \| 1.232\ \text{k}\Omega$$
$$\cong 1.1\ \text{k}\Omega$$

$$I_s = \frac{V_s}{R_{in(T)}} = \frac{10\ \text{mV}}{1.1\ \text{k}\Omega} \cong 9.1\ \mu\text{A}$$

Next we must determine I_c.

$$I_c = \frac{V_{out}}{R_C} = \frac{A_v V_s}{R_C} = \frac{(381.5)(10 \text{ mV})}{4.7 \text{ k}\Omega} = 0.812 \text{ mA}$$

$$A_i = \frac{I_c}{I_s} = \frac{812 \text{ }\mu\text{A}}{9.1 \text{ }\mu\text{A}} \cong 89.2$$

The power gain is

$$A_p = A_v A_i = (381.5)(89.2) = 34{,}029$$

The voltage gain and the power gain in decibels are as follows:

$$A_v = 20 \log(381.5) = 51.63 \text{ dB}$$
$$A_p = 10 \log(34{,}029) = 45.32 \text{ dB}$$

PRACTICE EXERCISE 4–3
Determine A_v, A_i, and A_p in this example taking into account R_{IN}.

4–1 REVIEW QUESTIONS

1. What is the purpose of the bypass capacitor in a CE amplifier?
2. How is the CE voltage gain determined?
3. If A_v is 50 and A_i is 200, what is the power gain?

4–2 COMMON-COLLECTOR AMPLIFIERS

The common-collector (CC) amplifier, commonly referred to as an emitter-follower, is the second of the three basic amplifier configurations. The input is applied to the base and the output is at the emitter. There is no collector resistor. The voltage gain of a CC amplifier is approximately 1, and its main advantage is its high input resistance.

Figure 4–7 shows a **common collector (CC)** circuit with a voltage-divider bias. Notice that the input is applied to the base and the output is taken off the emitter.

FIGURE 4–7
Typical common-collector (CC) or emitter-follower amplifier.

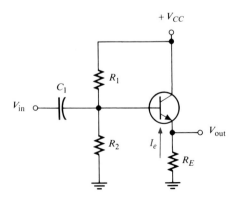

VOLTAGE GAIN

As in all amplifiers, the voltage gain in a CC amplifier is $A_v = V_{out}/V_{in}$. For the emitter-follower, $V_{out} = I_e R_E$, and $V_{in} = I_e(r_e + R_E)$. Therefore, the gain is

$$A_v = \frac{I_e R_E}{I_e(r_e + R_E)}$$

The currents cancel, and the gain expression simplifies to

$$A_v = \frac{R_E}{r_e + R_E} \qquad (4-8)$$

It is important to notice here that the gain is always less than 1. If r_e is much less than R_E, then a good approximation is $A_v \cong 1$.

Since the output voltage is the emitter voltage, it is in phase with the base or the input voltage. As a result, and because the voltage gain is close to 1, the output voltage follows the input voltage; thus the term **emitter-follower**.

INPUT RESISTANCE

The emitter-follower is characterized by a high input resistance, which makes it a very useful circuit. Because of the high input resistance, the emitter-follower can be used as a buffer to minimize loading effects when one circuit is driving another.

The derivation of the input resistance viewed from the base is similar to that for the CE amplifier. In this case, however, the emitter resistor is not bypassed.

$$R_{in} = \frac{V_b}{I_b} = \frac{I_e(r_e + R_E)}{I_b} \cong \frac{\beta_{ac} I_b(r_e + R_E)}{I_b} = \beta_{ac}(r_e + R_E)$$

If R_E is at least ten times larger than r_e, then the input resistance at the base is

$$R_{in} \cong \beta_{ac} R_E \qquad (4-9)$$

In Figure 4–7, the bias resistors appear in parallel with R_{in} to the input signal, just as in the CE amplifier with voltage-divider bias. The total ac input resistance is

$$R_{in(T)} = R_1 \| R_2 \| R_{in} \qquad (4-10)$$

Since R_{in} can be made large with the proper selection of R_E, a much higher input resistance results for this configuration than for the CE circuit.

CURRENT GAIN

The signal current gain for the emitter-follower is I_e/I_s where I_s is the signal current and can be calculated as $V_s/R_{in(T)}$. If the bias resistors are large enough to be neglected so that $I_s = I_b$, then the current gain of the amplifier is equal to the current gain of the transistor, β_{ac}. Of course, the same was also true for the CE amplifier. β_{ac} is the maximum achievable current gain in both amplifiers.

$$A_i = \frac{I_e}{I_s} \qquad (4-11)$$

Power Gain

The power gain for the CC amplifier is the product of the voltage gain and the current gain. For the emitter-follower, the power gain is approximately equal to the current gain because the voltage gain is approximately 1.

$$A_p \cong A_i \qquad (4-12)$$

EXAMPLE 4–4

Determine the input resistance of the emitter-follower in Figure 4–8. Also find the voltage gain, current gain, and power gain.

FIGURE 4–8

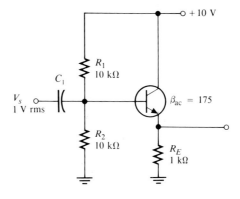

Solution
The approximate input resistance viewed from the base is

$$R_{in} \cong \beta_{ac}R_E = (175)(1\ k\Omega) = 175\ k\Omega$$

The total input resistance is

$$R_{in(T)} = R_1 \| R_2 \| R_{in} = 10\ k\Omega \| 10\ k\Omega \| 175\ k\Omega$$
$$= 4.86\ k\Omega$$

The voltage gain is, neglecting r_e,

$$A_v \cong 1$$

The current gain is

$$A_i = \frac{I_e}{I_s}$$

$$I_e = \frac{V_e}{R_E} = \frac{A_v V_b}{R_E} \cong \frac{1\ V}{1\ k\Omega} = 1\ mA$$

$$I_s = \frac{V_s}{R_{in(T)}} = \frac{1\ V}{4.86\ k\Omega} = 0.21\ mA$$

$$A_i = \frac{1\ mA}{0.21\ mA} = 4.76$$

The power gain is

$$A_p \cong A_i = 4.76$$

PRACTICE EXERCISE 4-4
If R_E in Figure 4-8 is decreased to 820 Ω, how is the power gain affected?

THE DARLINGTON PAIR

As you have seen, β_{ac} is a major factor in determining the input resistance. The β_{ac} of the transistor limits the maximum achievable input resistance you can get from a given emitter-follower circuit.

One way to boost input resistance is to use a **Darlington pair,** as shown in Figure 4-9. The collectors of two transistors are connected, and the emitter of the first drives the base of the second. This configuration achieves β_{ac} multiplication as shown in the following steps. The emitter current of the first transistor is

$$I_{e1} \cong \beta_{ac1} I_{b1}$$

This emitter current becomes the base current for the second transistor, producing a second emitter current of

$$I_{e2} \cong \beta_{ac2} I_{e1}$$
$$I_{e2} = \beta_{ac1} \beta_{ac2} I_{b1}$$

Therefore, the effective current gain of the Darlington pair is

$$\beta_{ac} = \beta_{ac1} \beta_{ac2} \qquad (4-13)$$

The input resistance is $\beta_{ac1}\beta_{ac2}R_E$.

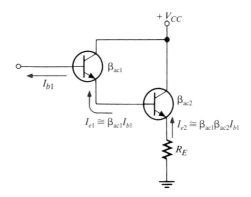

FIGURE 4-9
Darlington pair.

4-2 REVIEW QUESTIONS

1. What is a common-collector amplifier called?
2. What is the ideal maximum voltage gain of a CC amplifier?
3. What is the most important characteristic of the CC amplifier?

4–3 COMMON-BASE AMPLIFIERS

The third basic amplifier configuration is the common-base (CB). It provides high voltage gain with no current gain. Since it has a low input resistance, the CB amplifier is the most appropriate type for certain high-frequency applications where sources tend to have very low-resistance outputs.

A typical **common-base** circuit is pictured in Figure 4–10. The base is at signal (ac) ground, and the input is applied to the emitter. The output is taken off the collector.

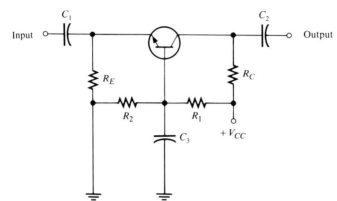

FIGURE 4–10
Typical common-base (CB) amplifier.

VOLTAGE GAIN

The input voltage is the emitter voltage V_e. The output voltage is the collector voltage V_c. With this in mind, we develop the voltage gain formula as follows:

$$A_v = \frac{V_c}{V_e} = \frac{I_c R_C}{I_e r_e} \cong \frac{I_e R_C}{I_e r_e}$$

$$A_v = \frac{R_C}{r_e} \tag{4–14}$$

Notice that the gain expression is the same as that for the CE amplifier (when R_E is bypassed).

INPUT RESISTANCE

The resistance viewed from the emitter appears to the input signal as follows:

$$R_{in} = \frac{V_{in}}{I_{in}} = \frac{V_e}{I_e} = \frac{I_e r_e}{I_e}$$

$$R_{in} = r_e \tag{4–15}$$

Viewed from the source, R_E appears in parallel with R_{in}. However, r_e is normally so small compared to R_E that Equation (4–15) is also valid for the total input resistance, $R_{in(T)}$.

CURRENT GAIN

The current gain is the output current I_c divided by the input current I_e. Since $I_c \cong I_e$, the signal current gain is approximately 1.

$$A_i \cong 1 \tag{4-16}$$

POWER GAIN

Since the current gain is approximately 1 for the CB amplifier, the power gain is approximately equal to the voltage gain.

$$A_p \cong A_v \tag{4-17}$$

■ **EXAMPLE 4-5** Find the input resistance, voltage gain, current gain, and power gain for the CB amplifier in Figure 4-11.

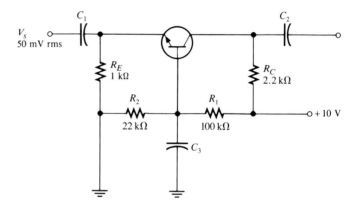

FIGURE 4-11

SOLUTION

First, let us find I_E so that we can determine r_e. Then $R_{in} = r_e$. Thus,

$$V_B \cong \left(\frac{R_2}{R_1 + R_2}\right)V_{CC} = \left(\frac{22\ k\Omega}{122\ k\Omega}\right)10\ V = 1.8\ V$$

$$V_E = V_B - 0.7\ V = 1.8\ V - 0.7\ V = 1.1\ V$$

$$I_E = \frac{V_E}{R_E} = \frac{1.1\ V}{1\ k\Omega} = 1.1\ mA$$

$$R_{in} = r_e \cong \frac{25\ mV}{I_E} = \frac{25\ mV}{1.1\ mA} = 22.73\ \Omega$$

The signal voltage gain is

$$A_v = \frac{R_C}{r_e} = \frac{2.2\ k\Omega}{22.73\ \Omega} = 96.8$$

Thus,

$$A_i \cong 1$$
$$A_p \cong 96.8$$

PRACTICE EXERCISE 4–5
If R_E is increased to 1.5 kΩ in Figure 4–11, what happens to the voltage gain?

SUMMARY

Table 4–1 summarizes the important characteristics of each of the three amplifier configurations. Also, relative values are indicated for general comparison of the amplifiers.

TABLE 4–1
Comparison of amplifier configurations. The current gains and the input resistance are the maximum achievable values, with the bias resistors neglected.

	CE	CC	CB
Voltage gain A_v	R_C/r_e High	$\cong 1$ Low	R_C/r_e High
Current gain $A_{i(max)}$	β_{ac} High	β_{ac} High	$\cong 1$ Low
Power gain A_p	$A_i A_v$ Very high	$\cong A_i$ High	$\cong A_v$ High
Input resistance $R_{in(max)}$	$\beta_{ac} r_e$ Low	$\beta_{ac} R_E$ High	r_e Very Low

4–3 REVIEW QUESTIONS

1. Can the same voltage gain be achieved with a CB as with a CE amplifier?
2. Is the input resistance of a CB amplifier very low or very high?

4–4 FET AMPLIFIERS

Field-effect transistors, both JFETs and MOSFETs, can be used as amplifiers in any of three circuit configurations similar to those for the bipolar transistor. The FET configurations are common-source, common-drain, and common-gate. These are similar to the bipolar configurations of common-emitter, common-collector, and common-base, respectively.

TRANSCONDUCTANCE OF AN FET

Recall from Chapter 3 that in a bipolar transistor, the base current controls the collector current, and the relationship between these two currents is expressed by the parameter β_{ac}

where $I_C = \beta_{ac}I_b$. In an FET, the gate voltage controls the drain current. An important FET parameter is the **transconductance**, g_m, which is defined as

$$g_m = \frac{I_d}{V_{gs}} \qquad (4-18)$$

The transconductance is one factor that determines the voltage gain of an FET amplifier. On data sheets, the transconductance is sometimes called the *forward transadmittance* and is designated y_{fs}.

COMMON-SOURCE (CS) AMPLIFIERS

Figure 4–12 shows a **common-source** amplifier with a self-biased n-channel JFET. An ac source is capacitively coupled to the gate. The resistor, R_G, serves two purposes: (1) It keeps the gate at approximately 0 V dc (because I_{GSS} is extremely small), and (2) its large value (usually several megohms) prevents loading of the ac signal source. The bias voltage is created by the drop across R_S. The bypass capacitor, C_2, keeps the source of the FET effectively at ac ground.

FIGURE 4–12

JFET common-source amplifier.

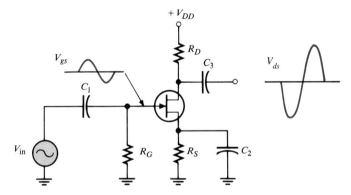

The signal voltage causes the gate-to-source voltage to swing above and below its Q-point value, causing a swing in drain current. As the drain current increases, the voltage drop across R_D also increases, causing the drain voltage (with respect to ground) to decrease.

The drain current swings above and below its Q-point value in-phase with the gate-to-source voltage. The drain-to-source voltage swings above and below its Q-point value 180° out-of-phase with the gate-to-source voltage, as illustrated in Figure 4–12.

D-MOSFET Figure 4–13 shows a zero-biased n-channel D-MOSFET with an ac source capacitively coupled to the gate. The gate is at approximately 0 V dc and the source terminal is at ground, thus making $V_{GS} = 0$ V.

The signal voltage causes V_{gs} to swing above and below its 0 value, producing a swing in I_d. The negative swing in V_{gs} produces the depletion mode, and I_d decreases. The positive swing in V_{gs} produces the enhancement mode, and I_d increases.

FIGURE 4–13
Zero-biased D-MOSFET common-source amplifier.

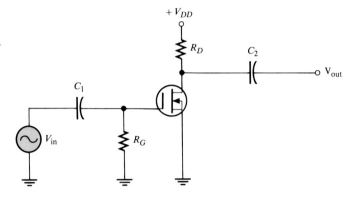

E-MOSFET Figure 4–14 shows a voltage divider-biased, n-channel E-MOSFET with an ac signal source capacitively coupled to the gate. The gate is biased with a positive voltage such that $V_{GS} > V_{GS(\text{th})}$.

FIGURE 4–14
Common-source E-MOSFET amplifier with voltage-divider bias.

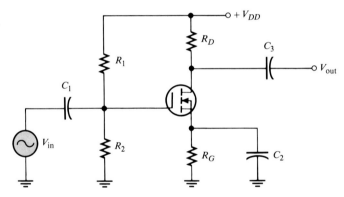

As with the JFET and D-MOSFET, the signal voltage produces a swing in V_{gs} above and below its Q-point value. This swing, in turn, causes a swing in I_d. Operation is entirely in the enhancement mode.

VOLTAGE GAIN Voltage gain, A_v, of an amplifier always equals V_{out}/V_{in}. In the case of the CS amplifier, V_{in} is equal to V_{gs}, and V_{out} is equal to the signal voltage developed across R_D, which is $I_d R_D$. Thus,

$$A_v = \frac{I_d R_D}{V_{gs}}$$

Since $g_m = I_d/V_{gs}$, the common-source voltage gain is

$$A_v = g_m R_D \tag{4-19}$$

AMPLIFIERS AND OSCILLATORS

INPUT RESISTANCE Because the input to a CS amplifier is at the gate, the input resistance is extremely high. Ideally, it approaches infinity and can be neglected. As you know, the high input resistance is produced by the reverse-biased pn junction in a JFET and by the insulated gate structure in a MOSFET.

The actual input resistance seen by the signal source is the gate-to-ground resistor, R_G, in parallel with the FET's input resistance at the gate, $R_{IN(gate)}$. The input resistance at the gate of a JFET is

$$R_{IN(gate)} = \frac{V_{GS}}{I_{GSS}}$$

The reverse leakage current, I_{GSS}, is typically given on the data sheet for a specific value of V_{GS} so that the input resistance of the device can be calculated.

EXAMPLE 4–6

(a) What is the total output voltage (dc + ac) of the amplifier in Figure 4–15? The g_m is 4500 µS, I_D is 2 mA, $V_{GS(off)}$ is −10 V, and I_{GSS} is 15 nA.
(b) What is the input resistance seen by the signal source?

FIGURE 4–15

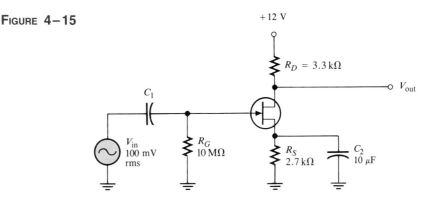

SOLUTION
(a) First, find the dc output voltage.

$$V_D = V_{DD} - I_D R_D = 12 \text{ V} - (2 \text{ mA})(3.3 \text{ k}\Omega) = 5.4 \text{ V}$$

Next, find the ac output voltage by using the gain formula:

$$A_v = \frac{V_{out}}{V_{in}} = g_m R_D$$

$$V_{out} = g_m R_D V_{in} = (4500 \text{ µS})(3.3 \text{ k}\Omega)(100 \text{ mV})$$
$$= 1.49 \text{ V rms}$$

The total output voltage is an ac signal with a peak-to-peak value of 1.49 V × 2.828 = 4.21 V, riding on a dc level of 5.4 V.

(b) The input resistance is determined as follows (since $V_G = 0$ V).

$$V_{GS} = I_D R_S = (2 \text{ mA})(2.7 \text{ k}\Omega) = 5.4 \text{ V}$$

The input resistance at the gate of the JFET is

$$R_{IN(gate)} = \frac{V_{GS}}{I_{GSS}} = \frac{5.4 \text{ V}}{15 \text{ nA}} = 360 \text{ M}\Omega$$

The input resistance seen by the signal source is

$$R_{in} = R_G \| R_{IN(gate)} = 10 \text{ M}\Omega \| 360 \text{ M}\Omega = 9.73 \text{ M}\Omega$$

PRACTICE EXERCISE 4–6
What will happen in the amplifier of Figure 4–15 if a transistor with $V_{GS(off)} = -2$ V is used? Assume the other parameters are the same.

COMMON-DRAIN (CD) AMPLIFIER

A **common-drain** JFET amplifier is shown in Figure 4–16 with voltages indicated. Self-biasing is used in this circuit. The input signal is applied to the gate through a coupling capacitor, and the output is at the source terminal. There is no drain resistor. This circuit, of course, is analogous to the bipolar emitter-follower and is sometimes called a *source-follower*.

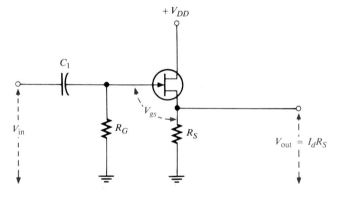

FIGURE 4–16
JFET common-drain amplifier (source-follower).

VOLTAGE GAIN As in all amplifiers, the voltage gain is $A_v = V_{out}/V_{in}$. For the source-follower, V_{out} is $I_d R_S$ and V_{in} is $V_{gs} + I_d R_S$, as shown in Figure 4–16. Therefore, the gate-to-source voltage gain is $I_d R_S/(V_{gs} + I_d R_S)$. Substituting $I_d = g_m V_{gs}$ into the expression gives the following result:

$$A_v = \frac{g_m V_{gs} R_S}{V_{gs} + g_m V_{gs} R_S}$$

Canceling V_{gs}, we get

$$A_v = \frac{g_m R_S}{1 + g_m R_S} \qquad (4-20)$$

Notice here that the gain is always slightly less than 1. If $g_m R_S \gg 1$, then a good approximation is $A_v \cong 1$. Since the output voltage is at the source, it is in phase with the gate (input) voltage.

AMPLIFIERS AND OSCILLATORS

INPUT RESISTANCE Because the input signal is applied to the gate, the input resistance seen by the input signal source is extremely high, just as in the CS amplifier configuration. The gate resistor R_G, in parallel with the input resistance looking in at the gate, is the total input resistance.

EXAMPLE 4–7

(a) Determine the voltage gain of the amplifier in Figure 4–17(a) using the data sheet information in Figure 4–17(b).

(b) Also determine the input resistance. Assume minimum data sheet values where available.

FIGURE 4–17

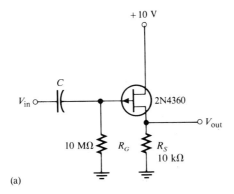

(a)

*ELECTRICAL CHARACTERISTICS (T_A = 25°C unless otherwise noted)

Characteristic	Symbol	Min	Max	Unit
OFF CHARACTERISTICS				
Gate-Source Breakdown Voltage (I_G = 10 µAdc, V_{DS} = 0)	$V_{(BR)GSS}$	20	–	Vdc
Gate-Source Cutoff Voltage (V_{DS} = -10 Vdc, I_D = 1.0 µAdc)	$V_{GS(off)}$	0.7	10	Vdc
Gate Reverse Current (V_{GS} = 15 Vdc, V_{DS} = 0)	I_{GSS}	–	10	nAdc
(V_{GS} = 15 Vdc, V_{DS} = 0, T_A = 65°C)		–	0.5	µAdc
ON CHARACTERISTICS				
Zero-Gate Voltage Drain Current (Note 1) (V_{DS} = -10 Vdc, V_{GS} = 0)	I_{DSS}	3.0	30	mAdc
Gate-Source Voltage (V_{DS} = -10 Vdc, I_D = 0.3 mAdc)	V_{GS}	0.4	9.0	Vdc
SMALL-SIGNAL CHARACTERISTICS				
Drain-Source "ON" Resistance (V_{GS} = 0, I_D = 0, f = 1.0 kHz)	$r_{ds(on)}$	–	700	Ohms
Forward Transadmittance (Note 1) (V_{DS} = -10 Vdc, V_{GS} = 0, f = 1.0 kHz)	$|y_{fs}|$	2000	8000	µmhos
Forward Transconductance (V_{DS} = -10 Vdc, V_{GS} = 0, f = 1.0 MHz)	$Re(y_{fs})$	1500	–	µmhos
Output Admittance (V_{DS} = -10 Vdc, V_{GS} = 0, f = 1.0 kHz)	$|y_{os}|$	–	100	µmhos
Input Capacitance (V_{DS} = -10 Vdc, V_{GS} = 0, f = 1.0 MHz)	C_{iss}	–	20	pF
Reverse Transfer Capacitance (V_{DS} = -10 Vdc, V_{GS} = 0, f = 1.0 MHz)	C_{rss}	–	5.0	pF
Common-Source Noise Figure (V_{DS} = -10 Vdc, I_D = 1.0 mAdc, R_G = 1.0 Megohm, f = 100 Hz)	NF	–	5.0	dB
Equivalent Short-Circuit Input Noise Voltage (V_{DS} = -10 Vdc, I_D = 1.0 mAdc, f = 100 Hz, BW = 15 Hz)	E_n	–	0.19	$\mu V/\sqrt{Hz}$

*Indicates JEDEC Registered Data.

Note 1: Pulse Test: Pulse Width ≤ 630 ms, Duty Cycle ≤ 10%.

(b)

SOLUTION
(a) From the data sheet, $g_m = y_{fs} = 2000 \ \mu S$ minimum. The gain is

$$A_v = \frac{g_m R_S}{1 + g_m R_S} = \frac{(2000 \ \mu S)(10 \ k\Omega)}{1 + (2000 \ \mu S)(10 \ k\Omega)} \cong 0.952$$

(b) From the data sheet, $I_{GSS} = 10$ nA maximum at $V_{GS} = 15$ V. Therefore,

$$R_{IN(gate)} = \frac{15 \ V}{10 \ nA} = 1500 \ M\Omega$$

$$R_{IN} = R_G \| R_{IN(gate)} = 10 \ M\Omega \| 1500 \ M\Omega = 9.93 \ M\Omega$$

PRACTICE EXERCISE 4–7
If the g_m of the JFET in the source-follower of Figure 4–17 is doubled, what is the voltage gain?

COMMON-GATE (CG) AMPLIFIER

A typical **common-gate amplifier** is shown in Figure 4–18. The gate is connected directly to ground. The input signal is applied at the source terminal through C_1. The output is coupled through C_2 from the drain terminal.

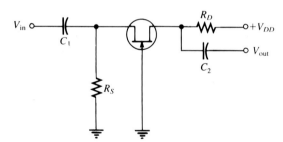

FIGURE 4–18
JFET common-gate amplifier.

VOLTAGE GAIN The voltage gain from source to drain is developed as follows.

$$A_v = \frac{V_{out}}{V_{in}} = \frac{V_d}{V_{gs}} = \frac{I_d R_D}{V_{gs}} = \frac{g_m V_{gs} R_D}{V_{gs}}$$

$$A_v = g_m R_D \qquad (4-21)$$

Notice that the gain expression is the same as for the CS JFET amplifier.

INPUT RESISTANCE As you have seen, both the CS and the CD configurations have extremely high input resistances because the gate is the input terminal. In contrast, the common-gate configuration has a low input resistance, as shown in the following steps. First, the input current (source current) is equal to the drain current.

$$I_{in} = I_d = g_m V_{gs}$$

Amplifiers and Oscillators

Second, the input voltage equals V_{gs}.

$$V_{in} = V_{gs}$$

Therefore, the input resistance at the source terminal is

$$R_{in(source)} = \frac{V_{in}}{I_{in}} = \frac{V_{gs}}{g_m V_{gs}}$$

$$R_{in(source)} = \frac{1}{g_m} \qquad (4-22)$$

EXAMPLE 4–8 (a) Determine the voltage gain of the amplifier in Figure 4–19.
(b) Determine the input resistance.

FIGURE 4–19

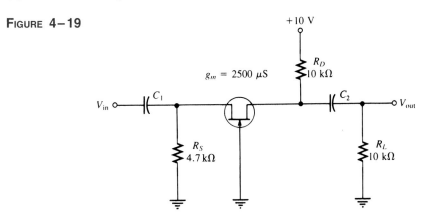

SOLUTION
(a) This CG amplifier has a load resistor effectively in parallel with R_D, so the effective drain resistance is $R_D \| R_L$ and the gain is

$$A_v = g_m(R_D \| R_L) = (2500 \ \mu S)(10 \ k\Omega \| 10 \ k\Omega) = 12.5$$

(b) The input resistance at the source terminal is

$$R_{in(source)} = \frac{1}{g_m} = \frac{1}{2500 \ \mu S} = 400 \ \Omega$$

The signal source actually sees R_S in parallel with $R_{in(source)}$, so the total input resistance is

$$R_{in} = 400 \ \Omega \| 4.7 \ k\Omega = 369 \ \Omega$$

PRACTICE EXERCISE 4–8
How much does the input resistance change in Figure 4–19 if R_S is changed to 10 kΩ?

SUMMARY

A summary of the gain and input resistance characteristics for the three FET amplifier configurations is given in Table 4–2.

TABLE 4–2
Comparison of FET amplifier configurations

	CS	CD	CG
Voltage gain, A_v	$g_m R_D$	$\dfrac{g_m R_S}{1 + g_m R_S}$	$g_m R_D$
Input resistance, R_{in}	$\left(\dfrac{V_{GS}}{I_{GSS}}\right) \| R_G$	$\left(\dfrac{V_{GS}}{I_{GSS}}\right) \| R_G$	$\left(\dfrac{1}{g_m}\right) \| R_G$

4–4 REVIEW QUESTIONS

1. What factors determine the voltage gain of a CS JFET amplifier?
2. A certain CS amplifier has an $R_D = 1$ kΩ. When a load resistance of 1 kΩ is capacitively coupled to the drain, how much does the gain change?
3. What is a major difference between a CG amplifier and the other two configurations?

4–5 MULTISTAGE AMPLIFIERS

Several amplifiers can be connected in a cascaded arrangement with the output of one amplifier driving the input of the next. Each amplifier in the cascaded arrangement is known as a stage. The purpose of a multistage arrangement is to increase the overall gain.

MULTISTAGE GAIN

The overall gain A'_v of cascaded amplifiers as in Figure 4–20 is the product of the individual gains.

$$A'_v = A_{v1} A_{v2} A_{v3} \cdots A_{vn} \qquad (4\text{–}23)$$

where n is the number of stages.

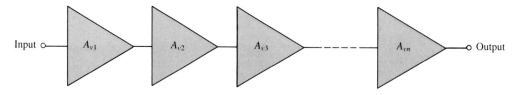

FIGURE 4–20
Cascaded amplifiers. Each triangle represents a separate amplifier.

DECIBEL VOLTAGE GAIN

Amplifier voltage gain is often expressed in decibels (dB) as follows:

$$A_v \text{ (dB)} = 20 \log A_v \qquad (4\text{–}24)$$

This formula is particularly useful in multistage systems because the overall dB voltage gain is the sum of the individual dB gains.

$$A'_v \text{ (dB)} = A_{v1} \text{ (dB)} + A_{v2} \text{ (dB)} + \cdots + A_{vn} \text{ (dB)} \qquad (4-25)$$

EXAMPLE 4-9

A given cascaded amplifier arrangement has the following voltage gains: $A_{v1} = 10$, $A_{v2} = 15$, $A_{v3} = 20$. What is the overall gain? Also express each gain in decibels and determine the total decibel voltage gain.

SOLUTION

$$A'_v = A_{v1}A_{v2}A_{v3} = (10)(15)(20) = 3000$$
$$A_{v1} \text{ (dB)} = 20 \log 10 = 20 \text{ dB}$$
$$A_{v2} \text{ (dB)} = 20 \log 15 = 23.52 \text{ dB}$$
$$A_{v3} \text{ (dB)} = 20 \log 20 = 26.02 \text{ dB}$$
$$A'_v \text{ (dB)} = 20 \text{ dB} + 23.52 \text{ dB} + 26.02 \text{ dB} = 69.54 \text{ dB}$$

PRACTICE EXERCISE 4-9

In a certain multistage amplifier, the individual stages have the following voltage gains: $A_{v1} = 25$, $A_{v2} = 5$, and $A_{v3} = 12$. What is the overall gain? Express each gain in dB and determine the total dB voltage gain.

MULTISTAGE ANALYSIS

The two-stage amplifier in Figure 4–21 illustrates multistage analysis. Notice that both stages are identical CE amplifiers with the output of the first stage capacitively coupled to the input of the second stage. Capacitive coupling prevents the dc bias of one stage from affecting that of the other. Also notice that the transistors are designated Q_1 and Q_2.

FIGURE 4–21
A two-stage common-emitter amplifier.

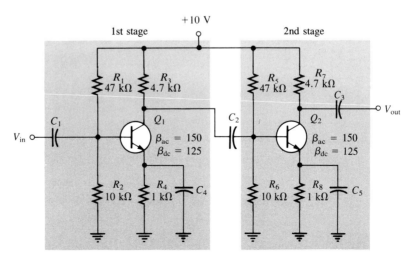

LOADING EFFECTS

In determining the gain of the first stage, we must consider the loading effect of the second stage. Because the coupling capacitor C_2 appears as a short to the signal frequency, the total input resistance of the second stage presents an ac load to the first stage.

Looking from the collector of Q_1, the two biasing resistors, R_5 and R_6, appear in parallel with the input resistance at the base of Q_2. In other words, the signal at the collector of Q_1 "sees" R_3 and R_5, R_6, and $R_{in(base2)}$ of the second stage all in parallel to ac ground. Thus, the effective ac collector resistance of Q_1 is the total of all these in parallel, as Figure 4–22 illustrates.

FIGURE 4–22
AC equivalent of first stage in Figure 4–21, showing loading from second stage.

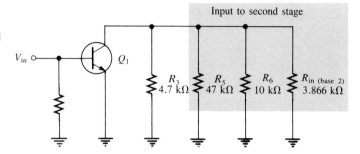

The voltage gain of the first stage is reduced by the loading of the second stage, because the effective ac collector resistance of the first stage is less than the actual value of its collector resistor, R_3. Remember that $A_v = R_C/r_e$ for an unloaded amplifier.

VOLTAGE GAIN OF THE FIRST STAGE

The ac collector resistance of the first stage is

$$R_{c1} = R_3 \| R_5 \| R_6 \| R_{in(base2)}$$

Keep in mind that lowercase subscripts denote ac quantities such as for R_c.

You can verify that $I_E = 1.05$ mA, $r_e = 23.8\ \Omega$, and $R_{in(base2)} = 3.57$ kΩ. The effective ac collector resistance of the first stage is as follows:

$$R_{c1} = 4.7\ \text{k}\Omega \| 47\ \text{k}\Omega \| 10\ \text{k}\Omega \| 3.57\ \text{k}\Omega = 1.63\ \text{k}\Omega$$

Therefore, the base-to-collector voltage gain of the first stage is

$$A_{v1} = \frac{R_{c1}}{r_e} = \frac{1.63\ \text{k}\Omega}{23.8\ \Omega} = 68.5$$

VOLTAGE GAIN OF THE SECOND STAGE

The second stage has no load resistor, so the ac collector resistance is R_7, as shown in Figure 4–21, and the gain is

$$A_{v2} = \frac{R_7}{r_e} = \frac{4.7\ \text{k}\Omega}{23.8\ \Omega} = 197.5$$

AMPLIFIERS AND OSCILLATORS

Compare this to the gain of the first stage, and notice how much the second-stage loading reduced the gain of the first stage.

OVERALL VOLTAGE GAIN

The overall amplifier gain with no load on the output is

$$A'_v = A_{v1}A_{v2} = (68.5)(197.5) \cong 13{,}529$$

If an input signal of, say, 100 μV is applied to the first stage and if the **attenuation** of the input base circuit is neglected, an output from the second stage of (100 μV)(13,529) = 1.3529 V will result. The overall gain can be expressed in decibels as follows:

$$A'_v \text{ (dB)} = 20 \log(13{,}529) = 82.63 \text{ dB}$$

4–5 REVIEW QUESTIONS

1. What does the term *stage* mean?
2. How is the overall gain of a multistage amplifier determined?
3. Express a voltage gain of 500 in decibels.

4–6 CLASS A AMPLIFIER OPERATION

When an amplifier, whether it is one of the three configurations of bipolar or an FET type, is biased such that it always operates in the linear region where the output signal is an amplified replica of the input signal, it is a **class A** *amplifier. The discussion and formulas in the previous sections apply to class A operation.*

When the output signal takes up only a small percentage of the total load line excursion, the amplifier is a small-signal amplifier. When the output signal is larger and approaches the limits of the load line, the amplifier is a large-signal type. Amplifiers are typically operated as large-signal devices when power amplification is the major objective. Figure 4–23 illustrates class A operation.

FIGURE 4–23
Class A operation.

WHY THE Q POINT MUST BE CENTERED FOR MAXIMUM OUTPUT SIGNAL

When the dc operating point (Q point) is at the center of the load line, a maximum class A signal can be obtained. You can see this concept by examining the graph of the load line for a given amplifier in Figure 4–24(a). This graph shows the load line with the Q point at its center. The collector current can vary from its Q-point value, I_{CQ}, up to its saturation value, $I_{C(sat)}$, and down to its cutoff value of zero. Likewise, the collector-to-emitter voltage can swing from its Q-point value, V_{CEQ}, up to its cutoff value, $V_{CE(cutoff)}$, and down to its saturation value of near zero. This operation is indicated in Figure 4–24(b). The peak value of the collector current equals I_{CQ}, and the peak value of the collector-to-emitter voltage equals V_{CEQ} in this case. This signal is the maximum that we can obtain from the class A amplifier. Actually, we cannot quite reach saturation or cutoff, so the practical maximum is slightly less.

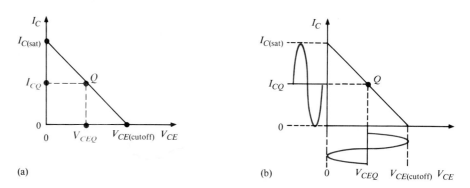

FIGURE 4–24
Maximum class A output occurs when the Q point is centered on the load line.

How a Noncentered Q Point Limits Output Swing

If the Q point is not centered, the output signal is limited. Figure 4–25 shows a load line with the Q point moved away from center toward cutoff. The output variation is limited by cutoff in this case. The collector current can only swing down to near zero and an equal amount above I_{CQ}. The collector-to-emitter voltage can only swing up to its cutoff value and an equal amount below V_{CEQ}. This situation is illustrated in Figure 4–25(a). If the amplifier is driven any further than this, it will "clip" at cutoff, as shown in Figure 4–25(b).

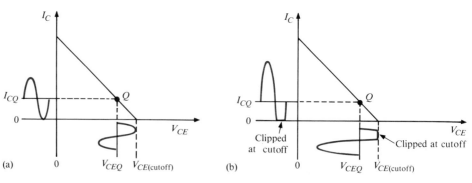

FIGURE 4–25
Q point closer to cutoff.

Amplifiers and Oscillators

Figure 4–26 shows a load line with the Q point moved away from center toward saturation. In this case, the output variation is limited by saturation. The collector current can only swing up to near saturation and an equal amount below I_{CQ}. The collector-to-emitter voltage can only swing down to its saturation value and an equal amount above V_{CEQ}. This situation is illustrated in Figure 4–26(a). If the amplifier is driven any further, it will "clip" at saturation, as shown in Figure 4–26(b).

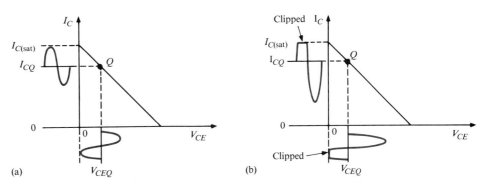

Figure 4–26
Q point closer to saturation.

Power Gain

The main purpose of a large-signal amplifier is to achieve power gain. If we assume that the large-signal current gain A_i is approximately equal to β_{dc}, then the power gain for a common-emitter amplifier is

$$A_p = A_i A_v = \beta_{dc} A_v$$

$$A_p = \beta_{dc}\left(\frac{R_C}{r_e}\right) \tag{4-26}$$

DC Quiescent Power

The power dissipation of a transistor with no signal input is the product of its Q-point current and voltage.

$$P_{DQ} = I_{CQ} V_{CEQ} \tag{4-27}$$

The quiescent power is the maximum power that the class A transistor must handle; therefore, its power rating should exceed this value.

Output Power

In general, for any Q-point location, the output power of a CE amplifier is the product of the rms collector current and the rms collector-to-emitter voltage.

$$P_{out} = V_{ce} I_c \tag{4-28}$$

Q POINT CENTERED When the Q point is centered, the maximum collector current swing is I_{CQ}, and the maximum collector-to-emitter voltage swing is V_{CEQ}, as was shown in Figure 4–24(b). The output power therefore is

$$P_{out} = (0.707V_{CEQ})(0.707I_{CQ})$$
$$P_{out} = 0.5V_{CEQ}I_{CQ} \qquad (4\text{–}29)$$

This is the maximum ac output power from a class A amplifier under signal conditions. Notice that it is one-half the quiescent power dissipation.

EFFICIENCY

Efficiency of an amplifier is the ratio of ac output power to dc input power. The dc input power is the dc supply voltage times the current drawn from the supply.

$$P_{dc} = V_{CC}I_{CC}$$

The average supply current I_{CC} equals I_{CQ}, and the supply voltage V_{CC} is twice V_{CEQ} when the Q point is centered. Therefore, the maximum efficiency is

$$\text{eff}_{max} = \frac{P_{out}}{P_{dc}} = \frac{0.5V_{CEQ}I_{CQ}}{V_{CC}I_{CC}} = \frac{0.5V_{CEQ}I_{CQ}}{2V_{CEQ}I_{CQ}} = \frac{0.5}{2}$$
$$\text{eff}_{max} = 0.25 \qquad (4\text{–}30)$$

Thus, 25% is the highest possible efficiency available from a class A amplifier and is approached only when the Q point is at the center of the load line.

■ EXAMPLE 4–10

Determine the following values for a class A amplifier operated with a centered Q point with $I_{CQ} = 50$ mA and $V_{CEQ} = 7.5$ V: **(a)** minimum transistor power rating, **(b)** ac output power, and **(c)** efficiency.

SOLUTION

(a) The maximum power that the transistor must be able to handle is the minimum rating that you would use. Thus,

$$P_{DQ} = I_{CQ}V_{CEQ} = (50 \text{ mA})(7.5 \text{ V}) = 0.375 \text{ W}$$

(b) $P_{out} = 0.5V_{CEQ}I_{CQ} = 0.5(50 \text{ mA})(7.5 \text{ V}) = 0.1875$ W

(c) Since the Q point is centered, the efficiency is at its maximum possible value of 25%.

PRACTICE EXERCISE 4–10

Explain what happens to the output voltage of the class A amplifier in this example if the Q point is shifted to $V_{CE} = 3$ V.

4–6 REVIEW QUESTIONS

1. What is the optimum Q-point location for class A amplifiers?
2. What is the maximum efficiency of a class A amplifier?
3. A certain amplifier has a centered Q point of $I_{CQ} = 10$ mA and $V_{CEQ} = 7$ V. What is the *maximum* ac output power?

4–7 CLASS B PUSH-PULL AMPLIFIER OPERATION

When an amplifier is biased such that it operates in the linear region for 180° of the input cycle and is in cutoff for 180°, it is a class B amplifier. The primary advantage of a class B amplifier over a class A is that the class B is more efficient; you can get more output power for a given amount of input power. A disadvantage of class B is that it is more difficult to implement the circuit in order to get a linear reproduction of the input waveform. As you will see in this section, the term **push-pull** *refers to a common type of class B amplifier circuit in which the input wave shape is approximately reproduced at the output.*

Class B amplifier operation is illustrated in Figure 4–27, where the output waveform is shown relative to the input.

FIGURE 4–27
Class B amplifier operation (noninverting).

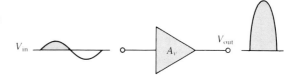

WHY THE Q POINT IS AT CUTOFF

The class B amplifier is biased at cutoff so that $I_{CQ} = 0$ and $V_{CEQ} = V_{CE(\text{cutoff})}$. It is brought out of cutoff and operates in its linear region when the input signal drives it into conduction. This is illustrated in Figure 4–28 with an emitter-follower circuit.

FIGURE 4–28
Common-collector class B amplifier.

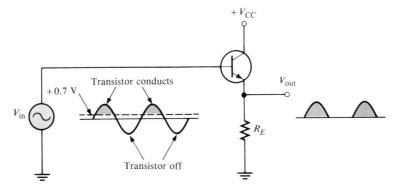

PUSH-PULL OPERATION

Figure 4–29 shows one type of push-pull class B amplifier using two emitter-followers. This is a complementary amplifier because one emitter-follower uses an npn transistor and the other a pnp, which conduct on *opposite* alternations of the input cycle. Notice that

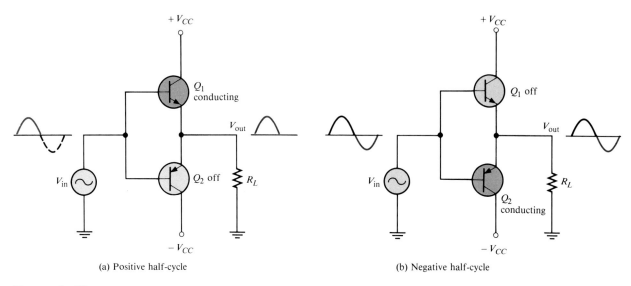

FIGURE 4–29
Class B push-pull operation.

there is no dc base bias voltage ($V_B = 0$). Thus, only the signal voltage drives the transistors into conduction. Q_1 conducts during the positive half of the input cycle, and Q_2 conducts during the negative half.

CROSSOVER DISTORTION

When the dc base voltage is zero, the input signal voltage must exceed V_{BE} before a transistor conducts. As a result, there is a time interval between the positive and negative alternations of the input when neither transistor is conducting, as shown in Figure 4–30. The resulting distortion in the output waveform is quite common and is called *crossover distortion*.

FIGURE 4–30
Illustration of crossover distortion in a class B push-pull amplifier.

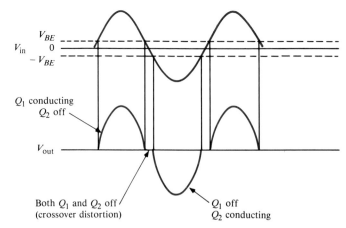

AMPLIFIERS AND OSCILLATORS

BIASING THE PUSH-PULL AMPLIFIER

To eliminate crossover distortion, both transistors in the push-pull arrangement must be biased slightly above cutoff when there is no signal. This can be done with a voltage divider and diode arrangement, as shown in Figure 4–31. When the diode characteristics of D_1 and D_2 are closely matched to the characteristics of the transistor base-emitter junctions, a stable bias is maintained.

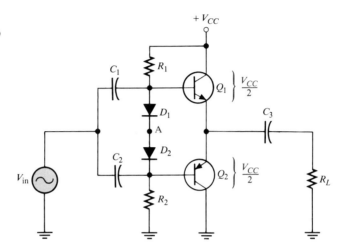

FIGURE 4–31

Biasing the push-pull amplifier to eliminate crossover distortion.

Since R_1 and R_2 are of equal value, the voltage with respect to ground at point A between the two diodes is $V_{CC}/2$. Assuming that both diodes and both transistors are identical, the drop across D_1 equals the V_{BE} of Q_1, and the drop across D_2 equals the V_{BE} of Q_2. As a result, the voltage at the emitters is also $V_{CC}/2$; therefore, $V_{CEQ1} = V_{CEQ2} = V_{CC}/2$, as indicated. Because both transistors are biased near cutoff, $I_{CQ} \cong 0$.

AC OPERATION

Under maximum conditions, transistors Q_1 and Q_2 are alternately driven from near cutoff to near saturation. During the positive alternation of the input signal, the Q_1 emitter is driven from its Q-point value of $V_{CC}/2$ to near V_{CC}, producing a positive peak voltage approximately equal to V_{CEQ}. At the same time, the Q_1 current swings from its Q-point value near zero to near-saturation value, as shown in Figure 4–32(a).

During the negative alternation of the input signal, the Q_2 emitter is driven from its Q-point value of $V_{CC}/2$ to near zero, producing a negative peak voltage approximately equal to V_{CEQ}. Also, the Q_2 current swings from near zero to near-saturation value, as shown in Figure 4–32(b).

Because the peak voltage across each transistor is V_{CEQ}, the ac saturation current is

$$I_{c(sat)} = \frac{V_{CEQ}}{R_L} \qquad (4\text{--}31)$$

Since $I_e \cong I_c$ and the output current is the emitter current, the peak output current is also V_{CEQ}/R_L.

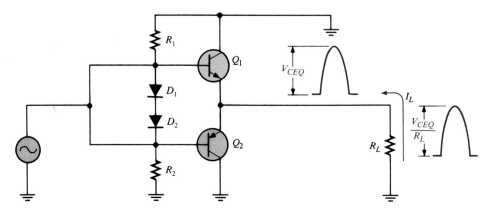

(a) Q_1 conducting with maximum signal output

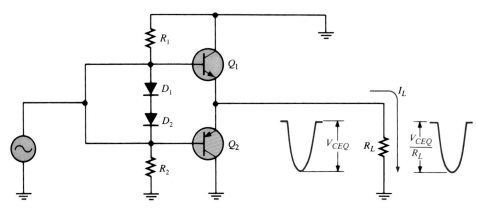

(b) Q_2 conducting with maximum signal output

FIGURE 4–32
AC push-pull operation. Capacitors are assumed to be shorts at the signal frequency, and the dc source is at ac ground.

■ **EXAMPLE 4–11**   Determine the maximum peak values for the output voltage and current in Figure 4–33.

SOLUTION
The maximum peak output is

$$V_{p(out)} \cong V_{CEQ} = \frac{V_{CC}}{2} = \frac{20 \text{ V}}{2} = 10 \text{ V}$$

The maximum peak output current is

$$I_{p(out)} \cong I_{c(sat)} = \frac{V_{CEQ}}{R_L} = \frac{10 \text{ V}}{5 \text{ }\Omega} = 2 \text{ A}$$

FIGURE 4-33

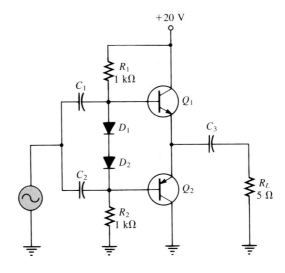

PRACTICE EXERCISE 4-11
Find the maximum peak values for the output voltage and current in Figure 4-33 if V_{CC} is lowered to 15 V and the load resistance is changed to 10 Ω.

MAXIMUM OUTPUT POWER

It has been shown that the maximum peak output current is approximately $I_{c(sat)}$, and the maximum peak output voltage is approximately V_{CEQ}. The maximum average output power is therefore

$$P_{out} = V_{(rms)out} I_{(rms)out}$$

Since

$$V_{(rms)out} = 0.707 V_{p(out)} = 0.707 V_{CEQ}$$

and

$$I_{(rms)out} = 0.707 I_{p(out)} = 0.707 I_{c(sat)}$$

then

$$P_{out} = 0.5 V_{CEQ} I_{c(sat)}$$

Substituting $V_{CC}/2$ for V_{CEQ}, we get

$$P_{out} = 0.25 V_{CC} I_{c(sat)} \qquad (4-32)$$

INPUT POWER

The input power comes from the V_{CC} supply and is

$$P_{dc} = V_{CC} I_{CC}$$

Since each transistor draws current for a half-cycle, the current is a half-wave signal with an average value of

$$I_{CC} = \frac{I_{c(sat)}}{\pi}$$

Thus,

$$P_{dc} = \frac{V_{CC}I_{c(sat)}}{\pi} \quad (4-33)$$

EFFICIENCY

The great advantage of push-pull class B amplifiers over class A is a much higher efficiency. This advantage usually overrides the difficulty of biasing the class B push-pull amplifier to eliminate crossover distortion. The efficiency is again defined as the ratio of ac output power to dc input power.

$$\text{eff} = \frac{P_{out}}{P_{dc}}$$

The maximum efficiency for a class B amplifier is designated eff_{max} and is developed as follows, starting with Equation (4-32).

$$P_{out} = 0.25 V_{CC} I_{c(sat)}$$

$$\text{eff}_{max} = \frac{P_{out}}{P_{dc}} = \frac{0.25 V_{CC} I_{c(sat)}}{V_{CC} I_{c(sat)}/\pi} = 0.25\pi$$

$$\text{eff}_{max} = 0.785 \quad (4-34)$$

Therefore, the maximum efficiency is 78.5%. Recall that the maximum efficiency for class A amplifiers is 0.25 (25%).

EXAMPLE 4–12

Find the maximum ac output power and the dc input power of the amplifier in Figure 4–33 of Example 4–11.

SOLUTION

In Example 4–11, $I_{c(sat)}$ was found to be 2 A. Thus,

$$P_{out} = 0.25 V_{CC} I_{c(sat)} = 0.25(20 \text{ V})(2 \text{ A}) = 10 \text{ W}$$

$$P_{dc} = \frac{V_{CC}I_{c(sat)}}{\pi} = \frac{(20 \text{ V})(2 \text{ A})}{\pi} = 12.73 \text{ W}$$

PRACTICE EXERCISE 4–12

What is the maximum ac output power if the load is 8 Ω?

4–7 REVIEW QUESTIONS

1. Where is the Q point for a class B amplifier?
2. What causes crossover distortion?
3. What is the maximum efficiency of a push-pull class B amplifier?
4. Explain the purpose of the push-pull configuration for class B.

4–8 CLASS C AMPLIFIER OPERATION

Class C amplifiers are biased so that conduction occurs for much less than 180°. Class C amplifiers are more efficient than either class A or push-pull class B, which means that more output power can be obtained from class C operation. Because the output waveform is severely distorted, class C amplifiers are normally limited to applications as tuned amplifiers at radio frequencies (RF) as you will see in this section.

BASIC OPERATION

Class C amplifier operation is illustrated in Figure 4–34. A basic common-emitter class C amplifier with a resistive load is shown in Figure 4–35(a). It is biased below cutoff with the $-V_{BB}$ supply. The ac source voltage has a peak value that is slightly greater than $V_{BB} + V_{BE}$ so that the base voltage exceeds the barrier potential of the base-emitter junction for a short time near the positive peak of each cycle, as illustrated in Figure 4–35(b). During this short interval, the transistor is turned on.

When the entire ac load is used, the maximum collector current is approximately $I_{C(\text{sat})}$, and the minimum collector voltage is approximately $V_{CE(\text{sat})}$.

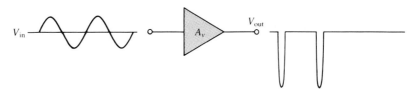

FIGURE 4–34
Class C amplifier operation (inverting).

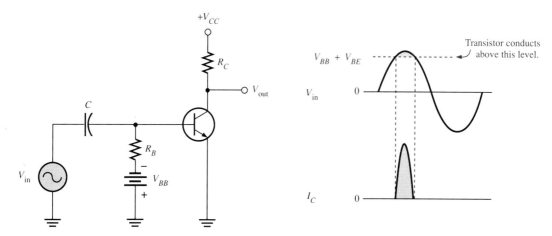

(a) Basic class C amplifier circuit

(b) Input voltage and output current waveforms

FIGURE 4–35
A class C amplifier circuit and waveforms.

POWER DISSIPATION

The power dissipation of the transistor in a class C amplifier is low because it is on for only a small percentage of the input cycle. Figure 4–36 shows the collector current pulses. The time between the pulses is the period (T) of the ac input voltage.

FIGURE 4–36
Collector current pulses in a class C amplifier.

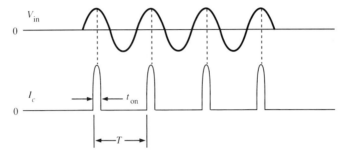

The transistor is on for a short time, t_{on}, and off for the rest of the input cycle. Since the power dissipation averaged over the entire cycle depends on the ratio of t_{on} to T and on the power dissipation during t_{on}, it is typically very low.

TUNED OPERATION

Because the collector voltage (output) is not a replica of the input, the resistively loaded class C amplifier is of no value in linear applications. Therefore, it is necessary to use a class C amplifier with a parallel resonant circuit (tank), as shown in Figure 4–37(a). The resonant frequency of the tank circuit is determined by the formula $f_r = 1/(2\pi\sqrt{LC})$. The short pulse of collector current on each cycle of the input initiates and sustains the oscillation of the tank circuit so that an output sine wave voltage is produced, as illustrated in Figure 4–37(b).

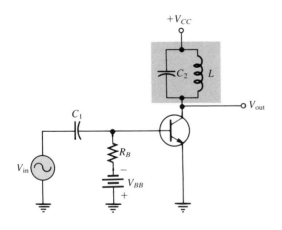

(a) Basic circuit

(b) Output waveforms

FIGURE 4–37
Tuned class C amplifier.

The amplitude of each successive cycle of the oscillation would be less than that of the previous cycle because of energy loss in the resistance of the tank circuit, as shown in Figure 4–38(a), and the oscillation would eventually die out. However, the regular recurrences of the collector current pulse re-energizes the resonant circuit and sustains the oscillations at a constant amplitude. When the tank circuit is tuned to the frequency of the input signal, re-energizing occurs on each cycle of the tank voltage, as shown in Figure 4–38(b).

FIGURE 4–38
Tank circuit oscillations.

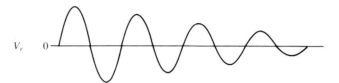

(a) Oscillation dies out due to energy loss.

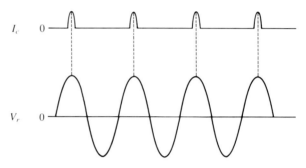

(b) Oscillation is sustained by short pulses of collector current.

MAXIMUM OUTPUT POWER AND EFFICIENCY

Since the voltage developed across the tank circuit has a peak-to-peak value of approximately $2V_{CC}$, the maximum output power can be expressed as

$$P_{out} = \frac{V_{rms}^2}{R_c} = \frac{(0.707V_{CC})^2}{R_c}$$

$$P_{out} = \frac{0.5V_{CC}^2}{R_c} \quad (4\text{–}35)$$

where R_c is the equivalent parallel resistance of the collector tank circuit and represents the parallel combination of the coil resistance and the load resistance. It usually has a low value.

The total power that must be supplied to the amplifier is

$$P_T = P_{out} + P_{D(avg)}$$

Therefore, the efficiency is

$$\text{eff} = \frac{P_{out}}{P_{out} + P_{D(avg)}} \qquad (4\text{--}36)$$

When $P_{out} \gg P_{D(avg)}$, the class C efficiency closely approaches 100%.

EXAMPLE 4–13

A certain class C amplifier has a $P_{D(avg)}$ of 2 mW, a V_{CC} equal to 24 V, and an R_c of 100 Ω. Determine the efficiency.

SOLUTION

$$P_{out} = \frac{0.5 V_{CC}^2}{R_c} = \frac{0.5(24 \text{ V})^2}{100 \text{ Ω}} = 2.88 \text{ W}$$

Therefore,

$$\text{eff} = \frac{P_{out}}{P_{out} + P_{D(avg)}} = \frac{2.88 \text{ W}}{2.88 \text{ W} + 2 \text{ mW}} = 0.9993$$

or

$$\%\text{eff} = 99.93\%$$

PRACTICE EXERCISE 4–13

What happens to the efficiency of the amplifier if R_c is increased?

4–8 REVIEW QUESTIONS

1. How is a class C amplifier normally biased?
2. What is the purpose of the tuned circuit in a class C amplifier?
3. A certain class C amplifier has a power dissipation of 100 mW and an output power of 1 W. What is its efficiency?

4–9 OSCILLATORS

Oscillators *are circuits that generate an output signal without having an externally applied input signal. They are used as signal sources in many applications. The oscillator is essentially an amplifier in which a portion of the output is fed back to the input. Its operation is based on the principle of* **positive feedback.**

The block diagram in Figure 4–39 shows an amplifier with gain *A*. The output drives a feedback circuit with gain *B*. The output of the feedback circuit provides the input to the amplifier. This is the basic form of an oscillator. If conditions are proper, the oscillator will continuously produce an output signal by amplifying the feedback signal.

CONDITIONS FOR OSCILLATION

There are two conditions that must be met in order for a circuit to oscillate. First, the phase shift through the amplifier and the feedback circuit must be 0°. There must be no phase shift so that the feedback signal will tend to reinforce itself rather than cancel.

AMPLIFIERS AND OSCILLATORS

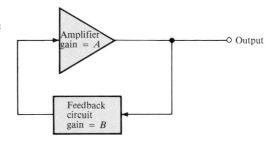

FIGURE 4–39
Block diagram of basic oscillator.

Second, the gain through the amplifier and the feedback circuit must be equal to or greater than 1 ($AB \geq 1$). This gain is called the *loop gain*. If the gain were less than 1, the output signal would decrease and die out.

THE RC OSCILLATOR

The basic *RC* oscillator shown in Figure 4–40 uses an *RC* network as its feedback circuit. In this case, three *RC* lag networks have a total phase shift of 180°. The common-emitter transistor contributes a 180° phase shift. The total phase shift through the amplifier and feedback circuit therefore is 360°, which is effectively 0° (no phase shift). The attenuation of the *RC* network and the gain of the amplifier must be such that the overall gain around the feedback loop is equal to 1 at the frequency of oscillation. This circuit will produce a continuous sine wave output.

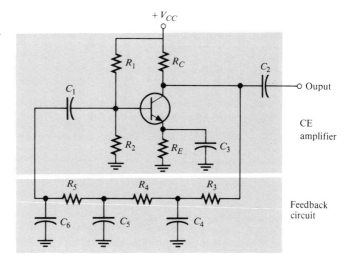

FIGURE 4–40
Basic *RC* oscillator.

THE COLPITTS OSCILLATOR

One basic type of tuned oscillator is the Colpitts, named after its inventor. As shown in Figure 4–41, this type of oscillator uses an *LC* circuit in the feedback loop to provide the necessary phase shift and to act as a filter that passes only the specified frequency of

FIGURE 4–41
A basic Colpitts oscillator.

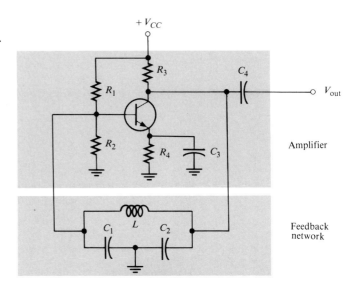

oscillation. The approximate frequency of oscillation is established by the values of C_1, C_2, and L according to the following familiar formula:

$$f_r \cong \frac{1}{2\pi\sqrt{LC_T}}$$

Because the capacitors effectively appear in series around the tank circuit, the total capacitance is

$$C_T = \frac{C_1 C_2}{C_1 + C_2}$$

THE HARTLEY OSCILLATOR

Another basic type of oscillator circuit is the Hartley, which is similar to the Colpitts except that the feedback network consists of two inductors and one capacitor, as shown in Figure 4–42.

The frequency of oscillation is

$$f_r \cong \frac{1}{2\pi\sqrt{L_T C}}$$

The total inductance is the series combination of L_1 and L_2.

THE CLAPP OSCILLATOR

The Clapp oscillator is similar to the Colpitts except that there is an additional capacitor in series with the inductor, as shown in Figure 4–43. C_1 and C_2 can be selected for optimum feedback, and C_3 can be adjusted to obtain the desired frequency of oscillation. Also, a capacitor having a negative temperature coefficient can be used for C_3 to stabilize the frequency of oscillation when there are temperature changes.

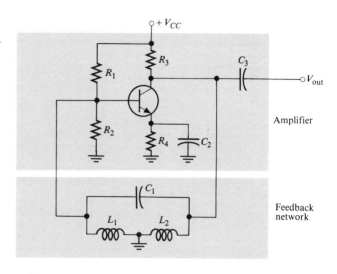

FIGURE 4–42
A basic Hartley oscillator.

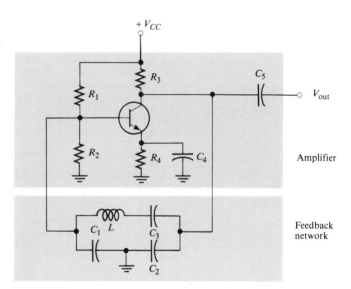

FIGURE 4–43
A basic Clapp oscillator.

THE CRYSTAL OSCILLATOR

A crystal oscillator is essentially a tuned-circuit oscillator that uses a quartz crystal as the resonant tank circuit. Other types of crystals can be used, but quartz is the most prevalent. Crystal oscillators offer greater frequency stability than other types.

Quartz is a substance found in nature that exhibits a property called the *piezoelectric effect*. When a changing mechanical stress is applied across the crystal to cause it to vibrate, a voltage is developed at the frequency of the mechanical vibration. Conversely, when an ac voltage is applied across the crystal, it vibrates at the frequency of the applied voltage.

The symbol for a crystal is shown in Figure 4–44(a), the electrical equivalent is shown in part (b), and a typical crystal is shown in part (c). In construction, a slab of quartz is mounted as shown in Figure 4–44(d). Series resonance occurs in the crystal when the reactances in the series branch are equal. Parallel resonance occurs, at a higher frequency, when the reactance of L_S equals the reactance of C_m.

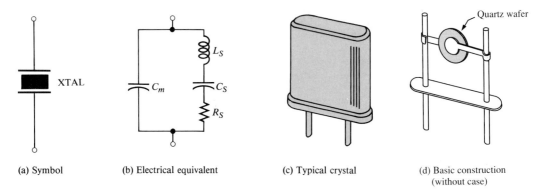

(a) Symbol   (b) Electrical equivalent   (c) Typical crystal   (d) Basic construction (without case)

FIGURE 4–44
Quartz crystal.

A crystal oscillator using the crystal as a series resonant tank circuit is shown in Figure 4–45(a). The impedance of the crystal is *minimum* at the series resonance, thus providing maximum feedback. The capacitor C_C is a tuning capacitor used to fine-tune the frequency. A modified Colpitts configuration, shown in Figure 4–45(b), uses the crystal

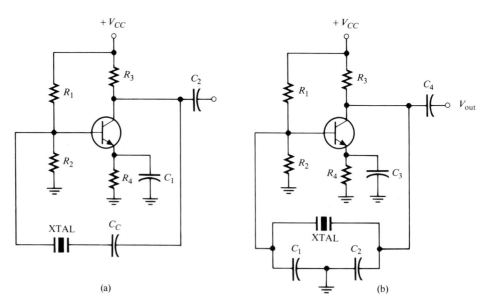

FIGURE 4–45
Basic crystal oscillators.

AMPLIFIERS AND OSCILLATORS

in its parallel resonant mode. The impedance of the crystal is maximum at parallel resonance, thus developing the maximum voltage across both C_1 and C_2. The voltage across C_1 is fed back to the input.

4–9 REVIEW QUESTIONS

1. Name four types of oscillators.
2. Describe the basic difference between the Colpitts and Hartley oscillators.

4–10 TROUBLESHOOTING

In working with any circuit, you must first know how it is supposed to work before you can troubleshoot it for a failure. The two-stage capacitively coupled amplifier discussed in Section 4–5 is used to illustrate a typical troubleshooting procedure.

The proper signal levels and dc voltage levels for the capacitively coupled two-stage amplifier are shown in Figure 4–46.

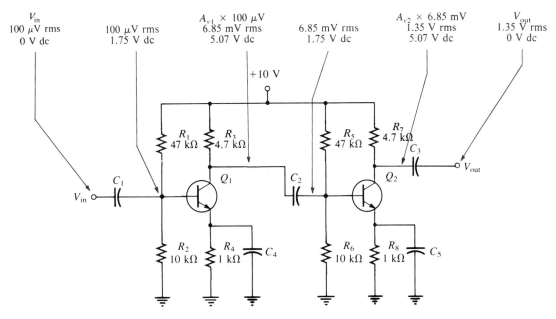

FIGURE 4–46
Two-stage amplifier with proper ac and dc voltage levels indicated.

TROUBLESHOOTING PROCEDURE

A basic procedure for troubleshooting called *signal tracing*, which is usually done with an oscilloscope, is illustrated in Figure 4–47 using the two-stage amplifier as an example. This general procedure can be expanded to any number of stages.

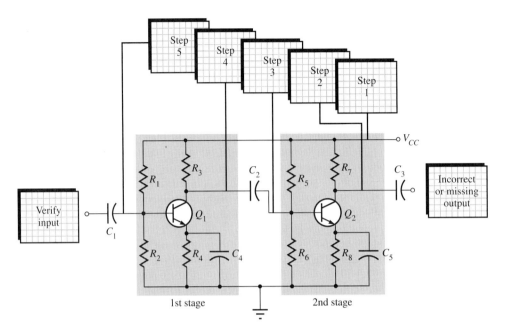

FIGURE 4–47
Basic troubleshooting procedure for a two-stage amplifier with no output signal.

We will begin by assuming that the final output signal of the amplifier has been determined to be missing. We will also assume that there is an input signal and that this has been verified. When troubleshooting, you generally start at the point where the signal is missing and work back point-by-point toward the input until a correct voltage is found. The fault then lies somewhere between the point of the first good voltage check and the missing or incorrect voltage point.

Before you begin checking the voltages, it is usually a good idea to visually check the circuit board or assembly for obvious problems such as broken or poor connections, solder splashes, wire clippings, or burned components.

STEP 1 Check the dc supply voltage. Often something as simple as a blown fuse or the power switch being in the *off* position may be the problem. The circuit will certainly not work without power.

STEP 2 Check the voltage at the collector of Q_2. If the correct signal is present at this point, the coupling capacitor, C_3, is open. If there is no signal at this point, proceed to Step 3.

STEP 3 Check the signal at the base of Q_2. If the signal is present at this point, the fault is in the second stage of the amplifier. First, do an in-circuit check of the transistor. If it is OK, then one of the biasing resistors may be open; if this is the case, the dc voltages will be incorrect. If there is no signal at this point, proceed to Step 4.

AMPLIFIERS AND OSCILLATORS

STEP 4 Check the signal at the collector of Q_1. If the signal is present at this point, the coupling capacitor, C_2, is open. If there is no signal at the point, proceed to Step 5.

STEP 5 Check the signal at the base of Q_1. If the signal is present at this point, the fault is in the first stage of the amplifier. First, do an in-circuit check of the transistor. If it is OK, then one of the biasing resistors may be open; if this is the case, the dc voltages will be incorrect. If there is no signal at this point, the coupling capacitor, C_1, is open because we started out by verifying that there is a correct signal at the input.

4–10 REVIEW QUESTIONS

1. If C_5 in Figure 4–47 were open, how would the output signal be affected? How would the dc level at the collector of Q_2 be affected?
2. If R_5 in Figure 4–47 were open, how would the output signal be affected?
3. If the coupling capacitor C_2 in Figure 4–47 shorted out, would any of the dc voltages in the amplifier be changed? If so, which ones?

4–11 A SYSTEM APPLICATION

The audio preamplifier in the receiver system that you saw at the opening of this chapter accepts a very small audio signal out of the detector circuit and amplifies it to a level for input to the power amplifier that provides audio power to the speaker. In this application, the transistors are used as common-emitter small-signal linear amplifiers. In this section, you will

- □ *See how a small-signal amplifier is used in a system application.*
- □ *See how a two-stage amplifier is used in a system application.*
- □ *See how a potentiometer is used for audio volume control.*
- □ *Translate between a printed circuit board and a schematic.*
- □ *Troubleshoot some common amplifier failures.*

A BRIEF DESCRIPTION OF THE SYSTEM

The block diagram for the AM radio receiver is shown in Figure 4–48. A very basic system description follows. The antenna picks up all radiated signals that pass by and feeds them into the RF amplifier. The voltages induced in the antenna by the electromagnetic radiation are extremely small. The RF amplifier is tuned to select and amplify a desired frequency within the AM broadcast band. Since it is a frequency-selective circuit, the RF amplifier eliminates essentially all but the selected frequency band. The output of the RF amplifier goes to the mixer where it is combined with the output of the local oscillator which is 455 kHz above the selected RF frequency. In the mixer, a nonlinear process called *heterodyning* takes place and produces one frequency that is the sum of the RF and local oscillator frequencies and another frequency that is the difference (always 455 kHz). The sum frequency is filtered out and only the 455 kHz difference frequency is used. This frequency is amplitude modulated just like the much higher RF frequency and, therefore, contains the audio signal. The IF (intermediate

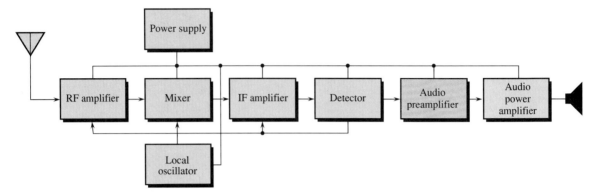

FIGURE 4–48

frequency) amplifier is tuned to 455 kHz and amplifies the mixer output. The detector takes the amplified 455 kHz AM signal and recovers the audio from it while eliminating the intermediate frequency. The output of the detector is a small audio signal that goes to the audio preamplifier and then to the power amplifier, which drives the speaker to convert the electrical audio signal into sound.

Now, so that you can take a closer look at the audio preamplifier, let's take it out of the system and put it on the test bench.

On the Test Bench

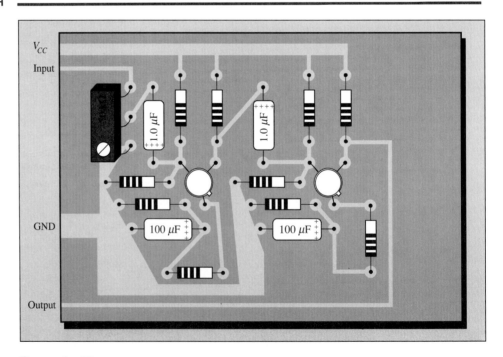

FIGURE 4–49

AMPLIFIERS AND OSCILLATORS

■ **ACTIVITY 1** **RELATE THE PC BOARD TO THE SCHEMATIC**

Notice that there are no component labels on the pc board shown in Figure 4–49. Label the components using the schematic in Figure 4–50 as a guide. That is, find and label R_1, R_2, and so forth.

FIGURE 4–50

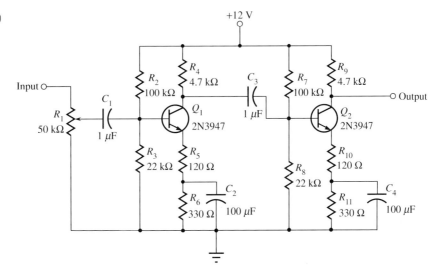

■ **ACTIVITY 2** **ANALYZE THE AMPLIFIER**

Refer to the 2N3947 data sheet in Appendix A and use minimum values of β when available.

STEP 1 Calculate all of the dc voltages in the audio preamp.

STEP 2 Determine the voltage gain of each stage and the overall voltage gain of the two-stage amplifier. Don't forget to take into account the loading effect of the input resistance of the second stage. Assume that the second stage is unloaded and all capacitive reactances are zero.

STEP 3 Specify what you would expect to happen as the frequency is reduced to a sufficiently low value. Why?

STEP 4 Repeat Step 2 with a 15 kΩ load resistor connected to the output of the second stage through a coupling capacitor.

■ **ACTIVITY 3** **WRITE A TECHNICAL REPORT**

Discuss the overall operation of the amplifier circuit. State the purpose of each component on the board. Make sure to explain how the potentiometer R_1 is used. Use the results of Activity 2 as appropriate.

■ **ACTIVITY 4** **TROUBLESHOOT THE CIRCUIT BOARD FOR EACH OF THE FOLLOWING PROBLEMS BY STATING THE PROBABLE CAUSE OR CAUSES IN EACH CASE**

1. No output signal when there is a verified input signal.
2. Proper voltages at base of Q_2, but signal voltage at collector is less than it should be.

3. Proper signal at collector of Q_1, but no signal at base of Q_2.
4. Amplitude of output signal much less than it should be.
5. Collector of Q_2 at $+12$ V dc and no signal voltage with a verified signal at the base.

4–11 REVIEW QUESTIONS

1. Explain why R_1 is in the circuit of Figure 4–50.
2. What happens if C_2 opens?
3. What happens if C_3 opens?
4. How can you reduce the gain of each stage without changing the dc voltages?

SUMMARY

- The three bipolar transistor amplifier configurations are common-emitter (CE), common-collector (CC), and common-base (CB).
- The general characteristics of a common-emitter amplifier are high voltage gain, high current gain, very high power gain, and low input resistance.
- The general characteristics of a common-collector amplifier are low voltage gain (≤ 1), high current gain, high power gain, and high input resistance.
- The general characteristics of a common-base amplifier are high voltage gain, low current gain (≤ 1), high power gain, and very low input resistance.
- The three FET amplifier configurations are common-source (CS), common-drain (CD), and common-gate (CG).
- The overall voltage gain of a multistage amplifier is the product of the gains of all the individual stages.
- Any of the bipolar or FET configurations can be operated as class A, class B, or class C amplifiers.
- The class A amplifier conducts for the entire 360° of the input cycle.
- The class B amplifier conducts for 180° of the input cycle.
- The class C amplifier conducts for a small portion of the input cycle.

GLOSSARY

Attenuation The reduction in the level of power, current, or voltage.

Class A A category of amplifier circuit that conducts for the entire input cycle and produces an output signal that is a replica of the input signal in terms of its waveshape.

Class B A category of amplifier circuit that conducts for half of the input cycle.

Class C A category of amplifier circuit that conducts for a very small portion of the input cycle.

Common-base A type of BJT amplifier configuration in which the base is the common (grounded) terminal.

Common-collector A type of BJT amplifier configuration in which the collector is the common (grounded) terminal.

Common-drain An FET amplifier configuration in which the drain is the grounded terminal.

Common-emitter A type of BJT amplifier configuration in which the emitter is the common (grounded) terminal.

Common-gate An FET amplifier configuration in which the gate is the grounded terminal.

Common-source An FET amplifier configuration in which the source is the grounded terminal.

Current gain The ratio of output current to input current.

Darlington pair A two-transistor arrangement that produces a multiplication of current gain.

Emitter-follower A popular term for a common-collector amplifier.

Feedback The process of returning a portion of a circuit's output signal to the input in such a way as to create certain specified operating conditions.

Gain The amount by which an electrical signal is increased or decreased; the ratio of output to input.

Oscillator An electronic circuit consisting of an amplifier and a phase-shift network connected in a feedback loop that produces a time-varying output signal without an external input signal using positive feedback.

Positive feedback The return of a portion of the output signal to the input such that it is in-phase with the input signal.

Power gain The ratio of output power to input power; the product of voltage gain and current gain.

Push-pull A type of class B amplifier in which one output transistor conducts for one half-cycle and the other conducts for the other half-cycle.

Transconductance, g_m The ratio of drain current to gate-to-source voltage in an FET.

Voltage gain The ratio of output voltage to input voltage.

FORMULAS

(4–1) $A_v = \dfrac{R_C}{r_e + R_E}$ CE voltage gain (unbypassed)

(4–2) $A_v = \dfrac{R_C}{r_e}$ CE voltage gain (bypassed)

(4–3) $r_e \cong \dfrac{25 \text{ mV}}{I_E}$ Internal emitter resistance

(4–4)	$R_{in} \cong \beta_{ac} r_e$	CE input resistance
(4–5)	$R_{in(T)} = R_1 \| R_2 \| R_{in}$	CE total input resistance
(4–6)	$A_i = \dfrac{I_c}{I_s}$	CE current gain
(4–7)	$A_p = A_v A_i$	CE power gain
(4–8)	$A_v = \dfrac{R_E}{r_e + R_E}$	CC voltage gain
(4–9)	$R_{in} \cong \beta_{ac} R_E$	CC input resistance
(4–10)	$R_{in(T)} = R_1 \| R_2 \| R_{in}$	CC total input resistance
(4–11)	$A_i = \dfrac{I_e}{I_s}$	CC current gain
(4–12)	$A_p \cong A_i$	CC power gain
(4–13)	$\beta_{ac} = \beta_{ac1} \beta_{ac2}$	Beta for a Darlington pair
(4–14)	$A_v = \dfrac{R_C}{r_e}$	CB voltage gain
(4–15)	$R_{in} = r_e$	CB input resistance
(4–16)	$A_i \cong 1$	CB current gain
(4–17)	$A_p \cong A_v$	CB power gain
(4–18)	$g_m = \dfrac{I_d}{V_{gs}}$	FET transconductance
(4–19)	$A_v = g_m R_D$	CS voltage gain
(4–20)	$A_v \cong \dfrac{g_m R_S}{1 + g_m R_S}$	CD voltage gain
(4–21)	$A_v = g_m R_D$	CG voltage gain
(4–22)	$R_{in(source)} = \dfrac{1}{g_m}$	CG input resistance
(4–23)	$A'_v = A_{v1} A_{v2} A_{v3} \cdots A_{vn}$	Multistage gain
(4–24)	$A_v \text{ (dB)} = 20 \log A_v$	Voltage gain in dB
(4–25)	$A'_v \text{(dB)} = A_{v1}\text{(dB)} + A_{v2}\text{(dB)} + \cdots + A_{vn}\text{(dB)}$	Multistage dB gain
(4–26)	$A_p = \beta_{dc} \left(\dfrac{R_C}{r_e} \right)$	CE large-signal power gain
(4–27)	$P_{DQ} = I_{CQ} V_{CEQ}$	Transistor power dissipation
(4–28)	$P_{out} = V_{ce} I_c$	CE output power

Amplifiers and Oscillators

(4–29)	$P_{out} = 0.5 V_{CEQ} I_{CQ}$	CE output power with centered Q point
(4–30)	$eff_{max} = 0.25$	Class A maximum efficiency
(4–31)	$I_{c(sat)} = \dfrac{V_{CEQ}}{R_L}$	Class B ac saturation current
(4–32)	$P_{out} = 0.25 V_{CC} I_{c(sat)}$	Class B output power (maximum)
(4–33)	$P_{dc} = \dfrac{V_{CC} I_{c(sat)}}{\pi}$	Class B input power
(4–34)	$eff_{max} = 0.785$	Class B maximum efficiency
(4–35)	$P_{out} = \dfrac{0.5 V_{CC}^2}{R_c}$	Class C output power (maximum)
(4–36)	$eff = \dfrac{P_{out}}{P_{out} + P_{D(avg)}}$	Class C efficiency

Self-Test

1. In a common-emitter (CE) amplifier, the capacitor from emitter to ground is called the
   (a) coupling capacitor (b) decoupling capacitor
   (c) bypass capacitor   (d) tuning capacitor
2. If the capacitor from emitter to ground in a CE amplifier is removed, the voltage gain
   (a) increases (b) decreases (c) is not affected (d) becomes erratic
3. When the collector resistor in a CE amplifier is increased in value, the voltage gain
   (a) increases (b) decreases (c) is not affected (d) becomes erratic
4. The input resistance of a CE amplifier is affected by
   (a) α and r_e (b) β and r_e (c) R_c and r_e (d) R_e, r_e, and β
5. The output signal of a CE amplifier is always
   (a) in phase with the input signal   (b) out of phase with the input signal
   (c) larger than the input signal   (d) equal to the input signal
6. The output signal of a common-collector amplifier is always
   (a) in phase with the input signal   (b) out of phase with the input signal
   (c) larger than the input signal   (d) exactly equal to the input signal
7. The largest *theoretical* voltage gain obtainable with a CC amplifier is
   (a) 100 (b) 10 (c) 1 (d) dependent on β

8. Using a Darlington pair has the advantage of
   (a) increasing the overall voltage gain   (b) less cost
   (c) decreasing the input resistance   (d) increasing the overall β
9. Compared to CE and CC amplifiers, the common-base (CB) amplifier has
   (a) a lower input resistance   (b) a much larger voltage gain
   (c) a larger current gain   (d) a higher input resistance
10. In terms of higher power gain, the amplifier of choice is
    (a) CB   (b) CC   (c) CE   (d) any of these
11. The most efficient amplifier configuration is
    (a) class A   (b) class B   (c) class C
12. When an FET with a lower transconductance is substituted into an FET amplifier circuit,
    (a) the voltage gain increases   (b) the voltage gain decreases
    (c) the input resistance decreases   (d) nothing changes
13. When amplifiers are cascaded,
    (a) the gain of each amplifier is increased   (b) each amplifier has to work less
    (c) a lower supply voltage is required   (d) the overall gain is increased
14. If three amplifiers, each with a voltage gain of 30, are connected in a multistage arrangement, the overall voltage gain is
    (a) 27,000   (b) 90   (c) 10   (d) 30
15. In a class A amplifier, the output signal is
    (a) distorted   (b) clipped
    (c) the same shape as the input   (d) smaller in amplitude than the input
16. Oscillators operate on the principle of
    (a) signal feedthrough   (b) positive feedback
    (c) negative feedback   (d) attenuation

PROBLEMS

SECTION 4–1 COMMON-EMITTER AMPLIFIERS

1. Determine the voltage gain for Figure 4–51.
2. Determine each of the dc voltages, V_B, V_C, and V_E, with respect to ground in Figure 4–51.
3. Determine the following dc values for the amplifier in Figure 4–52:
   (a) V_B (b) V_E (c) I_E (d) I_C (e) V_C (f) V_{CE}
4. Determine the following ac values for the amplifier in Figure 4–52:
   (a) R_{in} (b) $R_{in(T)}$ (c) A_v (d) A_i (e) A_p

FIGURE 4–51

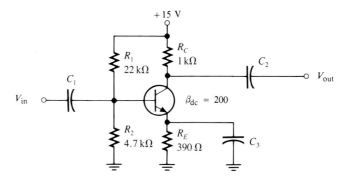

FIGURE 4–52

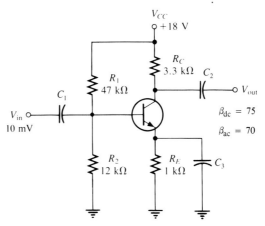

5. The amplifier in Figure 4–53 has a variable gain control, using a 100 Ω potentiometer for R_E with the wiper ac grounded. As the potentiometer is adjusted, more or less of R_E is bypassed to ground, thus varying the gain. The total R_E remains constant to dc, keeping the bias fixed. Determine the maximum and minimum gains for this amplifier.

FIGURE 4–53

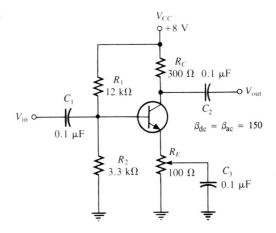

6. If a load resistance of 600 Ω is placed on the output of the amplifier in Figure 4–53, what is the maximum gain?

SECTION 4–2 COMMON-COLLECTOR AMPLIFIERS
7. Determine the *exact* voltage gain for the emitter-follower in Figure 4–54.
8. What is the total input resistance in Figure 4–54? What is the dc output voltage?
9. A load resistance is capacitively coupled to the emitter in Figure 4–54. In terms of signal operation, the load appears in parallel with R_E and reduces the effective emitter resistance. How does this affect the voltage gain?

FIGURE 4–54

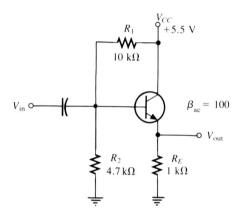

SECTION 4–3 COMMON-BASE AMPLIFIERS
10. What is the main disadvantage of the CB amplifier compared to the CE and the emitter-follower?
11. Find R_{in}, A_v, A_i, and A_p for the amplifier in Figure 4–55.

FIGURE 4–55

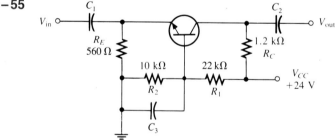

SECTION 4–4 FET AMPLIFIERS
12. Determine the voltage gain of each CS amplifier in Figure 4–56.
13. Find the gain of each amplifier in Figure 4–57.
14. Determine the gain of each amplifier in Figure 4–57 when a 10 kΩ load is capacitively coupled from source to ground.

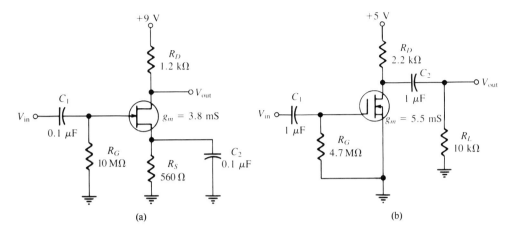

FIGURE 4–56

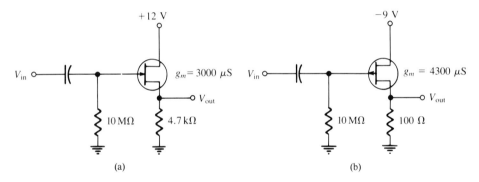

FIGURE 4–57

SECTION 4–5 MULTISTAGE AMPLIFIERS
15. Each of three cascaded amplifier stages has a decibel voltage gain of 10. What is the overall decibel voltage gain? What is the actual overall voltage gain?
16. For the two-stage, capacitively coupled amplifier in Figure 4–58, find the following values:
    (a) Voltage gain of each stage
    (b) Overall voltage gain
    (c) Express the gains found above in decibels

SECTION 4–6 CLASS A AMPLIFIER OPERATION
17. Determine the minimum power rating for each of the transistors in Figure 4–59.

SECTION 4–7 CLASS B PUSH-PULL AMPLIFIER OPERATION
18. Determine the dc voltages at the bases and emitters of Q_1 and Q_2 in Figure 4–60. Also determine V_{CEQ} for each transistor.

FIGURE 4–58

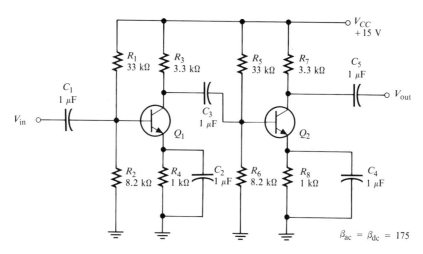

FIGURE 4–59

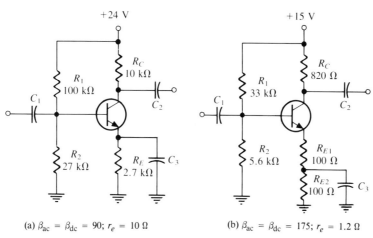

(a) $\beta_{ac} = \beta_{dc} = 90$; $r_e = 10\ \Omega$ (b) $\beta_{ac} = \beta_{dc} = 175$; $r_e = 1.2\ \Omega$

FIGURE 4–60

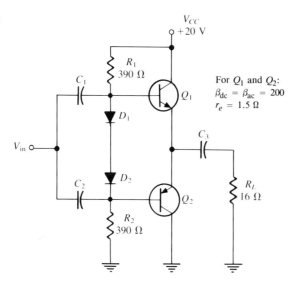

19. Determine the maximum peak output voltage and peak load current for the circuit in Figure 4–60.

20. The efficiency of a certain class B push-pull amplifier is 0.71, and the dc input power is 16.25 W. What is the ac output power?

SECTION 4–8 CLASS C AMPLIFIER OPERATION

21. What is the resonant frequency of the tank circuit in a class C amplifier with $L = 10$ mH and $C = 0.001$ μF?

22. Determine the efficiency of the class C amplifier when $P_{D(avg)} = 10$ mW, $V_{CC} = 15$ V, and the equivalent parallel resistance in the collector tank circuit is 50 Ω.

SECTION 4–9 OSCILLATORS

23. Calculate the frequency of oscillation for each circuit in Figure 4–61, and identify each type of oscillator.

FIGURE 4–61

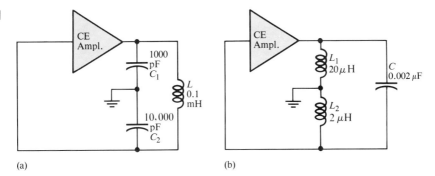

(a)   (b)

SECTION 4–10 TROUBLESHOOTING

24. Assume that the coupling capacitor C_2 is shorted in Figure 4–21. What dc voltage will appear at the collector of Q_1?

25. Assume that R_5 opens in Figure 4–21. Will Q_2 be in cutoff or in conduction? What dc voltage will you observe at the Q_2 collector?

26. Refer to Figure 4–58 and determine the general effect of each of the following failures:

   (a) C_2 opens (b) C_3 opens
   (c) C_4 opens (d) C_2 shorts
   (e) Base-collector junction of Q_1 opens (f) Base-emitter of Q_2 opens

27. Assume that you must troubleshoot the amplifier in Figure 4–58. Set up a table of test point values, input, output, and all transistor terminals that include both dc and rms values that you expect to observe when a 300 Ω test signal source with a 25 μV rms output is used.

28. What symptom(s) would indicate each of the following failures under signal conditions in Figure 4–62?

   (a) Q_1 open from drain to source (b) R_3 open (c) C_2 shorted
   (d) C_3 shorted (e) Q_2 open from drain to source

FIGURE 4–62

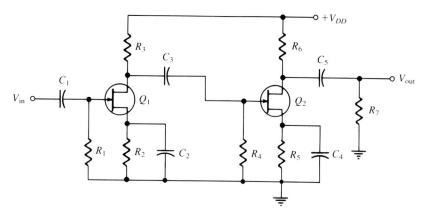

Answers to Review Questions

Section 4–1
1. To increase voltage gain
2. Ratio of collector resistance to total emitter resistance
3. 10,000

Section 4–2
1. Emitter-follower
2. 1
3. High input resistance

Section 4–3
1. Yes
2. Very low

Section 4–4
1. g_m and R_d
2. It is halved.
3. Common-gate has low input resistance.

Section 4–5
1. One amplifier in a cascaded arrangement
2. Product of individual gains
3. 53.98 dB

Section 4–6
1. Centered on the load line
2. 25%
3. 35 mW

Section 4–7
1. At cutoff
2. The barrier potential of the base-emitter junction

AMPLIFIERS AND OSCILLATORS   **221**

3. 78.5%
4. To reproduce both positive and negative alternations of the input signal with greater efficiency

SECTION 4–8
1. In cutoff
2. To produce a sine wave output
3. 90.9%

SECTION 4–9
1. RC phase-shift, Colpitts, Hartley, crystal
2. Colpitts uses two capacitors, center-tapped to ground in parallel with an inductor. Hartley uses two coils, center-tapped to ground in parallel with a capacitor.

SECTION 4–10
1. If C_5 opens, the gain drops. The dc level would not be affected.
2. Q_2 would be biased in cutoff.
3. The collector voltage of Q_1 and the base voltage of Q_2 would change. A change in V_{B2} will also cause V_{E2}, I_{E2}, and V_{C2} to change.

SECTION 4–11
1. R_1 adjusts the volume by increasing or decreasing the input voltage.
2. The gain of the first stage decreases if C_2 opens.
3. There will be no output if C_3 opens because the signal is not coupled through to the second stage.
4. The gain can be reduced by partially bypassing the emitter resistors.

ANSWERS TO PRACTICE EXERCISES

4–1	235	
4–2	7.83 kΩ	
4–3	$A_v = 338.4$; $A_i = 89.2$; $A_p = 30{,}185$	
4–4	A_p increases from 4.76 to 5.89.	
4–5	A_v decreases to 64.5.	
4–6	I_D will be 584 μA. V_D will increase to 10.1 V.	
4–7	0.976	
4–8	$R_{in} = 384.6$ Ω; $\Delta R_{in} = 15.6$ Ω	
4–9	$A'_v = 1500$, A_{v1} (dB) $= 27.96$ dB, A_{v2} (dB) $= 13.98$ dB, A_{v3} (dB) $= 21.58$ dB, A'_v (dB) $= 63.52$ dB	
4–10	The peak unclipped output voltage decreases to 3 V.	
4–11	7.5 V, 0.75 A	
4–12	6.25 W	
4–13	The efficiency decreases.	

5
OPERATIONAL AMPLIFIERS

5–1 INTRODUCTION TO OPERATIONAL AMPLIFIERS
5–2 THE DIFFERENTIAL AMPLIFIER
5–3 OP-AMP DATA SHEET PARAMETERS
5–4 NEGATIVE FEEDBACK
5–5 OP-AMP CONFIGURATIONS WITH NEGATIVE FEEDBACK
5–6 EFFECTS OF NEGATIVE FEEDBACK ON OP-AMP IMPEDANCES
5–7 BIAS CURRENT AND OFFSET VOLTAGE COMPENSATION
5–8 TROUBLESHOOTING
5–9 A SYSTEM APPLICATION

After completing this chapter, you should be able to

☐ Describe the basic operational amplifier.
☐ Compare ideal to practical op-amp characteristics.
☐ Explain the operation of a basic differential amplifier.
☐ Discuss single and double-ended operation in differential amplifiers.
☐ Define *open-loop gain*.
☐ Define and calculate the common-mode rejection ratio.
☐ Use op-amp data sheet parameters.
☐ Discuss negative feedback and how it is used in amplifiers.
☐ Explain how negative feedback affects the voltage gain of an op-amp.
☐ Explain how negative feedback affects input and output impedances.
☐ Recognize and analyze inverting, noninverting, and voltage follower op-amp configurations.

So far in this book, you have studied a number of important electronic devices. These devices, such as the diode and the transistor, are separate devices that are individually packaged and interconnected in a circuit with other devices to form a complete, functional unit. Such devices are referred to as *discrete* components.

Now we move into the area of linear integrated circuits, where many transistors, diodes, resistors, and capacitors are fabricated on a single tiny chip of semiconductor material and packaged in a single case to form a functional circuit. An integrated circuit, such as an operational amplifier (op-amp), is treated as a single device. This means that you will be concerned with what the circuit does more from an external viewpoint than from an internal, component-level viewpoint.

In this chapter, you will learn the basics of operational amplifiers, which are the most versatile and widely used of all linear integrated circuits.

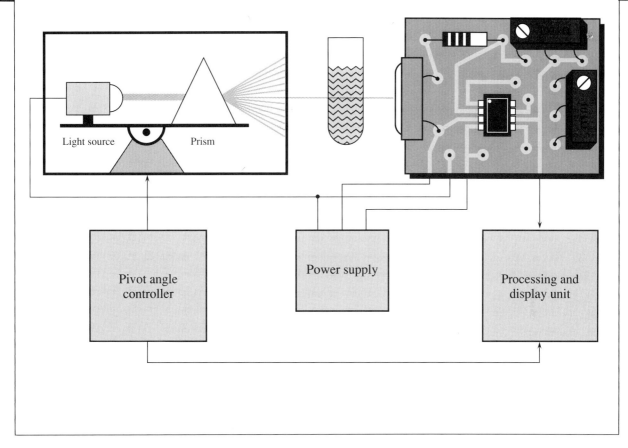

A SYSTEM APPLICATION

In medical laboratories, an instrument known as a spectrophotometer is used to analyze chemicals in solutions by determining how much absorption of light occurs over a range of wavelengths. A basic system is shown in the above diagram. Light is passed through a prism; and as the light source and prism are pivoted, different wavelengths of visible light pass through the slit. The wavelength coming through the slit at a given pivot angle passes through the solution and is detected by a photocell. The op-amp circuit is used to amplify the output of the photocell and send the signal to a processor and display instrument. Since every chemical and compound absorbs light in a different way, the output of the spectrophotometer can be used to accurately identify the contents of the solution.

For the system application in Section 5–9, in addition to the other topics, be sure you understand

☐ The functions of the inputs and outputs of an op-amp.
☐ How an op-amp works.
☐ The pin configurations of an op-amp.
☐ How to null an op-amp's output.

5-1 INTRODUCTION TO OPERATIONAL AMPLIFIERS

Early **operational amplifiers** *(op-amps) were used primarily to perform mathematical operations such as addition, subtraction, integration, and differentiation, hence the term operational. These early devices were constructed with vacuum tubes and worked with high voltages. Today's op-amps are linear integrated circuits that use relatively low supply voltages and are reliable and inexpensive.*

SYMBOL AND TERMINALS

The standard op-amp symbol is shown in Figure 5-1(a). It has two input terminals, called the *inverting input* ($-$) and the *noninverting input* ($+$), and one output terminal. The typical op-amp operates with two dc supply voltages, one positive and the other negative, as shown in Figure 5-1(b). Usually these dc voltage terminals are left off the schematic symbol for simplicity but are always understood to be there. Some typical op-amp IC packages are shown in Figure 5-1(c).

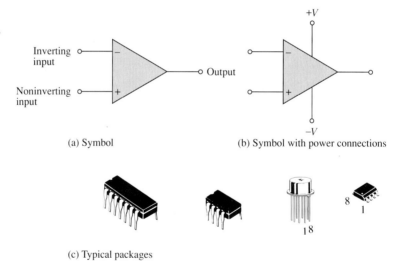

FIGURE 5-1
Op-amp symbols and packages. Part (c) copyright of Motorola, Inc. Used by permission.

(a) Symbol

(b) Symbol with power connections

(c) Typical packages

THE IDEAL OP-AMP

In order to get a concept of what an op-amp is, we will consider its ideal characteristics. A practical op-amp, of course, falls short of these ideal standards, but it is much easier to understand and analyze the device from an ideal point of view.

First, the ideal op-amp has infinite voltage gain and infinite bandwidth. Also, it has an infinite input impedance (open), so that it does not load the driving source. Finally, it has a zero output impedance. These characteristics are illustrated in Figure 5-2. The input voltage V_{in} appears between the two input terminals, and the output voltage is $A_v V_{in}$, as indicated by the internal voltage source symbol. The concept of infinite input impedance is a particularly valuable analysis tool for the various op-amp configurations covered later.

FIGURE 5–2
Ideal op-amp representation.

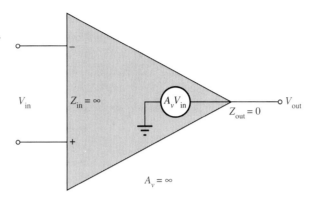

THE PRACTICAL OP-AMP

Although modern integrated circuit op-amps approach parameter values that can be treated as ideal in many cases, the ideal device has not been and probably will not be developed even though improvements continue to be made. Any device has limitations, and the integrated circuit op-amp is no exception. Op-amps have both voltage and current limitations. Peak-to-peak output voltage, for example, is usually limited to slightly less than the two supply voltages. Output current is also limited by internal restrictions such as power dissipation and component ratings. Characteristics of a practical op-amp are high voltage gain, high input impedance, low output impedance, and wide bandwidth. Some of these are illustrated in Figure 5–3.

FIGURE 5–3
Practical op-amp representation.

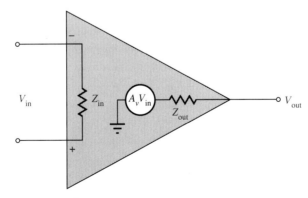

5–1 REVIEW QUESTIONS

1. What are the connections to a basic op-amp?
2. Describe some of the characteristics of a *practical* op-amp.

5–2 THE DIFFERENTIAL AMPLIFIER

The op-amp, in its basic form, typically consists of two or more differential amplifier stages. Because the differential amplifier (diff-amp) is fundamental to the op-amp's internal operation, it is useful to spend some time in acquiring a basic understanding of this type of circuit.

A basic **differential amplifier** circuit is shown in Figure 5–4(a) and its block symbol in Figure 5–4(b). The diff-amp stages that make up part of the op-amp provide high voltage gain and common-mode rejection (defined later).

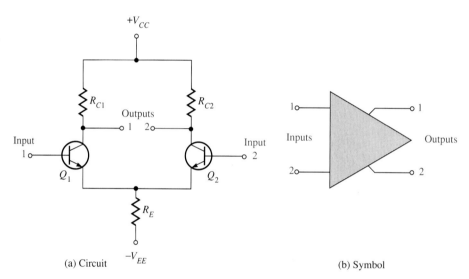

FIGURE 5–4
Basic differential amplifier.

(a) Circuit   (b) Symbol

BASIC OPERATION

Although an op-amp typically has more than one differential amplifier stage, we will use a single diff-amp to illustrate the basic operation. The following discussion is in relation to Figure 5–5 and consists of a basic dc analysis of the diff-amp's operation. First, when both inputs are grounded (0 V), the emitters are at -0.7 V, as indicated in Figure 5–5(a). It is assumed that the transistors are identically matched by careful process control during manufacturing so that their dc emitter currents are the same when there is no input signal.

$$I_{E1} = I_{E2}$$

Since both emitter currents combine through R_E,

$$I_{E1} = I_{E2} = \frac{I_{R_E}}{2} \qquad (5\text{–}1)$$

OPERATIONAL AMPLIFIERS

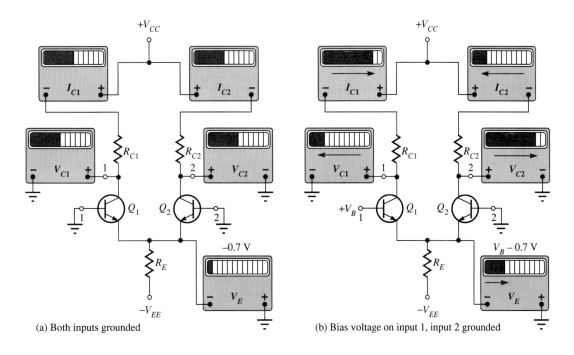

(a) Both inputs grounded

(b) Bias voltage on input 1, input 2 grounded

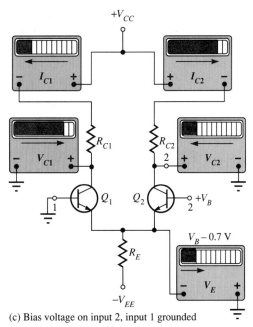

(c) Bias voltage on input 2, input 1 grounded

FIGURE 5–5
Basic operation of a differential amplifier (ground is zero volts).

where

$$I_{R_E} = \frac{V_E - V_{EE}}{R_E} \qquad (5-2)$$

Based on the approximation that $I_C \cong I_E$, it can be stated that

$$I_{C1} = I_{C2} \cong \frac{I_{R_E}}{2} \qquad (5-3)$$

Since both collector currents and both collector resistors are equal (when the input voltage is zero),

$$V_{C1} = V_{C2} = V_{CC} - I_{C1}R_{C1} \qquad (5-4)$$

This condition is illustrated in Figure 5–5(a). Next, input 2 is left grounded, and a positive bias voltage is applied to input 1, as shown in Figure 5–5(b). The positive voltage on the base of Q_1 increases I_{C1} and raises the emitter voltage to

$$V_E = V_B - 0.7 \text{ V} \qquad (5-5)$$

This action reduces the forward bias (V_{BE}) of Q_2 because its base is held at 0 V (ground), thus causing I_{C2} to decrease as indicated in part (b) of the diagram. The net result is that the increase in I_{C1} causes a decrease in V_{C1}, and the decrease in I_{C2} causes an increase in V_{C2}, as shown. Finally, input 1 is grounded and a positive bias voltage is applied to input 2, as shown in Figure 5–5(c).

The positive bias voltage causes Q_2 to conduct more, thus increasing I_{C2}. Also, the emitter voltage is raised. This reduces the forward bias of Q_1, since its base is held at ground, and causes I_{C1} to decrease. The result is that the increase in I_{C2} produces a decrease in V_{C2}, and the decrease in I_{C1} causes V_{C1} to increase, as shown in Figure 5–5(c).

MODES OF SIGNAL OPERATION

SINGLE-ENDED INPUT When a diff-amp is operated in this mode, one input is grounded and the signal voltage is applied only to the other input, as shown in Figure 5–6. In the case where the signal voltage is applied to input 1 as in part (a), an inverted, amplified signal voltage appears at output 1 as shown. Also, a signal voltage appears in-phase at the emitter of Q_1. Since the emitters of Q_1 and Q_2 are common, the emitter signal becomes an input to Q_2, which functions as a common-base amplifier. The signal is amplified by Q_2 and appears, noninverted, at output 2. This action is illustrated in part (a).

In the case where the signal is applied to input 2 with input 1 grounded, as in Figure 5–6(b), an inverted, amplified signal voltage appears at output 2. In this situation, Q_1 acts as a common-base amplifier, and a noninverted, amplified signal appears at output 1. This action is illustrated in part (b) of the figure.

DIFFERENTIAL INPUT In this mode, two opposite-polarity (out-of-phase) signals are applied to the inputs, as shown in Figure 5–7(a). This type of operation is also referred to as

FIGURE 5–6

Single-ended input operation of a differential amplifier.

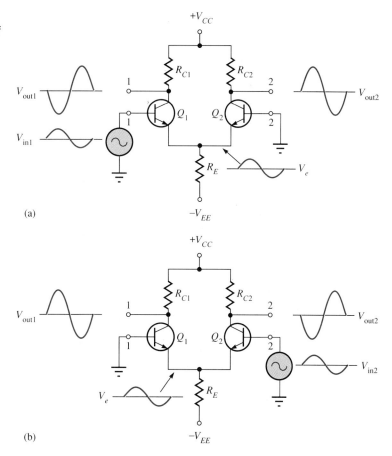

double-ended. Each input affects the outputs, as you will see in the following discussion. Figure 5–7(b) shows the output signals due to the signal on input 1 acting alone as a single-ended input. Figure 5–7(c) shows the output signals due to the signal on input 2 acting alone as a single-ended input. Notice, in parts (b) and (c), that the signals on output 1 are of the same polarity. The same is also true for output 2. By superimposing both output 1 signals and both output 2 signals, we get the total differential operation, as pictured in Figure 5–7(d).

COMMON-MODE INPUT One of the most important aspects of the operation of a differential amplifier can be seen by considering the **common-mode** condition where two signal voltages of the same phase, frequency, and amplitude are applied to the two inputs, as shown in Figure 5–8(a). Again, by considering each input signal as acting alone, the basic operation can be understood. Figure 5–8(b) shows the output signals due to the signal on only input 1, and Figure 5–8(c) shows the output signals due to the signal on only input 2. Notice that the corresponding signals on output 1 are of the opposite

FIGURE 5–7
Differential operation of a differential amplifier. (a) Differential inputs. (b) Outputs due to V_{in1}. (c) Outputs due to V_{in2}. (d) Total outputs due to differential inputs.

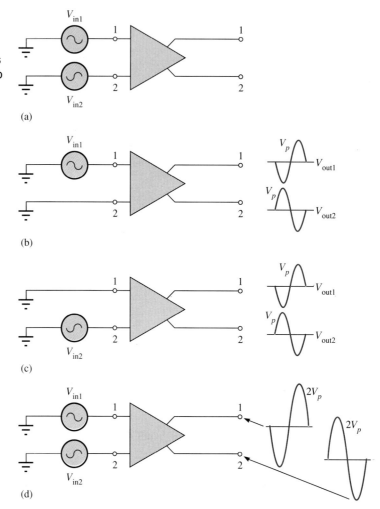

polarity, and so are the ones on output 2. When these are superimposed, they cancel, resulting in a zero output voltage, as shown in Figure 5–8(d).

This action is called *common-mode rejection*. Its importance lies in the situation where an unwanted signal appears commonly on both diff-amp inputs. Common-mode rejection means that this unwanted signal will not appear on the outputs to distort the desired signal. Common-mode signals (noise) generally are the result of the pick-up of radiated energy on the input lines, from adjacent lines, or the 60 Hz power line, or other sources.

In summary, desired signals appear on only one input or with opposite polarities on both input lines. These desired signals are amplified and appear on the outputs as previously discussed. Unwanted signals (noise) appearing with the same polarity on both input lines are essentially cancelled by the diff-amp and do not appear on the outputs. The measure of an amplifier's ability to reject common-mode signals is a parameter called the *common-mode rejection ratio* (CMRR) and is discussed next.

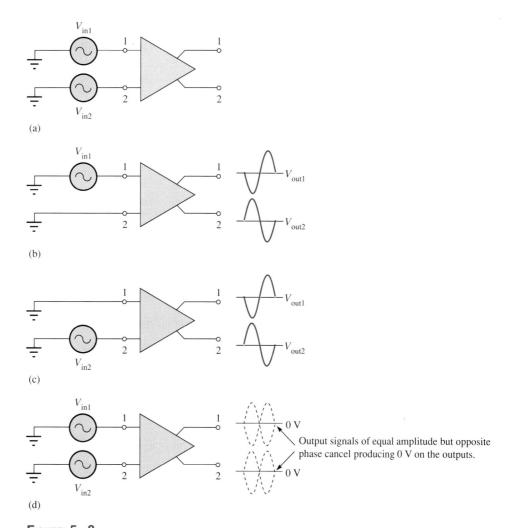

FIGURE 5–8
Common-mode operation of a differential amplifier. (a) Common-mode inputs. (b) Outputs due to V_{in1}. (c) Outputs due to V_{in2}. (d) Outputs cancel when common-mode signals are applied.

COMMON-MODE GAIN

Ideally, a differential amplifier provides a very high gain for desired signals (single-ended or differential), and zero gain for common-mode signals. Practical diff-amps, however, do exhibit a very small common-mode gain (usually much less than one), while providing a high differential voltage gain (usually several thousand). The higher the differential gain with respect to the common-mode gain, the better the performance of the diff-amp in terms of rejection of common-mode signals. This suggests that a good measure of the diff-amp's performance in rejecting unwanted common-mode signals is the ratio of the differential gain $A_{v(d)}$ to the common-mode gain, A_{cm}. This ratio is called the **common-mode rejection ratio,** CMRR.

$$\text{CMRR} = \frac{A_{v(d)}}{A_{cm}} \qquad (5\text{-}6)$$

The higher the CMRR, the better, as you can see. A very high value of CMRR means that the differential gain $A_{v(d)}$ is high and the common-mode gain A_{cm} is low. The CMRR is often expressed in decibels (dB) as

$$\text{CMRR} = 20 \log \frac{A_{v(d)}}{A_{cm}} \qquad (5\text{-}7)$$

■ **EXAMPLE 5-1** A certain differential amplifier has a differential voltage gain of 2000 and a common-mode gain of 0.2. Determine the CMRR and express it in dB.

SOLUTION
$A_{v(d)} = 2000$, and $A_{cm} = 0.2$. Therefore,

$$\text{CMRR} = \frac{A_{v(d)}}{A_{cm}} = \frac{2000}{0.2} = 10{,}000$$

In dB, CMRR = 20 log(10,000) = 80 dB.

PRACTICE EXERCISE 5-1
Determine the CMRR and express it in dB for an amplifier with a differential voltage gain of 8500 and a common-mode gain of 0.25.

A CMRR of 10,000, for example, means that the desired input signal (differential) is amplified 10,000 times more than the unwanted noise (common-mode). So, as an example, if the amplitudes of the differential input signal and the common-mode noise are equal, the desired signal will appear on the output 10,000 times greater in amplitude than the noise. Thus, the noise or interference has been essentially eliminated. Example 5-2 should help reinforce the idea of common-mode rejection and the general signal operation of the differential amplifier.

■ **EXAMPLE 5-2** The differential amplifier shown in Figure 5-9 has a differential voltage gain of 2500 and a CMRR of 30,000. In part (a) of the figure, a single-ended input signal of 500 μV rms is applied. At the same time a 1-V, 60 Hz interference signal appears on both inputs as a result of radiated pick-up from the ac power system. In part (b) of the figure, differential input signals of 500 μV rms each are applied to the inputs. The common-mode interference is the same as in part (a).

(a) Determine the common-mode gain.
(b) Express the CMRR in dB.
(c) Determine the rms output signal for parts (a) and (b) of the figure.
(d) Determine the rms interference voltage on the output.

FIGURE 5–9

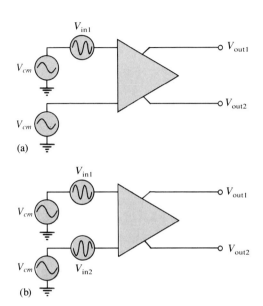

SOLUTION

(a) CMRR $= \dfrac{A_{v(d)}}{A_{cm}}$. Therefore,

$$A_{cm} = \dfrac{A_{v(d)}}{\text{CMRR}} = \dfrac{2500}{30{,}000} = 0.083$$

(b) CMRR $= 20 \log(30{,}000) = 89.5$ dB

(c) In Figure 5–9(a), the differential input voltage is the difference between the voltage on input 1 and that on input 2. Since input 2 is grounded, its voltage is zero. Therefore,

$$V_{in(diff)} = V_{in1} - V_{in2}$$
$$= 500 \ \mu V - 0 \ V$$
$$= 500 \ \mu V$$

The output signal voltage in this case is taken at output 1.

$$V_{out1} = A_{v(d)} V_{in(diff)}$$
$$= (2500)(500 \ \mu V)$$
$$= 1.25 \ V \ \text{rms}$$

In Figure 5–9(b), the differential input voltage is the difference between the two opposite-polarity, 500 μV signals.

$$V_{in(diff)} = V_{in1} - V_{in2}$$
$$= 500 \, \mu V - (-500 \, \mu V)$$
$$= 1000 \, \mu V$$
$$= 1 \, mV$$

The output voltage signal is

$$V_{out1} = A_{v(d)} V_{in(diff)}$$
$$= (2500)(1 \, mV)$$
$$= 2.5 \, V \, rms$$

This shows that a differential input (two opposite-polarity signals) results in a gain that is double that for a single-ended input.

(d) The common-mode input is 1 V rms. The common-mode gain A_{cm} is 0.083. The interference (common-mode) voltage on the output is therefore

$$A_{cm} = \frac{V_{out(cm)}}{V_{in(cm)}}$$
$$V_{out(cm)} = A_{cm} V_{in(cm)}$$
$$= (0.083)(1 \, V)$$
$$= 0.083 \, V$$

PRACTICE EXERCISE 5–2

The amplifier in Figure 5–9 has a differential voltage gain of 4200 and a CMRR of 25,000. For the same single-ended and differential input signals as described in the example: **(a)** Find A_{cm}. **(b)** Express the CMRR in dB. **(c)** Determine the rms output signal for parts (a) and (b) of the figure. **(d)** Determine the rms interference (common-mode) voltage appearing on the output.

SIMPLE OP-AMP ARRANGEMENT

Figure 5–10 shows two differential amplifier (diff-amp) stages and an emitter-follower connected to form a simple op-amp. The first stage can be used with a single-ended or a differential input. The differential outputs of the first stage are directly coupled into the differential inputs of the second stage. The output of the second stage is single-ended to drive an emitter-follower to achieve a relatively low output impedance. Both differential stages together provide a high voltage gain and a high CMRR.

5–2 REVIEW QUESTIONS

1. Distinguish between differential and single-ended inputs.
2. What is common-mode rejection?
3. For a given value of differential gain, does a higher CMRR result in a higher or lower common-mode gain?

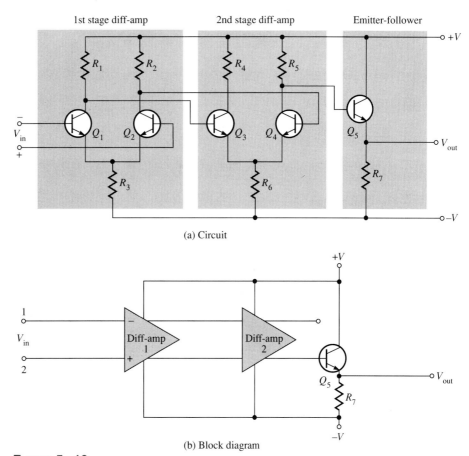

FIGURE 5–10
Simplified internal circuitry of a basic op-amp.

5–3 OP-AMP DATA SHEET PARAMETERS

In this section, several important op-amp parameters are defined. These are the input offset voltage, the input offset voltage drift, the input bias current, the input impedance, the input offset current, the output impedance, the common-mode range, the open-loop voltage gain, the common mode rejection ratio, the slew rate, and the frequency response. Also four popular IC op-amps are compared in terms of these parameters.

INPUT OFFSET VOLTAGE

The ideal op-amp produces zero volts out for zero volts in. In a practical op-amp, however, a small dc voltage appears at the output when no differential input voltage is applied. Its primary cause is a slight mismatch of the base-to-emitter voltages of the differential input stage, as illustrated in Figure 5–11(a). The output voltage of the

FIGURE 5–11

Input offset voltage, V_{OS}.

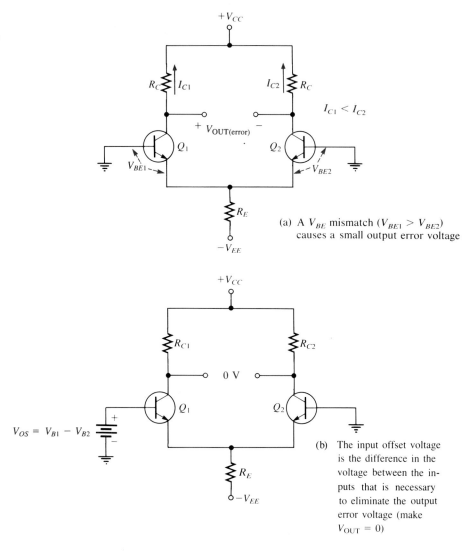

(a) A V_{BE} mismatch ($V_{BE1} > V_{BE2}$) causes a small output error voltage

(b) The input offset voltage is the difference in the voltage between the inputs that is necessary to eliminate the output error voltage (make $V_{OUT} = 0$)

differential input stage is expressed as

$$V_{OUT} = I_{C2}R_C - I_{C1}R_C \qquad (5-8)$$

A small difference in the base-to-emitter voltages of Q_1 and Q_2 causes a small difference in the collector currents. This results in a nonzero value of V_{OUT}. (The collector resistors are equal.) As specified on an op-amp data sheet, the *input offset voltage* V_{OS} is the differential dc voltage required between the inputs to force the differential output to zero volts, as demonstrated in Figure 5–11(b). Typical values of input offset voltage are in the range of 2 mV or less. In the ideal case, it is 0 V.

INPUT OFFSET VOLTAGE DRIFT WITH TEMPERATURE

The *input offset voltage drift* is a parameter related to V_{OS} that specifies how much change occurs in the input offset voltage for each degree change in temperature. Typical values range anywhere from about 5 μV per degree Celsius to about 50 μV per degree Celsius. Usually, an op-amp with a higher nominal value of input offset voltage exhibits a higher drift.

INPUT BIAS CURRENT

You have seen that the input terminals of a bipolar differential amplifier are the transistor bases and, therefore, the input currents are the base currents. The *input bias current* is the dc current required by the inputs of the amplifier to properly operate the first stage. By definition, the input bias current is the average of both input currents and is calculated as follows:

$$I_{BIAS} = \frac{I_1 + I_2}{2} \tag{5-9}$$

The concept of input bias current is illustrated in Figure 5–12.

FIGURE 5–12
Input bias current is the average of the two op-amp input currents.

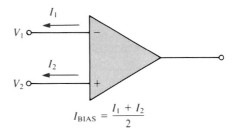

INPUT IMPEDANCE

Two basic ways of specifying the *input impedance* of an op-amp are the differential and the common mode. The differential input impedance is the total resistance between the inverting and the noninverting inputs and is illustrated in Figure 5–13(a). Differential impedance is measured by determining the change in bias current for a given change in differential input voltage. The common-mode input impedance is the resistance between each input and ground and is measured by determining the change in bias current for a given change in common-mode input voltage. It is depicted in Figure 5–13(b).

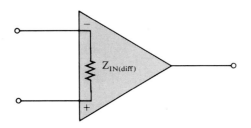

(a) Differential input impedance

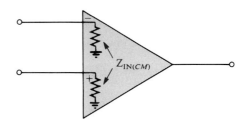

(b) Common-mode input impedance

FIGURE 5–13
Op-amp input impedance.

Input Offset Current

Ideally, the two input bias currents are equal, and thus their difference is zero. In a practical op-amp, however, the bias currents are not exactly equal. The *input offset current* is the difference of the input bias currents, expressed as

$$I_{OS} = |I_1 - I_2| \qquad (5\text{--}10)$$

Actual magnitudes of offset current are usually at least an order of magnitude (ten times) less than the bias current. In many applications, the offset current can be neglected. However, high-gain, high-input impedance amplifiers should have as little I_{OS} as possible, because the difference in currents through large input resistances develops a substantial offset voltage, as shown in Figure 5–14.

The offset voltage developed by the input offset current is

$$V_{OS} = I_1 R_{in} - I_2 R_{in}$$
$$= (I_1 - I_2) R_{in}$$
$$V_{OS} = I_{OS} R_{in} \qquad (5\text{--}11)$$

The error created by I_{OS} is amplified by the gain A_v of the op-amp and appears in the output as

$$V_{OUT(error)} = A_v I_{OS} R_{in} \qquad (5\text{--}12)$$

The change in offset current with temperature is often an important consideration. Values of temperature coefficient in the range of 0.5 nA per degree Celsius are common.

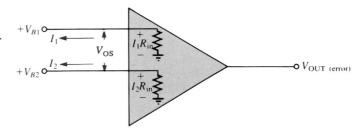

FIGURE 5–14
Effect of input offset current.

Output Impedance

Output impedance is the resistance viewed from the output terminal of the op-amp, as indicated in Figure 5–15.

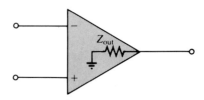

FIGURE 5–15
Op-amp output impedance.

OPERATIONAL AMPLIFIERS

COMMON-MODE RANGE

All op-amps have limitations on the range of voltages over which they will operate. The *common-mode range* is the range of input voltages which, when applied to both inputs, will not cause clipping or other output distortion. Many op-amps have common-mode ranges of ± 10 V with dc supply voltages of ± 15 V.

OPEN-LOOP VOLTAGE GAIN

Open-loop voltage gain is the gain of the op-amp without any external feedback from output to input. A good op-amp has a very high **open-loop** gain; 50,000 to 200,000 is typical.

COMMON-MODE REJECTION RATIO

The **common-mode rejection ratio** (CMRR), as discussed in conjunction with the diff-amp, is a measure of an op-amp's ability to reject common-mode signals. An infinite value of CMRR means that the output is zero when the same signal is applied to both inputs (common-mode).

An infinite CMRR is never achieved in practice, but a good op-amp does have a very high value of CMRR. As previously mentioned, common-mode signals are undesired interference voltages such as 60 Hz power-supply ripple and noise voltages due to pick-up of radiated energy. A high CMRR enables the op-amp to virtually eliminate these interference signals from the output.

The accepted definition of CMRR for an op-amp is the open-loop gain (A_{ol}) divided by the common-mode gain.

$$\text{CMRR} = \frac{A_{ol}}{A_{cm}} \qquad (5-13)$$

It is commonly expressed in decibels as follows:

$$\text{CMRR} = 20 \log \frac{A_{ol}}{A_{cm}} \qquad (5-14)$$

■ **EXAMPLE 5–3** A certain op-amp has an open-loop gain of 100,000 and a common-mode gain of 0.25. Determine the CMRR and express it in dB.

SOLUTION

$$\text{CMRR} = \frac{A_{ol}}{A_{cm}} = \frac{100{,}000}{0.25} = 400{,}000$$

$$\text{CMRR} = 20 \log(400{,}000) = 112 \text{ dB}$$

PRACTICE EXERCISE 5–3
If a particular op-amp has a CMRR of 90 dB and a common-mode gain of 0.4, what is the open-loop gain?

Slew Rate

The maximum rate of change of the output voltage in response to a step input voltage is the **slew rate** of an op-amp. The slew rate is dependent upon the high-frequency response of the amplifier stages within the op-amp. Slew rate is measured with an op-amp connected as shown in Figure 5–16(a). This particular op-amp connection is a unity-gain, noninverting configuration which will be discussed later. It gives a worst-case (slowest) slew rate. Recall that the high-frequency components of a voltage step are contained in the rising edge and that the upper critical frequency of an amplifier limits its response to a step input. The lower f_{ch} is, the more slope there is on the output for a step input.

FIGURE 5–16
Slew-rate measurement.

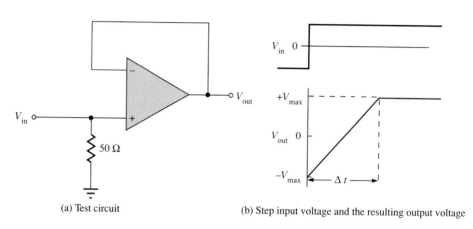

(a) Test circuit

(b) Step input voltage and the resulting output voltage

A pulse is applied to the input as shown, and the ideal output voltage is measured as indicated in Figure 5–16(b). The width of the input pulse must be sufficient to allow the output to "slew" from its lower limit to its upper limit, as shown. As you can see, a certain time interval, Δt, is required for the output voltage to go from its lower limit $-V_{max}$ to its upper limit $+V_{max}$, once the input step is applied. The slew rate is expressed as

$$\text{Slew rate} = \frac{\Delta V_{out}}{\Delta t} \qquad (5\text{–}15)$$

where $\Delta V_{out} = +V_{max} - (-V_{max})$. The unit of slew rate is volts per microsecond (V/μs).

■ **EXAMPLE 5–4** The output voltage of a certain op-amp appears as shown in Figure 5–17 in response to a step input. Determine the slew rate.

SOLUTION
The output goes from the lower to the upper limit in 1 μs. Since this is not an ideal response, the limits are taken at the 90 percent points, as indicated. So, the upper limit is +9 V and the lower limit is −9 V. The slew rate is

$$\frac{\Delta V}{\Delta t} = \frac{+9 \text{ V} - (-9 \text{ V})}{1 \text{ } \mu s} = 18 \text{ V}/\mu s$$

OPERATIONAL AMPLIFIERS

FIGURE 5–17

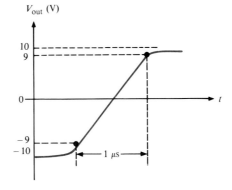

PRACTICE EXERCISE 5–4
When a pulse is applied to an op-amp, the output voltage goes from -8 V to $+7$ V in 0.75 μs. What is the slew rate?

FREQUENCY RESPONSE

The internal amplifier stages that make up an op-amp have voltage gains limited by junction capacitances. Although the differential amplifiers used in op-amps are somewhat different from the basic amplifiers discussed, the same principles apply. An op-amp has no internal coupling capacitors, however; therefore, the low-frequency response extends down to dc. Frequency-related characteristics will be discussed in the next chapter.

COMPARISON OF OP-AMP PARAMETERS

Table 5–1 provides a comparison of values of some of the parameters just described for four common integrated circuit op-amps. Any values not listed were not given on the manufacturer's data sheet. All values are typical at 25°C.

TABLE 5–1

Parameter	Op-Amp Type			
	741C	LM101A	LM108	LM218
Input offset voltage	1 mV	1 mV	0.7 mV	2 mV
Input bias current	80 nA	120 nA	0.8 nA	120 nA
Input offset current	20 nA	40 nA	0.05 nA	6 nA
Input impedance	2 MΩ	800 kΩ	70 MΩ	3 MΩ
Output impedance	75 Ω	—	—	—
Open-loop gain	200,000	160,000	300,000	200,000
Slew rate	0.5 V/μs	—	—	70 V/μs
CMRR	90 dB	90 dB	100 dB	100 dB

OTHER FEATURES

Most available op-amps have three important features—short-circuit protection, no latch-up, and input offset nulling. Short-circuit protection keeps the circuit from being damaged if the output becomes shorted, and the no latch-up feature prevents the op-amp from hanging up in one output state (high- or low-voltage level) under certain input conditions. Input offset nulling is achieved by an external potentiometer that sets the output voltage at precisely zero with zero input.

5-3 REVIEW QUESTIONS

1. List ten or more op-amp parameters.
2. Which two parameters, not including frequency response, are frequency dependent?

5-4 NEGATIVE FEEDBACK

Negative feedback is one of the most useful concepts in electronic circuits, particularly in op-amp applications. Negative feedback is the process whereby a portion of the output voltage of an amplifier is returned to the input with a phase angle that opposes (or subtracts from) the input signal.

Negative feedback is illustrated in Figure 5–18. The inverting input effectively makes the feedback signal 180° out of phase with the input signal.

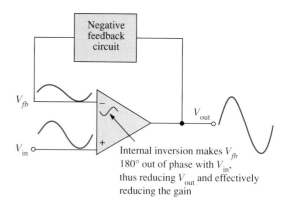

FIGURE 5–18
Illustration of negative feedback.

Internal inversion makes V_{fb} 180° out of phase with V_{in}, thus reducing V_{out} and effectively reducing the gain

WHY USE NEGATIVE FEEDBACK?

As you have seen, the inherent open-loop gain of a typical op-amp is very high (usually greater than 100,000). Therefore, an extremely small input voltage drives the op-amp into its saturated output states. In fact, even the input offset voltage of the op-amp can drive it into saturation. For example, assume $V_{in} = 1$ mV and $A_{ol} = 100,000$. Then,

$$V_{in}A_{ol} = (1 \text{ mV})(100,000) = 100 \text{ V}$$

Since the output level of an op-amp can never reach 100 V, it is driven deep into saturation and the output is limited to its maximum output levels, as illustrated in Figure 5–19 for both a positive and a negative input voltage of 1 mV.

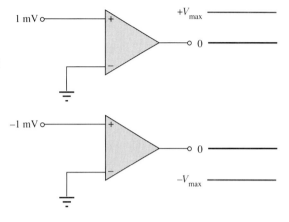

FIGURE 5–19
Without negative feedback, a small input voltage drives the op-amp to its output limits and it becomes nonlinear.

The usefulness of an op-amp operated in this manner is severely restricted and is generally limited to comparator applications (to be studied in a later chapter). With negative feedback, the overall voltage gain (A_{cl}) can be reduced and controlled so that the op-amp can function as a linear amplifier. In addition to providing a controlled, stable voltage gain, negative feedback also provides for control of the input and output impedances and amplifier bandwidth. (We will cover these topics in detail later in this chapter.) Table 5–2 summarizes the general effects of negative feedback on op-amp performance.

TABLE 5–2

	Voltage Gain	Input Z	Output Z	Bandwidth
Without negative feedback	A_{ol} is too high for linear amplifier applications	Relatively high (see Table 5–1)	Relatively low (see Table 5–1)	Relatively narrow
With negative feedback	A_{cl} is set by the feedback circuit to desired value	Can be increased or reduced to a desired value depending on type of circuit	Can be reduced to a desired value	Significantly wider

5–4 REVIEW QUESTIONS

1. What are the benefits of negative feedback in an op-amp circuit?
2. Why is it necessary to reduce the gain of an op-amp from its open-loop value?

5-5 OP-AMP CONFIGURATIONS WITH NEGATIVE FEEDBACK

In this section, we will discuss several basic ways in which an op-amp can be connected using negative feedback to stabilize the gain and increase frequency response. As mentioned, the extremely high open-loop gain of an op-amp creates an unstable situation because a small noise voltage on the input can be amplified to a point where the amplifier is driven out of its linear region. Also, unwanted oscillations can occur. In addition, the open-loop gain parameter of an op-amp can vary greatly from one device to the next. Negative feedback takes a portion of the output and applies it back out-of-phase with the input, creating an effective reduction in gain. This closed-loop gain is usually much less than the open-loop gain and independent of it.

NONINVERTING AMPLIFIER

An op-amp connected as a **noninverting amplifier** with a controlled amount of voltage gain is shown in Figure 5–20. The input signal is applied to the noninverting input. The output is applied back to the inverting input through the feedback network formed by R_i and R_f. This creates negative feedback as follows. R_i and R_f form a voltage-divider network, which reduces the output V_{out} and connects the reduced voltage V_f to the inverting input. The feedback voltage is expressed as

$$V_f = \left(\frac{R_i}{R_i + R_f}\right)V_{out} \qquad (5-16)$$

The difference of the input voltage V_{in} and the feedback voltage V_f is the differential input to the op-amp, as shown in Figure 5–21. This differential voltage is amplified by the open-loop gain of the op-amp (A_{ol}) and produces an output voltage expressed as

$$V_{out} = A_{ol}(V_{in} - V_f) \qquad (5-17)$$

Letting $R_i/(R_i + R_f) = B$ and then substituting BV_{out} for V_f in Equation (5–17), we get the following algebraic steps.

$$V_{out} = A_{ol}(V_{in} - BV_{out})$$
$$V_{out} = A_{ol}V_{in} - A_{ol}BV_{out}$$
$$V_{out} + A_{ol}BV_{out} = A_{ol}V_{in}$$
$$V_{out}(1 + A_{ol}B) = A_{ol}V_{in}$$

Since the total voltage gain of the amplifier in Figure 5–20 is V_{out}/V_{in}, it can be expressed as

$$\frac{V_{out}}{V_{in}} = \frac{A_{ol}}{1 + A_{ol}B} \qquad (5-18)$$

The product $A_{ol}B$ is usually much greater than 1, so Equation (5–18) simplifies to

$$\frac{V_{out}}{V_{in}} = \frac{A_{ol}}{A_{ol}B}$$

OPERATIONAL AMPLIFIERS

Since

$$A_{cl(NI)} = \frac{V_{out}}{V_{in}}$$

then

$$A_{cl(NI)} = \frac{1}{B} = \frac{R_i + R_f}{R_i} = 1 + \frac{R_f}{R_i} \qquad (5-19)$$

Equation (5–19) shows that the closed-loop gain, $A_{cl(NI)}$, of the noninverting (NI) amplifier is the reciprocal of the attenuation (B) of the feedback network (voltage-divider). It is interesting to note that the closed-loop gain is not at all dependent on the op-amp's open-loop gain under the condition $A_{ol}B \gg 1$. The closed-loop gain can be set by selecting values of R_i and R_f.

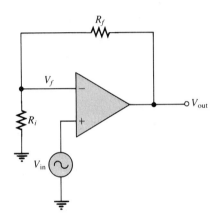

FIGURE 5–20
Noninverting amplifier.

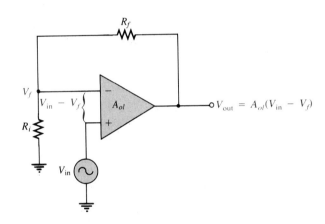

FIGURE 5–21
Differential input.

EXAMPLE 5–5

Determine the gain of the amplifier in Figure 5–22. The open-loop voltage gain is 100,000.

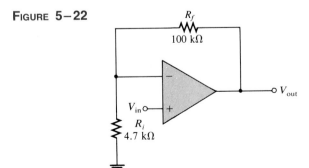

FIGURE 5–22

SOLUTION

This is a noninverting op-amp configuration. Therefore, the closed-loop gain is

$$A_{cl(NI)} = 1 + \frac{R_f}{R_i} = 1 + \frac{100 \text{ k}\Omega}{4.7 \text{ k}\Omega} = 22.3$$

PRACTICE EXERCISE 5-5

If the open-loop gain of the amplifier in Figure 5-22 is 150,000 and R_f is increased to 150 kΩ, determine the closed-loop gain.

VOLTAGE FOLLOWER

The **voltage-follower** configuration is a special case of the noninverting amplifier where all of the output voltage is fed back to the inverting input, as shown in Figure 5-23. As you can see, the straight feedback connection has a voltage gain of approximately one. The closed-loop voltage gain of a noninverting amplifier is $1/B$ as previously derived. Since $B = 1$, the closed-loop gain of the voltage follower is

$$A_{cl(VF)} = \frac{1}{B} = 1 \qquad (5-20)$$

The most important features of the voltage-follower configuration are its very high input impedance and its very low output impedance. These features make it a nearly ideal buffer amplifier for interfacing high-impedance sources and low-impedance loads. This is discussed further in the next section.

FIGURE 5-23
Op-amp voltage follower.

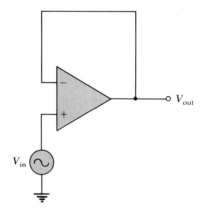

INVERTING AMPLIFIER

An op-amp connected as an **inverting amplifier** with a controlled amount of voltage gain is shown in Figure 5-24. The input signal is applied through a series input resistor R_i to the inverting input. Also, the output is fed back through R_f to the same input. The noninverting input is grounded.

OPERATIONAL AMPLIFIERS

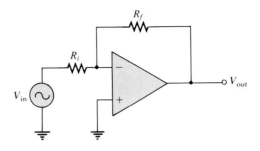

FIGURE 5–24
Inverting amplifier.

At this point, the ideal op-amp parameters mentioned earlier are very useful in simplifying the analysis of this circuit. In particular, the concept of infinite input impedance is of great value. An infinite input impedance implies that there is *no* current out of the inverting input. If there is no current through the input impedance, then there must be *no* voltage drop between the inverting and noninverting inputs. This means that the voltage at the inverting (−) input is zero because the other input (+) is grounded. This zero voltage at the inverting input terminal is referred to as *virtual ground*. This condition is illustrated in Figure 5–25(a).

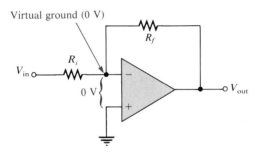

(a) Virtual ground

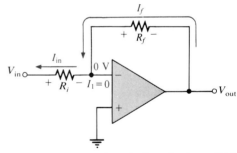

(b) $I_{in} = I_f$ and current into the inverting input (I_1) is 0

FIGURE 5–25
Virtual ground concept and closed-loop voltage gain development for the inverting amplifier.

Since there is no current out of the inverting input, the current through R_i and the current through R_f are equal, as shown in Figure 5–25(b).

$$I_{in} = I_f$$

The voltage across R_i equals V_{in} because of virtual ground on the other side of the resistor. Therefore,

$$I_{in} = \frac{V_{in}}{R_i}$$

Also, the voltage across R_f equals $-V_{out}$ because of virtual ground, and therefore,

$$I_f = \frac{-V_{out}}{R_f}$$

Since $I_f = I_{in}$,

$$\frac{-V_{out}}{R_f} = \frac{V_{in}}{R_i}$$

Rearranging the terms, we get

$$\frac{V_{out}}{V_{in}} = -\frac{R_f}{R_i}$$

Of course, you recognize V_{out}/V_{in} as the overall gain of the amplifier.

$$A_{cl(I)} = -\frac{R_f}{R_i} \qquad (5-21)$$

Equation (5–21) shows that the **closed-loop** voltage gain $A_{cl(I)}$ of the inverting amplifier is the ratio of the feedback resistance R_f to the resistance R_i. *The closed-loop gain is independent of the op-amp's internal open-loop gain.* Thus, the negative feedback stabilizes the voltage gain. The negative sign indicates inversion.

■ EXAMPLE 5–6

Given the op-amp configuration in Figure 5–26, determine the value of R_f required to produce a closed-loop voltage gain of 100.

FIGURE 5–26

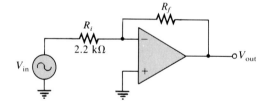

SOLUTION

$R_i = 2.2 \text{ k}\Omega$, and $A_{cl(I)} = 100$

$$A_{cl(I)} = \left|\frac{R_f}{R_i}\right|$$

$$R_f = A_{cl(I)}R_i$$
$$= (100)(2.2 \text{ k}\Omega)$$
$$= 220 \text{ k}\Omega$$

PRACTICE EXERCISE 5–6

If R_i is changed to 2.7 kΩ in Figure 5–26, what value of R_f is required to produce a closed-loop gain of 25?

5-5 REVIEW QUESTIONS

1. What is the main purpose of negative feedback?
2. The closed-loop voltage gain of each of the op-amp configurations discussed is dependent on the internal open-loop voltage gain of the op-amp (T or F).
3. The attenuation of the negative feedback network of a noninverting op-amp configuration is 0.02. What is the closed-loop gain of the amplifier?

5-6 EFFECTS OF NEGATIVE FEEDBACK ON OP-AMP IMPEDANCES

In this section, you will see how a negative feedback connection affects the input and output impedances of an op-amp. The effects on both inverting and noninverting amplifiers are examined.

INPUT IMPEDANCE OF THE NONINVERTING AMPLIFIER

The input impedance of this op-amp configuration is developed with the aid of Figure 5–27. For this analysis, a small differential voltage V_{diff} is assumed to exist between the two inputs, as indicated. This means that the op-amp's input impedance is not assumed to be infinite, nor the input current to be zero. The input voltage can be expressed as

$$V_{in} = V_{diff} + V_f$$

Substituting BV_{out} for V_f,

$$V_{in} = V_{diff} + BV_{out}$$

Since $V_{out} \cong A_{ol}V_{diff}$ (A_{ol} is the open-loop gain of the op-amp),

$$V_{in} = V_{diff} + A_{ol}BV_{diff}$$
$$= (1 + A_{ol}B)V_{diff}$$

Because $V_{diff} = I_{in}Z_{in}$,

$$V_{in} = (1 + A_{ol}B)I_{in}Z_{in}$$

where Z_{in} is the open-loop input impedance of the op-amp (without feedback connections).

$$\frac{V_{in}}{I_{in}} = (1 + A_{ol}B)Z_{in}$$

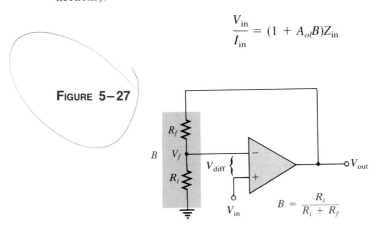

FIGURE 5–27

$$B = \frac{R_i}{R_i + R_f}$$

V_{in}/I_{in} is the overall input impedance of the closed-loop noninverting configuration.

$$Z_{in(NI)} = (1 + A_{ol}B)Z_{in} \qquad (5-22)$$

This equation shows that the input impedance of this amplifier configuration with negative feedback is much greater than the internal input impedance of the op-amp itself (without feedback).

OUTPUT IMPEDANCE OF THE NONINVERTING AMPLIFIER

An expression for output impedance is developed with the aid of Figure 5–28. By applying Kirchhoff's law to the output circuit, we get

$$V_{out} = A_{ol}V_{diff} - Z_{out}I_{out}$$

The differential input voltage is $V_{in} - V_f$; therefore, under the assumption that $A_{ol}V_{diff} \gg Z_{out}I_{out}$, the output voltage can be expressed as

$$V_{out} \cong A_{ol}(V_{in} - V_f)$$

Substituting BV_{out} for V_f, we get

$$V_{out} \cong A_{ol}(V_{in} - BV_{out})$$

Remember, B is the attenuation of the negative feedback network. Expanding and factoring, we get

$$V_{out} \cong A_{ol}V_{in} - A_{ol}BV_{out}$$
$$A_{ol}V_{in} \cong V_{out} + A_{ol}BV_{out}$$
$$\cong (1 + A_{ol}B)V_{out}$$

Since the output impedance of the noninverting configuration is $Z_{out(NI)} = V_{out}/I_{out}$, we can substitute $I_{out}Z_{out(NI)}$ for V_{out}.

$$A_{ol}V_{in} = (1 + A_{ol}B)I_{out}Z_{out(NI)}$$

Dividing both sides of the above expression by I_{out}, we get

$$\frac{A_{ol}V_{in}}{I_{out}} = (1 + A_{ol}B)Z_{out(NI)}$$

The term on the left is the internal output impedance of the op-amp (Z_{out}) because, without feedback, $A_{ol}V_{in} = V_{out}$. Therefore,

$$Z_{out} = (1 + A_{ol}B)Z_{out(NI)}$$

Thus,

$$Z_{out(NI)} = \frac{Z_{out}}{1 + A_{ol}B} \qquad (5-23)$$

This equation shows that the output impedance of this amplifier configuration with negative feedback is much less than the internal output impedance of the op-amp itself (without feedback).

FIGURE 5-28

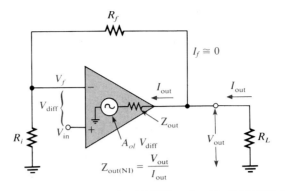

EXAMPLE 5-7   (a) Determine the input and output impedances of the amplifier in Figure 5-29. The op-amp data sheet gives $Z_{in} = 2\ M\Omega$, $Z_{out} = 75\ \Omega$, and $A_{ol} = 200{,}000$.
(b) Find the closed-loop voltage gain.

FIGURE 5-29

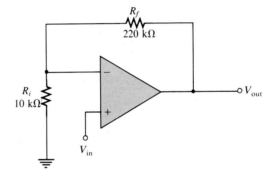

SOLUTION

(a) The attenuation of the feedback network is

$$B = \frac{R_i}{R_i + R_f} = \frac{10\ k\Omega}{230\ k\Omega} = 0.0435$$

$$\begin{aligned}
Z_{in(NI)} &= (1 + A_{ol}B)Z_{in} \\
&= [1 + (200{,}000)(0.0435)](2\ M\Omega) \\
&= (1 + 8700)(2\ M\Omega) \\
&= 17{,}402\ M\Omega
\end{aligned}$$

$$\begin{aligned}
Z_{out(NI)} &= \frac{Z_{out}}{1 + A_{ol}B} \\
&= \frac{75\ \Omega}{1 + 8700} \\
&= 0.0086\ \Omega
\end{aligned}$$

(b) $A_{cl(NI)} = \dfrac{1}{B} = \dfrac{1}{0.0435} \cong 23$

PRACTICE EXERCISE 5–7

(a) Determine the input and output impedances in Figure 5–29 for op-amp data sheet values of $Z_{in} = 3.5$ MΩ, $Z_{out} = 82$ Ω, and $A_{ol} = 135,000$.
(b) Find A_{cl}.

VOLTAGE-FOLLOWER IMPEDANCES

Since the voltage follower is a special case of the noninverting configuration, the same impedance formulas are used with $B = 1$.

$$Z_{in(VF)} = (1 + A_{ol})Z_{in} \quad (5\text{--}24)$$

$$Z_{out(VF)} = \frac{Z_{out}}{1 + A_{ol}} \quad (5\text{--}25)$$

As you can see, the voltage-follower input impedance is greater for a given A_{ol} and Z_{in} than for the noninverting configuration with the voltage-divider feedback network. Also, its output impedance is much smaller.

EXAMPLE 5–8

The same op-amp in Example 5–7 is used in a voltage-follower configuration. Determine the input and output impedances.

SOLUTION
Since $B = 1$,

$$Z_{in(VF)} = (1 + A_{ol})Z_{in}$$
$$= (1 + 200,000)(2 \text{ M}\Omega)$$
$$\cong 400,000 \text{ M}\Omega$$

$$Z_{out(VF)} = \frac{Z_{out}}{1 + A_{ol}}$$
$$= \frac{75 \text{ }\Omega}{1 + 200,000}$$
$$= 0.00038 \text{ }\Omega$$

Notice that $Z_{in(VF)}$ is much greater than $Z_{in(NI)}$, and $Z_{out(VF)}$ is much less than $Z_{out(NI)}$ from Example 5–7.

PRACTICE EXERCISE 5–8
If the op-amp in this example is replaced with one having a higher open-loop gain, how are the input and output impedances affected?

IMPEDANCES OF THE INVERTING AMPLIFIER

The input impedance of this op-amp configuration is developed with the aid of Figure 5–30. Because both the input signal and the negative feedback are applied, through

OPERATIONAL AMPLIFIERS

resistors, to the inverting terminal, Miller's theorem can be applied to this configuration. According to Miller's theorem, the effective input impedance of an amplifier with a feedback resistor from output to input as in Figure 5–30 is

$$Z_{in(Miller)} = \frac{R_f}{A_{ol} + 1} \qquad (5\text{–}26)$$

and

$$Z_{out(Miller)} = \left(\frac{A_{ol}}{A_{ol} + 1}\right) R_f \qquad (5\text{–}27)$$

Applying Miller's theorem to the circuit of Figure 5–30, we get the equivalent circuit of Figure 5–31. As indicated, the Miller input impedance appears in parallel with the internal input impedance of the op-amp, and R_i appears in series with this as follows:

$$Z_{in(I)} = R_i + \frac{R_f}{A_{ol} + 1} \| Z_{in} \qquad (5\text{–}28)$$

Typically, $R_f/(A_{ol} + 1)$ is much less than the Z_{in} of an open-loop op-amp; also, $A_{ol} \gg 1$. So Equation (5–28) simplifies to

$$Z_{in(I)} \cong R_i + \frac{R_f}{A_{ol}}$$

Since R_i appears in series with R_f/A_{ol} and if $R_i \gg R_f/A_{ol}$, $Z_{in(I)}$ reduces to

$$Z_{in(I)} \cong R_i \qquad (5\text{–}29)$$

The Miller output impedance appears in parallel with Z_{out} of the op-amp.

$$Z_{out(I)} = \left(\frac{A_{ol}}{A_{ol} + 1}\right) R_f \| Z_{out} \qquad (5\text{–}30)$$

Normally $A_{ol} \gg 1$ and $R_f \gg Z_{out}$, so $Z_{out(I)}$ simplifies to

$$Z_{out(I)} \cong Z_{out} \qquad (5\text{–}31)$$

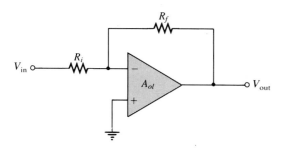

FIGURE 5–30
Inverting amplifier.

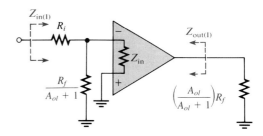

FIGURE 5–31
Miller equivalent for the inverting amplifier in Figure 5–30.

EXAMPLE 5–9

Find the values of the input and output impedances in Figure 5–32. Also, determine the closed-loop voltage gain. The op-amp has the following parameters: $A_{ol} = 50{,}000$; $Z_{in} = 4 \text{ M}\Omega$; and $Z_{out} = 50 \text{ }\Omega$.

FIGURE 5–32

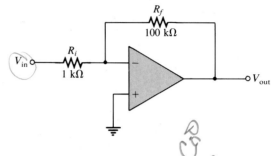

Solution

$$Z_{in(I)} \cong R_i = 1 \text{ k}\Omega$$
$$Z_{out(I)} \cong Z_{out} = 50 \text{ }\Omega$$

$$A_{cl(I)} = -\frac{R_f}{R_i} = -\frac{100 \text{ k}\Omega}{1 \text{ k}\Omega} = -100$$

Practice Exercise 5–9

Determine the input and output impedances and the closed-loop voltage gain in Figure 5–32. The op-amp parameters and circuit values are as follows: $A_{ol} = 100{,}000$; $Z_{in} = 5 \text{ M}\Omega$; $Z_{out} = 75 \text{ }\Omega$; $R_i = 560 \text{ }\Omega$; and $R_f = 82 \text{ k}\Omega$.

5–6 Review Questions

1. How does the input impedance of a noninverting amplifier configuration compare to the input impedance of the op-amp itself?
2. When an op-amp is connected in a voltage-follower configuration, does the input impedance increase or decrease?
3. Given that $R_f = 100 \text{ k}\Omega$; $R_i = 2 \text{ k}\Omega$; $A_{ol} = 120{,}000$; $Z_{in} = 2 \text{ M}\Omega$; and $Z_{out} = 60 \text{ }\Omega$, what are $Z_{in(I)}$ and $Z_{out(I)}$ for an inverting amplifier configuration?

5–7 BIAS CURRENT AND OFFSET VOLTAGE COMPENSATION

Up until now, the op-amp has been treated as an ideal device in many of our discussions. However, since it is not an ideal device, certain "flaws" in the op-amp must be recognized because of their effects on its operation. Transistors within the op-amp must be biased so that they have the correct values of base and collector current and collector-to-emitter voltages. The ideal op-amp has no input current at its terminals, but in fact, the practical op-amp has small input bias currents typically in the nA range. Also, small internal imbalances in the transistors effectively produce a small offset voltage between the inputs. These nonideal parameters were described in Section 5–3.

EFFECT OF AN INPUT BIAS CURRENT

Figure 5–33 is an inverting amplifier with zero input voltage. Ideally, the current through R_i is zero because the input voltage is zero and the voltage at the inverting (−) terminal is zero. The small input current I_1 flows to the output terminal through R_f. I_1 creates a voltage drop across R_f, as indicated. The positive side of R_f is the output terminal, and therefore, the output error voltage is I_1R_f when it should be zero.

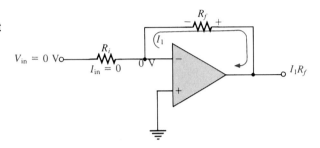

FIGURE 5–33
Input bias current creates output error voltage (I_1R_f) in inverting amplifier.

Figure 5–34 is a voltage follower with zero input voltage and a source resistance R_s. In this case, an input current I_2 creates an output voltage error (a path exists for I_2 through the negative voltage supply and back to ground). I_2 produces a drop across R_s, as shown. The voltage at the inverting input terminal decreases to $-I_2R_s$ because the negative feedback tends to maintain a differential voltage of zero, as indicated. Since the inverting terminal is connected directly to the output terminal, the output error voltage is $-I_2R_s$.

Figure 5–35 is a noninverting amplifier with zero input voltage. Ideally, the voltage at the inverting terminal is also zero, as indicated. The input current I_1 produces a voltage drop across R_f and thus creates an output error voltage of I_1R_f, just as with the inverting amplifier.

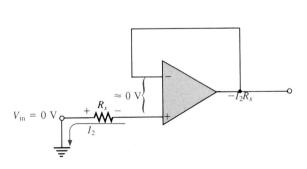

FIGURE 5–34
Input bias current creates output error voltage in voltage follower.

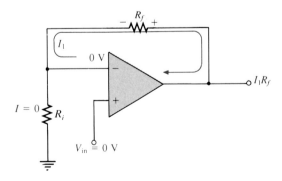

FIGURE 5–35
Input bias current creates output error voltage in noninverting amplifier.

BIAS CURRENT COMPENSATION IN A VOLTAGE FOLLOWER

The output error voltage due to bias currents in a voltage follower can be sufficiently reduced by adding a resistor equal to R_s in the feedback path, as shown in Figure 5–36. The voltage drop created by I_1 across the added resistor subtracts from the $-I_2R_s$ output error voltage. If $I_1 = I_2$, then the output voltage is zero. Usually I_1 does not quite equal I_2; but even in this case, the output error voltage is reduced as follows, because I_{OS} is less than I_2.

$$V_{OUT(error)} = |I_1 - I_2|R_s$$
$$V_{OUT(error)} = I_{OS}R_s \qquad (5\text{–}32)$$

where I_{OS} is the input offset current.

FIGURE 5–36
Bias current compensation in a voltage follower.

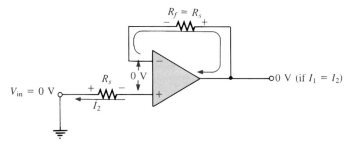

BIAS CURRENT COMPENSATION IN OTHER OP-AMP CONFIGURATIONS

To compensate for the effect of bias current in the noninverting amplifier, a resistor R_c is added, as shown in Figure 5–37(a). The compensating resistor value equals the parallel combination of R_i and R_f. The input current I_2 creates a voltage drop across R_c that offsets the voltage across the $R_i\text{-}R_f$ combination, thus sufficiently reducing the output error voltage. The inverting amplifier is similarly compensated, as shown in Figure 5–37(b).

FIGURE 5–37
Bias current compensation.

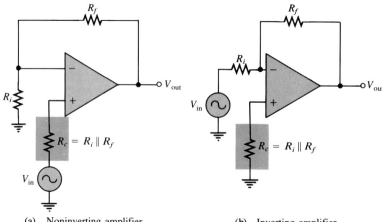

(a) Noninverting amplifier   (b) Inverting amplifier

Use of a BIFET Op-Amp to Eliminate the Need for Bias Current Compensation

A BIFET op-amp uses both bipolar junction transistors and JFETs in its internal circuitry. The JFETs are used as the input devices to achieve a higher input impedance than is possible with standard bipolar amplifiers. Because of their very high input impedance, BIFETs typically have input bias currents that are much smaller than in bipolar op-amps, thus reducing or eliminating the need for bias current compensation.

Effect of Input Offset Voltage

The output voltage of an op-amp should be zero when the differential input is zero. However, there is always a small output error voltage present whose value typically ranges from microvolts to millivolts. This is due to unavoidable imbalances within the internal op-amp transistors aside from the bias currents previously discussed. In a negative feedback configuration, the input offset voltage V_{IO} can be visualized as an equivalent small dc voltage source, as illustrated in Figure 5–38 for a voltage follower. The output error voltage due to the input offset voltage in this case is

$$V_{OUT(error)} = V_{IO} \tag{5-33}$$

Figure 5–38
Input offset voltage equivalent.

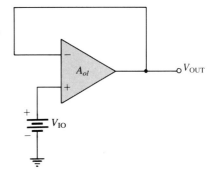

Input Offset Voltage Compensation

Most integrated circuit op-amps provide a means of compensating for offset voltage. This is usually done by connecting an external potentiometer to designated pins on the IC package, as illustrated in Figure 5–39(a) and (b) on page 258 for a 741 op-amp. The two terminals are labelled *offset null*. With no input, the potentiometer is simply adjusted until the output voltage reads 0, as shown in Figure 5–39(c).

5–7 Review Questions

1. What are two sources of dc output error voltages?
2. How do you compensate for bias current in a voltage follower?

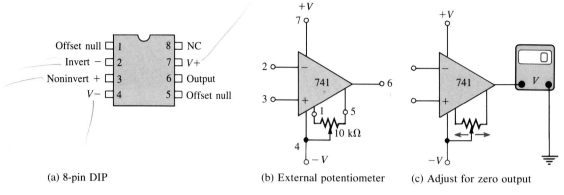

(a) 8-pin DIP (b) External potentiometer (c) Adjust for zero output

FIGURE 5–39
Input offset voltage compensation for a 741.

5–8 TROUBLESHOOTING

As a technician, you will no doubt encounter situations in which an op-amp or its associated circuitry has malfunctioned. The op-amp is a complex integrated circuit with many types of internal failures possible. However, since you cannot troubleshoot the op-amp internally, you treat it as a single device with only a few connections to it. If it fails, you replace it just as you would a resistor, capacitor, or transistor.

In the basic op-amp configurations, there are only a few external components that can fail. These are the feedback resistor, the input resistor, and the potentiometer used for offset voltage compensation. Also, of course, the op-amp itself can fail or there can be faulty contacts in the circuit. We will now examine the three basic configurations for possible faults and the associated symptoms.

FAULTS IN THE NONINVERTING AMPLIFIER

The first thing to do when you suspect a faulty circuit is to check for the proper supply voltage and ground. Having done that, several other possible faults are as follows.

OPEN FEEDBACK RESISTOR If the feedback resistor, R_f, in Figure 5–40 opens, the op-amp is operating with its very high open-loop gain, which causes the input signal to drive the device into nonlinear operation and results in a severely clipped output signal as shown in part (a).

OPEN INPUT RESISTOR In this case, we still have a closed-loop configuration. But, since R_i is open and effectively equal to infinity, the closed-loop gain from Equation (5–19) is

$$A_{cl(NI)} = 1 + \frac{R_f}{R_i} = 1 + \frac{R_f}{\infty} = 1 + 0 = 1$$

OPERATIONAL AMPLIFIERS

FIGURE 5–40
Faults in the noninverting amplifier.

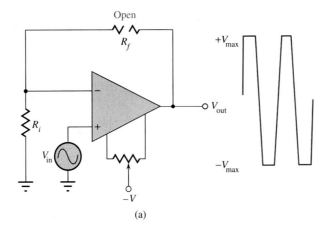
(a)

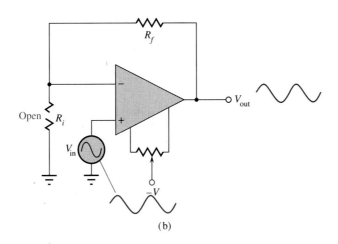
(b)

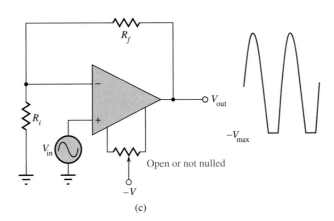
(c)

This shows that the amplifier acts like a voltage follower. You would observe an output signal that is the same as the input, as indicated in Figure 5–40(b).

OPEN OR INCORRECTLY ADJUSTED OFFSET NULL POTENTIOMETER In this situation, the output offset voltage will cause the output signal to begin clipping on only one peak as the input signal is increased to a sufficient amplitude. This is indicated in Figure 5–40(c).

FAULTY OP-AMP As mentioned, many things can happen to an op-amp. In general, an internal failure will result in a loss or distortion of the output signal. The best approach is to first make sure that there are no external failures or faulty conditions. If everything else is good, then the op-amp must be bad.

FAULTS IN THE VOLTAGE FOLLOWER

The voltage follower is a special case of the noninverting amplifier. Except for a bad op-amp, a bad external connection, or a problem with the offset null potentiometer, about the only thing that can happen in a voltage-follower circuit is an open feedback loop. This would have the same effect as an open feedback resistor as previously discussed.

FIGURE 5–41
Faults in the inverting amplifier.

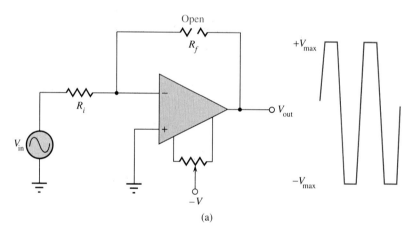

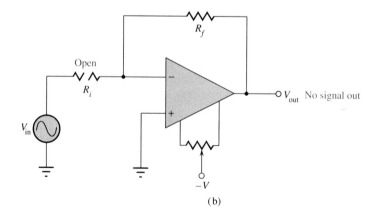

FAULTS IN THE INVERTING AMPLIFIER

OPEN FEEDBACK RESISTOR If R_f opens as indicated in Figure 5–41(a), the input signal still feeds through the input resistor and is amplified by the high open-loop gain of the op-amp. This forces the device to be driven into nonlinear operation, and you will see an output something like that shown. This is the same result as in the noninverting configuration.

OPEN INPUT RESISTOR This prevents the input signal from getting to the op-amp input, so there will be no output signal, as indicated in Figure 5–41(b).

Failures in the op-amp itself or the offset null potentiometer have the same effects as previously discussed for the noninverting amplifier.

5–8 REVIEW QUESTIONS

1. If you notice that the op-amp output signal is beginning to clip on one peak as you increase the input signal, what should you check?
2. If there is no op-amp output signal when there is a verified input signal, what would you suspect as being faulty?

5–9 A SYSTEM APPLICATION

The spectrophotometer system presented at the beginning of this chapter combines light optics with electronics to analyze the chemical makeup of various solutions. This type of system is common in medical laboratories as well as many other areas. It is another example of mixed systems in which electronic circuits interface with other types of systems, such as mechanical and optical, to accomplish a specific function. When you are a technician or technologist in industry, you will probably be working with different types of mixed systems from time to time. In this section, you will

- *See the role of electronics in a system that is not totally electronic.*
- *See how an op-amp is used in the system.*
- *See how an electronic circuit interfaces with an optical device.*
- *Translate between a printed circuit board and a schematic.*
- *Troubleshoot some common system problems.*

A BRIEF DESCRIPTION OF THE SYSTEM

The light source shown in Figure 5–42 produces a beam of visible light containing a wide spectrum of wavelengths. Each component wavelength in the beam of light is refracted at a different angle by the prism as indicated. Depending on the angle of the platform as set by the pivot angle controller, a certain wavelength passes through the narrow slit and is transmitted through the solution under analysis. By precisely pivoting the light source and prism, a selected wavelength can be transmitted. Every chemical and compound absorbs different wavelengths of light in different ways, so the resulting light coming through the solution has a unique "signature" that can be used to define the chemicals in the solution.

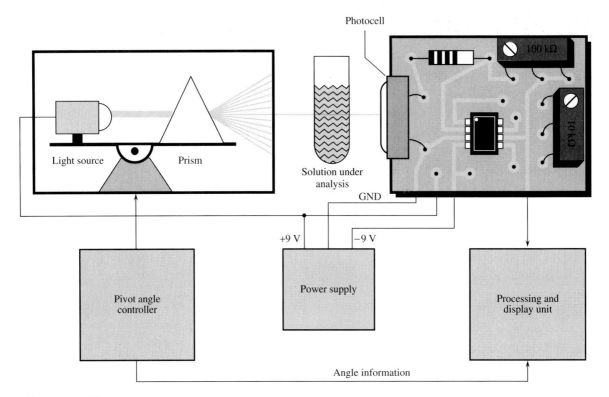

FIGURE 5–42

The photocell on the circuit board, produces a voltage that is proportional to the amount of light and wavelength. The op-amp circuit amplifies the photovoltaic cell output and sends the resulting signal to the processing and display unit where the type of chemical(s) in the solution is identified. This is usually a microprocessor-based digital system. Although these other system blocks are interesting, our focus in this section is the photocell and op-amp circuit board.

OPERATIONAL AMPLIFIERS

Now, so that you can take a closer look at the photocell and op-amp circuit, let's take it out of the system and put it on the test bench.

ON THE TEST BENCH

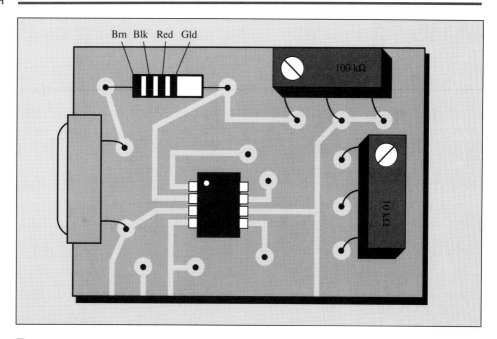

FIGURE 5–43

■ **ACTIVITY 1   RELATE THE PC BOARD TO THE SCHEMATIC**

Develop a complete schematic diagram by carefully following the conductive traces on the pc board shown in Figure 5–43 to see how the components are interconnected. Some of the interconnecting traces are on the reverse side of the board, but if you are familiar with basic op-amp configurations, you should have no trouble figuring out the connections. Refer to the chapter material or the 741 data sheet for the pin layout. This op-amp is housed in a surface-mount SO-8 package. A pad to which no component lead is connected represents a feedthrough to the other side.

■ **ACTIVITY 2   ANALYZE THE CIRCUIT**

STEP 1 Determine the resistance value to which the feedback rheostat must be adjusted for a voltage gain of 50.

STEP 2 Assume the maximum linear output of the op-amp is 1 V less than the supply voltage. Determine the voltage gain required and the value to which the feedback resistance must be set to achieve the maximum linear output. The maximum voltage from the photocell is 0.5 V.

STEP 3 The system light source produces wavelengths ranging from 400 nm to 700 nm, which is approximately the full range of visible light from violet to red. Determine the op-amp output voltage over this range of wavelengths in 50 nm intervals and plot a graph of the results. Refer to the photocell response characteristic in Figure 5–44.

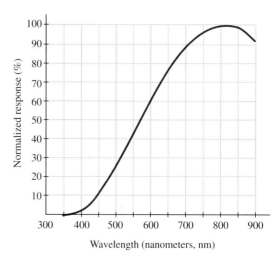

FIGURE 5–44
Photocell response curve.

■ **ACTIVITY 3   WRITE A TECHNICAL REPORT**

Describe the circuit operation. Be sure to identify the type of op-amp circuit configuration and explain the purpose of the two potentiometers. Use the results of Activity 2 to specify the performance of the circuit.

■ **ACTIVITY 4   TROUBLESHOOT THE CIRCUIT FOR EACH OF THE FOLLOWING PROBLEMS BY STATING THE PROBABLE CAUSE OR CAUSES**

1. No voltage at the op-amp output.
2. Output of op-amp stays at approximately −9 V.
3. A small dc voltage on the op-amp output under no-light conditions.
4. Zero output voltage as light source is pivoted with verified photocell output voltage.

5–9 REVIEW QUESTIONS

1. What is the purpose of the 100 kΩ potentiometer on the circuit board?
2. What is the purpose of the 10 kΩ potentiometer?
3. Explain why the light source and prism must be pivoted.

SUMMARY

- The basic op-amp has three terminals not including power and ground: inverting input (−), noninverting input (+), and output.
- Most op-amps require both a positive and a negative dc supply voltage.

OPERATIONAL AMPLIFIERS

- The ideal (perfect) op-amp has infinite input impedance, zero output impedance, infinite open-loop voltage gain, infinite bandwidth, and infinite CMRR.
- A good practical op-amp has high input impedance, low output impedance, high open-loop voltage gain, and a wide bandwidth.
- A differential amplifier is normally used for the input stage of an op-amp.
- A differential input voltage appears between the inverting and noninverting inputs of a differential amplifier.
- A single-ended input voltage appears between one input and ground (with the other input grounded).
- A differential output voltage appears between two output terminals of a diff-amp.
- A single-ended output voltage appears between the output and ground of a diff-amp.
- Common-mode occurs when equal in-phase voltages are applied to both input terminals.
- Input offset voltage produces an output error voltage (with no input voltage).
- Input bias current also produces an output error voltage (with no input voltage).
- Input offset current is the difference between the two bias currents.
- Open-loop voltage gain is the gain of the op-amp with no external feedback connections.
- The common-mode rejection ratio (CMRR) is a measure of an op-amp's ability to reject common-mode inputs.
- Slew rate is the rate in volts per microsecond at which the output voltage of an op-amp can change in response to a step input.
- There are three basic op-amp configurations—inverting, noninverting, and voltage follower.
- All op-amp configurations listed employ negative feedback. Negative feedback occurs when a portion of the output voltage is connected back to the inverting input such that it subtracts from the input voltage, thus reducing the voltage gain but increasing the stability and bandwidth.
- A noninverting amplifier configuration has a higher input impedance and a lower output impedance than the op-amp itself (without feedback).
- An inverting amplifier configuration has an input impedance approximately equal to the input resistor R_i and an output impedance approximately equal to the output impedance of the op-amp itself.
- The voltage follower has the highest input impedance and the lowest output impedance of the three configurations.
- All practical op-amps have small input bias currents and input offset voltages that produce small output error voltages.
- The input bias current effect can be compensated for with external resistors.
- The input offset voltage can be compensated for with an external potentiometer between the two offset null pins provided on the IC op-amp package and as recommended by the manufacturer.

Glossary

Closed-loop An op-amp connection in which the output is connected back to the input through a feedback circuit.

Common mode A condition characterized by the presence of the same signal on both op-amp inputs.

Common-mode rejection ratio (CMRR) The ratio of open-loop gain to common-mode gain; a measure of an op-amp's ability to reject common-mode signals.

Differential amplifier (diff-amp) An amplifier that produces an output voltage proportional to the difference of the two input voltages.

Inverting amplifier An op-amp closed-loop configuration in which the input signal is applied to the inverting input.

Negative feedback The process of returning a portion of the output signal to the input of an amplifier such that it is out of phase with the input signal.

Noninverting amplifier An op-amp closed-loop configuration in which the input signal is applied to the noninverting input.

Open-loop A condition in which an op-amp has no feedback.

Operational amplifier (op-amp) A type of amplifier that has very high voltage gain, very high input impedance, very low output impedance, and good rejection of common-mode signals.

Slew rate The rate of change of the output voltage of an op-amp in response to a step input.

Voltage follower A closed-loop, noninverting op-amp with a voltage gain of one.

Formulas

Differential Amplifiers

(5–1) $\quad I_{E1} = I_{E2} = \dfrac{I_{R_E}}{2}$ $\qquad$ Diff-amp emitter current

(5–2) $\quad I_{R_E} = \dfrac{V_E - V_{EE}}{R_E}$ $\qquad$ Combined emitter current

(5–3) $\quad I_{C1} = I_{C2} \cong \dfrac{I_{R_E}}{2}$ $\qquad$ Diff-amp collector current

(5–4) $\quad V_{C1} = V_{C2} = V_{CC} - I_{C1}R_{C1}$ $\qquad$ Diff-amp collector voltage

(5–5) $\quad V_E = V_B - 0.7 \text{ V}$ $\qquad$ Diff-amp emitter voltage

(5–6) $\quad \text{CMRR} = \dfrac{A_{v(d)}}{A_{cm}}$ $\qquad$ Common-mode rejection ratio (diff-amp)

$(5-7)$ $\text{CMRR} = 20 \log \dfrac{A_{v(d)}}{A_{cm}}$ Common-mode rejection ratio (dB)

Op-Amp Parameters

$(5-8)$ $V_{OUT} = I_{C2}R_C - I_{C1}R_C$ Differential output

$(5-9)$ $I_{BIAS} = \dfrac{I_1 + I_2}{2}$ Input bias current

$(5-10)$ $I_{OS} = |I_1 - I_2|$ Input offset current

$(5-11)$ $V_{OS} = I_{OS}R_{in}$ Offset voltage

$(5-12)$ $V_{OUT(error)} = A_v I_{OS} R_{in}$ Output error voltage

$(5-13)$ $\text{CMRR} = \dfrac{A_{ol}}{A_{cm}}$ Common-mode rejection ratio (op-amp)

$(5-14)$ $\text{CMRR} = 20 \log \dfrac{A_{ol}}{A_{cm}}$ Common-mode rejection ratio (dB)

$(5-15)$ $\text{Slew rate} = \dfrac{\Delta V_{out}}{\Delta t}$ Slew rate

Op-Amp Configurations

$(5-16)$ $V_f = \left(\dfrac{R_i}{R_i + R_f}\right) V_{out}$ Feedback voltage (noninverting)

$(5-17)$ $V_{out} = A_{ol}(V_{in} - V_f)$ Output voltage (noninverting)

$(5-18)$ $\dfrac{V_{out}}{V_{in}} = \dfrac{A_{ol}}{1 + A_{ol}B}$ Voltage gain (noninverting)

$(5-19)$ $A_{cl(NI)} = \dfrac{1}{B} = 1 + \dfrac{R_f}{R_i}$ Voltage gain (noninverting)

$(5-20)$ $A_{cl(VF)} = \dfrac{1}{B} = 1$ Voltage gain (voltage follower)

$(5-21)$ $A_{cl(I)} = -\dfrac{R_f}{R_i}$ Voltage gain (inverting)

Op-Amp Impedances

$(5-22)$ $Z_{in(NI)} = (1 + A_{ol}B)Z_{in}$ Input impedance (noninverting)

$(5-23)$ $Z_{out(NI)} = \dfrac{Z_{out}}{1 + A_{ol}B}$ Output impedance (noninverting)

$(5-24)$ $Z_{in(VF)} = (1 + A_{ol})Z_{in}$ Input impedance (voltage follower)

$(5-25)$ $Z_{out(VF)} = \dfrac{Z_{out}}{1 + A_{ol}}$ Output impedance (voltage follower)

(5–26) $Z_{in(Miller)} = \dfrac{R_f}{A_{ol} + 1}$ Miller input impedance (inverting)

(5–27) $Z_{out(Miller)} = \left(\dfrac{A_{ol}}{A_{ol} + 1}\right) R_f$ Miller output impedance (inverting)

(5–28) $Z_{in(I)} = R_i + \dfrac{R_f}{A_{ol} + 1} \| Z_{in}$ Input impedance (inverting)

(5–29) $Z_{in(I)} \cong R_i$ Input impedance (inverting)

(5–30) $Z_{out(I)} = \left(\dfrac{A_{ol}}{A_{ol} + 1}\right) R_f \| Z_{out}$ Output impedance (inverting)

(5–31) $Z_{out(I)} \cong Z_{out}$ Output impedance (inverting)

Error Voltage

(5–32) $V_{OUT(error)} = I_{OS} R_s$ Output error voltage

(5–33) $V_{OUT(error)} = V_{IO}$ Output error voltage (input offset)

Self-Test

1. An integrated circuit (IC) op-amp has
   (a) two inputs and two outputs
   (b) one input and one output
   (c) two inputs and one output
2. Which of the following characteristics does not *necessarily* apply to an op-amp?
   (a) High gain   (b) Low power
   (c) High input impedance (d) Low output impedance
3. A differential amplifier
   (a) is part of an op-amp   (b) has one input and one output
   (c) has two outputs   (d) a and c
4. When a differential amplifier is operated single-ended,
   (a) the output is grounded
   (b) one input is grounded and a signal is applied to the other
   (c) both inputs are connected together
   (d) the output is not inverted
5. In the differential mode,
   (a) opposite polarity signals are applied to the inputs
   (b) the gain is one

(c) the outputs are different amplitudes

(d) only one supply voltage is used

6. In the common mode,

(a) both inputs are grounded

(b) the outputs are connected together

(c) an identical signal appears on both inputs

(d) the output signals are in-phase

7. Common-mode gain is

(a) very high   (b) very low

(c) always unity   (d) unpredictable

8. Differential gain is

(a) very high   (b) very low

(c) dependent on the input voltage   (d) about 100

9. If $A_{v(d)} = 3500$ and $A_{cm} = 0.35$, the CMRR is

(a) 1225   (b) 10,000   (c) 80 dB   (d) b and c

10. With zero volts on both inputs, an op-amp ideally should have an output

(a) equal to the positive supply voltage

(b) equal to the negative supply voltage

(c) equal to zero

(d) equal to the CMRR

11. Of the values listed, the most *realistic* value for open-loop gain of an op-amp is

(a) 1   (b) 2000   (c) 80 dB   (d) 100,000

12. A certain op-amp has bias currents of 50 μA and 49.3 μA. The input offset current is

(a) 700 nA   (b) 99.3 μA   (c) 49.65 μA   (d) none of these

13. The output of a particular op-amp increases 8 V in 12 μs. The slew rate is

(a) 96 V/μs   (b) 0.67 V/μs   (c) 1.5 V/μs   (d) none of these

14. The purpose of offset nulling is to

(a) reduce the gain   (b) equalize the input signals

(c) zero the output error voltage   (d) b and c

15. For an op-amp with negative feedback, the output is

(a) equal to the input   (b) increased

(c) fed back to the inverting input   (d) fed back to the noninverting input

16. The use of negative feedback

(a) reduces the voltage gain of an op-amp

(b) makes the op-amp oscillate

(c) makes linear operation possible
(d) a and c

17. Negative feedback
    (a) increases the input and output impedances
    (b) increases the input impedance and the bandwidth
    (c) decreases the output impedance and the bandwidth
    (d) does not affect impedances or bandwidth

18. A certain noninverting amplifier has an R_i of 1 kΩ and an R_f of 100 kΩ. The closed-loop gain is
    (a) 100,000   (b) 1000   (c) 101   (d) 100

19. If the feedback resistor in Question 18 is open, the voltage gain
    (a) increases   (b) decreases   (c) is not affected   (d) depends on R_i

20. A certain inverting amplifier has a closed-loop gain of 25. The op-amp has an open-loop gain of 100,000. If another op-amp with an open-loop gain of 200,000 is substituted in the configuration, the closed-loop gain
    (a) doubles   (b) drops to 12.5
    (c) remains at 25   (d) increases slightly

21. A voltage follower
    (a) has a gain of one   (b) is noninverting
    (c) has no feedback resistor   (d) has all of these

PROBLEMS

SECTION 5–1 INTRODUCTION TO OPERATIONAL AMPLIFIERS

1. Compare a practical op-amp to the ideal.

2. Two IC op-amps are available to you. Their characteristics are listed below. Choose the one you think is more desirable.
   Op-amp 1: Z_{in} = 5 MΩ, Z_{out} = 100 Ω, A_{ol} = 50,000
   Op-amp 2: Z_{in} = 10 MΩ, Z_{out} = 75 Ω, A_{ol} = 150,000

SECTION 5–2 THE DIFFERENTIAL AMPLIFIER

3. Identify the type of input and output configuration for each basic differential amplifier in Figure 5–45.

4. The dc base voltages in Figure 5–46 are zero. Using your knowledge of transistor analysis, determine the dc differential output voltage. Assume that Q_1 has an α = 0.98 and Q_2 has an α = 0.975.

OPERATIONAL AMPLIFIERS

FIGURE 5–45

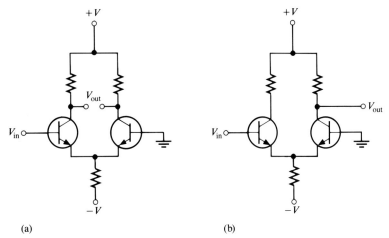

(a)　　　　　　　　　　　(b)

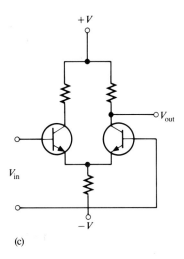　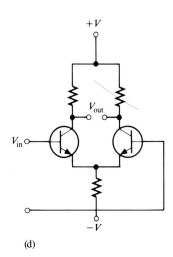

(c)　　　　　　　　　　　(d)

FIGURE 5–46

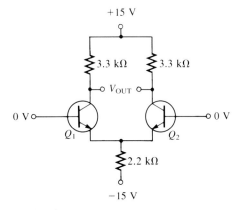

5. Identify the quantity being measured by each meter in Figure 5–47.

FIGURE 5–47

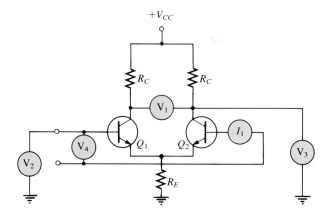

6. A differential amplifier stage has collector resistors of 5.1 kΩ each. If I_{C1} = 1.35 mA and I_{C2} = 1.29 mA, what is the differential output voltage?

SECTION 5–3   OP-AMP DATA SHEET PARAMETERS

7. Determine the bias current, I_{BIAS}, given that the input currents to an op-amp are 8.3 μA and 7.9 μA.

8. Distinguish between input bias current and input offset current, and then calculate the input offset current in Problem 7.

9. A certain op-amp has a CMRR of 250,000. Convert this to dB.

10. The open-loop gain of a certain op-amp is 175,000. Its common-mode gain is 0.18. Determine the CMRR in dB.

11. An op-amp data sheet specifies a CMRR of 300,000 and an A_{ol} of 90,000. What is the common-mode gain?

12. Figure 5–48 shows the output voltage of an op-amp in response to a step input. What is the slew rate?

FIGURE 5–48

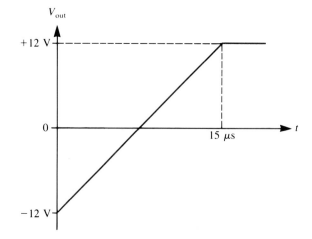

13. How long does it take the output voltage of an op-amp to go from -10 V to $+10$ V, if the slew rate is 0.5 V/μs?

SECTION 5–5 OP-AMP CONFIGURATIONS WITH NEGATIVE FEEDBACK

14. Identify each of the op-amp configurations in Figure 5–49.

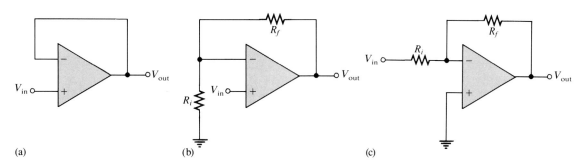

FIGURE 5–49

15. A noninverting amplifier has an R_i of 1 kΩ and an R_f of 100 kΩ. Determine V_f and B, if $V_{out} = 5$ V.

16. For the amplifier in Figure 5–50, determine the following:

    (a) $A_{cl(NI)}$ (b) V_{out} (c) V_f

17. Determine the closed-loop gain of each amplifier in Figure 5–51.

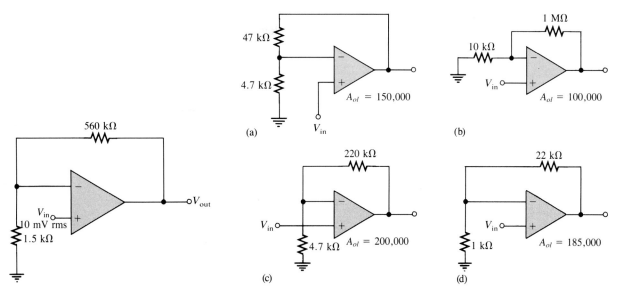

FIGURE 5–50

FIGURE 5–51

18. Find the value of R_f that will produce the indicated closed-loop gain in each amplifier in Figure 5–52.

FIGURE 5–52

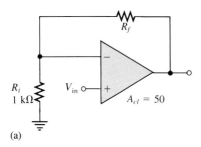

(a)

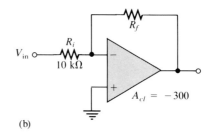

(b)

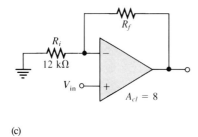

(c)

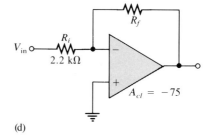

(d)

19. Find the gain of each amplifier in Figure 5–53.

FIGURE 5–53

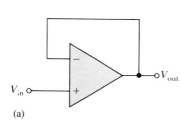

(a)

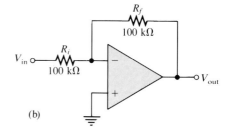

(b)

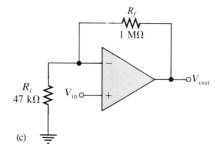

(c)

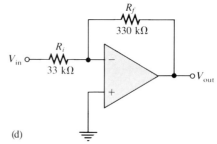

(d)

20. If a signal voltage of 10 mV rms is applied to each amplifier in Figure 5–53, what are the output voltages and what is their phase relationship with inputs?

21. Determine the approximate values for each of the following quantities in Figure 5–54.

   (a) I_{in} (b) I_f (c) V_{out} (d) Closed-loop gain

FIGURE 5–54

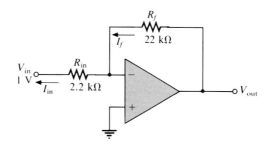

SECTION 5–6 EFFECTS OF NEGATIVE FEEDBACK ON OP-AMP IMPEDANCES

22. Determine the input and output impedances for each amplifier configuration in Figure 5–55.

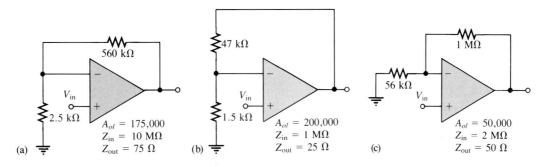

FIGURE 5–55

23. Repeat Problem 22 for each circuit in Figure 5–56.

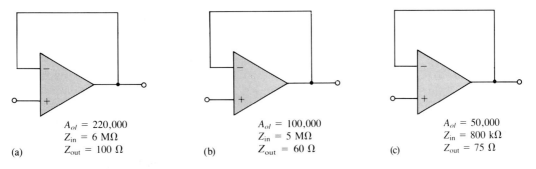

FIGURE 5–56

24. Repeat Problem 22 for each circuit in Figure 5–57.

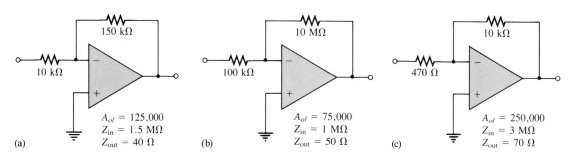

FIGURE 5–57

SECTION 5–7 BIAS CURRENT AND OFFSET VOLTAGE COMPENSATION

25. A voltage follower is driven by a voltage source with a source resistance of 75 Ω.

    (a) What value of compensating resistor is required for bias current, and where should the resistor be placed?

    (b) If the two input currents after compensation are 42 μA and 40 μA, what is the output error voltage?

26. Determine the compensating resistor value for each amplifier configuration in Figure 5–55, and indicate the placement of the resistor.

27. A particular op-amp has an input offset voltage of 2 nV and an open-loop gain of 100,000. What is the output error voltage?

28. What is the input offset voltage of an op-amp if a dc output voltage of 35 mV is measured when the input voltage is zero? The op-amp's open-loop gain is specified to be 200,000.

SECTION 5–8 TROUBLESHOOTING

29. Determine the most likely fault(s) for each of the following symptoms in Figure 5–58 with a 100 mV signal applied.

    (a) No output signal.

    (b) Output severely clipped on both positive and negative swings.

    (c) Clipping on only positive peaks when input signal is increased to a certain point.

FIGURE 5–58

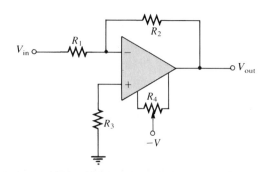

OPERATIONAL AMPLIFIERS

 30. On the circuit board in Figure 5–59, what happens if the middle lead (wiper) of the 100 kΩ potentiometer is broken?

FIGURE 5–59

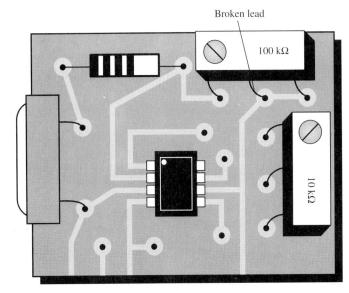

ANSWERS TO REVIEW QUESTIONS

SECTION 5–1
1. Inverting input, noninverting input, output, positive and negative supply voltages.
2. A practical op-amp has high imput impedance, low output impedance, high voltage gain, and wide bandwidth.

SECTION 5–2
1. Differential input is between two input terminals. Single-ended input is from one input terminal to ground (with other input grounded).
2. Common-mode rejection is the ability of an op-amp to produce very little output when the same signal is applied to both inputs.
3. A higher CMRR means less common-mode gain.

SECTION 5–3
1. Input bias current, input offset voltage, drift, input offset current, input impedance, output impedance, common-mode range, CMRR, open-loop voltage gain, slew rate, frequency response.
2. Slew rate and voltage gain are both frequency dependent.

SECTION 5–4
1. Negative feedback provides a stable controlled gain, control of impedances, and wider bandwidth.
2. The open-loop gain is so high that a very small signal on the input will drive the op-amp into saturation.

Section 5–5
1. The main purpose of negative feedback is to stabilize the gain.
2. False
3. $A_{cl} = 1/0.02 = 50$

Section 5–6
1. The noninverting configuration has a higher Z_{in} than the op-amp alone.
2. Z_{in} increases in a voltage follower.
3. $Z_{in(I)} \cong R_i = 2 \text{ k}\Omega$, $Z_{out(I)} \cong Z_{out} = 60 \text{ }\Omega$.

Section 5–7
1. Input bias current and input offset voltage are sources of output error.
2. Add a resistor in the feedback path equal to the input source resistance.

Section 5–8
1. Check the output null adjustment.
2. The op-amp is probably bad.

Section 5–9
1. The 100 kΩ potentiometer is the feedback resistor.
2. The 10 kΩ potentiometer is for nulling the output.
3. The light source and prism must be pivoted to allow different wavelengths of light to pass through the slit.

Answers to Practice Exercises

5–1 34,000; 90.6 dB
5–2 (a) 0.168   (b) 87.96 dB   (c) 2.1 V rms, 4.2 V rms   (d) 0.168 V
5–3 12,649
5–4 20 V/μs
5–5 32.9
5–6 67.5 kΩ
5–7 (a) 20,557 MΩ, 0.014 Ω (b) 23
5–8 Input Z increases, output Z decreases.
5–9 $Z_{in(I)} = 560 \text{ }\Omega$, $Z_{out(I)} = 75 \text{ }\Omega$, $A_{cl} = -146$

6
OP-AMP RESPONSES

6–1 BASIC CONCEPTS
6–2 OP-AMP OPEN-LOOP RESPONSE
6–3 OP-AMP CLOSED-LOOP RESPONSE
6–4 POSITIVE FEEDBACK AND STABILITY
6–5 OP-AMP COMPENSATION
6–6 A SYSTEM APPLICATION

After completing this chapter, you should be able to

☐ Distinguish the difference between open-loop voltage gain and closed-loop voltage gain of an op-amp.
☐ Compare an op-amp's frequency response for closed-loop and open-loop conditions.
☐ Discuss the effects of closed-loop operation on bandwidth.
☐ Define *gain-bandwidth product*.
☐ Describe what happens when positive feedback occurs in an op-amp.
☐ Define *phase margin*.
☐ Explain stability and what factors affect the stability of an op-amp.
☐ Use compensation to ensure stability in an op-amp.

In this chapter, you will learn more about frequency response, bandwidth, phase shift, and other frequency-related parameters. The effects of negative feedback will be further examined, and you will learn about stability requirements and how to compensate op-amp circuits to ensure stable operation.

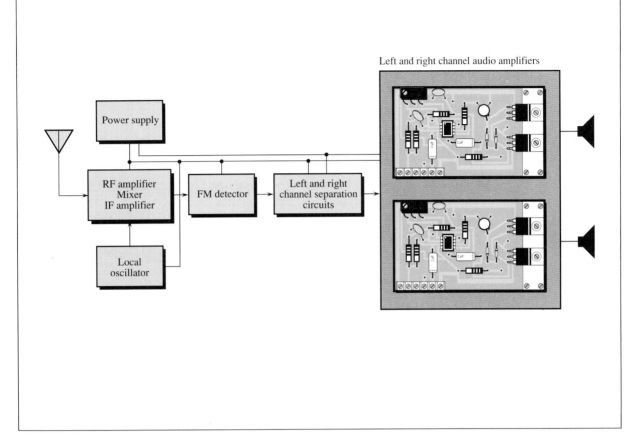

A System Application

The block diagram above shows an FM stereo receiver. Notice that the system is basically the same as the standard superheterodyne receiver up through the FM detector. Stereo systems use two separate frequency-modulated (FM) signals to reproduce sound as, for example, from the left and right sides of the stage in a concert performance. When the signal is processed by a stereo receiver, the sound comes out of both the left and right speaker, and you get the original sound effects in terms of direction and distribution. When a stereo broadcast is received by a single-speaker (monophonic) system, the sound from the speaker is actually the composite or sum of the left and right channel sounds so you get the original sound without separation. Op-amps can be used for many purposes in stereo systems such as this, but we will focus on the identical left and right channel audio amplifiers for this system application.

For the system application in Section 6-6, in addition to the other topics, be sure you understand

☐ The noninverting op-amp configuration.
☐ How capacitors can be connected for frequency compensation.
☐ Discrete transistor push-pull amplifier operation (from Chapter 4).

6-1 BASIC CONCEPTS

The last chapter demonstrated how closed-loop voltage gains of the basic op-amp configurations are determined, and the distinction between open-loop gain and closed-loop gain was established. Because of the importance of these two different types of gain, the definitions are restated in this section.

OPEN-LOOP GAIN

The **open-loop voltage gain** of an op-amp is the internal voltage gain of the device and represents the ratio of output voltage to input voltage, as indicated in Figure 6-1(a). Notice that there are no external components, so the open-loop gain is set entirely by the internal design. Open-loop gain can range up to 200,000 and is not a well-controlled parameter. Data sheets often refer to the open-loop gain as the *large-signal voltage gain*.

FIGURE 6-1
Open-loop and closed-loop op-amp configurations.

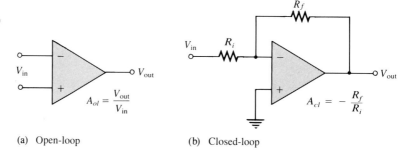

(a) Open-loop (b) Closed-loop

CLOSED-LOOP GAIN

The **closed-loop voltage gain** is for an entire amplifier configuration consisting of the op-amp and an external negative feedback circuit that connects the output to the inverting input. The closed-loop gain is determined by the external component values, as illustrated in Figure 6-1(b) for an inverting amplifier configuration. The closed-loop gain can be precisely controlled by external component values.

THE GAIN IS FREQUENCY DEPENDENT

In the last chapter, all of the gain expressions applied to the midrange gain and were considered independent of the frequency. The midrange open-loop gain of an op-amp extends from zero frequency (dc) up to a critical frequency at which the gain is 3 dB less than the midrange value. The difference here is that op-amps are dc amplifiers (no capacitive coupling between stages), and therefore, there is no lower critical frequency. This means that the midrange gain extends down to zero frequency, and dc voltages are amplified the same as midrange signal frequencies.

OP-AMP RESPONSES

An open-loop response curve (Bode plot) for a certain op-amp is shown in Figure 6–2. Most op-amp data sheets show this type of curve or specify the midrange open-loop gain. Notice that the curve rolls off at −20 dB per decade (−6 dB per octave). The midrange gain is 200,000, which is 106 dB, and the critical (cutoff) frequency is approximately 10 Hz.

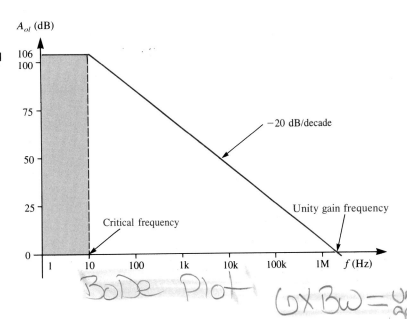

FIGURE 6–2
Ideal plot of open-loop voltage gain versus frequency for typical op-amp. The frequency scale is logarithmic.

3 dB OPEN-LOOP BANDWIDTH

Recall that the bandwidth of an ac amplifier is the frequency range between the points where the gain is 3 dB less than midrange, equalling the upper critical frequency minus the lower critical frequency.

$$BW = f_{c(high)} - f_{c(low)} \tag{6-1}$$

Since $f_{c(low)}$ for an op-amp is zero, the bandwidth is simply equal to the upper critical frequency.

$$BW = f_{c(high)} \tag{6-2}$$

From now on, $f_{c(high)}$ will be referred to simply as f_c. Open-loop (*ol*) or closed-loop (*cl*) subscript designators will be used.

UNITY-GAIN BANDWIDTH

Notice in Figure 6–2 that the gain steadily decreases to a point where it is equal to one (0 dB). The value of the frequency at which this unity gain occurs in the *unity-gain bandwidth*.

Gain-Versus-Frequency Analysis

The *RC* lag (low-pass) networks within an op-amp are responsible for the roll-off in gain as the frequency increases. From basic ac circuit theory, the attenuation of an *RC* lag network, such as in Figure 6–3, is expressed as

$$\frac{V_{out}}{V_{in}} = \frac{X_C}{\sqrt{R^2 + X_C^2}} \qquad (6\text{–}3)$$

Dividing both the numerator and denominator to the right of the equal sign by X_C, we get

$$\frac{V_{out}}{V_{in}} = \frac{1}{\sqrt{1 + R^2/X_C^2}} \qquad (6\text{–}4)$$

The critical frequency of an *RC* network is

$$f_c = \frac{1}{2\pi RC}$$

Dividing both sides by *f* gives

$$\frac{f_c}{f} = \frac{1}{2\pi RCf} = \frac{1}{(2\pi fC)R}$$

Since $X_C = 1/(2\pi f C)$, the above expression can be written as

$$\frac{f_c}{f} = \frac{X_C}{R} \qquad (6\text{–}5)$$

Substituting this result into Equation (6–4) produces the following expression for the attenuation of an *RC* lag network.

$$\frac{V_{out}}{V_{in}} = \frac{1}{\sqrt{1 + f^2/f_c^2}} \qquad (6\text{–}6)$$

If an op-amp is represented by a voltage gain element and a single *RC* lag network, as shown in Figure 6–4, then the total open-loop gain is the product of the midrange open-loop gain $A_{ol(mid)}$ and the attenuation of the *RC* network.

$$A_{ol} = \frac{A_{ol(mid)}}{\sqrt{1 + f^2/f_c^2}} \qquad (6\text{–}7)$$

As you can see from Equation (6–7), the open-loop gain equals the midrange value when the signal frequency *f* is much less than the critical frequency f_c and drops off as the frequency increases. Since f_c is part of the open-loop response of an op-amp, we will refer to it as $f_{c(ol)}$.

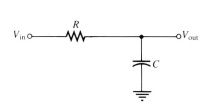

FIGURE 6–3
RC lag network.

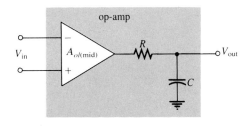

FIGURE 6–4
Op-amp represented by gain element and internal RC network.

EXAMPLE 6–1

Determine A_{ol} for the following values of f. Assume $f_{c(ol)} = 100$ Hz and $A_{ol(mid)} = 100{,}000$.

(a) $f = 0$ Hz
(b) $f = 10$ Hz
(c) $f = 100$ Hz
(d) $f = 1000$ Hz

SOLUTION

(a) $A_{ol} = \dfrac{A_{ol(mid)}}{\sqrt{1 + f^2/f_{c(ol)}^2}} = \dfrac{100{,}000}{\sqrt{1 + 0}} = 100{,}000$

(b) $A_{ol} = \dfrac{100{,}000}{\sqrt{1 + (0.1)^2}} = 99{,}503$

(c) $A_{ol} = \dfrac{100{,}000}{\sqrt{1 + (1)^2}} = \dfrac{100{,}000}{\sqrt{2}} = 70{,}710$

(d) $A_{ol} = \dfrac{100{,}000}{\sqrt{1 + (10)^2}} = 9950$

This exercise has demonstrated how the open-loop gain decreases as the frequency increases above $f_{c(ol)}$.

PRACTICE EXERCISE 6–1
Find A_{ol} for the following frequencies. Assume $f_{c(ol)} = 200$ Hz, $A_{ol(mid)} = 80{,}000$.
(a) $f = 2$ Hz   (b) $f = 10$ Hz   (c) $f = 2500$ Hz

PHASE SHIFT

As you know, an RC network causes a propagation delay from input to output, thus creating a **phase** difference between the input signal and the output signal. An RC lag network such as found in an op-amp stage causes the output signal voltage to lag the input, as shown in Figure 6–5. From basic ac circuit theory, the phase shift is

$$\phi = -\tan^{-1}\left(\frac{R}{X_C}\right) \quad (6\text{–}8)$$

Substituting the relationship in Equation (6–5), we get

$$\phi = -\tan^{-1}\left(\frac{f}{f_c}\right) \quad (6\text{–}9)$$

The negative sign indicates that the output lags the input. This equation shows that the phase shift increases with frequency and approaches $-90°$ as f becomes much greater than f_c.

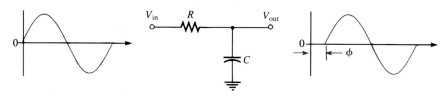

FIGURE 6–5
Output voltage lags input voltage.

EXAMPLE 6–2   Calculate the phase shift for an RC lag network for each of the following frequencies, and then plot the curve of phase shift versus frequency. Assume $f_c = 100$ Hz.

(a) $f = 1$ Hz
(b) $f = 10$ Hz
(c) $f = 100$ Hz
(d) $f = 1000$ Hz
(e) $f = 10,000$ Hz

SOLUTION

(a) $\phi = -\tan^{-1}\left(\dfrac{f}{f_c}\right) = -\tan^{-1}\left(\dfrac{1 \text{ Hz}}{100 \text{ Hz}}\right) = -0.573°$

(b) $\phi = -\tan^{-1}\left(\dfrac{10 \text{ Hz}}{100 \text{ Hz}}\right) = -5.71°$

(c) $\phi = -\tan^{-1}\left(\dfrac{100 \text{ Hz}}{100 \text{ Hz}}\right) = -45°$

(d) $\phi = -\tan^{-1}\left(\dfrac{1000 \text{ Hz}}{100 \text{ Hz}}\right) = -84.29°$

(e) $\phi = -\tan^{-1}\left(\dfrac{10,000 \text{ Hz}}{100 \text{ Hz}}\right) = -89.43°$

The phase shift-versus-frequency curve is plotted in Figure 6–6. Note that the frequency axis is logarithmic.

FIGURE 6–6

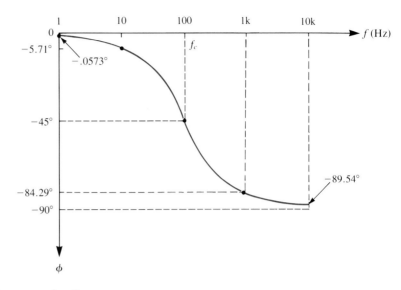

PRACTICE EXERCISE 6–2
At what frequency, in this example, is the phase shift 60°?

6–1 REVIEW QUESTIONS

1. How do the open-loop gain and the closed-loop gain of an op-amp differ?
2. The upper critical frequency of a particular op-amp is 100 Hz. What is its open-loop 3 dB bandwidth?
3. Does the open-loop gain increase or decrease with frequency above the critical frequency?

6–2   OP-AMP OPEN-LOOP RESPONSE

In this section, you will learn about the open-loop frequency response and the open-loop phase response of an op-amp. Open-loop responses relate to an op-amp with no external feedback. The frequency response indicates how the voltage gain changes with frequency, and the phase response indicates how the phase shift between the input and output signal changes with frequency. The open-loop gain, like the β of a transistor, varies greatly from one device to the next of the same type and cannot be depended upon to have a constant value.

FREQUENCY RESPONSE

In the previous section, we considered an op-amp to have a constant roll-off of -20 dB/decade above its critical frequency. Actually, the situation is often more complex than that. A typical IC operational amplifier may consist of two or more cascaded amplifier stages. The gain of each stage is frequency dependent and rolls off at -20 dB/decade above its critical frequency. Therefore, the total response of an op-amp is a composite of the individual responses of the internal stages. As an example, a three-stage op-amp is represented in Figure 6–7(a), and the frequency response of each stage is shown in Figure 6–7(b). As you know, dB gains are added so that the total op-amp frequency response is as shown in Figure 6–7(c). Since the roll-off rates are additive, the total roll-off rate increases by -20 dB/decade (-6 dB/octave) as each critical frequency is reached.

FIGURE 6–7
Op-amp open-loop frequency response.

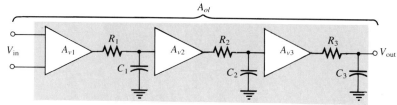

(a) Representation of an op-amp with three internal stages

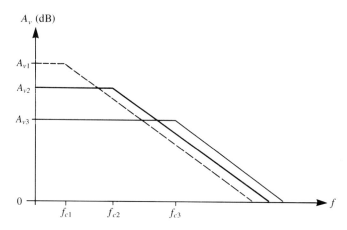

(b) Individual responses

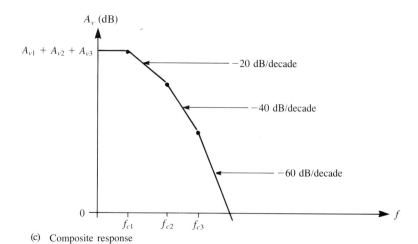

(c) Composite response

Op-Amp Responses

Phase Response

In a multistage amplifier, each stage contributes to the total phase lag. As you have seen, each RC lag network can produce up to a $-90°$ phase shift. Since each stage in an op-amp includes an RC lag network, a three-stage op-amp, for example, can have a maximum phase lag of $-270°$. Also, the phase lag of each stage is less than $-45°$ below the critical frequency, equal to $-45°$ at the critical frequency, and greater than $-45°$ above the critical frequency. The phase lags of the stages of an op-amp are added to produce a total phase lag, according to the following formula for three stages.

$$\phi_{tot} = -\tan^{-1}\left(\frac{f}{f_{c1}}\right) - \tan^{-1}\left(\frac{f}{f_{c2}}\right) - \tan^{-1}\left(\frac{f}{f_{c3}}\right) \qquad (6-10)$$

■ EXAMPLE 6–3

A certain op-amp has three internal amplifier stages with the following gains and critical frequencies:

Stage 1: $A_{v1} = 40$ dB, $f_{c1} = 2000$ Hz
Stage 2: $A_{v2} = 32$ dB, $f_{c2} = 40$ kHz
Stage 3: $A_{v3} = 20$ dB, $f_{c3} = 150$ kHz

Determine the open-loop midrange dB gain and the total phase lag when $f = f_{c1}$.

Solution

$$\begin{aligned} A_{ol(mid)} &= A_{v1} + A_{v2} + A_{v3} \\ &= 40 \text{ dB} + 32 \text{ dB} + 20 \text{ dB} \\ &= 92 \text{ dB} \end{aligned}$$

$$\begin{aligned} \phi_{tot} &= \tan^{-1}\left(\frac{f}{f_{c1}}\right) - \tan^{-1}\left(\frac{f}{f_{c2}}\right) - \tan^{-1}\left(\frac{f}{f_{c3}}\right) \\ &= -\tan^{-1}(1) - \tan^{-1}\left(\frac{2}{40}\right) - \tan^{-1}\left(\frac{2}{150}\right) \\ &= -45° - 2.86° - 0.76° \\ &= -48.62° \end{aligned}$$

Practice Exercise 6–3

The internal stages of a two-stage amplifier have the following characteristics: $A_{v1} = 50$ dB, $A_{v2} = 25$ dB, $f_{c1} = 1500$ Hz, and $f_{c2} = 3000$ Hz. Determine the open-loop midrange gain in dB and the total phase lag when $f = f_{c1}$. ■

6–2 Review Questions

1. If the individual stage gains of an op-amp are 20 dB and 30 dB, what is the total gain in dB?
2. If the individual phase lags are $-49°$ and $-5.2°$, what is the total phase lag?

6-3 OP-AMP CLOSED-LOOP RESPONSE

Op-amps are normally used in a closed-loop configuration with negative feedback in order to achieve precise control of the gain and bandwidth. In this section, you will see how feedback affects the gain and frequency response of an op-amp.

Recall from Chapter 5 that midrange gain is reduced by negative feedback, as indicated by the following closed-loop gain expressions for the three configurations previously covered, where B is the feedback attenuation. For noninverting,

$$A_{cl(NI)} = \frac{A_{ol}}{1 + A_{ol}B} \cong \frac{1}{B}$$

For inverting,

$$A_{cl(I)} \cong -\frac{R_f}{R_i}$$

For voltage follower,

$$A_{cl(VF)} \cong 1$$

EFFECT OF NEGATIVE FEEDBACK ON BANDWIDTH

You know how negative feedback affects the gain; now you will learn how it affects the amplifier's bandwidth. The closed-loop critical frequency is

$$f_{c(cl)} = f_{c(ol)}(1 + BA_{ol(mid)}) \qquad (6-11)$$

This expression shows that the closed-loop critical frequency, $f_{c(cl)}$, is higher than the open-loop critical frequency $f_{c(ol)}$ by the factor $1 + BA_{ol(mid)}$. A derivation of Equation (6-11) can be found in Appendix B.

Since $f_{c(cl)}$ equals the bandwidth for the closed-loop amplifier, the bandwidth is also increased by the same factor.

$$BW_{cl} = BW_{ol}(1 + BA_{ol(mid)}) \qquad (6-12)$$

■ **EXAMPLE 6-4** A certain amplifier has an open-loop midrange gain of 150,000 and an open-loop 3 dB bandwidth of 200 Hz. The attenuation of the feedback loop is 0.002. What is the closed-loop bandwidth?

SOLUTION
$$BW_{cl} = BW_{ol}(1 + BA_{ol(mid)})$$
$$= 200 \text{ Hz}[1 + (150,000)(0.002)]$$
$$= 60.2 \text{ kHz}$$

PRACTICE EXERCISE 6-4
If $A_{ol(mid)} = 200,000$ and $B = 0.05$, what is the closed loop bandwidth?

OP-AMP RESPONSES

Figure 6–8 graphically illustrates the concept of closed-loop response. When the open-loop gain of an op-amp is reduced by negative feedback, the bandwidth is increased. The closed-loop gain is independent of the open-loop gain up to the point of intersection of the two gain curves. This point of intersection is the critical frequency for the closed-loop response. Notice that the closed-loop gain has the same roll-off rate as the open-loop gain, beyond the closed-loop critical frequency.

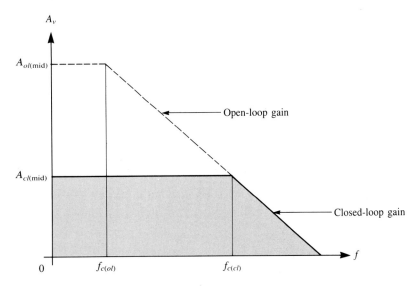

FIGURE 6–8
Closed-loop gain compared to open-loop gain.

GAIN-BANDWIDTH PRODUCT

An increase in closed-loop gain causes a decrease in the bandwidth and vice versa, such that the product of gain and bandwidth is a constant. This is true as long as the roll-off rate is fixed. Letting A_{cl} stand for the gain of any of the closed-loop configurations and $f_{c(cl)}$ for the closed-loop critical frequency (also the bandwidth), then

$$A_{cl}f_{c(cl)} = A_{ol}f_{c(ol)} \qquad (6\text{–}13)$$

The gain-bandwidth product is always equal to the frequency at which the op-amp's open-loop gain is unity (unity-gain bandwidth).

$$A_{cl}f_{c(cl)} = \text{unity-gain bandwidth} \qquad (6\text{–}14)$$

EXAMPLE 6–5

Determine the bandwidth of each of the amplifiers in Figure 6–9. Both op-amps have an open-loop gain of 100 dB and a unity-gain bandwidth of 3 MHz.

SOLUTION

(a) For the noninverting amplifier in part (a) of the figure, the closed-loop gain is

$$A_{cl} \cong \frac{1}{B} = \frac{1}{R_i/(R_i + R_f)} = \frac{1}{3.3 \text{ k}\Omega/223.3 \text{ k}\Omega} = 67.67$$

FIGURE 6-9

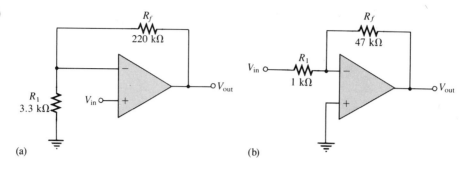

Using Equation (6–14) and solving for $f_{c(cl)}$, we get (where $f_{c(cl)} = BW_{cl}$)

$$f_{c(cl)} = BW_{cl} = \frac{\text{unity-gain }BW}{A_{cl}}$$

$$BW_{cl} = \frac{3 \text{ MHz}}{67.67} = 44.33 \text{ kHz}$$

(b) For the inverting amplifier in part (b) of the figure, the closed-loop gain is

$$A_{cl} = -\frac{R_f}{R_i} = -\frac{47 \text{ k}\Omega}{1 \text{ k}\Omega} = -47$$

The closed-loop bandwidth is

$$BW_{cl} = \frac{3 \text{ MHz}}{47} = 63.8 \text{ kHz}$$

PRACTICE EXERCISE 6–5
Determine the bandwidth of each of the amplifiers in Figure 6–9. Both op-amps have an A_{ol} of 90 dB and a unity-gain bandwidth of 2 MHz.

6–3 REVIEW QUESTIONS

1. Is the closed-loop gain always less than the open-loop gain?
2. A certain op-amp is used in a feedback configuration having a gain of 30 and a bandwidth of 100 kHz. If the external resistor values are changed to increase the gain to 60, what is the new bandwidth?
3. What is the unity-gain bandwidth of the op-amp in Question 2?

6–4 POSITIVE FEEDBACK AND STABILITY

Stability is a very important consideration when using op-amps. Stable operation means that the op-amp does not oscillate under any condition. Instability produces oscillations, which are unwanted voltage swings on the output when there is no signal present on the input, or in response to noise or transient voltages on the input.

POSITIVE FEEDBACK

To understand stability, instability and its causes must first be examined. As you know, with negative feedback, the signal fed back to the input is out of phase with the input signal, thus subtracting from it and effectively reducing the voltage gain. As long as the feedback is negative, the amplifier is stable. When the signal fed back from output to input is in phase with the input signal, a positive feedback condition exists. That is, positive feedback occurs when the total phase shift through the op-amp and feedback network is 360°, which is equivalent to no phase shift (0°).

LOOP GAIN

For instability to occur, (1) there must be positive feedback, and (2) the loop gain of the closed-loop amplifier must be greater than 1. The loop gain of a closed-loop amplifier is defined to be the op-amp's open-loop gain times the attenuation of the feedback network.

$$\text{Loop gain} = A_{ol}B \qquad (6-15)$$

PHASE MARGIN

Notice that for each amplifier configuration in Figure 6–10, the feedback loop is connected to the inverting input. There is an inherent phase shift of 180° because of the *inversion* between input and output. Additional phase shift (ϕ_{tot}) is produced by the *RC* lag networks within the amplifier. So, the total phase shift around the loop is $180° + \phi_{tot}$.

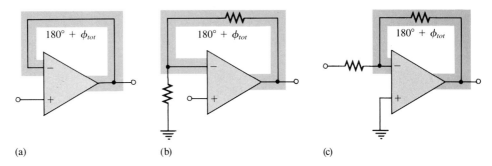

(a)   (b)   (c)

FIGURE 6–10
Feedback-loop phase shift.

The **phase margin**, θ_{pm}, is the amount of additional phase shift required to make the total phase shift around the loop 360°. (360° is equivalent to 0°.)

$$180° + \phi_{tot} + \theta_{pm} = 360°$$
$$\theta_{pm} = 180° - |\phi_{tot}| \qquad (6-16)$$

If the phase margin is positive, the total phase shift is less than 360° and the amplifier is stable. If the phase margin is zero or negative, then the amplifier is potentially unstable because the signal fed back can be in phase with the input. As you can see from Equation (6–16), when the total lag network phase shift ϕ_{tot} equals or exceeds 180°, then the phase margin is 0° or negative and an unstable condition exists.

STABILITY ANALYSIS

Since most op-amp configurations use a loop gain greater than 1 ($A_{ol}B > 1$), the criteria for stability are based on the phase angle of the internal lag networks. As previously mentioned, operational amplifiers are composed of multiple stages, each of which has a critical frequency. For purposes of illustrating the concept of **stability,** we will use a three-stage op-amp with an open-loop response as shown in the Bode plot of Figure 6–11. Notice that there are three different critical frequencies, which indicates three internal RC lag networks. At the first critical frequency, f_{c1}, the gain begins rolling off at -20 dB/decade; when the second critical frequency, f_{c2}, is reached, the gain decreases at -40 dB/decade; and when the third critical frequency, f_{c3}, is reached, the gain drops at -60 dB/decade.

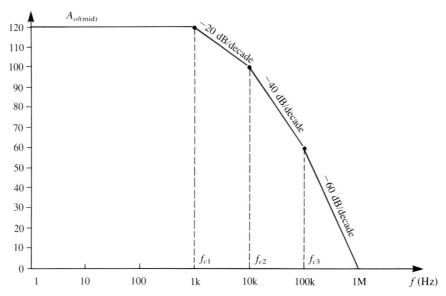

FIGURE 6–11
Bode plot of example of three-stage op-amp response.

To analyze a closed-loop amplifier for stability, the phase margin must be determined. A positive phase margin will indicate that the amplifier is stable for a given value of closed-loop gain. Three example cases will be considered in order to demonstrate the conditions for instability.

CASE 1 The closed-loop gain intersects the open-loop response on the -20 dB/decade slope, as shown in Figure 6–12. For this example, the midrange closed-loop gain is 106 dB, and the closed-loop critical frequency is 5 kHz. If we assume that the amplifier is not operated out of its midrange, the maximum phase shift for the 106 dB amplifier

OP-AMP RESPONSES

FIGURE 6–12
Case where closed-loop gain intersects open-loop gain on −20 dB/decade slope (stable operation).

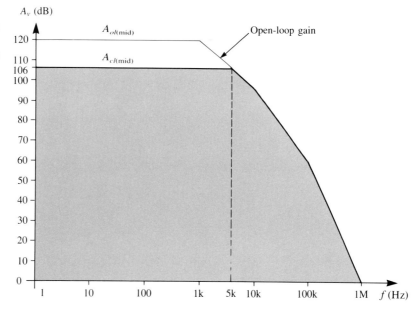

occurs at the highest midrange frequency (in this case, 5 kHz). The total phase shift at this frequency due to the three lag networks is calculated as follows:

$$\phi_{tot} = -\tan^{-1}\left(\frac{f}{f_{c1}}\right) - \tan^{-1}\left(\frac{f}{f_{c2}}\right) - \tan^{-1}\left(\frac{f}{f_{c3}}\right)$$

where
$$f = 5 \text{ kHz}$$
$$f_{c1} = 1 \text{ kHz}$$
$$f_{c2} = 10 \text{ kHz}$$
$$f_{c3} = 100 \text{ kHz}$$

Therefore,

$$\phi_{tot} = -\tan^{-1}\left(\frac{5 \text{ kHz}}{1 \text{ kHz}}\right) - \tan^{-1}\left(\frac{5 \text{ kHz}}{10 \text{ kHz}}\right) - \tan^{-1}\left(\frac{5 \text{ kHz}}{100 \text{ kHz}}\right)$$
$$= -78.69° - 26.57° - 2.86°$$
$$= -108.12°$$

The phase margin is

$$\theta_{pm} = 180° - |\phi_{tot}| = 180° - 108.12° = +71.88°$$

θ_{pm} is positive, so the amplifier is stable for all frequencies in its midrange. In general, an amplifier is stable for all midrange frequencies if its closed-loop gain intersects the open-loop response curve on a −20 dB/decade slope.

CASE 2 The closed-loop gain is lowered to where it intersects the open-loop response on the −40 dB/decade slope, as shown in Figure 6–13. The midrange closed-loop gain in

FIGURE 6–13
Case where closed-loop gain intersects open-loop gain on −40 dB/decade slope (marginally stable operation).

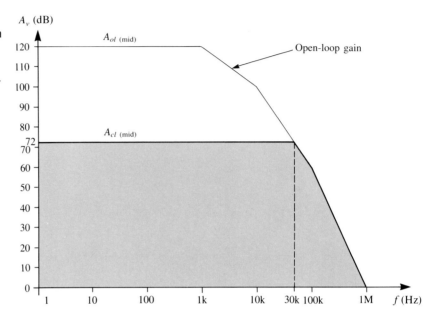

this case is 72 dB, and the closed-loop critical frequency is approximately 30 kHz. The total phase shift at $f = 30$ kHz due to the three lag networks is calculated as follows.

$$\phi_{tot} = -\tan^{-1}\left(\frac{30 \text{ kHz}}{1 \text{ kHz}}\right) - \tan^{-1}\left(\frac{30 \text{ kHz}}{10 \text{ kHz}}\right) - \tan^{-1}\left(\frac{30 \text{ kHz}}{100 \text{ kHz}}\right)$$
$$= -88.09° - 71.57° - 16.7°$$
$$= -176.36°$$

The phase margin is

$$\theta_{pm} = 180° - 176.36° = +3.64°$$

The phase margin is positive, so the amplifier is still stable for all frequencies in its midrange, but a very slight increase in frequency above f_c would cause it to oscillate. Therefore, it is marginally stable and very close to instability because instability occurs where $\theta_{pm} = 0°$ and, as a general rule, a minimum 45° phase margin is recommended to avoid marginal conditions.

CASE 3 The closed-loop gain is further decreased until it intersects the open-loop response on the −60 dB/decade slope, as shown in Figure 6–14. The midrange closed-loop gain in this case is 18 dB, and the closed-loop critical frequency is 500 kHz. The total phase shift at $f = 500$ kHz due to the three lag networks is

$$\phi_{tot} = -\tan^{-1}\left(\frac{500 \text{ kHz}}{1 \text{ kHz}}\right) - \tan^{-1}\left(\frac{500 \text{ kHz}}{10 \text{ kHz}}\right) - \tan^{-1}\left(\frac{500 \text{ kHz}}{100 \text{ kHz}}\right)$$
$$= -89.89° - 88.85° - 78.69° = -257.43°$$

The phase margin is

$$\theta_{pm} = 180° - 257.43° = -77.43°$$

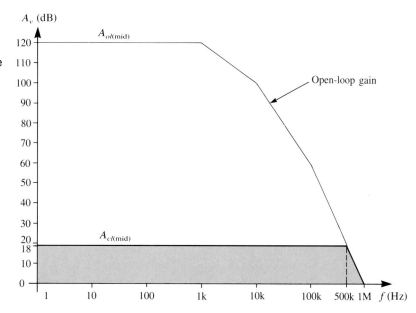

FIGURE 6–14
Case where closed-loop gain intersects open-loop gain on −60 dB/decade slope (unstable operation).

Here the phase margin is negative and the amplifier is unstable at the upper end of its midrange. Using the following BASIC program, you can determine the approximate midrange frequency at which instability occurs. The program computes the phase margin for each of a specified number of frequency values and determines whether the op-amp is stable or unstable. The display shows a list of frequencies with corresponding phase shifts, phase margins, and stability conditions. Instability occurs between the last two frequencies printed. Inputs required for this program are all of the critical frequencies in the open-loop response of the given op-amp, the highest frequency of operation, and the increments of frequency for which the parameters are to be computed.

```
10   CLS
20   PRINT"THIS PROGRAM COMPUTES THE PHASE MARGINS FOR EACH"
30   PRINT"FREQUENCY IN A SPECIFIED RANGE AND DETERMINES THE"
40   PRINT"MAXIMUM FREQUENCY FOR STABLE OPERATION OF THE
     OP-AMP"
50   PRINT:PRINT:PRINT
60   INPUT "TO CONTINUE PRESS 'ENTER'";X
70   CLS
80   INPUT "NUMBER OF OPEN-LOOP CRITICAL FREQUENCIES";N
90   CLS
100  FOR Y=1 TO N
110  INPUT "CRITICAL FREQUENCY IN HERTZ";FC(Y)
120  NEXT
130  INPUT "THE HIGHEST INPUT FREQUENCY";FH
140  INPUT "INCREMENTS OF FREQUENCY FROM ZERO TO THE
     HIGHEST";FI
150  CLS
160  PRINT"FREQUENCY","PHASE SHIFT","PHASE MARGIN","STABILITY"
170  FOR F=0 TO FH STEP FI
180  PH=0
190  FOR Y=1 TO N
```

```
200 PH=PH-ATN(F/FC(Y))*57.29578
210 NEXT Y
220 PM=180+PH
230 IF PM<=0 THEN S$="UNSTABLE" ELSE S$="STABLE"
240 PRINT F,PH,PM,S$
250 IF PM<=0 THEN END
260 NEXT F
```

SUMMARY OF STABILITY CRITERIA

This analysis has demonstrated that an amplifier's closed-loop gain must intersect the open-loop gain curve on a -20 dB/decade slope to ensure stability for all of its mid-range frequencies. If the closed-loop gain is lowered to a value that intersects on a -40 dB/decade slope, then marginal stability or complete instability can occur. In the previous analyses (Cases 1, 2, and 3), the closed-loop gain should be greater than 72 dB.

If the closed-loop gain intersects the open-loop response on a -60 dB/decade slope, definite instability will occur at some frequency within the amplifier's midrange. Therefore, to ensure stability for all of the midrange frequencies, an op-amp must be operated at a closed-loop gain such that the roll-off rate beginning at its dominant critical frequency does not exceed -20 dB/decade.

6–4 REVIEW QUESTIONS

1. Under what feedback condition can an amplifier oscillate?
2. How much can the phase shift of an amplifier's internal *RC* network be before instability occurs? What is the phase margin at the point where instability begins?
3. What is the maximum roll-off rate of the open-loop gain of an op-amp for which the device will still be stable?

6–5 OP-AMP COMPENSATION

The last section demonstrated that instability can occur when an op-amp's response has roll-off rates exceeding -20 dB/decade and the op-amp is operated in a closed-loop configuration having a gain curve that intersects a higher roll-off rate portion of the open-loop response. In situations like those examined in the last section, the closed-loop voltage gain is restricted to very high values. In many applications, lower values of closed-loop gain are necessary or desirable. To allow op-amps to be operated at low closed-loop gain, phase lag compensation is required.

PHASE LAG COMPENSATION

As you have seen, the cause of instability is excessive phase shift through an op-amp's internal lag networks. When these phase shifts equal or exceed 180°, the amplifier can oscillate. **Compensation** is used to either eliminate open-loop roll-off rates greater than -20 dB/decade or extend the -20 dB/decade rate to a lower gain. These concepts are illustrated in Figure 6–15.

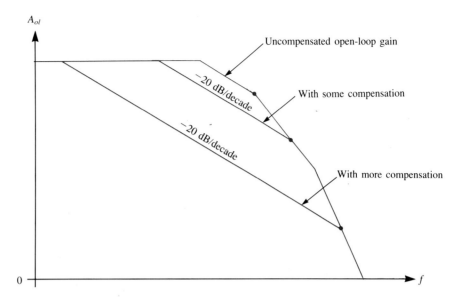

FIGURE 6–15
Bode plot illustrating effect of phase compensation on open-loop gain of typical op-amp.

COMPENSATING NETWORK

There are two basic methods of compensation for integrated circuit op-amps: internal and external. In either case an RC network is added. The basic compensating action is as follows. Consider first the RC network shown in Figure 6–16(a). At low frequencies where X_{C_c} is extremely large, the output voltage approximately equals the input voltage. When the frequency reaches its critical value, $f_c = 1/[2\pi(R_1 + R_2)C_c]$, the output voltage decreases at -20 dB/decade. This roll-off rate continues until $X_{C_c} \cong 0$, at which point the output voltage levels off to a value determined by R_1 and R_2, as indicated in Figure 6–16(b). This is the principle used in the phase compensation of an op-amp.

FIGURE 6–16
Basic compensating network action.

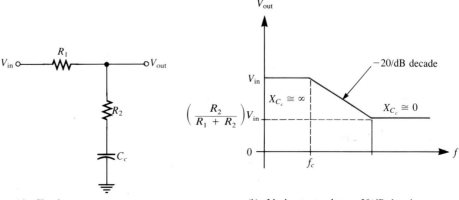

(a) Circuit

(b) Ideal output voltage -20/dB decade

CHAPTER 6

To see how a compensating network changes the open-loop response of an op-amp, refer to Figure 6–17. This diagram represents a two-stage op-amp. The individual stages are within the shaded blocks along with the associated lag networks. A compensating network is shown connected at point A on the output of stage 1.

The critical frequency of the compensating network is set to a value less than the dominant (lowest) critical frequency of the internal lag networks. This causes the

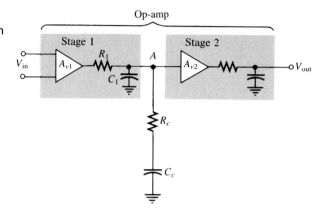

FIGURE 6–17
Representation of op-amp with compensation.

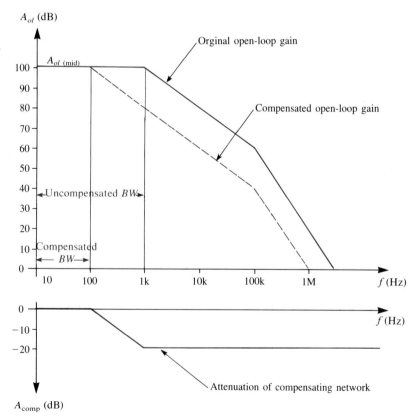

FIGURE 6–18
Example of compensated op-amp frequency response.

−20 dB/decade roll-off to begin at the compensating network's critical frequency. The roll-off of the compensating network continues up to the critical frequency of the dominant lag network. At this point, the response of the compensating network levels off, and the −20 dB/decade roll-off of the dominant lag network takes over. The net result is a shift of the open-loop response to the left, thus reducing the bandwidth, as shown in Figure 6–18. The response curve of the compensating network is shown in proper relation to the overall open-loop response.

■ **EXAMPLE 6–6**

A certain op-amp has the open-loop response in Figure 6–19. As you can see, the lowest closed-loop gain for which stability is assured is approximately 40 dB (where the closed-loop gain line still intersects the −20 dB/decade slope). In a particular application, a 20 dB closed-loop gain is required.

(a) Determine the critical frequency for the compensating network.
(b) Sketch the ideal response curve for the compensating network.
(c) Sketch the total ideal compensated open-loop response.

FIGURE 6–19
Original open-loop response.

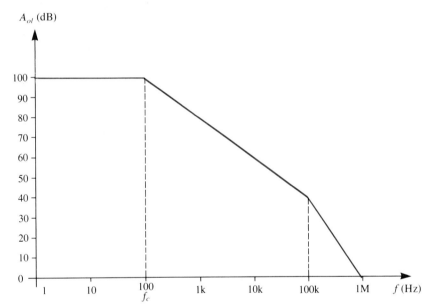

SOLUTION

(a) The gain must be dropped so that the −20 dB/decade roll-off extends down to 20 dB rather than to 40 dB. To achieve this, the midrange open-loop gain must be made to roll off a decade sooner. Therefore, the critical frequency of the compensating network must be 10 Hz.

(b) The roll-off of the compensating network must end at 100 Hz, as shown in Figure 6–20(a).

(c) The total open-loop response resulting from compensation is shown in Figure 6–20(b).

FIGURE 6-20

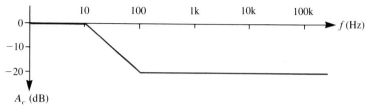

(a) Compensating network response

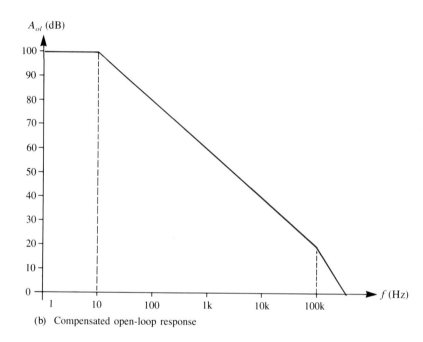

(b) Compensated open-loop response

PRACTICE EXERCISE 6-6

In this example, what is the uncompensated bandwidth? What is the compensated bandwidth?

EXTENT OF COMPENSATION

A larger compensating capacitor will cause the open-loop roll-off to begin at a lower frequency and thus extend the -20 dB/decade roll-off to lower gain levels, as shown in Figure 6-21(a). With a sufficiently large compensating capacitor, an op-amp can be made unconditionally stable, as illustrated in Figure 6-21(b), where the -20 dB/decade slope is extended all the way down to unity gain. This is normally the case when internal compensation is provided by the manufacturer. An internally, fully compensated op-amp can be used for any value of closed-loop gain and remain stable. The 741 is an example of an internally compensated device.

A disadvantage of fully compensated op-amps is that bandwidth is sacrificed; thus the slew rate is decreased. Therefore, many IC op-amps have provisions for external compensation. Figure 6-22 shows typical package layouts of an LM101A op-amp with

FIGURE 6–21
Extent of compensation.

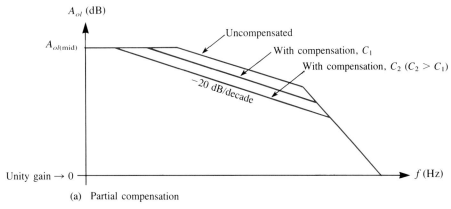
(a) Partial compensation

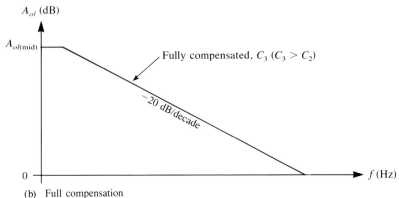
(b) Full compensation

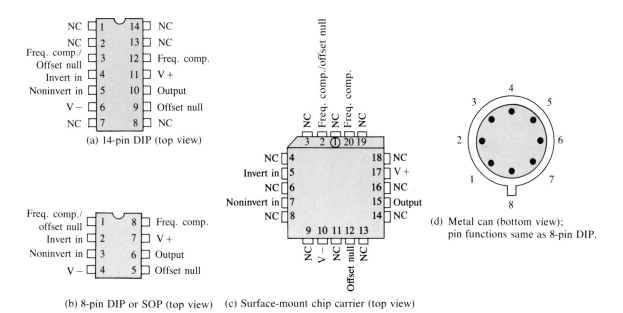

(a) 14-pin DIP (top view)

(b) 8-pin DIP or SOP (top view)

(c) Surface-mount chip carrier (top view)

(d) Metal can (bottom view); pin functions same as 8-pin DIP.

FIGURE 6–22
Typical op-amp packages.

pins available for external compensation with a small capacitor. With provisions for external connections, just enough compensation can be used for a given application without sacrificing more performance than necessary.

SINGLE-CAPACITOR COMPENSATION

As an example of compensating an IC op-amp, a capacitor C_1 is connected to pins 1 and 8 of an LM101A in an inverting amplifier configuration, as shown in Figure 6–23(a). Part (b) of the figure shows the open-loop frequency response curves for two values of C_1. The 3 pF compensating capacitor produces a unity-gain bandwidth approaching 10 MHz. Notice that the -20 dB/decade slope extends to a very low gain value. When C_1 is increased ten times to 30 pF, the bandwidth is reduced by a factor of ten. Notice that the -20 dB/decade slope now extends through unity gain.

When the op-amp is used in a closed-loop configuration, as in Figure 6–23(c), the useful frequency range depends on the compensating capacitor. For example, with a closed-loop gain of 40 dB as shown in part (c), the bandwidth is approximately 10 kHz for $C_1 = 30$ pF and increases to approximately 100 kHz when C_1 is decreased to 3 pF.

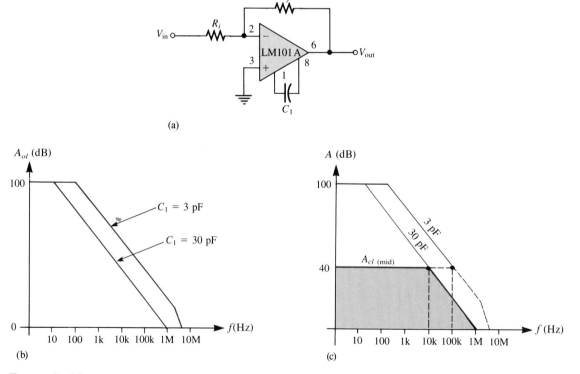

FIGURE 6–23

Example of single-capacitor compensation of an LM101A op-amp.

FEEDFORWARD COMPENSATION

Another method of phase compensation is called **feedforward.** This type of compensation results in less bandwidth reduction than the method previously discussed. The basic concept is to bypass the internal input stage of the op-amp at high frequencies and drive the higher-frequency second stage, as shown in Figure 6–24.

Feedforward compensation of an LM101A is shown in Figure 6–25(a). The feedforward capacitor C_1 is connected from the inverting input to the compensating terminal. A small capacitor is needed across R_f to ensure stability. The Bode plot in Figure 6–25(b) shows the feedforward compensated response and the standard compensated response that was discussed previously. The use of feedforward compensation is restricted to the inverting amplifier configuration. Other compensation methods are also used. Often, recommendations are provided by the manufacturer on the data sheet.

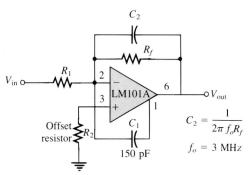

(a) Manufacturers' recommended configuration

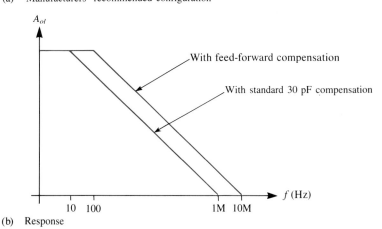

(b) Response

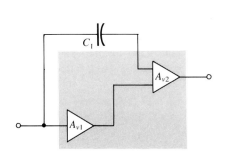

FIGURE 6–24
Feedforward compensation showing high-frequency bypassing of first stage.

FIGURE 6–25
Feedforward compensation of an LM101A op-amp and the response curves.

6-5 REVIEW QUESTIONS

1. What is the purpose of phase compensation?
2. What is the main difference between internal and external compensation?
3. When you compensate an amplifier, does the bandwidth increase or decrease?

6-6 A SYSTEM APPLICATION

In this system application, we are focusing on the audio amplifier boards in the FM stereo receiver presented at the beginning of the chapter. Both boards are identical except one is for the left channel sound and the other for the right channel sound. This circuit is a good example of a mixed use of an integrated circuit and discrete components. In this section, you will

☐ *See how an op-amp is used as an audio amplifier.*
☐ *Identify the functions of various components on the board.*
☐ *Analyze the circuit's operation.*
☐ *Translate between a printed circuit board and a schematic.*
☐ *Troubleshoot some common amplifier failures.*

A BRIEF DESCRIPTION OF THE SYSTEM

Some general information about the stereo system might be helpful before you concentrate on the audio amplifiers. When an FM stereo broadcast is received by a standard single-speaker system, the output to the speaker is equal to the sum of the left plus the right channel audio, so you get the original sound without separation. When a stereo receiver is used, the full stereo effect is reproduced by the two speakers. Stereo FM signals are transmitted on a carrier frequency of 88 MHz to 108 MHz. The complete stereo signal consists of three modulating signals. These are the sum of the left and right channel audio, the difference of the left and right channel audio, and a pilot subcarrier. These three signals are detected and are used to separate out the left and right channel audio by special circuits. The channel audio amplifiers then amplify each signal equally and drive the speakers. It is not necessary for you to understand this process for the purposes of this system application, although you may be interested in doing further study in this area on your own.

The two channel audio amplifiers are identical, so we will look at only one. The op-amp serves basically as a preamplifier that drives the power amplifier stage.

Op-Amp Responses

Now, so that you can take a closer look at one of the audio amplifier boards, let's take one out of the system and put it on the test bench.

On the Test Bench

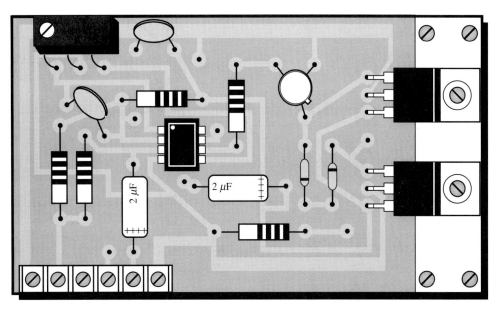

Figure 6–26

■ **Activity 1   Relate the PC Board to the Schematic**

The schematic for the audio amplifier board in Figure 6–26 is shown in Figure 6–27. Using this schematic, locate and label each component on the pc board. The board has several feed-through pads for connections that are on the back side, which you should locate and identify as you compare the board to the schematic.

Figure 6–27

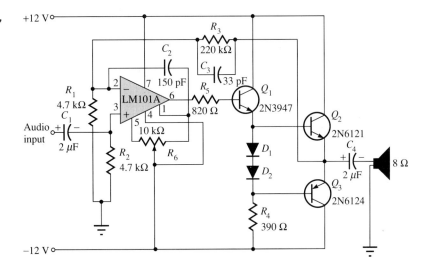

■ **ACTIVITY 2   ANALYZE THE CIRCUIT**

STEP 1   Determine the midrange voltage gain.

STEP 2   Determine the lower critical frequency. Given that the upper critical frequency is 15 kHz, what is the bandwidth?

STEP 3   Determine the maximum peak-to-peak input voltage that can be applied without producing a distorted output signal. Assume that the maximum output peaks are 1 V less than the supply voltages.

■ **ACTIVITY 3   WRITE A TECHNICAL REPORT**

Describe the overall operation of the circuit and the function of each component. In discussing the general operation and basic purpose of each component, make sure you identify the negative feedback loop, the type of op-amp configuration, which components determine the voltage gain, which components set the lower critical frequency, and the purpose of each of the capacitors. Use the results of Activity 2 when appropriate.

■ **ACTIVITY 4   TROUBLESHOOT THE AUDIO SECTION BOARDS FOR EACH OF THE FOLLOWING PROBLEMS BY STATING THE PROBABLE CAUSE OR CAUSES IN EACH CASE**

1. No final output signal when there is a verified input signal.
2. The positive half-cycle of the output voltage is severely distorted or missing.
3. Output severely clipped on both positive and negative cycles.

■ **ACTIVITY 5   TEST BENCH SPECIAL ASSIGNMENT**

Go to Test Bench 2 in the color insert section (which follows page 452) and carry out the assignment that is stated there.

6–6 REVIEW QUESTIONS

1. How can the lower critical frequency of the amplifier be reduced?
2. Which transistors form the class-B power amplifier?
3. What is the purpose of Q_1 and what type of circuit is it?
4. Calculate the power to the speaker for the maximum voltage output from Activity 2.

SUMMARY

□ *Open-loop gain* is the voltage gain of an op-amp without feedback.

□ *Closed-loop gain* is the voltage gain of an op-amp with negative feedback.

□ The closed-loop gain is always less than the open-loop gain.

□ The midrange gain of an op-amp extends down to dc.

□ The gain of an op-amp decreases as frequency increases above the critical frequency.

- The bandwidth of an op-amp equals the upper critical frequency.
- The internal *RC* lag networks that are inherently part of the amplifier stages cause the gain to roll off as frequency goes up.
- The internal *RC* lag networks also cause a phase shift between input and output signals.
- Negative feedback lowers the gain and increases the bandwidth.
- The product of gain and bandwidth is constant for a given op-amp.
- The gain-bandwidth product equals the frequency at which unity voltage gain occurs.
- Positive feedback occurs when the total phase shift through the op-amp (including 180° inversion) and feedback network is 0° (equivalent to 360°) or more.
- The phase margin is the amount of additional phase shift required to make the total phase shift around the loop 360°.
- When the closed-loop gain of an op-amp intersects the open-loop response curve on a -20 dB/decade (-6 dB/octave) slope, the amplifier is stable.
- When the closed-loop gain intersects the open-loop response curve on a slope greater than -20 dB/decade, the amplifier can be either marginally stable or unstable.
- A minimum phase margin of 45° is recommended to provide a sufficient safety factor for stable operation.
- A fully compensated op-amp has a -20 dB/decade roll-off all the way down to unity gain.
- Compensation reduces bandwidth and increases slew rate.
- Internally compensated op-amps such as the 741 are available. These are usually fully compensated with a large sacrifice in bandwidth.
- Externally compensated op-amps such as LM101A are available. External compensating networks can be connected to specified pins, and the compensation can be tailored to a specific application. In this way, bandwidth and slew rate are not degraded more than necessary.

Glossary

Closed-loop gain The overall voltage gain with external feedback.

Compensation The process of modifying the roll-off rate of an amplifier to ensure stability.

Feedforward A method of frequency compensation in op-amp circuits.

Open-loop gain The voltage gain of an op-amp without feedback.

Phase The relative angular displacement of a time-varying function relative to a reference.

Phase margin The difference between the total phase shift through an amplifier and 180°. The additional amount of phase shift that can be allowed before instability occurs.

Stability A condition in which an amplifier circuit does not oscillate.

FORMULAS

(6–1)	$BW = f_{c(high)} - f_{c(low)}$	General bandwidth
(6–2)	$BW = f_{c(high)}$	Op-amp bandwidth
(6–3)	$\dfrac{V_{out}}{V_{in}} = \dfrac{X_C}{\sqrt{R^2 + X_C^2}}$	RC attenuation, lag network
(6–4)	$\dfrac{V_{out}}{V_{in}} = \dfrac{1}{\sqrt{1 + R^2/X_C^2}}$	RC attenuation, lag network
(6–5)	$\dfrac{f_c}{f} = \dfrac{X_C}{R}$	Ratio of frequencies equals ratio of reactances
(6–6)	$\dfrac{V_{out}}{V_{in}} = \dfrac{1}{\sqrt{1 + f^2/f_c^2}}$	RC attenuation
(6–7)	$A_{ol} = \dfrac{A_{ol(mid)}}{\sqrt{1 + f^2/f_c^2}}$	Open-loop gain
(6–8)	$\phi = -\tan^{-1}\left(\dfrac{R}{X_C}\right)$	RC phase shift
(6–9)	$\phi = -\tan^{-1}\left(\dfrac{f}{f_c}\right)$	RC phase shift
(6–10)	$\phi_{tot} = -\tan^{-1}\left(\dfrac{f}{f_{c1}}\right) - \tan^{-1}\left(\dfrac{f}{f_{c2}}\right) - \tan^{-1}\left(\dfrac{f}{f_{c3}}\right)$	Total phase shift
(6–11)	$f_{c(cl)} = f_{c(ol)}(1 + BA_{ol(mid)})$	Closed-loop critical frequency
(6–12)	$BW_{cl} = BW_{ol}(1 + BA_{ol(mid)})$	Closed-loop bandwidth
(6–13)	$A_{cl}f_{c(cl)} = A_{ol}f_{c(ol)}$	Gain-bandwidth product
(6–14)	$A_{cl}f_{c(cl)} =$ unity-gain bandwidth	
(6–15)	Loop gain $= A_{ol}B$	
(6–16)	$\theta_{pm} = 180° - |\phi_{tot}|$	Phase margin

Self-Test

1. The open-loop gain of an op-amp is always
   (a) less than the closed-loop gain
   (b) equal to the closed-loop gain
   (c) greater than the closed-loop gain
   (d) a very stable and constant quantity for a given type of op-amp
2. The bandwidth of an ac amplifier having a lower critical frequency of 1 kHz and an upper critical frequency of 10 kHz is
   (a) 1 kHz (b) 9 kHz (c) 10 kHz (d) 11 kHz
3. The bandwidth of a dc amplifier having an upper critical frequency of 100 kHz is
   (a) 100 kHz (b) unknown (c) infinity (d) 0 kHz
4. The midrange open-loop gain of an op-amp
   (a) extends from the lower critical frequency to the upper critical frequency
   (b) extends from 0 Hz to the upper critical frequency
   (c) rolls off at 20 dB/decade beginning at 0 Hz
   (d) b and c
5. The frequency at which the open-loop gain is equal to one is called
   (a) the upper critical frequency (b) the cutoff frequency
   (c) the notch frequency (d) the unity-gain frequency
6. Phase shift through an op-amp is caused by
   (a) the internal RC networks (b) the external RC networks
   (c) the gain roll off (d) negative feedback
7. Each RC network in an op-amp
   (a) causes the gain to roll off at -6 dB/octave
   (b) causes the gain to roll off at -20 dB/decade
   (c) reduces the midrange gain by 3 dB
   (d) a and b
8. When negative feedback is used, the gain-bandwidth product of an op-amp
   (a) increases (b) decreases (c) stays the same (d) fluctuates
9. If a certain op-amp has a midrange open-loop gain of 200,000 and a unity-gain frequency of 5 MHz, the gain-bandwidth product is
   (a) 200,000 Hz (b) 5,000,000 Hz
   (c) 1×10^{12} Hz (d) not determinable from the information
10. If a certain op-amp has a closed-loop gain of 20 and an upper critical frequency of 10 MHz, the gain-bandwidth product is
    (a) 200 MHz (b) 10 MHz (c) the unity-gain frequency

11. Positive feedback occurs when
    (a) the output signal is fed back to the input in-phase with the input signal
    (b) the output signal is fed back to the input out-of-phase with the input signal
    (c) the total phase shift through the op-amp and feedback circuit is 360 degrees
    (d) a and c

12. For a closed-loop op-amp circuit to be unstable
    (a) there must be positive feedback
    (b) the loop gain must be greater than one
    (c) the loop gain must be less than one
    (d) a and b

13. The amount of additional phase shift required to make the total phase shift around a closed loop equal to zero is called
    (a) the unity-gain phase shift   (b) phase margin
    (c) phase lag   (d) phase bandwidth

14. For a given value of closed-loop gain, a positive phase margin indicates
    (a) an unstable condition   (b) too much phase shift
    (c) a stable condition   (d) nothing

15. The purpose of phase-lag compensation is to
    (a) make the op-amp stable at very high values of gain
    (b) make the op-amp stable at low values of gain
    (c) reduce the unity-gain frequency
    (d) increase the bandwidth

PROBLEMS

SECTION 6–1 BASIC CONCEPTS

1. The midrange open-loop gain of a certain op-amp is 120 dB. Negative feedback reduces this gain by 50 dB. What is the closed-loop gain?

2. The upper critical frequency of an op-amp's open-loop response is 200 Hz. If the midrange gain is 175,000, what is the ideal gain at 200 Hz? What is the actual gain? What is the op-amp's open-loop bandwidth?

3. An *RC* lag network has a critical frequency of 5 kHz. If the resistance value is 1 kΩ, what is X_C when $f = 3$ kHz?

4. Determine the attenuation of an *RC* lag network with $f_c = 12$ kHz for each of the following frequencies.
    (a) 1 kHz   (b) 5 kHz   (c) 12 kHz
    (d) 20 kHz   (e) 100 kHz

5. The midrange open-loop gain of a certain op-amp is 80,000. if the open-loop critical frequency is 1 kHz, what is the open-loop gain at each of the following frequencies?

   (a) 100 Hz   (b) 1 kHz   (c) 10 kHz   (d) 1 MHz

6. Determine the phase shift through each network in Figure 6–28 at a frequency of 2 kHz.

FIGURE 6–28

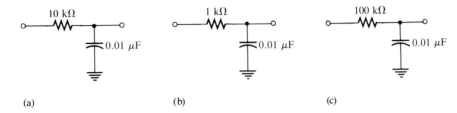

(a)   (b)   (c)

7. An RC lag network has a critical frequency of 8.5 kHz. Determine the phase for each frequency and plot a graph of its phase angle versus frequency.

   (a) 100 Hz   (b) 400 Hz   (c) 850 Hz
   (d) 8.5 kHz (e) 25 kHz   (f) 85 kHz

SECTION 6–2 OP-AMP OPEN-LOOP RESPONSE

8. A certain op-amp has three internal amplifier stages with midrange gains of 30 dB, 40 dB, and 20 dB. Each stage also has a critical frequency associated with it as follows: $f_{c1} = 600$ Hz, $f_{c2} = 50$ kHz, and $f_{c3} = 200$ kHz.

   (a) What is the midrange open-loop gain of the op-amp, expressed in dB?

   (b) What is the total phase shift through the amplifier, including inversion, when the signal frequency is 10 kHz?

9. What is the gain roll-off rate in Problem 8 between the following frequencies?

   (a) 0 Hz and 600 Hz   (b) 600 Hz and 50 kHz
   (c) 50 kHz and 200 kHz   (d) 200 kHz and 1 MHz

SECTION 6–3 OP-AMP CLOSED-LOOP RESPONSE

10. Determine the midrange gain in dB of each amplifier in Figure 6–29. Are these open- or closed-loop gains?

FIGURE 6–29

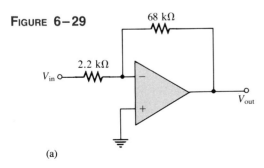

(a)

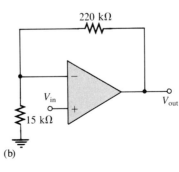

(b)

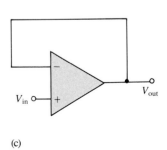

(c)

11. A certain amplifier has an open-loop gain in midrange of 180,000 and an open-loop critical frequency of 1500 Hz. If the attenuation of the feedback path is 0.015, what is the closed-loop bandwidth?
12. Given that $f_{c(ol)} = 750$ Hz, $A_{ol} = 89$ dB, and $f_{c(cl)} = 5.5$ kHz, determine the closed-loop gain in dB.
13. What is the unity-gain bandwidth in Problem 12?
14. For each amplifier in Figure 6–30, determine the closed-loop gain and bandwidth. The op-amps in each circuit exhibit an open-loop gain of 125 dB and a unity-gain bandwidth of 2.8 MHz.

FIGURE 6–30

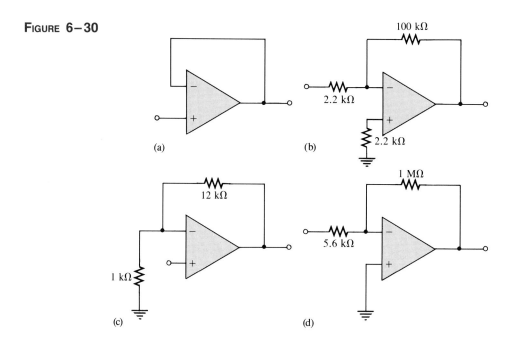

15. Which of the amplifiers in Figure 6–31 has the smaller bandwidth?

FIGURE 6–31

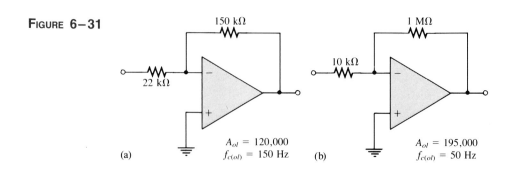

Section 6–4 Positive Feedback and Stability

16. It has been determined that the op-amp circuit in Figure 6–32 has three internal critical frequencies as follows: 1.2 kHz, 50 kHz, 250 kHz. If the midrange open-loop gain is 100 dB, is the amplifier configuration stable, marginally stable, or unstable?

Figure 6–32

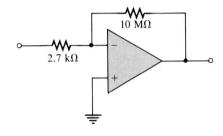

17. Determine the phase margin for each value of phase lag.
    (a) 30°   (b) 60°   (c) 120°   (d) 180°   (e) 210°

18. A certain op-amp has the following internal critical frequencies in its open-loop response: 125 Hz, 25 kHz, and 180 kHz. What is the total phase shift through the amplifier when the signal frequency is 50 kHz?

19. Each graph in Figure 6–33 shows both the open-loop and the closed-loop response of a particular op-amp configuration. Analyze each case for stability.

Figure 6–33

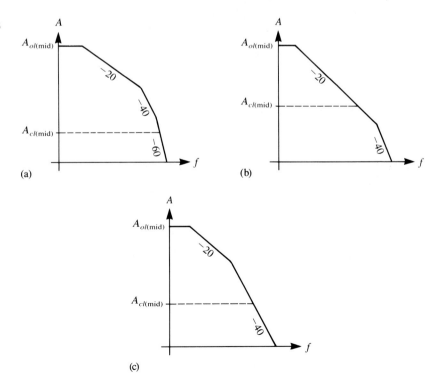

SECTION 6–5 OP-AMP COMPENSATION

20. A certain operational amplifier has an open-loop response curve as shown in Figure 6–34. A particular application requires a 30 dB closed-loop midrange gain. In order to achieve a 30 dB gain, compensation must be added because the 30 dB line intersects the uncompensated open-loop gain on the −40 dB/decade slope and, therefore, stability is not assured.

(a) Find the critical frequency of the compensating network such that the −20 dB/decade slope is lowered to a point where it intersects the 30 dB gain line.

(b) Sketch the ideal response curve for the compensating network.

(c) Sketch the total ideal compensated open-loop response.

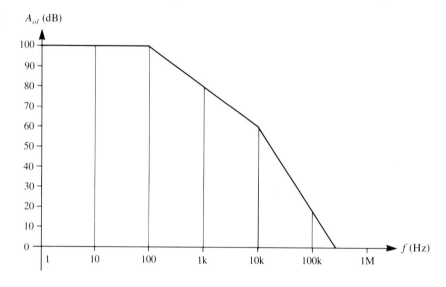

FIGURE 6–34

21. The open-loop gain of a certain op-amp rolls off at −20 dB/decade, beginning at $f = 250$ Hz. This roll-off rate extends down to a gain of 60 dB. If a 40 dB closed-loop gain is required, what is the critical frequency for the compensating network?

22. Repeat Problem 21 for a closed-loop gain of 20 dB.

ANSWERS TO REVIEW QUESTIONS

SECTION 6–1

1. Open-loop gain is without feedback, and closed-loop gain is with negative feedback. Open-loop gain is larger.
2. 100 Hz
3. Decrease

Section 6-2
1. 20 dB + 30 dB = 50 dB
2. $-49° + (-5.2°) = -54.2°$

Section 6-3
1. Yes
2. $BW = 3{,}000 \text{ kHz}/60 = 50 \text{ kHz}$
3. $3{,}000 \text{ kHz}/1 = 3 \text{ MHz}$

Section 6-4
1. Positive feedback
2. 180°, 0°
3. −20 dB/decade (−6 dB/octave)

Section 6-5
1. Phase compensation increases the phase margin at a given frequency.
2. Internal compensation is full compensation; external compensation can be tailored to maximize bandwidth.
3. Bandwidth decreases.

Section 6-6
1. f_{cl} can be reduced by increasing C_1 or R_2.
2. Q_2 and Q_3
3. Q_1 is an emitter-follower buffer stage
4. $P_{\text{avg}} = \dfrac{V_{\text{out(rms)}}^2}{R_L} \cong 4 \text{ W}$

Answers to Practice Exercises

6–1 (a) 79,996 (b) 79,900 (c) 6380
6–2 173.2 Hz
6–3 75 dB, −71.57°
6–4 2 MHz
6–5 (a) 29.56 kHz (b) 42.55 kHz
6–6 100 Hz, 10 Hz

7
BASIC OP-AMP CIRCUITS

7–1 COMPARATORS
7–2 SUMMING AMPLIFIERS
7–3 THE INTEGRATOR AND DIFFERENTIATOR
7–4 MORE OP-AMP CIRCUITS
7–5 TROUBLESHOOTING
7–6 A SYSTEM APPLICATION

After completing this chapter, you should be able to

☐ Use an op-amp as a comparator.
☐ Describe how hysteresis can be implemented in a comparator circuit and explain its purpose.
☐ Explain how bounded comparators work.
☐ Describe a basic window comparator.
☐ Show how comparators are used in a certain type of analog-to-digital converter and in an over-temperature sensing circuit.
☐ Use an op-amp as a summing amplifier.
☐ Explain how an averaging amplifier works.
☐ Explain how a scaling adder works.
☐ Show how a scaling adder can be used in a certain type of digital-to-analog converter.
☐ Discuss the operation of an op-amp integrator and a differentiator.
☐ Show how an op-amp can be used as a constant-current source.
☐ Show how an op-amp can be used as a current-to-voltage or voltage-to-current converter.
☐ Explain how a basic op-amp peak detector circuit works.
☐ Troubleshoot common op-amp circuit failures.

In the last two chapters, you learned about the principles, operation, and characteristics of the operational amplifier. Op-amps are used in such a wide variety of applications that it is impossible to cover all of them in one chapter, or even in one book. Therefore, in this chapter, we will examine some of the more fundamental applications to illustrate how versatile the op-amp is and to give you a foundation in basic op-amp circuits.

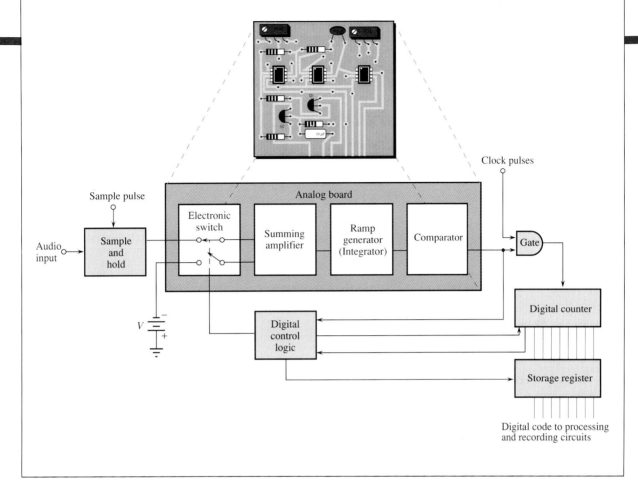

A SYSTEM APPLICATION

This system application illustrates a very interesting application of three types of op-amp circuits that will be studied in this chapter—the summing amplifier, integrator, and comparator. The system diagram above shows one basic type of analog-to-digital converter that takes an audio input, such as voice or music, and converts it to binary codes that can be recorded digitally. Analog-to-digital converters are covered thoroughly in Chapter 13.

Op-amps play a key role in this system, and we will be focusing on the analog board to see how these circuits are used in a representative application. The digital circuits are discussed just enough to allow you to understand what the overall system does. You do not need to have a background in digital circuits for our purposes here. However, this particular system application points out the fact, again, that many systems that you will be working with in industry will include combinations of both analog (linear) and digital circuits. You will, of course, get a thorough grounding in digital fundamentals in another course, if you haven't already.

For the system application in Section 7–6, in addition to the other topics, be sure you understand

☐ How a summing amplifier works.
☐ How an integrator works.
☐ How a comparator works.

7-1 COMPARATORS

Operational amplifiers are often used as nonlinear devices to compare the amplitude of one voltage with another. In this application, the op-amp is used in the open-loop configuration, with the input voltage on one input and a reference voltage on the other.

ZERO-LEVEL DETECTION

A basic application of the op-amp as a **comparator** is in determining when an input voltage exceeds a certain level. Figure 7–1(a) shows a zero-level detector. Notice that the inverting (−) input is grounded and that the input signal voltage is applied to the noninverting (+) input. Because of the high open-loop voltage gain, a very small difference voltage between the two inputs drives the amplifier into saturation, causing the output voltage to go to its limit. For example, consider an op-amp having $A_{ol} = 100,000$. A voltage difference of only 0.25 mV between the inputs could produce an output voltage of (0.25 mV)(100,000) = 25 V if the op-amp were capable. However, since most op-amps have output voltage limitations of less than ±15 V, the device would be driven into saturation.

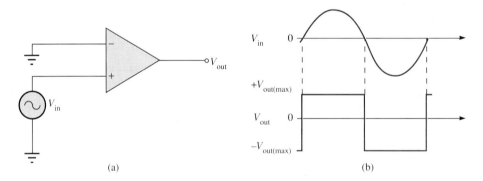

FIGURE 7–1
The op-amp as a zero-level detector.

Figure 7–1(b) shows the result of a sine wave input voltage applied to the noninverting input of the zero-level detector. When the sine wave is negative, the output is at its maximum negative level. When the sine wave crosses 0, the amplifier is driven to its opposite state and the output goes to its maximum positive level, as shown. As you can see, the zero-level detector can be used as a squaring circuit to produce a square wave from a sine wave.

Nonzero-Level Detection

The zero-level detector in Figure 7–1 can be modified to detect voltages other than zero by connecting a fixed reference voltage, as shown in Figure 7–2(a). A more practical arrangement is shown in Figure 7–2(b) using a voltage divider to set the reference voltage as follows:

$$V_{REF} = \frac{R_2}{R_1 + R_2}(+V) \tag{7-1}$$

where $+V$ is the positive op-amp supply voltage. The circuit in Figure 7–2(c) uses a zener diode to set the reference voltage ($V_{REF} = V_Z$). As long as the input voltage V_{in} is less than

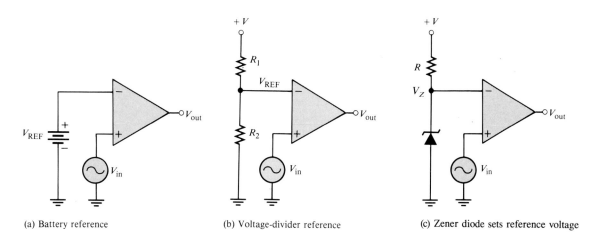

(a) Battery reference (b) Voltage-divider reference (c) Zener diode sets reference voltage

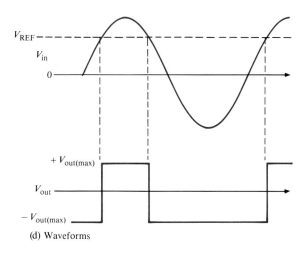

(d) Waveforms

FIGURE 7–2
Nonzero-level detectors.

V_{REF}, the output remains at the maximum negative level. When the input voltage exceeds the reference voltage, the output goes to its maximum positive state, as shown in Figure 7–2(d) with a sine wave input voltage.

■ **EXAMPLE 7–1** The input signal in Figure 7–3(a) is applied to the comparator circuit in Figure 7–3(b). Make a sketch of the output showing its proper relationship to the input signal. Assume the maximum output levels are ±12 V.

FIGURE 7–3

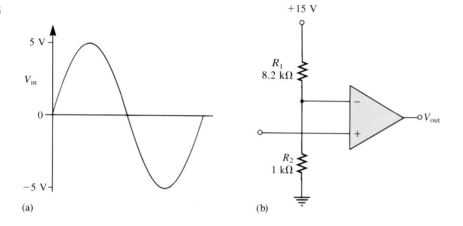

SOLUTION
The reference voltage is set by R_1 and R_2 as follows:

$$V_{REF} = \frac{R_2}{R_1 + R_2}(+V) = \frac{1 \text{ k}\Omega}{8.2 \text{ k}\Omega + 1 \text{ k}\Omega}(+15 \text{ V}) = 1.63 \text{ V}$$

Each time the input exceeds +1.63 V, the output voltage switches to its +12 V level. Each time the input goes below +1.63 V, the output switches back to its −12 V level, as shown in Figure 7–4.

FIGURE 7–4

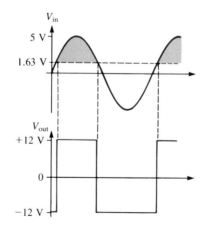

Practice Exercise 7–1
Determine the reference voltage in Figure 7–3 if $R_1 = 22$ kΩ and $R_2 = 3.3$ kΩ.

Effects of Input Noise on Comparator Operation

In many practical situations, unwanted voltage fluctuations (**noise**) appear on the input line. This noise voltage becomes superimposed on the input voltage, as shown in Figure 7–5, and can cause a comparator to erratically switch output states.

Figure 7–5
Sine wave with superimposed noise.

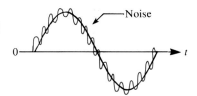

In order to understand the potential effects of noise voltage, consider a low-frequency sine wave applied to the input of an op-amp comparator used as a zero-level detector, as shown in Figure 7–6(a). Part (b) of the figure shows the input sine wave plus noise and the resulting output. As you can see, when the sine wave approaches 0, the fluctuations due to noise cause the total input to vary above and below 0 several times, thus producing an erratic output.

Figure 7–6
Effects of noise on comparator circuit.

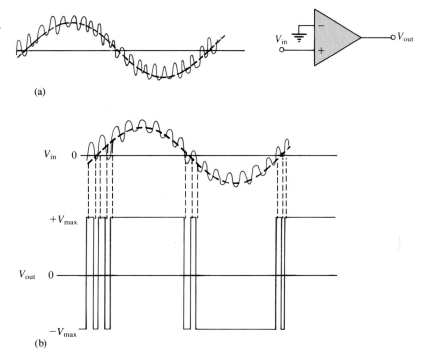

Reducing Noise Effects with Hysteresis

An erratic output voltage caused by noise on the input occurs because the op-amp comparator switches from its negative output state to its positive output state at the same input voltage level that causes it to switch in the opposite direction, from positive to negative. This unstable condition occurs when the input voltage hovers around the reference voltage, and any small noise fluctuations cause the comparator to switch first one way and then the other.

In order to make the comparator less sensitive to noise, a technique incorporating positive feedback, called **hysteresis,** is often employed. Basically, hysteresis means that there is a higher reference level when the input voltage goes from a lower to higher value than when it goes from a higher to a lower value. A good example of hysteresis is a common household thermostat that turns the furnace on at one temperature and off at another.

The two reference levels are referred to as the upper trigger point (UTP) and the lower trigger point (LTP). This two-level hysteresis is established with a positive feedback arrangement, as shown in Figure 7–7. Notice that the noninverting (+) input is connected to a resistive voltage divider such that a portion of the output voltage is fed back to the input. The input signal is applied to the inverting input (−) in this case.

FIGURE 7–7

Comparator with positive feedback for hysteresis.

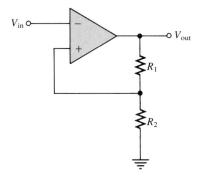

The basic operation of the comparator with hysteresis is as follows and is illustrated in Figure 7–8. Assume that the output voltage is at its positive maximum, $+V_{out(max)}$. The voltage fed back to the noninverting input is V_{UTP} and is expressed as

$$V_{UTP} = \frac{R_2}{R_1 + R_2}[+V_{out(max)}] \qquad (7-2)$$

When the input voltage V_{in} exceeds V_{UTP}, the output voltage drops to its negative maximum, $-V_{out(max)}$. Now the voltage fed back to the noninverting input is V_{LTP} and is expressed as

$$V_{LTP} = \frac{R_2}{R_1 + R_2}[-V_{out(max)}] \qquad (7-3)$$

Basic Op-Amp Circuits

The input voltage must now fall below V_{LTP} before the device will switch back to its other state. This means that a small amount of noise voltage has no effect on the output, as illustrated by Figure 7–8.

A comparator with hysteresis is sometimes known as a **Schmitt trigger.** The amount of hysteresis is defined by the difference of the two trigger levels.

$$V_{HYS} = V_{UTP} - V_{LTP} \qquad (7\text{–}4)$$

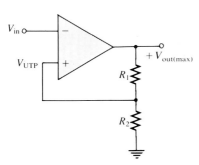

(a) Output in maximum positive state

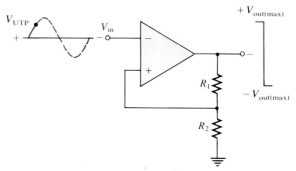

(b) Input exceeds UTP; output switches to maximum negative state

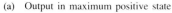

(c) Output in maximum negative state

(d) Input goes below LTP; output switches back to positive state

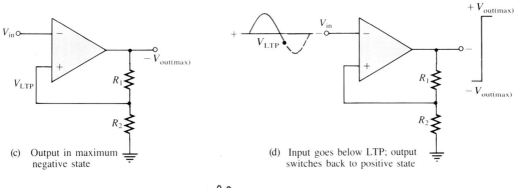

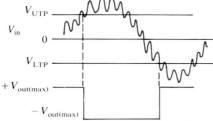

(e) Device triggers only once when UTP or LTP is reached; thus there is immunity to noise that is riding on the input signal.

FIGURE 7–8
Operation of a comparator with hysteresis.

EXAMPLE 7–2

Determine the upper and lower trigger points for the comparator circuit in Figure 7–9. Assume that $+V_{out(max)} = +5$ V and $-V_{out(max)} = -5$ V.

FIGURE 7–9

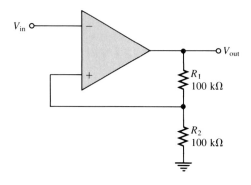

SOLUTION

$$V_{UTP} = \frac{R_2}{R_1 + R_2}[+V_{out(max)}] = 0.5(5 \text{ V}) = +2.5 \text{ V}$$

$$V_{LTP} = \frac{R_2}{R_1 + R_2}[-V_{out(max)}] = 0.5(-5 \text{ V}) = -2.5 \text{ V}$$

PRACTICE EXERCISE 7–2

Determine the upper and lower trigger points in Figure 7–9 for $R_1 = 68$ kΩ and $R_2 = 82$ kΩ. The maximum output levels are ± 7 V.

OUTPUT BOUNDING

In some applications, it is necessary to limit the output voltage levels of a comparator to a value less than that provided by the saturated op-amp. A single zener diode can be used as shown in Figure 7–10 to limit the output voltage swing to the zener voltage in one direction and to the forward diode drop in the other. This process of limiting the output range is called **bounding**.

FIGURE 7–10
Comparator with output bounding.

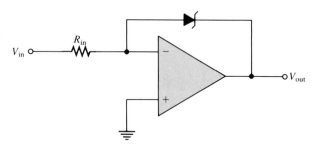

The operation is as follows. Since the anode of the zener is connected to the inverting input, it is at virtual ground ($\cong 0$ V). Therefore, when the output reaches a positive value equal to the zener voltage, it limits at that value, as illustrated in Figure 7–11. When the output switches negative, the zener acts as a regular diode and becomes forward-biased at 0.7 V, limiting the negative output swing to this value, as shown. Turning the zener around limits the output in the opposite direction.

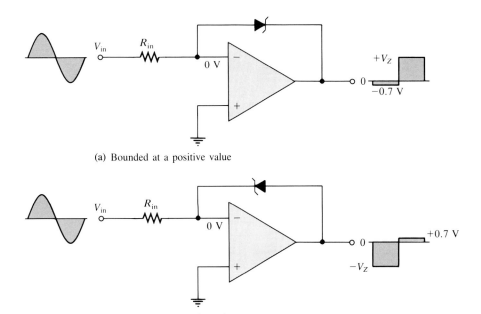

FIGURE 7–11
Operation of a bounded comparator.

Two zener diodes arranged as in Figure 7–12 limit the output voltage to the zener voltage plus the voltage drop (0.7 V) of the forward-biased zener, both positively and negatively, as shown in Figure 7–12. Op-amp comparators are available in IC form and are optimized for the comparison operation. Typical of these is the LM307.

FIGURE 7–12
Double-bounded comparator.

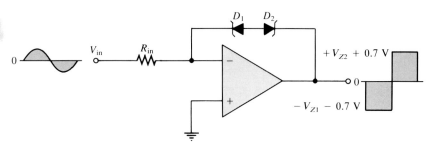

EXAMPLE 7-3

Determine the output voltage waveform for Figure 7–13.

FIGURE 7–13

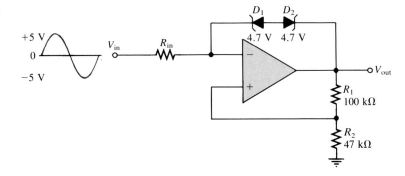

SOLUTION
This comparator has both hysteresis and zener bounding.

The voltage across D_1 and D_2 in either direction is 4.7 V + 0.7 V = 5.4 V. This is because one zener is always forward-biased with a drop of 0.7 V when the other one is in breakdown.

The voltage at the inverting (−) op-amp input is

$$V_{out} \pm 5.4 \text{ V}$$

Since the differential voltage is negligible, the voltage at the noninverting (+) op-amp input is also $V_{out} \pm 5.4$ V. Thus,

$$V_{R1} = V_{out} - (V_{out} \pm 5.4 \text{ V}) = \pm 5.4 \text{ V}$$

$$I_{R1} = \frac{V_{R1}}{R_1} = \frac{\pm 5.4 \text{ V}}{100 \text{ k}\Omega} = \pm 54 \text{ }\mu\text{A}$$

Since the current into the noninverting input is negligible,

$$I_{R2} = I_{R1} = \pm 54 \text{ }\mu\text{A}$$
$$V_{R2} = (47 \text{ k}\Omega)(\pm 54 \text{ }\mu\text{A}) = \pm 2.54 \text{ V}$$
$$V_{out} = V_{R1} + V_{R2} = \pm 5.4 \text{ V} \pm 2.54 \text{ V} = \pm 7.94 \text{ V}$$

The upper trigger point (UTP) and the lower trigger point (LTP) are as follows.

$$V_{UTP} = \left(\frac{R_2}{R_1 + R_2}\right)(+V_{out}) = \left(\frac{47 \text{ k}\Omega}{147 \text{ k}\Omega}\right)(+7.94 \text{ V}) = +2.54 \text{ V}$$

$$V_{LTP} = \left(\frac{R_2}{R_1 + R_2}\right)(-V_{out}) = \left(\frac{47 \text{ k}\Omega}{147 \text{ k}\Omega}\right)(-7.94 \text{ V}) = -2.54 \text{ V}$$

The output waveform for the given input voltage is shown in Figure 7–14.

BASIC OP-AMP CIRCUITS

FIGURE 7-14

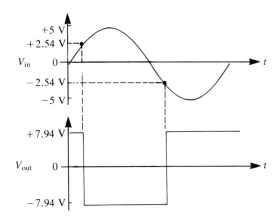

PRACTICE EXERCISE 7-3
Determine the upper and lower trigger points for Figure 7-13 if $R_1 = 150$ kΩ, $R_2 = 68$ kΩ, and the zener diodes are 3.3 V devices.

WINDOW COMPARATOR

Two individual op-amp comparators arranged as in Figure 7-15 form what is known as a *window comparator*. This circuit detects when an input voltage is between two limits, an upper and a lower, called the "window."

FIGURE 7-15
A basic window comparator.

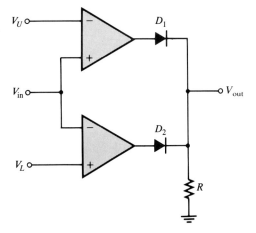

The upper and lower limits are set by reference voltages designated V_U and V_L. These voltages can be established with voltage dividers, zener diodes, or any type of voltage source. As long as V_{in} is within the window (less than V_U and greater than V_L), the output of each comparator is at its low saturated level. Under this condition, both diodes are reverse-biased and V_{OUT} is held at zero by the resistor to ground. When V_{in} goes above V_U or below V_L, the output of the associated comparator goes to its high saturated level. This action forward-biases the diode and produces a high-level V_{OUT}. This is illustrated in Figure 7-16 with V_{in} changing arbitrarily.

FIGURE 7–16
Example of window comparator operation.

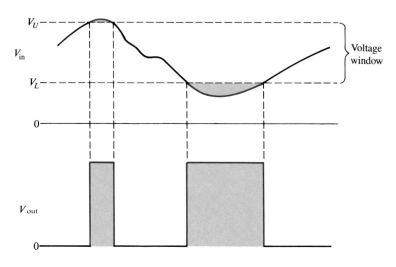

A COMPARATOR APPLICATION: OVER-TEMPERATURE SENSING CIRCUIT

Figure 7–17 shows an op-amp comparator used in a precision over-temperature sensing circuit. The circuit consists of a Wheatstone bridge with the op-amp used to detect when the bridge is balanced. One leg of the bridge contains a thermistor (R_1), which is a temperature-sensing resistor with a negative temperature coefficient (its resistance decreases as temperature increases). The potentiometer (R_2) is set at a value equal to the resistance of the thermistor at the critical temperature. At normal temperatures (below critical), R_1 is greater than R_2, thus creating an unbalanced condition that drives the op-amp to its low saturated output level and keeps transistor Q_1 off.

As the temperature increases, the resistance of the thermistor decreases. When the temperature reaches the critical value, R_1 becomes equal to R_2, and the bridge becomes

FIGURE 7–17
An over-temperature sensing circuit.

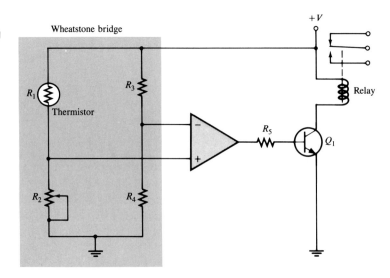

BASIC OP-AMP CIRCUITS

balanced (since $R_3 = R_4$). At this point the op-amp switches to its high saturated output level, turning Q_1 on. This energizes the relay, which can be used to activate an alarm or initiate an appropriate response to the over-temperature condition.

A COMPARATOR APPLICATION: ANALOG-TO-DIGITAL (A/D) CONVERSION

A/D conversion is a common interfacing process often used when a linear **analog** system must provide inputs to a **digital** system. Many methods for A/D conversion are available and some of these will be covered thoroughly in Chapter 13. However, in this discussion, only one type is used to demonstrate the concept.

The *simultaneous*, or *flash*, method of A/D conversion uses parallel comparators as shown in Figure 7–18 to compare the linear input signal with various reference voltages developed by a voltage divider. When the input voltage exceeds the reference voltage for

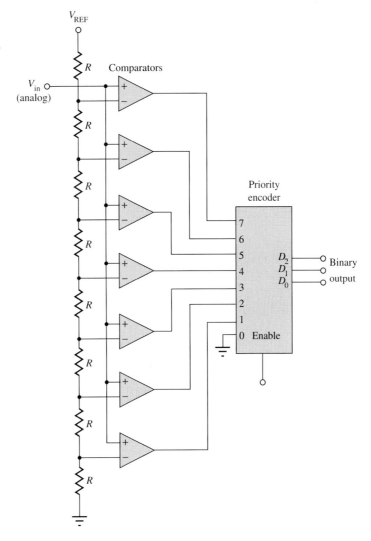

FIGURE 7–18
A simultaneous A/D converter using op-amps as comparators.

a given comparator, a high level is produced on that comparator's output. Figure 7–18 shows a converter that produces three-digit binary numbers on its output, which represent the values of the analog input voltage as it changes. This converter requires seven comparators. In general, $2^n - 1$ comparators are required for conversion to an n-digit binary number. The large number of comparators necessary for a reasonably sized binary number is one of the drawbacks of this type of A/D converter. Its chief advantage is that it provides a fast conversion time.

The reference voltage for each comparator is set by the resistive voltage-divider network and V_{REF}. The output of each comparator is connected to an input of the priority encoder. The *priority encoder* is a digital device that produces a binary number representing the highest value input.

The encoder *samples* its input when a pulse occurs on the enable line (sampling pulse), and a three-digit binary number proportional to the value of the analog input signal appears on the encoder's outputs. The sampling rate determines the accuracy with which the sequence of binary numbers represents the changing input signal.

7–1 REVIEW QUESTIONS

1. What is the reference voltage for each comparator in Figure 7–19?
2. What is the purpose of hysteresis in a comparator?
3. Define the term *bounding* in relation to a comparator's output.

FIGURE 7–19

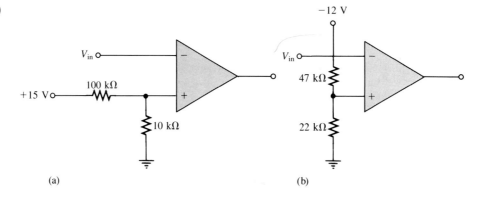

(a) (b)

7–2 SUMMING AMPLIFIERS

The summing amplifier is a variation of the inverting op-amp configuration covered in Chapter 5. The summing amplifier has two or more inputs, and its output voltage is proportional to the negative of the algebraic sum of its input voltages. In this section, you will see how a summing amplifier works, and you will learn about the averaging amplifier and the scaling amplifier, which are variations of the basic summing amplifier.

Basic Op-Amp Circuits

A two-input summing amplifier is shown in Figure 7-20, but any number of inputs can be used. The operation of the circuit and derivation of the output expression are as follows. Two voltages, V_{IN1} and V_{IN2}, are applied to the inputs and produce currents I_1 and I_2, as shown. Using the concepts of infinite input impedance and virtual ground, you can see that the inverting input of the op-amp is approximately 0 V, and there is no current from the input. This means that both input currents I_1 and I_2 combine at this summing point and form the total current, which is through R_f, as indicated.

$$I_T = I_1 + I_2$$

Since $V_{OUT} = -I_T R_f$, the following steps apply.

$$V_{OUT} = -(I_1 + I_2)R_f$$

$$= -\left(\frac{V_{IN1}}{R_1} + \frac{V_{IN2}}{R_2}\right)R_f$$

If all three of the resistors are equal to the same value R ($R_1 = R_2 = R_f = R$), then

$$V_{OUT} = -\left(\frac{V_{IN1}}{R} + \frac{V_{IN2}}{R}\right)R$$

$$V_{OUT} = -(V_{IN1} + V_{IN2}) \tag{7-5}$$

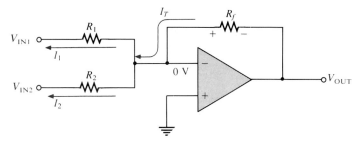

FIGURE 7-20
Two-input inverting summing amplifier.

Equation (7-5) shows that the output voltage is the sum of the two input voltages. A general expression is given in Equation (7-6) for a summing amplifier with n inputs, as shown in Figure 7-21 where all resistors are equal in value.

$$V_{OUT} = -(V_{IN1} + V_{IN2} + \cdots + V_{INn}) \tag{7-6}$$

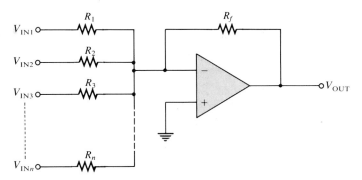

FIGURE 7-21
Summing amplifier with n inputs.

EXAMPLE 7–4

Determine the output voltage in Figure 7–22.

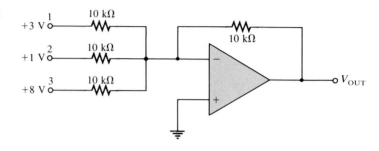

FIGURE 7–22

SOLUTION

$$V_{OUT} = -(V_{IN1} + V_{IN2} + V_{IN3})$$
$$= -(3\text{ V} + 1\text{ V} + 8\text{ V})$$
$$= -12\text{ V}$$

PRACTICE EXERCISE 7–4
If a fourth input of +0.5 V is added to Figure 7–22 with a 10 kΩ resistor, what is the output voltage?

SUMMING AMPLIFIER WITH GAIN GREATER THAN UNITY

When R_f is larger than the input resistors, the amplifier has a gain of R_f/R, where R is the value of each input resistor. The general expression for the output is

$$V_{OUT} = -\frac{R_f}{R}(V_{IN1} + V_{IN2} + \cdots + V_{INn}) \tag{7–7}$$

As you can see, the output is the sum of all the input voltages multiplied by a constant determined by the ratio R_f/R.

EXAMPLE 7–5

Determine the output voltage for the summing amplifier in Figure 7–23.

FIGURE 7–23

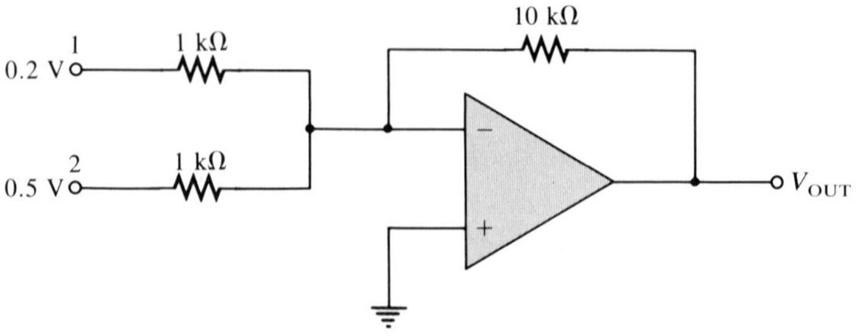

Basic Op-Amp Circuits

Solution

$R_f = 10$ kΩ, and $R = 1$ kΩ.

$$V_{OUT} = -\frac{R_f}{R}(V_{IN1} + V_{IN2})$$

$$= -\frac{10 \text{ k}\Omega}{1 \text{ k}\Omega}(0.2 \text{ V} + 0.5 \text{ V})$$

$$= -10(0.7 \text{ V})$$

$$= -7 \text{ V}$$

Practice Exercise 7-5

Determine the output voltage in Figure 7-23 if the two input resistors are 2.2 kΩ and the feedback resistor is 18 kΩ.

Averaging Amplifier

A summing amplifier can be made to produce the mathematical average of the input voltages. This is done by setting the ratio R_f/R equal to the reciprocal of the number of inputs. You obtain the average of several numbers by first adding the numbers and then dividing by the quantity of numbers you have. Examination of Equation (7-7) and a little thought will convince you that a summing amplifier will do this, as the next example illustrates.

Example 7-6

Show that the amplifier in Figure 7-24 produces an output whose magnitude is the mathematical average of the input voltages.

Figure 7-24

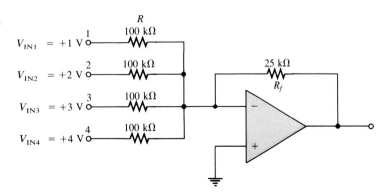

CHAPTER 7

SOLUTION
The output voltage is

$$V_{OUT} = -\frac{R_f}{R}(V_{IN1} + V_{IN2} + V_{IN3} + V_{IN4})$$

$$= -\frac{25 \text{ k}\Omega}{100 \text{ k}\Omega}(1 \text{ V} + 2 \text{ V} + 3 \text{ V} + 4 \text{ V})$$

$$= -\frac{1}{4}(10 \text{ V}) = -2.5 \text{ V}$$

It can easily be shown that the average of the input values is the same magnitude as V_{OUT} but of opposite sign.

$$\frac{1 \text{ V} + 2 \text{ V} + 3 \text{ V} + 4 \text{ V}}{4} = \frac{10 \text{ V}}{4} = 2.5 \text{ V}$$

PRACTICE EXERCISE 7-6
Specify the changes required in the averaging amplifier in Figure 7-24 in order to handle five inputs.

SCALING ADDER

A different weight can be assigned to each input of a summing amplifier by simply adjusting the values of the input resistors. As you have seen, the output voltage can be expressed as

$$V_{OUT} = -\left(\frac{R_f}{R_1}V_{IN1} + \frac{R_f}{R_2}V_{IN2} + \cdots + \frac{R_f}{R_n}V_{INn}\right) \quad (7\text{-}8)$$

The weight of a particular input is set by the ratio of R_f to the input resistance. For example, if an input voltage is to have a weight of 1, then $R = R_f$. Or, if a weight of 0.5 is required, $R = 2R_f$. The smaller the value of R, the greater the weight, and vice versa. (R is the input resistor.)

EXAMPLE 7-7 Determine the weight of each input voltage for the scaling adder in Figure 7-25 and find the output voltage.

FIGURE 7-25

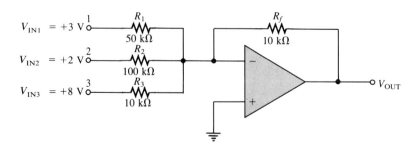

Basic Op-Amp Circuits

Solution

Weight of input 1: $\dfrac{R_f}{R_1} = \dfrac{10 \text{ k}\Omega}{50 \text{ k}\Omega} = 0.2$

Weight of input 2: $\dfrac{R_f}{R_2} = \dfrac{10 \text{ k}\Omega}{100 \text{ k}\Omega} = 0.1$

Weight of input 3: $\dfrac{R_f}{R_3} = \dfrac{10 \text{ k}\Omega}{10 \text{ k}\Omega} = 1$

The output voltage is

$$V_{OUT} = -\left(\dfrac{R_f}{R_1}V_{IN1} + \dfrac{R_f}{R_2}V_{IN2} + \dfrac{R_f}{R_3}V_{IN3}\right)$$
$$= -[0.2(3 \text{ V}) + 0.1(2 \text{ V}) + 1(8 \text{ V})]$$
$$= -(0.6 \text{ V} + 0.2 \text{ V} + 8 \text{ V})$$
$$= -8.8 \text{ V}$$

Practice Exercise 7-7

Determine the weight of each input voltage in Figure 7-25 if $R_1 = 22$ kΩ, $R_2 = 82$ kΩ, $R_3 = 56$ kΩ, and $R_f = 10$ kΩ. Also find V_{OUT}.

A Scaling Adder Application: Digital-to-Analog (D/A) Conversion

D/A conversion is an important interface process for converting digital signals to analog (linear) signals. An example is a voice signal that is digitized for storage, processing, or transmission and must be changed back into an approximation of the original audio signal in order to drive a speaker. Digital-to-analog converters will be covered thoroughly in Chapter 13.

One method of D/A conversion uses a scaling adder with input resistor values that represent the binary weights of the input code. Figure 7-26 shows a four-digit D/A converter of this type (called a *binary-weighted resistor D/A converter*). The switch symbols represent transistor switches for applying each of the four binary digits to the inputs.

Figure 7-26
A scaling adder as a four-digit digital-to-analog converter.

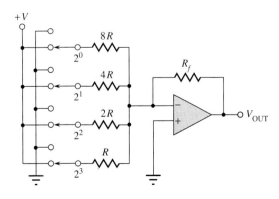

The inverting input is at virtual ground, so that the output is proportional to the current through the feedback resistor R_f (sum of input currents). The lowest value resistor R corresponds to the highest weighted binary input (2^3). All of the other resistors are multiples of R and correspond to the binary weights 2^2, 2^1, and 2^0.

7-2 Review Questions

1. Define *summing point*.
2. What is the value of R_f/R for a five-input averaging amplifier?
3. A certain scaling adder has two inputs, one having twice the weight of the other. If the resistor value for the lower weighted input is 10 kΩ, what is the value of the other input resistor?

7-3 The Integrator and Differentiator

An op-amp integrator simulates mathematical integration, which is basically a summing process that determines the total area under the curve of a function. An op-amp differentiator simulates mathematical differentiation, which is a process of determining the instantaneous rate of change of a function. It is not necessary for you to understand mathematical integration or differentiation, at this point, in order to learn how an integrator and differentiator work.

The Op-Amp Integrator

A basic **integrator** circuit is shown in Figure 7–27. Notice that the feedback element is a capacitor that forms an RC circuit with the input resistor.

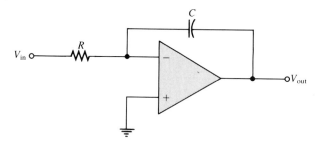

Figure 7–27
An op-amp integrator.

How a Capacitor Charges To understand how the integrator works, it is important to review how a capacitor charges. Recall that the charge Q on a capacitor is proportional to the charging current and the time.

$$Q = I_C t$$

Also, in terms of the voltage, the charge on a capacitor is

$$Q = CV_C$$

Basic Op-Amp Circuits

From these two relationships, the capacitor voltage can be expressed as

$$V_C = \left(\frac{I_C}{C}\right) t$$

You should recognize this expression as an equation for a straight line beginning at zero with a constant slope of I_C/C. (Remember from algebra that the general formula for a straight line is $y = mx + b$. In this case, $y = V_C$, $m = I_C/C$, $x = t$, and $b = 0$).

Recall that the capacitor voltage in a simple RC circuit is not linear but is exponential. This is because the charging current continuously decreases as the capacitor charges and causes the rate of change of the voltage to continuously decrease. The key thing about using an op-amp with an RC circuit to form an integrator is that the capacitor's charging current is made constant, thus producing a straight-line (linear) voltage rather than an exponential voltage. Now let's see why this is true.

In Figure 7–28, the inverting input of the op-amp is at virtual ground (0 V), so the voltage across R_i equals V_{in}. Therefore, the input current is

$$I_{in} = \frac{V_{in}}{R_i}$$

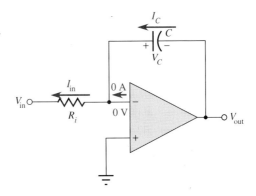

FIGURE 7–28
Currents in an integrator.

If V_{in} is a constant voltage, then I_{in} is also a constant because the inverting input always remains at 0 V, keeping a constant voltage across R_i. Because of the very high input impedance of the op-amp, there is negligible current from the inverting input. This makes all of the input current flow through the capacitor, as indicated in the figure, so

$$I_C = I_{in}$$

THE CAPACITOR VOLTAGE Since I_{in} is constant, so is I_C. The constant I_C charges the capacitor linearly and produces a linear voltage across C. The positive side of the capacitor is held at 0 V by the virtual ground of the op-amp. The voltage on the negative side of the capacitor decreases linearly from zero as the capacitor charges, as shown in Figure 7–29. This voltage is called a *negative ramp*.

FIGURE 7–29
A linear ramp voltage is produced across C by the constant charging current.

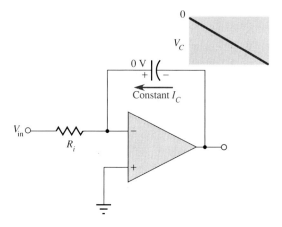

THE OUTPUT VOLTAGE V_{out} is the same as the voltage on the negative side of the capacitor. When a constant input voltage in the form of a step or pulse (a pulse has a constant amplitude when high) is applied, the output ramp decreases negatively until the op-amp saturates at its maximum negative level. This is indicated in Figure 7–30.

FIGURE 7–30
A constant input voltage produces a ramp on the output.

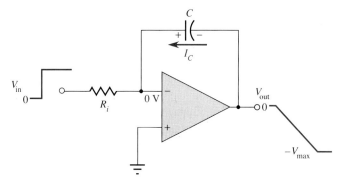

RATE OF CHANGE OF THE OUTPUT The rate at which the capacitor charges, and therefore the slope of the output ramp, is set by the ratio I_C/C, as you have seen. Since $I_C = V_{in}/R_i$, the rate of change or slope of the integrator's output voltage is

$$\frac{\Delta V_{out}}{\Delta t} = -\frac{V_{in}}{R_i C} \quad (7\text{–}9)$$

Integrators are especially useful in triangular-wave generators as you will see in Chapter 9.

■ **EXAMPLE 7–8**   (a) Determine the rate of change of the output voltage in response to a single pulse input, as shown for the integrator in Figure 7–31(a). The output voltage is initially zero.
(b) Draw the output waveform.

FIGURE 7–31

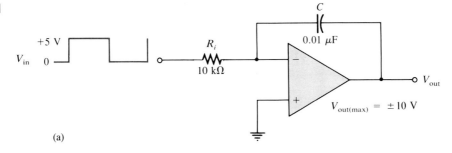

(a)

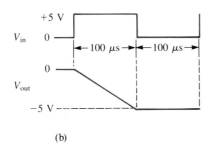

(b)

SOLUTION

(a) The rate of change of the output voltage is

$$\frac{\Delta V_{out}}{\Delta t} = \frac{-V_{IN}}{R_i C} = -\frac{5 \text{ V}}{(10 \text{ k}\Omega)(0.01 \text{ }\mu\text{F})}$$

$$= -50 \text{ kV/s} = -50 \text{ mV/}\mu\text{s}$$

(b) The rate of change was found to be -50 mV/μs in part (a). When the input is at $+5$ V, the output is a negative-going ramp. When the input is at 0 V, the output is a constant level. In 100 μs, the voltage decreases.

$$\Delta V_{out} = (-50 \text{ mV/}\mu\text{s})(100 \text{ }\mu\text{s}) = -5 \text{ V}$$

Therefore, the negative-going ramp reaches -5 V at the end of the pulse. The output voltage then remains constant at -5 V for the time that the input is zero. The waveforms are shown in Figure 7–31(b).

PRACTICE EXERCISE 7–8

Modify the integrator in Figure 7–31 to make the output change from 0 to -5 V in 50 μs with the same input. ∎

THE OP-AMP DIFFERENTIATOR

A basic **differentiator** is shown in Figure 7–32. Notice how the placement of the capacitor and resistor differ from the integrator. The capacitor is now the input element.

FIGURE 7–32
An op-amp differentiator.

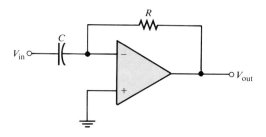

A differentiator produces an output that is proportional to the rate of change of the input voltage.

To see how the differentiator works, we will apply a positive-going ramp voltage to the input as indicated in Figure 7–33. In this case, $I_C = I_{in}$ and the voltage across the capacitor is equal to V_{in} at all times ($V_C = V_{in}$) because of virtual ground on the inverting input.

FIGURE 7–33
A differentiator with a ramp input.

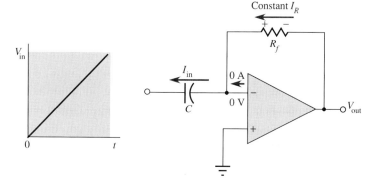

From the basic formula, $V_C = (I_C/C)t$, we get

$$I_C = \left(\frac{V_C}{t}\right)C$$

Since the current from the inverting input is negligible, $I_R = I_C$. Both currents are constant because the slope of the capacitor voltage (V_C/t) is constant. The output voltage is also constant and equal to the voltage across R_f because one side of the feedback resistor is always 0 V (virtual ground).

$$V_{out} = I_R R_f = I_C R_f$$

$$V_{out} = \left(\frac{V_C}{t}\right) R_f C \qquad (7\text{–}10)$$

The output is negative when the input is a positive-going ramp and positive when the input is a negative-going ramp as illustrated in Figure 7–34. During the positive slope of the input, the capacitor is charging from the input source and the constant current through the

Basic Op-Amp Circuits

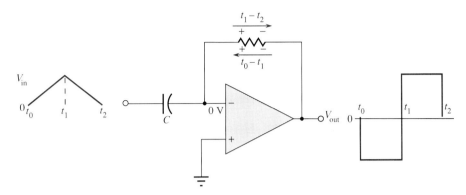

FIGURE 7–34
Output of a differentiator with a series of positive and negative ramps (triangle wave) on the input.

feedback resistor is in the direction shown. During the negative slope of the input, the current is in the opposite direction because the capacitor is discharging.

Notice in Equation (7–10) that the term V_C/t is the slope of the input. If the slope increases, V_{out} increases. If the slope decreases, V_{out} decreases. So, the output voltage is proportional to the slope (rate of change) of the input. The constant of proportionality is the time constant, $R_f C$.

EXAMPLE 7–9 Determine the output voltage of the op-amp differentiator in Figure 7–35 for the triangular-wave input shown.

FIGURE 7–35

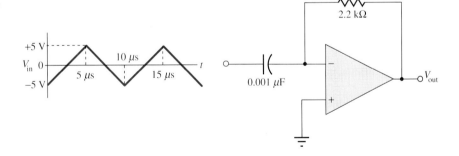

Solution
Starting at $t = 0$, the input voltage is a positive-going ramp ranging from -5 V to $+5$ V (a $+10$-V change) in 5 μs. Then it changes to a negative-going ramp ranging from $+5$ V to -5 V (a -10-V change) in 5 μs.

The time constant is

$$R_f C = (2.2 \text{ k}\Omega)(0.001 \text{ } \mu\text{F}) = 2.2 \text{ } \mu\text{s}$$

The slope or rate of change (V_C/t) of the positive-going ramp is determined, and the output voltage is calculated as follows.

$$\frac{V_C}{t} = \frac{10 \text{ V}}{5 \text{ }\mu\text{s}} = 2 \text{ V}/\mu\text{s}$$

$$V_{out} = (2 \text{ V}/\mu\text{s})2.2 \text{ }\mu\text{s} = +4.4 \text{ V}$$

Likewise, the slope of the negative-going ramp is -2 V/μs, and the output voltage is calculated as follows.

$$V_{out} = (-2 \text{ V}/\mu\text{s})2.2 \text{ }\mu\text{s} = -4.4 \text{ V}$$

Finally, the output voltage waveform is graphed relative to the input as shown in Figure 7–36.

FIGURE 7–36

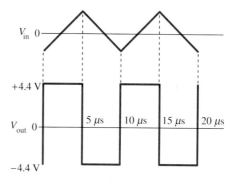

PRACTICE EXERCISE 7–9

What would the output voltage be if the feedback resistor in Figure 7–35 is changed to 3.3 kΩ? ■

7–3 REVIEW QUESTIONS

1. What is the feedback element in an op-amp integrator?
2. For a constant input voltage to an integrator, why is the voltage across the capacitor linear?
3. What is the feedback element in an op-amp differentiator?
4. How is the output of a differentiator related to the input?

7–4 MORE OP-AMP CIRCUITS

This section introduces a few more op-amp circuits that represent basic applications of the op-amp. You will learn about the constant-current source, the current-to-voltage converter, the voltage-to-current converter, and the peak detector. This is, of

BASIC OP-AMP CIRCUITS

course, not a comprehensive coverage of all possible op-amp circuits but is intended only to introduce you to some common and basic uses.

CONSTANT-CURRENT SOURCE

The purpose of a constant-current source is to deliver a load current that remains constant when the load resistance changes. Figure 7–37 shows a basic circuit in which a stable voltage source (V_{in}) provides a constant current (I_i) through the input resistor (R_i). Since the inverting (−) input of the op-amp is at virtual ground (0 V), the value of I_i is determined by V_{IN} and R_i as

$$I_i = \frac{V_{IN}}{R_i}$$

Now, since the internal input impedance of the op-amp is extremely high (ideally infinite), practically all of I_i flows through R_L, which is connected in the feedback path. Since $I_i = I_L$,

$$I_L = \frac{V_{IN}}{R_i} \qquad (7-11)$$

If R_L changes, I_L remains constant as long as V_{IN} and R_i are held constant.

FIGURE 7–37
A basic constant-current source.

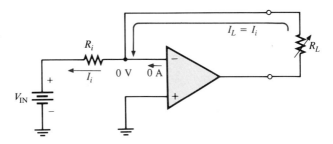

CURRENT-TO-VOLTAGE CONVERTER

The purpose of a current-to-voltage converter is to convert a variable input current to a proportional output voltage. A basic circuit that accomplishes this is shown in Figure 7–38(a). Since practically all of I_i flows through the feedback path, the voltage dropped across R_f is $I_i R_f$. Because the left side of R_f is at virtual ground (0 V), the output voltage equals the voltage across R_f, which is proportional to I_i.

$$V_{out} = I_i R_f \qquad (7-12)$$

A specific application of this circuit is illustrated in Figure 7–38(b), where a photoconductive cell is used to sense changes in light level. As the amount of light changes, the current through the photoconductive cell varies because of the cell's change in resistance. This change in resistance produces a proportional change in the output voltage ($\Delta V_{out} = \Delta I_i R_f$).

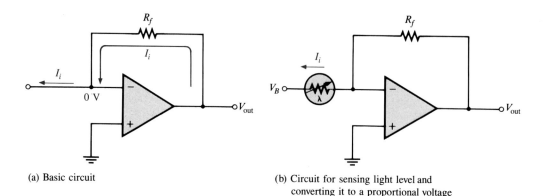

(a) Basic circuit

(b) Circuit for sensing light level and converting it to a proportional voltage

FIGURE 7–38
Current-to-voltage converter.

VOLTAGE-TO-CURRENT CONVERTER

A basic voltage-to-current converter is shown in Figure 7–39. This circuit is used in applications where it is necessary to have an output (load) current that is controlled by an input voltage.

FIGURE 7–39
Voltage-to-current converter.

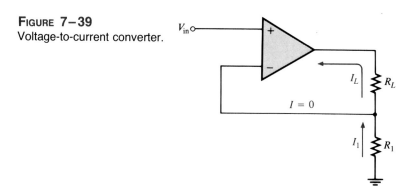

Neglecting the input offset voltage, both inverting and noninverting input terminals of the op-amp are at the same voltage, V_{in}. Therefore, the voltage across R_1 equals V_{in}. Since negligible current flows from the inverting input, the same current that flows through R_1 also flows through R_L; thus

$$I_L = \frac{V_{in}}{R_1} \qquad (7-13)$$

PEAK DETECTOR

An interesting application of the op-amp is in a peak detector circuit such as the one shown in Figure 7–40. In this case the op-amp is used as a comparator. The purpose of this circuit is to detect the peak of the input voltage and store that peak voltage on a

BASIC OP-AMP CIRCUITS

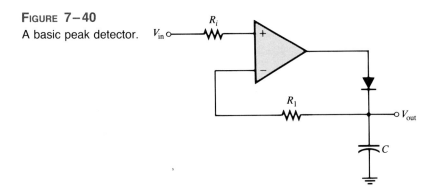

FIGURE 7–40
A basic peak detector.

capacitor. For example, this circuit can be used to detect and store the maximum value of a voltage surge; this value can then be measured at the output with a voltmeter or recording device. The basic operation is as follows. When a positive voltage is applied to the noninverting input of the op-amp through R_i, the high-level output voltage of the op-amp forward-biases the diode and charges the capacitor. The capacitor continues to charge until its voltage reaches a value equal to the input voltage and thus both op-amp inputs are at the same voltage. At this point, the op-amp comparator switches, and its output goes to the low level. The diode is now reverse-biased, and the capacitor stops charging. It has reached a voltage equal to the peak of V_{in} and will hold this voltage until the charge eventually leaks off. If a greater input peak occurs, the capacitor charges to the new peak.

7–4 REVIEW QUESTIONS

1. For the constant-current source in Figure 7–37, the input reference voltage is 6.8 V and R_i is 10 kΩ. What value of constant current does the circuit supply to a 1 kΩ load? To a 5 kΩ load?
2. What element determines the constant of proportionality that relates input current to output voltage in the current-to-voltage converter?

7–5 TROUBLESHOOTING

Although integrated circuit op-amps are extremely reliable and trouble-free, failures do occur from time to time. One type of internal failure mode is a condition where the op-amp output is in a saturated state resulting in a constant high or constant low level, regardless of the input. Also, external component failures will produce various types of failure modes in op-amp circuits. Some examples are presented in this section.

Figure 7–41 illustrates an internal failure of a comparator circuit that results in a failed output.

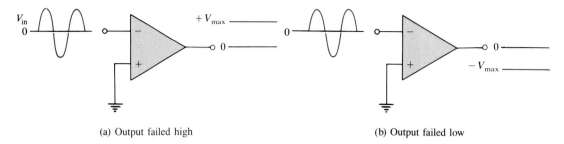

(a) Output failed high   (b) Output failed low

FIGURE 7–41
Typical internal comparator failures.

EXTERNAL COMPONENT FAILURES IN COMPARATOR CIRCUITS

Comparators with zener-bounding and hysteresis are shown in Figure 7–42. In addition to a failure of the op-amp itself, a zener diode or one of the resistors could go bad. Suppose one of the zener diodes opens. This effectively eliminates both zeners, and the circuit operates as an unbounded comparator, as indicated in Figure 7–43(a). With a shorted diode, the output is limited to the zener voltage (bounded) only in one direction depending on which diode remains operational, as illustrated in Figure 7–43(b). In the other direction, the output is held at the forward diode voltage.

FIGURE 7–42
A bounded comparator.

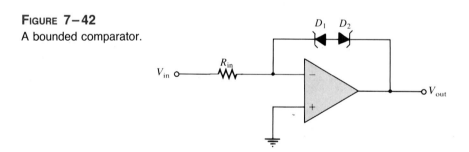

Recall that R_1 and R_2 set the UTP and LTP for the hysteresis comparator. Now, suppose that R_2 opens. Essentially all of the output voltage is fed back to the noninverting input, and, since the input voltage will never exceed the output, the device will remain in one of its saturated states. This symptom can also indicate a faulty op-amp, as mentioned before. Now, assume that R_1 opens. This leaves the noninverting input near ground potential and causes the circuit to operate as a zero-level detector. These conditions are shown in parts (c) and (d) of Figure 7–43.

BASIC OP-AMP CIRCUITS

FIGURE 7–43
Examples of comparator circuit failures and their effects.

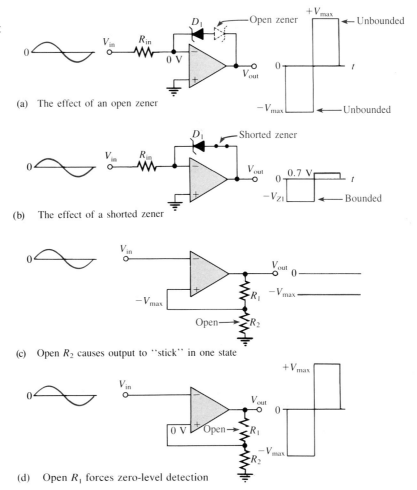

(a) The effect of an open zener

(b) The effect of a shorted zener

(c) Open R_2 causes output to "stick" in one state

(d) Open R_1 forces zero-level detection

■ **EXAMPLE 7–10**   One channel of a dual-trace oscilloscope is connected to the comparator output and the other channel to the input signal, as shown in Figure 7–44. From the observed waveforms, determine if the circuit is operating properly, and if not, what the most likely failure is.

SOLUTION
The output should be limited to ± 8.67 V. However, the positive maximum is $+0.88$ V and the negative maximum is -7.79 V. This indicates that D_2 is shorted. Refer to Example 7–3 for analysis.

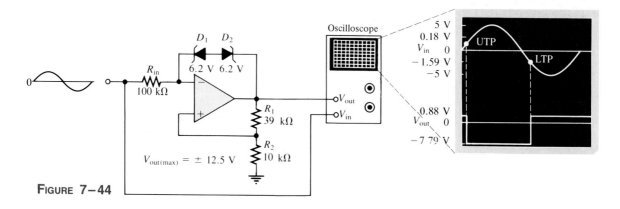

FIGURE 7–44

PRACTICE EXERCISE 7–10
What would the output voltage look like if D_1 shorted rather than D_2?

COMPONENT FAILURES IN SUMMING AMPLIFIERS

If one of the input resistors in a unity-gain summing amplifier opens, the output will be less than the normal value by the amount of the voltage applied to the open input. Stated another way, the output will be the sum of the remaining input voltages.

If the summing amplifier has a nonunity gain, an open input resistor causes the output to be less than normal by an amount equal to the gain times the voltage at the open input.

■ **EXAMPLE 7–11** (a) What is the normal output voltage in Figure 7–45? (b) What is the output voltage if R_2 opens? (c) What happens if R_5 opens?

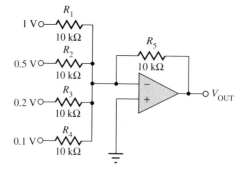

FIGURE 7–45

SOLUTION
(a) $V_{OUT} = -(1\text{ V} + 0.5\text{ V} + 0.2\text{ V} + 0.1\text{ V}) = -1.8\text{ V}$
(b) $V_{OUT} = -(1\text{ V} + 0.2\text{ V} + 0.1\text{ V}) = -1.3\text{ V}$
(c) If R_5 opens, the circuit becomes a comparator and the output goes to $-V_{max}$.

PRACTICE EXERCISE 7–11
In Figure 7–45, $R_5 = 47\text{ k}\Omega$. What is the output voltage if R_1 opens?

Basic Op-Amp Circuits

As another example, let's look at an averaging amplifier. An open input resistor will result in an output voltage that is the average of all the inputs with the open input averaged in as a zero.

EXAMPLE 7–12   (a) What is the normal output voltage for the averaging amplifier in Figure 7–46?
(b) If R_4 opens, what is the output voltage? What does the output voltage represent?

FIGURE 7–46

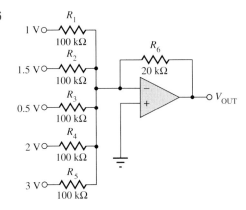

SOLUTION

(a) $V_{OUT} = -\dfrac{20 \text{ k}\Omega}{100 \text{ k}\Omega}(1 \text{ V} + 1.5 \text{ V} + 0.5 \text{ V} + 2 \text{ V} + 3 \text{ V}) = -\dfrac{1}{5}(8 \text{ V}) = -1.6 \text{ V}$

(b) $V_{OUT} = -\dfrac{20 \text{ k}\Omega}{100 \text{ k}\Omega}(1 \text{ V} + 1.5 \text{ V} + 0.5 + 3 \text{ V}) = -\dfrac{1}{5}(6 \text{ V}) = -1.2 \text{ V}$

1.2 V is the average of five voltages with the 2-V input replaced by 0 V. Notice that the output is not the average of the four remaining input voltages.

PRACTICE EXERCISE 7–12

If R_4 is open, as was the case in this example, what would you have to do to make the output equal to the average of the remaining four input voltages?

7–5 REVIEW QUESTIONS

1. Describe one type of internal op-amp failure.
2. If a certain malfunction is attributable to more than one possible component failure, what would you do to isolate the problem?

7–6   A SYSTEM APPLICATION

The system presented at the beginning of this chapter is a dual-slope analog-to-digital (A/D) converter. This is one of several methods for A/D conversion. You saw another type, called a simultaneous A/D converter as an example in the chapter.

The topic of data conversion including both A/D and D/A converters is covered thoroughly in Chapter 13. Although A/D conversion is used for many purposes, in this particular application the converter is used to change an audio signal into digital form for recording. Even though many parts of this system are digital, we are going to concentrate on the analog board, which includes op-amps used in several types of circuits that you have learned about in this chapter. In this section, you will

- *See one way in which a summing amplifier is used.*
- *See how the integrator is a key element in A/D conversion.*
- *See how a comparator is used.*
- *Translate between a printed circuit board and a schematic.*
- *Troubleshoot some common system problems.*

A BRIEF DESCRIPTION OF THE SYSTEM

The dual-slope A/D converter in Figure 7–47 accepts an audio signal and converts it to a series of digital codes for the purpose of recording. The audio signal voltage is applied to the sample-and-hold circuit. (Sample-and-hold circuits are covered in detail in Chapter 13.) At a given instant, a sample pulse causes the amplitude at that point on the audio waveform to be converted to a proportional dc level which is then processed by the rest of the circuits and represented by a digital code. The sample pulses occur at a much higher

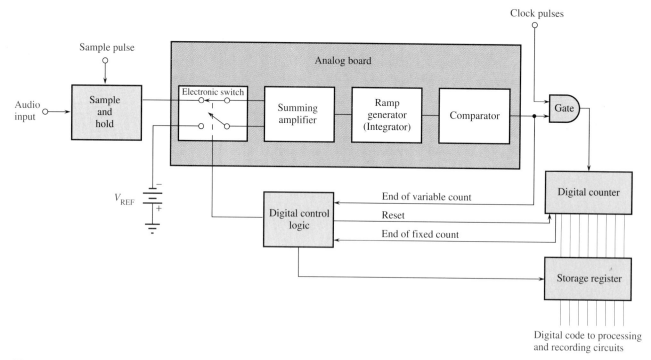

FIGURE 7–47

Basic dual-slope analog-to-digital converter.

rate than the audio frequency so that a sufficient number of points on the audio waveform are sampled and converted to obtain an accurate representation of the audio signal. A rough approximation of the sampling process is illustrated in Figure 7–48.

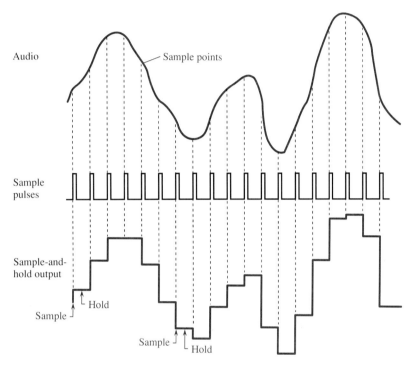

FIGURE 7–48
Sample-and-hold process. The sample-and-hold output is a rough approximation of the audio voltage for purposes of illustration. As the frequency of the sample pulses is increased, an increasingly accurate representation is achieved.

During the time between each sample pulse, the dc level from the sample-and-hold circuit is switched into the summing amplifier by the electronic switch under digital control and then applied to the integrator. At the same time, the digital counter starts counting up from zero. During the fixed interval of the counting sequence, the integrator produces a negative-going ramp whose slope depends on the level of the sampled audio voltage. At the end of the fixed counting sequence, the ramp voltage at the output of the integrator has reached a voltage that is proportional to the sampled audio voltage. At this time, the digital control logic switches from the sample-and-hold input to the reference voltage input and resets the digital counter to zero. The summing amplifier applies this negative dc reference to the integrator input, which starts a positive-going ramp on the output. This ramp has a slope that is fixed by the value of the dc reference voltage. At the same time, the digital counter begins to count up again and will continue to count up until the positive ramp output of the integrator reaches zero volts. At this point, the comparator switches to its negative saturated output voltage and disables the gate so that there are no additional clock pulses to the counter. At this time, the digital code in the counter is proportional to the time that it took for the positive-going ramp to reach zero and it will vary for each different sampled value. Recall that the positive-going ramp started at a negative voltage that was dependent on the sampled value of the audio signal. Therefore, the digital code in the counter is also proportional to, and represents, the amplitude of the sampled audio voltage. This code is then shifted out to the register and then processed and recorded.

This process is repeated many times during a typical audio cycle. The result is a sequence of digital codes that represent the audio voltage amplitude as it varies with time. Figure 7–49 illustrates this for several sampled values. As mentioned, you will focus on the analog board, which contains the electronic switch, summing amplifier, integrator, and comparator.

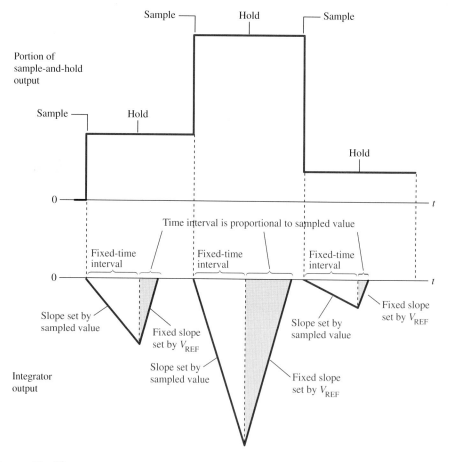

FIGURE 7–49
During the fixed-time interval of the negative-going ramp, the sampled audio input is applied to the integrator. During the variable-time interval of the positive-going ramp, the reference voltage is applied to the integrator. The counter sets the fixed-time interval and is then reset. Another count begins during the variable interval and the code in the counter at the end of this interval represents the sampled value.

Basic Op-Amp Circuits

Now, so you can take a closer look at the analog board, let's take it out of the system and put it on the test bench.

On the Test Bench

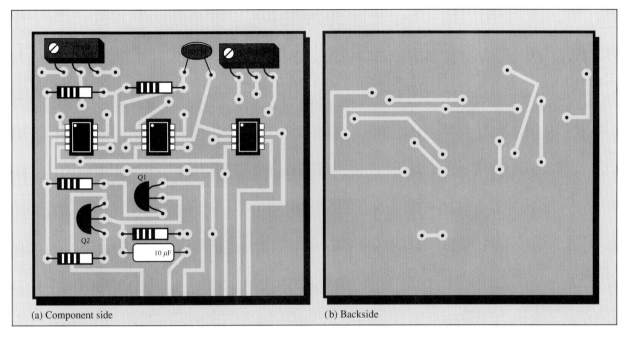

(a) Component side (b) Backside

FIGURE 7–50

■ **ACTIVITY 1**   **RELATE THE PC BOARD TO THE SCHEMATIC**

Identify each component on the circuit board in Figure 7–50 using the schematic in Figure 7–51. Also, identify each input and output on the board. Check to make sure the interconnections on both sides of the board correspond with the schematic. Locate

FIGURE 7–51

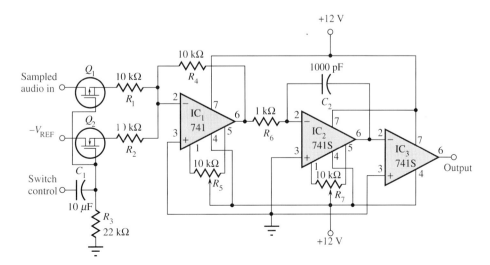

CHAPTER 7

and identify the components associated with the electronic switch, the summing amplifier, the integrator, and the comparator.

■ **ACTIVITY 2   ANALYZE THE CIRCUIT**

STEP 1   Determine the gain of the summing amplifier.

STEP 2   Determine the slope of the integrator ramp in volts per microsecond when a sampled audio voltage of $+2$ V is applied.

STEP 3   Determine the slope of the integrator ramp in volts per microsecond when the reference voltage of -8 V is applied.

STEP 4   Given that the reference voltage is -8 V and the fixed-time interval of the negative-going slope is 1 μs, sketch the dual-slope output of the integrator when an instantaneous audio voltage of $+3$ V is sampled.

STEP 5   Assuming that the maximum audio voltage to be sampled is $+6$ V, determine the maximum audio frequency that can be sampled by this particular system if there are to be 100 samples per cycle. What is the sample pulse rate in this case?

■ **ACTIVITY 3   WRITE A TECHNICAL REPORT**

Discuss the detailed operation of the analog board circuitry and explain how it interfaces with the overall system. Discuss the purpose of each component on the circuit board.

■ **ACTIVITY 4   TROUBLESHOOT THE CIRCUIT FOR EACH OF THE FOLLOWING PROBLEMS BY STATING THE PROBABLE CAUSE OR CAUSES**

1. Zero volts on the output of IC_1 when there are voltages on the sampled audio input and on the reference voltage input.
2. IC_1 goes back and forth between its saturated states as the positive audio voltage and the negative reference voltage are alternately switched in.
3. The inverting input of IC_1 never goes negative.
4. The output of IC_2 stays at zero volts under normal operating conditions.

7–6 REVIEW QUESTIONS

1. Identify the summing amplifier, the integrator, and the comparator by IC number.
2. The 741S op-amps are high slew-rate devices with a minimum slew rate of 10 V/μs. Why are these used in this application?
3. What is the purpose of R_5 and R_7 in the circuit of Figure 7–51?
4. What type of output voltage does the analog board produce and how is it used in the system?
5. If a sample pulse rate of 500 kHz is used, how long does each sampled audio voltage remain on the input to the analog board?
6. Although IC_1 is connected in the form of a summing amplifier, it does not actually perform a summing operation in this application. Why?

Basic Op-Amp Circuits

Summary

- In an op-amp comparator, when the input voltage exceeds a specified reference voltage, the output changes state.
- Hysteresis gives an op-amp noise immunity.
- A comparator switches to one state when the input reaches the upper trigger point (UTP) and back to the other state when the input drops below the lower trigger point (LTP).
- The difference between the UTP and the LTP is the hysteresis voltage.
- Bounding limits the output amplitude of a comparator.
- The output voltage of a summing amplifier is proportional to the sum of the input voltages.
- An averaging amplifier is a summing amplifier with a closed-loop gain equal to the reciprocal of the number of inputs.
- In a scaling adder, a different weight can be assigned to each input, thus making the input contribute more or contribute less to the output.
- Integration is a mathematical process for determining the area under a curve.
- Integration of a step produces a ramp with a slope proportional to the amplitude.
- Differentiation is a mathematical process for determining the rate of change of a function.
- Differentiation of a ramp produces a step with an amplitude proportional to the slope.

Glossary

A/D conversion A process whereby information in analog form is converted into digital form.

Analog Characterized by a linear process in which a variable takes on a continuous set of values.

Bounding The process of limiting the output range of an amplifier or other circuit.

Comparator A circuit which compares two input voltages and produces an output in either of two states indicating the greater than or less than relationship of the inputs.

Differentiator A circuit that produces an output which approximates the instantaneous rate of change of the input function.

Digital Characterized by a process in which a variable takes on either of two values.

Hysteresis Characteristic of a circuit in which two different trigger levels create an offset or lag in the switching action.

Integrator A circuit that produces an output which approximates the area under the curve of the input function.

Noise An unwanted signal.

Schmitt trigger A comparator with hysteresis.

Formulas

Comparators

(7–1) $\quad V_{REF} = \dfrac{R_2}{R_1 + R_2}(+V) \quad$ Comparator reference

(7–2) $\quad V_{UTP} = \dfrac{R_2}{R_1 + R_2}[+V_{out(max)}] \quad$ Upper trigger point

(7–3) $\quad V_{LTP} = \dfrac{R_2}{R_1 + R_2}[-V_{out(max)}] \quad$ Lower trigger point

(7–4) $\quad V_{HYS} = V_{UTP} - V_{LTP} \quad$ Hysteresis voltage

Summing Amplifier

(7–5) $\quad V_{OUT} = -(V_{IN1} + V_{IN2}) \quad$ Two-input adder

(7–6) $\quad V_{OUT} = -(V_{IN1} + V_{IN2} + \cdots + V_{INn}) \quad$ n-input adder

(7–7) $\quad V_{OUT} = -\dfrac{R_f}{R}(V_{IN1} + V_{IN2} + \cdots + V_{INn}) \quad$ Adder with gain

(7–8) $\quad V_{OUT} = -\left(\dfrac{R_f}{R_1}V_{IN1} + \dfrac{R_f}{R_2}V_{IN2} + \cdots + \dfrac{R_f}{R_n}V_{INn}\right) \quad$ Adder with gain

Integrator and Differentiator

(7–9) $\quad \dfrac{\Delta V_{out}}{\Delta t} = -\dfrac{V_{in}}{R_i C} \quad$ Integrator, rate of change

(7–10) $\quad V_{out} = \left(\dfrac{V_C}{t}\right) R_f C \quad$ Differentiator output with ramp input

Miscellaneous

(7–11) $\quad I_L = \dfrac{V_{IN}}{R_i} \quad$ Constant-current source

(7–12) $\quad V_{out} = I_i R_f \quad$ Current-to-voltage converter

(7–13) $\quad I_L = \dfrac{V_{in}}{R_1} \quad$ Voltage-to-current converter

Self-Test

1. In a zero-level detector, the output changes state when the input
    (a) is positive
    (b) is negative
    (c) crosses zero
    (d) has a zero rate of change

Basic Op-Amp Circuits

2. The zero-level detector is one application of a
   - (a) comparator
   - (b) differentiator
   - (c) summing amplifier
   - (d) diode
3. Noise on the input of a comparator can cause the output to
   - (a) hang up in one state
   - (b) go to zero
   - (c) change back and forth erratically between two states
   - (d) produce the amplified noise signal
4. The effects of noise can be reduced by
   - (a) lowering the supply voltage
   - (b) using positive feedback
   - (c) using negative feedback
   - (d) using hysteresis
   - (e) b and d
5. A comparator with hysteresis
   - (a) has one trigger point
   - (b) has two trigger points
   - (c) has a variable trigger point
   - (d) is like a magnetic circuit
6. In a comparator with hysteresis,
   - (a) a bias voltage is applied between the two inputs
   - (b) only one supply voltage is used
   - (c) a portion of the output is fed back to the inverting input
   - (d) a portion of the output is fed back to the noninverting input
7. Using output bounding in a comparator
   - (a) makes it faster
   - (b) keeps the output positive
   - (c) limits the output levels
   - (d) stabilizes the output
8. A window comparator detects when
   - (a) the input is between two specified limits
   - (b) the input is not changing
   - (c) the input is changing too fast
   - (d) the amount of light exceeds a certain value
9. A summing amplifier can have
   - (a) only one input
   - (b) only two inputs
   - (c) any number of inputs
10. If the voltage gain for each input of a summing amplifier with a 4.7 kΩ feedback resistor is unity, the input resistors must have a value of
    - (a) 4.7 kΩ
    - (b) 4.7 kΩ divided by the number of inputs
    - (c) 4.7 kΩ times the number of inputs
11. An averaging amplifier has five inputs. The ratio R_f/R_{in} must be
    - (a) 5
    - (b) 0.2
    - (c) 1

12. In a scaling adder, the input resistors are
    (a) all the same value
    (b) all of different values
    (c) each proportional to the weight of its inputs
    (d) related by a factor of two
13. In an integrator, the feedback element is a
    (a) resistor   (b) capacitor
    (c) zener diode   (d) voltage divider
14. For a step input, the output of an integrator is
    (a) a pulse   (b) a triangular waveform
    (c) a spike   (d) a ramp
15. The rate of change of an integrator's output voltage in response to a step input is set by
    (a) the RC time constant   (b) the amplitude of the step input
    (c) the current through the capacitor   (d) all of these
16. In a differentiator, the feedback element is a
    (a) resistor   (b) capacitor
    (c) zener diode   (d) voltage divider
17. The output of a differentiator is proportional to
    (a) the RC time constant   (b) the rate at which the input is changing
    (c) the amplitude of the input   (d) a and b
18. When you apply a triangular waveform to the input of a differentiator, the output is
    (a) a dc level   (b) an inverted triangular waveform
    (c) a square waveform (d) the first harmonic of the triangular waveform

Problems

Section 7–1 Comparators

1. A certain op-amp has an open-loop gain of 80,000. The maximum saturated output levels of this particular device are ±12 V when the dc supply voltages are ±15 V. If a differential voltage of 0.15 mV rms is applied between the inputs, what is the peak-to-peak value of the output?

2. Determine the output level (maximum positive or maximum negative) for each comparator in Figure 7–52.

BASIC OP-AMP CIRCUITS

FIGURE 7-52

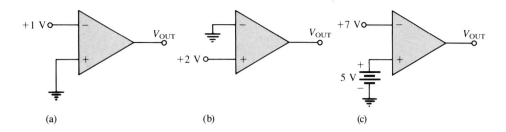

(a) (b) (c)

3. Calculate the V_{UTP} and V_{LTP} in Figure 7-53. $V_{out(max)} = -10$ V.
4. What is the hysteresis voltage in Figure 7-53?

FIGURE 7-53

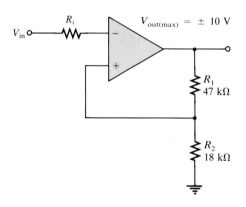

5. Sketch the output voltage waveform for each circuit in Figure 7-54 with respect to the input. Show voltage levels.

FIGURE 7-54

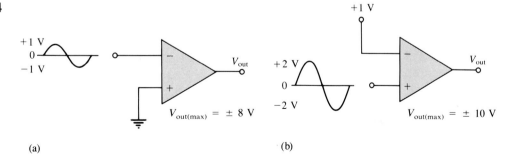

(a) (b)

6. Determine the hysteresis voltage for each comparator in Figure 7-55. The maximum output levels are ± 11 V.

FIGURE 7–55

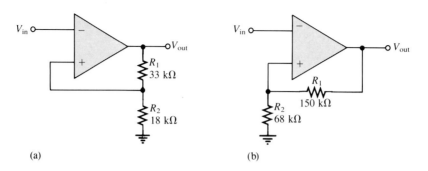

(a)   (b)

7. A 6.2 V zener diode is connected from the output to the inverting input in Figure 7–53 with the cathode at the output. What are the positive and negative output levels?

8. Determine the output voltage waveform in Figure 7–56.

FIGURE 7–56

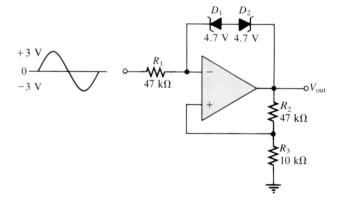

SECTION 7–2   SUMMING AMPLIFIERS

9. Determine the output voltage for each circuit in Figure 7–57.

FIGURE 7–57

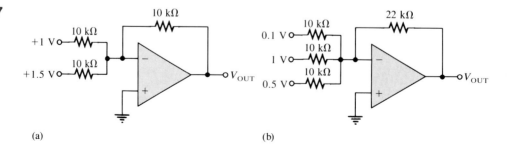

(a)   (b)

Basic Op-Amp Circuits

10. Refer to Figure 7–58. Determine the following:
    (a) V_{R1} and V_{R2}
    (b) Current through R_f
    (c) V_{OUT}

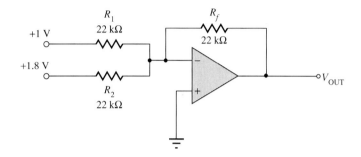

FIGURE 7–58

11. Find the value of R_f necessary to produce an output that is five times the sum of the inputs in Figure 7–58.
12. Design a summing amplifier that will average eight input voltages. Use input resistances of 10 kΩ each.
13. Find the output voltage when the input voltages shown in Figure 7–59 are applied to the scaling adder. What is the current through R_f?

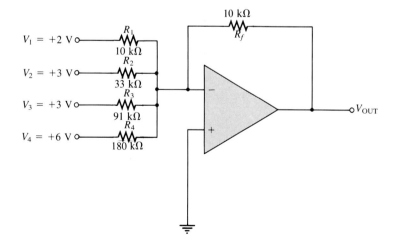

FIGURE 7–59

14. Determine the values of the input resistors required in a six-input scaling adder so that the lowest weighted input is 1 and each successive input has a weight twice the previous one. Use $R_f = 100$ kΩ.

Section 7-3 The Integrator and Differentiator

15. Determine the rate of change of the output voltage in response to the step input to the integrator in Figure 7–60.

Figure 7–60

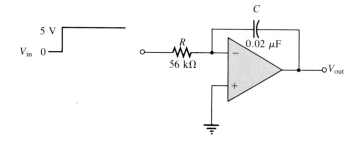

16. A triangular waveform is applied to the input of the circuit in Figure 7–61 as shown. Determine what the output should be and sketch its waveform in relation to the input.
17. What is the magnitude of the capacitor current in Problem 16?

Figure 7–61

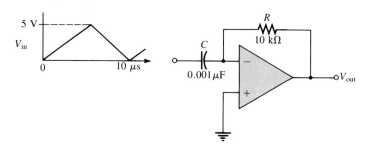

18. A triangular waveform with a peak-to-peak voltage of 2 V and a period of 1 ms is applied to the differentiator in Figure 7–62(a). What is the output voltage?
19. Beginning in position 1 in Figure 7–62(b), the switch is thrown into position 2 and held there for 10 ms, then back to position 1 for 10 ms, and so forth. Sketch the resulting output waveform. The saturated output levels of the op-amp are ± 12 V.

Figure 7–62

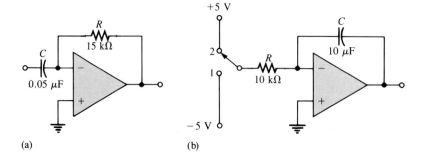

(a) (b)

Section 7-4 More Op-Amp Circuits

20. Determine the load current in each circuit of Figure 7-63.

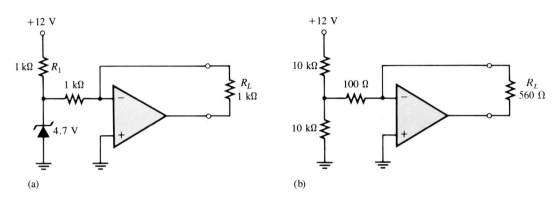

Figure 7-63

21. Devise a circuit for remotely sensing temperature and producing a proportional voltage that can then be converted to digital form for display. A thermistor can be used as the temperature-sensing element.

Section 7-5 Troubleshooting

22. The waveforms given in Figure 7-64(a) are observed at the indicated points in Figure 7-64(b). Is the circuit operating properly? If not, what is a likely fault?

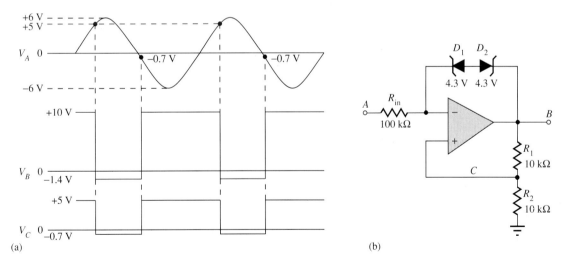

Figure 7-64

23. The waveforms shown for the window comparator in Figure 7-65 are measured. Determine if the output waveform is correct and, if not, specify the possible fault(s).

FIGURE 7–65

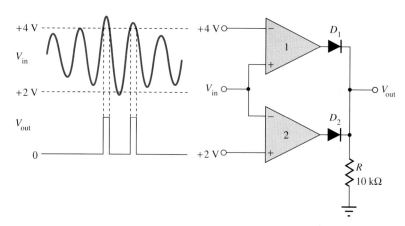

 24. The sequences of voltage levels shown in Figure 7–66 are applied to the summing amplifier and the indicated output is observed. First, determine if this output is correct. If it is not correct, determine the fault.

FIGURE 7–66

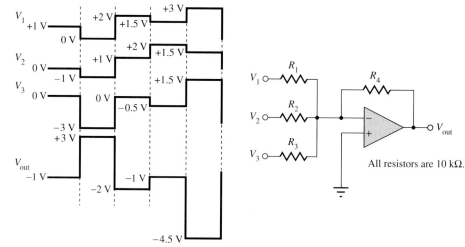

 25. The given ramp voltages are applied to the op-amp circuit in Figure 7–67. Is the given output correct? If it isn't, what is the problem?

26. The analog board, shown in Figure 7–68, for the A/D converter in the system application has just come off the assembly line and a pass/fail test indicates that it doesn't work. The board now comes to you for troubleshooting. What is the very first thing you should do? Can you isolate the problem(s) by this first step in this case?

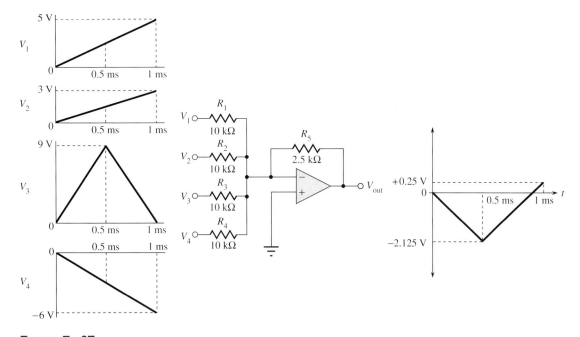

FIGURE 7–67

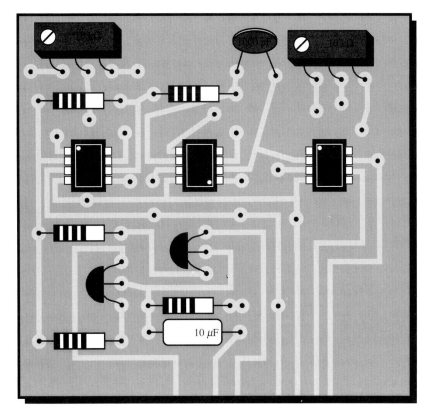

FIGURE 7–68

Answers to Review Questions

Section 7-1
1. (a) $V = (10 \text{ k}\Omega/110 \text{ k}\Omega)15 \text{ V} = 1.36 \text{ V}$
   (b) $V = (22 \text{ k}\Omega/69 \text{ k}\Omega)(-12 \text{ V}) = -3.83 \text{ V}$
2. Hysteresis makes the comparator noise free.
3. Bounding limits the output amplitude to a specified level.

Section 7-2
1. The summing point is the point where the input resistors are commonly connected.
2. $R_f/R = 1/5 = 0.2$
3. $5 \text{ k}\Omega$

Section 7-3
1. The feedback element in an integrator is a capacitor.
2. The capacitor voltage is linear because the capacitor current is constant.
3. The feedback element in a differentiator is a resistor.
4. The output of a differentiator is proportional to the rate of change of the input.

Section 7-4
1. $I_L = 6.8 \text{ V}/10 \text{ k}\Omega = 0.68 \text{ mA}$; same value to $5 \text{ k}\Omega$ load.
2. The feedback resistor is the constant of proportionality.

Section 7-5
1. An op-amp can fail with a shorted output.
2. Replace suspected components one by one.

Section 7-6
1. Summing amplifier—IC_1, integrator—IC_2, comparator—IC_3
2. A high slew-rate op-amp is used in the integrator to avoid slew-rate limitation of the output ramps. One is used as a comparator, to achieve a fast switching time.
3. R_4 and R_6 are for eliminating output offset (nulling).
4. The board output is the comparator output. The transition of the comparator output from its positive state to its negative state notifies the control logic of the end of the variable-time interval.
5. $1/500 \text{ kHz} = 2 \text{ }\mu\text{s}$
6. Because of the electronic switch, only one input voltage at a time is actually applied.

Answers to Practice Exercises

- **7–1** 1.96 V
- **7–2** +3.83 V, −3.83 V
- **7–3** +1.81 V, −1.81 V
- **7–4** −12.5 V
- **7–5** −5.73 V
- **7–6** Changes require an additional 100 kΩ input resistor and a change of R_f to 20 kΩ.
- **7–7** 0.45, 0.12, 0.18; $V_{OUT} = -3.03$ V
- **7–8** Change C to 5000 pF.
- **7–9** Same waveform with an amplitude of 6.6 V
- **7–10** A pulse from −0.88 V to +7.79 V
- **7–11** −3.76 V
- **7–12** Change R_6 to 25 kΩ.

8
ACTIVE FILTERS

8–1 BASIC FILTER RESPONSES
8–2 FILTER RESPONSE CHARACTERISTICS
8–3 ACTIVE LOW-PASS FILTERS
8–4 ACTIVE HIGH-PASS FILTERS
8–5 ACTIVE BAND-PASS FILTERS
8–6 ACTIVE BAND-STOP FILTERS
8–7 FILTER RESPONSE MEASUREMENTS
8–8 A SYSTEM APPLICATION

After completing this chapter, you should be able to

☐ Identify low-pass, high-pass, band-pass, and band-stop filter responses.
☐ Recognize Butterworth, Chebyshev, and Bessel response characteristics.
☐ Describe the effect of the damping factor on
☐ filter response.
Define the term *pole* in relation to filters.
☐ Determine the critical (cutoff) frequencies of specific filters.
☐ Explain how the roll-off rate of a filter is related to the number of poles.
☐ Discuss the frequency response of cascaded
☐ filters.
Implement the Butterworth response in a filter.
☐ Analyze Sallen-Key low-pass and high-pass filters.
☐ Analyze multiple-feedback and state-variable band-pass and band-stop filters.
☐ Measure filter response using two different methods.

Power supply filters were introduced in Chapter 2. In this chapter, we introduce active filters used for signal processing. Filters are circuits that are capable of passing input signals with certain selected frequencies through to the output while rejecting signals with other frequencies. This property is called *selectivity*.

Filters use active devices such as transistors or op-amps and passive *RC* networks. The active devices provide voltage gain and the passive networks provide frequency selectivity. In terms of general response, there are four basic categories of active filters: low-pass, high-pass, band-pass, and band-stop. In this chapter, we will concentrate on active filters using op-amps and *RC* networks.

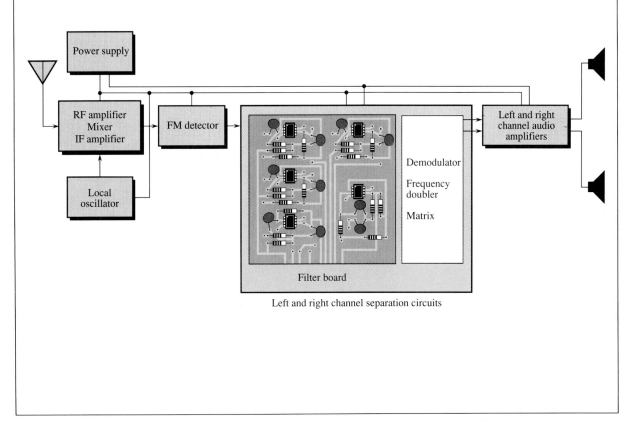

Left and right channel separation circuits

A System Application

In Chapter 6, you worked with an FM stereo multiplex receiver, concentrating on the audio amplifiers. In this chapter, we again take a look at this same system, but this time our focus is on the filters in the left and right channel separation circuits in which several types of active filters are used. The FM stereo multiplex signal that is received is quite complex and beyond the scope of our coverage to investigate the reasons it is transmitted in such a way. It is interesting, however, to see how filters, such as the ones studied in this chapter can be used to separate out the audio signals that go to the left and right speakers.

For the system application in Section 8–8, in addition to the other topics, be sure you understand

☐ Filter responses.
☐ How low-pass and band-pass filters work.

8–1 BASIC FILTER RESPONSES

*Filters are usually categorized by the manner in which the output voltage varies with the frequency of the input voltage. The categories of **active** filters are low-pass, high-pass, band-pass, and band-stop. We will examine each of these general responses in this section.*

LOW-PASS FILTER RESPONSE

The pass band of the basic **low-pass filter** is defined to be from 0 Hz (dc) up to the critical (cutoff) frequency, f_c, at which the output voltage is 70.7 percent of the pass-band voltage, as indicated in Figure 8–1(a). The ideal pass band, shown by the shaded region within the dashed lines, has an instantaneous roll-off at f_c. The bandwidth of this filter is equal to f_c.

$$BW = f_c$$

Although the ideal response is not achievable in practice, roll-off rates of −20 dB/decade and higher are obtainable. Figure 8–1(b) illustrates ideal low-pass filter response curves with several roll-off rates. The −20 dB/decade rate is obtained with a single *RC* network

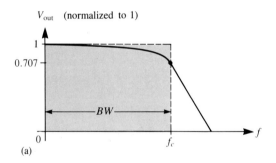

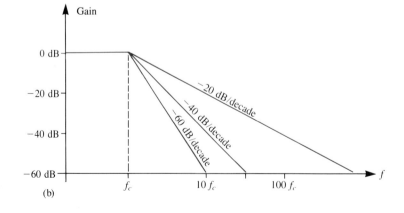

FIGURE 8–1
Low-pass filter responses.

consisting of one resistor and one capacitor. The higher roll-off rates require additional RC networks. Each network is called a **pole**.

The critical frequency of the RC low-pass filter occurs when $X_C = R$, where

$$f_c = \frac{1}{2\pi RC} \qquad (8-1)$$

HIGH-PASS FILTER RESPONSE

A **high-pass filter** response is one that significantly attenuates all frequencies below f_c and passes all frequencies above f_c. The critical frequency is, of course, the frequency at which the output voltage is 70.7 percent of the pass-band voltage, as shown in Figure 8–2(a). The ideal response, shown by the shaded region within the dashed lines, has an instantaneous drop at f_c, which, of course, is not achievable. Roll-off rates of 20 dB/decade/pole are realizable. Figure 8–2(b) illustrates high-pass filter responses with several roll-off rates.

FIGURE 8–2
High-pass filter responses.

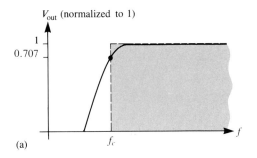

(a)

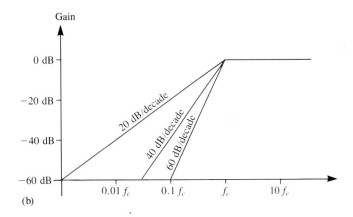

(b)

As with the RC low-pass filter, the high-pass critical frequency corresponds to the value where $X_C = R$ and is calculated with the formula $f_c = 1/2\pi RC$. The response of a high-pass filter extends from f_c up to a frequency that is determined by the limitations of the active element (transistor or op-amp) used.

BAND-PASS FILTER RESPONSE

A **band-pass filter** passes all signals lying within a band between a lower- and an upper-frequency limit and essentially rejects all other frequencies that are outside this specified band. A generalized band-pass response curve is shown in Figure 8–3. The *bandwidth (BW)* is defined as the difference between the upper critical frequency (f_{c2}) and the lower critical frequency (f_{c1}).

$$BW = f_{c2} - f_{c1} \qquad (8\text{--}2)$$

The critical frequencies are, of course, the points at which the response curve is 70.7 percent of its maximum. Recall from Chapter 6 that these critical frequencies are also called *3 dB frequencies*. The frequency about which the pass band is centered is called the *center frequency*, f_0, defined as the geometric mean of the critical frequencies.

$$f_0 = \sqrt{f_{c1}f_{c2}} \qquad (8\text{--}3)$$

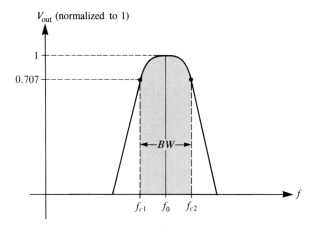

FIGURE 8–3
General band-pass response curve.

QUALITY FACTOR The **quality factor** (Q) of a band-pass filter is the ratio of the center frequency to the bandwidth.

$$Q = \frac{f_0}{BW} \qquad (8\text{--}4)$$

The value of Q is an indication of the selectivity of a band-pass filter. The higher the value of Q, the narrower the bandwidth and the better the selectivity for a given value of f_0. Band-pass filters are sometimes classified as narrow-band ($Q > 10$) or wide-band ($Q < 10$). The Q can also be expressed in terms of the damping factor (DF) of the filter as

$$Q = \frac{1}{DF} \qquad (8\text{--}5)$$

We will study the damping factor in Section 8–2.

EXAMPLE 8–1

A certain band-pass filter has a center frequency of 15 kHz and a bandwidth of 1 kHz. Determine the Q and classify the filter as narrow-band or wide-band.

SOLUTION

$$Q = \frac{f_0}{BW} = \frac{15 \text{ kHz}}{1 \text{ kHz}} = 15$$

Because $Q > 10$, this is a narrow-band filter.

PRACTICE EXERCISE 8–1
If the Q of the filter is doubled, what will the bandwidth be?

BAND-STOP FILTER RESPONSE

Another category of active filter is the band-stop, also known as *notch, band-reject,* or *band-elimination* filters. You can think of the operation as opposite to that of the band-pass filter because frequencies within a certain bandwidth are rejected, and frequencies outside the bandwidth are passed. A general response curve for a band-stop filter is shown in Figure 8–4. Notice that the bandwidth is the band of frequencies between the 3 dB points, just as in the case of the band-pass filter response.

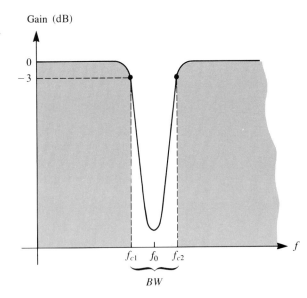

FIGURE 8–4
General band-stop filter response.

8–1 REVIEW QUESTIONS

1. What determines the bandwidth of a low-pass filter?
2. What limits the bandwidth of an active high-pass filter?
3. How are the Q and the bandwidth of a band-pass filter related? Explain how the selectivity is affected by the Q of a filter.

8-2 FILTER RESPONSE CHARACTERISTICS

Each type of filter response (low-pass, high-pass, band-pass, or band-stop) can be tailored by circuit component values to have either a Butterworth, Chebyshev, or Bessel characteristic. Each of these characteristics is identified by the shape of the response curve, and each has an advantage in certain applications.

THE BUTTERWORTH CHARACTERISTIC

The **Butterworth** characteristic provides a very flat amplitude response in the pass band and a roll-off rate of 20 dB/decade/pole. The phase response is not linear, however, and the phase shift (thus, time delay) of signals passing through the filter varies nonlinearly with frequency. Therefore, a pulse applied to a filter with a Butterworth response will cause overshoots on the output, because each frequency component of the pulse's rising and falling edges experiences a different time delay. Filters with the Butterworth response are normally used when all frequencies in the pass band must have the same gain. The Butterworth response is often referred to as a *maximally flat response*.

THE CHEBYSHEV CHARACTERISTIC

Filters with the **Chebyshev** response characteristic are useful when a rapid roll-off is required because it provides a roll-off rate greater than 20 dB/decade/pole. This is a greater rate than that of the Butterworth, so filters can be implemented with the Chebyshev response with fewer poles and less complex circuitry for a given roll-off rate. This type of filter response is characterized by overshoot or ripples in the pass band (depending on the number of poles) and an even less linear phase response than the Butterworth.

THE BESSEL CHARACTERISTIC

The **Bessel** response exhibits a linear phase characteristic, meaning that the phase shift increases linearly with frequency. The result is almost no overshoot on the output with a pulse input. For this reason, filters with the Bessel response are used for filtering pulse waveforms without distorting the shape of the waveform.

Butterworth, Chebyshev, or Bessel response characteristics can be realized with most active filter circuit configurations by proper selection of certain component values, as we will see later. A general comparison of the three response characteristics for a low-pass filter response curve is shown in Figure 8-5. High-pass and band-pass filters can also be designed to have any one of the characteristics.

THE DAMPING FACTOR

As mentioned, an active filter can be designed to have either a Butterworth, Chebyshev, or Bessel response characteristic regardless of whether it is a low-pass, high-pass, band-pass, or band-stop type. The **damping factor** (*DF*) of an active filter circuit determines which response characteristic the filter exhibits. To explain the basic concept, a generalized active filter is shown in Figure 8-6. It includes an amplifier, a negative

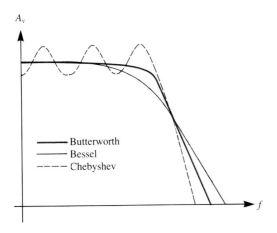

FIGURE 8-5
Comparative plots of three types of filter response characteristics.

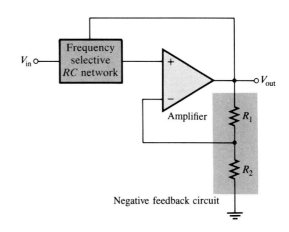

FIGURE 8-6
General diagram of an active filter.

feedback circuit, and a filter section. The amplifier and feedback are connected in a noninverting configuration. The damping factor is determined by the negative feedback circuit and is defined by the following equation:

$$DF = 2 - \frac{R_1}{R_2} \tag{8-6}$$

Basically, the damping factor affects the filter response by negative feedback action. Any attempted increase or decrease in the output voltage is offset by the opposing effect of the negative feedback. This tends to make the response curve flat in the pass band of the filter if the value for the damping factor is precisely set. By advanced mathematics, which we will not cover, values for the damping factor have been derived for various orders of filters to achieve the maximally flat response of the Butterworth characteristic.

The value of the damping factor required to produce a desired response characteristic depends on the *order* (number of poles) of the filter. A *pole,* for our purposes, is simply a circuit with one resistor and one capacitor. The more poles a filter has, the faster its roll-off rate is. To achieve a second-order Butterworth response, for example, the damping factor must be 1.414. To implement this damping factor, the feedback resistor ratio must be

$$\frac{R_1}{R_2} = 2 - DF = 2 - 1.414 = 0.586$$

This ratio gives the closed-loop gain of the noninverting filter amplifier, $A_{cl(\text{NI})}$, a value of 1.586, derived as follows:

$$A_{cl(\text{NI})} = \frac{1}{B} = \frac{R_1 + R_2}{R_2} = \frac{R_1}{R_2} + 1 = 0.586 + 1 = 1.586$$

■ **EXAMPLE 8–2** If resistor R_2 in the feedback circuit of an active two-pole filter of the type in Figure 8–6 is 10 kΩ, what value must R_1 be to obtain a maximally flat Butterworth response?

SOLUTION

$$\frac{R_1}{R_2} = 0.586$$

$$R_1 = 0.586 R_2 = 0.586(10 \text{ k}\Omega) = 5860 \text{ }\Omega$$

Using the nearest standard 5 percent value of 5600 Ω will get very close to the ideal Butterworth response.

PRACTICE EXERCISE 8–2
What is the damping factor for $R_2 = 10$ kΩ and $R_1 = 5.6$ kΩ?

■

CRITICAL FREQUENCY AND ROLL-OFF RATE

The critical frequency is determined by the values of the resistor and capacitors in the *RC* network, as shown in Figure 8–6. For a single-pole (first-order) filter, as shown in Figure 8–7, the critical frequency is

$$f_c = \frac{1}{2\pi RC}$$

Although we show a low-pass configuration, the same formula is used for the f_c of a single-pole high-pass filter. The number of poles determines the roll-off rate of the filter. A Butterworth response produces 20 dB/decade/pole. So, a first-order (one-pole) filter has a roll-off of 20 dB/decade; a second-order (two-pole) filter has a roll-off rate of 40 dB/decade; a third-order (three-pole) filter has a roll-off rate of 60 dB/decade; and so on.

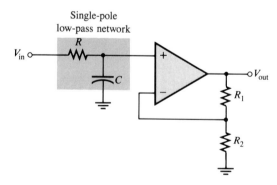

FIGURE 8–7
First-order (one-pole) low-pass filter.

Generally, to obtain a filter with three poles or more, one-pole or two-pole filters are cascaded, as shown in Figure 8–8. To obtain a third-order filter, for example, we cascade a second-order and a first-order filter; to obtain a fourth-order filter, we cascade two second-order filters; and so on. Each filter in a cascaded arrangement is called a *stage* or *section*.

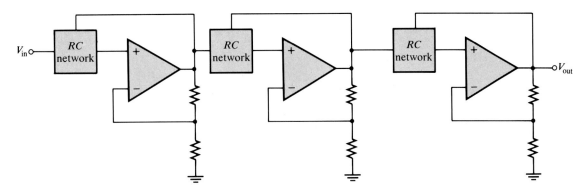

FIGURE 8–8
The number of filter poles can be increased by cascading.

Because of its maximally flat response, the Butterworth characteristic is the most widely used. Therefore, we will limit our coverage to the Butterworth response to illustrate basic filter concepts. Table 8–1 lists the roll-off rates, damping factors, and R_1/R_2 ratios for up to sixth-order Butterworth filters.

TABLE 8–1
Values for the Butterworth response

Order	Roll-off dB/decade	1st stage			2nd stage			3rd stage		
		Poles	DF	R_1/R_2	Poles	DF	R_1/R_2	Poles	DF	R_1/R_2
1	20	1	Optional							
2	40	2	1.414	0.586						
3	60	2	1.00	1	1	1.00	1			
4	80	2	1.848	0.152	2	0.765	1.235			
5	100	2	1.00	1	2	1.618	0.382	1	1.618	1.382
6	120	2	1.932	0.068	2	1.414	0.586	2	0.518	1.482

8–2 REVIEW QUESTIONS

1. Explain how Butterworth, Chebyshev, and Bessel responses differ.
2. What determines the response characteristic of a filter?
3. Name the basic parts of an active filter.

8–3 ACTIVE LOW-PASS FILTERS

Filters that use op-amps as the active element provide several advantages over passive filters (R, L, and C elements only). The op-amp provides gain, so that the signal is not attenuated as it passes through the filter. The high input impedance of the op-amp prevents excessive loading of the driving source, and the low output imped-

ance of the op-amp prevents the filter from being affected by the load that it is driving. Active filters are also easy to adjust over a wide frequency range without altering the desired response.

A SINGLE-POLE FILTER

Figure 8–9(a) shows an active filter with a single low-pass RC network that provides a roll-off of -20 dB/decade above the critical frequency, as indicated by the response curve in Figure 8–9(b). The critical frequency of the single-pole filter is $f_c = 1/2\pi RC$. The op-amp in this filter is connected as a noninverting amplifier with the closed-loop voltage gain in the pass band set by the values of R_1 and R_2 ($A_{cl} = R_1/R_2 + 1$).

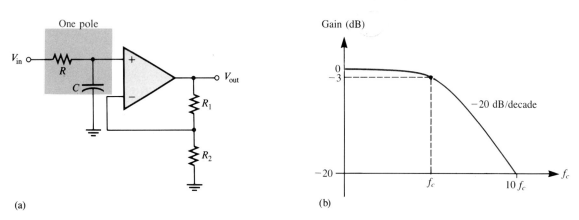

FIGURE 8–9
Single-pole active low-pass filter and response curve.

THE SALLEN-KEY LOW-PASS FILTER

The Sallen-Key is one of the most common configurations for a second-order (two-pole) filter. It is also known as a VCVS (voltage-controlled voltage source) filter. A low-pass version of the Sallen-Key filter is shown in Figure 8–10. Notice that there are two low-pass RC networks that provide a roll-off of -40 dB/decade above the critical frequency (assuming a Butterworth characteristic). One RC network consists of R_A and C_A, and the second network consists of R_B and C_B. A unique feature is the capacitor C_A that provides feedback for shaping the response near the edge of the pass band. The critical frequency for the second-order Sallen-Key filter is

$$f_c = \frac{1}{2\pi\sqrt{R_A R_B C_A C_B}} \tag{8-7}$$

For simplicity, the component values can be made equal so that $R_A = R_B = R$ and $C_A = C_B = C$. In this case, the expression for the critical frequency simplifies to $f_c = 1/2\pi RC$.

As in the single-pole filter, the op-amp in the second-order Sallen-Key filter acts as a noninverting amplifier with the negative feedback provided by the R_1/R_2 network. As you have learned, the damping factor is set by the values of R_1 and R_2, thus making the

FIGURE 8–10
Basic Sallen-Key second-order low-pass filter.

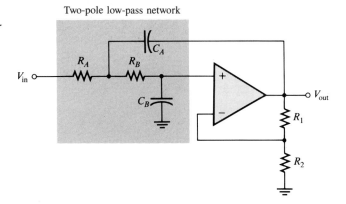

filter response either Butterworth, Chebyshev, or Bessel. For example, from Table 8–1, the R_1/R_2 ratio must be 0.586 to produce the damping factor of 1.414 required for a second-order Butterworth response.

EXAMPLE 8–3

Determine the critical frequency of the low-pass filter in Figure 8–11, and set the value of R_1 for an approximate Butterworth response.

FIGURE 8–11

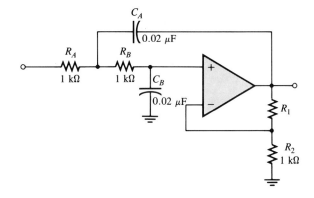

SOLUTION
Since $R_A = R_B = 1\ \text{k}\Omega$ and $C_A = C_B = 0.02\ \mu\text{F}$,

$$f_c = \frac{1}{2\pi RC} = \frac{1}{2\pi(1\ \text{k}\Omega)(0.02\ \mu\text{F})} = 7.958\ \text{kHz}$$

For a Butterworth response, $R_1/R_2 = 0.586$.

$$R_1 = 0.586 R_2 = 0.586(1\ \text{k}\Omega) = 586\ \Omega$$

Select a standard value as near as possible to this calculated value.

PRACTICE EXERCISE 8–3
Determine f_c for Figure 8–11 if $R_A = R_B = R_2 = 2.2\ \text{k}\Omega$ and $C_A = C_B = 0.01\ \mu\text{F}$. Also determine the value of R_1 for a Butterworth response.

Cascaded Low-Pass Filters Achieve a Higher Roll-Off Rate

A three-pole filter is required to get a third-order low-pass response (−60 dB/decade). This is done by cascading a two-pole low-pass filter and a single-pole low-pass filter, as shown in Figure 8–12(a). Figure 8–12(b) shows a four-pole configuration obtained by cascading two two-pole filters.

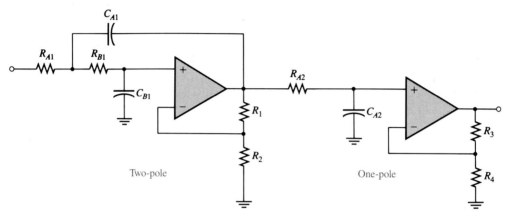

(a) Third-order

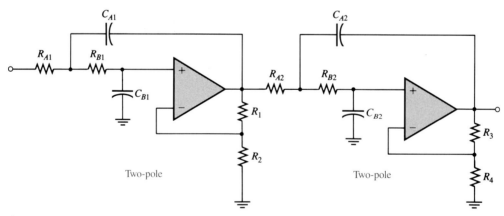

(b) Fourth-order

FIGURE 8–12
Cascaded low-pass filters.

EXAMPLE 8–4 For the four-pole filter in Figure 8–12(b), determine the capacitance values required to produce a critical frequency of 2680 Hz if all the resistors are 1.8 kΩ. Also select values for the feedback resistors to get a Butterworth response.

SOLUTION
Both stages must have the same f_c. Assuming equal-value capacitors,

$$f_c = \frac{1}{2\pi RC}$$

$$C = \frac{1}{2\pi R f_c} = \frac{1}{2\pi(1.8 \text{ k}\Omega)(2680 \text{ Hz})} = 0.032 \text{ }\mu\text{F}$$

$$C_{A1} = C_{B1} = C_{A2} = C_{B2} = 0.032 \text{ }\mu\text{F}$$

Select $R_2 = R_4 = 1.8$ kΩ for simplicity. Refer to Table 8–1.
For a Butterworth response in the first stage,

$$DF = 1.848, \quad \frac{R_1}{R_2} = 0.152$$

$$R_1 = 0.152 R_2 = 0.152(1800 \text{ }\Omega) = 273.6 \text{ }\Omega$$

Choose $R_1 = 270$ Ω.
In the second stage,

$$DF = 0.765, \quad \frac{R_3}{R_4} = 1.235$$

$$R_3 = 1.235 R_4 = 1.235(1800 \text{ }\Omega) = 2.223 \text{ k}\Omega$$

Choose $R_3 = 2.2$ kΩ.

PRACTICE EXERCISE 8–4
For the filter in Figure 8–12(b), determine the capacitance values for $f_c = 1$ kHz if all the filter resistors are 680 Ω. Also specify the values for the feedback resistors to produce a Butterworth response. ∎

8–3 REVIEW QUESTIONS

1. How many poles does a second-order low-pass filter have? How many resistors and how many capacitors are used in the frequency-selective network?
2. Why is the damping factor of a filter important?
3. What is the primary purpose of cascading low-pass filters?

8–4 ACTIVE HIGH-PASS FILTERS

In high-pass filters, the roles of the capacitor and resistor are reversed in the RC networks. Otherwise, the basic considerations are the same as for the low-pass filters.

A SINGLE-POLE FILTER

A high-pass active filter with a 20 dB/decade roll-off is shown in Figure 8–13(a). Notice that the input circuit is a single high-pass RC network. The negative feedback circuit is the same as for the low-pass filters previously discussed. The high-pass response curve is shown in Figure 8–13(b).

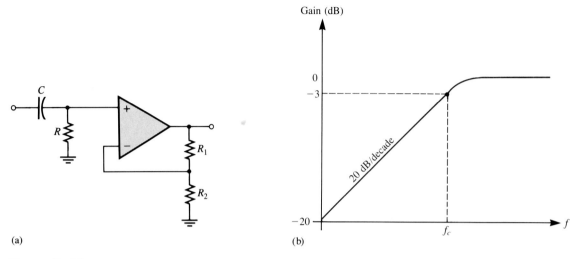

FIGURE 8–13
Single-pole active high-pass filter and response curve.

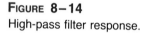

FIGURE 8–14
High-pass filter response.

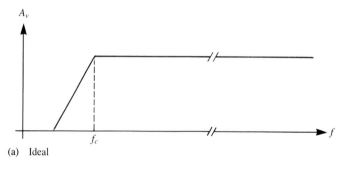

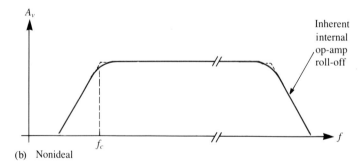

Ideally, a high-pass filter passes all frequencies above f_c without limit, as indicated in Figure 8–14(a), although in practice, this is not the case. As you have learned, all op-amps inherently have internal RC networks that limit the amplifier's response at high frequencies. Therefore, there is an upper-frequency limit on the high-pass filter's response which, in effect, makes it a band-pass filter with a very wide bandwidth. In the majority

THE SALLEN-KEY HIGH-PASS FILTER

A high-pass second-order Sallen-Key configuration is shown in Figure 8–15. The components R_A, C_A, R_B, and C_B form the two-pole frequency-selective network. Notice that the positions of the resistors and capacitors in the frequency-selective network are opposite to those in the low-pass configuration. As with the other filters, the response characteristic can be optimized by proper selection of the feedback resistors, R_1 and R_2.

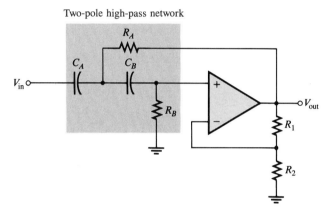

FIGURE 8–15
Basic Sallen-Key second-order high-pass filter.

■ EXAMPLE 8–5 Choose values for the Sallen-Key high-pass filter in Figure 8–15 to implement an equal-value second-order Butterworth response with a critical frequency of approximately 10 kHz.

SOLUTION
Start by selecting a value for R_A and R_B (R_1 or R_2 can also be the same value as R_A and R_B for simplicity).

$$R = R_A = R_B = R_2 = 3.3 \text{ k}\Omega \quad \text{(an arbitrary selection)}$$

Next, calculate the capacitance value from $f_c = 1/2\pi RC$.

$$C = C_A = C_B = \frac{1}{2\pi R f_c} = \frac{1}{2\pi(3.3 \text{ k}\Omega)(10 \text{ kHz})} = 0.004 \text{ }\mu\text{F}$$

For a Butterworth response, the damping factor must be 1.414 and $R_1/R_2 = 0.586$.

$$R_1 = 0.586 R_2 = 0.586(3.3 \text{ k}\Omega) = 1.93 \text{ k}\Omega$$

If we had let $R_1 = 3.3 \text{ k}\Omega$, then

$$R_2 = \frac{R_1}{0.586} = \frac{3.3 \text{ k}\Omega}{0.586} = 5.63 \text{ k}\Omega$$

Either way, an approximate Butterworth response is realized by choosing the nearest standard value.

PRACTICE EXERCISE 8–5
Select values for all the components in the high-pass filter of Figure 8–15 to obtain an $f_c = 300$ Hz. Use equal-value components and optimize for a Butterworth response.

CASCADING HIGH-PASS FILTERS

As with the low-pass configuration, first- and second-order high-pass filters can be cascaded to provide three or more poles and thereby create faster roll-off rates. Figure 8–16 shows a six-pole high-pass filter consisting of three two-pole stages. With this configuration optimized for a Butterworth response, a roll-off of 120 dB/decade is achieved.

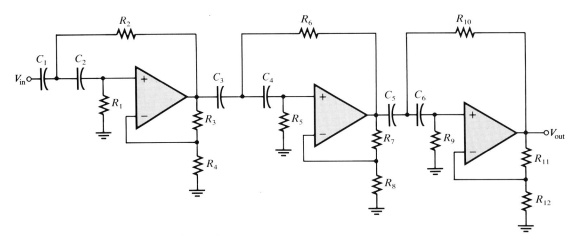

FIGURE 8–16
Sixth-order high-pass filter.

8–4 REVIEW QUESTIONS

1. How does a high-pass Sallen-Key filter differ from the low-pass configuration?
2. To increase the critical frequency of a high-pass filter, would you increase or decrease the resistor values?
3. If three two-pole high-pass filters and one single-pole high-pass filter are cascaded, what is the resulting roll-off?

8–5 ACTIVE BAND-PASS FILTERS

As mentioned, band-pass filters pass all frequencies bounded by a lower- and an upper-frequency limit and reject all others lying outside this specified band. A band-pass response can be thought of as the overlapping of a low-frequency response curve and a high-frequency response curve.

Cascaded Low-Pass and High-Pass Filters Achieve a Band-Pass Response

One way to implement a band-pass filter is a cascaded arrangement of a high-pass filter and a low-pass filter, as shown in Figure 8–17(a), as long as the critical frequencies are sufficiently separated. Each of the filters shown is a two-pole Sallen-Key Butterworth configuration so that the roll-off rates are ±40 dB/decade, indicated in the composite response curve of Figure 8–17(b). The critical frequency of each filter is chosen so that the response curves overlap sufficiently, as indicated. The critical frequency of the high-pass filter must be sufficiently lower than that of the low-pass stage.

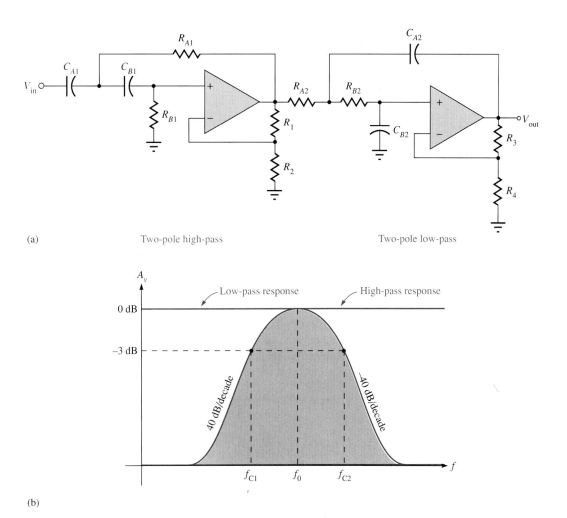

Figure 8–17
Band-pass filter formed by cascading a two-pole high-pass and a two-pole low-pass filter (it does not matter in which order the filters are cascaded).

The lower frequency f_{c1} of the pass band is the critical frequency of the high-pass filter. The upper frequency f_{c2} is the critical frequency of the low-pass filter. Ideally, as discussed earlier, the center frequency f_0 of the pass band is the geometric mean of f_{c1} and f_{c2}. The following formulas express the three frequencies of the band-pass filter in Figure 8–17.

$$f_{c1} = \frac{1}{2\pi\sqrt{R_{A1}R_{B1}C_{A1}C_{B1}}}$$

$$f_{c2} = \frac{1}{2\pi\sqrt{R_{A2}R_{B2}C_{A2}C_{B2}}}$$

$$f_0 = \sqrt{f_{c1}f_{c2}}$$

Of course, if equal-value components are used in implementing each filter, the critical frequency equations simplify to the form $f_c = 1/2\pi RC$.

MULTIPLE-FEEDBACK BAND-PASS FILTER

Another type of filter configuration, shown in Figure 8–18, is a multiple-feedback band-pass filter. The two feedback paths are through R_2 and C_1. Components R_1 and C_1 provide the low-pass response, and R_2 and C_2 provide the high-pass response. The maximum gain, A_0, occurs at the center frequency. Q values of less than 10 are typical in this type of filter. An expression for the center frequency is developed as follows, recognizing that R_1 and R_3 appear in parallel as viewed from the C_1 feedback path (with the V_{in} source replaced by a short).

$$f_0 = \frac{1}{2\pi\sqrt{(R_1\|R_3)R_2C_1C_2}}$$

Making $C_1 = C_2 = C$ yields

$$f_0 = \frac{1}{2\pi\sqrt{(R_1\|R_3)R_2C^2}}$$

$$= \frac{1}{2\pi C\sqrt{(R_1\|R_3)R_2}}$$

$$= \frac{1}{2\pi C}\sqrt{\frac{1}{R_2(R_1\|R_3)}}$$

$$= \frac{1}{2\pi C}\sqrt{\left(\frac{1}{R_2}\right)\left(\frac{1}{R_1R_3/(R_1+R_3)}\right)}$$

$$f_0 = \frac{1}{2\pi C}\sqrt{\frac{R_1+R_3}{R_1R_2R_3}} \qquad (8-8)$$

A convenient value for the capacitors is chosen, then the three resistor values are calculated based on the desired values for f_0, BW, and A_0. As you know, the Q can be

FIGURE 8–18
Multiple-feedback band-pass filter.

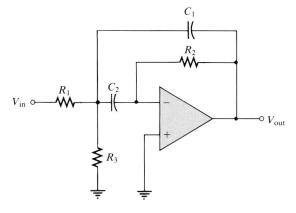

determined from the relation $Q = f_0/BW$, and the resistors are found using the following formulas (stated without derivation).

$$R_1 = \frac{Q}{2\pi f_0 C A_0}$$

$$R_2 = \frac{Q}{\pi f_0 C}$$

$$R_3 = \frac{Q}{2\pi f_0 C(2Q^2 - A_0)}$$

To develop a gain expression, we solve for Q in the first two equations above.

$$Q = 2\pi f_0 A_0 C R_1$$
$$Q = \pi f_0 C R_2$$

Then,

$$2\pi f_0 A_0 C R_1 = \pi f_0 C R_2$$

Cancelling, we get

$$2A_0 R_1 = R_2$$

$$A_0 = \frac{R_2}{2R_1} \qquad (8\text{–}9)$$

In order for the denominator of the equation $R_3 = Q/2\pi f_0 C(2Q^2 - A_0)$ to be positive, $A_0 < 2Q^2$, which imposes a limitation on the gain.

EXAMPLE 8–6 Determine the center frequency, maximum gain, and bandwidth for the filter in Figure 8–19.

FIGURE 8–19

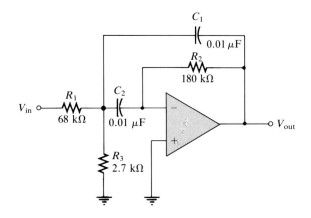

SOLUTION

$$f_0 = \frac{1}{2\pi C}\sqrt{\frac{R_1 + R_3}{R_1 R_2 R_3}}$$

$$= \frac{1}{2\pi(0.01\ \mu F)}\sqrt{\frac{68\ k\Omega + 2.7\ k\Omega}{(68\ k\Omega)(180\ k\Omega)(2.7\ k\Omega)}}$$

$$= 736\ Hz$$

$$A_0 = \frac{R_2}{2R_1} = \frac{180\ k\Omega}{2(68\ k\Omega)} = 1.32$$

$$Q = \pi f_0 C R_2$$
$$= \pi(736\ Hz)(0.01\ \mu F)(180\ k\Omega)$$
$$= 4.16$$

$$BW = \frac{f_0}{Q} = \frac{736\ Hz}{4.16} = 176.9\ Hz$$

PRACTICE EXERCISE 8–6
If R_2 in Figure 8–19 is increased to 330 kΩ, how does this affect the gain, center frequency, and bandwidth of the filter?

STATE-VARIABLE BAND-PASS FILTER

The state-variable or universal active filter is widely used for band-pass applications. As shown in Figure 8–20, it consists of a summing amplifier and two op-amp integrators (which act as single-pole low-pass filters) that are combined in a cascaded arrangement to form a second-order filter. Although used primarily as a band-pass filter, the state-variable configuration also provides low-pass (*LP*) and high-pass (*HP*) outputs. The center frequency is set by the *RC* networks in both integrators. When used as a band-pass filter, the critical frequencies of the integrators are usually made equal, thus setting the center frequency of the pass band.

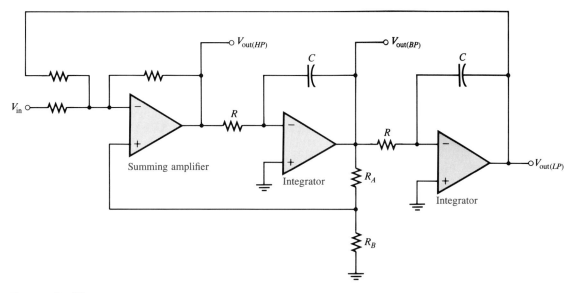

FIGURE 8–20
State-variable band-pass filter.

BASIC OPERATION At input frequencies below f_c, the input signal passes through the summing amplifier and integrators and is fed back 180° out-of-phase. Thus, the feedback signal and input signal cancel for all frequencies below approximately f_c. As the low-pass response of the integrators rolls off, the feedback signal diminishes, thus allowing the input to pass through to the band-pass output. Above f_c, the low-pass response disappears, thus preventing the input signal from passing through the integrators. As a result, the band-pass output peaks sharply at f_c, as indicated in Figure 8–21. Stable Qs up to 100 can be obtained with this type of filter. The Q is set by the feedback resistors R_A and R_B according to the following equation.

$$Q = \frac{1}{3}\left(\frac{R_A}{R_B} + 1\right) \tag{8-10}$$

FIGURE 8–21
General state-variable response curves.

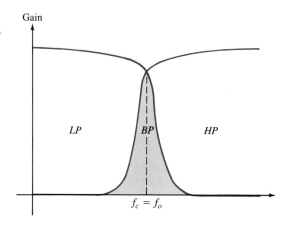

The state-variable filter cannot be optimized for low-pass, high-pass, and band-pass performance simultaneously for this reason: To optimize for a low-pass or a high-pass Butterworth response, DF must equal 1.414. Since $Q = 1/DF$, a Q of 0.707 will result. Such a low Q provides a very poor band-pass response (large BW and poor selectivity). For optimization as a band-pass filter, the Q must be set high.

EXAMPLE 8–7

Determine the center frequency, Q, and BW for the band-pass output of the state-variable filter in Figure 8–22.

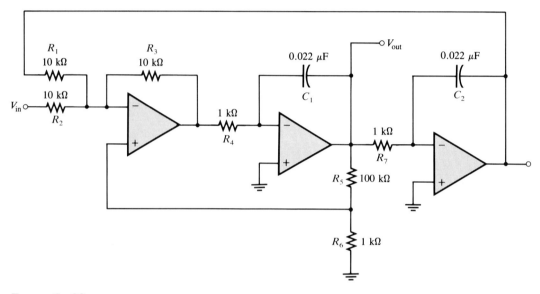

FIGURE 8–22

SOLUTION
For each integrator,

$$f_c = \frac{1}{2\pi R_4 C_1} = \frac{1}{2\pi R_7 C_2} = \frac{1}{2\pi (1 \text{ k}\Omega)(0.022 \text{ }\mu\text{F})} = 7.23 \text{ kHz}$$

The center frequency is approximately equal to the critical frequencies of the integrators.

$$f_0 = f_c = 7.23 \text{ kHz}$$

$$Q = \frac{1}{3}\left(\frac{R_5}{R_6} + 1\right) = \frac{1}{3}\left(\frac{100 \text{ k}\Omega}{1 \text{ k}\Omega} + 1\right) = 33.67$$

$$BW = \frac{f_0}{Q} = \frac{7.23 \text{ kHz}}{33.67} = 214.7 \text{ Hz}$$

PRACTICE EXERCISE 8–7
Determine f_0, Q, and BW for the filter in Figure 8–22 if $R_4 = R_6 = R_7 = 330 \text{ }\Omega$ with all other component values the same as shown on the schematic.

8-5 REVIEW QUESTIONS

1. What determines selectivity in a band-pass filter?
2. One filter has a $Q = 5$ and another has a $Q = 25$. Which has the narrower bandwidth?
3. List the elements that make up a state-variable filter.

8-6 ACTIVE BAND-STOP FILTERS

Band-stop filters reject a specified band of frequencies and pass all others. The response is opposite to that of a band-pass filter.

MULTIPLE-FEEDBACK BAND-STOP FILTER

Figure 8–23 shows a multiple-feedback band-stop filter. Notice that this configuration is similar to the band-pass version except that R_3 has been omitted and R_A and R_B have been added.

FIGURE 8–23
Multiple-feedback band-stop filter.

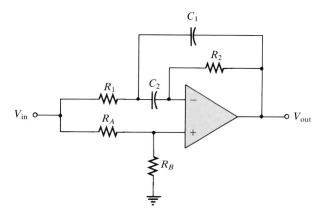

STATE-VARIABLE BAND-STOP FILTER

Summing the low-pass and the high-pass responses of the state-variable filter covered in Section 8–5 creates a band-stop response as shown in Figure 8–24. One important

FIGURE 8–24
State-variable band-stop filter.

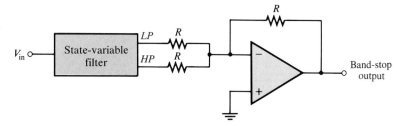

application of this filter is minimizing the 60 Hz "hum" in audio systems by setting the center frequency to 60 Hz.

EXAMPLE 8–8 Verify that the band-stop filter in Figure 8–25 has a center frequency of 60 Hz, and optimize it for a Q of 30.

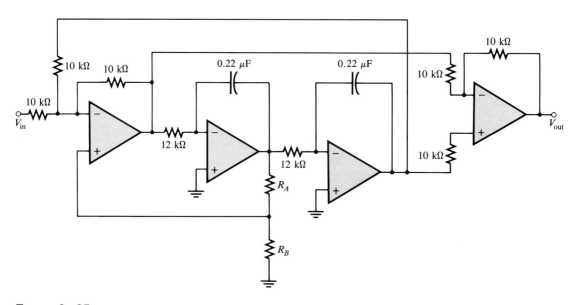

FIGURE 8–25

SOLUTION
f_0 equals the f_c of the integrator stages.

$$f_0 = \frac{1}{2\pi RC} = \frac{1}{2\pi(12\ \text{k}\Omega)(0.22\ \mu\text{F})} = 60\ \text{Hz}$$

We obtain a $Q = 30$ by choosing R_B and then calculating R_A.

$$Q = \frac{1}{3}\left(\frac{R_A}{R_B} + 1\right)$$

$$R_A = (3Q - 1)R_B$$

Choose $R_B = 1\ \text{k}\Omega$. Then

$$R_A = [3(30) - 1]1\ \text{k}\Omega = 89\ \text{k}\Omega$$

PRACTICE EXERCISE 8–8
How would you change the center frequency to 120 Hz in Figure 8–25?

8–6 Review Questions

1. How does a band-stop response differ from a band-pass response?
2. How is a state-variable band-pass filter converted to a band-stop filter?

8–7 Filter Response Measurements

In this section, we discuss two methods of determining a filter's response by measurement—discrete point measurement and swept frequency measurement.

Discrete Point Measurement

Figure 8–26 shows an arrangement for taking filter output voltage measurements at discrete values of input frequency using common laboratory instruments. The general procedure is as follows:

1. Set the amplitude of the sine wave generator to a desired voltage level.
2. Set the frequency of the sine wave generator to a value well below the expected critical frequency of the filter under test. For a low-pass filter, set the frequency as near as possible to 0 Hz. For a band-pass filter, set the frequency well below the expected lower critical frequency.
3. Increase the frequency in predetermined steps sufficient to allow enough data points for an accurate response curve.
4. Maintain a constant input voltage amplitude while varying the frequency.
5. Record the output voltage at each value of frequency.
6. After recording a sufficient number of points, plot a graph of output voltage versus frequency.

If the frequencies to be measured exceed the response of the DMM, an oscilloscope may have to be used instead.

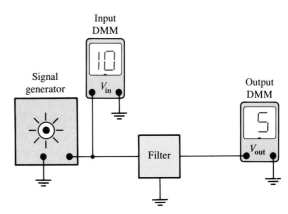

Figure 8–26
Test set-up for discrete point measurement of the filter response. (Readings are arbitrary and for display only.)

Swept Frequency Measurement

The swept frequency method requires more elaborate test equipment than does the discrete point method, but it is much more efficient and can result in a more accurate response curve. A general test set-up is shown in Figure 8–27 using a swept frequency generator and a spectrum analyzer.

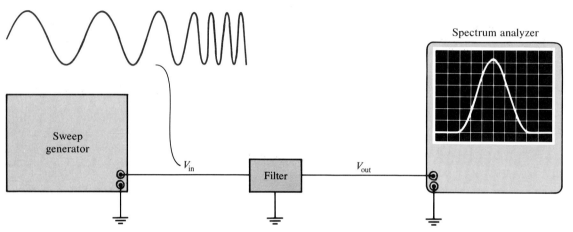

Figure 8–27
Test set-up for swept frequency measurement of the filter response.

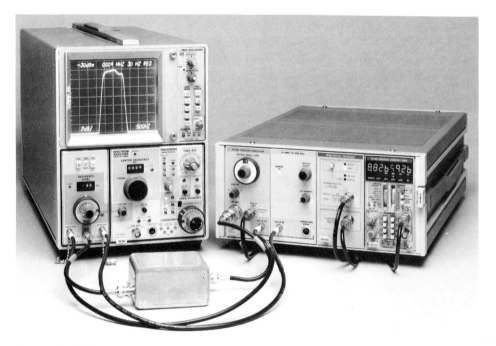

Figure 8–28
Test set-up with actual equipment (courtesy of Tektronix, Inc.).

ACTIVE FILTERS   397

The swept frequency generator produces a constant amplitude output signal whose frequency increases linearly between two preset limits, as indicated in Figure 8–27. The spectrum analyzer is essentially an elaborate oscilloscope that can be calibrated for a desired *frequency span/division* rather than for the usual *time/division* setting. Therefore, as the input frequency to the filter sweeps through a preselected range, the response curve is traced out on the screen of the spectrum analyzer. An actual swept frequency test set-up using typical equipment is shown in Figure 8–28.

8–7 REVIEW QUESTIONS

1. What is the purpose of the two tests discussed in this section?
2. Name one disadvantage and one advantage of each test method.

8–8 A SYSTEM APPLICATION

In this system application, the focus is on the filter board, which is part of the channel separation circuits in the FM stereo receiver. In addition to the active filters, the left and right channel separation circuit includes a demodulator, a frequency doubler, and a stereo matrix. Except for mentioning their purpose, we will not deal specifically with the demodulator, doubler, or matrix. However, the matrix is an interesting application of summing amplifiers, which were studied in Chapter 7 and these will be shown in detail on the schematic although we will not concentrate on them. In this section, you will

- *See how low-pass and band-pass active filters are used.*
- *Use the schematic to locate and identify the components on the pc board.*
- *Analyze the operation of the filters.*
- *Troubleshoot some common amplifier failures.*

A BRIEF DESCRIPTION OF THE SYSTEM

Stereo FM **(frequency modulation)** signals are transmitted on a **carrier** frequency of 88 MHz to 108 MHz. The standard transmitted stereo signal consists of three modulating signals. These are the sum of the left and right channel audio (L + R), the difference of the left and right channel audio (L − R), and a 19 kHz pilot subcarrier. The L + R audio extends from 30 Hz to 15 kHz and the L − R signal is contained in two sidebands extending from 23 kHz to 53 kHz as indicated in Figure 8–29. These frequencies come from the FM detector and go into the filter circuits where they are separated. The frequency doubler and demodulator are used to extract the audio signal from the 23 kHz to 53 kHz sidebands after which the 30 Hz to 15 kHz L − R signal is passed through a filter. The L + R and L − R audio signals are then sent to the matrix where they are applied to the summing circuits to produce the left and right channel audio (−2L and −2R). Our focus is on the filters.

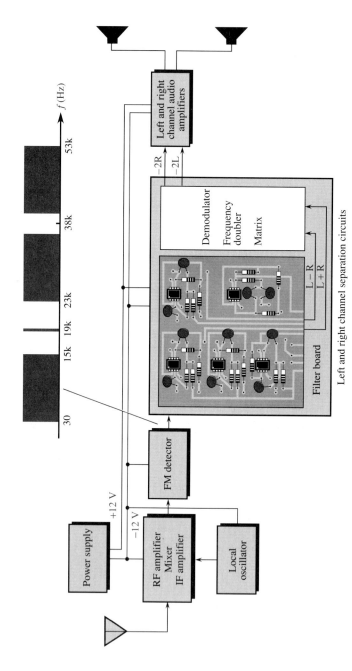

FIGURE 8–29
FM stereo receiver system.

ACTIVE FILTERS 399

Now, so that you can take a closer look at the filter board, let's take it out of the system and put it on the test bench.

On the Test Bench

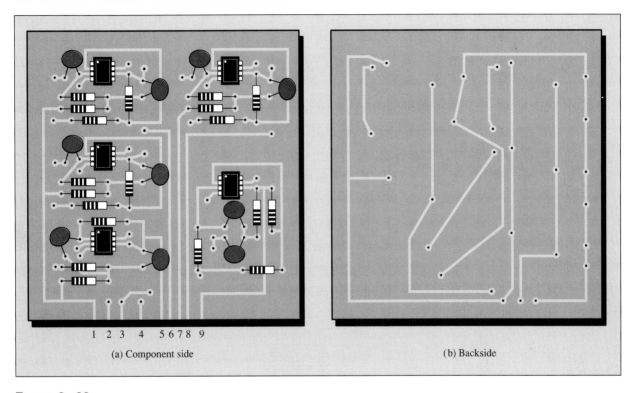

1 2 3 4 5 6 7 8 9
(a) Component side   (b) Backside

FIGURE 8–30

■ **ACTIVITY 1 RELATE THE PC BOARD TO THE SCHEMATIC**

Locate and identify all components on the pc board in Figure 8–30, using the schematic diagram in Figure 8–31. Label the pc board components to correspond with the schematic. Identify all inputs and outputs. Trace out the pc board to verify that it corresponds with the schematic. The shaded areas on the schematic are the filter circuits contained on the board. The other blocks and circuits are on the demodulator, frequency doubler, and matrix board located elsewhere.

■ **ACTIVITY 2 ANALYZE THE FILTER CIRCUITS**

  STEP 1 Using the component values, determine the critical frequencies of each Sallen-Key-type filter

  STEP 2 Using the component values, determine the center frequency of the multiple-feedback filter.

  STEP 3 Determine the bandwidth of each filter.

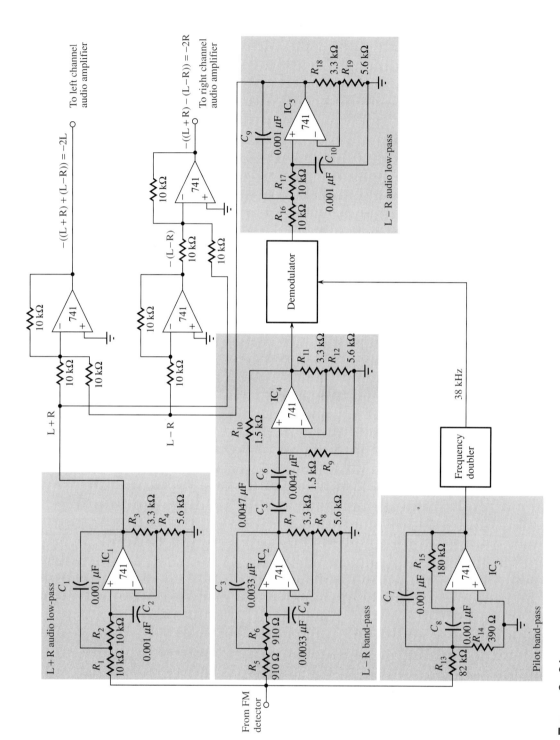

FIGURE 8–31 Left and right channel separation circuits.

ACTIVE FILTERS

STEP 4 Determine the voltage gain of each filter.

STEP 5 Verify that the Sallen-Key filters have an approximate Butterworth response characteristic.

■ **ACTIVITY 3   WRITE A TECHNICAL REPORT**

Describe each filter in detail specifying the type of filter, the frequency responses, and the function of the filter within the overall circuitry. Also, describe the overall operation of the complete channel separation circuitry.

■ **ACTIVITY 4   TROUBLESHOOT THE FILTER BOARD**

The filter board is plugged into a test fixture that permits access to each input and output, as shown in Figure 8-32, where the socket numbers correspond to the board pin numbers. The text instruments to be used are a sweep generator, a spectrum analyzer, and a dual power supply. For the sweep generator, a minimum and a maximum frequency is selected and the instrument produces an output that repetitively sweeps through all frequencies between the minimum and maximum setting. The spectrum analyzer is a type of oscilloscope that will plot out a frequency response curve.

Develop a basic test procedure for completely testing the board in the fixture, using general references to instrument inputs, outputs, and settings. Include a diagram of a complete test set-up.

FIGURE 8-32
Filter board in a test fixture.

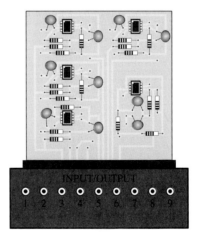

8-8 REVIEW QUESTIONS

1. What is the purpose of the filter board in this system?
2. What is the bandwidth of the L + R low-pass filter?
3. What is the bandwidth of the L − R low-pass filter?
4. Which filters on the board have approximate Butterworth responses?
5. What is the purpose of the stereo matrix circuitry?

Summary

- The bandwidth in a low-pass filter equals the critical frequency because the response extends to 0 Hz.
- The bandwidth in a high-pass filter extends above the critical frequency and is limited only by the inherent frequency limitation of the active circuit.
- A band-pass filter passes all frequencies within a band between a lower and an upper critical frequency and rejects all others outside this band.
- The bandwidth of a band-pass filter is the difference between the upper critical frequency and the lower critical frequency.
- A band-stop filter rejects all frequencies within a specified band and passes all those outside this band.
- Filters with the Butterworth response characteristic have a very flat response in the pass band, exhibit a roll-off of 20 dB/decade/pole, and are used when all the frequencies in the pass band must have the same gain.
- Filters with the Chebyshev characteristic have ripples or overshoot in the pass band and exhibit a faster roll-off per pole than filters with the Butterworth characteristic.
- Filters with the Bessel characteristic are used for filtering pulse waveforms. Their linear phase characteristic results in minimal waveshape distortion. The roll-off rate per pole is slower than for the Butterworth.
- In filter terminology, a single RC network is called a *pole*.
- Each pole in a Butterworth filter causes the output to roll off at a rate of 20 dB/decade.
- The quality factor Q of a band-pass filter determines the filter's selectivity. The higher the Q, the narrower the bandwidth and the better the selectivity.
- The damping factor determines the filter response characteristic (Butterworth, Chebyshev, or Bessel).

Glossary

Active filter A frequency-selective circuit consisting of active devices such as transistors or op-amps coupled with reactive components.

Band-pass filter A type of filter that passes a range of frequencies lying between a certain lower frequency and a certain higher frequency.

Band-stop filter A type of filter that blocks or rejects a range of frequencies lying between a certain lower frequency and a certain higher frequency.

Bessel A type of filter response having a linear phase characteristic and less than 20 dB/decade/pole roll-off.

Butterworth A type of filter response characterized by flatness in the pass band and a 20 dB/decade/pole roll-off.

Carrier The high frequency (RF) signal that carries modulated information in AM, FM, or other systems.

Chebyshev A type of filter response characterized by ripples in the pass band and a greater than 20 dB/decade/pole roll-off.

Damping factor A filter characteristic that determines the type of response.

Frequency modulation (FM) A communication method in which a lower frequency intelligence-carrying signal modulates (varies) the frequency of a higher frequency signal.

High-pass filter A type of filter that passes frequencies above a certain frequency while rejecting lower frequencies.

Low-pass filter A type of filter that passes frequencies below a certain frequency while rejecting higher frequencies.

Pole A network containing one resistor and one capacitor that contributes 20 dB/decade to a filter's roll-off rate.

Quality factor (Q) The ratio of a band-pass filter's center frequency to its bandwidth.

FORMULAS

(8–1) $\quad f_c = \dfrac{1}{2\pi RC}\quad$ Filter critical frequency

(8–2) $\quad BW = f_{c2} - f_{c1}\quad$ Filter bandwidth

(8–3) $\quad f_0 = \sqrt{f_{c1}f_{c2}}\quad$ Center frequency of a band-pass filter

(8–4) $\quad Q = \dfrac{f_0}{BW}\quad$ Quality factor of a band-pass filter

(8–5) $\quad Q = \dfrac{1}{DF}\quad$ Q in terms of damping factor

(8–6) $\quad DF = 2 - \dfrac{R_1}{R_2}\quad$ Damping factor

(8–7) $\quad f_c = \dfrac{1}{2\pi\sqrt{R_A R_B C_A C_B}}\quad$ Critical frequency for a second-order Sallen-Key filter

(8–8) $\quad f_0 = \dfrac{1}{2\pi C}\sqrt{\dfrac{R_1 + R_3}{R_1 R_2 R_3}}\quad$ Center frequency of a multiple-feedback filter

(8–9) $\quad A_0 = \dfrac{R_2}{2R_1}\quad$ Gain of a multiple-feedback filter

(8–10) $\quad Q = \dfrac{1}{3}\left(\dfrac{R_A}{R_B} + 1\right)\quad$ Q of a state-variable filter

Self-Test

1. The term *pole* in filter terminology refers to
   - (a) a high-gain op-amp
   - (b) one complete active filter
   - (c) a single *RC* network
   - (d) the feedback circuit
2. An *RC* circuit produces a roll-off rate of
   - (a) −20 dB/decade
   - (b) −40 dB/decade
   - (c) −6 dB/octave
   - (d) a and c
3. A band-pass response has
   - (a) two critical frequencies
   - (b) one critical frequency
   - (c) a flat curve in the pass band
   - (d) a wide bandwidth
4. The lowest frequency passed by a low-pass filter is
   - (a) 1 Hz
   - (b) 0 Hz
   - (c) 10 Hz
   - (d) dependent on the critical frequency
5. The Q of a band-pass filter depends on
   - (a) the critical frequencies
   - (b) only the bandwidth
   - (c) the center frequency and the bandwidth
   - (d) only the center frequency
6. The damping factor of an active filter determines
   - (a) the voltage gain
   - (b) the critical frequency
   - (c) the response characteristic
   - (d) the roll-off rate
7. A maximally flat frequency response is known as
   - (a) Chebyshev
   - (b) Butterworth
   - (c) Bessel
   - (d) Colpitts
8. The damping factor of a filter is set by
   - (a) the negative feedback circuit
   - (b) the positive feedback circuit
   - (c) the frequency-selective circuit
   - (d) the gain of the op-amp
9. The number of poles in a filter affect the
   - (a) voltage gain
   - (b) bandwidth
   - (c) center frequency
   - (d) roll-off rate
10. Sallen-Key filters are
    - (a) single-pole filters
    - (b) second-order filters
    - (c) Butterworth filters
    - (d) band-pass filters
11. When filters are cascaded, the roll-off rate
    - (a) increases
    - (b) decreases
    - (c) does not change

ACTIVE FILTERS

12. When a low-pass and a high-pass filter are cascaded to get a band-pass filter, the critical frequency of the low-pass filter must be
    (a) equal to the critical frequency of the high-pass filter
    (b) less than the critical frequency of the high-pass filter
    (c) greater than the critical frequency of the high-pass filter
13. A state-variable filter consists of
    (a) one op-amp with multiple-feedback paths
    (b) a summing amplifier and two integrators
    (c) a summing amplifier and two differentiators
    (d) three Butterworth stages
14. When the gain of a filter is minimum at its center frequency, it is a
    (a) band-pass filter   (b) a band-stop filter
    (c) a notch filter   (d) b and c

PROBLEMS

SECTION 8–1 BASIC FILTER RESPONSES

1. Identify each type of filter response (low-pass, high-pass, band-pass, or band-stop) in Figure 8–33.

FIGURE 8–33

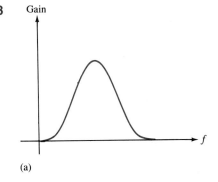

(a)

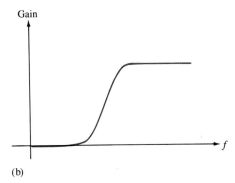

(b)

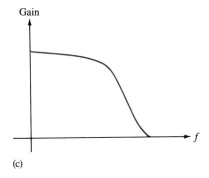

(c)

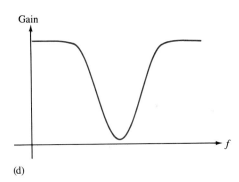

(d)

2. A certain low-pass filter has a critical frequency of 800 Hz. What is its bandwidth?

3. A single-pole high-pass filter has a frequency-selective circuit with $R = 2.2$ kΩ and $C = 0.0015$ μF. What is the critical frequency? Can you determine the bandwidth from the available information?

4. What is the roll-off rate of the filter described in Problem 3?

5. What is the bandwidth of a band-pass filter whose critical frequencies are 3.2 kHz and 3.9 kHz? What is the Q of this filter?

6. What is the center frequency of a filter with a Q of 15 and a bandwidth of 1 kHz?

SECTION 8–2 FILTER RESPONSE CHARACTERISTICS

7. What is the damping factor in each active filter shown in Figure 8–34? Which filters are approximately optimized for a Butterworth response characteristic?

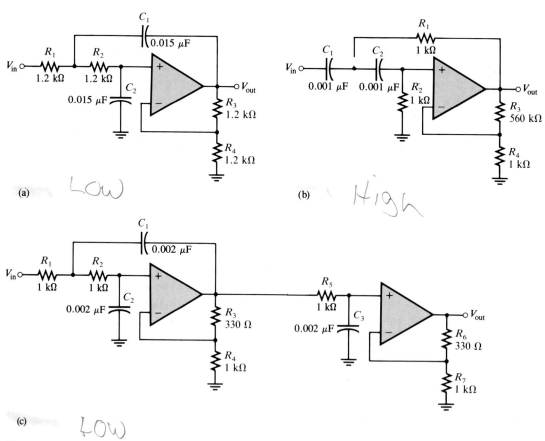

FIGURE 8–34

Active Filters

8. For the filters in Figure 8–34 that do not have a Butterworth response, specify the changes necessary to convert them to Butterworth responses. (Use nearest standard values.)

9. Response curves for high-pass second-order filters are shown in Figure 8–35. Identify each as Butterworth, Chebyshev, or Bessel.

FIGURE 8–35

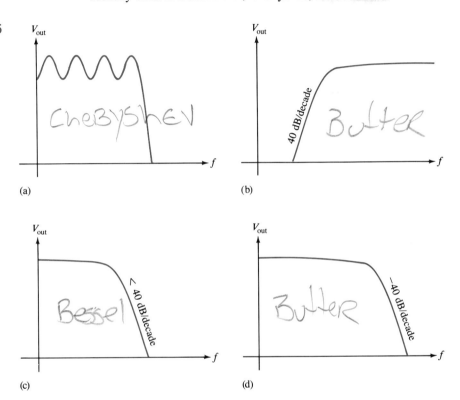

SECTION 8–3 ACTIVE LOW-PASS FILTERS

10. Is the four-pole filter in Figure 8–36 a low-pass or a high-pass type? Is it approximately optimized for a Butterworth response? What is the roll-off rate?

FIGURE 8–36

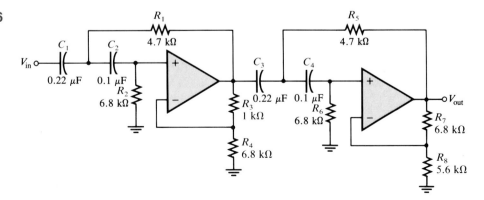

11. Determine the critical frequency in Figure 8–36.
12. Without changing the response curve, adjust the component values in the filter of Figure 8–36 to make it an equal-value filter.
13. Modify the filter in Figure 8–36 to increase the roll-off rate to −120 dB/decade while maintaining an approximate Butterworth response.
14. Using a block diagram format, show how to implement the following roll-off rates using single-pole and two-pole low-pass filters with Butterworth responses.
    (a) −40 dB/decade (b) −20 dB/decade
    (c) −60 dB/decade (d) −100 dB/decade
    (e) −120 dB/decade

SECTION 8–4 ACTIVE HIGH-PASS FILTERS

15. Convert the filter in Problem 12 to a low-pass with the same critical frequency and response characteristic.
16. Make the necessary circuit modification to reduce by half the critical frequency in Problem 15.
17. For the filter in Figure 8–37, (a) How would you increase the critical frequency? (b) How would you increase the gain?

FIGURE 8–37

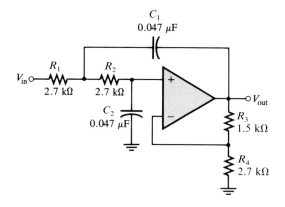

SECTION 8–5 ACTIVE BAND-PASS FILTERS

18. Identify each band-pass filter configuration in Figure 8–38.
19. Determine the center frequency and bandwidth for each filter in Figure 8–38.
20. Optimize the state-variable filter in Figure 8–39 for $Q = 50$. What bandwidth is achieved?

ACTIVE FILTERS

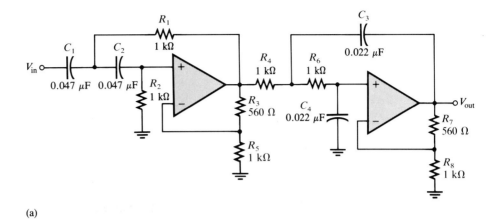

(a)

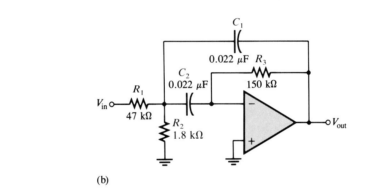

(b)

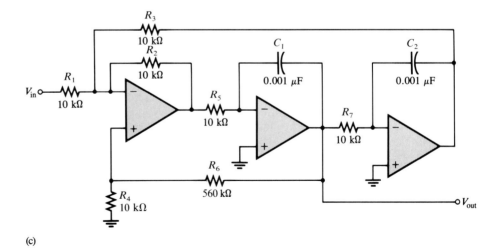

(c)

FIGURE 8–38

FIGURE 8–39

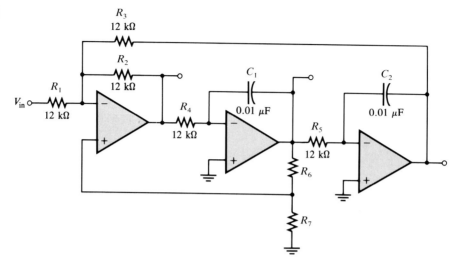

SECTION 8–6 ACTIVE BAND-STOP FILTERS
21. Show how to make a notch (band-stop) filter using the basic circuit in Figure 8–39.
22. Modify the band-stop filter in Problem 21 for a center frequency of 120 Hz.

ANSWERS TO REVIEW QUESTIONS

SECTION 8–1
1. The critical frequency determines the bandwidth.
2. The inherent frequency limitation of the op-amp limits the bandwidth.
3. Q and BW are inversely related. The higher the Q, the better the selectivity, and vice versa.

SECTION 8–2
1. Butterworth is very flat in the pass band and has a 20 dB/decade/pole roll-off. Chebyshev has ripples in the pass band and has greater than 20 dB/decade/pole roll-off.
Bessel has a linear phase characteristic and less than 20 dB/decade/pole roll-off.
2. The damping factor determines the response characteristic.
3. Frequency-selection network, gain element, and negative feedback circuit are the parts of an active filter.

SECTION 8–3
1. A second-order filter has two poles. Two resistors and two capacitors make up the frequency-selective circuit.
2. The damping factor sets the response characteristic.
3. Cascading increases the roll-off rate.

ACTIVE FILTERS

SECTION 8–4
1. The positions of the Rs and Cs in the frequency-selection circuit are opposite for low-pass and high-pass configurations.
2. Decrease the R values to increase f_c.
3. 140 dB/decade

SECTION 8–5
1. Q determines selectivity.
2. $Q = 25$. Higher Q gives narrower BW.
3. A summing amplifier and two integrators make up a state-variable filter.

SECTION 8–6
1. A band-stop rejects frequencies within the stop band. A band-pass passes frequencies within the pass band.
2. The low-pass and high-pass outputs are summed.

SECTION 8–7
1. To check the frequency response of a filter
2. Discrete point measurement—tedious and less complete; simpler equipment. Swept frequency measurement—uses more expensive equipment; more efficient, can be more accurate and complete.

SECTION 8–8
1. The filter board takes the detected FM signal and separates the L + R and L − R audio signals.
2. $BW = 15.9$ kHz
3. $BW = 15.9$ kHz
4. The L + R low-pass, the L − R low-pass, and the L − R band-pass
5. The stereo matrix combines the L + R and L − R signals and produces the separate left and right channel audio signals.

ANSWERS TO PRACTICE EXERCISES

8–1 500 Hz

8–2 1.44

8–3 7.234 kHz, 1.29 kΩ

8–4 $C_{A1} = C_{A2} = C_{B1} = C_{B2} = 0.234$ μF, $R_2 = R_4 = 680$ Ω, $R_1 = 103$ Ω, $R_3 = 840$ Ω

8–5 $R_A = R_B = R_2 = 10$ kΩ, $C_A = C_B = 0.053$ μF, $R_1 = 586$ Ω

8–6 Gain increases to 2.43, frequency decreases to 544 Hz, and bandwidth decreases to 96.5 Hz.

8–7 $f_0 = 21.922$ kHz, $Q = 101$, $BW = 217$ Hz

8–8 Decrease the input resistors or the feedback capacitors of the two integrator stages by half.

9
SIGNAL GENERATORS AND TIMERS

9–1 DEFINITION OF AN OSCILLATOR
9–2 OSCILLATOR PRINCIPLES
9–3 SINE WAVE OSCILLATORS
9–4 NONSINUSOIDAL OSCILLATORS
9–5 THE 555 TIMER AS AN OSCILLATOR
9–6 THE 555 TIMER AS A ONE-SHOT
9–7 A SYSTEM APPLICATION

After completing this chapter, you should be able to

□ Describe the basic concept of an oscillator.
□ Define positive feedback.
□ List the conditions for oscillation.
□ Recognize and explain the operation of the Wien-bridge, phase-shift, and twin-T oscillators.
□ Discuss the conditions for oscillator start-up.
□ Explain the operation of triangular-wave, sawtooth, and square-wave oscillators.
□ Explain how a voltage-controlled oscillator (VCO) works.
□ Apply a 555 timer IC in oscillator applications.
□ Apply a 555 timer in one-shot and time delay applications.

Oscillators are circuits that generate an output signal without an input signal. They are used as signal generators in all sorts of applications. Different types of oscillators produce various types of outputs including sine waves, square waves, triangular waves, and sawtooth waves. In this chapter, several types of basic oscillator circuits using an op-amp as the gain element are introduced. Also, a very popular integrated circuit, called the 555 timer, is discussed.

Oscillator operation is based on the principle of positive feedback, where a portion of the output signal is fed back to the input in a way that causes it to reinforce itself and thus sustain a continuous output signal. Oscillators are widely used in most communications systems as well as in digital systems, including computers, to generate required frequencies and timing signals. Also, oscillators are found in many types of test instruments like those used in the laboratory.

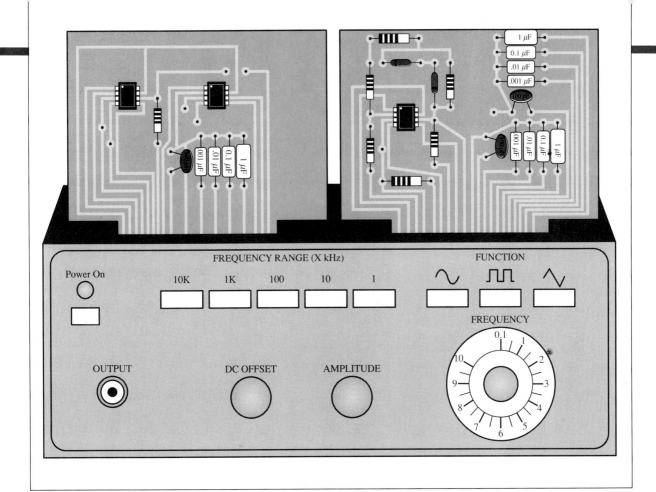

A SYSTEM APPLICATION

The function generator shown in the diagram is a good illustration of a system application for oscillators. The oscillator is a major part of this particular system. No doubt, you are already familiar with the use of the signal or function generator in your lab. As with most types of systems, a function generator can be implemented in more than one way. The system in this chapter uses circuits with which you are already familiar without some of the refinements and features found in many commercial instruments. The system reinforces what you have studied and lets you see these circuits "at work" in a specific application.

For the system application in Section 9–7, in addition to the other topics, be sure you understand

☐ How *RC* oscillators work.
☐ How a zero-level detector works.
☐ How an integrator works.

9-1 Definition of an Oscillator

An oscillator is a circuit that produces a repetitive waveform on its output with only the dc supply voltage as an input. A repetitive input signal is not required. The output voltage can be either sinusoidal or nonsinusoidal, depending on the type of oscillator.

The basic oscillator concept is illustrated in Figure 9–1. Essentially, an oscillator converts electrical energy in the form of dc to electrical energy in the form of ac. A basic oscillator consists of an amplifier for gain (either discrete transistor or op-amp) and a positive feedback circuit that produces phase shift and provides attenuation, as shown in Figure 9–2.

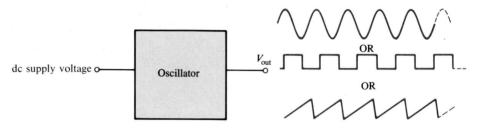

FIGURE 9–1
The basic oscillator concept showing three possible types of output waveforms.

FIGURE 9–2
Basic elements of an oscillator.

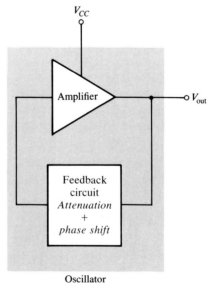

9–1 REVIEW QUESTIONS

1. What is an oscillator?
2. What type of feedback does an oscillator use?
3. What is the purpose of the feedback circuit?

9–2 OSCILLATOR PRINCIPLES

With the exception of the relaxation oscillator, which we will cover in Section 9–4, oscillator operation is based on the principle of positive feedback. In this section, we will examine this concept and look at the general conditions required for oscillation to occur.

POSITIVE FEEDBACK

Positive feedback is characterized by the condition wherein a portion of the output voltage of an amplifier is fed back to the input with no net phase shift, resulting in a reinforcement of the output signal. This basic idea is illustrated in Figure 9–3. As you can see, the in-phase feedback voltage is amplified to produce the output voltage, which in turn produces the feedback voltage. That is, a loop is created in which the signal sustains itself and a continuous sine wave output is produced. This phenomenon is called *oscillation*.

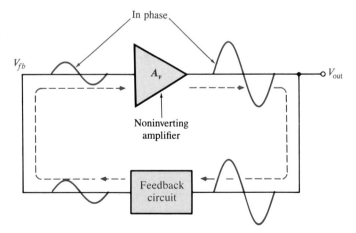

FIGURE 9–3
Positive feedback produces oscillation.

CONDITIONS FOR OSCILLATION

Two conditions are required for a sustained state of oscillation:

1. The phase shift around the feedback loop must be 0°.
2. The voltage gain around the closed feedback loop must equal 1 (unity).

The voltage gain around the closed feedback loop (A_{cl}) is the product of the amplifier gain (A_v) and the attenuation of the feedback circuit (B).

$$A_{cl} = A_v B$$

For example, if the amplifier has a gain of 100, the feedback circuit must have an attenuation of 0.01 to make the loop gain equal to 1 ($A_v B = 100 \times 0.01 = 1$). These conditions for oscillation are illustrated in Figure 9–4.

FIGURE 9–4
Conditions for oscillation.

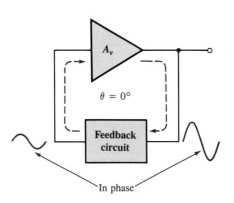
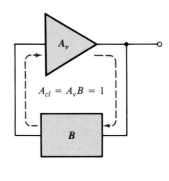

(a) The phase shift around the loop is 0°.   (b) The closed loop gain is 1.

START-UP CONDITIONS

So far, we have seen what it takes for an oscillator to produce a continuous sine wave output. Now we examine the requirements for the oscillation to start when the dc supply voltage is turned on. As you have seen, the unity-gain condition must be met for oscillation to be sustained. For oscillation to begin, the voltage gain around the positive feedback loop must be greater than 1 so that the amplitude of the output can build up to

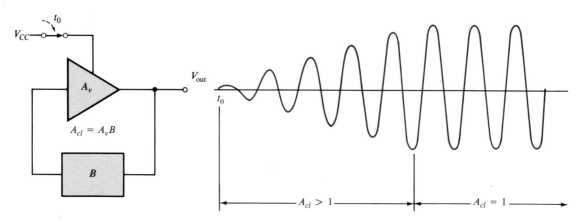

FIGURE 9–5
When oscillation starts at t_0, the condition $A_{cl} > 1$ causes the sinusoidal output voltage amplitude to build up to a desired level, where A_{cl} decreases to 1 and maintains the desired amplitude.

a desired level. The gain must then decrease to 1 so that the output stays at the desired level. (Several ways to achieve this reduction in gain after start-up are discussed later.) The conditions for both starting and sustaining oscillation are illustrated in Figure 9–5.

A question that normally arises is this: If the oscillator is off (no dc voltage) and there is no output voltage, how does a feedback signal originate to start the positive feedback build-up process? Initially, a small positive feedback voltage develops from thermally produced broad-band noise in the resistors or other components or from turn-on transients. The feedback circuit permits only a voltage with a frequency equal to the selected oscillation frequency to appear in-phase on the amplifier's input. This initial feedback voltage is amplified and continually reinforced, resulting in a buildup of the output voltage as previously discussed.

9–2 REVIEW QUESTIONS

1. What are the conditions required for a circuit to oscillate?
2. Define positive feedback.
3. What are the start-up conditions for an oscillator?

9–3 SINE WAVE OSCILLATORS

In this section you will learn about three types of RC oscillator circuits that produce sinusoidal outputs—the Wien-bridge oscillator, the phase-shift oscillator, and the twin-T oscillator. Generally, RC oscillators are used for frequencies up to about 1 MHz. The Wien-bridge is by far the most widely used type of RC oscillator for this range of frequencies.

THE WIEN-BRIDGE OSCILLATOR

One type of sine wave oscillator is the *Wien-bridge* oscillator. A fundamental part of the Wien-bridge oscillator is a lead-lag network like that shown in Figure 9–6(a). R_1 and C_1 together form the lag portion of the network; R_2 and C_2 form the lead portion. The operation of this circuit is as follows. At lower frequencies, the lead network dominates due to the high reactance of C_2. As the frequency increases, X_{C2} decreases, thus allowing the output voltage to increase. At some specified frequency, the response of the lag network takes over, and the decreasing value of X_{C1} causes the output voltage to decrease.

FIGURE 9–6
A lead-lag network.

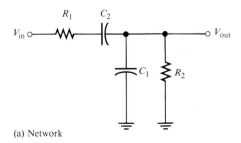

(a) Network

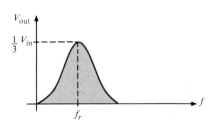

(b) Response curve

418　CHAPTER 9

So, we have a response curve like that shown in Figure 9–6(b) where the output voltage peaks at a frequency f_r. At this point, the attenuation (V_{out}/V_{in}) of the network is ⅓ if $R_1 = R_2$ and $X_{C1} = X_{C2}$ as stated by the following equation, which is derived in Appendix B.

$$\frac{V_{out}}{V_{in}} = \frac{1}{3} \tag{9-1}$$

The formula for the resonant frequency is also derived in Appendix B and is

$$f_r = \frac{1}{2\pi RC} \tag{9-2}$$

To summarize, the lead-lag network has a resonant frequency f_r at which the phase shift through the network is 0° and the attenuation is ⅓. Below f_r, the lead network dominates and the output leads the input. Above f_r, the lag network dominates and the output lags the input.

The lead-lag network is used in the positive feedback loop of an op-amp, as shown in Figure 9–7(a). A voltage divider is used in the negative feedback loop.

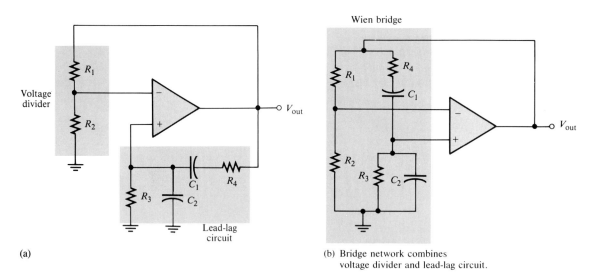

FIGURE 9–7
Two ways to draw the schematic of a Wien-bridge oscillator.

THE BASIC CIRCUIT

The Wien-bridge oscillator circuit can be viewed as a noninverting amplifier configuration with the input signal fed back from the output through the lead-lag network. Recall that the closed-loop gain of the amplifier is determined by the voltage divider.

$$A_{cl} = \frac{1}{B} = \frac{1}{R_2/(R_1 + R_2)} = \frac{R_1 + R_2}{R_2}$$

Signal Generators and Timers

The circuit is redrawn in Figure 9–7(b) to show that the op-amp is connected across the Wien bridge. One leg of the bridge is the lead-lag network, and the other is the voltage divider.

Positive Feedback Conditions for Oscillation

As you know, for the circuit to produce a sustained sine wave output (oscillate), the phase shift around the positive feedback loop must be 0° and the gain around the loop must be at least unity (1). The 0° phase-shift condition is met when the frequency is f_r, because the phase shift through the lead-lag network is 0° and there is no inversion from the noninverting input (+) of the op-amp to the output. This is shown in Figure 9–8(a).

The unity-gain condition in the feedback loop is met when

$$A_{cl} = 3$$

This offsets the ⅓ attenuation of the lead-lag network, thus making the gain around the positive feedback loop equal to 1, as depicted in Figure 9–8(b). To achieve a closed-loop gain of 3,

$$R_1 = 2R_2$$

Then

$$A_{cl} = \frac{R_1 + R_2}{R_2} = \frac{2R_2 + R_2}{R_2} = \frac{3R_2}{R_2} = 3$$

FIGURE 9–8
Conditions for oscillation.

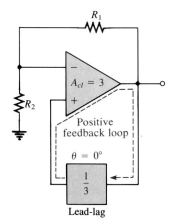 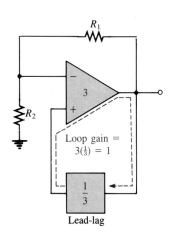

(a) The phase shift around the loop is 0°. (b) The voltage gain around the loop is 1.

Start-Up Conditions

Initially, the closed-loop gain of the amplifier must be more than 1 ($A_{cl} > 3$) until the output signal builds up to a desired level. The gain must then decrease to 1 so that the output signal stays at the desired level. This is illustrated in Figure 9–9.

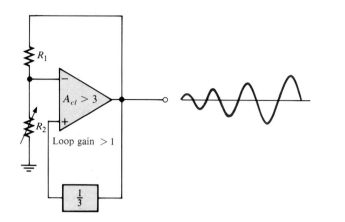
(a) Loop gain greater than 1 causes output to build up.

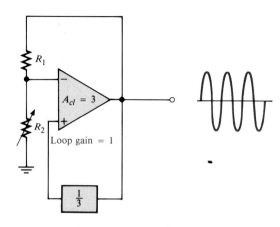
(b) Loop gain of 1 causes a sustained constant output.

FIGURE 9–9
Oscillator start-up conditions.

The circuit in Figure 9–10 illustrates a basic method for achieving the condition described above. Notice that the voltage-divider network has been modified to include an additional resistor R_3 in parallel with a back-to-back zener diode arrangement. When dc power is first applied, both zener diodes appear as opens. This places R_3 in series with R_1, thus increasing the closed-loop gain as follows ($R_1 = 2R_2$).

$$A_{cl} = \frac{R_1 + R_2 + R_3}{R_2} = \frac{3R_2 + R_3}{R_2} = 3 + \frac{R_3}{R_2}$$

FIGURE 9–10
Self-starting Wien-bridge oscillator using back-to-back zener diodes.

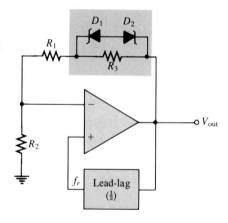

Initially, a small positive feedback signal develops from noise or turn-on transients. The lead-lag network permits only a signal with a frequency equal to f_r to appear in-phase on the noninverting input. This feedback signal is amplified and continually reinforced, resulting in a buildup of the output voltage. When the output signal reaches the zener breakdown voltage, the zeners conduct and effectively short out R_3. This lowers the amplifier's closed-loop gain to 3. At this point the output signal levels off and the oscillation is sustained. (Incidentally, the frequency of oscillation can be adjusted by using gang-tuned capacitors in the lead-lag network.)

Another method sometimes used to ensure self-starting employs a tungsten lamp in the voltage divider, as shown in Figure 9–11. When the power is first turned on, the resistance of the lamp is lower than its nominal value. This keeps the negative feedback small and makes the closed-loop gain of the amplifier greater than 3. As the output voltage builds up, the voltage across the tungsten lamp—and thus its current—increases. As a result, the lamp resistance increases until it reaches a value equal to one-half the feedback resistance. At this point the closed-loop gain is 3, and the output is sustained at a constant level.

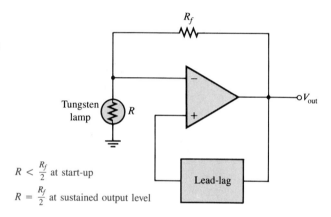

FIGURE 9–11
Self-starting Wien-bridge oscillator using a tungsten lamp in the negative feedback loop.

EXAMPLE 9–1

Determine the frequency of oscillation for the Wien-bridge oscillator in Figure 9–12. Also, verify that oscillations will start and then continue when the output signal reaches 5.4 V.

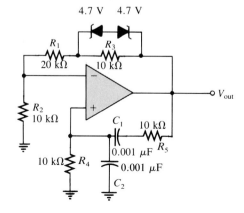

FIGURE 9–12

SOLUTION

For the lead-lag network, $R_4 = R_5 = R = 10 \text{ k}\Omega$ and $C_1 = C_2 = C = 0.001 \text{ }\mu\text{F}$. The frequency is

$$f_r = \frac{1}{2\pi RC} = \frac{1}{2\pi(10 \text{ k}\Omega)(0.001 \text{ }\mu\text{F})} = 15.92 \text{ kHz}$$

Initially, the closed-loop gain is

$$A_{cl} = \frac{R_1 + R_2 + R_3}{R_2} = \frac{40 \text{ k}\Omega}{10 \text{ k}\Omega} = 4$$

Since $A_{cl} > 3$, the start-up condition is met.

When the output reaches 5.4 V (4.7 V + 0.7 V), the zeners conduct (their forward resistance is assumed small, compared to 10 kΩ), and the closed-loop gain is reached. Thus, oscillation is sustained.

$$A_{cl} = \frac{R_1 + R_2}{R_2} = \frac{30 \text{ k}\Omega}{10 \text{ k}\Omega} = 3$$

PRACTICE EXERCISE 9–1

What change is required in the oscillator in Figure 9–12 to produce an output with an amplitude of 6.8 V?

THE PHASE-SHIFT OSCILLATOR

Figure 9–13 shows a type of sine-wave oscillator called the *phase-shift oscillator*. Each of the three RC networks in the feedback loop can provide a maximum phase shift approaching 90°. Oscillation occurs at the frequency where the total phase shift through the three RC networks is 180°. The inversion of the op-amp, itself, provides the additional 180° to meet the requirement for oscillation.

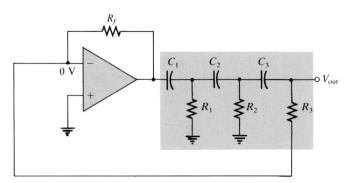

FIGURE 9–13
Op-amp phase-shift oscillator.

The attenuation B of the three-section RC feedback network is

$$B = \frac{1}{29} \tag{9-3}$$

Signal Generators and Timers

The derivation of this unusual result is given in Appendix B. To meet the greater-than-unity loop gain requirement, the closed-loop voltage gain of the op-amp must be greater than 29 (set by R_f and R_3). The frequency of oscillation is also derived in Appendix B and stated in the following equation, where $R_1 = R_2 = R_3 = R$ and $C_1 = C_2 = C_3 = C$.

$$f_r = \frac{1}{2\pi\sqrt{6}RC} \qquad (9\text{-}4)$$

EXAMPLE 9–2

(a) Determine the value of R_f necessary for the circuit in Figure 9–14 to operate as an oscillator.

(b) Determine the frequency of oscillation.

FIGURE 9–14

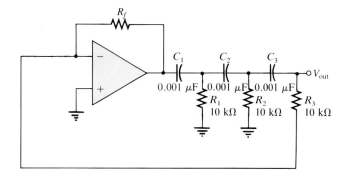

SOLUTION

(a) $A_{cl} = 29$, and $A_{cl} = \dfrac{R_f}{R_3}$. Therefore,

$$\frac{R_f}{R_3} = 29$$

$$R_f = 29R_3 = 29(10 \text{ k}\Omega) = 290 \text{ k}\Omega$$

(b) $f_r = \dfrac{1}{2\pi\sqrt{6}RC} = \dfrac{1}{2\pi\sqrt{6}(10 \text{ k}\Omega)(0.001\ \mu\text{F})} \cong 6.5 \text{ kHz}$

PRACTICE EXERCISE 9–2

(a) If R_1, R_2, and R_3 in Figure 9–14 are changed to 8.2 kΩ, what value must R_f be for oscillation?

(b) What is the value of f_r?

TWIN-T OSCILLATOR

Another type of RC oscillator is called the *twin-T* because of the two T-type RC filters used in the negative feedback loop, as shown in Figure 9–15(a). One of the twin-T filters has a low-pass response, and the other has a high-pass response. The combined parallel filters produce a band-stop or notch response with a center frequency equal to the desired frequency of oscillation, f_r, as shown in Figure 9–15(b).

FIGURE 9–15

Twin-T oscillator and twin-T filter response.

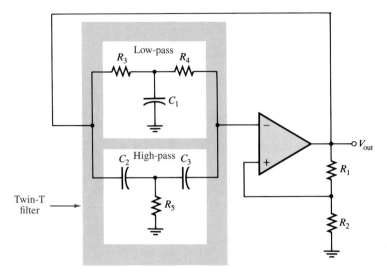

(a) Oscillator circuit

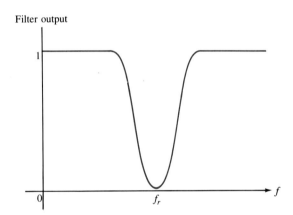

(b) Twin-T filter's frequency response curve

Oscillation cannot occur at frequencies above or below f_r because of the negative feedback through the filters. At f_r, however, there is negligible negative feedback, and thus, the positive feedback through the voltage divider (R_1 and R_2) allows the circuit to oscillate. Self-starting can be achieved by using a tungsten lamp in the place of R_1.

9–3 REVIEW QUESTIONS

1. There are two feedback loops in the Wien-bridge oscillator. What is the purpose of each?
2. A certain lead-lag network has $R_1 = R_2$ and $C_1 = C_2$. An input voltage of 5 V rms is applied. The input frequency equals the resonant frequency of the network. What is the rms output voltage?

3. Why must the phase shift through the *RC* feedback circuit in a phase-shift oscillator equal 180°?

9–4 NONSINUSOIDAL OSCILLATORS

In this section, several types of op-amp oscillator circuits that produce triangular, sawtooth, or square waveforms are discussed. Some of these types of oscillators are frequently referred to as signal generators and multivibrators depending on the particular circuit implementation.

A TRIANGULAR-WAVE OSCILLATOR

The op-amp integrator covered in the last chapter can be used as the basis for a triangular wave generator. The basic idea is illustrated in Figure 9–16(a) where a dual-polarity, switched input is used. We use the switch only to introduce the concept; it is not a practical way to implement this circuit. When the switch is in position 1, the negative voltage is applied, and the output is a positive-going ramp. When the switch is thrown into position 2, a negative-going ramp is produced. If the switch is thrown back and forth at fixed intervals, the output is a triangular wave consisting of alternating positive-going and negative-going ramps, as shown in Figure 9–16(b).

FIGURE 9–16

Basic triangular-wave generator.

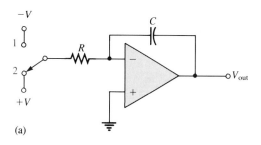

(a)

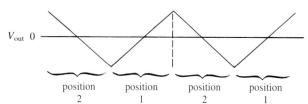

(b) Output voltage as the switch is thrown back and forth at regular intervals

A PRACTICAL CIRCUIT

One practical implementation of a triangular-wave generator utilizes an op-amp comparator to perform the switching function, as shown in Figure 9–17. The operation is as follows. To begin, assume that the output voltage of the comparator is at its maximum negative level. This output is connected to the inverting input of the integrator through R_1,

FIGURE 9–17
A triangular-wave generator using two op-amps.

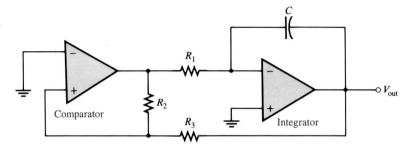

producing a positive-going ramp on the output of the integrator. When the ramp voltage reaches the upper trigger point (UTP), the comparator switches to its maximum positive level. This positive level causes the integrator ramp to change to a negative-going direction. The ramp continues in this direction until the lower trigger point (LTP) of the comparator is reached. At this point, the comparator output switches back to the maximum negative level and the cycle repeats. This action is illustrated in Figure 9–18.

Since the comparator produces a square-wave output, the circuit in Figure 9–17 can be used as both a triangular-wave generator and a square-wave generator. Devices of this type are commonly known as *function generators* because they produce more than one output function. The output amplitude of the square wave is set by the output swing of the comparator, and the resistors R_2 and R_3 set the amplitude of the triangular output by establishing the UTP and LTP voltages according to the following formulas:

$$V_{UTP} = +V_{max}\left(\frac{R_3}{R_2}\right) \quad (9-5)$$

$$V_{LTP} = -V_{max}\left(\frac{R_3}{R_2}\right) \quad (9-6)$$

where the comparator output levels, $+V_{max}$ and $-V_{max}$, are equal. The frequency of both waveforms depends on the R_1C time constant as well as the amplitude-setting resistors, R_2

FIGURE 9–18
Waveforms for the circuit in Figure 9–17.

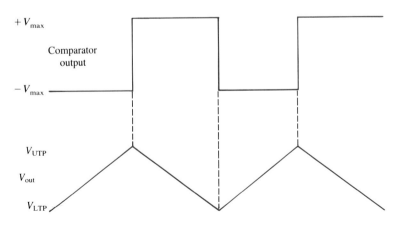

and R_3. By varying R_1, the frequency of oscillation can be adjusted without changing the output amplitude.

$$f = \frac{1}{4R_1C}\left(\frac{R_2}{R_3}\right) \tag{9-7}$$

■ **EXAMPLE 9-3** Determine the frequency of the circuit in Figure 9-19. To what value must R_1 be changed to make the frequency 20 kHz?

FIGURE 9-19

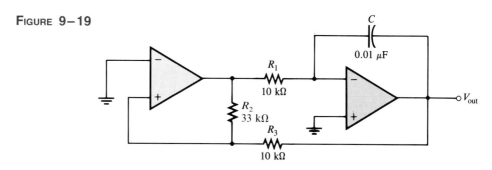

SOLUTION

$$f = \frac{1}{4R_1C}\left(\frac{R_2}{R_3}\right) = \left(\frac{1}{4(10\ \text{k}\Omega)(0.01\ \mu\text{F})}\right)\left(\frac{33\ \text{k}\Omega}{10\ \text{k}\Omega}\right) = 8.25\ \text{kHz}$$

To make $f = 20$ kHz,

$$R_1 = \frac{1}{4fC}\left(\frac{R_2}{R_3}\right) = \left(\frac{1}{4(20\ \text{kHz})(0.01\ \mu\text{F})}\right)\left(\frac{33\ \text{k}\Omega}{10\ \text{k}\Omega}\right) = 4.13\ \text{k}\Omega$$

PRACTICE EXERCISE 9-3

What is the amplitude of the triangular wave in Figure 9-19 if the comparator output is ± 10 V?

■

A VOLTAGE-CONTROLLED SAWTOOTH OSCILLATOR (VCO)

The voltage-controlled oscillator (VCO) is an oscillator whose frequency can be changed by a variable dc control voltage. VCOs can be either sinusoidal or nonsinusoidal. One way to build a voltage-controlled sawtooth oscillator is with an op-amp integrator that uses a switching device (PUT) in parallel with the feedback capacitor to terminate each ramp at a prescribed level and effectively "reset" the circuit. Figure 9-20(a) shows the implementation.

The PUT is a programmable unijunction transistor with an anode, a cathode, and a gate terminal. The gate is always biased positively with respect to the cathode. When the anode voltage exceeds the gate voltage by approximately 0.7 V, the PUT turns on and acts as a forward-biased diode. When the anode voltage falls below this level, the PUT turns off. Also, the current must be above the holding value to maintain conduction.

FIGURE 9–20

Voltage-controlled sawtooth oscillator operation.

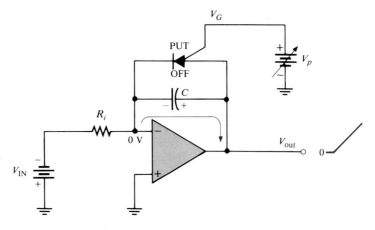

(a) Initially, the capacitor charges, the output ramp begins, and the PUT is off.

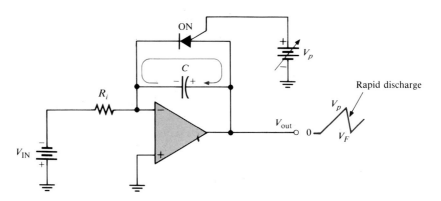

(b) The capacitor rapidly discharges when the PUT momentarily turns on.

The operation of the sawtooth generator begins when the negative dc input voltage, $-V_{IN}$, produces a positive-going ramp on the output. During the time that the ramp is increasing, the circuit acts as a regular integrator. The PUT triggers on when the output ramp (at the anode) exceeds the gate voltage by 0.7 V. The gate is set to the approximate desired sawtooth peak voltage. When the PUT turns on, the capacitor rapidly discharges, as shown in Figure 9–20(b). The capacitor does not discharge completely to zero because of the PUT's forward voltage, V_F. Discharge continues until the PUT current falls below the holding value. At this point, the PUT turns off and the capacitor begins to charge again, thus generating a new output ramp. The cycle continually repeats, and the resulting output is a repetitive sawtooth waveform, as shown. The sawtooth amplitude and period can be adjusted by varying the PUT gate voltage.

The frequency is determined by the R_iC time constant of the integrator and the peak voltage set by the PUT. Recall that the charging rate of the capacitor is V_{IN}/R_iC. The time

Signal Generators and Timers

it takes the capacitor to charge from V_F to V_p is the period, T, of the sawtooth (neglecting the rapid discharge time).

$$T = \frac{V_p - V_F}{|V_{IN}|/R_iC} \quad (9-8)$$

From $f = 1/T$, we get

$$f = \frac{|V_{IN}|}{R_iC}\left(\frac{1}{V_p - V_F}\right) \quad (9-9)$$

EXAMPLE 9-4

(a) Find the amplitude and frequency of the sawtooth output in Figure 9–21. Assume that the forward PUT voltage V_F is approximately 1 V.
(b) Sketch the output waveform.

FIGURE 9–21

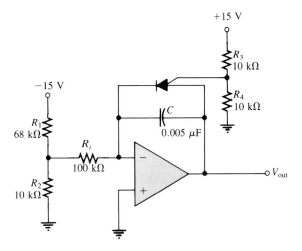

SOLUTION

(a) First, find the gate voltage in order to establish the approximate voltage at which the PUT turns on.

$$V_G = \frac{R_4}{R_3 + R_4}(+V) = \frac{10\ k\Omega}{20\ k\Omega}(15\ V) = 7.5\ V$$

This voltage sets the approximate maximum peak value of the sawtooth output (neglecting the 0.7 V).

$$V_p \cong 7.5\ V$$

The minimum peak value (low point) is

$$V_F \cong 1\ V$$

The period is determined as follows:

$$V_{IN} = \frac{R_2}{R_1 + R_2}(-V) = \frac{10 \text{ k}\Omega}{78 \text{ k}\Omega}(-15 \text{ V}) = -1.92 \text{ V}$$

$$T = \frac{V_p - V_F}{|V_{IN}|/R_iC} = \frac{7.5 \text{ V} - 1 \text{ V}}{1.92 \text{ V}/(100 \text{ k}\Omega)(0.005 \text{ }\mu\text{F})} = 1.69 \text{ ms}$$

$$f = \frac{1}{1.69 \text{ ms}} \cong 592 \text{ Hz}$$

(b) The output waveform is shown in Figure 9–22.

FIGURE 9–22
Output of the circuit in Figure 9–21.

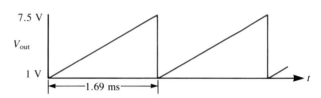

PRACTICE EXERCISE 9–4
If R_i is changed to 56 kΩ in Figure 9–21, what is the frequency?

A SQUARE-WAVE RELAXATION OSCILLATOR

The basic square-wave generator shown in Figure 9–23 is a type of relaxation oscillator because its operation is based on the charging and discharging of a capacitor. Notice that the op-amp's inverting input is the capacitor voltage and the noninverting input is a portion of the output fed back through resistors R_2 and R_3. When the circuit is first turned on, the capacitor is uncharged, and thus the inverting input is at 0 V. This makes the output a positive maximum, and the capacitor begins to charge toward V_{out} through R_1.

FIGURE 9–23
A square-wave relaxation oscillator.

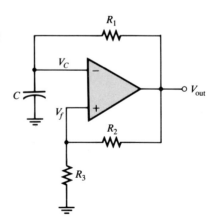

When the capacitor voltage reaches a value equal to the feedback voltage on the noninverting input, the op-amp switches to the maximum negative state. At this point, the capacitor begins to discharge from $+V_f$ toward $-V_f$. When the capacitor voltage reaches $-V_f$, the op-amp switches back to the maximum positive state. This action continues to repeat, as shown in Figure 9–24, and a square-wave output voltage is obtained.

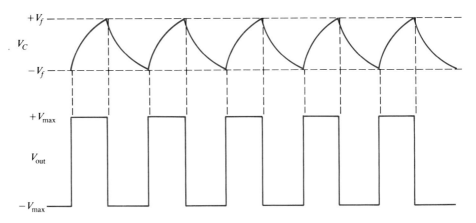

FIGURE 9–24
Waveforms for the square-wave relaxation oscillator.

9–4 REVIEW QUESTIONS

1. What is a VCO, and basically, what does it do?
2. Upon what principle does a relaxation oscillator operate?

9–5 THE 555 TIMER AS AN OSCILLATOR

The 555 timer is a versatile integrated circuit with many applications. In this section, you will see how the 555 is configured as an astable or free-running multivibrator, which is essentially a square-wave oscillator. We will also discuss the use of the 555 timer as a voltage-controlled oscillator (VCO).

The 555 timer consists basically of two comparators, a flip-flop, a discharge transistor, and a resistive voltage divider, as shown in Figure 9–25. The flip-flop (bistable multivibrator) is a digital device that is perhaps unfamiliar to you at this point unless you already have taken a digital fundamentals course. Briefly, it is a two-state device whose output can be at either a high voltage level (set) or a low voltage level (reset). The state of the output can be changed with proper input signals.

The resistive voltage divider is used to set the voltage comparator levels. All three resistors are of equal value; therefore, the upper comparator has a reference of $\frac{2}{3}V_{CC}$, and the lower comparator has a reference of $\frac{1}{3}V_{CC}$. The comparators' outputs control the state of the flip-flop. When the trigger voltage goes below $\frac{1}{3}V_{CC}$, the flip-flop sets and the

FIGURE 9–25
Internal diagram of a 555 integrated circuit timer. (IC pin numbers are in parentheses.)

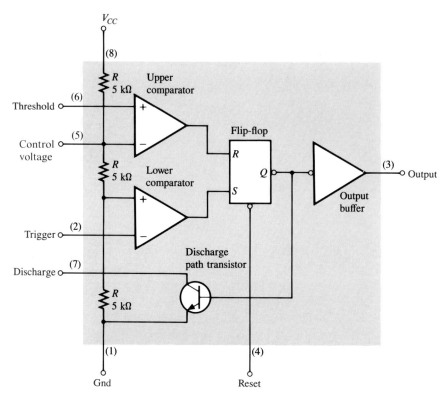

output jumps to its high level. The threshold input is normally connected to an external RC timing network. When the external capacitor voltage exceeds $\frac{2}{3}V_{CC}$, the upper comparator resets the flip-flop, which in turn switches the output back to its low level. When the device output is low, the discharge transistor Q_d is turned on and provides a path for rapid discharge of the external timing capacitor. This basic operation allows the timer to be configured with external components as an oscillator, a one-shot, or a time-delay element.

ASTABLE OPERATION

A 555 timer connected to operate as an astable **multivibrator,** which is a free-running nonsinusoidal oscillator, is shown in Figure 9–26. Notice that the threshold input (THRESH) is now connected to the trigger input (TRIG). The external components R_1, R_2, and C_{ext} form the timing network that sets the frequency of oscillation. The 0.01 μF capacitor connected to the control (CONT) input is strictly for decoupling and has no effect on the operation; in some cases it can be left off.

Initially, when the power is turned on, the capacitor C_{ext} is uncharged and thus the trigger voltage (2) is at 0 V. This causes the output of the lower comparator to be high and the output of the upper comparator to be low, forcing the output of the flip-flop, and thus the base of Q_d, low and keeping the transistor off. Now, C_{ext} begins charging through R_1 and R_2 as indicated in Figure 9–27. When the capacitor voltage reaches $\frac{1}{3}V_{CC}$, the lower

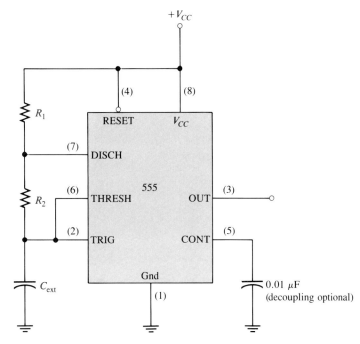

FIGURE 9–26
The 555 timer connected as an astable multivibrator.

comparator switches to its low output state, and when the capacitor voltage reaches $\frac{2}{3}V_{CC}$, the upper comparator switches to its high output state. This resets the flip-flop, causes the base of Q_d to go high, and turns on the transistor. This sequence creates a discharge path for the capacitor through R_2 and the transistor, as indicated. The capacitor now begins to discharge, causing the upper comparator to go low. At the point where the capacitor discharges down to $\frac{1}{3}V_{CC}$, the lower comparator switches high, setting the flip-flop, which makes the base of Q_d low and turns off the transistor. Another charging cycle begins, and the entire process repeats. The result is a rectangular wave output whose duty cycle depends on the values of R_1 and R_2. The frequency of oscillation is given by the formula or it can be found using the graph in Figure 9–28.

$$f = \frac{1.44}{(R_1 + 2R_2)C_{ext}} \qquad (9\text{--}10)$$

By selecting R_1 and R_2, the duty cycle of the output can be adjusted. Since C_{ext} charges through $R_1 + R_2$ and discharges only through R_2, duty cycles approaching a minimum of 50 percent can be achieved if $R_2 \gg R_1$ so that the charging and discharging times are approximately equal. An expression to calculate the duty cycle is developed as follows.

The time that the output is high is how long it takes C_{ext} to charge from $\frac{1}{3}V_{CC}$ to $\frac{2}{3}V_{CC}$. It is expressed as

$$t_H = 0.693(R_1 + R_2)C_{ext}$$

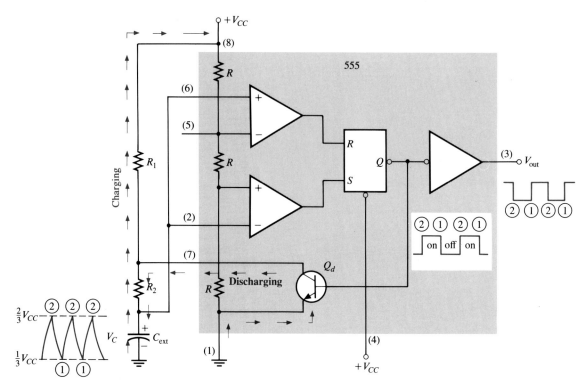

FIGURE 9–27
Operation of the 555 timer in the astable mode.

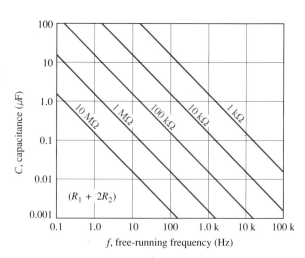

FIGURE 9–28
Frequency of oscillation (free-running frequency) as a function of C_{ext} and $R_1 + 2R_2$. The sloped lines are values of $R_1 + 2R_2$.

The time that the output is low is how long it takes C_{ext} to discharge from $\tfrac{2}{3}V_{CC}$ to $\tfrac{1}{3}V_{CC}$. It is expressed as

$$t_L = 0.693 R_2 C_{ext}$$

The period, T, of the output waveform is the sum of t_H and t_L.

$$T = t_H + t_L = 0.693(R_1 + 2R_2)C_{ext}$$

This is the reciprocal of f in Equation (9–10). Finally, the duty cycle is

$$\text{Duty cycle} = \frac{t_H}{T} = \frac{t_H}{t_H + t_L}$$

$$\text{Duty cycle} = \frac{R_1 + R_2}{R_1 + 2R_2} \times 100\% \quad (9\text{–}11)$$

To achieve duty cycles of less than 50 percent, the circuit in Figure 9–26 can be modified so that C_{ext} charges through only R_1 and discharges through R_2. This is achieved with a diode D_1 placed as shown in Figure 9–29. The duty cycle can be made less than 50 percent by making R_1 less than R_2. Under this condition, the expression for the duty cycle is

$$\text{Duty cycle} = \frac{R_1}{R_1 + R_2} \times 100\% \quad (9\text{–}12)$$

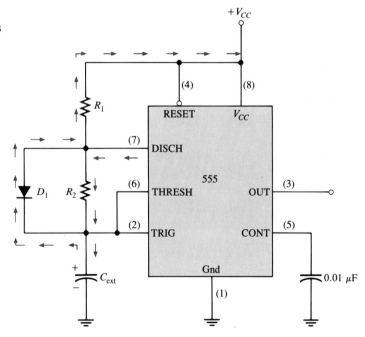

FIGURE 9–29
The addition of diode D_1 allows the duty cycle of the output to be adjusted to less than 50 percent by making $R_1 < R_2$.

EXAMPLE 9–5 A 555 timer configured to run in the astable mode (oscillator) is shown in Figure 9–30. Determine the frequency of the output and the duty cycle.

FIGURE 9–30

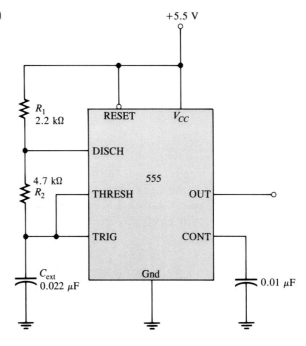

SOLUTION

$$f = \frac{1.44}{(R_1 + 2R_2)C_{ext}} = \frac{1.44}{(2.2\ k\Omega + 9.4\ k\Omega)0.022\ \mu F} = 5.64\ kHz$$

$$\text{Duty cycle} = \frac{R_1 + R_2}{R_1 + 2R_2} \times 100\% = \frac{2.2\ k\Omega + 4.7\ k\Omega}{2.2\ k\Omega + 9.4\ k\Omega} \times 100\% = 59.5\%$$

PRACTICE EXERCISE 9–5
Determine the duty cycle in Figure 9–30 if a diode is connected across R_2 as indicated in Figure 9–29.

OPERATION AS A VOLTAGE-CONTROLLED OSCILLATOR (VCO)

A 555 timer can be configured to operate as a VCO by using the same external connections as for astable operation, with the exception that a variable control voltage is applied to the CONT input (pin 5), as indicated in Figure 9–31.

As shown in Figure 9–32, the control voltage (V_{CONT}) changes the threshold values of $\frac{1}{3}V_{CC}$ and $\frac{2}{3}V_{CC}$ for the internal comparators. With the control voltage, the upper value is V_{CONT} and the lower value is $\frac{1}{2}V_{CONT}$, as you can see by examining the internal diagram of the 555 timer. When the control voltage is varied, the output frequency also varies. An increase in V_{CONT} increases the charging and discharging time of the external capacitor and causes the frequency to decrease. A decrease in V_{CONT} decreases the charging and discharging time of the capacitor and causes the frequency to increase.

SIGNAL GENERATORS AND TIMERS

FIGURE 9–31
The 555 timer connected as a voltage-controlled oscillator (VCO). Note the variable control voltage input on pin 5.

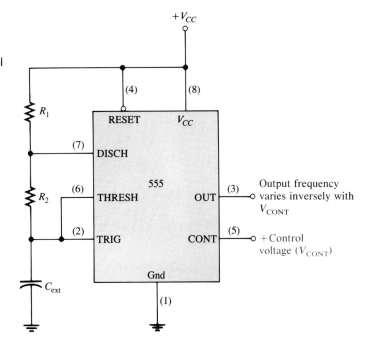

FIGURE 9–32
The VCO output frequency varies inversely with V_{CONT} because the charging and discharging time of C_{ext} is directly dependent on the control voltage.

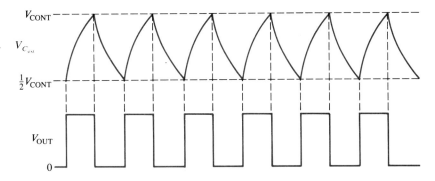

An interesting application of the VCO is in phase-locked loops, which are used in various types of communications receivers to track variations in the frequency of incoming signals. We will cover the basic operation of a phase-locked loop in Chapter 12.

9–5 REVIEW QUESTIONS

1. Name the five basic elements in a 555 timer IC.
2. When the 555 timer is configured as an astable multivibrator, how is the duty cycle determined?

9–6 THE 555 TIMER AS A ONE-SHOT

A one-shot is a monostable multivibrator that produces a single output pulse for each input trigger pulse. The term monostable means that the device has only one stable state. When a one-shot is triggered, it temporarily goes to its unstable state but it always returns to its stable state. The time that it remains in its unstable state establishes the width of the output pulse and is set by the values of an external resistor and capacitor.

A 555 timer connected for **monostable** operation is shown in Figure 9–33. Compare this configuration to the one used for **astable** operation in Figure 9–26 and note the difference in the external circuit.

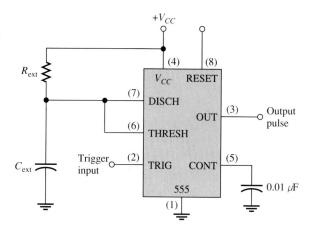

FIGURE 9–33
The 555 timer connected as a monostable multivibrator (one-shot).

MONOSTABLE OPERATION

As you can see in Figure 9–33, a single resistor (R_{ext}) and a capacitor (C_{ext}) are connected externally to the 555 timer. These components form a timing circuit that sets the width of the output pulse. In the nonactive state, the trigger input is at its high level making the output of the lower comparator in Figure 9–34 high. This keeps the flip-flop output at the high level which, in turn, creates a discharge path for C_{ext} by keeping Q_d turned on.

At t_0, when the trigger input falls below $0.33V_{CC}$, the lower comparator changes state and causes the output of the flip-flop to go low. This turns Q_d off and causes the output of the timer to go high, thus beginning the output pulse. This action is illustrated in Figure 9–34.

As soon as Q_d turns off, C_{ext} begins to charge exponentially through R_{ext}. The rate of charge of the capacitor is based on the time constant, $R_{ext}C_{ext}$. At t_1, when the capacitor voltage reaches $0.67V_{CC}$, the upper comparator causes the flip-flop output to go back high. This action turns Q_d on and causes the timer output to go low, thus ending the output pulse. C_{ext} immediately discharges back to zero volts through Q_d. This action is illustrated in Figure 9–34.

Signal Generators and Timers

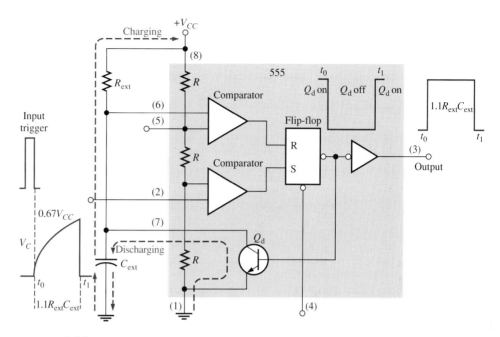

FIGURE 9-34
One-shot action in the 555 timer.

Once triggered, the one-shot cannot be retriggered until it completely times out; that is, until the capacitor charges to $0.67V_{CC}$ and then discharges, completing a full output pulse. Once it times out, the one-shot can then be triggered again to produce another output pulse. A low level on the reset input can be used to prematurely terminate the output pulse. The width of the output pulse is determined by the following formula:

$$t_W = 1.1 R_{ext} C_{ext} \qquad (9\text{--}13)$$

The graph in Figure 9–35 shows various combinations of R_{ext} and C_{ext} and the associated output pulse widths. This graph can be used to select component values for a desired pulse width.

FIGURE 9–35
555 one-shot timing.

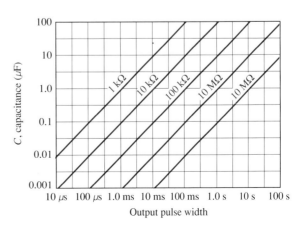

EXAMPLE 9–6

A 555 timer is connected as a one-shot with $R_{ext} = 10\ \text{k}\Omega$ and $C_{ext} = 0.1\ \mu\text{F}$. What is the pulse width of the output?

SOLUTION
You can determine the pulse width in two ways. You can use either Equation (9–13) or the graph in Figure 9–35. Using the formula,

$$t_W = 1.1 R_{ext} C_{ext}$$
$$= 1.1(10\ \text{k}\Omega)(0.1\ \mu\text{F}) = 1.1\ \text{ms}$$

To use the graph, move along the $C = 0.1\ \mu\text{F}$ line until it intersects with the sloped line corresponding to $R = 10\ \text{k}\Omega$. At that point, project down to the horizontal axis and you get a pulse width of 1.1 ms as illustrated in Figure 9–36.

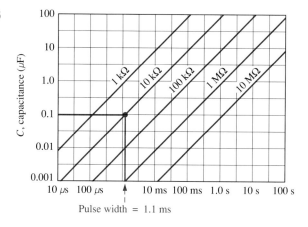

FIGURE 9–36

PRACTICE EXERCISE 9–6
To what value must R_{ext} be changed to increase the one-shot's output pulse width to 5 ms?

USING ONE-SHOTS FOR TIME DELAY

In many applications, it is necessary to have a fixed time delay between certain events. Figure 9–37(a) shows two 555 timers connected as one-shots. The output of the first goes to the input of the second. When the first one-shot is triggered, it produces an output pulse whose width establishes a time delay. At the end of this pulse, the second one-shot is triggered. Therefore, we have an output pulse from the second one-shot that is delayed from the input trigger to the first one-shot by a time equal to the pulse width of the first one-shot, as indicated in the timing diagram in Figure 9–37(b).

Signal Generators and Timers

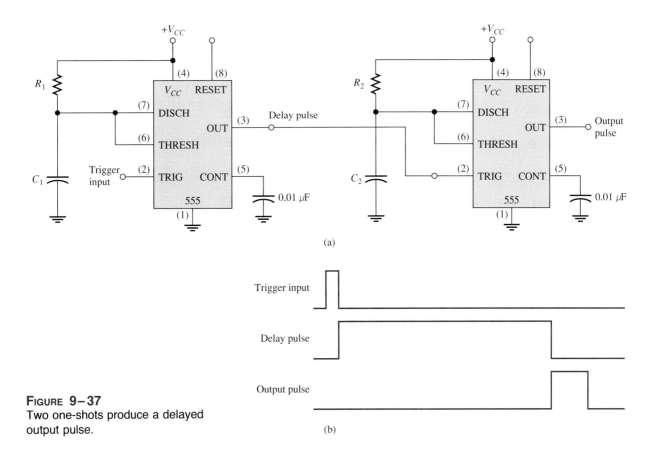

FIGURE 9–37
Two one-shots produce a delayed output pulse.

■ **EXAMPLE 9–7** Show the timing diagram (relationships of the input and output pulses) for the circuit in Figure 9–38.

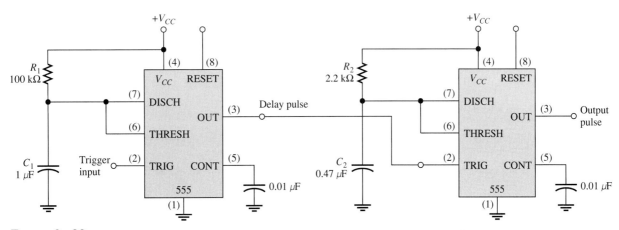

FIGURE 9–38

SOLUTION

The time relationship of the inputs and outputs are shown in Figure 9–39. The pulse widths for the two one-shots are

$$t_{W1} = 1.1R_1C_1 = 1.1(100 \text{ k}\Omega)(1\mu\text{F}) = 110 \text{ ms}$$
$$t_{W2} = 1.1R_2C_2 = 1.1(2.2 \text{ k}\Omega)(0.47\mu\text{F}) = 1.14 \text{ ms}$$

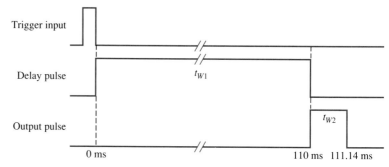

FIGURE 9–39

PRACTICE EXERCISE 9–7

Suggest a way that the circuit in Figure 9–38 can be modified so that the delay can be made adjustable from 10 ms to 200 ms.

9–6 REVIEW QUESTIONS

1. How many stable states does a one-shot have?
2. A certain 555 one-shot circuit has a time constant of 5 ms. What is the output pulse width?
3. How can you decrease the pulse width of a one-shot?

9–7 A SYSTEM APPLICATION

The function generator presented at the beginning of the chapter is a laboratory instrument used as a source for sine waves, square waves, and triangular waves. In this section, you will

☐ *See how an oscillator is used as a signal source.*
☐ *See how the frequency and amplitude of the generated signal are varied.*
☐ *Translate between printed circuit boards and a schematic.*
☐ *Interconnect the front panel controls and two pc boards.*
☐ *Troubleshoot some common system problems.*

A Brief Description of the System

The function generator in this system application produces either a sine wave, a square wave, or a triangular wave depending on the function selected by the front panel switches. The frequency of the selected waveform can be varied from less than 1 Hz to greater than 80 kHz using the range switches and the frequency dial. The amplitude of the output waveform can be adjusted up to approximately +10 V. Also, any dc offset voltage can be nulled out.

The system block diagram is shown in Figure 9–40. The concept of this particular function generator is very simple. The oscillator produces a sine wave that drives a zero-level detector (comparator) to produce a square wave of the same frequency as the sine wave. The level detector output goes to an integrator, which generates a triangular output voltage also with the same frequency as the sine wave.

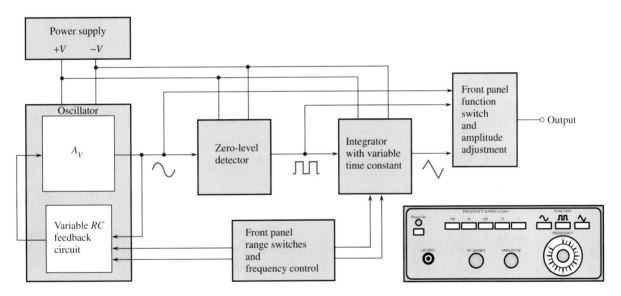

FIGURE 9–40
Function generator block diagram with front panel inset.

The frequency of the sine wave oscillator is controlled by the selection of any one of five capacitor values in the oscillator feedback circuit. These capacitors produce the five frequency ranges indicated on the front panel switches. Variation of the frequency within each range is accomplished by adjusting the resistances in the feedback circuit. These resistances are in the form of ganged potentiometers directly linked to the front panel frequency control knob. The integrator time constant is adjusted in step with the frequency by selection of capacitor and input resistor values.

On the Test Bench

Now, so that you can take a closer look at the oscillator boards, let's take them out of the system and put them on the test bench.

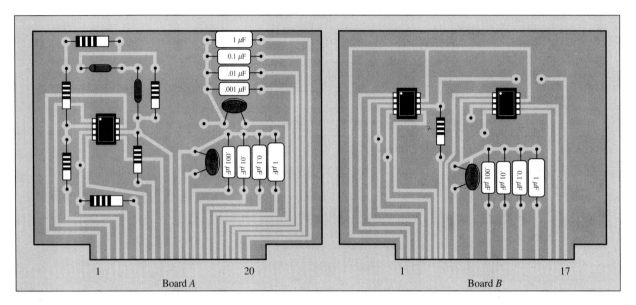

FIGURE 9–41

■ **ACTIVITY 1** **RELATE THE PC BOARDS TO THE SCHEMATIC**

Locate and identify each component on the pc boards, shown in Figure 9–41, using the system schematic in Figure 9–42. The portions shaded in color are front panel components and are not on the boards. The range switches are mechanically linked and the frequency rheostats are mechanically linked.

Develop a board-to-board wiring list specifying which pins on the two pc boards connect to each other and also indicate which pins go to the front panel.

■ **ACTIVITY 2** **ANALYZE THE SYSTEM**

STEP 1 Determine the maximum frequency of the oscillator for each range switch position ($\times 1$, $\times 10$, and so on). Only one set of three switches corresponding to a given range setting can be closed at a time. There is a set of three switches for $\times 1$, a set of three switches for $\times 10$, and so on.

STEP 2 Determine the minimum frequency of the oscillator for each range switch.

STEP 3 Determine the approximate maximum peak-to-peak output voltages for each function. The dc supply voltages are $+15$ V and -15 V.

Signal Generators and Timers

Figure 9-42

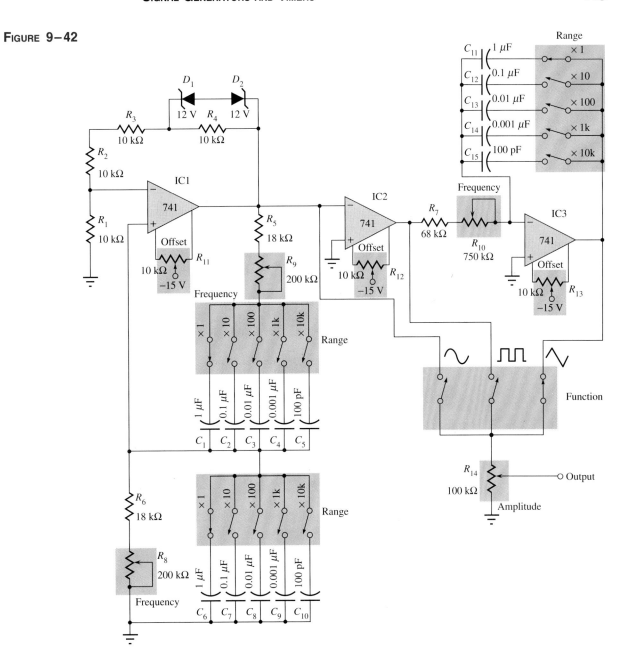

■ Activity 3   Write a Technical Report

Describe the overall operation of the function generator. Specify how each circuit works and what its purpose is. Identify the type of oscillator circuit used. Explain how the function, frequency, and amplitude are selected. Use the results of Activity 2 where appropriate.

■ Activity 4   Troubleshoot the System for Each of the Following Problems by Stating the Probable Cause or Causes

1. There is a square wave output when a triangular wave output is selected and only when the ×1k range is selected.
2. There is no output on any function setting.
3. There is no output when the square or triangular function is selected, but the sine wave output is OK.
4. Both the sine wave and the square wave outputs are OK, but there is no triangular wave output.

9–7 Review Questions

1. What type of oscillator is used in this function generator?
2. How many frequency ranges are available?
3. List the components that determine the output frequency.
4. What is the purpose of the zener diodes in the oscillator circuit?

Summary

- Oscillators operate with positive feedback.
- The two conditions for positive feedback are the phase shift around the feedback loop must be 0° and the voltage gain around the feedback loop must equal 1.
- For initial start-up, the loop gain must be greater than 1.
- Sinusoidal RC oscillators include the Wien-bridge, phase-shift, and twin-T.
- The frequency in a voltage-controlled oscillator (VCO) can be varied with a dc control voltage.
- The 555 timer is an integrated circuit that can be used as an oscillator or as a one-shot by proper connection of external components.

Glossary

Astable   Characterized by having no stable states; a type of oscillator.

Monostable   Characterized by having one stable state.

Multivibrator   A type of circuit that can operate as an oscillator or as a one-shot.

One-Shot   A monostable multivibrator.

Oscillator   An electronic circuit that operates with positive feedback and produces a time-varying output signal without an external input signal.

SIGNAL GENERATORS AND TIMERS

Positive feedback The return of a portion of the output signal to the input such that it sustains the output.

FORMULAS

(9–1)	$\dfrac{V_{out}}{V_{in}} = \dfrac{1}{3}$	Wien-bridge positive feedback attenuation
(9–2)	$f_r = \dfrac{1}{2\pi RC}$	Wien-bridge frequency
(9–3)	$B = \dfrac{1}{29}$	Phase-shift feedback attenuation
(9–4)	$f_r = \dfrac{1}{2\pi\sqrt{6}RC}$	Phase-shift oscillator frequency
(9–5)	$V_{UTP} = +V_{max}\left(\dfrac{R_3}{R_2}\right)$	Triangular wave generator upper trigger point
(9–6)	$V_{LTP} = -V_{max}\left(\dfrac{R_3}{R_2}\right)$	Triangular wave generator lower trigger point
(9–7)	$f = \dfrac{1}{4R_1 C}\left(\dfrac{R_2}{R_3}\right)$	Triangular wave generator frequency
(9–8)	$T = \dfrac{V_p - V_F}{\lvert V_{IN}\rvert / R_i C}$	Sawtooth VCO period
(9–9)	$f = \dfrac{\lvert V_{IN}\rvert}{R_i C}\left(\dfrac{1}{V_p - V_F}\right)$	Sawtooth VCO frequency
(9–10)	$f = \dfrac{1.44}{(R_1 + 2R_2)C_{ext}}$	555 astable frequency
(9–11)	Duty cycle $= \dfrac{R_1 + R_2}{R_1 + 2R_2} \times 100\%$	555 astable
(9–12)	Duty cycle $= \dfrac{R_1}{R_1 + R_2} \times 100\%$	555 astable (duty cycle < 50%)
(9–13)	$t_W = 1.1 R_{ext}C_{ext}$	555 one-shot pulse width

SELF-TEST

1. An oscillator differs from an amplifier because
   (a) it has more gain
   (b) it requires no input signal
   (c) it requires no dc supply
   (d) it always has the same output

2. All oscillators are based on
   (a) positive feedback (b) negative feedback
   (c) the piezoelectric effect (d) high gain

3. One condition for oscillation is
   (a) a phase shift around the feedback loop of 180°
   (b) a gain around the feedback loop of one-third
   (c) a phase shift around the feedback loop of 0°
   (d) a gain around the feedback loop of less than one

4. A second condition for oscillation is
   (a) no gain around the feedback loop
   (b) a gain of one around the feedback loop
   (c) the attenuation of the feedback circuit must be one-third
   (d) the feedback circuit must be capacitive

5. In a certain oscillator, $A_v = 50$. The attenuation of the feedback circuit must be
   (a) 1 (b) 0.01 (c) 10 (d) 0.02

6. For an oscillator to properly start, the gain around the feedback loop must initially be
   (a) 1 (b) less than 1 (c) greater than 1 (d) equal to B

7. In a Wien-bridge oscillator, if the resistances in the feedback circuit are decreased, the frequency
   (a) decreases (b) increases (c) remains the same

8. The Wien-bridge oscillator's positive feedback circuit is
   (a) an RL network (b) an LC network
   (c) a voltage divider (d) a lead-lag network

9. A phase-shift oscillator has
   (a) three RC networks (b) three LC networks
   (c) a T-type network (d) a π-type network

10. An oscillator whose frequency is changed by a variable dc voltage is known as
    (a) a Wien-bridge oscillator (b) a VCO
    (c) a phase-shift oscillator (d) an astable multivibrator

11. Which one of the following is not an input or output of the 555 timer?
    (a) Threshold (b) Control voltage (c) Clock
    (d) Trigger (e) Discharge (f) Reset

12. An astable multivibrator is
    (a) an oscillator (b) a one-shot (c) a time-delay circuit
    (d) characterized by having no stable states (e) a and d

Signal Generators and Timers

13. The output frequency of a 555 timer connected as an oscillator is determined by
    - (a) the supply voltage
    - (b) the frequency of the trigger pulses
    - (c) the external RC time constant
    - (d) the internal RC time constant
    - (e) a and d

14. The term *monostable* means
    - (a) one output
    - (b) one frequency
    - (c) one time constant
    - (d) one stable state

15. A 555 timer connected as a one-shot has $R_{ext} = 2$ kΩ and $C_{ext} = 2$ μF. The output pulse has a width of
    - (a) 1.1 ms
    - (b) 4 ms
    - (c) 4 μs
    - (d) 4.4 ms

Problems

Section 9–1 Definition of an Oscillator

1. What type of input is required for an oscillator?
2. What are the basic components of an oscillator circuit?

Section 9–2 Oscillator Principles

3. If the voltage gain of the amplifier portion of an oscillator is 75, what must be the attenuation of the feedback circuit to sustain the oscillation?
4. Generally describe the change required in the oscillator of Problem 3 in order for oscillation to begin when the power is initially turned on.

Section 9–3 Sine Wave Oscillators

5. A certain lead-lag network has a resonant frequency of 3.5 kHz. What is the rms output voltage if an input signal with a frequency equal to f_r and with an rms value of 2.2 V is applied to the input?
6. Calculate the resonant frequency of a lead-lag network with the following values: $R_1 = R_2 = 6.2$ kΩ, and $C_1 = C_2 = 0.02$ μF.
7. Determine the necessary value of R_2 in Figure 9–43 so that the circuit will oscillate. Neglect the forward resistance of the zener diodes.
8. Explain the purpose of R_3 in Figure 9–43.
9. What is the initial closed-loop gain in Figure 9–44? At what value of output voltage does A_{cl} change and to what value does it change? (The value of R_2 was found in Problem 7).
10. Find the frequency of oscillation for the Wien-bridge oscillator in Figure 9–43.
11. What value of R_f is required in Figure 9–44? What is f_r?

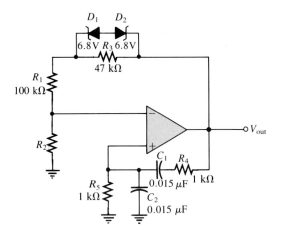

FIGURE 9–43   **FIGURE 9–44**

SECTION 9–4 NONSINUSOIDAL OSCILLATORS

12. What type of signal does the circuit in Figure 9–45 produce? Determine the frequency of the output.

13. Show how to change the frequency of oscillation in Figure 9–45 to 10 kHz.

14. Determine the amplitude and frequency of the output voltage in Figure 9–46. Use 1 V as the forward PUT voltage.

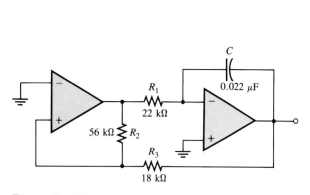
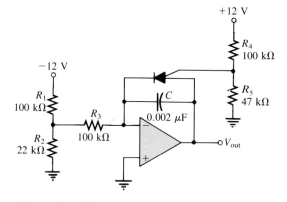

FIGURE 9–45   **FIGURE 9–46**

15. Modify the sawtooth generator in Figure 9–46 so that its peak-to-peak output is 4 V.

16. A certain sawtooth generator has the following parameter values: $V_{IN} = 3$ V, $R = 4.7$ kΩ, $C = 0.001$ μF, and V_F for the PUT is 1.2 V. Determine its peak-to-peak output voltage if the period is 10 μs.

SECTION 9–5 THE 555 TIMER AS AN OSCILLATOR

17. What are the two comparator reference voltages in a 555 timer when $V_{CC} = 10$ V?

18. Determine the frequency of oscillation for the 555 astable oscillator in Figure 9–47.

FIGURE 9-47

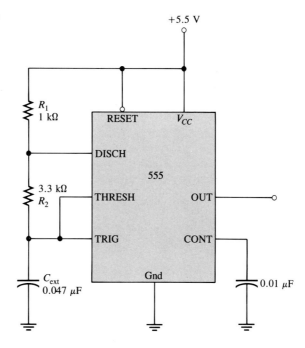

19. To what value must C_{ext} be changed in Figure 9-47 to achieve a frequency of 25 kHz?
20. In an astable 555 configuration, the external resistor $R_1 = 3.3$ kΩ. What must R_2 equal to produce a duty cycle of 75 percent?

SECTION 9-6 THE 555 TIMER AS A ONE-SHOT

21. A 555 timer connected in the monostable has a 56 kΩ external resistor and a 0.22 μF external capacitor. What is the pulse width of the output?
22. The output pulse width of a certain 555 one-shot is 12 ms. If $C_{ext} = 2.2$ μF, what is R_{ext}?
23. Suppose that you need to hook up a 555 timer as a one-shot in the lab to produce an output pulse with a width of 100 μs. Select the appropriate values for the external components.
24. Devise a circuit to produce two sequential 50 μs pulses. The first pulse must occur 100 ms after an intial trigger and the second pulse must occur 300 ms after the first pulse.

ANSWERS TO REVIEW QUESTIONS

SECTION 9-1

1. An oscillator is a circuit that produces a repetitive output waveform with only the dc supply voltage as an input.
2. Positive feedback
3. The feedback circuit provides attenuation and phase shift.

Section 9-2

1. Zero phase shift and unity voltage gain around the closed feedback
2. Positive feedback is when a portion of the output signal is fed back to the input of the amplifier such that it reinforces itself.
3. Loop gain greater than 1; zero phase shift and unity voltage gain

Section 9-3

1. The negative feedback loop sets the closed-loop gain; the positive feedback loop sets the frequency of oscillation.
2. 1.67 V
3. The three RC networks each contribute 60°.

Section 9-4

1. A voltage-controlled oscillator exhibits a frequency that can be varied with a dc control voltage.
2. The basis of a relaxation oscillator is the charging and discharging of a capacitor.

Section 9-5

1. Two comparators, a flip-flop, a discharge transistor, and a resistive voltage divider
2. The duty cycle is set by the external resistors and the external capacitor.

Section 9-6

1. A one-shot has one stable state.
2. t_W = 5.5 ms
3. The pulse width can be decreased by decreasing the external resistance or capacitance.

Section 9-7

1. A Wien-bridge oscillator
2. There are five frequency ranges.
3. R_5, R_6, R_8, R_9, C_1–C_5, C_6–C_{10}
4. To limit the oscillator output amplitude and to help ensure start-up

Answers to Practice Exercises

9–1 Change the zener diodes to 6.1 V devices.
9–2 (a) 238 kΩ (b) 7.92 kHz
9–3 6.06 V peak-to-peak
9–4 1055 Hz
9–5 31.9%
9–6 45.5 kΩ
9–7 Replace R_1 with a potentiometer with a maximum resistance of at least 182 kΩ.

TEST BENCH
SPECIAL ASSIGNMENTS

The three Test Bench assignments in this color section are each related to a different System Application as indicated in these graphics. These are representative of the circuit boards that you will be working with. Each Test Bench in this section contains a circuit board page and an instrumentation page. The assignment is stated on the circuit board page of each Test Bench.

Use this section only when directed by the special assigment activity in the System Application section of a chapter. You may use the special worksheets available from your instructor to facilitate performing these assignments as well as the other activities in the System Application.

Refer to the back page of this section for useful information.

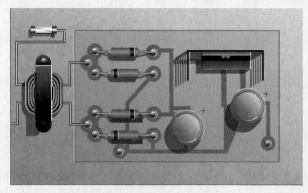

TEST BENCH 1 *Chapter* 2
DC Power Supply Board

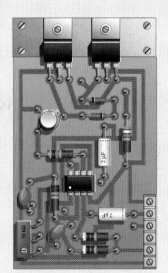

TEST BENCH 2 *Chapter* 6
Stereo Amplifer Board

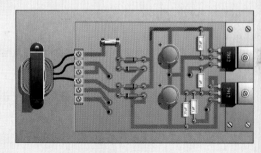

TEST BENCH 3 *Chapter* 10
Dual Power Supply Board

Assignment:
Evaluate the instrument settings and readings for this test setup and determine if the circuit is operating properly. If it is not operating properly, isolate the fault(s). The colored circled numbers indicate connections between the pc board and the instruments: 1(blue) goes to 1(blue), 2 (red) goes to 2 (red), etc. The scope probe is X1.

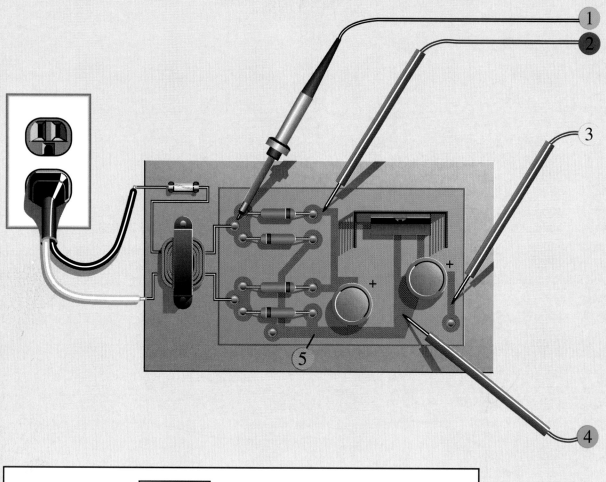

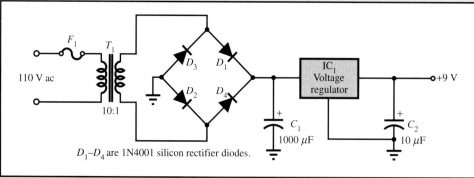

D_1–D_4 are 1N4001 silicon rectifier diodes.

TEST BENCH 1

DMM1

DMM2

DMM3

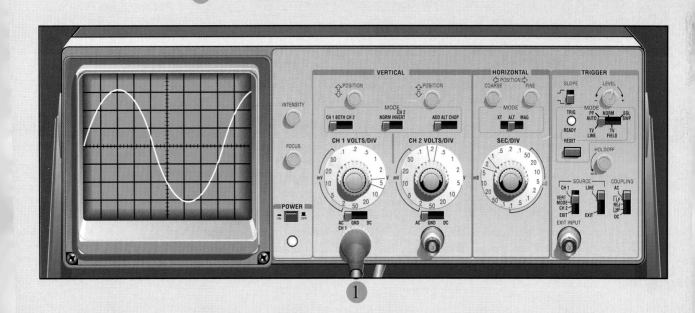

For Chapter 2
System Application

Assignment:
Evaluate the instrument settings and readings for this test setup and determine if the circuit is operating properly. If it is not operating properly, determine the most probable fault(s). The color circled numbers indicate connections between the pc board and the instruments: 1(blue) goes to 1(blue), 2 (red) goes to 2 (red), etc. The top waveform on the scope is CH 1.

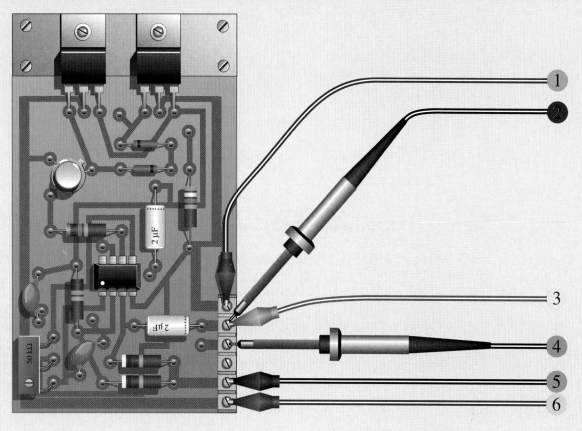

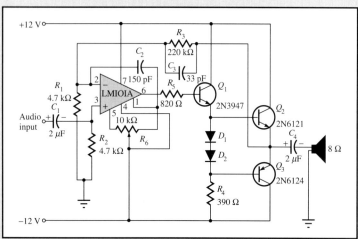

TEST BENCH 2

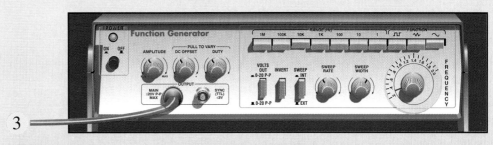
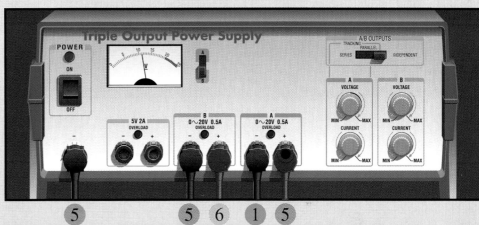
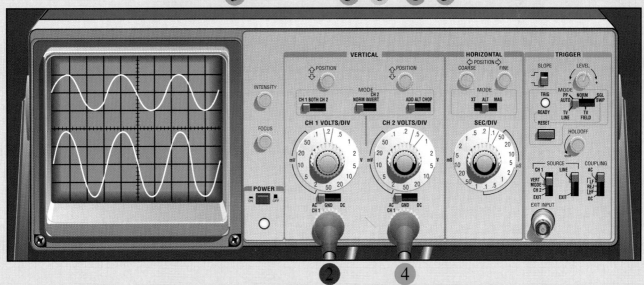

For Chapter 6
System Application

Assignment:
1. Set up the range and function controls on the instruments for observing the voltages at the probed points. Existing settings may not be correct.
2. Indicate the readings and displays you should observe if the power supply is working properly.
3. Indicate the readings and displays you should observe if diode D1 is open.
4. Indicate the readings and displays you should observe if capacitor C2 is open.

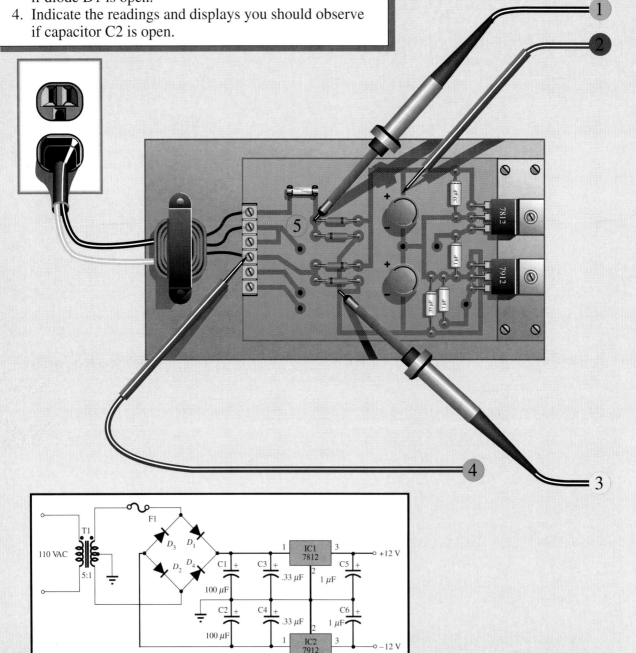

TEST BENCH 3

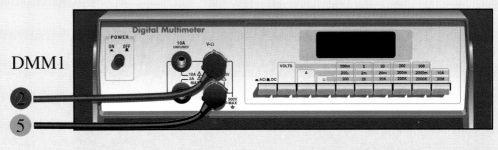

DMM1

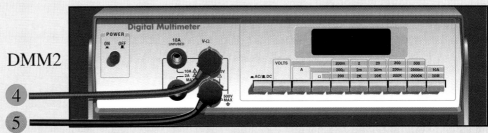

DMM2

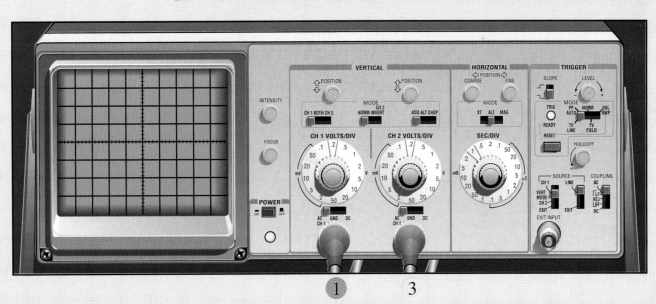

For Chapter 10
System Application

TEST BENCH INFORMATION

There are three Test Bench assignments in this color section. Each Test Bench contains a two-page spread and is related to one of the System Application sections in the text. The left page contains the assignment, the circuit board with connections and/or labeled points, and a schematic of the board. The right page contains certain instruments that are connected to the board or are to be connected to the board.

Connections between the circuit board and the instruments are indicated by corresponding numbers in color-coded circles. For example, a probe attached to the board with a *red circle 2* goes to the instrument input(s) or output(s) also labeled with a *red circle 2*.

The assignments in this section are of two basic types:
1. Evaluation of indicated instrument settings and readings for the purpose of troubleshooting the board.
2. Selection of instrument setting and proper readings for specified inputs to circuit boards.

Instrument readings that involve circular dial, rotary switch, push-button switch, or slide switch settings should be evident. Examples are the *Time/Div* switch on the oscilloscope and the *frequency* dial on the function generator.

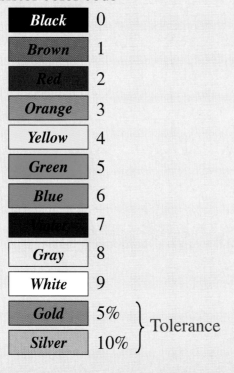

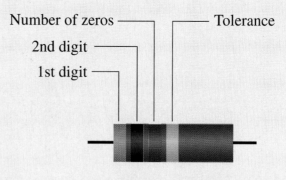

10
POWER SUPPLY CIRCUITS

10–1 VOLTAGE REGULATION
10–2 BASIC SERIES REGULATORS
10–3 BASIC SHUNT REGULATORS
10–4 BASIC SWITCHING REGULATORS
10–5 INTEGRATED CIRCUIT VOLTAGE REGULATORS
10–6 APPLICATIONS OF IC VOLTAGE REGULATORS
10–7 A SYSTEM APPLICATION

After completing this chapter, you should be able to

- Explain the basic concept of voltage regulation.
- Define line and load regulation and discuss the difference.
- Describe how a basic series voltage regulator works.
- Explain the need for overload protection.
- Describe how a basic shunt voltage regulator works.
- Relate the advantages of switching regulators and explain how switching regulators work.
- Discuss current limiting in regulators.
- Explain what fold-back current limiting is.
- Select specific positive or negative three-terminal IC regulators for a specific application.
- Use an external pass transistor with a three-terminal IC regulator to increase the current capability.
- Use a current-limiting circuit with a three-terminal regulator.
- Use a three-terminal regulator as a constant-current source.
- Configure an IC switching regulator for step-down or step-up operation.

A voltage regulator provides a constant dc output voltage that is practically independent of the input voltage, output load current, and temperature. The voltage regulator is one part of a power supply. Its input voltage comes from the filtered output of a rectifier derived from an ac voltage or from a battery in the case of portable systems.

Most voltage regulators fall into two broad categories—linear regulators and switching regulators. In the linear regulator category, two general types are the linear series regulator and the linear shunt regulator. These are normally available for either positive or negative output voltages. A dual regulator provides both positive and negative outputs. In the switching regulator category, three general configurations are step-down, step-up, and inverting.

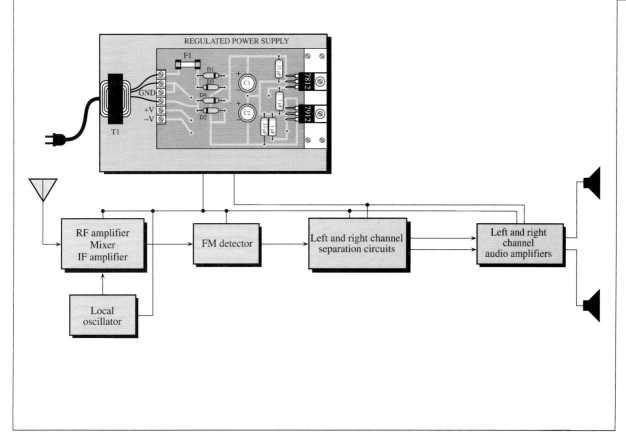

Many types of integrated circuit (IC) regulators are available. The most popular types of linear regulator are the three-terminal fixed voltage regulator and the three-terminal adjustable voltage regulator. Switching regulators are also widely used. In this chapter, specific IC devices are introduced as representative of the wide range of available devices.

A System Application

A dual polarity regulated power supply is used for the FM stereo system that you worked with in Chapter 8. Two regulators, one positive and the other negative, provide the positive voltage required for the receiver circuits and the dual polarity voltages for the op-amp circuits. The regulator input voltages come from a full-wave rectifier with filtered outputs.

For the system application in Section 10–7, in addition to the other topics, be sure you understand

- How three-terminal fixed-voltage regulators are used.
- The basic operation of a power supply rectifier and filter (review Chapter 2).
- How to set the current limit of a regulator.
- How to determine power dissipation in a pass transistor.

10-1 VOLTAGE REGULATION

Two basic categories of voltage regulation are line regulation and load regulation. Line regulation maintains a nearly constant output voltage when the input voltage varies. Load regulation maintains a nearly constant output voltage when the load varies.

LINE REGULATION

When the dc input (line) voltage changes, the voltage **regulator** must maintain a nearly constant output voltage, as illustrated in Figure 10–1.

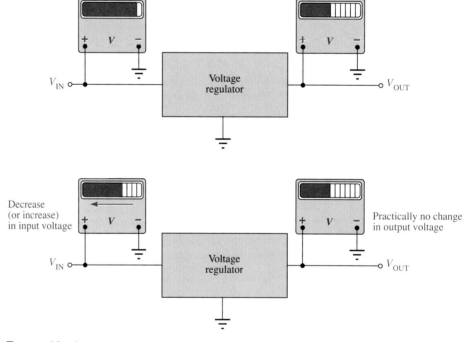

FIGURE 10–1
Line regulation. A change in input (line) voltage does not significantly affect the output voltage of a regulator (within certain limits).

Line regulation can be defined as the percentage change in the output voltage for a given change in the input (line) voltage. It is usually expressed in units of %/V. For example, a line regulation of 0.05%/V means that the output voltage changes 0.05 percent when the input voltage increases or decreases by one volt. Line regulation can be calculated using the following formula (Δ means "a change in"):

$$\text{Line regulation} = \frac{(\Delta V_{OUT}/V_{OUT})100\%}{\Delta V_{IN}} \qquad (10\text{--}1)$$

EXAMPLE 10–1

When the input to a particular voltage regulator decreases by 5 V, the output decreases by 0.25 V. The nominal output is 15 V. Determine the line regulation in %/V.

SOLUTION
The line regulation is

$$\frac{(\Delta V_{OUT}/V_{OUT})100\%}{\Delta V_{IN}} = \frac{(0.25 \text{ V}/15 \text{ V})100\%}{5 \text{ V}} = 0.333\%\text{V}$$

PRACTICE EXERCISE 10–1
The input of a certain regulator increases by 3.5 V. As a result, the output voltage increases by 0.42 V. The nominal output is 20 V. Determine the regulation in %/V.

LOAD REGULATION

When the amount of current through a load changes due to a varying load resistance, the voltage regulator must maintain a nearly constant output voltage across the load, as illustrated in Figure 10–2.

Load regulation can be defined as the percentage change in output voltage for a given change in load current. It can be expressed as a percentage change in output voltage from no-load (NL) to full-load (FL) as follows.

$$\text{Load regulation} = \frac{(V_{NL} - V_{FL})100\%}{V_{FL}} \quad (10\text{–}2)$$

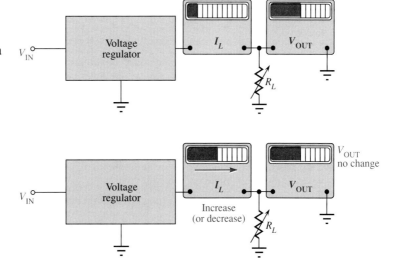

FIGURE 10–2
Load regulation. A change in load current has practically no effect on the output voltage of a regulator (within certain limits).

Alternately, the load regulation can be expressed as a percentage change in output voltage for each mA change in load current. For example, a load regulation of 0.01%/mA means that the output voltage changes 0.01 percent when the load current increases or decreases 1 mA.

EXAMPLE 10–2

A certain voltage regulator has a 12 V output when there is no load ($I_L = 0$). When there is a full-load current of 10 mA, the output voltage is 11.95 V. Express the voltage regulation as a percentage change from no-load to full-load and also as a percentage change for each mA change in load current.

SOLUTION
The no-load output voltage is

$$V_{NL} = 12 \text{ V}$$

The full-load output voltage is

$$V_{FL} = 11.95 \text{ V}$$

The load regulation is

$$\left(\frac{V_{NL} - V_{FL}}{V_{FL}}\right)100\% = \left(\frac{12 \text{ V} - 11.95 \text{ V}}{11.95 \text{ V}}\right)100\%$$
$$= 0.418\%$$

The load regulation can also be expressed as

$$\frac{0.418\%}{10 \text{ mA}} = 0.0418\%/\text{mA}$$

where the change in load current from no-load to full-load is 10 mA.

PRACTICE EXERCISE 10–2
A regulator has a no-load output voltage of 18 V and a full-load output of 17.85 V at a load current of 50 mA. Determine the voltage regulation as a percentage change from no-load to full-load and also as a percentage change for each mA change in load current.

10–1 REVIEW QUESTIONS

1. Define *line regulation*.
2. Define *load regulation*.

10–2 BASIC SERIES REGULATORS

The two fundamental classes of voltage regulators are linear regulators and switching regulators. Both of these are available in integrated circuit form. There are two basic types of linear regulator. One is the series regulator and the other is the shunt regulator. In this section, we will look at the series regulator. The shunt and switching regulators are covered in the next two sections.

A simple representation of a series type of linear regulator is shown in Figure 10–3(a), and the basic components are shown in the block diagram in Figure 10–3(b). Notice that

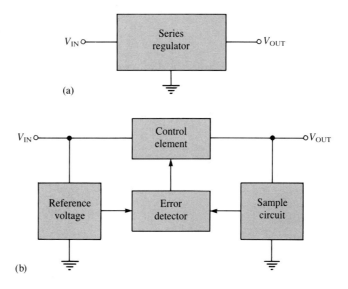

FIGURE 10–3
Simple series voltage regulator block diagram.

the control element is in series with the load between input and output. The output sample circuit senses a change in the output voltage. The error detector compares the sample voltage with a reference voltage and causes the control element to compensate in order to maintain a constant output voltage.

REGULATING ACTION

A basic op-amp series regulator circuit is shown in Figure 10–4. The operation of the series regulator is illustrated in Figure 10–5. The resistive voltage divider formed by R_2 and R_3 senses any change in the output voltage. When the output tries to decrease because of a decrease in V_{IN} or because of an increase in I_L, as indicated in parts (a) and (b), a proportional voltage decrease is applied to the op-amp's inverting input by the voltage divider. Since the zener diode holds the other op-amp input at a nearly fixed reference voltage V_{REF}, a small difference voltage (error voltage) is developed across the op-amp's inputs. This difference voltage is amplified, and the op-amp's output voltage increases.

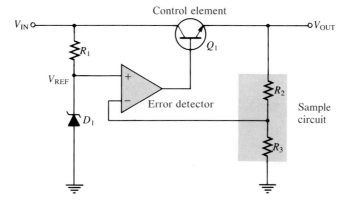

FIGURE 10–4
Basic op-amp series regulator.

This increase is applied to the base of Q_1, causing the emitter voltage V_{OUT} to increase until the voltage to the inverting input again equals the reference (zener) voltage. This action offsets the attempted decrease in output voltage, thus keeping it nearly constant. Q_1 is a power transistor and is often used with a heat sink because it must handle all of the load current.

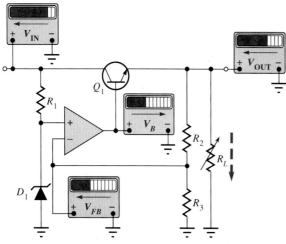

(a) When V_{IN} or R_L decreases, V_{OUT} attempts to decrease, V_{FB} also attempts to decrease, and as a result, V_B attempts to increase, thus compensating for the attempted decrease in V_{OUT}.

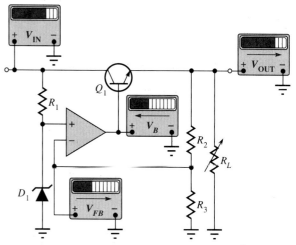

(b) When V_{IN} (or R_L) stabilizes at its new lower value, the voltages are at their original values, thus keeping V_{OUT} constant as a result of the negative feedback.

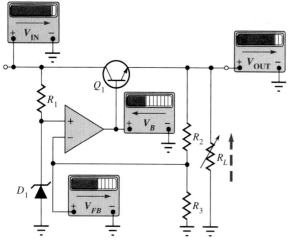

(c) When V_{IN} or R_L increases, V_{OUT} attempts to increase. The feedback voltage, V_{FB}, also attempts to increase, and, as a result, the op-amp's output voltage, V_B, applied to the base of the control transistor, attempts to decrease, thus compensating for the attempted increase in V_{OUT}.

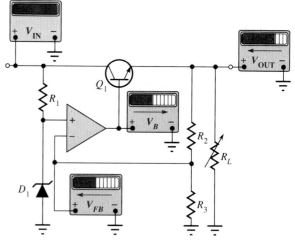

(d) When V_{IN} (or R_L) stabilizes at its new higher value, the voltages are at their original values, thus keeping V_{OUT} constant as result of the negative feedback.

FIGURE 10–5

Illustration of series regulator action that keeps V_{OUT} constant when V_{IN} or R_L changes.

The opposite action occurs when the output tries to increase, as indicated in Figure 10–5(c) and (d). The op-amp in the series regulator is actually connected as a noninverting amplifier where the reference voltage V_{REF} is the input at the noninverting terminal, and the R_2/R_3 voltage divider forms the negative feedback network. The closed-loop voltage gain is

$$A_{cl} = 1 + \frac{R_2}{R_3} \tag{10-3}$$

Therefore, the regulated output voltage (neglecting the base-emitter voltage of Q_1) is

$$V_{OUT} \cong \left(1 + \frac{R_2}{R_3}\right) V_{REF} \tag{10-4}$$

From this analysis, you can see that the output voltage is determined by the zener voltage and the resistors R_2 and R_3. It is relatively independent of the input voltage, and therefore, regulation is achieved (as long as the input voltage and load current are within specified limits).

EXAMPLE 10–3

Determine the output voltage for the regulator in Figure 10–6.

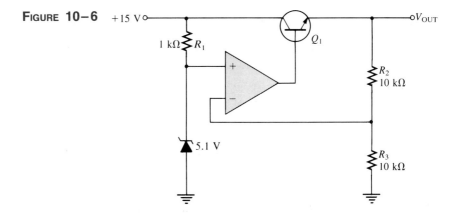

FIGURE 10–6

SOLUTION

$V_{REF} = 5.1$ V

$$V_{OUT} = \left(1 + \frac{R_2}{R_3}\right) V_{REF} = \left(1 + \frac{10\ k\Omega}{10\ k\Omega}\right) 5.1\ V = (2)5.1\ V = 10.2\ V$$

PRACTICE EXERCISE 10–3

The following changes are made in the circuit in Figure 10–6: $V_Z = 3.3$ V, $R_1 = 1.8$ kΩ, $R_2 = 22$ kΩ, and $R_3 = 18$ kΩ. What is the output voltage?

SHORT-CIRCUIT OR OVERLOAD PROTECTION

If an excessive amount of load current is drawn, the series-pass transistor can be quickly damaged or destroyed. Most regulators employ some type of excess current protection in the form of a current-limiting mechanism. Figure 10–7 shows one method of current limiting to prevent overloads called *constant current limiting*. The current-limiting circuit consists of transistor Q_2 and resistor R_4.

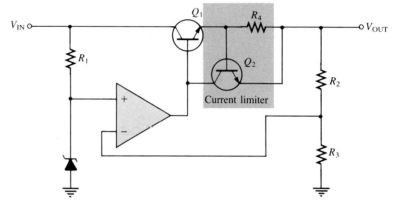

FIGURE 10–7
Series regulator with constant current limiting.

The load current through R_4 creates a voltage from base to emitter of Q_2. When I_L reaches a predetermined maximum value, the voltage drop across R_4 is sufficient to forward-bias the base-emitter junction of Q_2, thus causing it to conduct. Enough Q_1 base current is diverted into the collector of Q_2 so that I_L is limited to its maximum value $I_{L(max)}$. Since the base-to-emitter voltage of Q_2 cannot exceed about 0.7 V for a silicon transistor, the voltage across R_4 is held to this value, and the load current is limited to

$$I_{L(max)} = \frac{0.7 \text{ V}}{R_4} \qquad (10\text{–}5)$$

EXAMPLE 10–4 Determine the maximum current that the regulator in Figure 10–8 can provide to a load.

SOLUTION

$$I_{L(max)} = \frac{0.7 \text{ V}}{R_4} = \frac{0.7 \text{ V}}{1 \text{ }\Omega} = 0.7 \text{ A}$$

FIGURE 10–8

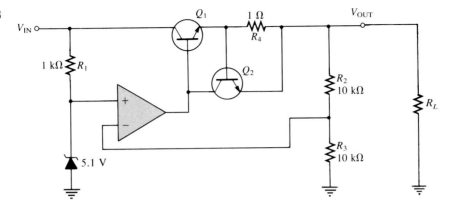

PRACTICE EXERCISE 10–4
If the output of the regulator in Figure 10–8 is shorted, what is the current?

REGULATOR WITH FOLD-BACK CURRENT LIMITING

In the previous current-limiting technique, the current is restricted to a maximum constant value.

Fold-back current limiting is a method used particularly in high-current regulators whereby the output current under overload conditions drops to a value well below the peak load current capability to prevent excessive power dissipation.

BASIC IDEA The basic concept of fold-back current limiting is as follows, with reference to Figure 10–9. The circuit is similar to the constant current-limiting arrangement in Figure 10–7, with the exception of resistors R_5 and R_6. The voltage drop developed across R_4 by the load current must not only overcome the base-emitter voltage required to turn on Q_2, but it must overcome the voltage across R_5. That is, the voltage across R_4 must be

$$V_{R4} = V_{R5} + V_{BE}$$

FIGURE 10–9
Series regulator with fold-back current limiting.

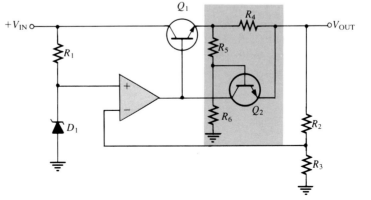

In an overload or short-circuit condition the load current increases to a value $I_{L(max)}$ that is sufficient to cause Q_2 to conduct. At this point the current can increase no further. The decrease in output voltage results in a proportional decrease in the voltage across R_5; thus less current through R_4 is required to maintain the forward-biased condition of Q_1. So, as V_{OUT} decreases, I_L decreases, as shown in the graph of Figure 10–10.

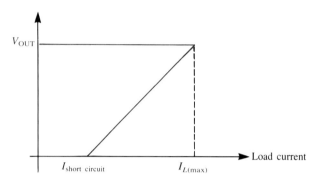

FIGURE 10–10
Fold-back current limiting (output voltage versus load current).

The advantage of this technique is that the regulator is allowed to operate with peak load current up to $I_{L(max)}$; but when the output becomes shorted, the current drops to a lower value to prevent overheating of the device.

10–2 REVIEW QUESTIONS

1. What are the basic components in a series regulator?
2. A certain series regulator has an output voltage of 8 V. If the op-amp's closed loop gain is 4, what is the value of the reference voltage?

10–3 BASIC SHUNT REGULATORS

The second basic type of linear voltage regulator is the shunt regulator. As you have learned, the control element in the series regulator is the series-pass transistor. In the shunt regulator, the control element is a transistor in parallel (shunt) with the load.

A simple representation of a shunt type of linear regulator is shown in Figure 10–11(a), and the basic components are shown in the block diagram in part (b) of the figure.

FIGURE 10–11
Simple shunt regulator block diagrams.

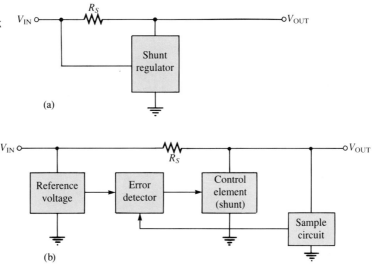

In the basic shunt regulator, the control element is a transistor Q_1 in parallel with the load, as shown in Figure 10–12. A resistor, R_1 is in series with the load. The operation of the circuit is similar to that of the series regulator, except that regulation is achieved by controlling the current through the parallel transistor Q_1.

FIGURE 10–12
Basic op-amp shunt regulator.

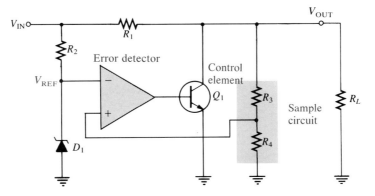

When the output voltage tries to decrease due to a change in input voltage or load current, as shown in Figure 10–13(a), the attempted decrease is sensed by R_3 and R_4 and applied to the op-amp's noninverting input. The resulting difference voltage reduces the op-amp's output, driving Q_1 less, thus reducing its collector current (shunt current) and increasing its internal collector-to-emitter resistance r_{ce}. Since r_{ce} acts as a voltage divider with R_1, this action offsets the attempted decrease in V_{OUT} and maintains it at an almost constant level. The opposite action occurs when the output tries to increase, as indicated in Figure 10–13(b).

FIGURE 10–13
Sequence of responses when V_{OUT} tries to decrease as a result of a decrease in R_L or V_{IN} (opposite responses for an attempted increase).

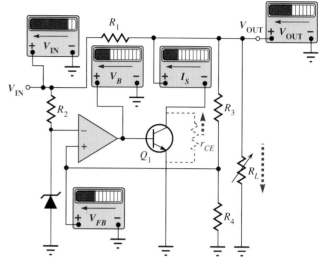

(a) Initial response to a decrease in V_{IN} or R_L

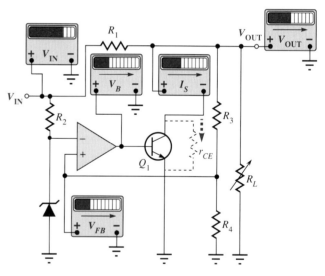

(b) V_{OUT} held constant by the feedback action.

With I_L and V_{OUT} constant, a change in the input voltage produces a change in shunt current (I_S) as follows (Δ means "a change in").

$$\Delta I_S = \frac{\Delta V_{IN}}{R_1} \qquad (10\text{–}6)$$

Power Supply Circuits

With a constant V_{IN} and V_{OUT}, a change in load current causes an opposite change in shunt current.

$$\Delta I_S = -\Delta I_L \qquad (10\text{--}7)$$

This formula says that if I_L increases, I_S decreases, and vice versa. The shunt regulator is less efficient than the series type but offers inherent short-circuit protection. If the output is shorted ($V_{OUT} = 0$), the load current is limited by the series resistor R_1 to a maximum value as follows ($I_S = 0$).

$$I_{L(max)} = \frac{V_{IN}}{R_1} \qquad (10\text{--}8)$$

EXAMPLE 10–5

In Figure 10–14, what power rating must R_1 have if the maximum input voltage is 12.5 V?

FIGURE 10–14

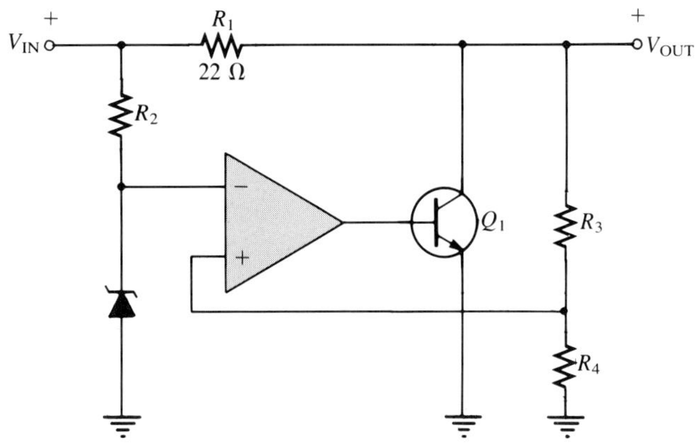

SOLUTION
The worst-case power dissipation in R_1 occurs when the output is short-circuited. $V_{OUT} = 0$, and when $V_{IN} = 12.5$ V, the voltage dropped across R_1 is $V_{IN} - V_{OUT} = 12.5$ V. The power dissipation in R_1 is

$$P_{R1} = \frac{V_{R1}^2}{R_1} = \frac{(12.5 \text{ V})^2}{22 \ \Omega} = 7.1 \text{ W}$$

Therefore, a resistor of at least 10 W should be used.

PRACTICE EXERCISE 10–5
In Figure 10–14, R_1 is changed to 33 Ω. What must be the power rating of R_1 if the maximum input voltage is 24 V?

10–3 REVIEW QUESTIONS

1. How does the control element in a shunt regulator differ from that in a series regulator?
2. What is one advantage of a shunt regulator over a series type? What is a disadvantage?

10-4 BASIC SWITCHING REGULATORS

The two types of linear regulators—series and shunt—have control elements (transistors) that are conducting all the time, with the amount of conduction varied as demanded by changes in the output voltage or current. The switching regulator is different; the control element operates as a switch. A greater efficiency can be realized with this type of voltage regulator than with the linear types because the transistor is not always conducting. Therefore, switching regulators can provide greater load currents at low voltage than linear regulators because the control transistor doesn't dissipate as much power. Three basic configurations of switching regulators are step-down, step-up, and inverting.

STEP-DOWN CONFIGURATION

In the step-down configuration, the output voltage is always less than the input voltage. A basic step-down switching regulator is shown in Figure 10–15(a), and its simplified equivalent is shown in Figure 10–15(b). Transistor Q_1 is used to switch the input voltage at a duty cycle that is based on the regulator's load requirement. The *LC* filter is then used

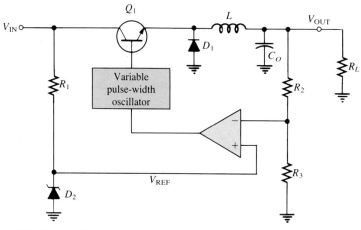

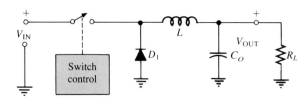

FIGURE 10–15
Basic step-down switching regulator.

(a) Typical circuit

(b) Simplified equivalent circuit

POWER SUPPLY CIRCUITS

to average the switched voltage. Since Q_1 is either *on* (saturated) or *off*, the power lost in the control element is relatively small. Therefore, the switching regulator is useful primarily in higher power applications or in applications where efficiency is of utmost concern.

The on and off intervals of Q_1 are shown in the waveform of Figure 10–16(a). The capacitor charges during the on-time (t_{on}) and discharges during the off-time (t_{off}). When the on-time is increased relative to the off-time, the capacitor charges more, thus increasing the output voltage, as indicated in Figure 10–16(b). When the on-time is decreased relative to the off-time, the capacitor discharges more, thus decreasing the output voltage, as in Figure 10–16(c). Therefore, by adjusting the duty cycle $t_{on}/(t_{on} + t_{off})$ of Q_1, the output voltage can be varied. The inductor further smooths the fluctuations of the output voltage caused by the charging and discharging action.

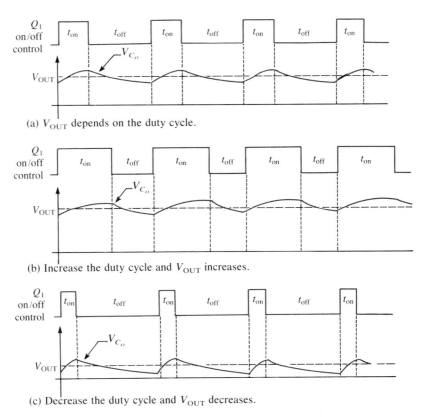

FIGURE 10–16
Switching regulator waveforms. The V_{C_O} waveform is for *no* inductive filtering to illustrate the charge and discharge action. L and C smooth V_{C_O} to a nearly constant level, as indicated by the dashed line for V_{OUT}.

The regulating action is as follows and is illustrated in Figure 10–17. When V_{OUT} tries to decrease, the on-time of Q_1 is increased, causing an additional charge on C_O to offset the attempted decrease. When V_{OUT} tries to increase, the on-time of Q_1 is decreased, causing C_O to discharge enough to offset the attempted increase.

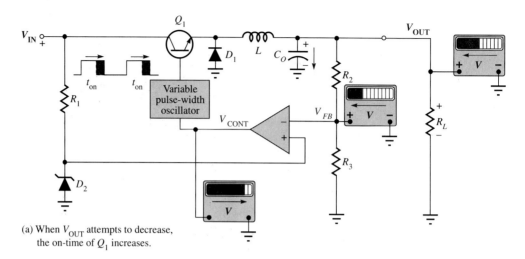

(a) When V_{OUT} attempts to decrease, the on-time of Q_1 increases.

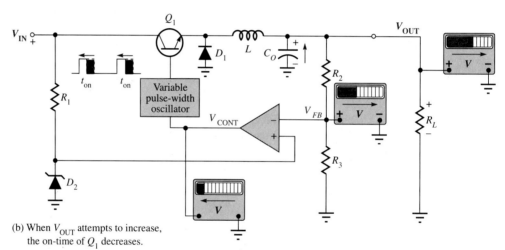

(b) When V_{OUT} attempts to increase, the on-time of Q_1 decreases.

FIGURE 10–17
Regulating action of the basic step-down switching regulator.

The output voltage is expressed as

$$V_{OUT} = \left(\frac{t_{on}}{T}\right)V_{IN} \qquad (10\text{–}9)$$

T is the period of the on-off cycle of Q_1 and is related to the frequency by $T = 1/f$. The period is the sum of the on-time and the off-time.

$$T = t_{on} + t_{off} \qquad (10\text{–}10)$$

The ratio t_{on}/T is called the *duty cycle*.

STEP-UP CONFIGURATION

A basic step-up type of switching regulator is shown in Figure 10–18. When Q_1 turns on, voltage across L increases instantaneously to $V_{IN} - V_{CE(sat)}$, and the inductor's magnetic field expands quickly, as indicated in Figure 10–19(a). During the on-time (t_{on}) of Q_1, V_L decreases from its initial maximum, as shown. The longer Q_1 is on, the smaller V_L becomes. When Q_1 turns off, the inductor's magnetic field collapses; and its polarity reverses so that its voltage adds to V_{IN}, thus producing an output voltage greater than the input, as indicated in Figure 10–19(b). During the off-time (t_{off}) of Q_1, the diode is forward-biased, allowing the capacitor to charge. The variations in the output voltage due to the charging and discharging action are sufficiently smoothed by the filtering action of L and C_O.

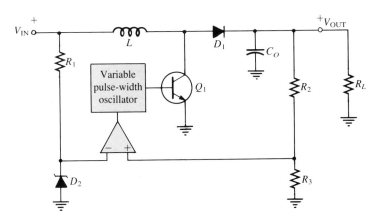

FIGURE 10–18
Basic step-up switching regulator.

The shorter the on-time of Q_1, the greater the inductor voltage is, and thus the greater the output voltage is (greater V_L adds to V_{IN}). The longer the on-time of Q_1, the smaller are the inductor voltage and the output voltage (small V_L adds to V_{IN}). When V_{OUT} tries to decrease because of increasing load or decreasing input voltage, t_{on} decreases and the attempted decrease in V_{OUT} is offset. When V_{OUT} tries to increase, t_{on} increases and the attempted increase in V_{OUT} is offset. This regulating action is illustrated in Figure 10–20. As you can see, the output voltage is inversely related to the duty cycle of Q_1 and can be expressed as follows.

$$V_{OUT} = \left(\frac{T}{t_{on}}\right) V_{IN} \qquad (10\text{–}11)$$

where $T = t_{on} + t_{off}$.

FIGURE 10–19
Step-up action of the basic switching regulator.

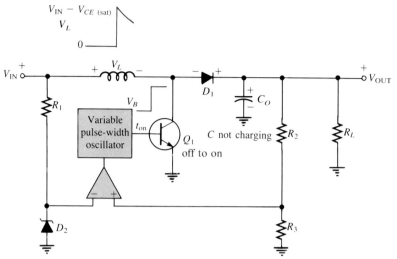

(a) When Q_1 is on

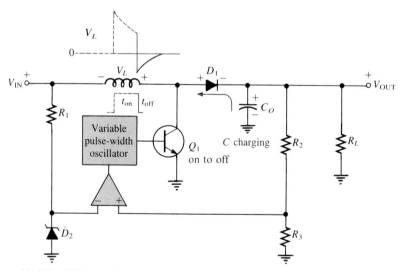

(b) When Q_1 turns off

FIGURE 10–20
Regulating action of the basic step-up switching regulator.

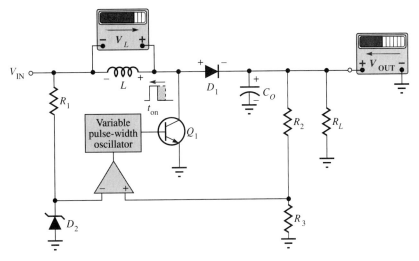

(a) When V_{OUT} tries to decrease, t_{on} decreases, causing V_L to increase. This compensates for the attempted decrease in V_{OUT}.

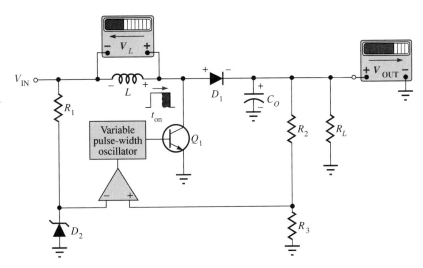

(b) When V_{OUT} tries to increase, t_{on} increases, causing V_L to decrease. This compensates for the attempted increase in V_{OUT}.

VOLTAGE-INVERTER CONFIGURATION

A third type of switching regulator produces an output voltage that is opposite in polarity to the input. A basic diagram is shown in Figure 10–21.

When Q_1 turns on, the inductor voltage jumps to $V_{IN} - V_{CE(sat)}$ and the magnetic field rapidly expands, as shown in Figure 10–22(a). While Q_1 is on, the diode is reverse-biased and the inductor voltage decreases from its initial maximum. When Q_1 turns off, the

FIGURE 10–21
Basic inverting switching regulator.

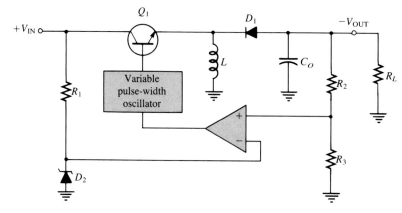

FIGURE 10–22
Inverting action of the basic inverting switching regulator.

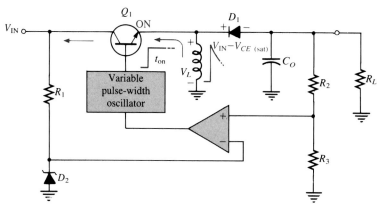

(a) When Q_1 is on, D_1 is reverse-biased.

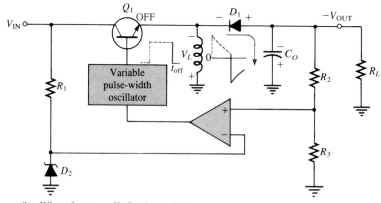

(b) When Q_1 turns off, D_1 forward biases.

magnetic field collapses and the inductor's polarity reverses, as shown in Figure 10–22(b). This forward-biases the diode, charges C_O, and produces a negative output voltage, as indicated. The repetitive on-off action of Q_1 produces a repetitive charging and discharging that is smoothed by the LC filter action.

Power Supply Circuits

As with the step-up regulator, the less time Q_1 is on, the greater the output voltage is, and vice versa. This regulating action is illustrated in Figure 10–23. Switching regulator efficiencies can be greater than 90 percent.

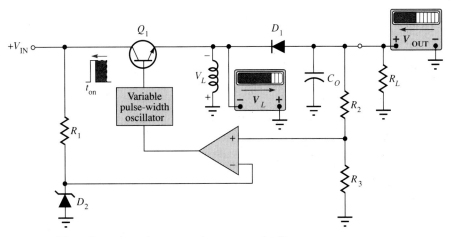

(a) When $-V_{OUT}$ tries to decrease, t_{on} decreases, causing V_L to increase. This compensates for the attempted decrease in $-V_{OUT}$.

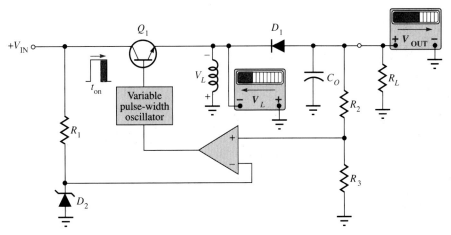

(b) When $-V_{OUT}$ tries to increase, t_{on} increases, causing V_L to decrease. This compensates for the attempted increase in $-V_{OUT}$.

FIGURE 10–23
Regulating action of the basic inverting switching regulator.

10–4 Review Questions

1. What are three types of switching regulators?
2. What is the primary advantage of switching regulators over linear regulators?
3. How are changes in output voltage compensated for in the switching regulator?

10–5 INTEGRATED CIRCUIT VOLTAGE REGULATORS

In the previous sections, we presented the basic voltage regulator configurations. Several types of both linear and switching regulators are available in integrated circuit (IC) form. Generally, the linear regulators are three-terminal devices that provide either positive or negative output voltages that can be either fixed or adjustable. In this section, typical linear and switching IC regulators are introduced.

FIXED POSITIVE LINEAR VOLTAGE REGULATORS

Although many types of IC regulators are available, the 7800 series of IC regulators is representative of three-terminal devices that provide a fixed positive output voltage. The three terminals are input, output, and ground as indicated in the standard fixed voltage configuration in Figure 10–24(a). The last two digits in the part number designate the output voltage. For example, the 7805 is a +5.0 V regulator. Other available output voltages are given in Figure 10–24(b) and common packages are shown in part (c) of Figure 10–24.

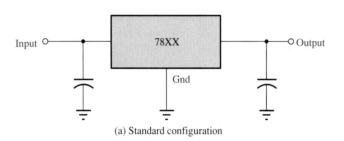

(a) Standard configuration

Type number	Output voltage
7805	+5.0 V
7806	+6.0 V
7808	+8.0 V
7809	+9.0 V
7812	+12.0 V
7815	+15.0 V
7818	+18.0 V
7824	+24.0 V

(b) The 7800 series

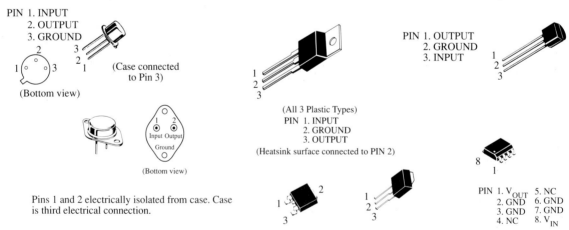

(c) Typical metal and plastic packages

FIGURE 10–24
The 7800 series three-terminal fixed positive voltage regulators. Part (c) copyright of Motorola, Inc. Used by permission.

Capacitors, although not always necessary, are sometimes used on the input and output as indicated. The output capacitor acts basically as a line filter to improve transient response. The input capacitor is used to prevent unwanted oscillations when the regulator is some distance from the power supply filter such that the line has a significant inductance.

The 7800 series can produce output current in excess of 1 A when used with an adequate heat sink. The 78L00 series can provide up to 100 mA, the 78M00 series can provide up to 500 mA, and the 78T00 series can provide in excess of 3 A. These devices are available with either a 2% or 4% output voltage tolerance.

The input voltage must be at least 2 V above the output voltage in order to maintain regulation. The circuits have internal thermal overload protection and short-circuit current-limiting features. **Thermal overload** occurs when the internal power dissipation becomes excessive and the temperature of the device exceeds a certain value.

FIXED NEGATIVE LINEAR VOLTAGE REGULATORS

The 7900 series is typical of three-terminal IC regulators that provide a fixed negative output voltage. This series is the negative-voltage counterpart of the 7800 series and shares most of the same features and characteristics. Figure 10–25 indicates the standard configuration and part numbers with corresponding output voltages that are available.

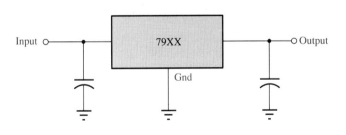

Type number	Output voltage
7905	−5.0 V
7905.2	−5.2 V
7906	−6.0 V
7908	−8.0 V
7912	−12.0 V
7915	−15.0 V
7918	−18.0 V
7924	−24.0 V

(a) Standard configuration (b) The 7900 series

FIGURE 10–25
The 7900 series three-terminal fixed negative voltage regulators.

ADJUSTABLE POSITIVE LINEAR VOLTAGE REGULATORS

The LM317 is an excellent example of a three-terminal positive regulator with an adjustable output voltage. A data sheet for this device is given in Appendix A. The standard configuration is shown in Figure 10–26. Input and output capacitors, although not shown, are often used for the reasons discussed previously. Notice that there is an input, an output, and an adjustment terminal. The external fixed resistor R_1 and the external variable resistor R_2 provide the output voltage adjustment. V_{OUT} can be varied from 1.2 V to 37 V depending on the resistor values. The LM317 can provide over 1.5 A of output current to a load.

FIGURE 10–26
The LM317 three-terminal adjustable positive voltage regulator.

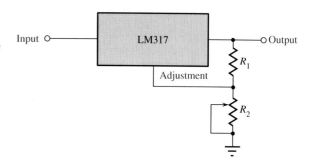

The LM317 is operated as a "floating" regulator because the adjustment terminal is not connected to ground, but floats to whatever voltage is across R_2. This allows the output voltage to be much higher than that of a fixed-voltage regulator.

BASIC OPERATION As indicated in Figure 10–27, a constant 1.25-V reference voltage (V_{REF}) is maintained by the regulator between the output terminal and the adjustment terminal. This constant reference voltage produces a constant current (I_{REF}) through R_1, regardless of the value of R_2. I_{REF} also flows through R_2.

$$I_{REF} = \frac{V_{REF}}{R_1} = \frac{1.25 \text{ V}}{R_1}$$

Also, there is a very small constant current into the adjustment terminal of approximately 50 μA called I_{ADJ}, which flows through R_2. An expression for the output voltage is developed as follows.

$$\begin{aligned}
V_{OUT} &= V_{R1} + V_{R2} \\
&= I_{REF}R_1 + I_{REF}R_2 + I_{ADJ}R_2 \\
&= I_{REF}(R_1 + R_2) + I_{ADJ}R_2 \\
&= \frac{V_{REF}}{R_1}(R_1 + R_2) + I_{ADJ}R_2
\end{aligned}$$

$$V_{OUT} = V_{REF}\left(1 + \frac{R_2}{R_1}\right) + I_{ADJ}R_2 \qquad (10\text{–}12)$$

As you can see, the output voltage is a function of both R_1 and R_2. Once the value of R_1 is set, the output voltage is adjusted by varying R_2.

FIGURE 10–27
Operation of the LM317 adjustable voltage regulator.

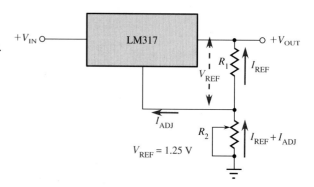

EXAMPLE 10–6

Determine the minimum and maximum output voltages for the voltage regulator in Figure 10–28. Assume $I_{ADJ} = 50 \ \mu A$.

FIGURE 10–28

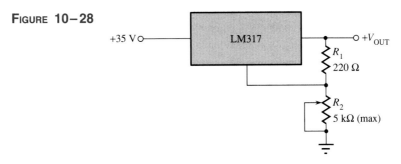

SOLUTION

$$V_{R1} = V_{REF} = 1.25 \text{ V}$$

When R_2 is set at its maximum of 5 kΩ,

$$V_{OUT} = V_{REF}\left(1 + \frac{R_2}{R_1}\right) + I_{ADJ}R_2 = 1.25 \text{ V}\left(1 + \frac{5 \text{ k}\Omega}{220 \ \Omega}\right) + (50 \ \mu A)5 \text{ k}\Omega$$
$$= 29.66 \text{ V} + 0.25 \text{ V} = 29.91 \text{ V}$$

When R_2 is set at its minimum of 0 Ω,

$$V_{OUT} = V_{REF}\left(1 + \frac{R_2}{R_1}\right) + I_{ADJ}R_2 = 1.25 \text{ V}(1) = 1.25 \text{ V}$$

PRACTICE EXERCISE 10–6

What is the output voltage of the regulator if R_2 is set at 2 kΩ?

ADJUSTABLE NEGATIVE LINEAR VOLTAGE REGULATORS

The LM337 is the negative output counterpart of the LM317 and is a good example of this type of IC regulator. Like the LM317, the LM337 requires two external resistors for output voltage adjustment as shown in Figure 10–29. The output voltage can be adjusted from -1.2 V to -37 V, depending on the external resistor values.

FIGURE 10–29
The LM337 three-terminal adjustable negative voltage regulator.

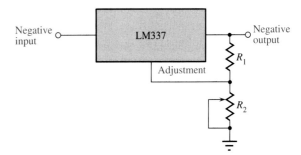

Switching Voltage Regulators

As an example of an IC switching voltage regulator, we will look at the 78S40. This is a universal device that can be used with external components to provide step-up, step-down, and inverting operation.

The internal circuitry of the 78S40 is shown in Figure 10–30. This circuit can be compared to the basic switching regulators that were covered in Section 10–4. For example, look back at Figure 10–15(a). The oscillator and comparator functions are directly comparable. The gate and flip-flop in the 78S40 were not included in the basic circuit of Figure 10–15(a), but they provide additional regulating action. Transistors Q_1 and Q_2 effectively perform the same function as Q_1 in the basic circuit. The 1.25-V reference block in the 78S40 has the same purpose as the zener diode in the basic circuit, and diode D_1 in the 78S40 corresponds to D_1 in the basic circuit.

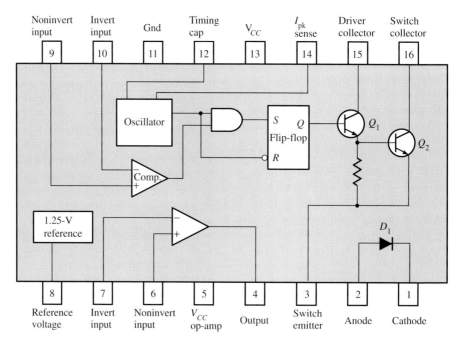

Figure 10–30
The 78S40 switching regulator.

The 78S40 also has an "uncommitted" op-amp thrown in for good measure. It is not used in any of the regulator configurations. External circuitry is required to make this device operate as a regulator, as you will see in Section 10–6.

10–5 REVIEW QUESTIONS

1. What are the three terminals of a fixed-voltage regulator?
2. What is the output voltage of a 7809? Of a 7915?
3. What are the three terminals of an adjustable-voltage regulator?
4. What external components are required for a basic LM317 configuration?

10–6 APPLICATIONS OF IC VOLTAGE REGULATORS

In the last section, you saw several devices that are representative of the general types of IC voltage regulators. Now, we will examine several different ways these devices can be modified with external circuitry to improve or alter their performance.

USING AN EXTERNAL PASS TRANSISTOR

As you know, an IC voltage regulator is capable of delivering only a certain amount of output current to a load. For example, the 7800 series regulators can handle a maximum output current of at least 1.3 A and typically 2.5 A. If the load current exceeds the maximum allowable value, there will be thermal overload and the regulator will shut down. A thermal overload condition means that there is excessive power dissipation inside the device.

If an application requires more than the maximum current that the regulator can deliver, an external pass transistor can be used. Figure 10–31 illustrates a three-terminal regulator with an external pass transistor for handling currents in excess of the output current capability of the basic regulator.

FIGURE 10–31

A 7800-series three-terminal regulator with an external pass transistor.

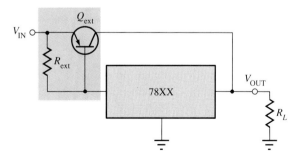

FIGURE 10–32
Operation of the regulator with an external pass transistor.

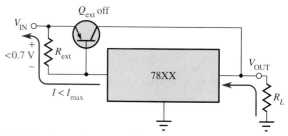

(a) When the regulator current is less than I_{max}, the external pass transistor is off and the regulator is handling all of the current.

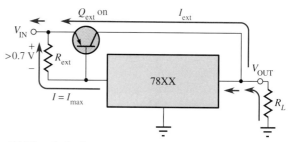

(b) When the load current exceeds I_{max}, the drop across R_{ext} turns Q_{ext} on and it conducts the excess current.

The value of the external current-sensing resistor R_{ext} determines the value of current at which Q_{ext} begins to conduct because it sets the V_{BE} voltage of the transistor. As long as the current is less than the value set by R_{ext}, the transistor Q_{ext} is off, and the regulator operates normally as shown in Figure 10–32(a). This is because the voltage drop across R_{ext} is less than the 0.7-V base-to-emitter voltage required to turn Q_{ext} on. R_{ext} is determined by the following formula, where I_{max} is the highest current that the voltage regulator is to handle internally.

$$R_{ext} = \frac{0.7 \text{ V}}{I_{max}} \tag{10-13}$$

When the current is sufficient to produce at least a 0.7 V drop across R_{ext}, the external pass transistor Q_{ext} turns on and conducts any current in excess of I_{max}, as indicated in Figure 10–32(b). Q_{ext} will conduct more or less, depending on the load requirements. For example, if the total load current is 3 A and I_{max} was selected to be 1 A, the external pass transistor will conduct 2 A, which is the excess over the current I_{max}, flowing internally through the voltage regulator.

EXAMPLE 10–7

What value is R_{ext} if the maximum current to be handled internally by the voltage regulator in Figure 10–31 is set at 700 mA?

SOLUTION

$$R_{ext} = \frac{0.7 \text{ V}}{I_{max}} = \frac{0.7 \text{ V}}{0.7 \text{ A}} = 1 \, \Omega$$

PRACTICE EXERCISE 10–7
If R_{ext} is changed to 1.5 Ω, at what current value will Q_{ext} turn on?

The external pass transistor is typically a power transistor with heat sink which must be capable of handling a maximum power of

$$P_{ext} = I_{ext}(V_{IN} - V_{OUT}) \qquad (10\text{–}14)$$

EXAMPLE 10–8

What must be the minimum power rating for the external pass transistor used with a 7824 regulator in a circuit such as that shown in Figure 10–31? The input voltage is 30 V and the load resistance is 10 Ω. The maximum internal current is to be 700 mA. Assume that there is no heat sink for this calculation. Keep in mind that the use of a heat sink increases the effective power rating of the transistor and you can use a lower rated transistor.

SOLUTION
The load current is

$$I_L = \frac{V_{OUT}}{R_L} = \frac{24 \text{ V}}{10 \, \Omega} = 2.4 \text{ A}$$

The current through Q_{ext} is

$$I_{ext} = I_L - I_{max} = 2.4 \text{ A} - 0.7 \text{ A} = 1.7 \text{ A}$$

The power dissipated by Q_{ext} is

$$\begin{aligned} P_{ext(min)} &= I_{ext}(V_{IN} - V_{OUT}) \\ &= 1.7 \text{ A}(30 \text{ V} - 24 \text{ V}) \\ &= 1.7 \text{ A}(6 \text{ V}) = 10.2 \text{ W} \end{aligned}$$

For a safety margin, choose a power transistor with a rating greater than 10.2 W, say at least 15 W.

PRACTICE EXERCISE 10–8
Rework this example using a 7815 regulator.

CURRENT LIMITING

A drawback of the circuit in Figure 10–31 is that the external transistor is not protected from excessive current, such as would result from a shorted output. An additional current-limiting circuit ($Q_{\lim}$ and $R_{\lim}$) can be added as shown in Figure 10–33 to protect Q_{ext} from excessive current and possible burn out.

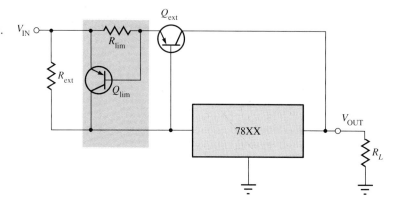

FIGURE 10–33
Regulator with current limiting.

The following describes the way the current-limiting circuit works. The current-sensing resistor $R_{\lim}$ sets the V_{BE} of transistor $Q_{\lim}$. The base-to-emitter voltage of Q_{ext} is now determined by $V_{R_{ext}} - V_{R_{\lim}}$ because they have opposite polarities. So, for normal operation, the drop across R_{ext} must be sufficient to overcome the opposing drop across $R_{\lim}$. If the current through Q_{ext} exceeds a certain maximum ($I_{ext(max)}$) because of a shorted output or a faulty load, the voltage across $R_{\lim}$ reaches 0.7 V and turns $Q_{\lim}$ on. $Q_{\lim}$ now conducts current away from Q_{ext} and through the regulator, forcing a thermal overload to occur and shut down the regulator. Remember, the IC regulator is internally protected from thermal overload as part of its design.

This action is illustrated in Figure 10–34. In part (a), the circuit is operating normally with Q_{ext} conducting less than the maximum current that it can handle with $Q_{\lim}$ off. Part (b) shows what happens when there is a short across the load. The current through Q_{ext} suddenly increases and causes the voltage drop across $R_{\lim}$ to increase, which turns $Q_{\lim}$ on. The current is now diverted through the regulator, which causes it to shut down due to thermal overload.

A CURRENT REGULATOR

The three-terminal regulator can be used as a current source when an application requires that a constant current be supplied to a variable load. The basic circuit is shown in Figure 10–35 where R is the current-setting resistor. The regulator provides a fixed constant voltage, V_{OUT}, between the ground terminal (not connected to ground in this case) and the output terminal. This determines the constant current supplied to the load.

$$I_L = \frac{V_{OUT}}{R} + I_G \qquad (10\text{–}15)$$

The current from the ground terminal I_G is very small compared to the output current and can often be neglected.

FIGURE 10–34
The current-limiting action of the regulator circuit.

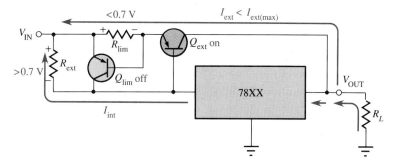

(a) During normal operation, when the load current is not excessive, Q_{lim} is off.

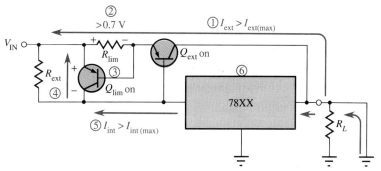

(b) When short occurs ①, the external current becomes excessive and the voltage across R_{lim} increases ② and turns on Q_{lim} ③, which then conducts current away from Q_{ext} and routes it through the regulator ④, causing the internal regulator current to become excessive ⑤ and to force the regulator into thermal shut down ⑥.

FIGURE 10–35
The three-terminal regulator as a current source.

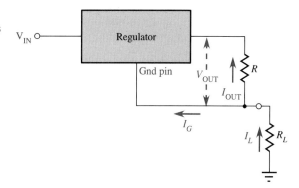

EXAMPLE 10–9

Use a 7805 regulator to provide a constant current of 1 A to a variable load. The input must be at least 2 V greater than the output and $I_G = 1.5$ mA.

SOLUTION
First, 1 A is within the limits of the 7805's capability (remember, it can handle at least 1.3 A without an external pass transistor).

The 7805 produces 5 V between its ground terminal and its output terminal. Therefore, if we want 1 A of current, the current-setting resistor must be (neglecting I_G)

$$R = \frac{V_{OUT}}{I_L} = \frac{5\text{ V}}{1\text{ A}} = 5\text{ }\Omega$$

The circuit is shown in Figure 10–36.

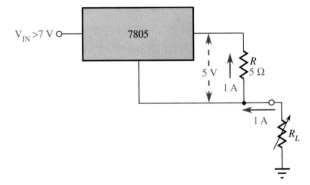

FIGURE 10–36
A 1-A constant-current source.

PRACTICE EXERCISE 10–9
If a 7812 regulator is used instead of the 7805, to what value would you change R to maintain a constant current of 1 A?

SWITCHING REGULATOR CONFIGURATIONS

In Section 10–5, the 78S40 was introduced as an example of an IC switching voltage regulator. Figure 10–37 shows the external connections for a step-down configuration where the output voltage is less than the input voltage and Figure 10–38 shows a step-up configuration in which the output voltage is greater than the input voltage. An inverting configuration is also possible, but it is not shown here.

The capacitor C_T is the timing capacitor that controls the pulse width and frequency of the oscillator and thus establishes the on-time of transistor Q_2. The voltage across the current-sensing resistor R_{cs} is used internally by the oscillator to vary the duty cycle based on the desired peak load current. The voltage divider, made up of R_1 and R_2, reduces the output voltage to a nominal value equal to the reference voltage. If V_{OUT} exceeds its set value, the output of the comparator switches to its low state, disabling the gate to turn Q_2 off until the output decreases. This regulating action is in addition to that produced by the duty cycle variation of the oscillator as described in Section 10–4 in relation to the basic switching regulator.

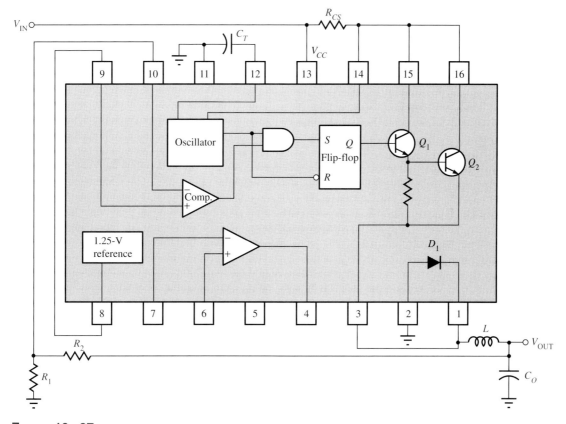

FIGURE 10–37
The step-down configuration of the 78S40 switching regulator.

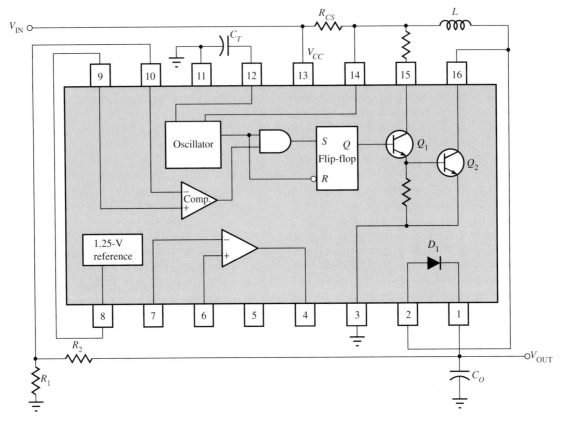

FIGURE 10–38
The step-up configuration of the 78S40 switching regulator.

10–6 REVIEW QUESTIONS

1. What is the purpose of using an external pass transistor with an IC voltage regulator?
2. What is the advantage of current limiting in a voltage regulator?
3. What does *thermal overload* mean?

10–7   A SYSTEM APPLICATION

In this system application, the focus is on the regulated power supply which provides the FM stereo receiver with dual polarity dc voltages. Recall from previous system applications that the op-amps in the channel separation circuits and the audio amplifiers operate from ± 12 V. Both positive and negative voltage regulators are used to regulate the rectified and filtered voltages from a bridge rectifier. In this section, you will

□ *See how dual supply voltages are produced by a rectifier.*
□ *See how positive and negative three-terminal IC regulators are used in a power supply.*

Power Supply Circuits

- Relate a schematic to a pc board.
- Analyze the operation of the power supply circuit.
- Troubleshoot some common power supply failures.

About the Power Supply

This power supply utilizes a full-wave bridge **rectifier** with both the positive and negative rectified voltages taken off the bridge at the appropriate points and filtered by electrolytic capacitors. A 7812 and a 7912 provide regulation.

Now, so that you can take a closer look at the dual power supply, let's take it out of the system and put it on the test bench.

On the Test Bench

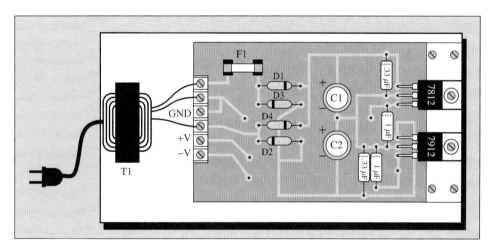

Figure 10–39

■ **Activity 1**   Relate the PC Board to the Schematic

Develop a schematic for the power supply in Figure 10–39. Add any missing labels and include the IC pin numbers by referring to the voltage regulator data sheets in Appendix A. The rectifier diodes are 1N4001s, the filter capacitors C1 and C2 are 100 μF, and the transformer has a turns ratio of 5:1. Determine the backside pc board connections as you develop the schematic.

■ **Activity 2**   Analyze the Power Supply Circuits

Step 1   Determine the approximate voltage at each of the four "corners" of the bridge with respect to ground.

Step 2   Calculate the peak inverse voltage of the rectifier diodes.

Step 3   Determine the voltage at the inputs of the voltage regulators.

Step 4   In this stereo system, assume that op-amps are used only in the channel separation circuits and the channel audio amplifiers. If all of the other circuits in the receiver use +12 V and draw an average dc current of 500 mA, determine how much total current each regulator must supply. Refer to the system applications in Chapters 6 and 8. Use the appropriate data sheets.

STEP 5 Based on the results in Step 4, do the IC regulators have to be attached to the heat sink or is this just for a safety margin?

■ **ACTIVITY 3**   **WRITE A TECHNICAL REPORT**

Describe the operation of the power supply with an emphasis on how both positive and negative voltages are obtained. State the purpose of each component. Use the results of Activity 2 where appropriate.

■ **ACTIVITY 4**   **TROUBLESHOOT THE POWER SUPPLY BY STATING THE PROBABLE CAUSE OR CAUSES IN EACH CASE**

1. Both positive and negative output voltages are zero.
2. Positive output voltage is zero and the negative output voltage is −12 V.
3. Negative output voltage is zero and the positive output voltage is +12 V.
4. Radical voltage fluctuations on output of positive regulator.

■ **ACTIVITY 5**   **TEST BENCH SPECIAL ASSIGNMENT**

Go to Test Bench 3 in the color insert section (which follows page 452) and carry out the assignment that is stated there.

10–7 REVIEW QUESTIONS

1. What should be the rating of the power supply fuse?
2. What purpose do the 0.33 μF capacitors serve?
3. Which regulator provides the negative voltage?
4. Would you recommend that an external pass transistor be used with the regulators in this power supply? Why?

SUMMARY

- Voltage regulators keep a constant dc output voltage when the input or load varies within limits.
- A basic voltage regulator consists of a reference voltage source, an error detector, a sampling element, and a control device. Protection circuitry is also found in most regulators.
- Two basic categories of voltage regulators are linear and switching.
- Two basic types of linear regulators are series and shunt.
- In a series linear regulator, the control element is a transistor in series with the load.
- In a shunt linear regulator, the control element is a transistor in parallel with the load.
- Three configurations for switching regulators are step-down, step-up, and inverting.

POWER SUPPLY CIRCUITS

- Switching regulators are more efficient than linear regulators and are particularly useful in low-voltage, high-current applications.
- Three-terminal linear IC regulators are available for either fixed output or variable output voltages of positive or negative polarities.
- An external pass transistor increases the current capability of a regulator.
- The 7800 series are three-terminal IC regulators with fixed positive output voltage.
- The 7900 series are three-terminal IC regulators with fixed negative output voltage.
- The LM317 is a three-terminal IC regulator with a positive variable output voltage.
- The LM337 is a three-terminal IC regulator with a negative variable output voltage.
- The 78S40 is a switching voltage regulator.

GLOSSARY

Fold-back current limiting A method of current limiting in voltage regulators.

Line regulation The percentage change in output voltage for a given change in line (input) voltage.

Load regulation The percentage change in output voltage for a given change in load current.

Rectifier An electronic circuit that converts ac into pulsating dc.

Regulator An electronic circuit that maintains an essentially constant output voltage with a changing input voltage or load current.

Thermal overload A condition in a rectifier where the internal power dissipation of the circuit exceeds a certain maximum due to excessive current.

FORMULAS

VOLTAGE REGULATION

(10–1) $\quad \text{Line regulation} = \dfrac{(\Delta V_{\text{OUT}}/V_{\text{OUT}})100\%}{\Delta V_{\text{IN}}} \quad$ Percent line regulation

(10–2) $\quad \text{Load regulation} = \dfrac{(V_{\text{NL}} - V_{\text{FL}})100\%}{V_{\text{FL}}} \quad$ Percent load regulation

BASIC SERIES REGULATOR

(10–3) $\quad A_{cl} = 1 + \dfrac{R_2}{R_3} \quad$ Closed-loop voltage gain

(10–4) $\quad V_{\text{OUT}} \cong \left(1 + \dfrac{R_2}{R_3}\right) V_{\text{REF}} \quad$ Regulator output

(10–5) $\quad I_{L(\max)} = \dfrac{0.7 \text{ V}}{R_4} \quad$ For constant current limiting

Basic Shunt Regulator

(10–6) $\quad \Delta I_S = \dfrac{\Delta V_{IN}}{R_1}$ $\qquad$ Change in shunt current

(10–7) $\quad \Delta I_S = -\Delta I_L$ $\qquad$ Change in shunt current

(10–8) $\quad I_{L(max)} = \dfrac{V_{IN}}{R_1}$ $\qquad$ Maximum load current

Basic Switching Regulators

(10–9) $\quad V_{OUT} = \left(\dfrac{t_{on}}{T}\right) V_{IN}$ $\qquad$ For step-down switching regulator

(10–10) $\quad T = t_{on} + t_{off}$ $\qquad$ Switching period

(10–11) $\quad V_{OUT} = \left(\dfrac{T}{t_{on}}\right) V_{IN}$ $\qquad$ For step-up switching regulator

IC Voltage Regulators

(10–12) $\quad V_{OUT} = V_{REF}\left(1 + \dfrac{R_2}{R_1}\right) + I_{ADJ}R_2$ $\qquad$ IC regulator

(10–13) $\quad R_{ext} = \dfrac{0.7\ V}{I_{max}}$ $\qquad$ For external pass circuit

(10–14) $\quad P_{ext} = I_{ext}(V_{IN} - V_{OUT})$ $\qquad$ For external pass transistor

(10–15) $\quad I_L = \dfrac{V_{OUT}}{R} + I_G$ $\qquad$ Regulator as a current source

Self-Test

1. In the case of line regulation,
   - (a) when the temperature varies, the output voltage stays constant
   - (b) when the output voltage changes, the load current stays constant
   - (c) when the input voltage changes, the output voltage stays constant
   - (d) when the load changes, the output voltage stays constant
2. In the case of load regulation,
   - (a) when the temperature varies, the output voltage stays constant
   - (b) when the input voltage changes, the load current stays constant
   - (c) when the load changes, the load current stays constant
   - (d) when the load changes, the output voltage stays constant
3. All of the following are parts of a basic voltage regulator *except*
   - (a) control element
   - (b) sampling circuit
   - (c) voltage follower
   - (d) error detector
   - (e) reference voltage

Power Supply Circuits

4. The basic difference between a series regulator and a shunt regulator is
   - (a) the amount of current that can be handled
   - (b) the position of the control element
   - (c) the type of sample circuit
   - (d) the type of error detector

5. In a basic series regulator, V_{OUT} is determined by
   - (a) the control element
   - (b) the sample circuit
   - (c) the reference voltage
   - (d) b and c

6. The main purpose of current limiting in a regulator is
   - (a) protection of the regulator from excessive current
   - (b) protection of the load from excessive current
   - (c) to keep the power supply transformer from burning up
   - (d) to maintain a constant output voltage

7. In a linear regulator, the control transistor is conducting
   - (a) a small part of the time
   - (b) half the time
   - (c) all of the time
   - (d) only when the load current is excessive

8. In a switching regulator, the control transistor is conducting
   - (a) part of the time
   - (b) all of the time
   - (c) only when the input voltage exceeds a set limit
   - (d) only when there is an overload

9. The LM317 is an example of an IC
   - (a) three-terminal negative voltage regulator
   - (b) fixed positive voltage regulator
   - (c) switching regulator
   - (d) linear regulator
   - (e) variable positive voltage regulator
   - (f) b and d only
   - (g) d and e only

10. An external pass transistor is used for
    - (a) increasing the output voltage
    - (b) improving the regulation
    - (c) increasing the current that the regulator can handle
    - (d) short-circuit protection

PROBLEMS

SECTION 10–1 VOLTAGE REGULATION

1. The nominal output voltage of a certain regulator is 8 V. The output changes 2 mV when the input voltage goes from 12 V to 18 V. Determine the line regulation and express it as a percentage change over the entire range of V_{IN}.

2. Express the line regulation found in Problem 1 in units of %/V.

3. A certain regulator has a no-load output voltage of 10 V and a full-load output voltage of 9.90 V. What is the percent load regulation?

4. In Problem 3, if the full-load current is 250 mA, express the load regulation in %/mA.

SECTION 10–2 BASIC SERIES REGULATORS

5. Label the functional blocks for the voltage regulator in Figure 10–40.

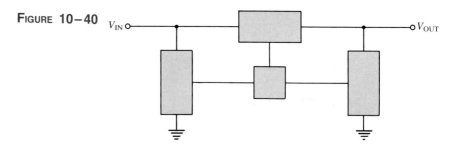

FIGURE 10–40

6. Determine the output voltage for the regulator in Figure 10–41.

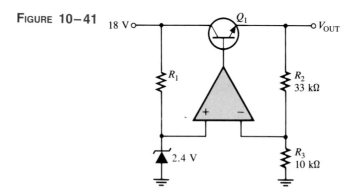

FIGURE 10–41

7. Determine the output voltage for the series regulator in Figure 10–42.

8. If R_3 in Figure 10–42 is increased to 4.7 kΩ, what happens to the output voltage?

9. If the zener voltage is 2.7 V instead of 2.4 V in Figure 10–42, what is the output voltage?

Power Supply Circuits

Figure 10–42

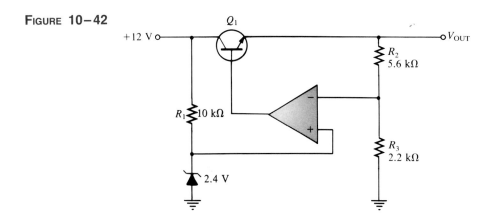

10. A series voltage regulator with constant current limiting is shown in Figure 10–43. Determine the value of R_4 if the load current is to be limited to a maximum value of 250 mA. What power rating must R_4 have?

Figure 10–43

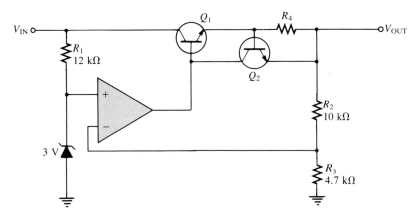

11. If the R_4 determined in Problem 10 is halved, what is the maximum load current?

Section 10–3 Basic Shunt Regulators

12. In the shunt regulator of Figure 10–44, when the load current increases, does Q_1 conduct more or less? Why?

13. Assume I_L remains constant and V_{IN} changes by 1 V in Figure 10–44. What is the change in the collector current of Q_1?

14. With a constant input voltage of 17 V, the load resistance in Figure 10–44 is varied from 1 kΩ to 1.2 kΩ. Neglecting any change in output voltage, how much does the shunt current through Q_1 change?

15. If the maximum allowable input voltage in Figure 10–44 is 25 V, what is the maximum possible output current when the output is short-circuited? What power rating should R_1 have?

FIGURE 10-44

SECTION 10-4 BASIC SWITCHING REGULATORS

16. A basic switching regulator is shown in Figure 10–45. If the switching frequency of the transistor is 100 Hz with an off-time of 6 ms, what is the output voltage?

FIGURE 10-45

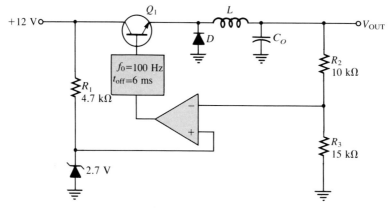

17. What is the duty cycle of the transistor in Problem 16?

18. Determine the output voltage for the switching regulator in Figure 10–46 when the duty cycle is 40 percent.

19. If the on-time of Q_1 in Figure 10–46 is decreased, does the output voltage increase or decrease?

FIGURE 10–46

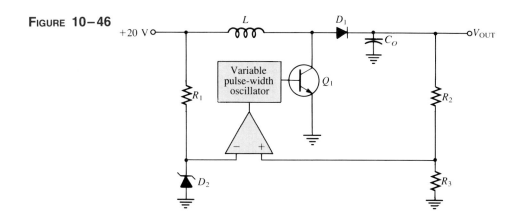

SECTION 10–5 INTEGRATED CIRCUIT VOLTAGE REGULATORS

20. What is the output voltage of each of the following IC regulators?

   (a) 7806   (b) 7905.2
   (c) 7818   (d) 7924

21. Determine the output voltage of the regulator in Figure 10–47.

FIGURE 10–47

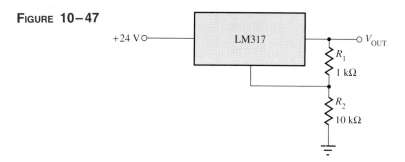

22. Determine the minimum and maximum output voltages for the circuit in Figure 10–48.

FIGURE 10–48

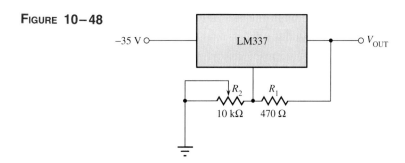

23. With no load connected, how much current is there through the regulator in Figure 10–47? Neglect the adjustment terminal current.

24. Select the values for the external resistors to be used in an LM317 circuit that is required to produce an output voltage of 12 V with an input of 18 V. The maximum regulator current with no load is to be 2 mA. There is no external pass transistor.

Section 10–6 Applications of IC Voltage Regulators

25. In the regulator circuit of Figure 10–49, determine R_{ext} if the maximum internal regulator current is to be 250 mA.

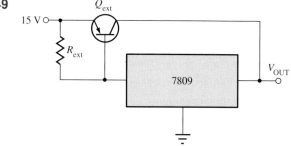

FIGURE 10–49

26. Using a 7812 voltage regulator and a 10 Ω load in Figure 10–49, how much power will the external pass transistor have to dissipate? The maximum internal regulator current is set at 500 mA by R_{ext}.

27. Show how to include current limiting in the circuit of Figure 10–49. What should the value of the limiting resistor be if the external current is to be limited to 2 A?

28. Using an LM317, design a circuit that will provide a constant current of 500 mA to a load.

29. Repeat Problem 28 using a 7909.

30. If a 78S40 switching regulator is to be used to regulate a 12-V input down to a 6-V output, calculate the values of the external voltage-divider resistors.

Answers to Review Questions

Section 10–1
1. The percentage change in the output voltage for a given change in input voltage.
2. The percentage change in output voltage for a given change in load current.

Section 10–2
1. Control element, error detector, sampling element, reference source
2. 2 V

POWER SUPPLY CIRCUITS

SECTION 10–3
1. In a shunt regulator, the control element is in parallel with the load rather than in series.
2. A shunt regulator has inherent current limiting. A disadvantage is that a shunt regulator is less efficient than a series regulator.

SECTION 10–4
1. Step-down, step-up, inverting
2. Switching regulators operate at a higher efficiency.
3. The duty cycle varies to regulate the output.

SECTION 10–5
1. Input, output, and ground
2. A 7809 has a +9-V output; A 7915 has a −15-V output.
3. Input, output, adjustment
4. A two-resistor voltage divider

SECTION 10–6
1. A pass transistor increases the current that can be handled.
2. Current limiting prevents excessive current and prevents damage to the regulator.
3. Thermal overload occurs when the internal power dissipation becomes excessive.

SECTION 10–7
1. 1 A
2. Those optional capacitors on the regulator inputs prevent oscillations.
3. The 7909 is a negative-voltage regulator.
4. No. The current that either regulator must supply is less than 1 A.

ANSWERS TO PRACTICE EXERCISES

10–1   0.6% V
10–2   0.84%, 0.0168%/mA
10–3   7.33 V
10–4   0.7 A
10–5   17.45 W
10–6   12.6 V
10–7   467 mA
10–8   12 W
10–9   12 Ω

11
SPECIAL AMPLIFIERS

11–1 INSTRUMENTATION AMPLIFIERS
11–2 ISOLATION AMPLIFIERS
11–3 OPERATIONAL TRANSCONDUCTANCE AMPLIFIERS (OTAs)
11–4 LOG AND ANTILOG AMPLIFIERS
11–5 ANALOG MULTIPLIERS AND DIVIDERS
11–6 A SYSTEM APPLICATION

After completing this chapter, you should be able to

☐ Describe the basic instrumentation amplifier.
☐ Set the voltage gain of an instrumentation amplifier.
☐ Discuss general applications of instrumentation amplifiers.
☐ Set up an AD521 instrumentation amplifier.
☐ Describe the basic isolation amplifier and discuss its operation.
☐ Explain how the sections of an isolation amplifier are electrically isolated.
☐ Discuss general applications of the isolation amplifier.
☐ Set up the AD295 isolation amplifier as a unity gain, noninverting device.
☐ Set up the AD295 for nonunity gain.
☐ Describe the basic operational transconductance amplifier (OTA).
☐ Explain the significance of the transconductance characteristic in the operation of an OTA.
☐ Set up a CA3080 OTA as a fixed-gain or variable-gain inverting amplifier.
☐ Use the OTA as an amplitude-modulating circuit.
☐ Use the OTA as a Schmitt-trigger circuit.
☐ Discuss the basic operation of a log amplifier.
☐ Discuss the basic operation of an antilog amplifier.
☐ Explain how logarithmic amplifiers are used to implement analog multipliers and dividers.

A general-purpose op-amp, such as the 741, is an extremely versatile and widely used device. However, some specialized IC amplifiers have been designed with certain types of applications in mind or with certain special features or characteristics. Most of these devices are actually derived from the basic op-amp. These special amplifiers include the instrumentation amplifier that is used in high-noise environments, the isolation amplifier that is used in high-voltage and medical applications, the operational transconductance amplifier (OTA) that is used as a

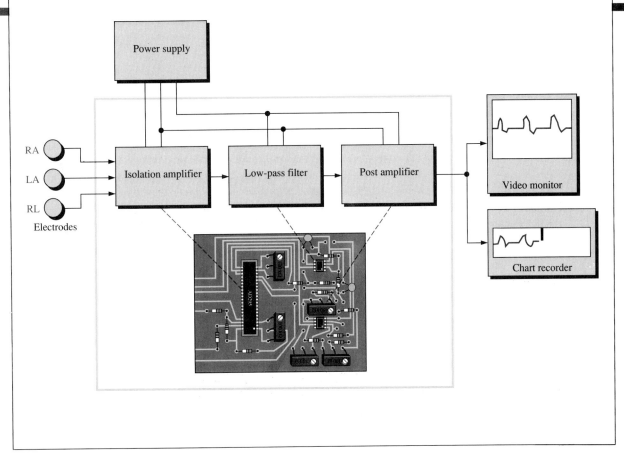

voltage-to-current amplifier, and the logarithmic amplifiers that are used for linearizing certain types of inputs and for mathematical operations. In this chapter, you will learn about each of these devices and some of their basic applications.

A SYSTEM APPLICATION

Medical electronics is a very important application area for electronic devices and, without doubt, one of the most beneficial. The electrocardiograph (ECG), one of the most common and important instruments in use for medical purposes, is used to monitor the heart function of patients in order to detect any irregularities or abnormalities in the heartbeat. Sensors called electrodes are placed at points on the body to pick up the small electrical signal produced by the heart. This signal goes through an amplifica-

tion process and is fed to a video monitor or chart recorder for viewing. Because of the safety hazards related to electrical equipment, it is very important that the patient be protected from the possibility of damaging or fatal electrical shock. For this reason, the isolation amplifier is used in medical equipment that comes in contact with the human body. The diagram above shows a basic block diagram for a simplified ECG system. Our focus in this system application is on the amplifier section.

For the system application in Section 11–6, in addition to the other topics, be sure you understand

☐ Basic op-amp operation.
☐ Isolation amplifiers.

11-1 INSTRUMENTATION AMPLIFIERS

An instrumentation amplifier is a differential voltage-gain device that amplifies the difference between the voltages existing at its two input terminals. The main purpose of an instrumentation amplifier is to amplify small signals riding on large common-mode voltages. The key characteristics are high input impedance, high common-mode rejection, low output offset, and low output impedance. A basic instrumentation amplifier is made up of three operational amplifiers and several resistors. The voltage gain is set with an external resistor. Instrumentation amplifiers are commonly used in environments with high common-mode noise such as in data acquisition systems where remote sensing of input variables is required.

THE BASIC INSTRUMENTATION AMPLIFIER

A basic **instrumentation** amplifier is shown in Figure 11–1. Op-amps 1 and 2 are noninverting configurations that provide high input impedance and voltage gain. Op-amp 3 is used as a unity-gain differential amplifier.

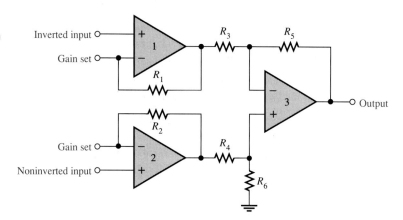

FIGURE 11–1
The basic instrumentation amplifier.

The gain-setting resistor, R_G, is connected externally as shown in Figure 11–2. Op-amp 1 receives the differential input signal V_{in1} on its noninverting input and amplifies this signal with a gain of

$$A_v = 1 + \frac{R_1}{R_G}$$

Op-amp 1 also receives the input signal V_{in2} through op-amp 2 and the path formed by R_2, R_G, and R_1. V_{in2} effectively appears on the inverting input of op-amp 1 and is amplified by a gain of

$$A_v = \frac{R_1}{R_G}$$

SPECIAL AMPLIFIERS

Also, the common-mode voltage on the noninverting input is amplified by the small common-mode gain of op-amp 1. (A_{cm} is typically less than 1). The total output voltage of op-amp 1 is

$$V_{out1} = \left(1 + \frac{R_1}{R_G}\right)V_{in1} - \left(\frac{R_1}{R_G}\right)V_{in2} + V_{cm} \qquad (11\text{–}1)$$

A similar analysis can be applied to op-amp 2 and results in the following output expression:

$$V_{out2} = \left(1 + \frac{R_2}{R_G}\right)V_{in2} - \left(\frac{R_2}{R_G}\right)V_{in1} + V_{cm} \qquad (11\text{–}2)$$

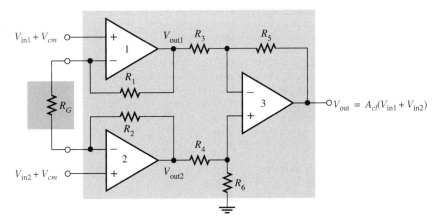

FIGURE 11–2
The instrumentation amplifier with the external gain-setting resistor R_G. Differential and common-mode signals are indicated.

Op-amp 3 has V_{out1} on one of its inputs and V_{out2} on the other. Therefore, the differential input voltage to op-amp 3 is $V_{out2} - V_{out1}$.

$$V_{out2} - V_{out1} = \left(1 + \frac{R_2}{R_G} + \frac{R_1}{R_G}\right)V_{in2} - \left(1 + \frac{R_2}{R_G} + \frac{R_1}{R_G}\right)V_{in1} + V_{cm} - V_{cm}$$

For $R_1 = R_2 = R$,

$$V_{out2} - V_{out1} = \left(1 + \frac{2R}{R_G}\right)V_{in2} - \left(1 + \frac{2R}{R_G}\right)V_{in1} + V_{cm} - V_{cm}$$

Notice that, since the common-mode voltages (V_{cm}) are equal, they cancel each other. Factoring out the differential gain gives the following expression for the differential input to op-amp 3.

$$V_{out2} - V_{out1} = \left(1 + \frac{2R}{R_G}\right)(V_{in2} - V_{in1})$$

Op-amp 3 has unity gain because $R_3 = R_5 = R_4 = R_6$ and $A_v = R_5/R_3 = R_6/R_4$. Therefore, the final output of the instrumentation amplifier (the output of op-amp 3) is

$$V_{out} = 1(V_{out2} - V_{out1})$$

$$V_{out} = \left(1 + \frac{2R}{R_G}\right)(V_{in2} - V_{in1}) \quad (11\text{--}3)$$

The closed-loop gain is

$$A_{cl} = 1 + \frac{2R}{R_G} \quad (11\text{--}4)$$

where $R_1 = R_2 = R$. Equation (11–4) shows that the gain of the instrumentation amplifier can be set by the value of the external resistor R_G when R_1 and R_2 have known fixed values.

The external gain-setting resistor R_G can be calculated for a desired voltage gain by using the following formula:

$$R_G = \frac{2R}{A_{cl} - 1} \quad (11\text{--}5)$$

■ EXAMPLE 11–1

Determine the value of the external gain-setting resistor R_G for a certain IC instrumentation amplifier with $R_1 = R_2 = 25$ kΩ. The voltage gain is to be 500.

SOLUTION

$$R_G = \frac{2R}{A_{cl} - 1} = \frac{50 \text{ k}\Omega}{500 - 1} \cong 100 \text{ }\Omega$$

PRACTICE EXERCISE 11–1

What value of external gain-setting resistor is required for an instrumentation amplifier with $R_1 = R_2 = 39$ kΩ to produce a gain of 325?

APPLICATIONS

As mentioned in the introduction to this section, the instrumentation amplifier is normally used to measure small differential signal voltages that are superimposed on a common-mode voltage often much larger than the signal voltage. Applications include situations where a quantity is sensed by a remote device, such as a temperature or pressure-sensitive transducer, and the resulting small electrical signal is sent over a long line subject to electrical noise that produces common-mode voltages in the line. The instrumentation amplifier at the end of the line must amplify the small signal from the remote sensor and reject the large common-mode voltage. Figure 11–3 illustrates this.

A SPECIFIC INSTRUMENTATION AMPLIFIER

Now that you have the basic idea of how an instrumentation amplifier works, let's take a look at a specific device. A representative device, the AD521, is shown in Figure 11–4 where IC pin numbers are given for reference. As you can see, there are some additional inputs and outputs that did not appear on the basic circuit. These provide additional features that are typical of many IC instrumentation amplifiers on the market.

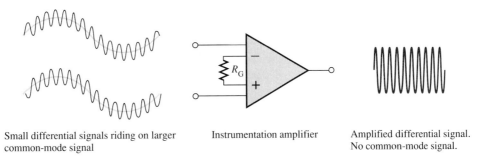

Small differential signals riding on larger common-mode signal

Instrumentation amplifier

Amplified differential signal. No common-mode signal.

FIGURE 11–3
Illustration of the rejection of large common-mode voltages and the amplification of smaller signal voltages by an instrumentation amplifier.

FIGURE 11–4
The AD521 instrumentation amplifier.

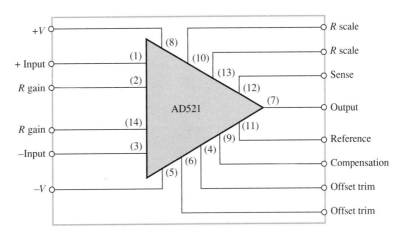

Some of the features of the AD521 are as follows. The voltage gain can be adjusted from 0.1 to 1000 with two external resistors. The input impedance is 3000 MΩ. The common-mode rejection ratio (CMRR) has a minimum value of 110 dB. Recall that a higher CMRR means better rejection of common-mode voltages. The AD521 has a gain-bandwidth product of 40 MHz. There is also an external provision for limiting the bandwidth, and the device is protected against excessive input voltages.

SETTING THE GAIN For the AD521, two external resistors must be used to set the voltage gain as indicated in Figure 11–5. Resistor R_G is connected between the *R*-gain terminals (pins 2 and 14). Resistor R_S is connected between the *R*-scale terminals (pins 10 and 13). R_S must be within ±15% of 100 kΩ, and R_G is selected for the desired gain based on the formula

$$A_v = \frac{R_S}{R_G} \qquad (11\text{–}6)$$

FIGURE 11–5
The AD521 with gain-setting resistors and output offset adjustment.

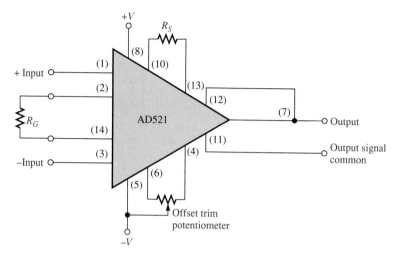

Don't be concerned about the difference in this gain expression and the closed-loop gain for the basic circuit stated in Equation (11–4). This difference is due to the subtle differences in design. If you were constructing an instrumentation amplifier from separate op-amps and discrete resistors, you would use the formulas discussed earlier.

OFFSET TRIM The offset **trim** adjustment (pins 4 and 6) is used to zero any output offset voltage caused by an input offset voltage multiplied by the gain. A potentiometer connected between pins 4 and 6 as shown in Figure 11–5 can be used to adjust the offset.

BANDWIDTH CONTROL When you need to set the amplifier's bandwidth to a desired value, the compensation input (pin 9) can be used in conjunction with an external RC network. A recommended configuration is shown in Figure 11–6. The values of the two resistors

FIGURE 11–6
The AD521 with a compensation circuit for controlling the bandwidth.

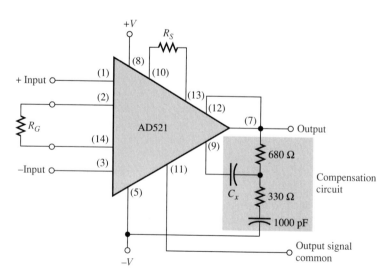

SPECIAL AMPLIFIERS

and one of the capacitors are set as recommended by the manufacturer and then the value of the capacitor C_x is selected for the desired bandwidth according to the following formula:

$$C_x = \frac{1}{100\pi f_c} \tag{11-7}$$

where $BW = f_c$ is in kilohertz (kHz) and C_x is in microfarads (μF).

EXAMPLE 11–2 Determine the gain and the bandwidth for the instrumentation amplifier in Figure 11–7.

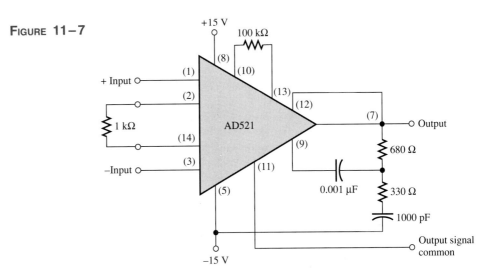

FIGURE 11–7

SOLUTION
The voltage gain is determined by R_S and R_G as follows.

$$A_v = \frac{R_S}{R_G} = \frac{100 \text{ k}\Omega}{1 \text{ k}\Omega} = 100$$

The bandwidth is determined as follows.

$$C_x = \frac{1}{100\pi f_c}$$

$$BW = f_c = \frac{1}{100\pi C_x} = \frac{1}{100\pi(0.001)} = 3.18 \text{ kHz}$$

Notice that the number of microfarads (0.001) is substituted into the formula, not the number of farads (0.001×10^{-6}).

PRACTICE EXERCISE 11–2
Modify the circuit in Figure 11–7 for a gain of approximately 45 and a bandwidth of approximately 10 kHz.

11–1 REVIEW QUESTIONS

1. What is the main purpose of an instrumentation amplifier and what are three of its key characteristics?
2. What components do you need to construct a basic instrumentation amplifier?
3. How is the gain determined in a basic instrumentation amplifier?
4. In a certain AD521 configuration, $R_S = 91$ kΩ and $R_G = 56$ kΩ. Is the voltage gain less than or greater than unity?

11–2 ISOLATION AMPLIFIERS

An isolation amplifier provides dc isolation between input and output for the protection of human life or sensitive equipment in those applications where hazardous power-line leakage or high-voltage transients are possible. The principal areas of application for isolation amplifiers are in medical instrumentation, power plant instrumentation, industrial processing, and automated testing.

THE BASIC ISOLATION AMPLIFIER

In some ways, the isolation amplifier can be viewed as an elaborate op-amp or instrumentation amplifier. The isolation amplifier has an input circuit that is electrically isolated from the output and power supply circuits using transformers or optical coupling. Transformer coupling is more common. The circuits are in IC form, but the transformers are not integrated. Although the packages that contain both the circuits and transformers are somewhat larger than standard IC packages, they are generally pin compatible for easy circuit board assembly.

The typical three-port isolation amplifier has three basic isolated sections that are transformer coupled. As shown in the block diagram of Figure 11–8, the sections are the input circuit, the output circuit, and the power source. The input section contains an instrumentation amplifier or an op-amp (in this case, it is an instrumentation amplifier), a power supply, and a modulator. The output section contains an op-amp, a power supply, and a demodulator. The power section contains an oscillator.

OPERATION An external dc supply voltage is applied to the power section where the internal oscillator is energized and converts the dc input power to ac power. The frequency of the oscillator is fairly high in order to keep the size of the transformers small; for example, it is 80 kHz in the AD295 isolation amplifier. The ac power signal from the oscillator is coupled to both the input and output sections through transformer T2. In the input and output sections, the ac power signal is rectified and filtered by the power-supply circuits to provide dc power for the amplifiers.

The ac power signal is also sent to the modulator in the input section where it is modulated by the output signal from the input amplifier. The modulated signal is coupled through transformer T1 to the output section where it is demodulated. The demodulation process recovers the original signal from the ac power signal. The output amplifier provides further gain before the signal goes to the final output.

Although the isolation amplifier is a fairly complex system in itself, in terms of its overall function, it is still simply an amplifier. You apply a dc voltage, put a signal in, and you get an amplified signal out. The isolation function is an unseen process.

Special Amplifiers

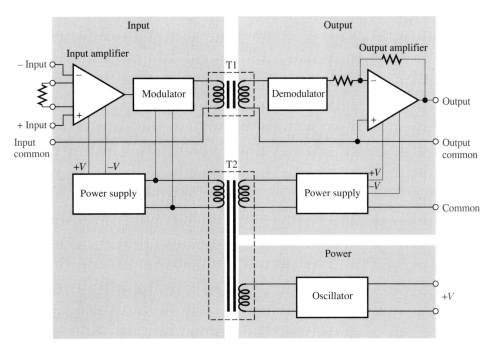

Figure 11–8
Basic isolation amplifier block diagram.

Applications

As previously mentioned, the isolation amplifier is mainly used in medical equipment and for remote sensing in high-noise industrial environments where interfacing to sensitive equipment is required. In medical applications where body functions such as heart and blood pressure are monitored, the very small monitored signals are combined with large common-mode signals, such as 60 Hz power line pickup from the skin. In these situations, without isolation, dc leakage or equipment failure could be fatal. In chemical, nuclear, and metal-processing industries, for example, millivolt signals typically exist in the presence of common-mode voltages that can be in the kilovolt range. In this type of environment, the isolation amplifier can amplify small signals from very noisy equipment and provide a safe output to sensitive equipment such as computers.

Figure 11–9 shows a simplified diagram of an isolation amplifier in a cardiac monitoring application. In this situation, we have heart signals, which are very small, combined with much larger common-mode signals caused by muscle noise, electrochemical noise, residual electrode voltages, and 60 Hz line pickup from the skin. The monitoring of fetal heartbeat, as illustrated, is the most demanding type of cardiac monitoring because in addition to the fetal heartbeat that typically generates 50 μV, there is also the mother's heartbeat that typically generates 1 mV. The common-mode voltages can run from about 1 mV to about 100 mV. The CMR (common-mode rejection) of the isolation amplifier separates the signal of the fetal heartbeat from that of the mother's heartbeat and from those signals of the common-mode voltages. So, the signal from the fetal heartbeat is essentially all that the amplifier sends to the monitoring equipment.

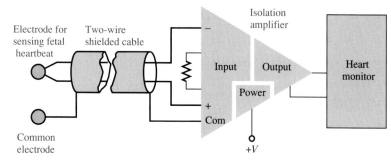

FIGURE 11–9
Fetal heartbeat monitoring using an isolation amplifier.

A Specific Isolation Amplifier

Now that you have learned basically what an isolation amplifier is and what it does, let's take a look at a representative device, the AD295, for a good introduction to practical IC isolation amplifiers. As you can see in Figure 11–10, the AD295 is similar to the basic isolation amplifier in Figure 11–8 except that the input amplifier is an op-amp and, also, it has quite a few more inputs and outputs. These additional pins provide for gain adjustments, offset adjustments, isolated dc voltage outputs, and other functions.

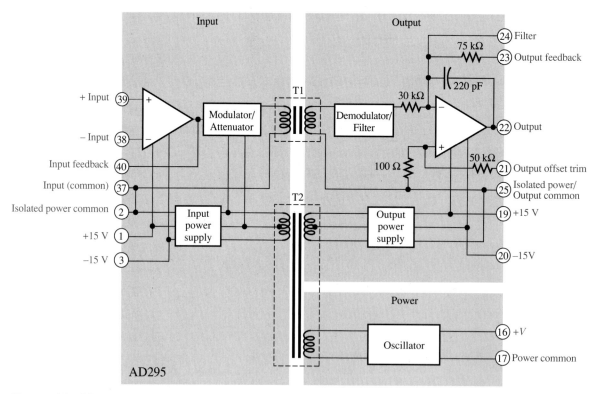

FIGURE 11–10
The AD295 isolation amplifier.

Special Amplifiers

Isolated Power Outputs The AD295 isolation amplifier provides ±15 V from both isolated power supplies. These voltages are available for powering associated external circuits such as preamplifiers, transducers, and the like.

Voltage Gain The gains of both the input and the output amplifiers can be set with external resistors. The overall gain of the device can be set at any value from 1 to 1000. Figure 11–11 shows the circuit connected for unity gain. In the input circuit, the amplifier output is connected directly back to the inverting input (pin 40 to pin 38) creating a voltage-follower configuration with a gain of 1. The attenuation circuit within the modulator/attenuator block has an inherent fixed attenuation of 0.4. To overcome this attenuation, the output amplifier has a gain of 2.5 (A_v = 75 kΩ/30 kΩ = 2.5) when the output is connected directly to the 75 kΩ feedback resistor (pin 22 to pin 23). This produces a combined gain of 1 (0.4 × 2.5 = 1).

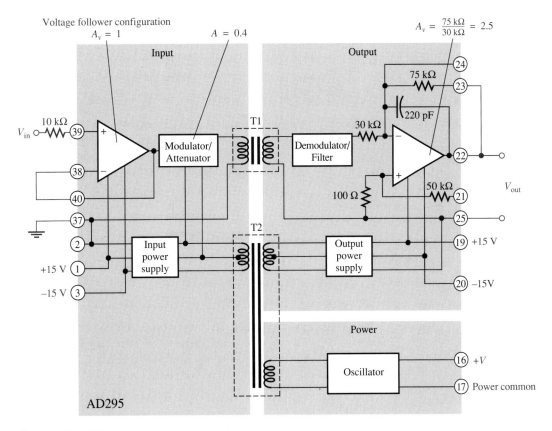

Figure 11–11
Unity-gain connections for the AD295 isolation amplifier.

Gains up to 1000 can be achieved by connecting external resistors as shown in Figure 11–12. Although the connection for the input amplifier is for a noninverting configuration, it can also be connected in an inverting configuration. The voltage gain of

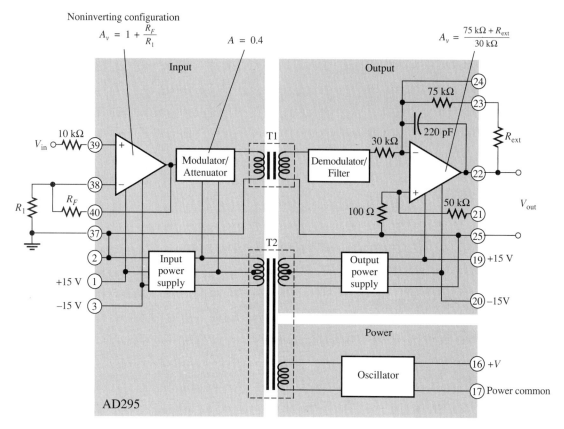

FIGURE 11–12
Nonunity gain connections for the AD295 isolation amplifier.

the input amplifier is

$$A_{v(\text{input})} = 1 + \frac{R_F}{R_1} \quad (11\text{–}8)$$

For the AD295, there must also be a 10 kΩ resistor in series with the input as indicated in the figure. Also, $R_F + R_1$ must be equal to or greater than 10 kΩ. R_F is the feedback resistor and R_1 is the input resistor for the op-amp configuration.

The gain of the output amplifier can be increased above 2.5 by adding an external resistor in series with the internal 75 kΩ resistor as shown in Figure 11–12. For this case, the voltage gain of the output amplifier is

$$A_{v(\text{output})} = \frac{75 \text{ k}\Omega + R_{\text{ext}}}{30 \text{ k}\Omega} \quad (11\text{–}9)$$

OFFSET ADJUSTMENTS The external connections for adjustment of the input and output offset voltages are shown in Figure 11–13 in conjunction with a unity-gain configuration. The resistor values shown are recommended by the manufacturer for this particular device.

SPECIAL AMPLIFIERS

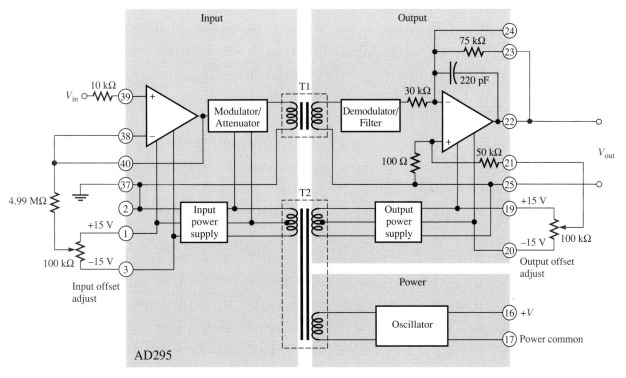

FIGURE 11–13
Offset voltage adjustments for the AD295 isolation amplifier.

■ **EXAMPLE 11–3**   Determine the overall voltage gain of the AD295 isolation amplifier in Figure 11–14.

FIGURE 11–14

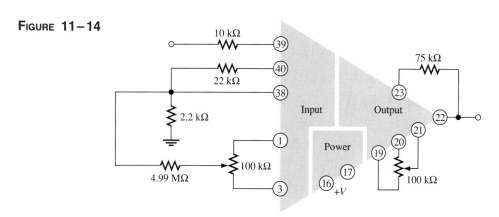

SOLUTION
The gain of the input amplifier is

$$A_{v(\text{input})} = 1 + \frac{R_F}{R_1} = 1 + \frac{22 \text{ k}\Omega}{2.2 \text{ k}\Omega} = 1 + 10 = 11$$

The gain of the output amplifier is

$$A_{v(\text{output})} = \frac{75 \text{ k}\Omega + R_{\text{ext}}}{30 \text{ k}\Omega} = \frac{75 \text{ k}\Omega + 75 \text{ k}\Omega}{30 \text{ k}\Omega} = 5$$

Since the fixed-internal attenuation is 0.4 for the AD295, the overall gain of the isolation amplifier is

$$A_v = A_{v(\text{input})} \times A_{v(\text{output})} \times \text{Atten} = (11)(5)(0.4) = 22$$

Practice Exercise 11–3
Select resistor values and specify the connections in Figure 11–14 that will produce an overall gain of approximately 10.

11–2 Review Questions

1. In what types of applications are isolation amplifiers used?
2. What are the three sections in a typical isolation amplifier?
3. How are the sections in an isolation amplifier connected?
4. What is the purpose of the oscillator in an isolation amplifier?

11–3 Operational Transconductance Amplifiers (OTAs)

Conventional op-amps are, as you know, primarily voltage amplifiers in which the output voltage equals the gain times the input voltage. The OTA is primarily a voltage-to-current amplifier in which the output current equals the gain times the input voltage.

Figure 11–15 shows the symbol for an operational transconductance amplifier (OTA). The double circle symbol at the output represents an output current source that is dependent on a bias current. Like the conventional op-amp, the OTA has two differential input terminals, a high input impedance, and a high CMRR. Unlike the conventional op-amp, the OTA has a bias-current input terminal, a high output impedance, and no fixed open-loop voltage gain.

Figure 11–15
Symbol for an operational transconductance amplifier (OTA).

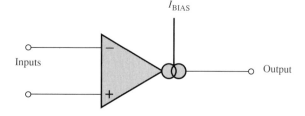

The Transconductance Is the Gain of an OTA

By definition, the transconductance of an electronic device is the ratio of the output current to the input voltage. For an OTA, voltage is the input variable and current is the output variable; therefore, the ratio of output current to input voltage is its gain. Consequently, the voltage-to-current gain of an OTA is the transconductance, g_m.

$$A = g_m = \frac{I_{out}}{V_{in}} \quad (11-10)$$

In an OTA, the transconductance is dependent on a constant (K) times the bias current (I_{BIAS}) as indicated in Equation (11–11). The value of the constant is dependent on the internal circuit design.

$$g_m = KI_{BIAS} \quad (11-11)$$

The output current is controlled by the input voltage and the bias current as shown by the following formulas:

$$I_{out} = g_m V_{in}$$
$$I_{out} = KI_{BIAS} V_{in}$$

THE TRANSCONDUCTANCE IS A FUNCTION OF BIAS CURRENT

The relationship of the transconductance and the bias current in an OTA is a very important characteristic. The graph in Figure 11–16 illustrates a typical relationship. Notice that the transconductance increases linearly with the bias current. The constant of proportionality, K, is the slope of the line.

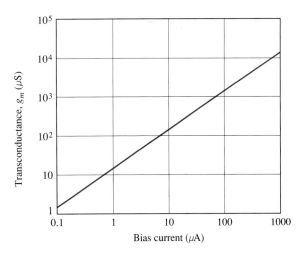

FIGURE 11–16
Graph of transconductance versus bias current for a typical OTA.

EXAMPLE 11–4 If an OTA has a g_m = 1000 μS, what is the output current when the input voltage is 50 mV?

SOLUTION

$$I_{out} = g_m V_{in} = (1000 \ \mu S)(50 \ mV) = 50 \ \mu A$$

Practice Exercise 11-4
From the graph in Figure 11-16, determine the bias current required to produce $g_m = 1000 \ \mu S$.

Basic OTA Circuits

Figure 11-17 shows the OTA used as an inverting amplifier with fixed-voltage gain. The voltage gain is set by the transconductance and the load resistance as follows.

$$V_{out} = I_{out}R_L$$

$$I_{out} = \frac{V_{out}}{R_L}$$

$$g_m = \frac{I_{out}}{V_{in}} = \frac{(V_{out}/R_L)}{V_{in}} = \left(\frac{V_{out}}{V_{in}}\right)\left(\frac{1}{R_L}\right)$$

$$g_m R_L = \frac{V_{out}}{V_{in}}$$

Since V_{out}/V_{in} is the voltage gain,

$$A_v = g_m R_L$$

The transconductance of the amplifier in Figure 11-17 is determined by the amount of bias current, which is set by the dc supply voltages and the bias resistor R_{BIAS}.

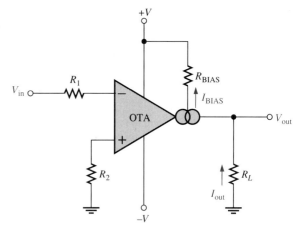

FIGURE 11-17
An OTA as an inverting amplifier with a fixed-voltage gain.

One of the most useful features of an OTA is that the voltage gain can be controlled by the amount of bias current. This can be done manually, as shown in Figure 11-18(a), by using a variable resistor in series with R_{BIAS} in the circuit of Figure 11-17. By changing the resistance, we can produce a change in I_{BIAS}, which changes the transconductance. A change in the transconductance changes the voltage gain. The voltage gain can also be controlled with an externally applied variable voltage as shown in Figure 11-18(b). A variation in the applied bias voltage causes a change in the bias current.

Special Amplifiers

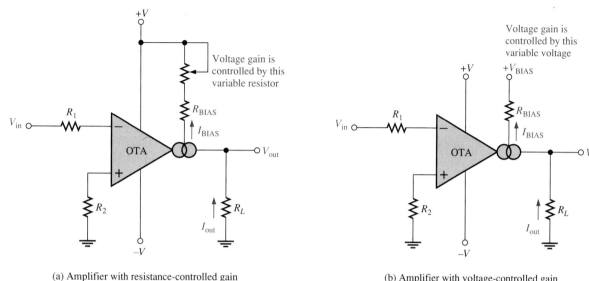

(a) Amplifier with resistance-controlled gain

(b) Amplifier with voltage-controlled gain

FIGURE 11–18
An OTA as an inverting amplifier with a variable-voltage gain.

A Specific OTA

The CA3080 is a typical OTA and serves as a representative device. Figure 11–19 shows its pin configuration for an eight-pin DIP. The maximum dc supply voltage is ±15 V, and its transconductance characteristic happens to be the same as indicated by the graph in Figure 11–16. For a CA3080, the bias current is determined by the following formula:

$$I_{BIAS} = \frac{(+V) - (-V) - 0.7 \text{ V}}{R_{BIAS}} \tag{11-12}$$

The 0.7 V is due to the internal circuit where a base-emitter junction connects the external R_{BIAS} with the negative supply voltage $(-V)$. The positive supply voltage is $+V$.

FIGURE 11–19
The CA3080 OTA.

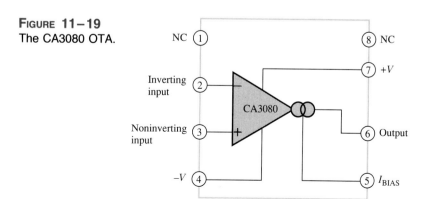

CHAPTER 11

Not only does the transconductance of an OTA vary with bias current, but so does the input and output resistances. Both the input and output resistances decrease as the bias current increases, as shown in Figure 11–20 for a CA3080.

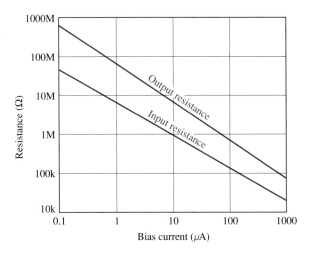

FIGURE 11–20
Input and output resistances versus bias current.

■ **EXAMPLE 11–5** The OTA in Figure 11–21 is connected as an inverting fixed-gain amplifier. Determine the voltage gain.

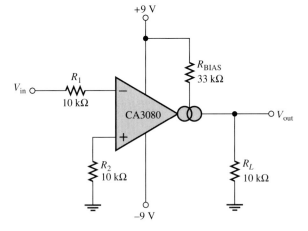

FIGURE 11–21

SOLUTION
The bias current is calculated as follows.

$$I_{BIAS} = \frac{(+V) - (-V) - 0.7 \text{ V}}{R_{BIAS}} = \frac{9 \text{ V} - (-9 \text{ V}) - 0.7 \text{ V}}{33 \text{ k}\Omega} = 524 \text{ } \mu\text{A}$$

SPECIAL AMPLIFIERS

From the graph in Figure 11–16, the value of transconductance corresponding to $I_{BIAS} = 524 \ \mu A$ is approximately 10,000 μS or 10 mS. Using this value of g_m, the voltage gain is calculated.

$$A_v = g_m R_L = (10 \text{ mS})(10 \text{ k}\Omega) = 100$$

PRACTICE EXERCISE 11–5

If the OTA in Figure 11–21 is operated with dc supply voltages of ±12 V, will this change the voltage gain and, if so, to what value?

TWO OTA APPLICATIONS

AMPLITUDE MODULATOR Figure 11–22 illustrates an OTA connected as an amplitude modulator. The voltage gain is varied by applying a modulation voltage to the bias input. When a constant-amplitude input signal is applied, the amplitude of the output signal will vary according to the modulation voltage on the bias input. The gain is dependent on bias current, and bias current is related to the modulation voltage by the following relationship:

$$I_{BIAS} = \frac{V_{MOD} - (-V) - 0.7 \text{ V}}{R_{BIAS}}$$

This modulating action is shown in Figure 11–22 for a higher frequency sine-wave input voltage and a lower frequency sine-wave modulating voltage.

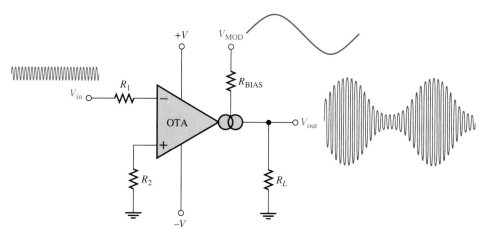

FIGURE 11–22
The OTA as an amplitude modulator.

EXAMPLE 11–6

The input to the OTA amplitude modulator in Figure 11–23 is a 50 mV peak-to-peak, 1 MHz sine wave. Determine the output signal, given the modulation voltage shown is applied to the bias input.

FIGURE 11–23

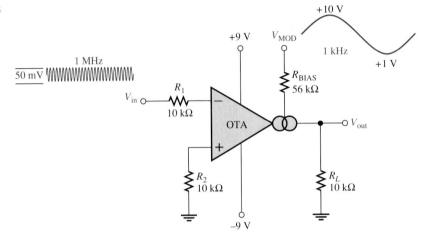

SOLUTION
The maximum voltage gain is when I_{BIAS}, and thus g_m, is maximum. This occurs at the maximum peak of the modulating voltage, V_{MOD}.

$$I_{BIAS(max)} = \frac{V_{MOD(max)} - (-V) - 0.7\text{ V}}{R_{BIAS}} = \frac{10\text{ V} - (-9\text{ V}) - 0.7\text{ V}}{56\text{ k}\Omega} = 327\ \mu\text{A}$$

From the graph in Figure 11–16, the constant K is approximately 16.

$$g_m = KI_{BIAS(max)} = 16(327\ \mu\text{A}) = 5.23\text{ mS}$$
$$A_{v(max)} = g_m R_L = (5.23\text{ mS})(10\text{ k}\Omega) = 52.3$$
$$V_{out(max)} = A_{v(max)} V_{in} = (52.3)(50\text{ mV}) = 2.62\text{ V}$$

The minimum bias current is

$$I_{BIAS(min)} = \frac{V_{MOD(min)} - (-V) - 0.7\text{ V}}{R_{BIAS}} = \frac{1\text{ V} - (-9\text{ V}) - 0.7\text{ V}}{56\text{ k}\Omega} = 166\ \mu\text{A}$$

$$g_m = KI_{BIAS(min)} = 16(166\ \mu\text{A}) = 2.66\text{ mS}$$
$$A_{v(min)} = g_m R_L = (2.66\text{ mS})(10\text{ k}\Omega) = 26.6$$
$$V_{out(min)} = A_{v(min)} V_{in} = (26.6)(50\text{ mV}) = 1.33\text{ V}$$

The resulting output voltage is shown in Figure 11–24.

FIGURE 11–24

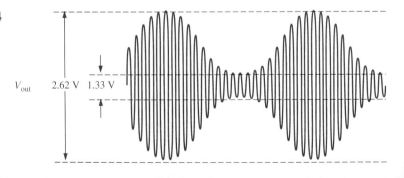

Special Amplifiers

Practice Exercise 11–6
Repeat this example with the sine-wave modulating signal replaced by a square wave with the same maximum and minimum levels and a bias resistor of 39 kΩ.

Schmitt Trigger Figure 11–25 shows an OTA used in a Schmitt-trigger configuration. Basically, a Schmitt trigger is a comparator with hysteresis where the input voltage is large enough to drive the device into its saturated states. When the input voltage exceeds a certain threshold value or trigger point, the device switches to one of its saturated output states. When the input falls back below another threshold value, the device switches back to its other saturated output state.

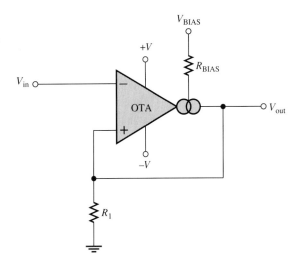

Figure 11–25
The OTA as a Schmitt trigger.

In the case of the OTA Schmitt trigger, the threshold levels are set by the current through resistor R_1. The maximum output current in an OTA equals the bias current. Therefore, in the saturated output states, $I_{out} = I_{BIAS}$. The maximum positive output voltage is $I_{out}R_1$, and this voltage is the positive threshold value or upper trigger point. When the input voltage exceeds this value, the output switches to its maximum negative voltage, which is $-I_{out}R_1$. Since $I_{out} = I_{BIAS}$, the trigger points can be controlled by the bias current. Figure 11–26 illustrates this operation.

11–3 Review Questions

1. What does OTA stand for?
2. If the bias current in an OTA is increased, does the transconductance increase or decrease?
3. What happens to the voltage gain if the OTA is connected as a fixed-voltage amplifier and the supply voltages are increased?
4. What happens to the voltage gain if the OTA is connected as a variable-gain voltage amplifier and the voltage at the bias terminal is decreased?

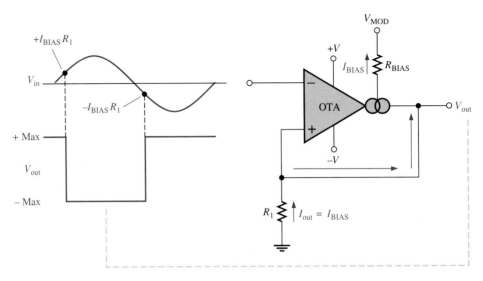

FIGURE 11–26
Basic operation of the OTA Schmitt trigger.

11–4   LOG AND ANTILOG AMPLIFIERS

A logarithmic amplifier produces an output that is proportional to the logarithm of the input and antilogarithmic amplifiers take the antilog or inverse log of the input. Log amplifiers are used in applications that require compression of analog input data, linearization of transducers that have exponential outputs, and analog multiplication and division. In this section, we will discuss the principles of logarithmic amplifiers.

THE BASIC LOGARITHMIC AMPLIFIER

The key element in a log amplifier is a device that exhibits a logarithmic characteristic that, when placed in the feedback loop of an op-amp, produces a logarithmic response. This means that the output voltage is a function of the logarithm of the input voltage, as expressed by the following general equation. K is a constant, and ln is the **natural logarithm** to the base e.

$$V_{out} = -K \ln(V_{in}) \tag{11–13}$$

Although we will use natural logarithms in the formulas in this section, each expression can be converted to a logarithm to the base 10 ($\log_{10}$) using this relationship, $\ln x = 2.3 \log_{10} x$.

The semiconductor pn junction in the form of either a diode or the base-emitter junction of a bipolar transistor provides a logarithmic characteristic. You may recall that a diode has a nonlinear characteristic up to a forward voltage of approximately 0.7 V. Figure 11–27 shows the characteristic curve, where I_D is the forward diode current and V_D is the forward diode voltage.

FIGURE 11–27
A portion of a diode (pn junction) characteristic curve (I_D versus V_D).

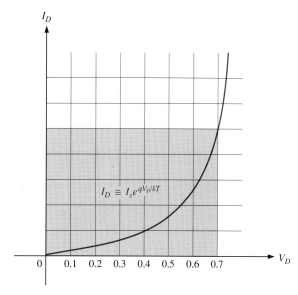

As you can see on the graph, the diode curve is nonlinear. Not only is the characteristic curve nonlinear, it is logarithmic and is specifically defined by the following formula:

$$I_D \cong I_s e^{qV_D/kT} \tag{11-14}$$

where I_s is the reverse leakage current, q is the charge on an electron, k is Boltzmann's constant, and T is the absolute temperature in degrees Kelvin. Equation (11–14) can be solved for the diode forward voltage, V_D, as follows. Take the natural logarithm (ln is the logarithm to the base e) of both sides.

$$\ln I_D = \ln I_s e^{qV_D/kT}$$

The ln of a product of two terms equals the ln of each term.

$$\ln I_D = \ln I_s + \ln e^{qV_D/kT} = \ln I_s + \frac{qV_D}{kT}$$

$$\ln I_D - \ln I_s = \frac{qV_D}{kT}$$

The difference of two ln terms equals the ln of the quotient of the terms.

$$\ln\left(\frac{I_D}{I_s}\right) = \frac{qV_D}{kT}$$

Solving for V_D,

$$V_D = \left(\frac{kT}{q}\right)\ln\left(\frac{I_D}{I_s}\right) \tag{11-15}$$

LOG AMPLIFIER WITH A DIODE When a diode is placed in the feedback loop of an op-amp circuit, as shown in Figure 11–28, we have a basic log amplifier. Since the inverting input is at virtual ground (0 V), the output is at $-V_D$ when the input is positive. Since V_D is logarithmic, so is V_{out}. The output is limited to a maximum value of approximately -0.7 V because the diode's logarithmic characteristic is restricted to voltages below 0.7 V. Also, the input must be positive when the diode is connected in the direction shown in the figure. To handle negative inputs, the diode must be turned around.

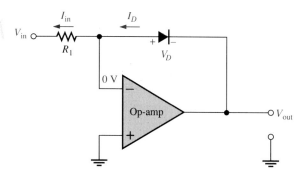

FIGURE 11–28
A basic log amplifier using a diode as the feedback element.

An analysis of the circuit in Figure 11–28 is as follows, beginning with the fact that $V_{out} = -V_D$ and the fact that $I_D = I_{in}$ because there is no current at the inverting input.

$$V_{out} = -V_D$$

$$I_D = I_{in} = \frac{V_{in}}{R_1}$$

Substituting into the formula for V_D,

$$V_{out} = -\left(\frac{kT}{q}\right)\ln\left(\frac{V_{in}}{I_s R_1}\right) \tag{11-16}$$

The term kT/q is a constant equal to approximately 25 mV at 25°C. Therefore, the output voltage can be expressed as

$$V_{out} \cong -(0.025 \text{ V})\ln\left(\frac{V_{in}}{I_s R_1}\right) \tag{11-17}$$

From Equation (11–17), you can see that the output voltage is the negative of a logarithmic function of the input voltage. The value of the output is controlled by the value of the input voltage and the value of the resistor R_1. The other factor, I_s, is a constant for a given diode.

EXAMPLE 11–7 Determine the output voltage for the log amplifier in Figure 11–29. Assume $I_s = 50$ nA.

Special Amplifiers

Figure 11-29

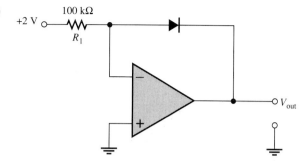

Solution
The input voltage and the resistor value are given in Figure 11–29.

$$V_{out} = -(0.025 \text{ V})\ln\left(\frac{V_{in}}{I_s R_1}\right)$$

$$= -(0.025 \text{ V})\ln\left(\frac{2 \text{ V}}{50 \text{ nA} \times 100 \text{ k}\Omega}\right)$$

$$= -(0.025 \text{ V})\ln(400)$$
$$= -(0.025 \text{ V})(5.99)$$
$$= -0.1498 \text{ V}$$

Practice Exercise 11-7
Calculate the output voltage of the log amplifier with a +4 V input.

Log Amplifier with a BJT The base-emitter junction of a bipolar transistor exhibits the same type of logarithmic characteristic as a diode because it is also a pn junction. A log amplifier with a BJT connected in a common-base form in the feedback loop is shown in Figure 11–30. Notice that V_{out} with respect to ground is equal to V_{BE}.

Figure 11-30
A basic log amplifier using a transistor as the feedback element.

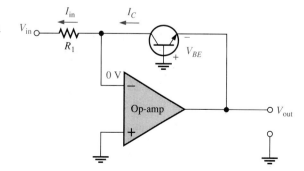

The analysis for this circuit is the same as for the diode log amplifier except that V_{BE} replaces V_D, I_C replaces I_D, and I_{EBO} replaces I_s. The expression for the I_C versus V_{BE} characteristic is

$$I_C = I_{EBO} e^{qV_{BE}/kT} \tag{11-18}$$

where I_{EBO} is the emitter-to-base leakage current. The expression for the output voltage is

$$V_{out} = -(0.025 \text{ V}) \ln\left(\frac{V_{in}}{I_{EBO} R_1}\right) \tag{11-19}$$

EXAMPLE 11-8

What is V_{out} for a transistor log amplifier with $V_{in} = 3$ V and $R_1 = 68$ kΩ? Assume $I_{EBO} = 40$ nA.

SOLUTION

$$V_{out} = -(0.025 \text{ V}) \ln\left(\frac{V_{in}}{I_{EBO} R_1}\right)$$

$$= -(0.025 \text{ V}) \ln\left(\frac{3 \text{ V}}{40 \text{ nA} \times 68 \text{ k}\Omega}\right)$$

$$= -(0.025 \text{ V}) \ln(1102.94)$$

$$= -0.1751 \text{ V}$$

PRACTICE EXERCISE 11-8

Calculate V_{out} if R_1 is changed to 33 kΩ.

THE BASIC ANTILOG AMPLIFIER

The **logarithm** of a number is the power to which the base must be raised to get that number. The antilogarithm of a number is the result obtained when the base is raised to a power equal to the logarithm of that number. To get the antilogarithm, you must take the exponential of the logarithm (antilogarithm of $x = e^{\ln x}$).

An antilog amplifier is formed by connecting a transistor (or diode) as the input element as shown in Figure 11-31. The exponential formula in Equation (11-18) still applies to the base-emitter pn junction. The output voltage is determined by the current (equal to the collector current) through the feedback resistor.

$$V_{out} = -R_F I_C$$

The characteristic equation of the pn junction is

$$I_C = I_{EBO} e^{qV_{BE}/kT}$$

Substituting into the equation for V_{out}, we get

$$V_{out} = -R_F I_{EBO} e^{qV_{BE}/kT}$$

As you can see in Figure 11-31, $V_{in} = V_{BE}$.

$$V_{out} = -R_F I_{EBO} e^{qV_{in}/kT}$$

Special Amplifiers

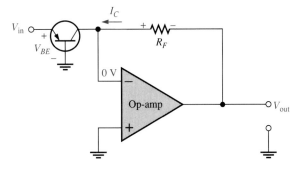

Figure 11–31
A basic antilog amplifier.

The exponential term can be expressed as an antilogarithm as follows.

$$V_{out} = -R_F I_{EBO}\,\text{antilog}\left(\frac{V_{in}q}{kT}\right)$$

Since kT/q is approximately 25 mV,

$$V_{out} = -R_F I_{EBO}\,\text{antilog}\left(\frac{V_{in}}{25\text{ mV}}\right) \qquad (11\text{–}20)$$

EXAMPLE 11–9

For the antilog amplifier in Figure 11–32, find the output voltage. Assume $I_{EBO} = 40$ nA.

FIGURE 11–32

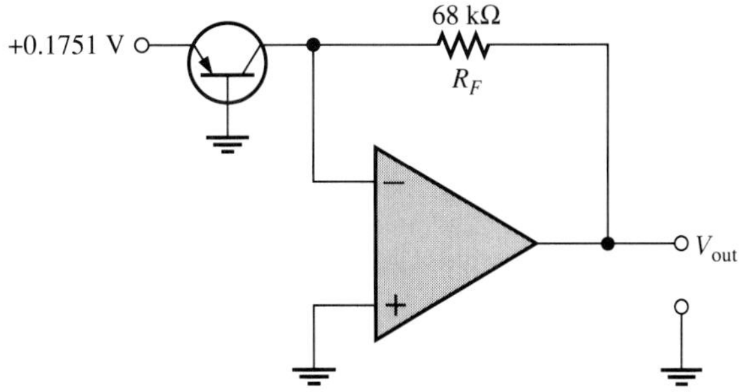

+0.1751 V, 68 kΩ R_F

SOLUTION
First of all, notice that the input voltage in Figure 11–32 is the same as the output voltage of the log amplifier in Example 11–8, where the output voltage is proportional to the logarithm of the input voltage. In this case, the antilog amplifier reverses the process and produces an output that is proportional to the antilog of the input. Stated another way, the input of an antilog amplifier is proportional to the logarithm of the output. So, the output voltage of the antilog amplifier in Figure 11–32 should have the same magnitude as the input voltage of the log amplifier in Example 11–8 because all the constants are the same. Let's see if it does.

$$V_{out} = -R_F I_{EBO} \text{antilog}\left(\frac{V_{in}}{25 \text{ mV}}\right)$$
$$= -(68 \text{ k}\Omega)(40 \text{ nA})\text{antilog}\left(\frac{0.1751 \text{ V}}{25 \text{ mV}}\right)$$
$$= -(68 \text{ k}\Omega)(40 \text{ nA})(1101)$$
$$= -3 \text{ V}$$

PRACTICE EXERCISE 11–9
Determine V_{out} for the amplifier in Figure 11–32 if the feedback resistor is changed to 100 kΩ.

SIGNAL COMPRESSION WITH LOGARITHMIC AMPLIFIERS

In certain applications, a signal may be too large in magnitude for a particular system to handle. The term *dynamic range* is often used to describe the range of voltages contained in a signal. In these cases, the signal voltage must be scaled down by a process called **signal compression** so that it can be properly handled by the system. If a linear circuit is used to scale a signal down in amplitude, the lower voltages are reduced by the same percentage as the higher voltages. Linear signal compression often results in the lower voltages becoming obscured by noise and difficult to accurately distinguish, as illustrated in Figure 11–33(a). To overcome this problem, a signal with a large dynamic range can be compressed using a logarithmic response, as shown in Figure 11–33(b). In logarithmic

FIGURE 11–33
The basic concept of signal compression with a logarithmic amplifier.

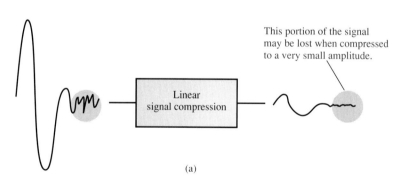

(a)

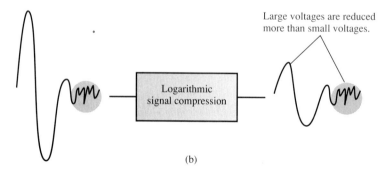

(b)

Special Amplifiers

signal compression, the higher voltages are reduced more than the lower voltages, thus keeping the lower voltage signals from being lost in noise.

11-4 Review Questions

1. What purpose does the diode or transistor perform in the feedback loop of a log amplifier?
2. Why is the output of a log amplifier limited to about 0.7 V?
3. What are the factors that determine the output voltage of a basic log amplifier?
4. In terms of implementation, how does a basic antilog amplifier differ from a basic log amplifier?

11-5 Analog Multipliers and Dividers

Multipliers and dividers are used in data acquisition and communications applications for modulation, demodulation, phase detection, analog computing, and many other applications. In this section, you will learn how the logarithmic amplifiers studied in Section 11-4 can be used in multiplication and division of analog signals. More applications are covered in Chapter 12.

An Analog Multiplier

Multipliers are based on the fundamental logarithmic relationship that states that the product of two terms equals the sum of the logarithms of each term. This relationship is shown in the following equation:

$$\ln(a \times b) = \ln a + \ln b \quad (11\text{-}21)$$

Equation (11–21) shows that two signal voltages are effectively multiplied if the logarithms of the signal voltages are added.

You know how to get the logarithm of a signal voltage by using a log amplifier. By summing the outputs of two log amplifiers, you get the logarithm of the product of the two original input voltages. Then, by taking the antilogarithm, you get the product of the two input voltages as indicated in the following equations:

$$\ln V_1 + \ln V_2 = \ln(V_1 V_2)$$
$$\text{antilog}[\ln(V_1 V_2)] = V_1 V_2$$

The block diagram in Figure 11–34 shows how the functions are connected to multiply two input voltages. Constant terms are omitted for simplicity.

Figure 11–35 shows the basic multiplier circuitry. The outputs of the log amplifiers were developed in Section 11–4 and are stated as follows:

$$V_{\text{out(log1)}} = -K_1 \ln\left(\frac{V_{\text{in1}}}{K_2}\right)$$

$$V_{\text{out(log2)}} = -K_1 \ln\left(\frac{V_{\text{in2}}}{K_2}\right)$$

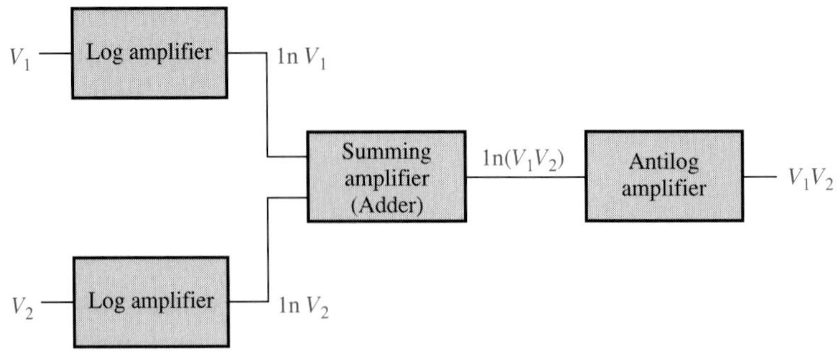

FIGURE 11–34
Basic block diagram of an analog multiplier.

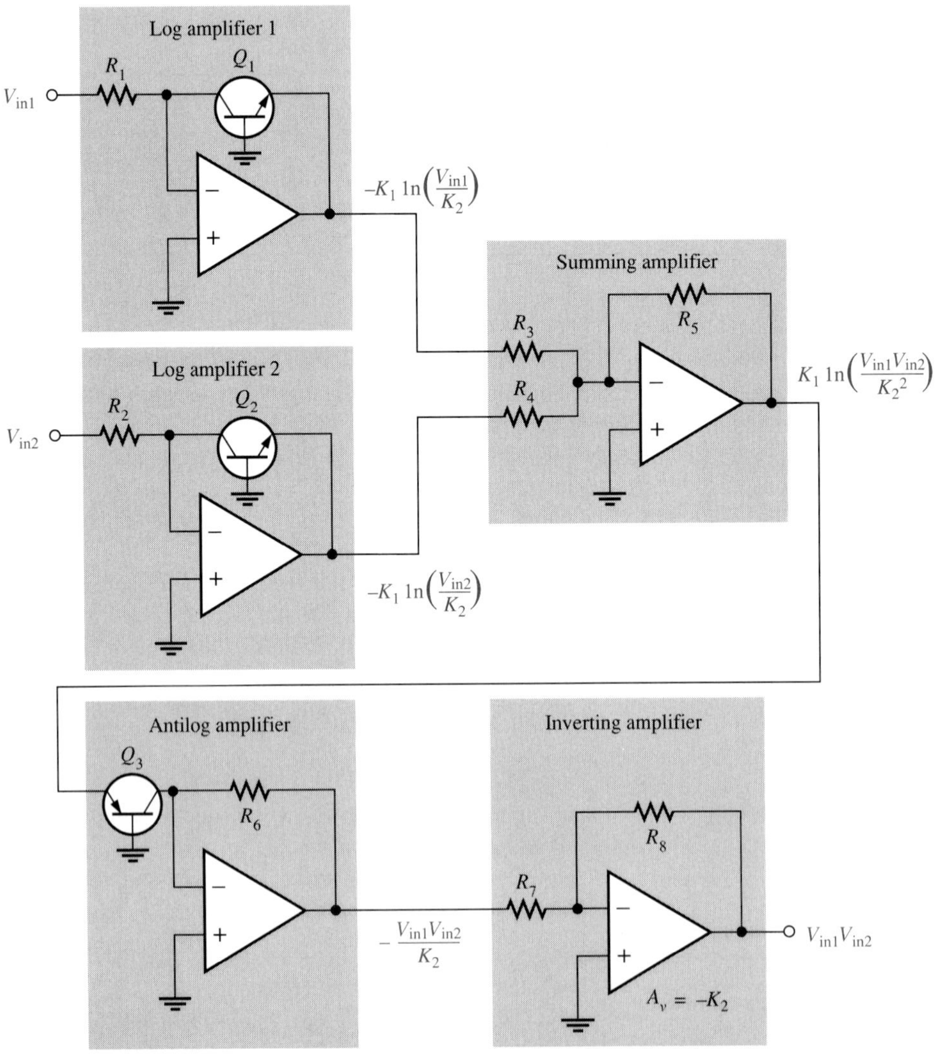

V_1 → Log amplifier → $\ln V_1$ → Summing amplifier (Adder) → $\ln(V_1 V_2)$ → Antilog amplifier → $V_1 V_2$

V_2 → Log amplifier → $\ln V_2$

Log amplifier 1: R_1, Q_1, V_{in1} → $-K_1 \ln\left(\dfrac{V_{in1}}{K_2}\right)$

Log amplifier 2: R_2, Q_2, V_{in2} → $-K_1 \ln\left(\dfrac{V_{in2}}{K_2}\right)$

Summing amplifier: R_3, R_4, R_5 → $K_1 \ln\left(\dfrac{V_{in1} V_{in2}}{K_2{}^2}\right)$

Antilog amplifier: Q_3, R_6 → $-\dfrac{V_{in1} V_{in2}}{K_2}$

Inverting amplifier: R_7, R_8, $A_v = -K_2$ → $V_{in1} V_{in2}$

FIGURE 11–35
A basic multiplier.

Special Amplifiers

where $K_1 = 0.025$ V, $K_2 = RI_{EBO}$, and $R = R_1 = R_2 = R_6$. The two output voltages from the log amplifiers are added and inverted by the unity-gain summing amplifier to produce the following result:

$$V_{out(sum)} = K_1 \ln\left[\left(\frac{V_{in1}}{K_2}\right) + \ln\left(\frac{V_{in2}}{K_2}\right)\right] = K_1 \ln\left(\frac{V_{in1} V_{in2}}{K_2^2}\right)$$

This expression is then applied to the antilog amplifier, and we get the expression for the multiplier output voltage as follows.

$$V_{out(antilog)} = -K_2 \text{antilog}\left(\frac{V_{out(sum)}}{K_1}\right) = -K_2 \text{antilog}\left[\frac{K_1 \ln\left(\frac{V_{in1} V_{in2}}{K_2^2}\right)}{K_1}\right]$$

$$= -K_2\left(\frac{V_{in1} V_{in2}}{K_2^2}\right) = -\frac{V_{in1} V_{in2}}{K_2}$$

As you can see, the output of the antilog amplifier is a constant ($1/K_2$) times the *product* of the input voltages. The final output is developed by an inverting amplifier with a voltage gain of $-K_2$.

$$V_{out} = -K_2\left(-\frac{V_{in1} V_{in2}}{K_2}\right) = V_{in1} V_{in2}$$

EXAMPLE 11–10 Determine the voltages at each indicated point in Figure 11–36 and verify that the output is the product of the two input voltages. Assume the $I_{EBO} = 40$ nA for each transistor.

SOLUTION

$$K_1 = 0.025 \text{ V}$$
$$R = R_1 = R_2 = R_3 = 68 \text{ k}\Omega$$
$$K_2 = RI_{EBO} = (68 \text{ k}\Omega)(40 \text{ nA}) = 0.00272 \text{ V}$$

The output of log amplifier 1 is

$$V_{out(log1)} = -K_1 \ln\left(\frac{V_{in1}}{K_2}\right) = -(0.025 \text{ V})\ln\left(\frac{0.1 \text{ V}}{0.00272 \text{ V}}\right)$$
$$= -(0.025 \text{ V})\ln(36.76) = -0.09 \text{ V}$$

The output of log amplifier 2 is

$$V_{out(log2)} = -K_1 \ln\left(\frac{V_{in2}}{K_2}\right) = -(0.025 \text{ V})\ln\left(\frac{0.12 \text{ V}}{0.00272 \text{ V}}\right)$$
$$= -(0.025 \text{ V})\ln(44.12) = -0.095 \text{ V}$$

The output of the summing amplifier is

$$V_{out(sum)} = -(V_{out(log1)} + V_{out(log2)}) = -[-0.09 \text{ V} + (-0.095 \text{ V})] = 0.185 \text{ V}$$

The output of the antilog amplifier is

$$V_{out(antilog)} = -\left(\frac{V_{in1} V_{in2}}{K_2}\right) = -\left(\frac{(0.1 \text{ V})(0.12 \text{ V})}{0.00272 \text{ V}}\right) = -4.41 \text{ V}$$

FIGURE 11-36

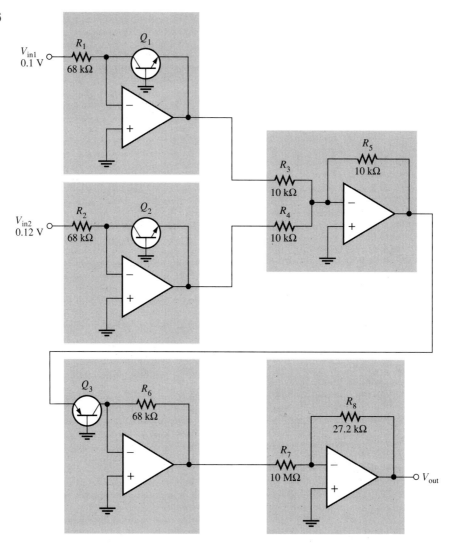

The gain of the inverting amplifier is

$$A_{v(I)} = -\frac{R_8}{R_7} = -\frac{27.2 \text{ k}\Omega}{10 \text{ M}\Omega} = -0.00272$$

The final output of the multiplier is

$$V_{out} = A_v V_{out(antilog)} = -0.00272(-4.41 \text{ V}) = 0.012 \text{ V}$$

To verify that this result is correct, the product of the two input voltages is

$$V_{in1} V_{in2} = (0.1 \text{ V})(0.12 \text{ V}) = 0.012 \text{ V}$$

SPECIAL AMPLIFIERS

PRACTICE EXERCISE 11-10
Would changing R_1, R_2, and R_6 to 100 kΩ require making any other changes in the circuit of Figure 11-36?

AN ANALOG DIVIDER

Dividers are based on the fundamental logarithmic relationship that states that the quotient of two terms equals the difference of the logarithms of each term.

$$\ln\left(\frac{a}{b}\right) = \ln a - \ln b \tag{11-22}$$

As shown in Equation (11-22), this relationship tells us that two signal voltages are effectively divided if the logarithms of the signal voltages are subtracted.

By subtracting the outputs of two log amplifiers, you get the logarithm of the quotient of the two input voltages. Then, by taking the antilogarithm, you get the quotient of the two input voltages.

$$\ln V_1 - \ln V_2 = \ln\left(\frac{V_1}{V_2}\right)$$

$$\text{antilog}\left[\ln\left(\frac{V_1}{V_2}\right)\right] = \frac{V_1}{V_2}$$

Two voltages can be subtracted by applying one of the voltages directly to the input of a summing amplifier and by inverting the other voltage that is to be subtracted from the first before applying it to the second summing amplifier input ($V_1 + (-V_2) = V_1 - V_2$).

Figure 11-37 shows the basic divider circuitry and indicates the expressions for the voltages at various points. The operation is basically the same as the multiplier circuit except that an inverting unity-gain amplifier follows log amplifier 2. This effectively makes the unity-gain summing amplifier a subtracter.

$$V_{out(sum)} = K_1 \ln\left(\frac{V_{in1}}{K_2} - \frac{V_{in2}}{K_2}\right) = K_1 \ln\frac{\left(\frac{V_{in1}}{K_2}\right)}{\left(\frac{V_{in2}}{K_2}\right)} = K_1 \ln\left(\frac{V_{in1}}{V_{in2}}\right)$$

where $K_1 = 0.025$, $K_2 = RI_{EBO}$, and $R = R_1 = R_2 = R_8$. This expression is then applied to the antilog amplifier.

$$V_{out(antilog)} = -K_2 \text{antilog}\left(\frac{V_{out(sum)}}{K_1}\right) = -K_2 \text{antilog}\left[\frac{K_1 \ln\left(\frac{V_{in1}}{V_{in2}}\right)}{K_1}\right]$$

$$= -K_2\left(\frac{V_{in1}}{V_{in2}}\right)$$

As you can see, the output of the antilog amplifier is a constant (K_2) times the *quotient* of the input voltages. The final output is developed by an inverting amplifier with a voltage gain of $-1/K_2$.

$$V_{out} = -\left(\frac{1}{K_2}\right)\left[-K_2\left(\frac{V_{in1}}{V_{in2}}\right)\right] = \frac{V_{in1}}{V_{in2}}$$

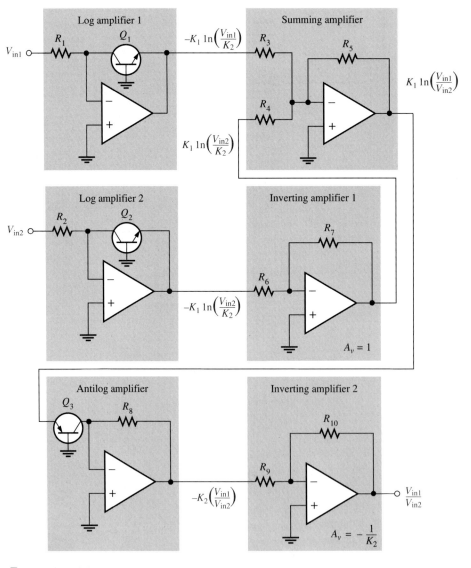

FIGURE 11–37
A basic divider.

11-5 REVIEW QUESTIONS

1. What circuits make up a basic analog multiplier?
2. State the mathematical relationship that is the basis for the analog multiplier.
3. What circuits make up a basic analog divider and how does a divider differ from a multiplier?
4. State the mathematical relationship that is the basis for the analog divider.

11-6 A SYSTEM APPLICATION

The electrocardiograph system, presented at the beginning of the chapter, is a medical instrument used for monitoring heart signals of patients. From the output waveform of an ECG, the doctor can detect abnormalities in the heartbeat. In this section, you will

- *See how an isolation amplifier is used in medical instrumentation.*
- *See how other op-amp circuits are also used as part of the ECG system.*
- *Translate between a printed circuit board and a schematic.*
- *Analyze the isolation amplifier board.*
- *Troubleshoot some common problems.*

A BRIEF DESCRIPTION OF THE SYSTEM

The human heart produces an electrical signal that can be picked up by electrodes in contact with the skin. When the heart signal is displayed on a chart recorder or on a video monitor, it is called an electrocardiograph or ECG. Typically, the heart signal picked up by the electrode is about 1 mV and has significant frequency components from less than 1 Hz to about 100 Hz.

As indicated in the block diagram in Figure 11–38, an ECG system has at least three electrodes. There is a right-arm (RA) electrode, a left-arm (LA) electrode, and a right-leg

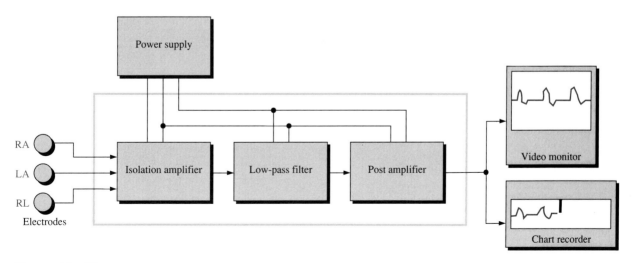

FIGURE 11-38
ECG block diagram.

(RL) electrode that is used as the common. The purposes of the isolation amplifier are to provide for differential inputs from the electrode sensors, provide a high CMR to eliminate the relatively high common-mode noise voltages associated with heart signals, and to provide electrical isolation for protection of the patient. The purpose of the active filter is to reject frequencies above those contained in the heart signal. The purpose of the post amplifier is to provide most of the amplification in the system and to drive a video monitor and/or a chart recorder. All three of of these circuits—the isolation amplifier, the low-pass filter, and the post amplifier—are on a single pc board called the amplifier board.

Now, so that you can take a closer look at the isolation amplifier board, let's take it out of the system and put it on the test bench.

On the Test Bench

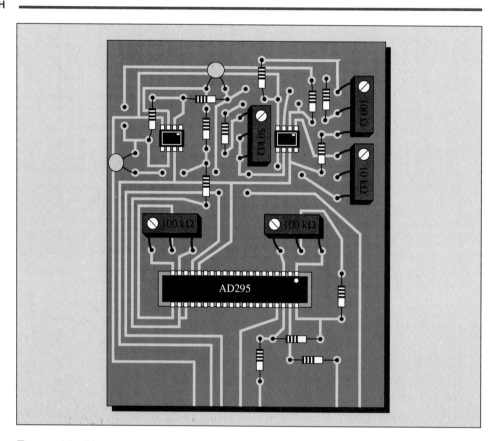

FIGURE 11–39

A Brief Description of the Circuits

The inputs from the electrode sensors come into the amplifier board shown in Figure 11–39 via a shielded cable to prevent noise pickup. The schematic for the amplifier board is shown in Figure 11–40. The shielded cable is basically a twisted pair of wires

SPECIAL AMPLIFIERS

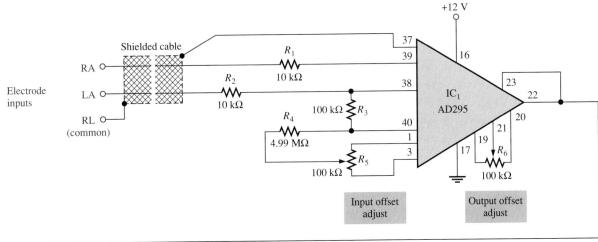

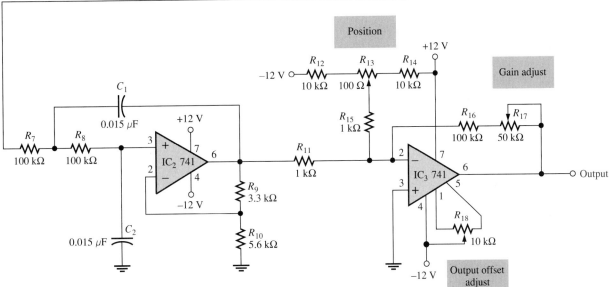

FIGURE 11–40

surrounded by a braided metal sheathing that is covered by an insulated sheathing. The metal shield serves as the conduit for the common connection. The incoming differential signal is amplified by the fixed gain of the AD295 isolation amplifier. The AD295 is housed in a package that is significantly larger than standard IC DIPs or SOPs. The larger configuration is required to accommodate the coupling transformers used in the isolation circuitry. The particular package in this system is a 40-pin package beginning with pin 1 at the dot. Pin 40 is directly across from pin 1.

The low-pass filter is a Sallen-Key two-pole filter, and the post amplifier is an inverting amplifier with adjustable gain. The inverting input of the post amplifier also serves as a summing point for the signal voltage and a dc voltage used for adding a dc level to the output for purposes of adjusting the vertical position of the display.

■ **ACTIVITY 1** **RELATE THE PC BOARD TO THE SCHEMATIC**

Locate and identify each component and each input/output pin on the pc board in Figure 11–39 using the schematic in Figure 11–40. Find and identify the pc connections on the back side of the board. Verify that the board and the schematic agree.

■ **ACTIVITY 2** **ANALYZE THE SYSTEM**

STEP 1 Determine the voltage gain of the isolation amplifier.

STEP 2 Determine the bandwidth of the active filter.

STEP 3 Determine the minimum and maximum voltage gain of the post amplifier.

STEP 4 Determine the overall gain range of the amplifier board.

STEP 5 Determine the voltage range at the wiper of the position adjustment potentiometer.

■ **ACTIVITY 3** **WRITE A TECHNICAL REPORT**

Describe the overall operation of the amplifier board. Specify how each circuit works and what its purpose is. Explain how the gain is adjusted and how the dc level of the output can be changed. Use the results of Activity 2 as appropriate.

■ **ACTIVITY 4** **TROUBLESHOOT THE SYSTEM FOR EACH OF THE FOLLOWING PROBLEMS BY STATING THE PROBABLE CAUSE OR CAUSES**

1. There is no final output voltage when there is a verified 1 mV input signal.
2. There is a 10 mV signal at the output of IC_1, but no signal at pin 3 of IC_2.
3. There is a 15 mV signal at the output of IC_2, but no signal at pin 2 of IC_3.
4. With a valid input signal, IC_3 is being driven into its saturated states and is basically acting as a comparator.

11–6 REVIEW QUESTIONS

1. Which resistors determine the voltage gain of the isolation amplifier?
2. What is the voltage gain of the output section of IC_1?
3. What are the lower and upper critical frequencies of the active filter?
4. What is the output voltage of the post amplifier if R_{17} is set at its midpoint resistance?

SUMMARY

☐ A basic instrumentation amplifier is formed by three op-amps and seven resistors, including the gain-setting resistor, R_G.

☐ An instrumentation amplifier has high input impedance, high CMRR, low output offset, and low output impedance.

☐ The voltage gain of a basic instrumentation amplifier is set by a single external resistor.

☐ An instrumentation amplifier is useful in applications where small signals are embedded in large common-mode noise.

Special Amplifiers

- A basic isolation amplifier has three electrically isolated sections: input, output, and power.
- Most isolation amplifiers use transformer coupling for isolation.
- Isolation amplifiers are used to interface sensitive equipment with high-voltage environments and to provide protection from electrical shock in certain medical applications.
- The operational transconductance amplifier (OTA) is a voltage-to-current amplifier.
- The output current of an OTA is the input voltage times the transconductance.
- In an OTA, transconductance varies with bias current; therefore, the gain of an OTA can be varied with a bias voltage.
- The operation of log and antilog amplifiers is based on the nonlinear (logarithmic) characteristic of a pn junction.
- A log amplifier has a BJT in the feedback loop.
- An antilog amplifier has a BJT in series with the input.
- Logarithmic amplifiers are used for analog multiplication and division.
- An analog multiplier is based on the mathematical principle that states the logarithm of the product of two variables equals the sum of the logarithms of the variables.
- An analog divider is based on the mathematical principle that states the logarithm of the quotient of two variables equals the difference of the logarithms of the variables.

Glossary

Instrumentation Related to an arrangement of instruments for the purpose of monitoring or measuring certain quantities.

Logarithm An exponent; the logarithm of a quantity is the exponent or power to which a given number called the base must be raised in order to equal the quantity.

Natural logarithm The exponent to which the base e ($e = 2.71828$) must be raised in order to equal a given quantity.

Signal compression The process of scaling down the amplitude of a signal voltage.

Trim To precisely adjust or fine tune a value.

Formulas

Instrumentation Amplifier

$$(11\text{--}1) \quad V_{out1} = \left(1 + \frac{R_1}{R_G}\right)V_{in1} - \left(\frac{R_1}{R_G}\right)V_{in2} + V_{cm}$$

$$(11\text{--}2) \quad V_{out2} = \left(1 + \frac{R_2}{R_G}\right)V_{in2} - \left(\frac{R_2}{R_G}\right)V_{in1} + V_{cm}$$

$$(11\text{-}3) \quad V_{out} = \left(1 + \frac{2R}{R_G}\right)(V_{in2} - V_{in1})$$

$$(11\text{-}4) \quad A_{cl} = 1 + \frac{2R}{R_G}$$

$$(11\text{-}5) \quad R_G = \frac{2R}{A_{cl} - 1}$$

$$(11\text{-}6) \quad A_v = \frac{R_S}{R_G}$$

$$(11\text{-}7) \quad C_x = \frac{1}{100\pi f_c}$$

ISOLATION AMPLIFIER

$$(11\text{-}8) \quad A_{v(input)} = 1 + \frac{R_F}{R_1}$$

$$(11\text{-}9) \quad A_{v(output)} = \frac{75\text{ k}\Omega + R_{ext}}{30\text{ k}\Omega}$$

OPERATIONAL TRANSCONDUCTANCE AMPLIFIER (OTA)

$$(11\text{-}10) \quad A = g_m = \frac{I_{out}}{V_{in}}$$

$$(11\text{-}11) \quad g_m = KI_{BIAS}$$

$$(11\text{-}12) \quad I_{BIAS} = \frac{(+V) - (-V) - 0.7\text{ V}}{R_{BIAS}}$$

LOGARITHMIC AMPLIFIER

$$(11\text{-}13) \quad V_{out} = -K \ln(V_{in})$$

$$(11\text{-}14) \quad I_D \cong I_s e^{qV_D/kT}$$

$$(11\text{-}15) \quad V_D = \left(\frac{kT}{q}\right)\ln\left(\frac{I_D}{I_s}\right)$$

$$(11\text{-}16) \quad V_{out} = -\left(\frac{kT}{q}\right)\ln\left(\frac{V_{in}}{I_s R_1}\right)$$

$$(11\text{-}17) \quad V_{out} \cong -(0.025\text{ V})\ln\left(\frac{V_{in}}{I_s R_1}\right)$$

$$(11\text{-}18) \quad I_C = I_{EBO} e^{qV_{BE}/kT}$$

$$(11\text{-}19) \quad V_{out} = -(0.025\text{ V})\ln\left(\frac{V_{in}}{I_{EBO} R_1}\right)$$

$$(11\text{-}20) \quad V_{out} = -R_F I_{EBO} \text{antilog}\left(\frac{V_{in}}{25\text{ mV}}\right)$$

Analog Multipliers and Dividers

$$\ln(a \times b) = \ln a + \ln b \qquad (11-21)$$

$$\ln\left(\frac{a}{b}\right) = \ln a - \ln b \qquad (11-22)$$

Self-Test

1. To make a basic instrumentation amplifier, it takes
   - (a) one op-amp with a certain feedback arrangement
   - (b) two op-amps and seven resistors
   - (c) three op-amps and seven capacitors
   - (d) three op-amps and seven resistors
2. Typically, an instrumentation amplifier has an external resistor used for
   - (a) establishing the input impedance
   - (b) setting the voltage gain
   - (c) setting the current gain
   - (d) for interfacing with an instrument
3. Instrumentation amplifiers are used primarily in
   - (a) high-noise environments
   - (b) medical equipment
   - (c) test instruments
   - (d) filter circuits
4. Isolation amplifiers are used primarily in
   - (a) remote, isolated locations
   - (b) systems that isolate a single signal from many different signals
   - (c) applications where there are high voltages and sensitive equipment
   - (d) applications where human safety is a concern
   - (e) c and d
5. The three sections of a basic isolation amplifier are
   - (a) amplifier, filter, and power
   - (b) input, output, and coupling
   - (c) input, output, and power
   - (d) gain, attenuation, and offset
6. The sections of most isolation amplifiers are connected by
   - (a) copper strips
   - (b) transformers
   - (c) microwave links
   - (d) current loops
7. The characteristic that allows an isolation amplifier to amplify small signal voltages in the presence of much greater noise voltages is its
   - (a) CMRR
   - (b) high gain
   - (c) high input impedance
   - (d) magnetic coupling between input and output
8. The term *OTA* means
   - (a) operational transistor amplifier
   - (b) operational transformer amplifier

(c) operational transconductance amplifier

(d) output transducer amplifier

9. In an OTA, the transconductance is controlled by
   (a) the dc supply voltage
   (b) the input signal voltage
   (c) the manufacturing process
   (d) a bias current

10. The voltage gain of an OTA circuit is set by
    (a) a feedback resistor
    (b) the transconductance only
    (c) the transconductance and the load resistor
    (d) the bias current and supply voltage

11. An OTA is basically a
    (a) voltage-to-current amplifier
    (b) current-to-voltage amplifier
    (c) current-to-current amplifier
    (d) voltage-to-voltage amplifier

12. The operation of a logarithmic amplifier is based on
    (a) the nonlinear operation of an op-amp
    (b) the logarithmic characteristic of a pn junction
    (c) the reverse breakdown characteristic of a pn junction
    (d) the logarithmic charge and discharge of an *RC* circuit

13. If the input to a log amplifier is x, the output is proportional to
    (a) e^x (b) $\ln x$ (c) $\log_{10} x$ (d) $2.3 \log_{10} x$
    (e) a and c   (f) b and d

14. If the input to an antilog amplifier is x, the output is proportional to
    (a) $e^{\ln x}$ (b) e^x (c) $\ln x$ (d) e^{-x}

15. The logarithm of the product of two numbers is equal to the
    (a) sum of the two numbers
    (b) sum of the logarithms of each of the numbers
    (c) difference of the logarithms of each of the numbers
    (d) ratio of the logarithms of the numbers

16. If you subtract $\ln y$ from $\ln x$, you get
    (a) $\ln x / \ln y$ (b) $(\ln x)(\ln y)$ (c) $\ln(x/y)$ (d) $\ln(y/x)$

PROBLEMS

SECTION 11–1 INSTRUMENTATION AMPLIFIERS

1. Determine the voltage gains of op-amps 1 and 2 for the instrumentation amplifier configuration in Figure 11–41.

2. Find the overall voltage gain of the instrumentation amplifier in Figure 11–41.

SPECIAL AMPLIFIERS

FIGURE 11–41

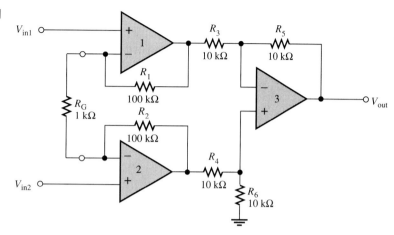

3. The following voltages are applied to the instrumentation amplifier in Figure 11–41: $V_{in1} = 5$ mV, $V_{in2} = 10$ mV, and $V_{cm} = 225$ mV. Determine the final output voltage.

4. What value of R_G must be used to change the gain of the instrumentation amplifier in Figure 11–41 to 1000?

5. What is the voltage gain of the AD521 instrumentation amplifier in Figure 11–42?

6. Determine the bandwidth of the amplifier in Figure 11–42.

7. Specify what you must do to change the gain of the amplifier in Figure 11–42 to approximately 50.

8. Specify what you must do to change the bandwidth of the amplifier in Figure 11–42 to approximately 15 kHz.

FIGURE 11–42

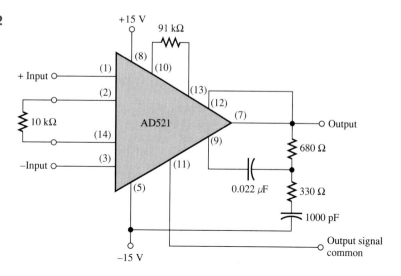

CHAPTER 11

SECTION 11-2 ISOLATION AMPLIFIERS

9. The op-amp in the input section of a certain isolation amplifier has a voltage gain of 30 and the attenuation network has an attenuation of 0.75. The output section is set for a gain of 10. What is the overall voltage gain of this device?

10. Determine the overall voltage gain of each AD295 in Figure 11–43.

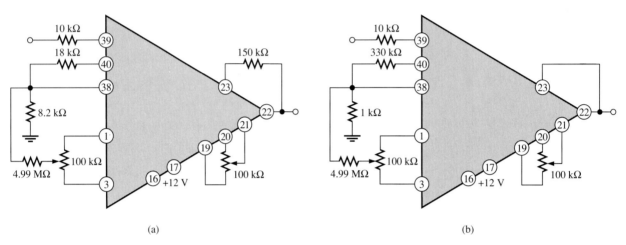

FIGURE 11–43

11. Specify how you would change the overall gain of the amplifier in Figure 11–43(a) to approximately 100 by changing only the gain of the input section.

12. Specify how you would change the overall gain in Figure 11–43(b) to approximately 100 by changing only the gain of the output section.

13. Specify how you would connect each amplifier in Figure 11–43 for unity gain.

SECTION 11-3 OPERATIONAL TRANSCONDUCTANCE AMPLIFIERS (OTAs)

14. A certain OTA has an input voltage of 10 mV and an output current of 10 μA. What is the transconductance?

15. A certain OTA with a transconductance of 5000 μS has a load resistance of 10 kΩ. If the input voltage is 100 mV, what is the output current? What is the output voltage?

16. The output voltage of a certain OTA with a load resistance is determined to be 3.5 V. If its transconductance is 4000 μS and the input voltage is 100 mV, what is the value of the load resistance?

17. Determine the voltage gain of the OTA in Figure 11–44. Use the graph in Figure 11–45.

18. If a 10 kΩ rheostat is added in series with the bias resistor in Figure 11–44, what are the minimum and maximum voltage gains?

FIGURE 11–44

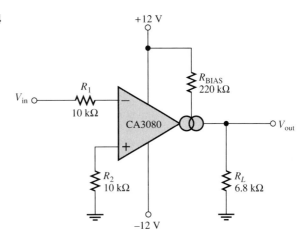

FIGURE 11–45

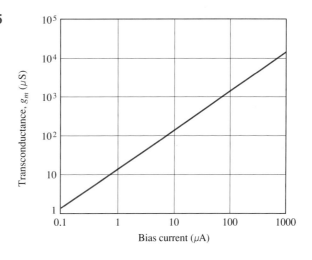

19. The OTA in Figure 11–46 functions as an amplitude modulation circuit. Determine the output voltage waveform for the given input waveforms.

20. Determine the trigger points for the Schmitt-trigger circuit in Figure 11–47.

21. Determine the output voltage waveform for the Schmitt trigger in Figure 11–47 in relation to a 1 kHz sine wave with peak values of ±10 V.

SECTION 11–4 LOG AND ANTILOG AMPLIFIERS

22. Using your calculator, find the natural logarithm (ln) of each of the following numbers:

    (a) 0.5 **(b)** 2 **(c)** 50 **(d)** 130

23. Repeat Problem 22 for $\log_{10}$.

24. What is the antilog of 1.6?

FIGURE 11–46

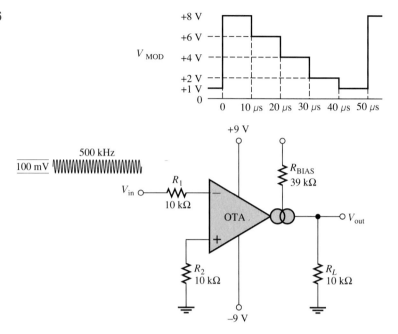

FIGURE 11–47

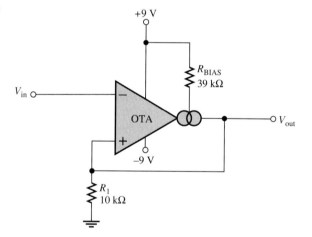

25. Explain why the output of a log amplifier is limited to approximately 0.7 V.

26. What is the output voltage of a certain log amplifier with a diode in the feedback path when the input voltage is 3 V? The input resistor is 82 kΩ and the reverse leakage current is 100 nA.

27. Determine the output voltage for the amplifier in Figure 11–48. Assume $I_{EBO} = 60$ nA.

FIGURE 11–48

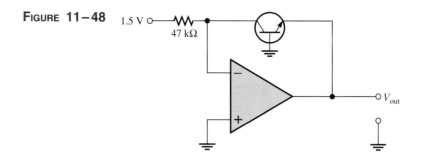

28. Determine the output voltage for the amplifier in Figure 11–49. Assume I_{EBO} = 60 nA.

FIGURE 11–49

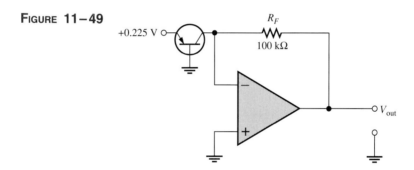

29. Signal compression is one application of logarithmic amplifiers. Suppose an audio signal with a maximum voltage of 1 V and a minimum voltage of 100 mV is applied to the log amplifier in Figure 11–48. What will be the maximum and minimum output voltages? What conclusion can you draw from this result?

Section 11–5 Analog Multipliers and Dividers

30. Determine the voltage at the output of each op-amp in the multiplier circuit of Figure 11–50. Assume I_{EBO} = 40 nA. Is the final output voltage equal to the product of the input voltages?

31. Determine the voltage at the output of each op-amp in the divider circuit of Figure 11–51. Assume I_{EBO} = 50 nA. Is the final output voltage equal to the quotient of the input voltages?

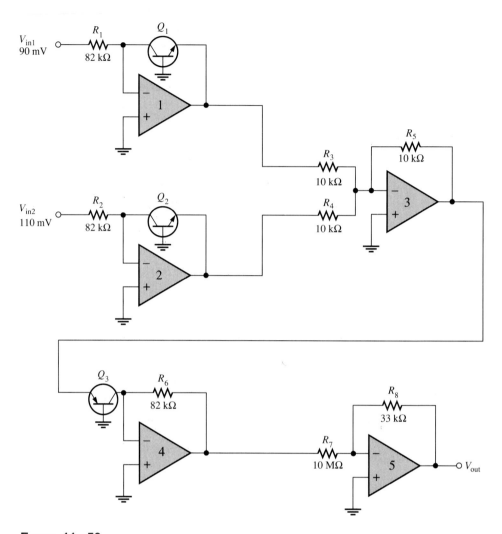

FIGURE 11–50

SPECIAL AMPLIFIERS

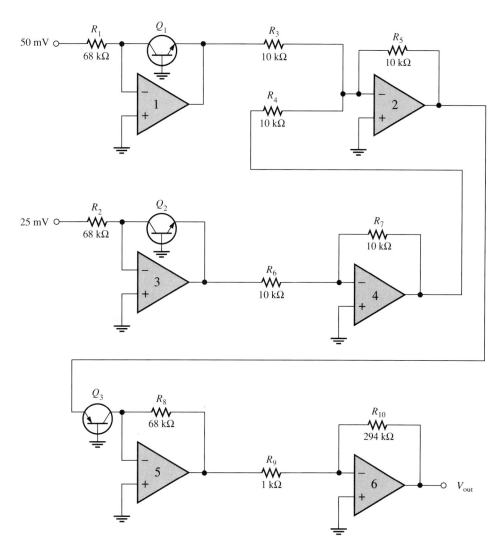

FIGURE 11–51

SECTION 11–6 A SYSTEM APPLICATION

32. With a 1 mV, 50 Hz signal applied to the ECG amplifier board in Figure 11–52, what voltage would you expect to see at each of the probed points? Assume that all offset voltages are nulled out and the position control is adjusted for zero deflection.

33. Repeat Problem 32 for a 2 mV, 1 kHz input signal.

FIGURE 11–52

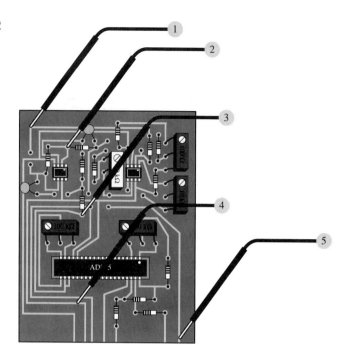

Answers to Review Questions

Section 11–1
1. The main purpose of an instrumentation amplifier is to amplify small signals that occur on large common-mode voltages. The key characteristics are high input impedance, high CMRR, low output impedance, and low output offset.
2. Three op-amps and seven resistors are required to construct a basic instrumentation amplifier.
3. The gain is set by the internal feedback resistors and an external resistor.
4. The gain is greater than unity.

Section 11–2
1. Isolation amplifiers are used in medical equipment, power plant instrumentation, industrial processing, and automated testing.
2. The three sections of an isolation amplifier are input, output, and power.
3. The sections are connected by transformer coupling and in some devices by optical coupling.
4. The oscillator acts as a dc to ac converter so that the dc power can be ac coupled to the input and output sections.

Section 11–3
1. OTA stands for Operational Transconductance Amplifier.
2. Transconductance increases with bias current.

Special Amplifiers

3. Assuming that the bias input is connected to the supply voltage, the voltage gain increases when the supply voltage is increased because this increases the bias current.
4. The gain decreases as the bias voltage decreases.

Section 11–4
1. A diode or transistor in the feedback loop provides the exponential (nonlinear) characteristic.
2. The output of a log amplifier is limited to the barrier potential of the pn junction (about 0.7 V).
3. The output voltage is determined by the input voltage, the input resistor, and the emitter-to-base leakage current.
4. The transistor in an antilog amplifier is in series with the input rather than in the feedback loop.

Section 11–5
1. A multiplier is made of two log amplifiers, a summing amplifier, an antilog amplifier, and an inverting amplifier.
2. $\ln(ab) = \ln a + \ln b$
3. A divider is made of two log amplifiers, a subtracting amplifier (a summing amplifier and an inverter), an antilog amplifier, and an inverting amplifier. The divider has a subtracter instead of an adder.
4. $\ln(a/b) = \ln a - \ln b$

Section 11–6
1. The gain is set by R_2 and R_3.   2. 75 kΩ/30 kΩ = 2.5
3. Lower: 0 Hz; Upper: $1/2\pi RC = 1/2\pi(0.015\ \mu\text{F})(100\ \text{k}\Omega) = 106$ Hz
4. 125 kΩ/1 kΩ = 125

Answers to Practice Exercises

11–1	240 Ω
11–2	Leave $R_S = 100$ kΩ and make $R_G = 2.2$ kΩ, $C_x = 0.00032\ \mu$F.
11–3	Many combinations are possible. Here is one: Connect the input for unity gain (pin 38 to 40, no R_1 or R_F). Connect a 680 kΩ resistor from pin 22 to pin 23.
11–4	Approximately 50 μA
11–5	Yes, the voltage gain will equal 200. (Answers may vary somewhat, depending on how accurately the graph is read.)
11–6	The output is a square-wave modulated signal with a maximum amplitude of 4.69 V and a minimum amplitude of 2.38 V. (Answers may vary somewhat, depending on how accurately the graph is read.)
11–7	−0.167 V 11–8 −0.193 V 11–9 −4.4 V
11–10	Yes. Change R_8 to 40 kΩ.

12

COMMUNICATIONS CIRCUITS

12–1 BASIC RECEIVERS
12–2 THE LINEAR MULTIPLIER
12–3 AMPLITUDE MODULATION
12–4 THE MIXER
12–5 AM DEMODULATION
12–6 IF AND AUDIO AMPLIFIERS
12–7 FREQUENCY MODULATION
12–8 THE PHASE-LOCKED LOOP (PLL)
12–9 A SYSTEM APPLICATION

After completing this chapter, you should be able to

- List the components in an AM and FM receiver system.
- Describe the purpose of each circuit in a receiver.
- Explain multiplier quadrants.
- Define the scale factor of a multiplier.
- Connect a linear multiplier as a squaring circuit.
- Connect a linear multiplier as a dividing circuit.
- Connect a linear multiplier as a square root circuit.
- Connect a linear multiplier as a mean square circuit.
- Show that amplitude modulation is a multiplication process.
- Explain amplitude modulation in terms of the frequency spectra.
- Explain the difference between balanced and standard AM.
- Describe the function of the mixer in a receiver system.
- Define *demodulation*.
- Implement an AM demodulator.
- Discuss the purpose of IF and audio amplifiers in a receiver.
- Describe the process of frequency modulation.
- Explain how a basic FM transmitter works.
- List the components of a phase-locked loop (PLL).
- Discuss the basic operation of a voltage-controlled oscillator (VCO).
- Explain how a PLL can be used as an FM demodulator.

Communications electronics encompasses a wide range of systems, including both analog (linear) and digital. Any system that sends information from one point to another over relatively long distances can be classified as a communications system. Some of the categories of communications systems are radio (broadcast, ham, CB, marine), television, telephony, radar, navigation, satellite, data (digital), and telemetry.

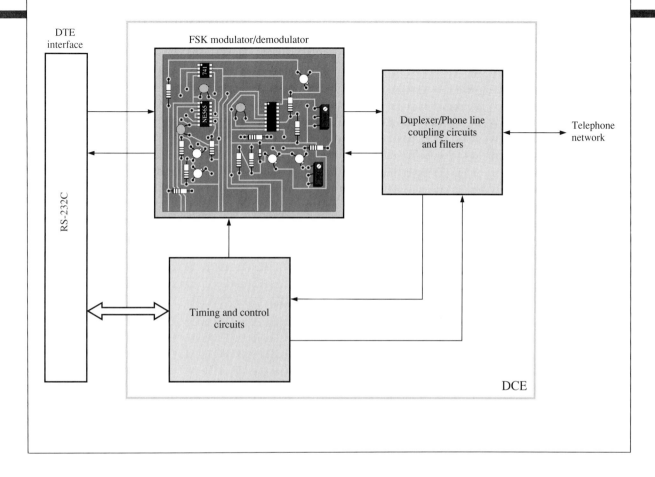

Many communications systems use either amplitude modulation (AM) or frequency modulation (FM) to send information. Other modulation methods include pulse modulation, phase modulation, and frequency shift keying (FSK) as well as more specialized techniques. By necessity, the scope of this chapter is limited and is intended to introduce you to basic AM and FM communications systems and circuits. You will cover communications electronics more thoroughly in another course.

A System Application

Digital data consisting of a series of binary digits (1s and 0s) are commonly sent from one computer to another over the telephone lines. Two voltage levels are used to represent the two types of bits, a high-voltage level and a low-voltage level. The data stream is made up of time intervals when the voltage has a constant high value or a constant low value with very fast transitions from one level to the other. In other words, the data stream contains very low frequencies (constant voltage intervals) and very high frequencies (transitions). Since the telephone system has a bandwidth of approximately 300 Hz to 3000 Hz, it cannot handle the very low and the very high frequencies that make up a typical data stream without losing most of the information. Because of the bandwidth limitation of the telephone system, it is necessary to modify digital data before they are sent out; and one method of doing this is with frequency shift keying (FSK), which is a form of frequency modulation. A simplified block diagram of a digital communications equipment (DCE) system for interfacing digital terminal equipment (DTE), such

553

as a computer, to the telephone network is shown on the previous page. The system FSK-modulates digital data before they are transmitted over the phone line and demodulates FSK signals received from another computer. Because the DTE's basic function is to *mod*ulate and *dem*odulate, it is called a *modem.* Although the modem performs many associated functions, as indicated by the different blocks, in this system application our focus will be on the modulation and demodulation circuit.

For the system application in Section 12–9, in addition to the other topics, be sure you understand

□ The basic operation of a VCO.
□ The basic operation of a PLL.
□ How to use an NE565 PLL.

12–1 BASIC RECEIVERS

Receivers based on the superheterodyne principle are standard in one form or another in most types of communications systems and are found in familiar systems such as standard broadcast radio, stereo, and television. In several of the system applications in previous chapters, we presented the superheterodyne receiver in order to focus on a given circuit; now we cover it from a system viewpoint. This section provides a basic introduction to amplitude and frequency modulation and an overview of the complete AM and FM receiver.

AMPLITUDE MODULATION (AM)

Amplitude modulation (AM) is a method for sending audible information, such as voice and music, by electromagnetic waves that are broadcast through the atmosphere. In AM, the amplitude of a signal with a specific frequency, called the **carrier,** is varied according to a modulating signal or audio signal (such as voice or music), as shown in Figure 12–1. The carrier frequency permits the receiver to be tuned to a specific known frequency. The resulting AM waveform contains the carrier frequency, an upper-side frequency equal to the carrier frequency plus the modulation frequency ($f_c + f_m$), and a lower-side frequency

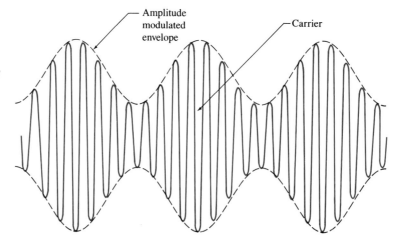

FIGURE 12–1
An example of an amplitude modulated signal. In this case, the higher-frequency carrier is modulated by a lower-frequency sine wave.

equal to $f_c - f_m$. Harmonics of these frequencies are also present. For example, if a 1 MHz carrier is amplitude modulated with a 5 kHz audio signal, the frequency components in the AM waveform are 1 MHz (carrier), 1 MHz + 5 kHz = 1,005,000 Hz (upper side), and 1 MHz − 5 kHz = 995,000 Hz (lower side).

The frequency band for AM broadcast receivers is 540 kHz to 1640 kHz. This means that an AM receiver can be tuned to any frequency in the broadcast band to pick up a specific carrier frequency that lies in the band. Each AM radio station transmits at a specific carrier frequency, so you can tune the receiver to pick up any desired station in your area.

THE SUPERHETERODYNE AM RECEIVER

A block diagram of a superheterodyne AM receiver is shown in Figure 12–2. The receiver consists of an antenna, an RF (radio frequency) amplifier (some AM receivers do not have this), a mixer, a local oscillator (LO), an IF (intermediate frequency) amplifier, a detector, an audio amplifier and a power amplifier, and a speaker.

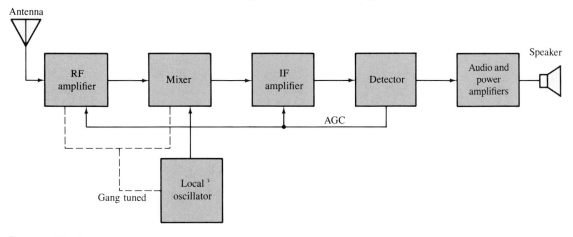

FIGURE 12–2

Superheterodyne AM receiver block diagram.

ANTENNA The antenna picks up all radiated signals and feeds them into the RF amplifier. These signals are very small (usually only a few microvolts).

RF AMPLIFIER This circuit can be adjusted (tuned) to select and amplify any frequency within the AM broadcast band. Only the selected frequency and its two side bands pass through the amplifier.

LOCAL OSCILLATOR This circuit generates a steady sine wave at a frequency 455 kHz above the selected RF frequency.

MIXER This circuit accepts two inputs, the amplitude modulated RF signal from the output of the RF amplifier (or the antenna when there is no RF amplifier) and the sine wave output of the local oscillator. These two signals are then "mixed" by a nonlinear process called *heterodyning* to produce sum and difference frequencies. For example, if the RF carrier has a frequency of 1000 kHz, the LO frequency is 1455 kHz and the sum and difference frequencies out of the mixer are 2455 kHz and 455 kHz, respectively. The difference frequency is always 455 kHz no matter what the RF carrier frequency.

IF Amplifier The input to the IF amplifier is the 455 kHz AM signal, a replica of the original AM carrier signal except that the frequency has been lowered to 455 kHz. The IF amplifier significantly increases the level of this signal.

Detector This circuit recovers the modulating signal (audio signal) from the 455 kHz IF. At this point the IF is no longer needed, so the output of the detector consists of only the audio signal.

Audio and Power Amplifiers This circuit amplifies the detected audio signal and drives the speaker to produce sound.

AGC The automatic gain control (**AGC**) provides a dc level out of the detector that is proportional to the strength of the received signal. This level is fed back to the IF amplifier, and sometimes to the mixer and RF amplifier, to adjust the gains so as to maintain constant signal levels throughout the system over a wide range of incoming carrier signal strengths.

Figure 12–3 shows the signal flow through an AM superheterodyne receiver. The receiver can be tuned to accept any frequency in the AM band. The RF amplifier, mixer, and local oscillator are tuned simultaneously so that the LO frequency is always 455 kHz above the incoming RF signal frequency. This is called *gang tuning*.

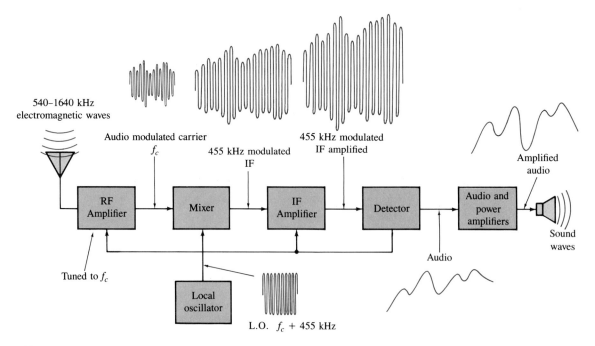

FIGURE 12–3
Illustration of signal flow through an AM receiver.

Frequency Modulation (FM)

In this method of **modulation,** the modulating signal (audio) varies the frequency of a carrier as opposed to the amplitude, as in the case of AM. Figure 12–4 illustrates basic **frequency modulation.** The standard FM broadcast band consists of carrier frequencies from 88 MHz to 108 MHz, which is significantly higher than AM. The FM receiver is similar to the AM receiver in many ways, but there are several differences.

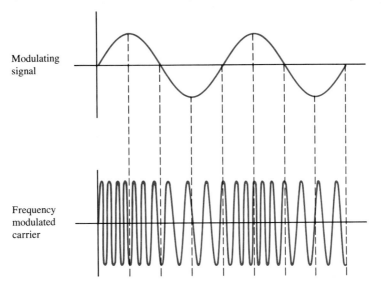

FIGURE 12–4
An example of frequency modulation.

The Superheterodyne FM Receiver

A block diagram of a superheterodyne FM receiver is shown in Figure 12–5. Notice that it includes an RF amplifier, mixer, local oscillator, and IF amplifier just as in the AM receiver. These circuits must, however, operate at higher frequencies than in the AM system. A significant difference in FM is the way the audio signal must be recovered from the modulated IF. This is accomplished by the limiter, discriminator, and deemphasis network. Figure 12–6 depicts the signal flow through an FM receiver.

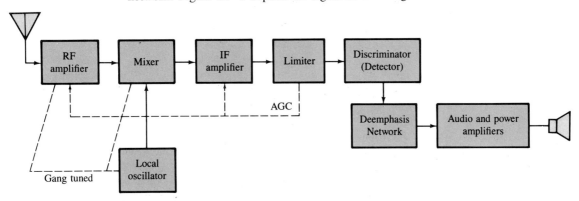

FIGURE 12–5
Superheterodyne FM receiver block diagram.

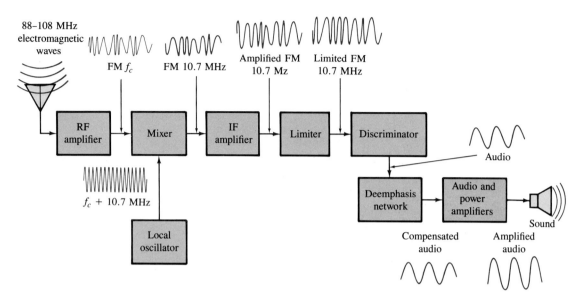

FIGURE 12–6
Example of signal flow through an FM receiver.

RF AMPLIFIER This circuit must be capable of amplifying any frequency between 88 MHz and 108 MHz. It is highly selective so that it passes only the selected carrier frequency and significant side-band frequencies that contain the **audio.**

LOCAL OSCILLATOR This circuit produces a sine wave at a frequency 10.7 MHz above the selected RF frequency.

MIXER This circuit performs the same function as in the AM receiver, except that its output is a 10.7 MHz FM signal regardless of the RF carrier frequency.

IF AMPLIFIER This circuit amplifies the 10.7 MHz FM signal.

LIMITER The limiter removes any unwanted variations in the amplitude of the FM signal as it comes out of the IF amplifier and produces a constant amplitude FM output at the 10.7 MHz intermediate frequency.

DISCRIMINATOR This circuit performs the equivalent function of the detector in an AM system and is often called a detector rather than a discriminator. The **discriminator** recovers the audio from the FM signal.

DEEMPHASIS NETWORK For certain reasons, the higher modulating frequencies are amplified more than the lower frequencies at the transmitting end of an FM system by a process called *preemphasis*. The deemphasis circuit in the FM receiver brings the high-frequency audio signals back to the proper amplitude relationship with the lower frequencies.

AUDIO AND POWER AMPLIFIERS This circuit is the same as in the AM system and can be shared when there is a dual AM/FM configuration.

12–1 REVIEW QUESTIONS

1. What do AM and FM mean?
2. How do AM and FM differ?
3. What are the standard broadcast frequency bands for AM and FM?

12–2 THE LINEAR MULTIPLIER

The linear multiplier is a key circuit in many types of communications systems. In this section, you will examine the basic principles of integrated circuit linear multipliers and look at a few applications that are found in communications as well as other areas. In the following sections, we will concentrate on multiplier applications in AM and FM systems.

MULTIPLIER QUADRANTS

There are one-quadrant, two-quadrant, and four-quadrant **multipliers.** The quadrant classification indicates the number of input polarity combinations that the multiplier can handle. A graphical representation of the quadrants is shown in Figure 12–7. A four-quadrant multiplier can accept any of the four possible input polarity combinations and produce an output with the corresponding polarity.

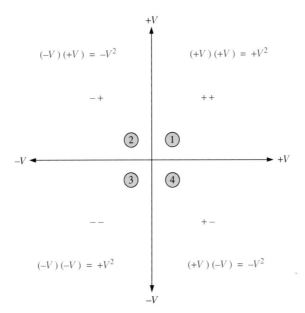

FIGURE 12–7
Four-quadrant polarities and their products.

The Multiplier Transfer Characteristic

Figure 12–8 shows the transfer characteristic for a typical IC linear multiplier. To find the output voltage from the transfer characteristic graph, you find the intersection of the two input voltages V_X and V_Y. Values of V_X run along the horizontal axis and values of V_Y are the sloped lines. The output voltage is found by projecting the point of intersection over to the vertical axis. An example will illustrate this.

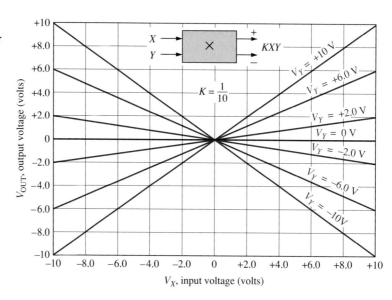

FIGURE 12–8
A four-quadrant multiplier transfer characteristic.

EXAMPLE 12–1

Determine the output voltage for a four-quadrant linear multiplier whose transfer characteristic is given in Figure 12–8. The input voltages are $V_X = -4$ V and $V_Y = +10$ V.

Solution
The output voltage is -4 V as illustrated in Figure 12–9. For this transfer characteristic, the output voltage is a factor of ten smaller than the actual product of the two input voltages. This is due to the *scale factor* of the multiplier, which is discussed next.

Practice Exercise 12–1
Find V_{OUT} if $V_X = -6$ V and $V_Y = +6$ V.

The Scale Factor, K

The scale factor, K, is basically an internal attenuation that reduces the output by a fixed amount. The scale factor on most IC multipliers is adjustable and has a typical value of

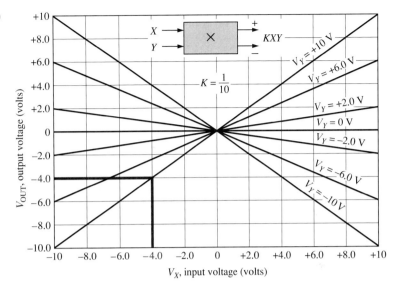

FIGURE 12–9

0.1. Figure 12–10 shows an MC1595 configured as a basic multiplier. The scale factor is determined by external resistors according to the formula

$$K = \frac{2R_L}{R_X R_Y I_2} \qquad (12\text{–}1)$$

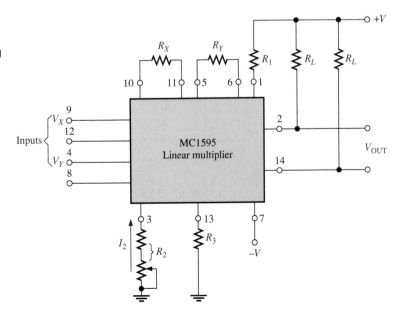

FIGURE 12–10
Basic MC1595 linear multiplier with external circuitry for setting the scale factor.

The current I_2 is set by internal and external parameters according to the formula

$$I_2 = \frac{|-V| - 0.7 \text{ V}}{R_2 + 500 \text{ }\Omega} \quad (12\text{--}2)$$

The potentiometer provides for fine adjustment by controlling I_2.

The expression for the output voltage of the IC linear multiplier includes the scale factor as indicated in Equation (12–3).

$$V_{\text{OUT}} = K V_X V_Y \quad (12\text{--}3)$$

■ **EXAMPLE 12–2** Determine the scale factor for the basic MC1595 multiplier in Figure 12–11. Assume the 5 kΩ potentiometer portion of R_2 is set to 2.5 kΩ. Also, determine the output voltage for the given inputs.

FIGURE 12–11

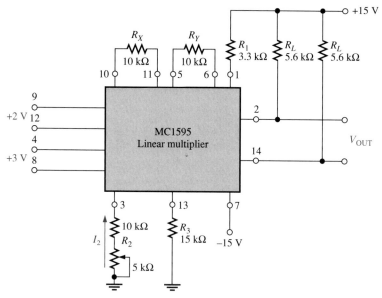

SOLUTION
I_2 is calculated as follows.

$$I_2 = \frac{|-V| - 0.7 \text{ V}}{R_2 + 500 \text{ }\Omega} = \frac{15 \text{ V} - 0.7 \text{ V}}{12.5 \text{ k}\Omega + 500 \text{ }\Omega} = \frac{14.3 \text{ V}}{13 \text{ k}\Omega} = 1.1 \text{ mA}$$

COMMUNICATIONS CIRCUITS

The scale factor is

$$K = \frac{2R_L}{R_X R_Y I_2} = \frac{2(5.6 \text{ k}\Omega)}{(10 \text{ k}\Omega)(10 \text{ k}\Omega)(1.1 \text{ mA})} = 0.1$$

The output voltage is

$$V_{OUT} = KV_X V_Y = 0.1(+2 \text{ V})(+3 \text{ V}) = 0.6 \text{ V}$$

PRACTICE EXERCISE 12-2

What is the output voltage in Figure 12-11 if the 5 kΩ potentiometer is set to its maximum resistance?

OFFSET ADJUSTMENT

Due to internal mismatches, generally small offset voltages are at the inputs and the output of an IC linear multiplier. External circuits to null out the offset voltages are shown in Figure 12-12. The resistive voltage dividers on the inputs allow the actual input voltages to be greater than the recommended maximum for the device. For example, the MC1595 has a maximum input voltage of 5 V. The voltage dividers allow a maximum of 10 V to be applied if the resistors are of equal value. The zener diodes in the input offset adjust circuit keep the inputs on pins 8 and 12 from exceeding the maximum of 5 V.

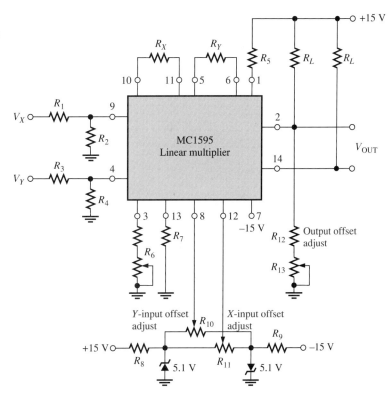

FIGURE 12-12
Basic MC1595 multiplier with both scale factor and offset circuitry.

BASIC APPLICATIONS OF THE MULTIPLIER

Applications of linear multipliers are almost endless. Some of the more basic applications are now presented.

MULTIPLIER The most obvious application of a linear multiplier is, of course, to multiply two voltages as indicated in Figure 12–13.

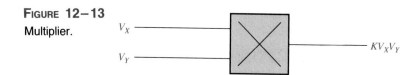

FIGURE 12–13
Multiplier.

SQUARING CIRCUIT A special case of the multiplier is a squaring circuit that is realized by simply applying the same variable to both inputs by connecting the inputs together as shown in Figure 12–14.

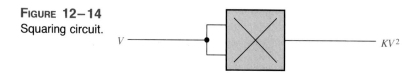

FIGURE 12–14
Squaring circuit.

DIVIDE CIRCUIT The circuit in Figure 12–15 shows the multiplier placed in the feedback loop of an op-amp. The basic operation is as follows. There is a virtual ground at the inverting input of the op-amp and the current from the inverting input is negligible. Therefore, I_1 and I_2 are equal. Since the inverting input is 0 V, the voltage across R_1 is KV_YV_{OUT} and the current through R_1 is

$$I_1 = \frac{KV_YV_{OUT}}{R_1}$$

FIGURE 12–15
Divide circuit.

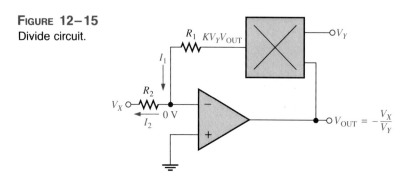

Communications Circuits

The voltage across R_2 is V_X, so the current through R_2 is

$$I_2 = \frac{V_X}{R_2}$$

Since $I_1 = -I_2$,

$$\frac{KV_Y V_{OUT}}{R_1} = -\frac{V_X}{R_2}$$

Solving for V_{OUT},

$$V_{OUT} = -\frac{V_X R_1}{KV_Y R_2}$$

If $R_1 = KR_2$,

$$V_{OUT} = -\frac{V_X}{V_Y}$$

SQUARE ROOT CIRCUIT The square root circuit is a special case of the divide circuit where V_{OUT} is applied to both inputs of the multiplier as shown in Figure 12–16.

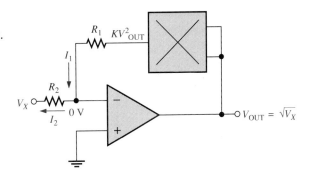

FIGURE 12–16
Square root circuit.

MEAN SQUARE CIRCUIT In this application, the multiplier is used as a squaring circuit with its output connected to an op-amp integrator as shown in Figure 12–17. The integrator produces the average or mean value of the squared input over time, as indicated by the integration sign ($\int$).

FIGURE 12–17
Mean square circuit.

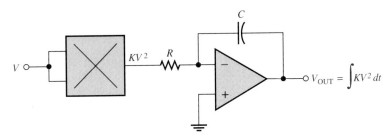

12–2 REVIEW QUESTIONS

1. Compare a four-quadrant multiplier to a one-quadrant multiplier in terms of the inputs that can be handled.
2. If 5 V and 1 V are applied to the inputs of a multiplier and its output is 0.5 V, what is the scale factor? What must the scale factor be for an output of 5 V?
3. How do you convert a basic multiplier to a squaring circuit?

12–3 AMPLITUDE MODULATION

As you learned in Section 12–1, amplitude modulation (AM) is an important method for transmitting information. Of course, the AM superheterodyne receiver is designed to receive transmitted AM signals. In this section we further define amplitude modulation and show how the linear multiplier can be used as an amplitude-modulated device.

As you learned in Section 12–1, amplitude modulation is the process of varying the amplitude of a signal of a given frequency (carrier) with another signal of much lower frequency (modulating signal). The higher-frequency carrier signal is necessary because audio or other signals with relatively low frequencies cannot be transmitted with antennas of a practical size. The basic concept of standard amplitude modulation is visually illustrated in Figure 12–18.

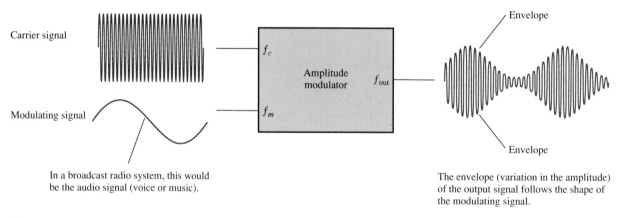

FIGURE 12–18
Basic concept of amplitude modulation.

A MULTIPLICATION PROCESS

If a signal is applied to the input of a variable-gain device, the resulting output is an amplitude-modulated signal because $V_{out} = A_v \times V_{in}$. The output voltage is the input voltage multiplied by the gain. For example, if the gain of an amplifier is made to vary sinusoidally at a certain frequency and an input signal is applied at a higher frequency, the output signal will have the higher frequency. However, its amplitude will vary according to the variation in gain as illustrated in Figure 12–19. Amplitude modulation is basically a multiplication process (input voltage multiplied by a variable gain).

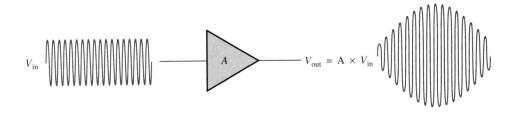

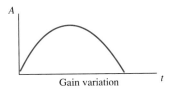

FIGURE 12–19
The amplitude of the output voltage varies according to the gain and is the product of gain and input voltage.

SUM AND DIFFERENCE FREQUENCIES

If the expressions for two sine waves of different frequencies are multiplied mathematically, a term containing both the difference and the sum of the two frequencies is produced. Recall from ac circuit theory that a sine-wave voltage can be expressed as

$$v = V_p \sin 2\pi ft \quad (12\text{–}4)$$

where V_p is the peak voltage and f is the frequency. Two different sine waves can be expressed as follows:

$$v_1 = V_{1(p)} \sin 2\pi f_1 t$$
$$v_2 = V_{2(p)} \sin 2\pi f_2 t$$

Multiplying these two sine-wave terms, we get

$$v_1 v_2 = (V_{1(p)} \sin 2\pi f_1 t)(V_{2(p)} \sin 2\pi f_2 t)$$
$$v_1 v_2 = V_{1(p)} V_{2(p)} (\sin 2\pi f_1 t)(\sin 2\pi f_2 t) \quad (12\text{–}5)$$

The basic trigonometric identity for the product of two sine functions is

$$(\sin A)(\sin B) = \frac{1}{2}[\cos(A - B) - \cos(A + B)] \quad (12\text{–}6)$$

Applying this identity to Equation (12–5),

$$v_1 v_2 = \frac{V_{1(p)} V_{2(p)}}{2}[(\cos 2\pi f_1 t - 2\pi f_2 t) - (\cos 2\pi f_1 t + 2\pi f_2 t)]$$

$$v_1 v_2 = \frac{V_{1(p)} V_{2(p)}}{2}[(\cos 2\pi (f_1 - f_2)t) - (\cos 2\pi (f_1 + f_2)t)]$$

$$v_1 v_2 = \frac{V_{1(p)} V_{2(p)}}{2}\cos 2\pi (f_1 - f_2)t - \frac{V_{1(p)} V_{2(p)}}{2}\cos 2\pi (f_1 + f_2)t \quad (12\text{–}7)$$

You can see in Equation (12–7) that the product of the two sine-wave voltages V_1 and V_2 contains a difference frequency $(f_1 - f_2)$ and a sum frequency $(f_1 + f_2)$. The fact that the product terms are cosine simply indicates a 90° phase shift in the multiplication process.

ANALYSIS OF BALANCED MODULATION

Since amplitude modulation is simply a multiplication process, the preceding analysis is now applied to carrier and modulating signals. The expression for the sinusoidal carrier signal can be written as

$$v_c = V_{c(p)} \sin 2\pi f_c t$$

Assuming a sinusoidal modulating signal, it can be expressed as

$$v_m = V_{m(p)} \sin 2\pi f_m t$$

Multiplying these two signals and then applying the trigonometric identity in Equation (12–6), we get

$$v_c v_m = \frac{V_{c(p)} V_{m(p)}}{2}\cos 2\pi (f_c - f_m)t - \frac{V_{c(p)} V_{m(p)}}{2}\cos 2\pi (f_c + f_m)t \quad (12\text{–}8)$$

An outout signal described by this expression for the product of two sine waves is produced by a linear multiplier. Notice that there is a difference frequency term $(f_c - f_m)$ and a sum frequency term $(f_c + f_m)$, but the original frequencies, f_c and f_m, do not appear alone in the expression. Thus, the product of two sinusoidal signals contains no signal with the carrier frequency, f_c, or with the modulating frequency, f_m. This is called **balanced modulation** because there is no carrier frequency in the output. The carrier frequency is "balanced out."

THE FREQUENCY SPECTRA OF A BALANCED MODULATOR

A graphical picture of the frequency content of a signal is called its frequency spectrum. A frequency spectrum shows voltage on a frequency base rather than on a time base as a waveform diagram does. The frequency spectra of the product of two sine waves as expressed in Equation (12–8) is shown in Figure 12–20. Part (a) shows the two input frequencies and part (b) shows the output frequencies. In communications terminology, the sum frequency is called the *upper-side frequency* and the difference frequency is called the *lower-side frequency* because the frequencies appear on each side of the missing carrier frequency.

COMMUNICATIONS CIRCUITS

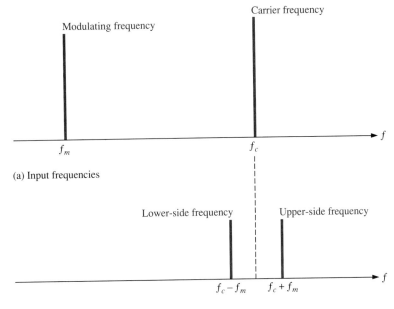

FIGURE 12–20
Illustration of the input and output frequency spectra for a linear multiplier.

THE LINEAR MULTIPLIER AS A BALANCED MODULATOR

As mentioned, the linear multiplier acts as a balanced modulator when a carrier signal and a modulating signal are applied to its inputs as illustrated in Figure 12–21. A balanced modulator produces an upper-side and a lower-side frequency, but it does not produce a carrier frequency. Since there is no carrier signal, balanced modulation is sometimes known as *suppressed-carrier modulation*. Balanced modulation is used in certain types of communications such as single side-band systems, but it is not used in standard AM broadcast systems.

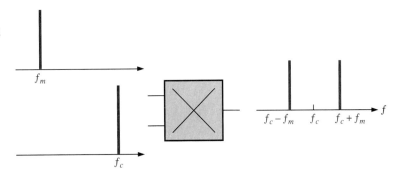

FIGURE 12–21
The linear multiplier as a balanced modulator.

■ **EXAMPLE 12–3** Determine the frequencies contained in the output signal of the balanced modulator in Figure 12–22.

CHAPTER 12

FIGURE 12–22

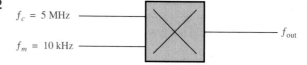

$f_c = 5$ MHz
$f_m = 10$ kHz
f_{out}

SOLUTION
The upper-side frequency is

$$f_c + f_m = 5 \text{ MHz} + 10 \text{ kHz} = 5.01 \text{ MHz}$$

The lower-side frequency is

$$f_c - f_m = 5 \text{ MHz} - 10 \text{ kHz} = 4.99 \text{ MHz}$$

PRACTICE EXERCISE 12–3
Explain how the separation between the side frequencies can be increased using the same carrier frequency.

STANDARD AMPLITUDE MODULATION (AM)

In standard AM systems, the output signal contains the carrier frequency as well as the sum and difference side frequencies. Standard amplitude modulation is illustrated by the frequency spectrum in Figure 12–23.

FIGURE 12–23
The output frequency spectrum of a standard amplitude modulator.

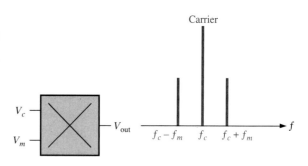

The expression for a standard amplitude-modulated signal is

$$V_{out} = \frac{V_{c(p)}^2}{2}\sin 2\pi f_c t + \frac{V_{c(p)}V_{m(p)}}{2}\cos 2\pi(f_c - f_m)t - \frac{V_{c(p)}V_{m(p)}}{2}\cos 2\pi(f_c + f_m)t \quad (12\text{–}9)$$

Notice in Equation (12–9) that the first term is for the carrier frequency and the other two terms are for the side frequencies. Let's see how the carrier-frequency term gets in the equation.

If a dc voltage equal to the peak of the carrier voltage is added to the modulating signal before the modulating signal is multiplied by the carrier signal, a carrier-signal term appears in the final result as shown in the following steps. Add the peak carrier voltage to the modulating signal, which results in the following expression:

$$V_{c(p)} + V_{m(p)}\sin 2\pi f_m t$$

Multiply by the carrier signal.

$$V_{out} = (V_{c(p)}\sin 2\pi f_c t)(V_{c(p)} + V_{m(p)}\sin 2\pi f_m t)$$
$$= \underbrace{V_{c(p)}^2 \sin 2\pi f_c t}_{\text{carrier term}} + \underbrace{V_{c(p)}V_{m(p)}(\sin 2\pi f_c t)(\sin 2\pi f_m t)}_{\text{product term}}$$

Apply the trigonometric identity in Equation (12–6) to the product term.

$$V_{out} = \frac{V_{c(p)}^2}{2}\sin 2\pi f_c + \frac{V_{c(p)}V_{m(p)}}{2}\cos 2\pi(f_c - f_m)t - \frac{V_{c(p)}V_{m(p)}}{2}\cos 2\pi(f_c + f_m)t$$

This result shows that the output of the multiplier contains a carrier term and two side-frequency terms. Figure 12–24 illustrates how a standard amplitude modulator can be implemented by a summing circuit followed by a linear multiplier. Figure 12–25 shows a possible implementation of the summing circuit.

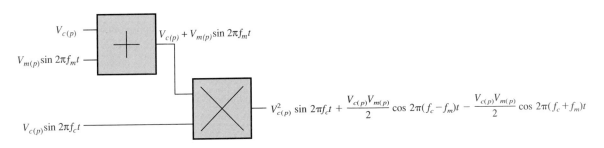

FIGURE 12–24
Basic block diagram of an amplitude modulator.

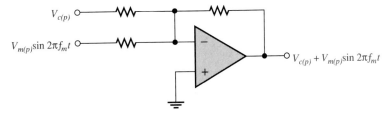

FIGURE 12–25
Implementation of the summing circuit in the amplitude modulator.

EXAMPLE 12–4

A carrier frequency of 1200 kHz is modulated by a sine wave with a frequency of 25 kHz by a standard amplitude modulator. Determine the output frequency spectrum.

SOLUTION
The lower-side frequency is

$$f_c - f_m = 1200 \text{ kHz} - 25 \text{ kHz} = 1175 \text{ kHz}$$

The upper-side frequency is

$$f_c + f_m = 1200 \text{ kHz} + 25 \text{ kHz} = 1225 \text{ kHz}$$

The output contains the carrier frequency and the two side frequencies as shown in Figure 12–26.

FIGURE 12–26

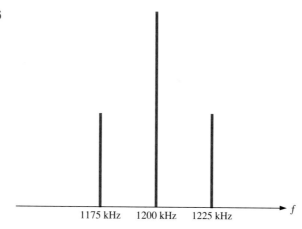

PRACTICE EXERCISE 12–4

Compare the output frequency spectrum in this example to that of a balanced modulator having the same inputs.

AMPLITUDE MODULATION WITH VOICE OR MUSIC

To this point in our discussion, we have considered the modulating signal to be a pure sine wave just to keep things fairly simple. If you receive an AM signal modulated by a pure sine wave in the audio frequency range, you will hear a single tone from the receiver's speaker.

A voice or music signal consists of many sinusoidal components within a range of frequencies from about 20 Hz to 20 kHz. For example, if a carrier frequency is amplitude modulated with voice or music with frequencies from 100 Hz to 10 kHz, the frequency spectrum is as shown in Figure 12–27. Instead of one lower-side and one upper-side

FIGURE 12–27
Example of a frequency spectrum for a voice or music signal.

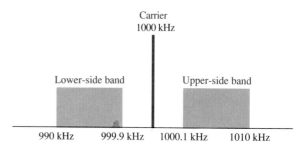

frequency as in the case of a single-frequency modulating signal, a band of lower-side frequencies and a band of upper-side frequencies correspond to the sum and difference frequencies of each sinusoidal component of the voice or music signal.

12–3 REVIEW QUESTIONS

1. What is amplitude modulation?
2. What is the difference between balanced modulation and standard AM?
3. What two input signals are used in amplitude modulation? Explain the purpose of each signal.
4. What are the upper-side frequency and the lower-side frequency?
5. How can a balanced modulator be changed to a standard amplitude modulator?

12–4 THE MIXER

The mixer in the receiver system discussed in Section 12–1 can be implemented with a linear multiplier as you will see in this section. The basic principles of linear multiplication of sine waves are covered, and you will see how sum and difference frequencies are produced. The difference frequency is a critical part of the operation of many types of receiver systems.

The **mixer** is basically a frequency converter because it changes the frequency of a signal to another value. The mixer in a receiver system takes the incoming modulated RF signal (which is sometimes amplified by an RF amplifier and sometimes not) along with the signal from the local oscillator and produces a modulated signal with a frequency equal to the difference of its two input frequencies (RF and LO). The mixer also produces a frequency equal to the sum of the input frequencies. The mixer function is illustrated in Figure 12–28.

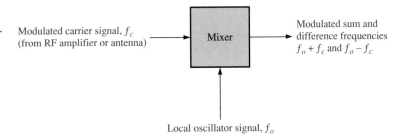

FIGURE 12–28
The mixer function.

THE MIXER IS A LINEAR MULTIPLIER

In the case of receiver applications, the mixer must produce an output that has a frequency component equal to the difference of its input frequencies. From the mathematical analysis in the preceding section, you can see that if two sine waves are multiplied, the product contains the difference frequency and the sum frequency. Thus, the mixer is actually a linear multiplier as indicated in Figure 12–29.

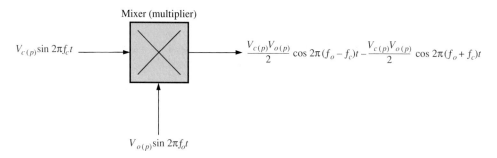

FIGURE 12–29
The mixer as a linear multiplier.

EXAMPLE 12–5

Determine the output expression for a multiplier with one sine-wave input having a peak voltage of 5 mV and a frequency of 1200 kHz and the other input having a peak voltage of 10 mV and a frequency of 1655 kHz.

SOLUTION
The two input expressions are

$$v_1 = (5 \text{ mV})\sin 2\pi(1200 \text{ kHz})t$$
$$v_2 = (10 \text{ mV})\sin 2\pi(1655 \text{ kHz})t$$

Multiplying, we get

$$v_1 v_2 = (5 \text{ mV})(10 \text{ mV})[\sin 2\pi(1200 \text{ kHz})t][\sin 2\pi(1655 \text{ kHz})t]$$

Applying the trigonometric identity in Equation (12–6), we get

$$V_{out} = \frac{(5 \text{ mV})(10 \text{ mV})}{2}\cos 2\pi(1655 \text{ kHz} - 1200 \text{ kHz})t$$
$$- \frac{(5 \text{ mV})(10 \text{ mV})}{2}\cos 2\pi(1655 \text{ kHz} + 1200 \text{ kHz})t$$

$$V_{out} = (25 \ \mu\text{V})\cos 2\pi(455 \text{ kHz})t - (25 \ \mu\text{V})\cos 2\pi(2855 \text{ kHz})t$$

PRACTICE EXERCISE 12–5
What is the value of the peak amplitude and frequency of the difference frequency component in this example?

In the receiver system, both the sum and difference frequencies from the mixer are applied to the IF (intermediate frequency) amplifier. The IF amplifier is actually a tuned amplifier that is designed to respond to the difference frequency while rejecting the sum frequency. You can think of the IF amplifier section of a receiver as a band-pass filter plus an amplifier because it uses resonant circuits to provide the frequency selectivity. This is illustrated in Figure 12–30.

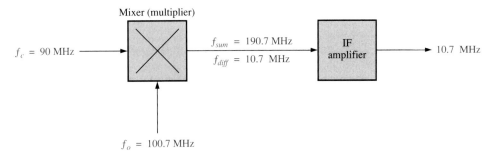

FIGURE 12–30
Example of frequencies in the mixer and IF portion of a receiver.

EXAMPLE 12–6 Determine the output frequency of the IF amplifier for the conditions shown in Figure 12–31.

FIGURE 12–31

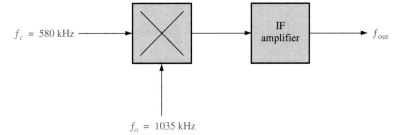

SOLUTION
The IF amplifier produces only the difference frequency signal on its output.

$$f_{\text{out}} = f_{\text{diff}} = f_o - f_c = 1035 \text{ kHz} - 580 \text{ kHz} = 455 \text{ kHz}$$

PRACTICE EXERCISE 12–6
Based on your basic knowledge of the superheterodyne receiver from Section 12–1, determine the IF output frequency when the incoming RF signal changes to 1550 kHz.

12–4 REVIEW QUESTIONS

1. What is the purpose of the mixer in a superheterodyne receiver?
2. How does the mixer produce its output?
3. If a mixer has 1000 kHz on one input and 350 kHz on the other, what frequencies appear on the output?

12-5 AM DEMODULATION

The linear multiplier can be used to demodulate or detect an AM signal as well as to perform the modulation process that was discussed in Section 12–3. **Demodulation** *can be thought of as reverse modulation. The purpose is to get back the original modulating signal (voice or music in the case of standard AM receivers). The detector in the AM receiver can be implemented using a multiplier, although another method using peak envelope detection is common.*

THE BASIC AM DEMODULATOR

An AM demodulator can be implemented with a linear multiplier followed by a low-pass filter, as shown in Figure 12–32. The critical frequency of the filter is the highest audio frequency that is required for a given application (15 kHz, for example).

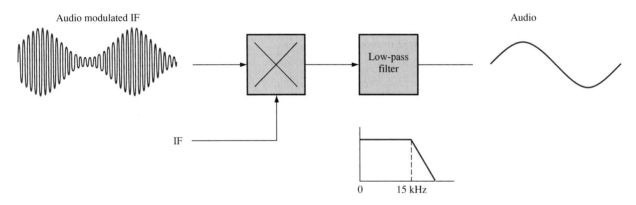

FIGURE 12–32
Basic AM demodulator.

OPERATION IN TERMS OF THE FREQUENCY SPECTRA

Let's assume a carrier modulated by a single tone with a frequency of 10 kHz is received and converted to a modulated intermediate frequency of 455 kHz, as indicated by the frequency spectra in Figure 12–33. Notice that the upper-side and lower-side frequencies are separated from both the carrier and the IF by 10 kHz.

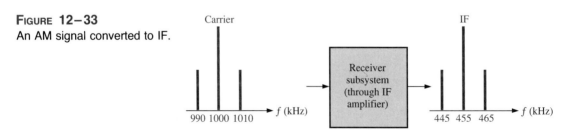

FIGURE 12–33
An AM signal converted to IF.

When the modulated output of the IF amplifier is applied to the demodulator along with the IF, sum and difference frequencies for each input frequency are produced as shown in Figure 12–34. Only the 10 kHz audio frequency is passed by the filter. A drawback to this type of AM detection is that a pure IF must be produced to mix with the modulated IF.

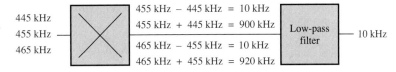

FIGURE 12–34
Example of demodulation.

12–5 REVIEW QUESTIONS

1. What is the purpose of the filter in the linear multiplier demodulator?
2. If a 455 kHz IF modulated by a 1 kHz audio frequency is demodulated, what frequency or frequencies appear on the output of the demodulator?

12–6   IF AND AUDIO AMPLIFIERS

In this section, integrated circuit amplifiers for intermediate and audio frequencies are introduced. A typical IF amplifier is discussed and audio preamplifiers and power amplifiers are covered. As you know, the IF amplifier in a communications receiver provides amplification of the modulated IF signal out of the mixer before it is applied to the detector. After the audio signal is recovered by the detector, it goes to the audio preamp where it is amplified and applied to the power amplifier that drives the speaker.

THE BASIC FUNCTION OF THE IF AMPLIFIER

The IF amplifier in a receiver is a tuned amplifier with a specified bandwidth operating at a center frequency of 455 kHz for AM and 10.7 kHz for FM. The IF amplifier is one of the key features of a superheterodyne receiver because it is set to operate at a single resonant frequency that remains the same over the entire band of carrier frequencies that can be received. Figure 12–35 illustrates the basic function of an IF amplifier in terms of the frequency spectra.

Assume, for example, that the received carrier frequency of $f_c = 1$ MHz is modulated by an audio signal with a maximum frequency of $f_m = 5$ kHz, indicated in Figure 12–35 by the frequency spectrum on the input to the mixer. For this frequency, the local oscillator is at a frequency of

$$f_o = 1 \text{ MHz} + 455 \text{ kHz} = 1.455 \text{ MHz}$$

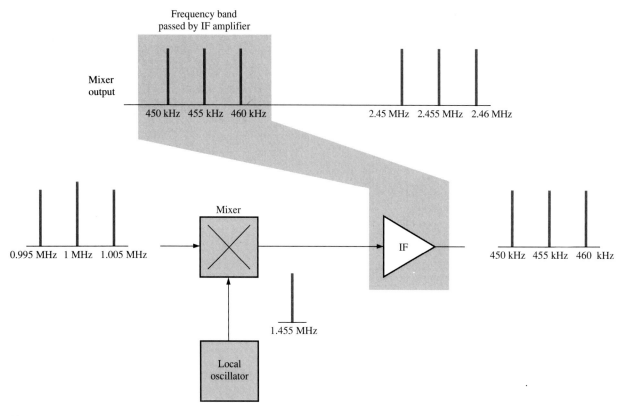

FIGURE 12–35
An illustration of the basic function of the IF amplifier in an AM receiver.

The mixer produces the following sum and difference frequencies as indicated in Figure 12–35.

$$f_o + f_c = 1.455 \text{ MHz} + 1 \text{ MHz} = 2.455 \text{ MHz}$$
$$f_o - f_c = 1.455 \text{ MHz} - 1 \text{ MHz} = 455 \text{ kHz}$$
$$f_o + (f_c + f_m) = 1.455 \text{ MHz} + 1.005 \text{ MHz} = 2.46 \text{ MHz}$$
$$f_o + (f_c - f_m) = 1.455 \text{ MHz} + 0.995 \text{ MHz} = 2.45 \text{ MHz}$$
$$f_o - (f_c + f_m) = 1.455 \text{ MHz} - 1.005 \text{ MHz} = 450 \text{ kHz}$$
$$f_o - (f_c - f_m) = 1.455 \text{ MHz} - 0.995 \text{ MHz} = 460 \text{ kHz}$$

Since the IF amplifier is a frequency-selective circuit, it responds only to 455 kHz and any side frequencies lying in the 10 kHz band centered at 455 kHz. So, all of the frequencies out of the mixer are rejected except the 455 kHz IF, all lower-side frequencies down to 450 kHz, and all upper-side frequencies up to 460 kHz. This frequency spectrum is the audio modulated IF.

A Basic IF Amplifier

Although the detailed circuitry of the IF amplifier may differ from one system to another, it always has a tuned (resonant) circuit on the input or on the output or on both. Figure 12–36(a) shows a basic IF amplifier with tuned transformer coupling at the input and output. The general frequency response curve is shown in Figure 12–36(b).

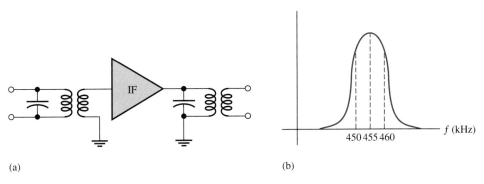

Figure 12–36
A basic IF amplifier with a tuned circuit on the input and output.

The MC1350 This device is representative of integrated circuit IF amplifiers. It can be used in either AM or FM systems and has a typical power gain of 62 dB at 455 kHz. Figure 12–37 shows packaging and a typical circuit diagram for application in an AM receiver. This configuration has a single-tuned transformer-coupled output. The AGC input is normally fed back from the detector in an AM receiver and is used to keep the IF gain at a constant level so that variations in the strength of the incoming RF signal does not cause the audio output to vary significantly. When the AGC voltage increases, the IF gain decreases and when the AGC voltage decreases, the IF gain increases.

Figure 12–37
A typical circuit configuration using the MC1350 IF amplifier.

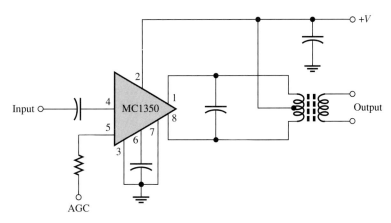

Audio Amplifiers

Audio amplifiers are used in a receiver system following the detector to provide amplification of the recovered audio signal and audio power to drive the speaker(s), as indicated in Figure 12–38. Audio amplifiers typically have bandwidths of 3 kHz to 15 kHz depending on the requirements of the system. Integrated circuit audio amplifiers are available with a range of capabilities.

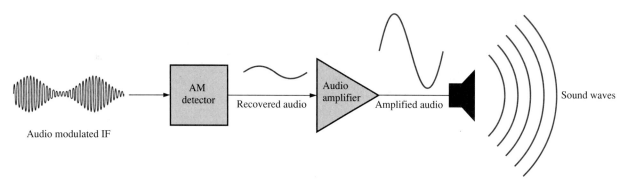

Figure 12–38
The audio amplifier in a receiver system.

The LM386 Audio-Power Amplifier This device is an example of a low-power audio amplifier that is capable of providing several hundred milliwatts to an 8 Ω speaker. It operates from any dc supply voltage in the 4 V to 12 V range. The pin configuration of the LM386 is shown in Figure 12–39(a). The voltage gain of the LM386 is 20 without external connections to the gain terminals, as shown in Figure 12–39(b). A gain of 200 is achieved by connecting a 10 μF capacitor from pin 1 to pin 8, as shown in Figure 12–39(c). Gains between 20 and 200 can be realized by a resistor and capacitor connected

Figure 12–39
Pin configuration and gain connections for the LM386 audio amplifier.

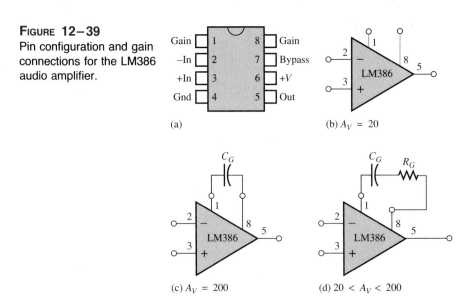

in series from pin 1 to pin 8 as shown in Figure 12–39(d). These external components are effectively placed in parallel with an internal gain-setting resistor.

A typical application of the LM386 as a power amplifier in a radio receiver is shown in Figure 12–40. Here the detected AM signal is fed to the inverting input through the volume control potentiometer R_1 and resistor R_2. C_1 is the input coupling capacitor and C_2 is the power supply decoupling capacitor. R_2 and C_3 filter out any residual RF or IF signal that may be on the output of the detector. R_3 and C_6 provide additional filtering before the audio signal is applied to the speaker through the coupling capacitor C_7.

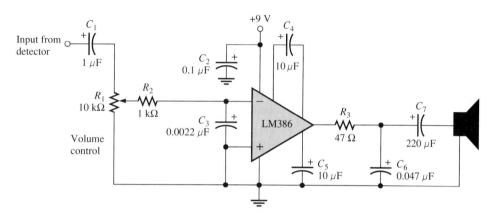

FIGURE 12–40
The LM386 as an AM audio power amplifier.

There are many IC audio amplifiers on the market of which the LM386 is just one example. Others, to mention a few, include the LM388, which is a 1.5 W device; the LM380, which is a 2 W device; and the LM384, which can handle up to 5 W of audio power.

12–6 REVIEW QUESTIONS

1. What is the purpose of the IF amplifier in an AM receiver?
2. What is the center frequency of an AM IF amplifier?
3. Why is the bandwidth of an IF amplifier 10 kHz?
4. Why must the audio amplifier follow the detector in a receiver system?
5. Compare the frequency response of the IF amplifier to that of the audio amplifier.

12–7 FREQUENCY MODULATION

As you have seen, modulation is the process of varying a parameter of a carrier signal with an information signal. Recall that in amplitude modulation the parameter of amplitude is varied. In frequency modulation (FM), the frequency of a carrier is varied above and below its normal or at-rest value by a modulating signal. This section provides a basic introduction to FM and discusses the differences between an AM and an FM receiver.

In a frequency-modulated (FM) signal, the carrier frequency is increased or decreased according to the modulating signal. The amount of deviation above or below the carrier frequency depends on the amplitude of the modulating signal. The rate at which the frequency deviation occurs depends on the frequency of the modulating signal.

Figure 12–41 illustrates both a square wave and a sine wave modulating the frequency of a carrier. The carrier frequency is highest when the modulating signal is at its maximum positive amplitude and is lowest when the modulating signal is at its maximum negative amplitude.

FIGURE 12–41
Examples of frequency modulation.

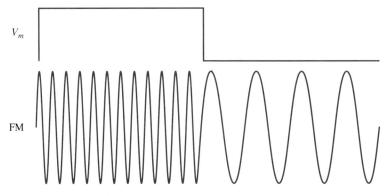

(a) Frequency modulation with a square wave

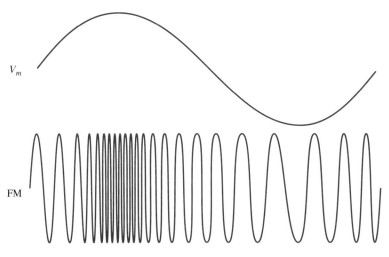

(b) Frequency modulation with a sine wave.

A Basic Frequency Modulator

Frequency modulation is achieved by varying the frequency of an oscillator with the modulating signal. A **voltage-controlled oscillator (VCO)** is typically used for this purpose, as illustrated in Figure 12–42.

FIGURE 12–42

Frequency modulation with a voltage-controlled oscillator.

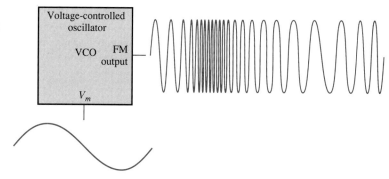

Generally, a variable-reactance type of VCO is used in FM applications. The variable-reactance VCO uses the varactor diode as a voltage-variable capacitance as illustrated in Figure 12–43, where the capacitance is varied with the modulating voltage, V_m.

FIGURE 12–43

Basic variable-reactance VCO.

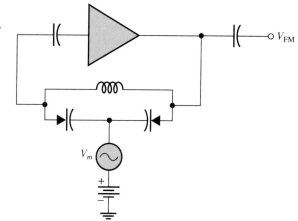

An Integrated Circuit FM Transmitter

An example of a single-chip FM transmitter is the MC2833, which is designed for cordless telephone and other FM communications equipment. The 16-pin package configuration showing the basic functional blocks is in Figure 12–44. The microphone amplifier (mic amp) amplifies the low-level input from a microphone and feeds its output to the variable-reactance circuit which connects to the RF oscillator. The reference voltage circuit (V_{REF}) provides stable bias to the reactance circuit. The two individual transistors can be connected as tuned amplifiers to boost the power output.

FIGURE 12–44
The MC2833 FM modulator.

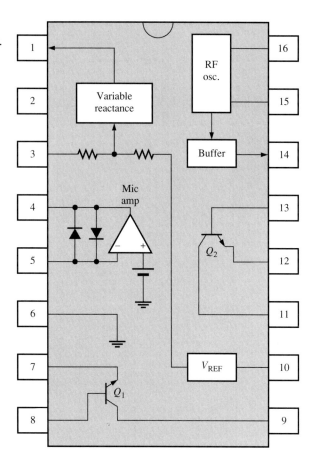

A typical VHF narrow-band FM transmitter using the MC2833 appropriate external circuitry is shown in Figure 12–45. This particular implementation has an output frequency of 49.7 MHz. The frequency of the oscillator is set by the external 16.5667 MHz crystal. The reactance circuit, of course, deviates this frequency with the amplified audio input to produce an FM signal. The 16.5667 MHz output of the oscillator goes to the buffer and is then applied to the input of Q_2, which is operated as a frequency tripler (16.5667 MHz $\times$ 3 = 49.7 MHz). A frequency tripler is basically a class C amplifier with the output tuned to a resonant frequency equal to three times the input. The signal from the Q_2 frequency tripler goes to the input of Q_1, which functions as a linear amplifier that drives the transmitting antenna from its resonant output circuit.

FM DEMODULATION

Except for the higher frequencies, the standard broadcast FM receiver is basically the same as the AM receiver up through the IF amplifier. The main difference between an FM receiver and an AM receiver is the method used to recover the audio signal from the modulated IF.

COMMUNICATIONS CIRCUITS

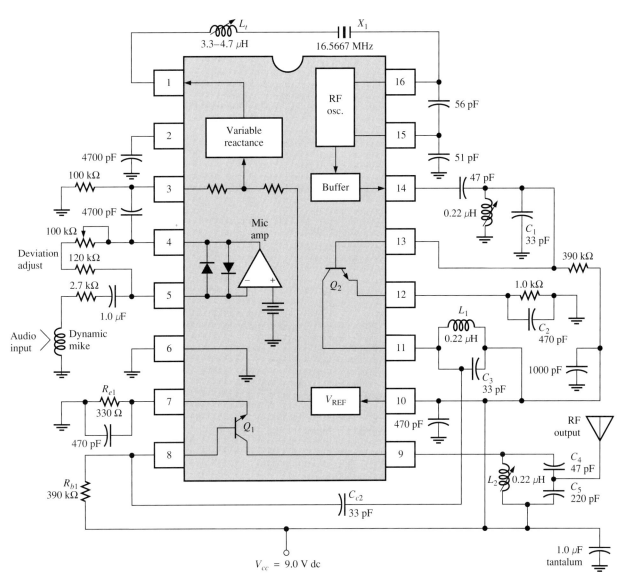

FIGURE 12–45
The MC2833 connected as a 49.7 MHz FM transmitter.

There are several methods for demodulating an FM signal. These include slope detection, phase-shift discrimination, ratio detection, quadrature detection, and phase-locked loop demodulation. Most of these methods are covered in detail in communications courses. However, because of its importance in many types of applications, we will cover the phase-locked loop (PLL) demodulation in the next section.

12-7 Review Questions

1. How does an FM signal carry information?
2. What does VCO stand for?
3. On what principle are most VCOs used in FM based?

12-8 The Phase-Locked Loop (PLL)

In the last section, the PLL was mentioned as a way to demodulate an FM signal. In addition to FM demodulation, PLLs are used in a wide variety of communications applications, which include TV receivers, tone decoders, telemetry receivers, modems, and data synchronizers, to name a few. Many of these applications are covered in an electronic communications course. In fact, entire books have been written on the finer points of PLL operation, analysis, and applications. The approach in this section is intended only to present the basic concept and give you an intuitive idea of how PLLs work and how they are used in FM demodulation. A specific PLL integrated circuit is also introduced.

The Basic PLL Concept

The **phase-locked loop (PLL)** is a feedback circuit consisting of a phase detector, a low-pass filter, and a voltage-controlled oscillator (VCO). Some PLLs also include an amplifier in the loop, and in some applications the filter is not used.

The PLL is capable of locking onto or synchronizing with an incoming signal. When the phase of the incoming signal changes, indicating a change in frequency, the phase detector's output increases or decreases just enough to keep the VCO frequency the same as the frequency of the incoming signal. A basic PLL block diagram is shown in Figure 12-46.

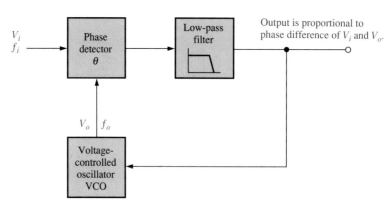

FIGURE 12-46
Basic PLL block diagram.

COMMUNICATIONS CIRCUITS

The general operation of the PLL is as follows. The phase detector compares the phase difference between the incoming signal, V_i, and the VCO signal, V_o. When the frequency of the incoming signal, f_i, is different from that of the VCO frequency, f_o, the phase angle between the two signals is also different. The output of the phase detector and the filter is proportional to the phase difference of the two signals. This proportional voltage is fed to the VCO, forcing its frequency to move toward the frequency of the incoming signal until the two frequencies are equal. At this point, the PLL is locked onto the incoming frequency. If f_i changes, the phase difference also changes, forcing the VCO to track the incoming frequency.

THE PHASE DETECTOR

The phase-detector circuit in a PLL is basically a linear multiplier. The following analysis illustrates how it works in a PLL application. The incoming signal, V_i, and the VCO signal, V_o, applied to the phase detector can be expressed as

$$v_i = V_i \sin(2\pi f_i t + \theta_i)$$
$$v_o = V_o \sin(2\pi f_o t + \theta_o)$$

where θ_i and θ_o are the relative phase angles of the two signals. The phase detector multiplies these two signals and produces a sum and difference frequency output, V_d, as follows.

$$V_d = V_i \sin(2\pi f_i t + \theta_i) \times V_o \sin(2\pi f_o t + \theta_o)$$
$$= \frac{V_i V_o}{2}\cos[(2\pi f_i t + \theta_i) - (2\pi f_o t + \theta_o)] - \frac{V_i V_o}{2}\cos[(2\pi f_i t + \theta_i) + (2\pi f_o t + \theta_o)]$$

When the PLL is locked,

$$f_i = f_o$$

and

$$2\pi f_i t = 2\pi f_o t$$

Therefore, the detector output voltage is

$$V_d = \frac{V_i V_o}{2}[\cos(\theta_i - \theta_o) - \cos(4\pi f_i t + \theta_i + \theta_o)] \quad (12\text{--}10)$$

The second cosine term in Equation (12–10) is a second harmonic term ($2 \times 2\pi f_i t$) and is filtered out by the low-pass filter. The voltage on the output of the filter is expressed as

$$V_c = \frac{V_i V_o}{2}\cos\theta_e \quad (12\text{--}11)$$

where $\theta_e = \theta_i - \theta_o$. θ_e is called the *phase error*. The filter output voltage is proportional to the phase difference between the incoming signal and the VCO signal and is used as the control voltage for the VCO. This operation is illustrated in Figure 12–47.

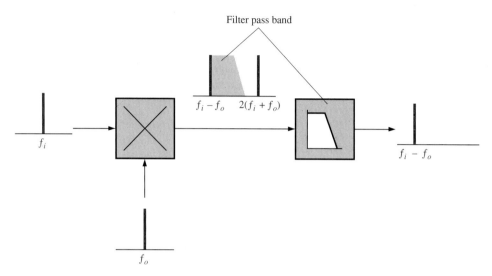

FIGURE 12–47
Basic phase detector/filter operation.

■ **EXAMPLE 12–7** A PLL is locked onto an incoming signal with a frequency of 1 MHz at a phase angle of 50°. The VCO signal is at a phase angle of 20°. The peak amplitude of the incoming signal is 0.5 V and that of the VCO output signal is 0.7 V.
(a) What is the VCO frequency?
(b) What is the value of the control voltage being fed back to the VCO at this point?

SOLUTION
(a) Since the PLL is in lock, $f_i = f_o = $ 1 MHz.
(b)
$$\theta_e = \theta_i - \theta_o = 50° - 20° = 30°$$
$$V_c = \frac{V_i V_o}{2} \cos \theta_e = \frac{(0.5 \text{ V})(0.7 \text{ V})}{2} \cos 30° = (0.175 \text{ V})\cos 30° = 0.152 \text{ V}$$

PRACTICE EXERCISE 12–7
If the phase angle of the incoming signal changes instantaneously to 30°, indicating a change in frequency, what is the instantaneous VCO control voltage?

THE VOLTAGE-CONTROLLED OSCILLATOR (VCO)

Voltage-controlled oscillators can take many forms. A VCO can be some type of *LC* or crystal oscillator as was shown in the last section or it can be some type of *RC* oscillator or multivibrator. No matter the exact type, most VCOs employed in PLLs operate on the principle of *variable reactance* using the varactor diode as a voltage-variable capacitor.

Recall from Chapter 2 that the capacitance of a varactor diode varies inversely with reverse-bias voltage. The capacitance decreases as reverse voltage increases and vice versa.

Communications Circuits

In a PLL, the control voltage fed back to the VCO is applied as a reverse-bias voltage to the varactor diode within the VCO. The frequency of oscillation is inversely related to capacitance by the formula

$$f_o = \frac{1}{2\pi RC}$$

for an *RC* type oscillator and

$$f_o = \frac{1}{2\pi\sqrt{LC}}$$

for an *LC* type oscillator. These formulas show that frequency increases as capacitance decreases and vice versa.

Capacitance decreases as reverse voltage (control voltage) increases. Therefore, an increase in control voltage to the VCO causes an increase in frequency and vice versa. Basic VCO operation is illustrated in Figure 12–48. The graph in part (b) shows that at the nominal control voltage, $V_{c(nom)}$, the oscillator is running at its nominal or free-running frequency, f_o. An increase in V_c above the nominal value forces the oscillator frequency to increase, and a decrease in V_c below the nominal value forces the oscillator frequency to decrease. There are, of course, limits on the operation as indicated by the minimum and maximum points. The transfer function or conversion gain, K, of the VCO is normally expressed as a certain frequency deviation per unit change in control voltage.

$$K = \frac{\Delta f_o}{\Delta V_c} \qquad (12\text{–}12)$$

FIGURE 12–48
Basic VCO operation.

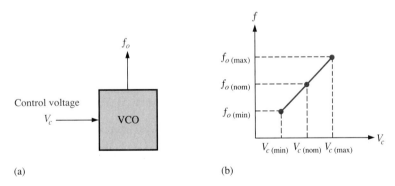

■ **EXAMPLE 12–8** The output frequency of a certain VCO changes from 50 kHz to 65 kHz when the control voltage increases from 0.5 V to 1 V. What is the conversion gain?

SOLUTION

$$K = \frac{\Delta f_o}{\Delta V_c} = \frac{65 \text{ kHz} - 50 \text{ kHz}}{1 \text{ V} - 0.5 \text{ V}} = \frac{15 \text{ kHz}}{0.5 \text{ V}} = 30 \text{ kHz/V}$$

PRACTICE EXERCISE 12-8

If the conversion gain of a certain VCO is 20 kHz/V, how much frequency deviation does a change in control voltage from 0.8 V to 0.5 V produce? If the VCO frequency is 250 kHz at 0.8 V, what is the frequency at 0.5 V?

BASIC PLL OPERATION

When the PLL is locked, the incoming frequency, f_i, and the VCO frequency, f_o, are equal. However, there is always a phase difference between them called the *static phase error*. The phase error, θ_e, is the parameter that keeps the PLL locked in. As you have seen, the filtered voltage from the phase detector is proportional to θ_e (Equation 12–11). This voltage controls the VCO frequency and is always just enough to keep $f_o = f_i$.

Figure 12–49 shows two sinusoidal signals of the same frequency but with a phase difference, θ_e. For this condition the PLL is in lock and the VCO control voltage is constant. If f_i decreases, θ_e increases to θ_{e1} as illustrated in Figure 12–50. This increase in θ_e is sensed by the phase detector causing the VCO control voltage to decrease, thus decreasing f_o until $f_o = f_i$ and keeping the PLL in lock. If f_i increases, θ_e decreases to θ_{e1} as illustrated in Figure 12–51. This decrease in θ_e causes the VCO control voltage to increase, thus increasing f_o until $f_o = f_i$ and keeping the PLL in lock.

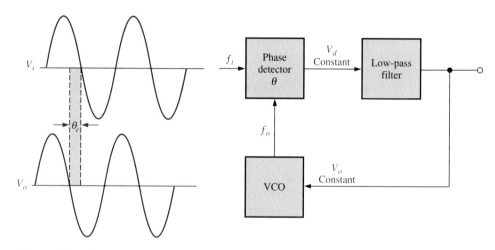

FIGURE 12-49
PLL in lock under static condition ($f_o = f_i$ and constant θ_e).

LOCK RANGE Once the PLL is locked, it will track frequency changes in the incoming signal. The range of frequencies over which the PLL can maintain lock is called the *lock* or *tracking range*. Limitations on the hold-in range are the maximum frequency

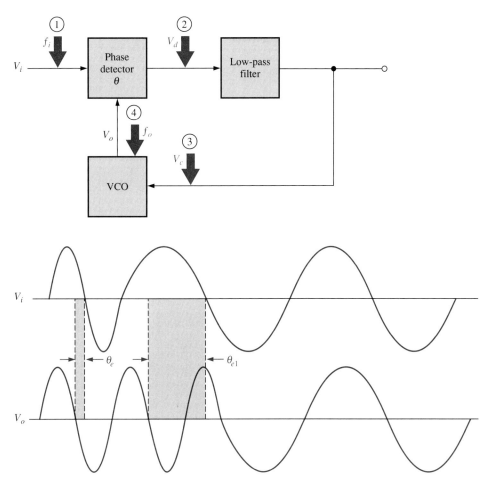

FIGURE 12–50
PLL action when f_i decreases.

deviations of the VCO and the output limits of the phase detector. The hold-in range is independent of the bandwidth of the low-pass filter because, when the PLL is in lock, the difference frequency $(f_i - f_o)$ is zero or a very low instantaneous value that falls well within the bandwidth. The hold-in range is usually expressed as a percentage of the VCO frequency.

CAPTURE RANGE Assuming the PLL is not in lock, the range of frequencies over which it can acquire lock with an incoming signal is called the *capture range*. Two basic conditions are required for a PLL to acquire lock. First, the difference frequency $(f_o - f_i)$ must be low enough to fall within the filter's bandwidth. This means that the incoming frequency must not be separated from the nominal or free-running frequency of the VCO

FIGURE 12–51
PLL action when f_i increases.

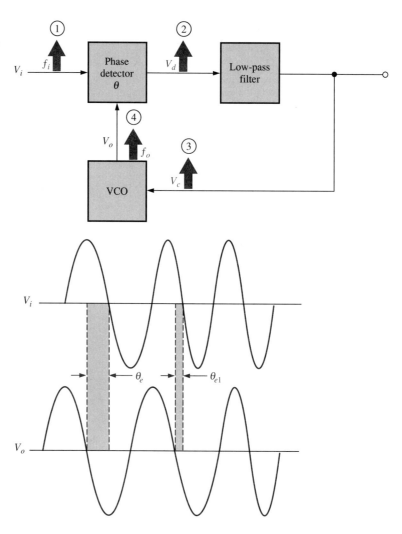

by more than the bandwidth of the low-pass filter. Second, the maximum deviation, Δf_{max}, of the VCO frequency must be sufficient to allow f_o to increase or decrease to a value equal to f_i. These conditions are illustrated in Figure 12–52; and when they exist, the PLL will "pull" the VCO frequency toward the incoming frequency until $f_o = f_i$.

THE NE565 PHASE-LOCKED LOOP

The NE565 is a good example of an integrated circuit PLL. The circuit consists of a VCO, phase detector, a low-pass filter formed by an internal resistor and an external capacitor, and an amplifier. The free-running VCO frequency can be set with external components. A block diagram is shown in Figure 12–53. The NE565 can be used for the frequency range from 0.001 Hz to 500 kHz.

COMMUNICATIONS CIRCUITS

FIGURE 12–52
Illustration of the conditions for a PLL to acquire lock.

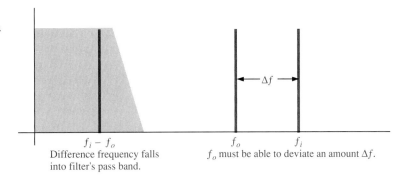

(a)

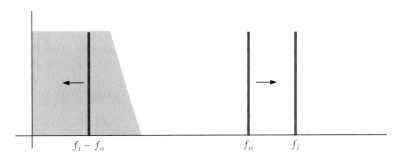

(b) $f_i - f_o$ decreases as f_o deviates towards f_i.

FIGURE 12–53
Block diagram of the NE565 PLL.

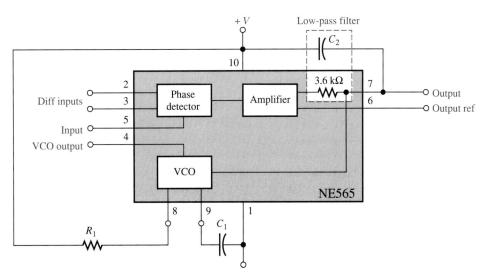

The free-running frequency of the VCO is set by the values of R_1 and C_1 in Figure 12–53 according to the following formula. The result is in hertz when the resistance is in ohms and the capacitance is in farads.

$$f_o \cong \frac{1.2}{4R_1C_1} \tag{12–13}$$

The lock range is given by

$$f_{lock} = \pm\frac{8f_o}{V_{CC}} \tag{12–14}$$

where V_{CC} is the sum of the positive and negative supply voltages.

The capture range is given by

$$f_{cap} \cong \pm\frac{1}{2\pi}\sqrt{\frac{2\pi f_{lock}}{(3600)C_2}} \tag{12–15}$$

The 3600 is the value of the internal filter resistor in ohms. You can see that the capture range is dependent on the filter bandwidth as determined by the internal resistor and the external capacitor C_2.

THE PLL AS AN FM DEMODULATOR

As you have seen, the VCO control voltage in a PLL depends on the deviation of the incoming frequency. The PLL will produce a voltage proportional to the frequency of the incoming signal which, in the case of FM, is the original modulating signal.

Figure 12–54 shows a typical connection for the NE565 as an FM demodulator. If the IF input is frequency modulated by a sine wave, we get a sine wave on the output as

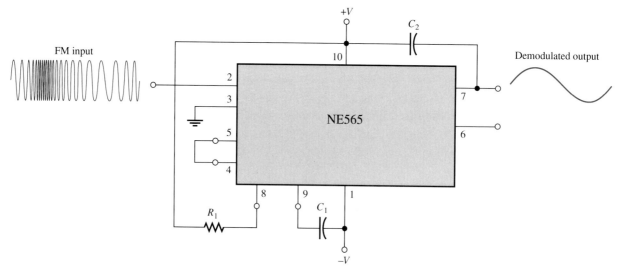

FIGURE 12–54
The NE565 as an FM demodulator.

Communications Circuits

indicated. Since the maximum operating frequency is 500 kHz, this device must be used in double-conversion FM receivers. A double-conversion FM receiver is one in which essentially two mixers are used to first convert the RF to a 10.7 MHz IF and then convert this to a 455 kHz IF.

The free-running frequency of the VCO is adjusted to approximately 455 kHz, which is the center of the modulated IF range. C_1 can be any value, but R_1 should be in the range from 2 kΩ to 20 kΩ. The input can be directly coupled as long as there is no dc voltage difference between pins 2 and 3. The VCO is connected to the phase detector by an external wire between pins 4 and 5. The capacitor between pins 7 and 8 is for eliminating possible oscillations. C_2 is the filter capacitor.

EXAMPLE 12–9

Determine the values for R_1, C_1, and C_2 for the NE565 in Figure 12–54 for a free-running frequency of 455 kHz and a capture range of ± 10 kHz. The dc supply voltages are ± 6 V.

SOLUTION

C_1 is calculated by using Equation (12–13). Choose $R_1 = 4.7$ kΩ.

$$f_o \cong \frac{1.2}{4R_1C_1}$$

$$C_1 \cong \frac{1.2}{4R_1f_o} = \frac{1.2}{4(4700 \;\Omega)(455 \times 10^3 \;\text{Hz})} = 140 \times 10^{-12} \;\text{F} = 140 \;\text{pF}$$

The lock range and capture range must be determined before C_2 can be calculated. The lock range is

$$f_{\text{lock}} = \pm\frac{8f_o}{V_{CC}} = \pm\frac{8(455 \;\text{kHz})}{12} = \pm 303 \;\text{kHz}$$

The capture range is

$$f_{\text{cap}} \cong \pm\frac{1}{2\pi}\sqrt{\frac{2\pi f_{\text{lock}}}{(3600)C_2}}$$

$$f_{\text{cap}}^2 \cong \left(\frac{1}{2\pi}\right)^2 \frac{2\pi f_{\text{lock}}}{(3600)C_2}$$

Therefore,

$$C_2 \cong \left(\frac{1}{2\pi}\right)^2 \frac{2\pi f_{\text{lock}}}{(3600)f_{\text{cap}}^2} = \left(\frac{1}{2\pi}\right)^2 \frac{2\pi (303 \times 10^3 \;\text{Hz})}{(3600)(10 \times 10^3 \;\text{Hz})^2} = 0.134 \times 10^{-6} \;\text{F} = 0.134 \;\mu\text{F}$$

PRACTICE EXERCISE 12–9

What can you do to increase the capture range from ± 10 kHz to ± 15 kHz?

12-8 REVIEW QUESTIONS

1. List the three basic components in a phase-locked loop.
2. What is another circuit used in some PLLs other than the three listed in Question 1?
3. What is the basic function of a PLL?
4. What is the difference between the lock range and the capture range of a PLL?
5. Basically, how does a PLL track the incoming frequency?

12-9 A SYSTEM APPLICATION

The DCE (data communications equipment) system introduced at the opening of this chapter includes an FSK (frequency shift keying) modem (modulator/demodulator). FSK is one method for modulating digital data for transmission over voice phone lines and is basically a form of frequency modulation. In this system application, the focus is on the low-speed modulator/demodulator board, which is implemented with a VCO for transmitting FSK signals and a PLL for receiving FSK signals. In this section, you will

- *See how a VCO and a PLL can be used in a communications system.*
- *See how FSK is used to send digital information over phone lines.*
- *Translate between a printed circuit board and a schematic.*
- *Analyze the modem circuitry.*
- *Troubleshoot some common problems.*

A Brief Description of the System

The FSK modem interfaces a computer with the telephone network so that digital data, which are incompatible with the phone system because of bandwidth limitations, can be transmitted and received over regular phone lines, thus allowing two computers to communicate with each other. Figure 12–55 shows a diagram of a simple data communications system in which a modem at each end of the phone line provides interfacing for a computer.

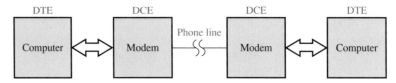

FIGURE 12–55
A data communications system.

The modem (DCE) consists of three basic functional blocks as shown in Figure 12–56: the FSK modulator/demodulator circuits, the phone line interface circuits, and the timing and control circuits. The dual polarity power supply is not shown. Although the focus of this system application is the FSK modulator/demodulator board, we will briefly look at each of the other parts to give you a basic idea of the overall system function. A complete and thorough treatment of the entire modem is beyond the scope of this limited coverage.

COMMUNICATIONS CIRCUITS

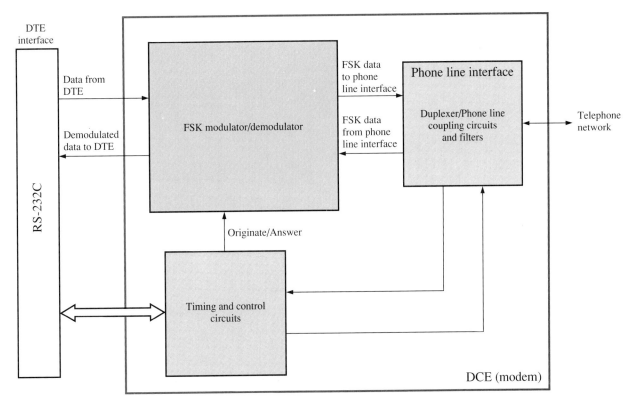

FIGURE 12–56
Basic block diagram of a modem.

THE PHONE LINE INTERFACE The main purposes of this circuitry are to couple the phone line to the modem by proper impedance matching, to provide necessary filtering, and to accommodate full-duplex transmission of data. *Full-duplex* means essentially that information can be going both ways on a single phone line at the same time. This allows a computer, connected to a modem, to be sending data and receiving data simultaneously without the transmitted data interfering with the received data. Full-duplexing is implemented by assigning the transmitted data one bandwidth and the received data another separate bandwidth within the 300 Hz to 3 kHz overall bandwidth of the phone network.

TIMING AND CONTROL One basic function of the timing and control circuits is to determine the proper mode of operation for the modem. The two modes are the originate mode and the answer mode. Another function is to provide a standard interface (such as RS-232C) with the DTE (computer). The RS-232C standard requires certain defined command and control signals, data signals, and voltage levels for each signal. We will not cover the RS-232C interface standard in detail here.

DIGITAL DATA Before we get into FSK, let's briefly review digital data. You may refer to Chapter 7; however, a detailed knowledge of binary numbers is not necessary for this system application. Information is represented in digital form by 1s and 0s, which are the binary digits or bits. In terms of voltage waveforms, a 1 is generally represented by a high level and a 0 by a low level. A stream of serial data consists of a sequence of bits as illustrated by an example in Figure 12–57(a).

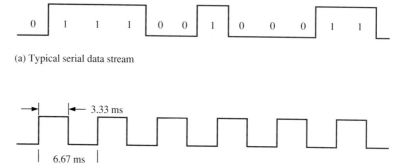

FIGURE 12–57
A serial stream of digital data.

(a) Typical serial data stream

(b) Period for 300 baud square wave

BAUD RATE A low-speed modem, such as the one we are focusing on, sends and receives digital data at a rate of 300 bits/s or 300 baud. For example, if we have an alternating sequence of 1s and 0s (highs and lows), as indicated in Figure 12–57(b), each bit takes 3.33 ms. Since it takes two bits, a 1 and a 0, to make up the period of this particular waveform, the fundamental frequency of this format is 1/6.67 ms = 150 Hz. This is the maximum frequency of a 300 baud data stream because normally there may be several consecutive 1s and/or several consecutive zeros in a sequence, thus reducing the frequency. As mentioned earlier, the telephone network has a 300 Hz minimum frequency response, so the fundamental frequency of the 300 baud data stream will fall outside of the telephone bandwidth. This prevents sending digital data in its pure form over the phone lines.

FREQUENCY-SHIFT KEYING (FSK) FSK is one method used to overcome the bandwidth limitation of the telephone system so that digital data can be sent over the phone lines. The basic idea of FSK is to represent 1s and 0s by two different frequencies within the telephone bandwidth. By the way, any frequency within the telephone bandwidth is an audible tone. The standard frequencies for a full-duplex 300 baud modem in the originate mode are 1070 Hz for a 0 (called a space) and 1270 Hz for a 1 (called a mark). In the answer mode, 2025 Hz is a 0 and 2225 Hz is a 1. The relationship of these FSK frequencies and the telephone bandwidth is illustrated in Figure 12–58. Signals in both the originate and answer bands can exist at the same time on the phone line and not interfere with each other because of the frequency separation.

An example of a digital data stream converted to FSK by a modem is shown in Figure 12–59.

Communications Circuits

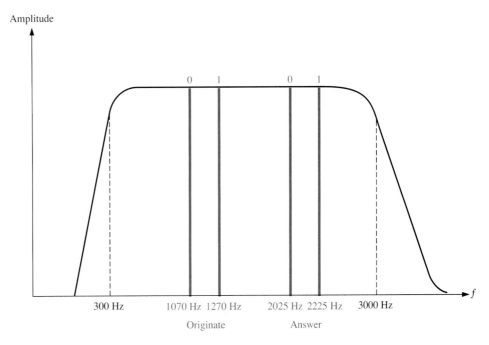

Figure 12–58
Frequencies for 300 baud, full-duplex data transmission.

Figure 12–59
Example of FSK data.

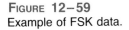

(a) Data stream

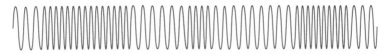

(b) Corresponding FSK signal (frequency relationships are not exact)

MODULATOR/DEMODULATOR CIRCUIT OPERATION

The FSK modulator/demodulator circuits, shown in Figure 12–60, contain an NE565 PLL and a VCO integrated circuit. The VCO can be a device such as the 4046 (not covered specifically in this chapter), which is a PLL device in which the VCO portion can be used by itself because all of the necessary inputs and outputs are available. The VCO in the NE565 cannot be used independent of the PLL because there is no input pin for the control voltage.

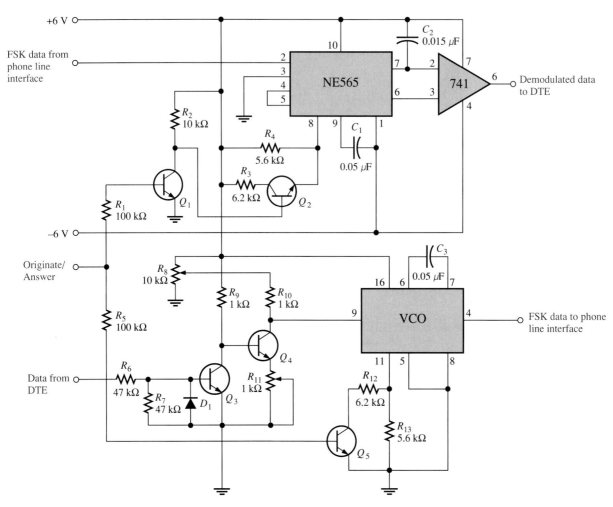

FIGURE 12–60
FSK modulator/demodulator circuit.

The function of the VCO is to accept digital data from a DTE and provide FSK modulation. The VCO is always the transmitting device. The digital data come in on the control voltage input (pin 9) of the VCO via a level-shifting circuit formed by Q_3 and Q_4. This circuit is used because the data from the RS-232C interface are dual polarity. Potentiometer R_8 is for adjusting the high level of the control voltage and R_{11} is for adjusting the low level for the purpose of fine tuning the frequency. Transistor Q_5 provides for originate/answer mode frequency selection by changing the value of the frequency-selection resistance from pin 11 to ground. Transistors Q_1 and Q_2 perform a similar function for the PLL.

When the digital data are at low levels, corresponding to 0s, the VCO oscillates at 1070 Hz in the originate mode and 2025 Hz in the answer mode. When the digital data

COMMUNICATIONS CIRCUITS

are at high levels, corresponding to 1s, the VCO oscillates at 1270 Hz in the originate mode and 2225 Hz in the answer mode. An example of the originate mode is when a DTE issues a request for data and transmits that request to another DTE. An example of the answer mode is when the receiving DTE responds to a request and sends data back to the originating DTE.

The function of the PLL is to accept incoming FSK-modulated data and convert it to a digital data format for use by the DTE. The PLL is always a receiving device. When the modem is in the originate mode, the PLL is receiving answer-mode data from the other modem. When the modem is in the answer mode, the PLL is receiving originate-mode data from the other modem. The 741 is connected as a comparator that changes the data levels from the PLL to a dual-polarity format for compatibility with the RS-232C interface.

Now, so that you can take a closer look at the FSK modulator/demodulator board, let's take it out of the system and put it on the test bench.

ON THE TEST BENCH

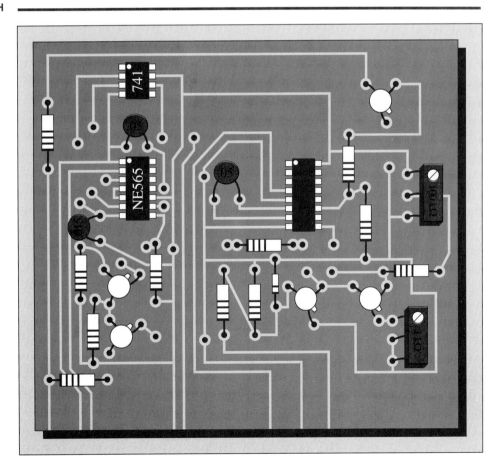

FIGURE 12–61

■ **ACTIVITY 1**   **RELATE THE PC BOARD TO THE SCHEMATIC**

Locate and identify each component and each input/output pin on the pc board in Figure 12–61 using the schematic in Figure 12–60. Find and identify the pc connections on the back side of the board. Verify that the board and the schematic agree.

■ **ACTIVITY 2**   **ANALYZE THE CIRCUITS**

For this application, the free-running frequencies of both the PLL and the VCO circuits are determined by the formula in Equation (12–13).

STEP 1   Verify that the free-running frequency for the PLL IC is approximately 1070 Hz in the originate mode and approximately 1270 Hz in the answer mode.

STEP 2   Repeat Step 1 for the VCO.

STEP 3   Determine the approximate minimum and maximum output voltages for the 741 comparator.

STEP 4   Determine the maximum high-level voltage on pin 9 of the VCO.

STEP 5   If a 300 Hz square wave that varies from +5 V to −5 V is applied to the data from the DTE input, what should you observe on pin 4 of the VCO?

STEP 6   When the data from the DTE are low, pin 9 of the VCO is at approximately 0 V. At this level, the VCO oscillates at 1070 Hz or 2025 Hz. When the data from the DTE go high, to what value should the voltage at pin 9 be adjusted to produce a 1270 Hz or 2225 Hz frequency if the transfer gain of the VCO is 50 Hz/V?

■ **ACTIVITY 3**   **WRITE A TECHNICAL REPORT**

Describe the overall operation of the FSK modem board. Specify how each circuit works and what its purpose is. Identify the function of each component. Use the results of Activity 2 as appropriate.

■ **ACTIVITY 4**   **TROUBLESHOOT THE SYSTEM FOR EACH OF THE FOLLOWING PROBLEMS BY STATING THE PROBABLE CAUSE OR CAUSES**

1. There is no demodulated data output voltage when there are verified FSK data from the phone line interface.
2. The NE565 properly demodulates 1070 Hz and 1270 Hz FSK data but does not properly demodulate 2025 Hz and 2225 Hz data.
3. The VCO produces no FSK output.
4. The VCO produces a continuous 1070 Hz tone in the originate mode and a continuous 2025 Hz tone in the answer mode when there are proper data from the DTE.

12–9 REVIEW QUESTIONS

1. The originate/answer input to the modem is *low*. In what mode is the system?
2. What is the purpose of diode D_1?
3. The VCO is transmitting 1070 Hz and 1270 Hz FSK signals. To what frequencies do the PLL respond?

4. If the VCO is transmitting a constant 2225 Hz tone, what does this correspond to in terms of digital data? In what mode is the modem?

Summary

- In amplitude modulation (AM), the amplitude of a higher-frequency carrier signal is varied by a lower-frequency modulating signal (usually an audio signal).
- A basic superheterodyne AM receiver consists of an RF amplifier (not always), a mixer, a local oscillator, an IF (intermediate frequency) amplifier, an AM detector, and audio and power amplifiers.
- The IF in a standard AM receiver is 455 kHz.
- The AGC (automatic gain control) in a receiver tends to keep the signal strength constant within the receiver to compensate for variations in the received signal.
- In frequency modulation (FM), the frequency of a carrier signal is varied by a modulating signal.
- A superheterodyne FM receiver is basically the same as an AM receiver except that it requires a limiter to keep the IF amplitude constant, a different kind of detector or discriminator, and a deemphasis network. The IF is 10.7 MHz.
- A four-quadrant linear multiplier can handle any combination of voltage polarities on its inputs.
- Amplitude modulation is basically a multiplication process.
- The multiplication of sine waves produces sum and difference frequencies.
- The output spectrum of a balanced modulator includes upper-side and lower-side frequencies, but no carrier frequency.
- The output spectrum of a standard amplitude modulator includes upper-side and lower-side frequencies and the carrier frequency.
- A linear multiplier is used as the mixer in receiver systems.
- A mixer converts the RF signal down to the IF signal. The radio frequency varies over the AM or FM band. The intermediate frequency is constant.
- One type of AM demodulator consists of a multiplier followed by a low-pass filter.
- The audio and power amplifiers boost the output of the detector or discriminator and drive the speaker.
- A voltage-controlled oscillator (VCO) produces an output frequency that can be varied by a control voltage. Its operation is based on a variable reactance.
- A VCO is a basic frequency modulator when the modulating signal is applied to the control voltage input.
- A phase-locked loop (PLL) is a feedback circuit consisting of a phase detector, a low-pass filter, a VCO, and sometimes an amplifier.
- The purpose of a PLL is to lock onto and track incoming frequencies.
- A linear multiplier can be used as a phase detector.

- ☐ A modem is a modulator/demodulator.
- ☐ DTE stands for digital terminal equipment.
- ☐ DCE stands for digital communications equipment.

Glossary

AGC Automatic gain control.

Amplitude modulation (AM) A communication method in which a lower-frequency signal modulates (varies) the amplitude of a higher-frequency signal (carrier).

Audio Related to the range of frequencies that can be heard by the human ear and generally considered to be in the 200 Hz to 20 kHz range.

Balanced modulation A form of amplitude modulation in which the carrier is suppressed; sometimes known as *suppressed-carrier modulation*.

Carrier The high frequency (RF) signal that carries modulated information in AM, FM, and other communications systems.

Demodulation The process in which the information signal is recovered from the IF carrier signal; the reverse of modulation.

Discriminator A type of FM demodulator.

Frequency modulation (FM) A communication method in which a lower-frequency signal modulates (varies) the frequency of a higher-frequency signal.

Mixer A device for down-converting frequencies in a receiver system.

Modulation The process in which a signal containing information is used to modify the amplitude, frequency, or phase of a much higher-frequency signal called the carrier.

Multiplier A linear device that produces an output voltage proportional to the product of two input voltages.

Phase-locked loop (PLL) A device for locking onto and tracking the frequency of an incoming signal.

Voltage-controlled oscillator (VCO) An oscillator for which the output frequency is dependent on a controlling input voltage.

Formulas

(12–1) $K = \dfrac{2R_L}{R_X R_Y I_2}$ Multiplier scale factor for MC1595

(12–2) $I_2 = \dfrac{|-V| - 0.7 \text{ V}}{R_2 + 500 \, \Omega}$ Scale factor current for MC1595

(12–3) $V_{out} = K V_X V_Y$ Multiplier output

(12–4) $v = V_p \sin 2\pi f t$ Sinusoidal voltage

(12–5) $v_1 v_2 = V_{1(p)} V_{2(p)} (\sin 2\pi f_1 t)(\sin 2\pi f_2 t)$ Product of two sine waves

Communications Circuits

(12–6) $(\sin A)(\sin B) = \frac{1}{2}[\cos(A - B) - \cos(A + B)]$ Trigonometric identity

(12–7) $v_1 v_2 = \dfrac{V_{1(p)} V_{2(p)}}{2} \cos 2\pi(f_1 - f_2)t$ Sum and difference

$\qquad\qquad - \dfrac{V_{1(p)} V_{2(p)}}{2} \cos 2\pi(f_1 + f_2)t$

(12–8) $v_c v_m = \dfrac{V_{c(p)} V_{m(p)}}{2} \cos 2\pi(f_c - f_m)t$ Sum and difference

$\qquad\qquad - \dfrac{V_{c(p)} V_{m(p)}}{2} \cos 2\pi(f_c + f_m)t$

(12–9) $V_{out} = \dfrac{V_{c(p)}^2}{2} \sin 2\pi f_c t$ Standard AM

$\qquad\qquad + \dfrac{V_{c(p)} V_{m(p)}}{2} \cos 2\pi(f_c - f_m)t$

$\qquad\qquad - \dfrac{V_{c(p)} V_{m(p)}}{2} \cos 2\pi(f_c + f_m)t$

(12–10) $V_d = \dfrac{V_i V_o}{2}[\cos(\theta_i - \theta_o) - \cos(4\pi f_i t + \theta_i + \theta_o)]$ PLL detector output

(12–11) $V_c = \dfrac{V_i V_o}{2} \cos \theta_e$ PLL control voltage

(12–12) $K = \dfrac{\Delta f_o}{\Delta V_c}$ VCO conversion gain

(12–13) $f_o \cong \dfrac{1.2}{4 R_1 C_1}$ Output frequency NE565

(12–14) $f_{lock} = \pm \dfrac{8 f_o}{V_{CC}}$ Lock range NE565

(12–15) $f_{cap} \cong \pm \dfrac{1}{2\pi} \sqrt{\dfrac{2\pi f_{lock}}{(3600)C_2}}$ Capture range NE565

Self-Test

1. In amplitude modulation, the pattern produced by the peaks of the carrier signal is called the
   (a) index (b) envelope (c) audio signal (d) upper-side frequency
2. Which of the following is not a part of an AM superheterodyne receiver?
   (a) Mixer (b) IF amplifier (c) DC restorer
   (d) Detector (e) Audio amplifier (f) Local oscillator

3. In an AM receiver, the local oscillator always produces a frequency that is above the incoming RF by
   (a) 10.7 kHz   (b) 455 MHz   (c) 10.7 MHz   (d) 455 kHz
4. An FM receiver has an IF frequency that is
   (a) in the 88 MHz to 108 MHz range
   (b) in the 540 kHz to 1640 kHz range
   (c) 455 kHz
   (d) greater than the IF in an AM receiver
5. The detector or discriminator in an AM or an FM receiver
   (a) detects the difference frequency from the mixer
   (b) changes the RF to IF
   (c) recovers the audio signal
   (d) maintains a constant IF amplitude
6. In order to handle all combinations of input voltage polarities, a multiplier must have
   (a) four-quadrant capability   (b) three-quadrant capability
   (c) four inputs   (d) dual-supply voltages
7. The internal attenuation of a multiplier is called the
   (a) transconductance   (b) scale factor   (c) reduction factor
8. When the two inputs of a multiplier are connected together, the device operates as a
   (a) voltage doubler   (b) square root circuit
   (c) squaring circuit   (d) averaging circuit
9. Amplitude modulation is basically
   (a) a summing of two signals   (b) a multiplication of two signals
   (c) a subtraction of two signals   (d) a nonlinear process
10. The frequency spectrum of a balanced modulator contains
    (a) a sum frequency   (b) a difference frequency   (c) a carrier frequency
    (d) a, b, and c   (e) a and b   (f) b and c
11. The IF in a receiver is the
    (a) sum of the local oscillator frequency and the RF carrier frequency
    (b) local oscillator frequency
    (c) difference of the local oscillator frequency and the carrier RF frequency
    (d) difference of the carrier frequency and the audio frequency
12. When a receiver is tuned from one RF frequency to another,
    (a) the IF changes by an amount equal to the LO (local oscillator) frequency
    (b) the IF stays the same

(c) the LO frequency changes by an amount equal to the audio frequency
(d) both the LO and the IF frequencies change

13. The output of the AM detector goes directly to the
    (a) IF amplifier   (b) mixer   (c) audio amplifier   (d) speaker

14. If the control voltage to a VCO increases, the output frequency
    (a) decreases   (b) does not change   (c) increases

15. A PLL maintains lock by
    (a) comparing the phase of two signals
    (b) comparing the frequency of two signals
    (c) comparing the amplitude of two signals

Problems

Section 12–1   Basic Receivers

1. Label each block in the AM receiver in Figure 12–62.

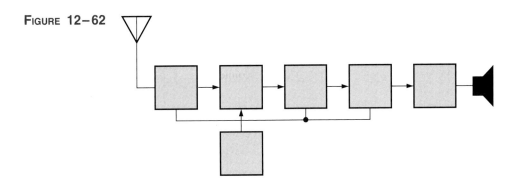

FIGURE 12–62

2. Label each block in the FM receiver in Figure 12–63.

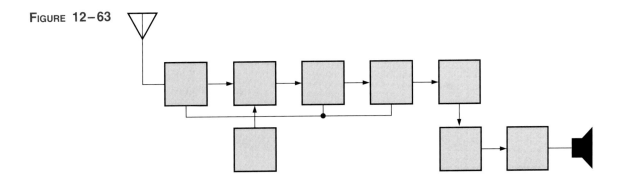

FIGURE 12–63

3. An AM receiver is tuned to a transmitted frequency of 680 kHz. What is the local oscillator (LO) frequency?
4. An FM receiver is tuned to a transmitted frequency of 97.2 MHz. What is the LO frequency?
5. The LO in an FM receiver is running at 101.9 MHz. What is the incoming RF? What is the IF?

Section 12–2 The Linear Multiplier

6. From the graph in Figure 12–64, determine the multiplier output voltage for each of the following pairs of input voltages.
   (a) $V_X = -4$ V, $V_Y = +6$ V   (b) $V_X = +8$ V, $V_Y = -2$ V
   (c) $V_X = -5$ V, $V_Y = -2$ V   (d) $V_X = +10$ V, $V_Y = +10$ V

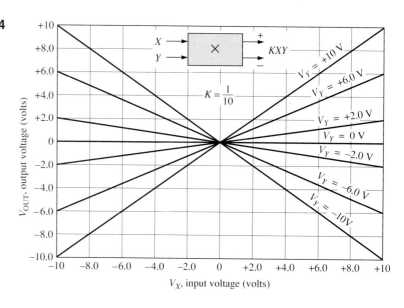

FIGURE 12–64

7. How much pin 3 current is there for the multiplier in Figure 12–65? The potentiometer is set at 2.8 kΩ.
8. Determine the scale factor for the multiplier in Figure 12–65.
9. If a certain multiplier has a scale factor of 0.8 and the inputs are +3.5 V and −2.9 V, what is the output voltage?
10. Show the connections for the multiplier in Figure 12–65 in order to implement a squaring circuit.
11. Determine the output voltage for each circuit in Figure 12–66.

FIGURE 12–65

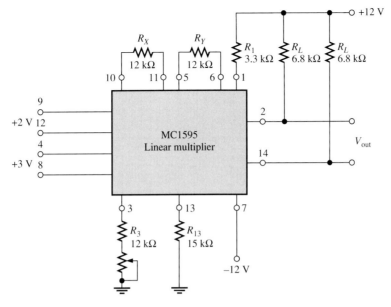

FIGURE 12–66

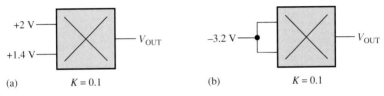
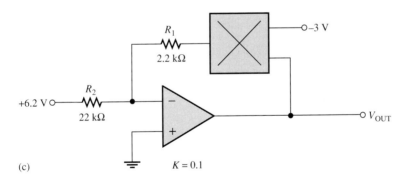
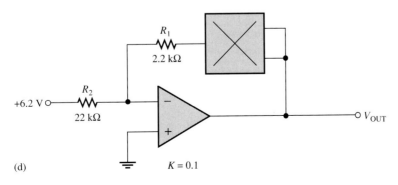

Section 12-3 Amplitude Modulation

12. If a 100 kHz signal and a 30 kHz signal are applied to a balanced modulator, what frequencies will appear on the output?

13. What are the frequencies on the output of the balanced modulator in Figure 12-67?

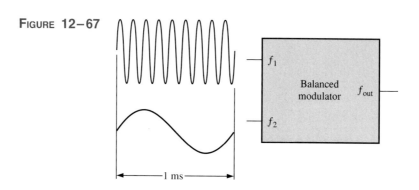

FIGURE 12-67

14. If a 1000 kHz signal and a 3 kHz signal are applied to a standard amplitude modulator, what frequencies will appear on the output?

15. What are the frequencies on the output of the standard amplitude modulator in Figure 12-68?

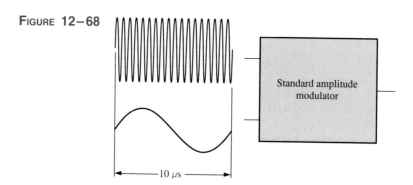

FIGURE 12-68

16. The frequency spectrum in Figure 12-69 is for the output of a standard amplitude modulator. Determine the carrier frequency and the modulating frequency.

17. The frequency spectrum in Figure 12-70 is for the output of a balanced modulator. Determine the carrier frequency and the modulating frequency.

18. A voice signal ranging from 300 Hz to 3 kHz amplitude modulates a 600 kHz carrier. Develop the frequency spectrum.

FIGURE 12–69

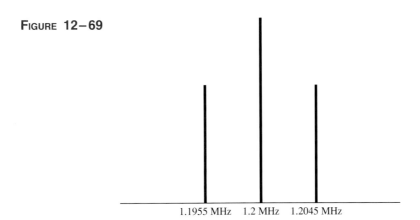

FIGURE 12–70

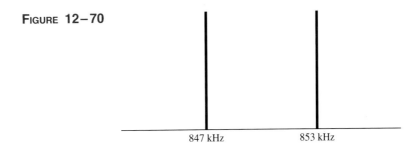

SECTION 12–4 THE MIXER

19. Determine the output expression for a multiplier with one sine-wave input having a peak voltage of 0.2 V and a frequency of 2200 kHz and the other input having a peak voltage of 0.15 V and a frequency of 3300 kHz.

20. Determine the output frequency of the IF amplifier for the frequencies shown in Figure 12–71.

FIGURE 12–71

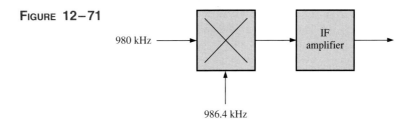

SECTION 12–5 AM DEMODULATION

21. The input to a certain AM receiver consists of a 1500 kHz carrier and two side frequencies separated from the carrier by 20 kHz. Determine the frequency spectrum at the output of the mixer amplifier.

22. For the same conditions stated in Problem 21, determine the frequency spectrum at the output of the IF amplifier.

23. For the same conditions stated in Problem 21, determine the frequency spectrum at the output of the AM detector (demodulator).

SECTION 12–6 IF AND AUDIO AMPLIFIERS

24. For a carrier frequency of 1.2 MHz and a modulating frequency of 8.5 kHz, list all of the frequencies on the output of the mixer in an AM receiver.

25. In a certain AM receiver, one amplifier has a bandpass from 450 kHz to 460 kHz and another has a bandpass from 10 Hz to 5 kHz. Identify these amplifiers.

26. Determine the maximum and minimum output voltages for the audio power amplifier in Figure 12–72.

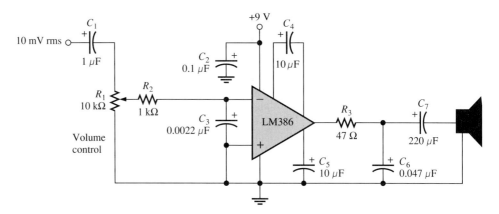

FIGURE 12–72

SECTION 12–7 FREQUENCY MODULATION

27. Explain how a VCO is used as a frequency modulator.

28. How does an FM signal differ from an AM signal?

29. What is the variable reactance element shown in the MC2833 diagram in Figure 12–44?

SECTION 12–8 THE PHASE-LOCKED LOOP (PLL)

30. Label each block in the PLL diagram of Figure 12–73.

31. A PLL is locked onto an incoming signal with a peak amplitude of 250 mV and a frequency of 10 MHz at a phase angle of 30°. The 400 mV peak VCO signal is at a phase angle of 15°.

   (a) What is the VCO frequency?

FIGURE 12–73

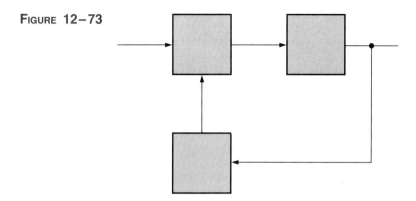

(b) What is the value of the control voltage being fed back to the VCO at this point?

32. What is the conversion gain of a VCO if a 0.5 V increase in the control voltage causes the output frequency to increase by 3.6 kHz?

33. If the conversion gain of a certain VCO is 1.5 kHz per volt, how much does the frequency change if the control voltage increases 0.67 V?

34. Name two conditions for a PLL to acquire lock.

35. Determine the free-running frequency, the lock range, and the capture range for the PLL in Figure 12–74.

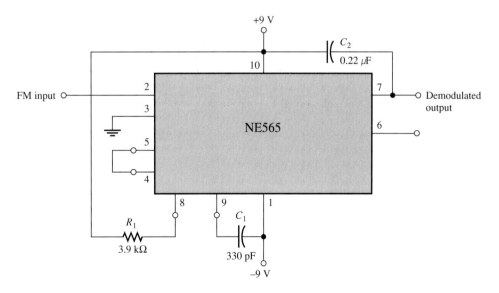

FIGURE 12–74

Answers to Review Questions

Section 12–1
1. AM is amplitude modulation. FM is frequency modulation.
2. In AM, the modulating signal varies the amplitude of a carrier. In FM, the modulating signal varies the frequency of a carrier.
3. AM: 540 kHz to 1640 kHz; FM: 88 MHz to 108 MHz

Section 12–2
1. A four-quadrant multiplier can handle any combination (4) of positive and negative inputs. A one-quadrant multiplier can only handle two positive inputs, for example.
2. $K = 0.1$. K must be 1 for an output of 5 V.
3. Connect the two inputs together and apply a single input variable.

Section 12–3
1. Amplitude modulation is the process of varying the amplitude of a carrier signal with a modulating signal.
2. Balanced modulation produces no carrier frequency on the output, whereas standard AM does.
3. The carrier signal is the modulated signal and has a sufficiently high frequency for transmission. The modulating signal is a lower-frequency signal that contains information and varies the carrier amplitude according to its waveshape.
4. The upper-side frequency is the sum of the carrier frequency and the modulating frequency. The lower-side frequency is the difference of the carrier frequency and the modulating frequency.
5. By summing the peak carrier voltage and the modulating signal before mixing with the carrier signal

Section 12–4
1. The mixer produces (among other frequencies) a signal representing the difference between the incoming carrier frequency and the local oscillator frequency. This is called the intermediate frequency.
2. The mixer multiplies the carrier and the local oscillator signals.
3. 1000 kHz + 350 kHz = 1350 kHz, 1000 kHz − 350 kHz = 650 kHz

Section 12–5
1. The filter removes all frequencies except the audio.
2. Only the 1 kHz

Section 12–6
1. To amplify the 455 kHz amplitude modulated IF coming from the mixer
2. The IF center frequency is 455 kHz.
3. The 10 kHz bandwidth allows the upper-side and lower-side frequencies that contain the information to pass.
4. The audio amplifier follows the detector because the detector is the circuit that recovers the audio from the modulated IF.

COMMUNICATIONS CIRCUITS   615

5. The IF has a response of approximately 455 kHz ± 5 kHz. The typical audio amplifier has a maximum bandwidth from tens of hertz up to about 15 kHz although for many amplifiers, the bandwidth can be much less than this typical maximum.

SECTION 12–7
1. The frequency variation of an FM signal bears the information.
2. VCO is voltage-controlled oscillator.
3. VCOs are based on the principle of voltage-variable reactance.

SECTION 12–8
1. Phase detector, low-pass filter, and VCO
2. Sometimes a PLL uses an amplifier in the loop.
3. A PLL locks onto and tracks a variable incoming frequency.
4. The lock range specifies how much a lock-on frequency can deviate without the PLL losing lock. The capture range specifies how close the incoming frequency must be from the free-running VCO frequency in order for the PLL to lock.
5. The PLL detects a change in the phase of the incoming signal compared to the VCO signal that indicates a change in frequency. The positive feedback then causes the VCO frequency to change along with the incoming frequency.

SECTION 12–9
1. A *low* on the originate/answer input puts the modem in the originate mode.
2. The diode clips excess negative voltage to protect the base-emitter junction of the transistor.
3. The PLL responds to 2025 Hz and 2225 Hz because the other modem is transmitting in the answer mode.
4. A constant 2225 Hz represents a continuous string of 1s; answer mode

ANSWERS TO PRACTICE EXERCISES

12–1 −3.6 V from the graph in Figure 12–9
12–2 0.732 V
12–3 Modulate the carrier with a higher-frequency signal.
12–4 The balanced modulator output has the same side frequencies but does not have a carrier frequency.
12–5 $V_p = 0.025$ mV, $f = 455$ kHz
12–6 455 kHz
12–7 0.172 V
12–8 A decrease of 6 kHz; 244 kHz
12–9 Decrease C_2 to 0.06 μF.

13

DATA CONVERSION CIRCUITS

13–1 ANALOG SWITCHES
13–2 SAMPLE-AND-HOLD AMPLIFIERS
13–3 INTERFACING THE ANALOG AND DIGITAL WORLDS
13–4 DIGITAL-TO-ANALOG (D/A) CONVERSION
13–5 BASIC CONCEPTS OF ANALOG-TO-DIGITAL (A/D) CONVERSION
13–6 ANALOG-TO-DIGITAL (A/D) CONVERSION METHODS
13–7 VOLTAGE-TO-FREQUENCY (V/F) AND FREQUENCY-TO-VOLTAGE (F/V) CONVERTERS
13–8 TROUBLESHOOTING
13–9 A SYSTEM APPLICATION

After completing this chapter, you should be able to

☐ List the types of analog switches.
☐ Describe how a basic analog switch works.
☐ Explain how analog switches are combined to form analog multiplexers.
☐ Discuss the basic purpose of an analog multiplexer.
☐ Describe a sample-and-hold amplifier and explain how it works.
☐ Explain the meaning of tracking in a sample-and-hold amplifier.
☐ Define the basic sample-and-hold performance parameters of aperture time, aperture jitter, acquisition time, droop, and feedthrough.
☐ Explain the purpose of analog-to-digital conversion.
☐ Distinguish an analog quantity from a digital quantity.
☐ Explain the need for analog-to-digital and digital-to-analog conversions and give examples.
☐ Analyze and compare the binary-weighted-input and the *R/2R* ladder types of D/A converters.
☐ Define resolution, accuracy, linearity, monotonicity, and settling time in relation to a D/A converter.
☐ Discuss resolution, conversion time, and quantization error in relation to A/D conversion.
☐ Discuss sampling and the meaning of the Nyquist rate.
☐ Describe the operation of several types of A/D converters including flash, single-slope, dual-slope, and successive-approximation.
☐ Explain why the flash A/D converter has the shortest conversion time.
☐ Explain the basic operation of a voltage-to-frequency converter.
☐ Explain the basic operation of a frequency-to-voltage converter.
☐ Test D/A converters for nonmonotonicity, nonlinearity, low or high gain, and offset error.
☐ Test A/D converters for a missing code, incorrect codes, and offset.

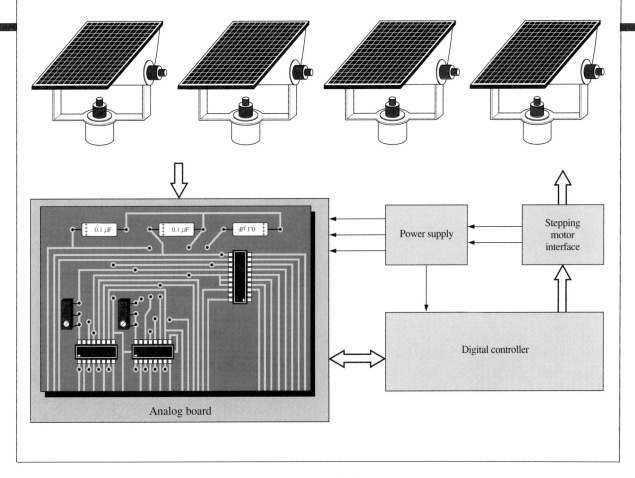

Data conversion circuits make interfacing between analog and digital systems possible. Most things in nature occur in analog form. For example, your voice is analog, time is analog, temperature and pressure are analog, and the speed of your vehicle is analog. These quantities and others are first sensed or measured with analog (linear) circuits and are then frequently converted to digital form to facilitate storage, processing, or display.

Also, in many applications, information in digital form must be converted back to analog form. An example of this is digitized music that is stored on a CD. Before you can hear the sounds, the digital information must be converted to its original analog form.

In this chapter, we will study several basic types of circuits found in applications requiring data conversion.

A System Application

A system for the angular control of solar cell arrays is depicted here. This particular system consists of four large solar cell arrays that can be individually positioned for proper orientation to the sun's rays. Both the azimuth and the elevation of each array are controlled by stepping motors. The angular position in both azimuth and elevation is sensed by position potentiometers that produce voltages proportional to the angular positions. Many systems use synchros or resolvers as angular transducers, but our system uses potentiometers for simplicity.

The electronic control circuits utilize an analog multiplexer to obtain the analog position from each solar array. The output of the analog multiplexer is then converted to digital form by the A/D converter for processing by the digital controller. The computer in the digital controller determines how much each array must be rotated in both azimuth and elevation based on stored information about the sun's position. The digital controller then sends the appropriate number of pulses to step the motors to the proper position. Basically, the system controls the solar arrays so that they track the sun each day to maintain an approximate 90° angle to the sun's rays.

This chapter's system application focuses on the analog multiplexer and A/D converter circuits.

For the system application in Section 13–9, in addition to the other topics, be sure you understand

☐ The basic operation of an analog switch.
☐ The basic operation of an analog multiplexer.
☐ The basic operation of A/D converters.

13–1 ANALOG SWITCHES

Analog switches are important in many types of electronic systems where it is necessary to switch signals on and off electronically. Major applications of analog switches are in signal selection, routing, and processing. Analog switches usually incorporate an FET as the basic switching element.

FIGURE 13–1
Basic types of analog switches.

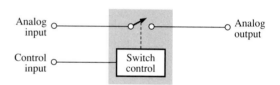

(a) Single pole–single throw (SPST)

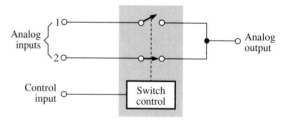

(b) Single pole–double throw (SPDT)

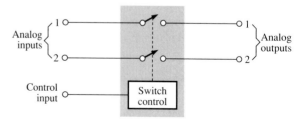

(c) Double pole–double throw (DPDT)

Data Conversion Circuits

Types of Analog Switches

Three basic types of analog switches in terms of their functional operation are

- Single pole–single throw (SPST)
- Single pole–double throw (SPDT)
- Double pole–single throw (DPST)

Figure 13–1 illustrates these three basic types of analog switches. As you can see, the analog switch consists of a control element and one or more input-to-output paths called *switch channels*.

An example of an analog switch is the ADG202A. This integrated circuit device has four independently operated SPST switches as shown in Figure 13–2(a). Typical packages are shown in Figure 13–2(b).

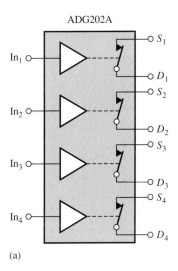

(a)

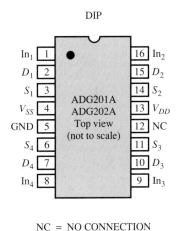

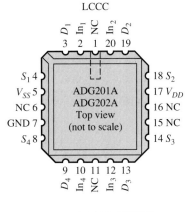

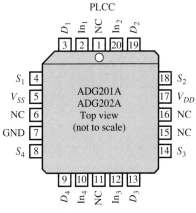

(b)

FIGURE 13–2
The ADG202A Quad SPST switches.

When the control input is at a high-level voltage (at least 2.4 V for the ADG202A), the switch is closed (on). When the control input is at a low-level voltage (no greater than 0.8 V for this device), the switch is open (off). The switches themselves are typically implemented with MOSFETs.

■ EXAMPLE 13–1

Determine the output waveform of the analog switch in Figure 13–3(a) for the control voltage and analog input voltage shown.

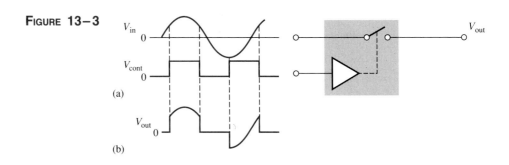

FIGURE 13–3

SOLUTION
When the control voltage is high, the switch is closed and the analog input passes through to the output. When the control voltage is low, the switch is open and there is no output voltage. The output waveform is shown in Figure 13–3(b) in relation to the other voltages.

PRACTICE EXERCISE 13–1
What will be the output waveform in Figure 13–3 if the frequency of the control voltage is doubled but keeping the same duty cycle?

MULTIPLE CHANNEL ANALOG SWITCHES

In data acquisition systems where inputs from several different sources must be independently converted to digital form for processing, a technique called *multiplexing* is used. A separate analog switch is used for each analog source as illustrated in Figure 13–4 for a four-channel system. In this type of application, all of the outputs of the analog switches are connected together to form a common output and only one switch can be closed at a given time. The common switch outputs are connected to the input of a voltage follower as indicated.

A good example of an IC analog multiplexer is the AD9300 shown in Figure 13–5. This device contains four analog switches that are controlled by a channel decoder. The

FIGURE 13–4
A four-channel analog multiplexer.

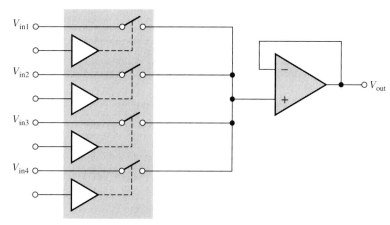

FIGURE 13–5
The AD9300 analog multiplexer.

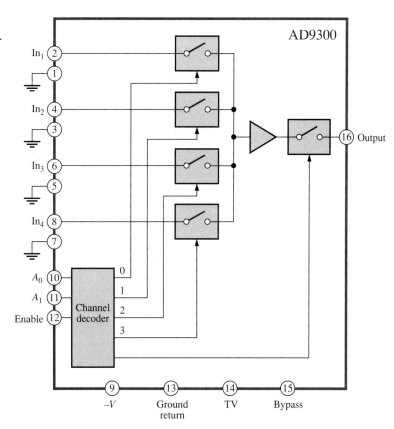

inputs A_0 and A_1 determine which one of the four switches is on. If A_0 and A_1 are both low, input In_1 is selected. If A_0 is high and A_1 is low, input In_2 is selected. If A_0 is low and A_1 is high, input In_3 is selected. If A_0 and A_1 are both high, input In_4 is selected. The enable input controls the switch that connects or disconnects the output.

EXAMPLE 13-2

Determine the output waveform of the analog multiplexer in Figure 13–6 for the control inputs and the analog inputs shown.

FIGURE 13–6

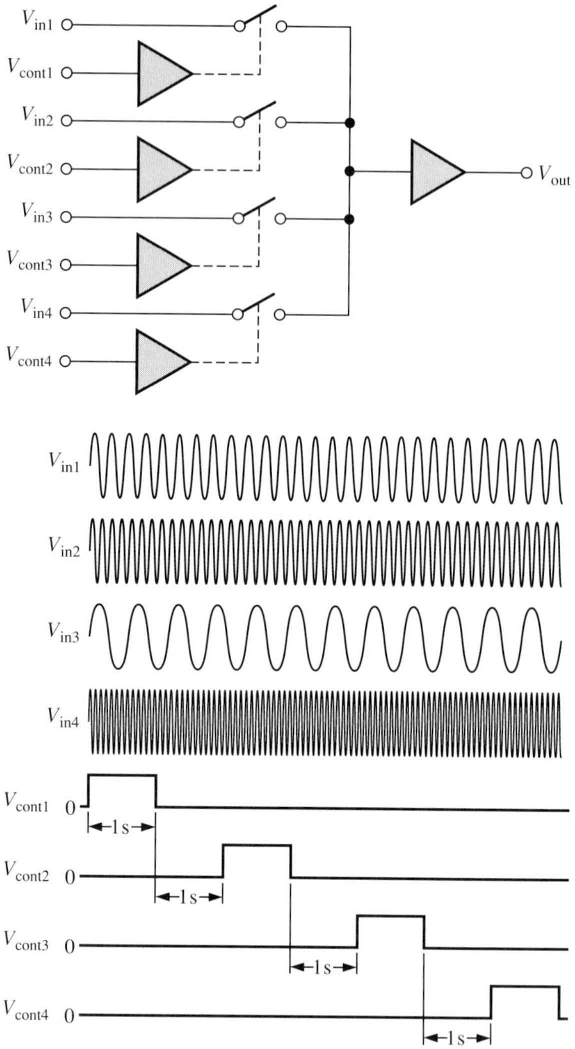

SOLUTION

When a control input is a high level, the corresponding switch is closed and the analog voltage on its input is switched through to the output. Notice that only one control voltage is high at a time. The inputs to the switches are sine waves, each having a different frequency. The resulting output is a sequence of different sine waves that last for one second and that are separated by a one-second interval, as indicated in Figure 13–7.

FIGURE 13–7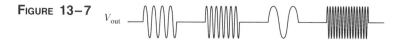

Practice Exercise 13-2
How is the output waveform in Figure 13-7 affected if the time interval between the control voltage pulses is decreased? ■

13-1 Review Questions

1. What is the purpose of an analog switch?
2. What is the basic function of an analog multiplexer?

13-2 Sample-and-Hold Amplifiers

A sample-and-hold amplifier samples an analog input voltage at a certain point in time and retains or holds the sampled voltage for an extended time after the sample is taken. The sample-and-hold process keeps the sampled analog voltage constant for the length of time necessary to allow an analog-to-digital (A/D) converter to convert the voltage to digital form.

A Basic Sample-and-Hold Circuit

A basic sample-and-hold circuit consists of an analog switch, a capacitor, and input and output buffer amplifiers as shown in Figure 13-8. The analog switch **samples** the analog input voltage through the input buffer amplifier, the capacitor (C_H) stores or holds the sampled voltage for a period of time, and the output buffer amplifier provides a high input impedance to prevent the capacitor from discharging quickly.

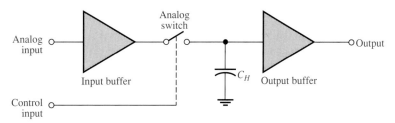

Figure 13-8
A basic sample-and-hold circuit.

As illustrated in Figure 13-9, a relatively narrow control voltage pulse closes the analog switch and allows the capacitor to charge to the value of the input voltage. The switch then opens, and the capacitor holds the voltage for a long period of time because of the very high impedance discharge path through the op-amp input. So, basically, the sample-and-hold circuit converts an instantaneous value of the analog input voltage to a dc voltage.

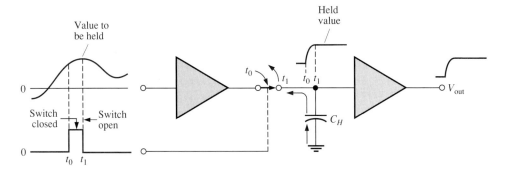

FIGURE 13–9
Basic action of a sample-and-hold.

TRACKING DURING SAMPLE TIME

Perhaps a more appropriate designation for a sample-and-hold amplifier is sample/track-and-hold because the circuit actually tracks the input voltage during the sample interval. As indicated in Figure 13–10, the output follows the input during the time that the control

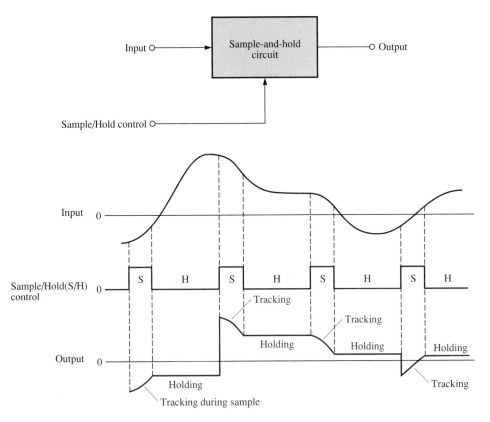

FIGURE 13–10
Example of tracking during a sample-and-hold sequence.

DATA CONVERSION CIRCUITS

voltage is high; and when the control voltage goes low, the last voltage is held until the next sample interval.

EXAMPLE 13–3 Determine the output voltage waveform for the sample/track-and-hold amplifier in Figure 13–11, given the input and control voltage waveforms.

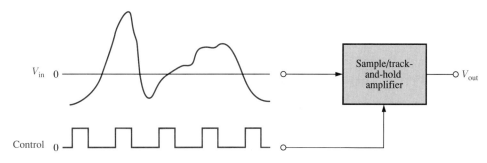

FIGURE 13–11

SOLUTION
During the time that the control voltage is high, the analog switch is closed and the circuit is tracking the input. When the control voltage goes low, the analog switch opens; and the last voltage value is held at a constant level until the next time the control voltage goes high. This is shown in Figure 13–12.

FIGURE 13–12

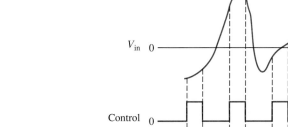

PRACTICE EXERCISE 13–3
Sketch the output voltage waveform for Figure 13–11 if the control voltage frequency is reduced by half while maintaining the same duty cycle.

Performance Specifications

In addition to specifications similar to those of a closed-loop op-amp that were discussed in Chapter 5, several specifications are peculiar to sample-and-hold amplifiers. These include the aperture time, aperture jitter, acquisition time, droop, and feedthrough.

- **Aperture time**—the time for the analog switch to fully open after the control voltage switches from its sample level to its hold level. Aperture time produces a delay in the effective sample point.
- **Aperture jitter**—the uncertainty in the aperture time.
- **Acquisition time**—the time required for the device to reach its final value when the control voltage switches from its hold level to its sample level.
- **Droop**—the change in voltage from the sampled value during the hold interval because of charge leaking off of the hold capacitor.
- **Feedthrough**—the component of the output voltage that follows the input signal after the analog switch is opened. The inherent capacitance from the input to the output of the switch causes feedthrough.

Each of these parameters is illustrated in Figure 13–13 for an example input voltage waveform.

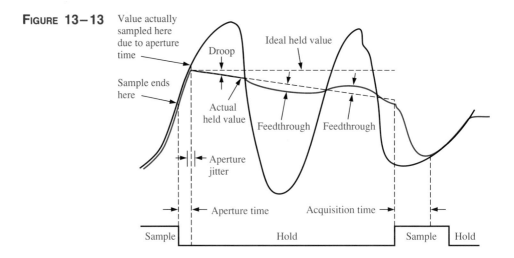

FIGURE 13–13

A Specific Device

A good example of a basic sample-and-hold amplifier is the AD582. The circuit and pin configuration are shown in Figure 13–14. As shown in the figure, this particular device consists of two buffer amplifiers and an analog switch that is controlled by a logic gate. The hold capacitor must be connected externally to pin 6 and its value is selected depending on the application requirements.

The control voltage for establishing the sample/hold intervals is applied between pins 11 and 12. The input signal to be sampled is applied to pin 1. A potentiometer for nulling

FIGURE 13–14

The AD582 sample-and-hold amplifier.

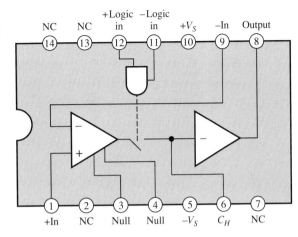

out the offset voltage can be connected between pins 3 and 4, and the overall gain of the device can be set with external feedback connections. Two typical configurations are shown in Figure 13–15 on page 628.

13–2 REVIEW QUESTIONS

1. What is the basic function of a sample-and-hold amplifier?
2. In reference to the output of a sample-and-hold amplifier, what does droop mean?
3. Define aperture time.
4. What is acquisition time?

13–3 INTERFACING THE ANALOG AND DIGITAL WORLDS

Analog quantities are sometimes called real-world quantities because most physical quantities are analog in nature. Many applications of computers and other digital systems require the input of real-world quantities, such as temperature, speed, position, pressure, and force. Real-world quantities can even include graphic images. Also, digital systems are often used to control real-world quantities. A basic familiarity with the binary number system is assumed for this and the next sections.

DIGITAL AND ANALOG SIGNALS

An **analog** quantity is one that has a continuous set of values over a given range, as contrasted with discrete values for the digital case. Almost any measurable quantity is analog in nature, such as temperature, pressure, speed, and time. To further illustrate the difference between an analog and a digital representation of a quantity, let's take the case of a voltage that varies over a range from 0 V to +15 V. The analog representation of this quantity takes in all values between 0 and +15 V of which there is an infinite number.

628   CHAPTER 13

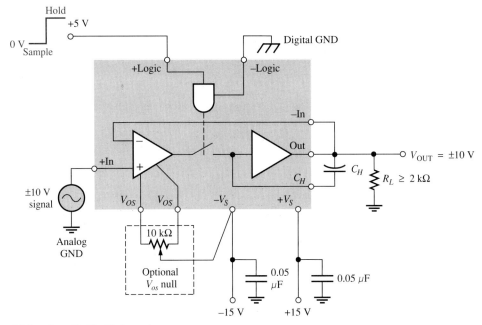

(a) Sample and hold with $A = +1$

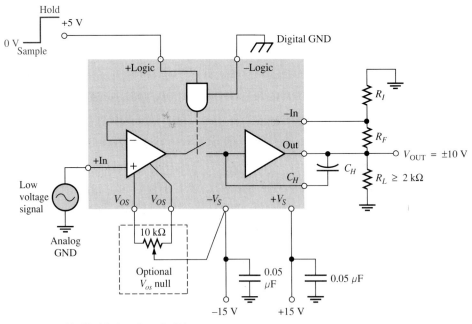

(b) Sample and hold with $A = (1 + R_F/R_I)$

FIGURE 13–15
Two configurations of the AD582 sample-and-hold amplifier.

In the case of a **digital** representation using a four-bit binary code, only sixteen values can be defined. More values between 0 and +15 can be represented by using more bits in the digital code. So an analog quantity can be represented to some degree of accuracy with a digital code that specifies discrete values within the range. This concept is illustrated in Figure 13–16, where the analog function shown is a smoothly changing curve that takes on values between 0 V and +15 V. If a four-bit code is used to represent this curve, each binary number represents a discrete point on the curve.

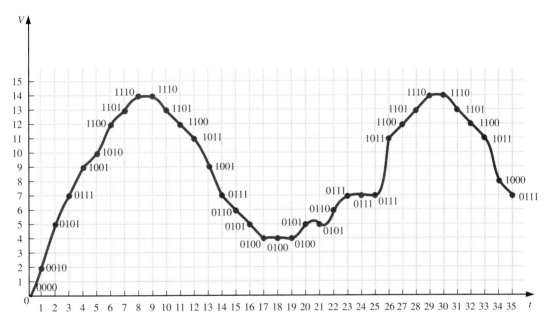

FIGURE 13–16
Discrete (digital) points on an analog curve.

In Figure 13–16 the voltage on the analog curve is measured, or sampled, at each of thirty-five equal intervals. The voltage at each of these intervals is represented by a four-bit code as indicated. At this point, we have a series of binary numbers representing various voltage values along the analog curve. This is the basic idea of analog-to-digital (A/D) conversion.

An approximation of the analog function in Figure 13–16 can be reconstructed from the sequence of digital numbers that has been generated. Obviously, there will be some error in the reconstruction because only certain values are represented (thirty-six in this example) and not the continuous set of values. If the digital values at all of the thirty-six points are graphed as shown in Figure 13–17, we have a reconstructed function. As you can see, the graph only approximates the original curve because values between the points are not known.

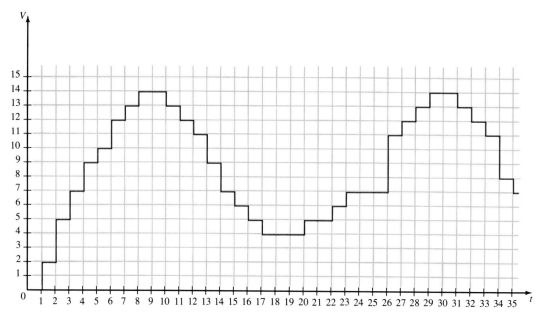

FIGURE 13–17
A rough digital reproduction of an analog curve.

REAL-WORLD INTERFACING

To interface between the digital and analog worlds, two basic processes are required. These are analog-to-digital (A/D) conversion and digital-to-analog (D/A) conversion. The following two system examples illustrate the application of these conversion processes.

AN ELECTRONIC THERMOSTAT A simplified block diagram of a microprocessor-based electronic thermostat is shown in Figure 13–18. The room temperature sensor produces an analog voltage that is proportional to the temperature. This voltage is increased by the linear amplifier and applied to the A/D converter, where it is converted to a digital code and periodically sampled by the microprocessor. For example, suppose the room temperature is 67°F. A specific voltage value corresponding to this temperature appears on the A/D converter input and is converted to an eight-bit binary number, 01000011.

Internally, the microprocessor compares this binary number with a binary number representing the desired temperature (say 01001000 for 72°F). This desired value has been previously entered from the keypad and stored in a register. The comparison shows that the actual room temperature is less than the desired temperature. As a result, the microprocessor instructs the unit control circuit to turn the furnace on. As the furnace runs, the microprocessor continues to monitor the actual temperature via the A/D converter (ADC). When the actual temperature equals or exceeds the desired temperature, the microprocessor turns the furnace off.

A DIGITAL AUDIO TAPE (DAT) PLAYER/RECORDER Another system example that includes both A/D and D/A conversion is the DAT player/recorder. A basic block diagram is shown in Figure 13–19.

DATA CONVERSION CIRCUITS

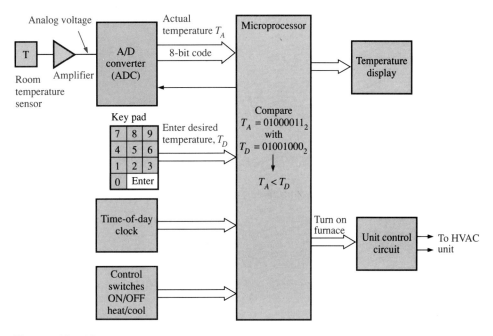

FIGURE 13–18
An electronic thermostat that uses an A/D converter.

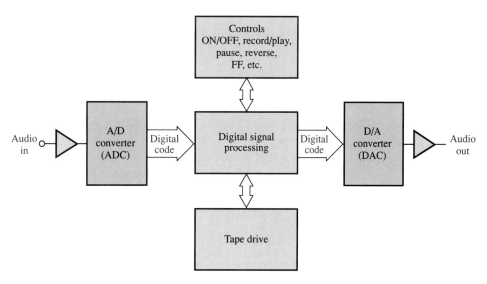

FIGURE 13–19
Basic block diagram of a DAT system.

An audio signal, of course, is an analog quantity. In the record mode, sound is picked up, amplified, and converted to digital form by the A/D converter. The digital codes representing the audio signal are processed and recorded on the tape.

In the play mode, the digitized audio signal is read from the tape, processed, and converted back to analog form by the D/A converter (DAC). It is then amplified and sent to the speaker system.

13-3 REVIEW QUESTIONS

1. In the real world, quantities normally appear in what form?
2. Explain the basic purpose of A/D conversion.
3. Explain the basic purpose of D/A conversion.

13-4 DIGITAL-TO-ANALOG (D/A) CONVERSION

D/A conversion is an important part of many systems. In this section, we will examine two basic types of D/A converters and learn about their performance characteristics. The binary-weighted-input D/A converter was introduced in Chapter 7 as an example of a scaling adder application and is covered more thoroughly in this section. Also, a more commonly used configuration called the R/2R ladder D/A converter is introduced.

BINARY-WEIGHTED-INPUT D/A CONVERTER

The binary-weighted-input D/A converter uses a resistor network with resistance values that represent the binary weights of the input bits of the digital code. Figure 13–20 shows a four-bit D/A converter of this type. Each of the input resistors will either have current or have no current, depending on the input voltage level. If the input voltage is zero (binary 0), the current is also zero. If the input voltage is high (binary 1), the amount of current depends on the input resistor value and is different for each input resistor, as indicated in the figure.

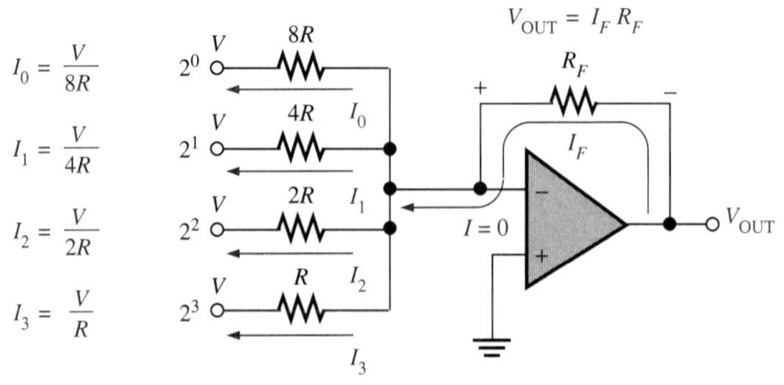

FIGURE 13–20

A four-bit D/A converter with binary-weighted inputs.

$I_0 = \dfrac{V}{8R}$

$I_1 = \dfrac{V}{4R}$

$I_2 = \dfrac{V}{2R}$

$I_3 = \dfrac{V}{R}$

DATA CONVERSION CIRCUITS

Since there is practically no current from the op-amp inverting input, all of the input currents sum together and flow through R_F. Since the inverting input is at 0 V (virtual ground), the drop across R_F is equal to the output voltage, so $V_{OUT} = I_F R_F$.

The values of the input resistors are chosen to be inversely proportional to the binary weights of the corresponding input bits. The lowest-value resistor (R) corresponds to the highest binary-weighted input (2^3). The other resistors are multiples of R—$2R$, $4R$, and $8R$—and correspond to the binary weights 2^2, 2^1, and 2^0, respectively. The input currents are also proportional to the binary weights. Thus, the output voltage is proportional to the sum of the binary weights because the sum of the currents flows through R_F.

One of the disadvantages of this type of D/A converter is the number of different resistor values. For example, an eight-bit converter requires eight resistors, ranging from some value of R to $128R$ in binary-weighted steps. This range of resistors requires tolerances of one part in 255 (less than 0.5%) to accurately convert the input, making this type of D/A converter very difficult to mass-produce.

■ **EXAMPLE 13–4** Determine the output of the D/A converter in Figure 13–21(a) if the waveforms representing a sequence of four-bit binary numbers in Figure 13–21(b) are applied to the inputs. Input D_0 is the least significant bit (LSB).

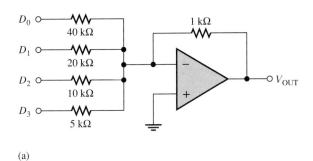
(a)

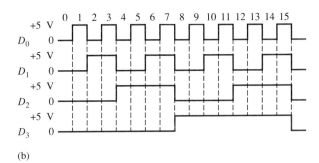
(b)

FIGURE 13–21

SOLUTION
First, we determine the current for each of the weighted inputs. Since the inverting input ($-$) of the op-amp is at 0 V (virtual ground) and a binary 1 corresponds to +5 V, the current through any of the input resistors is 5 V divided by the resistance value.

$$I_0 = \frac{5 \text{ V}}{40 \text{ k}\Omega} = 0.125 \text{ mA}$$

$$I_1 = \frac{5 \text{ V}}{20 \text{ k}\Omega} = 0.25 \text{ mA}$$

$$I_2 = \frac{5 \text{ V}}{10 \text{ k}\Omega} = 0.5 \text{ mA}$$

$$I_3 = \frac{5 \text{ V}}{5 \text{ k}\Omega} = 1.0 \text{ mA}$$

There is essentially no current from the inverting op-amp input because of its extremely high impedance. Therefore, we assume that all of the current goes through the feedback resistor R_F. Since one end of R_F is at 0 V (virtual ground), the drop across R_F equals the output voltage, which is negative with respect to virtual ground.

$$V_{OUT(D0)} = (1 \text{ k}\Omega)(-0.125 \text{ mA}) = -0.125 \text{ V}$$
$$V_{OUT(D1)} = (1 \text{ k}\Omega)(-0.25 \text{ mA}) = -0.25 \text{ V}$$
$$V_{OUT(D2)} = (1 \text{ k}\Omega)(-0.5 \text{ mA}) = -0.5 \text{ V}$$
$$V_{OUT(D3)} = (1 \text{ k}\Omega)(-1.0 \text{ mA}) = -1.0 \text{ V}$$

From Figure 13–21(b), the first binary input code is 0000, which produces an output voltage of 0 V. The next input code is 0001, which produces an output voltage of −0.125 V. The next code is 0010, which produces an output voltage of −0.25 V. The next code is 0011, which produces an output voltage of −0.125 V + −0.25 V = −0.375 V. Each successive binary code increases the output voltage by −0.125 V, so for this particular straight binary sequence on the inputs, the output is a stairstep waveform going from 0 V to −1.875 V in −0.125 V steps. This is shown in Figure 13–22.

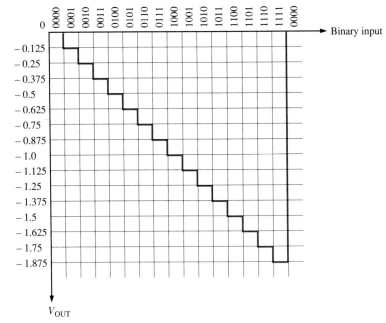

FIGURE 13–22
Output of D/A converter in Figure 13–21.

PRACTICE EXERCISE 13–4
What size are the output steps of the D/A converter if the feedback resistance is changed to 2 kΩ?

THE R/2R LADDER D/A CONVERTER

Another method of D/A conversion is the *R*/2*R* ladder, as shown in Figure 13–23 for four bits. It overcomes one of the problems in the binary-weighted-input D/A converter in that it requires only two resistor values.

FIGURE 13–23
An *R*/2*R* ladder D/A converter.

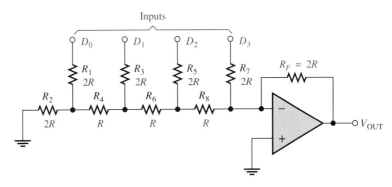

Start by assuming that the D_3 input is at a high level (+5 V) and the others are at a low level (ground, 0 V). This condition represents the binary number 1000. A circuit analysis will show that this reduces to the equivalent form shown in Figure 13–24(a). Essentially no current goes through the 2*R* equivalent resistance because the inverting input is at virtual ground. Thus, all of the current ($I = 5$ V/2*R*) through R_7 also goes through R_F, and the output voltage is -5 V.

Figure 13–24(b) shows the equivalent circuit when the D_2 input is at +5 V and the others are at ground. This condition represents 0100. If we thevenize looking from R_8, we get 2.5 V in series with *R*, as shown in part (b). This results in a current through R_F of $I = 2.5$ V/2*R*, which gives an output voltage of -2.5 V. Keep in mind that there is no current into the op-amp inverting input and that there is no current through the equivalent resistance to ground because it has 0 V across it, due to the virtual ground.

Figure 13–24(c) shows the equivalent circuit when the D_1 input is at +5 V and the others are at ground. Again thevenizing looking from R_8, we get 1.25 V in series with *R* as shown. This results in a current through R_F of $I = 1.25$ V/2*R*, which gives an output voltage of -1.25 V.

In part (d) of Figure 13–24, the equivalent circuit representing the case where D_0 is at +5 V and the other inputs are at ground is shown. Thevenizing from R_8 gives an equivalent of 0.625 V in series with *R* as shown. The resulting current through R_F is $I = 0.625$ V/2*R*, which gives an output voltage of -0.625 V.

Notice that each successively lower-weighted input produces an output voltage that is halved, so that the output voltage is proportional to the binary weight of the input bits.

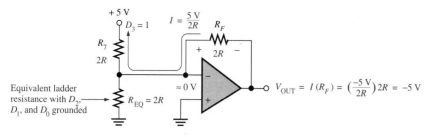

(a) Equivalent circuit for $D_3 = 1, D_2 = 0, D_1 = 0, D_0 = 0$

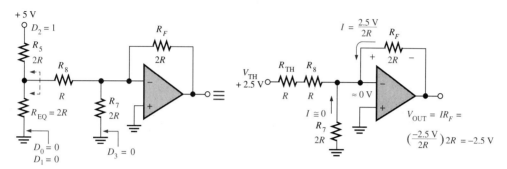

(b) Equivalent circuit for $D_3 = 0, D_2 = 1, D_1 = 0, D_0 = 0$

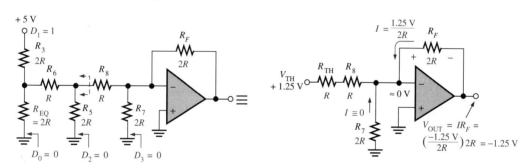

(c) Equivalent circuit for $D_3 = 0, D_2 = 0, D_1 = 1, D_0 = 0$

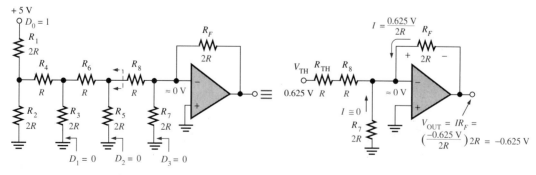

(d) Equivalent circuit for $D_3 = 0, D_2 = 0, D_1 = 0, D_0 = 1$

FIGURE 13–24

Analysis of the R/2R ladder D/A converter.

DATA CONVERSION CIRCUITS

PERFORMANCE CHARACTERISTICS OF D/A CONVERTERS

The performance characteristics of a D/A converter include resolution, accuracy, linearity, monotonicity, and settling time, each of which is discussed in the following list.

- *Resolution.* The resolution of a D/A converter is the reciprocal of the maximum number of discrete steps in the D/A output. Resolution, of course, is dependent on the number of input bits. For example, a four-bit D/A converter has a resolution of one part in $2^4 - 1$ (one part in fifteen). Expressed as a percentage, this is $(1/15)100 = 6.67\%$ The total number of discrete steps equals $2^n - 1$, where n is the number of bits. Resolution can also be expressed as the number of bits that are converted.
- *Accuracy.* Accuracy is a comparison of the actual output of a D/A converter with the expected output. It is expressed as a percentage of a full-scale, or maximum, output voltage. For example, if a converter has a full-scale output of 10 V and the accuracy is $\pm 0.1\%$, then the maximum error for any output voltage is $(10\text{ V})(0.001) = 10\text{ mV}$. Ideally, the accuracy should be, at most, $\pm\frac{1}{2}$ of a least significant bit (LSB). For an eight-bit converter, 1 LSB is $1/256 = 0.0039$ (0.39% of full scale). The accuracy should be approximately $\pm 0.2\%$.
- *Linearity.* A linear error is a deviation from the ideal straight-line output of a D/A converter. A special case is an offset error, which is the amount of output voltage when the input bits are all zeros.
- *Monotonicity.* A D/A converter is **monotonic** if it does not miss any steps when it is sequenced over its entire range of input bits.
- *Settling time.* Settling time is normally defined as the time it takes a D/A converter to settle within $\pm\frac{1}{2}$ LSB of its final value when a change occurs in the input code.

EXAMPLE 13–5

Determine the resolution, expressed as a percentage, of (a) an 8-bit and (b) a 12-bit D/A converter.

SOLUTION
(a) For the 8-bit converter,

$$\frac{1}{2^8 - 1} \times 100 = \frac{1}{255} \times 100 = 0.392\%$$

(b) For the 12-bit converter,

$$\frac{1}{2^{12} - 1} \times 100 = \frac{1}{4095} \times 100 = 0.0244\%$$

PRACTICE EXERCISE 13–5
Determine the percent resolution for an 18-bit converter.

13–4 REVIEW QUESTIONS

1. What is the disadvantage of the D/A converter with binary-weighted inputs?
2. What is the resolution of a four-bit D/A converter?

13-5 BASIC CONCEPTS OF ANALOG-TO-DIGITAL (A/D) CONVERSION

As you have seen, analog-to-digital conversion is the process by which an analog quantity is converted to digital form. A/D conversion is necessary when measured quantities must be in digital form for processing in a computer or for display or storage. Basic concepts of A/D conversion including resolution, conversion time, sampling theory, and quantization error are introduced in this section.

RESOLUTION

An analog-to-digital converter translates a continuous analog signal into a series of binary numbers. Each binary number represents the value of the analog signal at the time of conversion. The **resolution** of an A/D converter can be expressed as the number of bits (binary digits) used to represent each value of the analog signal. A 4-bit A/D converter can represent sixteen different values of an analog signal because $2^4 = 16$. An 8-bit A/D converter can represent 256 different values of an analog signal because $2^8 = 256$. A 12-bit A/D converter can represent 4096 different values of the analog signal because $2^{12} = 4096$. The more bits, the more accurate is the conversion and the greater is the resolution because more values of a given analog signal can be represented.

Resolution is basically illustrated in Figure 13–25 using the analog voltage ramp in part (a). For the case of 3-bit resolution as shown in part (b), only eight values of the voltage ramp can be represented by binary numbers. D/A reconstruction of the ramp using the eight binary values results in the stair-step approximation shown. For the case of 4-bit resolution as shown in part (c), sixteen values can be represented and D/A reconstruction results in a more accurate 16-step approximation as shown. For the case of 5-bit resolution as shown in part (d), D/A reconstruction produces an even more accurate 32-step approximation of the ramp.

CONVERSION TIME

In addition to resolution, another important characteristic of A/D converters is conversion time. The conversion of a value on an analog waveform into a digital quantity is not an instantaneous event, but it is a process that takes a certain amount of time. The conversion time can range from microseconds for fast converters to milliseconds for slower devices. Conversion time is illustrated in a basic way in Figure 13–26. As you can see, the value of the analog voltage to be converted occurs at time t_0 but the conversion is not complete until time t_1.

SAMPLING THEORY

In analog-to-digital conversion, an analog waveform is sampled at a given point and the sampled value is then converted to a binary number. Since it takes a certain interval of time to accomplish the conversion, the number of samples of an analog waveform during a given period of time is limited. For example, if a certain A/D converter can make one conversion in 1 ms, it can make 1000 conversions in one second. That is, it can convert 1000 different analog values to digital form in a one-second interval.

FIGURE 13–25
Illustration of the effect of resolution on the representation of an analog signal (a ramp in this case).

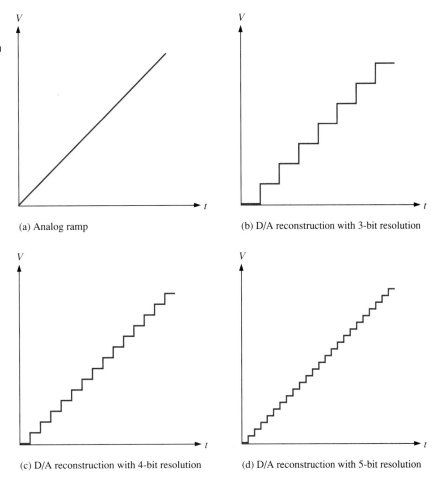

(a) Analog ramp

(b) D/A reconstruction with 3-bit resolution

(c) D/A reconstruction with 4-bit resolution

(d) D/A reconstruction with 5-bit resolution

FIGURE 13–26
An illustration of A/D conversion time.

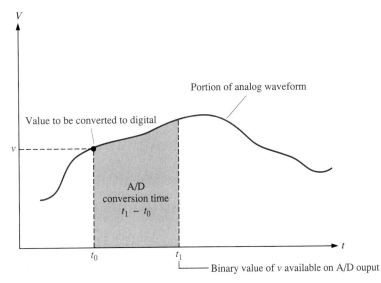

In order to represent an analog waveform with reasonable accuracy, the minimum sample rate must be twice the maximum frequency component of the analog signal. This minimum sampling rate is known as the **Nyquist rate.** At the Nyquist rate, an analog signal is sampled and converted two times per cycle, which establishes the fundamental frequency of the analog signal. Filtering can be used to obtain a reasonable facsimile of the original signal after D/A conversion. Obviously, a greater number of conversions per cycle of the analog signal results in a more accurate representation of the analog signal. This is illustrated in Figure 13–27 where the upper waveforms in each case are the original analog signals and the lower waveforms are the D/A reconstructions for various sample rates.

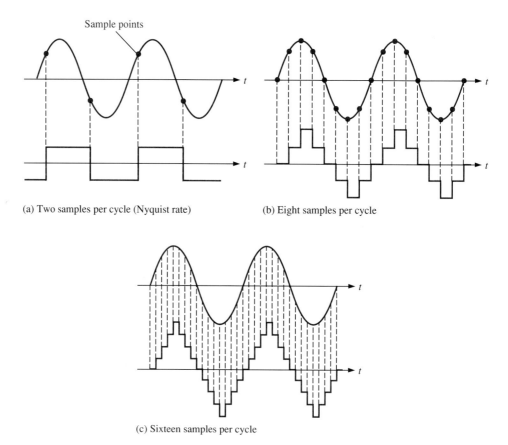

FIGURE 13–27
Illustration of various sampling rates.

Quantization Error

The term **quantization** in this context refers to determining a value for an analog quantity. Ideally, we would like to determine a value at a given instant and convert it immediately

DATA CONVERSION CIRCUITS

to digital form. This is, of course, impossible because of the conversion time of A/D converters. Since an analog signal may change during a conversion time, its value at the end of the conversion time may not be the same as it was at the beginning (unless the input is a constant dc). This change in the value of the analog signal during the conversion time produces what is called the **quantization error,** as illustrated in Figure 13–28.

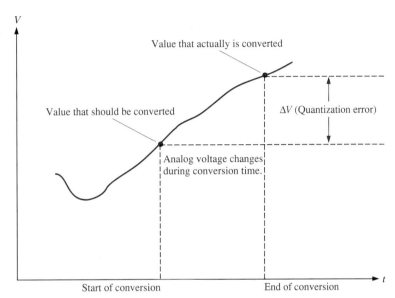

FIGURE 13–28
Illustration of quantization error in A/D conversion.

One way to avoid or at least minimize quantization error is to use a sample-and-hold circuit at the input to the A/D converter. As you learned in Section 13–2, a sample-and-hold amplifier quickly samples the analog input and then holds the sampled value for a certain time. When used in conjunction with an A/D converter, the sample-and-hold is held constant for the duration of the conversion time. This allows the A/D converter to convert a constant value to digital form and avoids the quantization error. A basic illustration of this process is shown in Figure 13–29 on page 642. When compared to the conversion in Figure 13–28, you can see that a more accurate representation of the analog input at the desired sample point is achieved.

13–5 REVIEW QUESTIONS

1. Which is more accurate, an 8-bit or a 12-bit A/D converter?
2. What is conversion time?
3. According to sampling theory, what is the minimum sampling rate for a 100 Hz sine wave?
4. Basically, how does a sample-and-hold circuit avoid quantization error in A/D conversion?

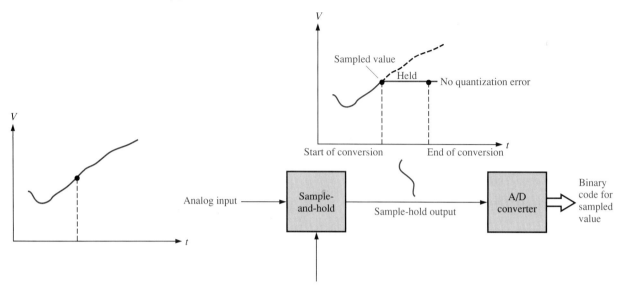

FIGURE 13-29
Using a sample-and-hold amplifier to avoid quantization error.

13-6   ANALOG-TO-DIGITAL (A/D) CONVERSION METHODS

Now that you are familiar with some basic A/D conversion concepts, we will look at several methods for A/D conversion. These methods are flash (simultaneous), stair-step ramp, tracking, single-slope, dual-slope, and successive approximation. The flash and dual-slope methods were introduced in Chapter 7 as examples of op-amp applications. Some of that material is reviewed and expanded upon in this section.

FLASH (SIMULTANEOUS) A/D CONVERTER

The **flash** (simultaneous) method utilizes comparators that compare reference voltages with the analog input voltage. When the analog voltage exceeds the reference voltage for a given comparator, a high-level output is generated. Figure 13-30 shows a three-bit converter that uses seven comparator circuits; a comparator is not needed for the all-0s condition. A four-bit converter of this type requires fifteen comparators. In general, $2^n - 1$ comparators are required for conversion to an *n*-bit binary code. The large number of comparators necessary for a reasonable-sized binary number is one of the disadvantages of the flash A/D converter. Its chief advantage is that it provides a fast conversion time.

The reference voltage for each comparator is set by the resistive voltage-divider network. The output of each comparator is connected to an input of the priority encoder. The encoder is sampled by a pulse on the Enable input, and a three-bit binary code representing the value of the analog input appears on the encoder's outputs. The binary code is determined by the highest-order input having a high level.

DATA CONVERSION CIRCUITS

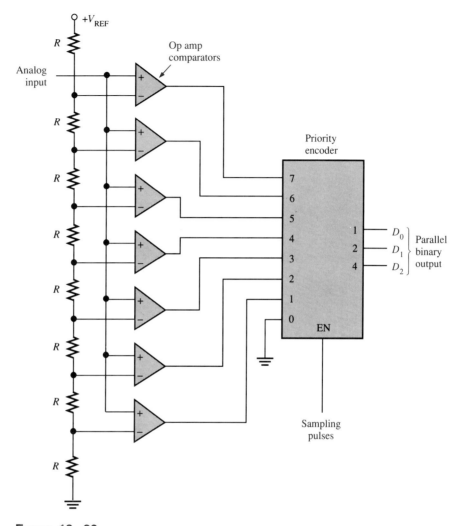

FIGURE 13-30
A three-bit flash A/D converter.

The sampling rate determines the accuracy with which the sequence of digital codes represents the analog input of the A/D converter. The more samples taken in a given unit of time, the more accurately the analog signal is represented in digital form.

The following example illustrates the basic operation of the flash A/D converter in Figure 13-30.

EXAMPLE 13-6 Determine the binary code output of the three-bit flash A/D converter for the analog input signal in Figure 13-31 and the sampling pulses (encoder Enable) shown. For this example, $V_{REF} = +8$ V.

FIGURE 13–31
Sampling of values on an analog waveform for conversion to digital form.

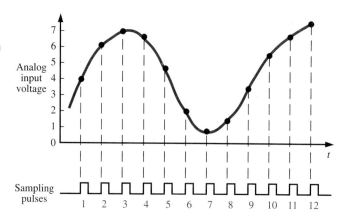

SOLUTION
The resulting A/D output sequence is listed as follows and shown in the waveform diagram of Figure 13–32 in relation to the sampling pulses.

100, 110, 110, 110, 100, 010, 000, 001, 011, 101, 110, 111

FIGURE 13–32
Resulting digital outputs for sampled values. Output D_0 is the least significant bit (LSB).

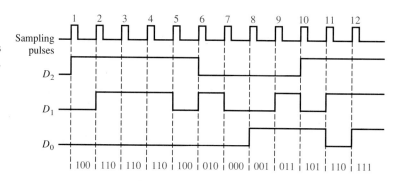

PRACTICE EXERCISE 13–6
If the amplitude of the analog voltage in Figure 13–31 is reduced by half, what will the A/D output sequence be?

STAIRSTEP-RAMP A/D CONVERTER

The stairstep-ramp method of A/D conversion is also known as the *digital-ramp* or the *counter* method. It employs a D/A converter and a binary counter to generate the digital value of an analog input. Figure 13–33 shows a diagram of this type of converter.

Assume that the counter begins in the reset state (all 0s) and the output of the D/A converter is zero. Now assume that an analog voltage is applied to the input. When it exceeds the reference voltage (output of D/A), the comparator switches to a high-level output state and enables the AND gate. The clock pulses begin advancing the counter through its binary states, producing a stairstep reference voltage from the D/A converter.

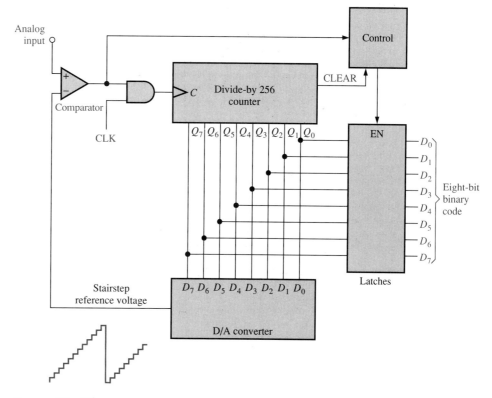

FIGURE 13–33
Stairstep-ramp A/D converter (eight bits).

The counter continues to advance from one binary state to the next, producing successively higher steps in the reference voltage. When the stairstep reference voltage reaches the analog input voltage, the comparator output will go to its low level and disable the AND gate, thus cutting off the clock pulses to stop the counter. The binary state of the counter at this point equals the number of steps in the reference voltage required to make the reference equal to or greater than the analog input. This binary number, of course, represents the value of the analog input. The control logic loads the binary count into the latches and resets the counter, thus beginning another count sequence to sample the input value.

The stairstep-ramp method is slower than the flash method because, in the worst case of maximum input, the counter must sequence through its maximum number of states before a conversion occurs. For an eight-bit conversion, this means a maximum of 256 counter states. Figure 13–34 illustrates a conversion sequence for a four-bit conversion. Notice that for each sample, the counter must count from zero up to the point at which the stairstep reference voltage reaches the analog input voltage. The conversion time varies, depending on the analog voltage.

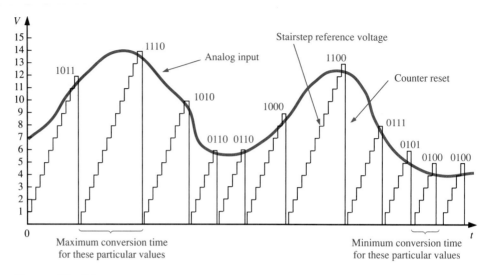

FIGURE 13–34
Example of a four-bit conversion, showing an analog input and the stairstep reference voltage.

TRACKING A/D CONVERTER

The tracking method uses an up/down counter (a counter that can go either way in a binary sequence) and is faster than the stairstep-ramp method because the counter is not reset after each sample, but rather tends to track the analog input. Figure 13–35 shows a typical eight-bit tracking A/D converter.

As long as the D/A output reference voltage is less than the analog input, the comparator output level is high, putting the counter in the up mode, which causes it to produce an up sequence of binary counts. This causes an increasing stairstep reference voltage out of the D/A converter, which continues until the stairstep reaches the value of the input voltage.

When the reference voltage equals the analog input, the comparator's output switches to its low level and puts the counter in the down mode, causing it to back up one count. If the analog input is decreasing, the counter will continue to back down in its sequence and effectively track the input. If the input is increasing, the counter will back down one count after the comparison occurs and then will begin counting up again. When the input is constant, the counter backs down one count when a comparison occurs. The reference output is now less than the analog input, and the comparator output goes to its high level, causing the counter to count up. As soon as the counter increases one state, the reference voltage becomes greater than the input, switching the comparator to its low-output state. This causes the counter to back down one count. This back-and-forth action continues as long as the analog input is a constant value, thus causing an oscillation between two binary states in the A/D output. This is a disadvantage of this type of converter.

Figure 13–36 illustrates the tracking action of this type of A/D converter for a four-bit conversion.

Data Conversion Circuits

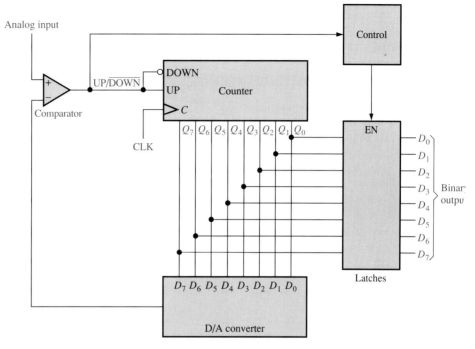

Figure 13-35
An eight-bit tracking A/D converter.

Figure 13-36
Tracking action of an A/D converter.

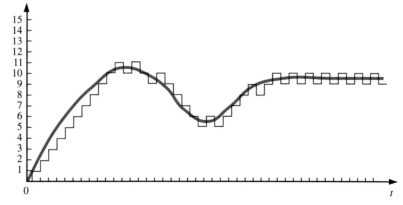

Single-Slope A/D Converter

Unlike the previous two methods, the single-slope converter does not require a D/A converter. It uses a linear ramp generator to produce a constant-slope reference voltage. A diagram is shown in Figure 13-37.

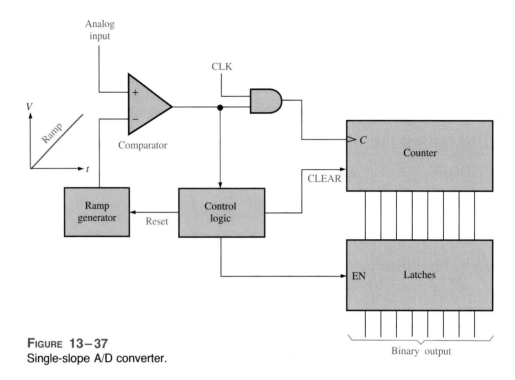

FIGURE 13–37
Single-slope A/D converter.

At the beginning of a conversion cycle, the counter is in the reset state and the ramp generator output is 0 V. The analog input is greater than the reference voltage at this point and therefore produces a high-level output from the comparator. This high level enables the clock to the counter and starts the ramp generator.

Assume that the slope of the ramp is 1 V/ms. The ramp will increase until it equals the analog input; at this point the ramp is reset, and the binary count is stored in the latches by the control logic. Let's assume that the analog input is 2 V at the point of comparison. This means that the ramp is also 2 V and has been running for 2 ms. Since the comparator output has been at its high level for 2 ms, 200 clock pulses have been allowed to pass through the gate to the counter (assuming a clock frequency of 100 kHz). At the point of comparison, the counter is in the binary state that represents decimal 200. With proper scaling and decoding, this binary number can be displayed as 2.00 V. This basic concept is used in some digital voltmeters.

DUAL-SLOPE A/D CONVERTER

The operation of the dual-slope A/D converter is similar to that of the single-slope type except that a variable-slope ramp and a fixed-slope ramp are both used. This type of converter is common in digital voltmeters and other types of measurement instruments.

A ramp generator (integrator), A_1, is used to produce the dual-slope characteristic. A block diagram of a dual-slope A/D converter is shown in Figure 13–38 for reference.

We will start by assuming that the counter is reset and the output of the integrator is zero. Now assume that a positive input voltage is applied to the input through the switch

DATA CONVERSION CIRCUITS

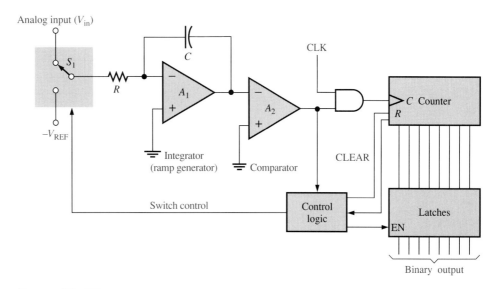

FIGURE 13–38
Dual-slope A/D converter.

(S_1) as selected by the control logic. Since the inverting input of A_1 is at virtual ground, and assuming that V_{IN} is constant for a period of time, there will be constant current through the input resistor R and therefore through the capacitor C. Capacitor C will charge linearly because the current is constant, and as a result, there will be a negative-going linear voltage ramp on the output of A_1, as illustrated in Figure 13–39(a).

When the counter reaches a specified count, it will be reset, and the control logic will switch the negative reference voltage ($-V_{REF}$) to input A_1 as shown in Figure 13–39(b). At this point the capacitor is charged to a negative voltage ($-V$) proportional to the input analog voltage.

Now the capacitor discharges linearly because of the constant current from the $-V_{REF}$ as shown in Figure 13–39(c). This linear discharge produces a positive-going ramp on the A_1 output, starting at $-V$ and having a constant slope that is independent of the charge voltage.

As the capacitor discharges, the counter advances from its reset state. The time it takes the capacitor to discharge to zero depends on the initial voltage $-V$ (proportional to V_{IN}) because the discharge rate (slope) is constant. When the integrator (A_1) output voltage reaches zero, the comparator (A_2) switches to its low state and disables the clock to the counter. The binary count is latched, thus completing one conversion cycle. The binary count is proportional to V_{IN} because the time it takes the capacitor to discharge depends only on $-V$, and the counter records this interval of time.

SUCCESSIVE-APPROXIMATION A/D CONVERTER

Perhaps the most widely used method of A/D conversion is **successive approximation.** It has a much faster conversion time than the other methods with the exception of the flash

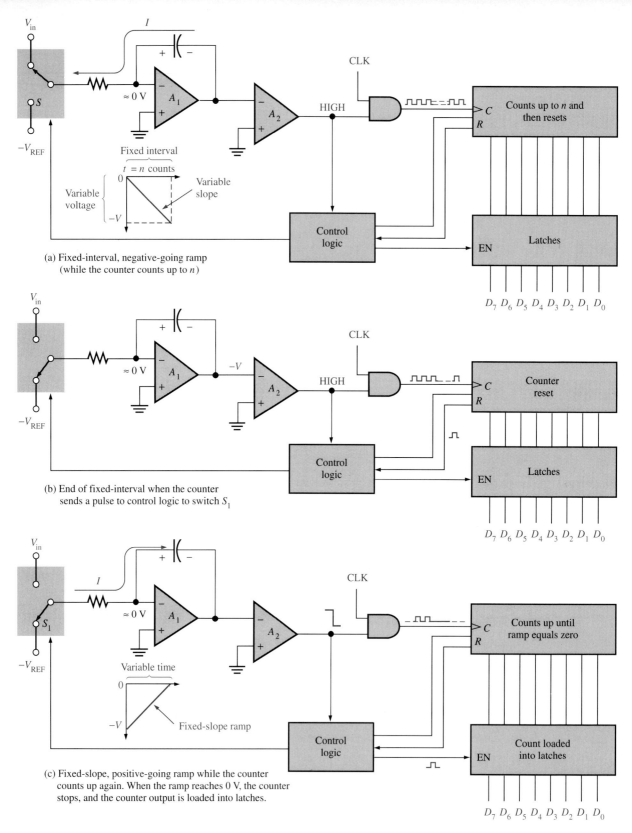

FIGURE 13–39
Dual-slope conversion.

method. It also has a fixed conversion time that is the same for any value of the analog input.

Figure 13–40 shows a basic block diagram of a four-bit successive-approximation A/D converter. It consists of a D/A converter, a successive-approximation register (SAR), and a comparator. The basic operation is as follows: The bits of the D/A converter are enabled one at a time, starting with the most significant bit (MSB). As each bit is enabled, the comparator produces an output that indicates whether the analog input voltage is greater or less than the output of the D/A. If the D/A output is greater than the analog input, the comparator's output is low, causing the bit in the register to reset. If the D/A output is less than the analog input, the bit is retained in the register.

FIGURE 13–40
Successive-approximation A/D converter.

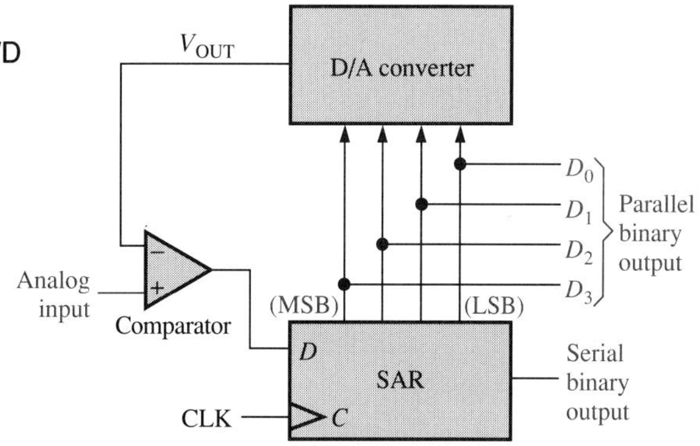

The system does this with the MSB first, then the next most significant bit, then the next, and so on. After all the bits of the D/A have been tried, the conversion cycle is complete.

In order to better understand the operation of the successive-approximation A/D converter, we will take a specific example of a four-bit conversion. Figure 13–41 illustrates the step-by-step conversion of a given analog input voltage (5 V in this case). We will assume that the D/A converter has the following output characteristic: $V_{OUT} = 8$ V for the 2^3 bit (MSB), $V_{OUT} = 4$ V for the 2^2 bit, $V_{OUT} = 2$ V for the 2^1 bit, and $V_{OUT} = 1$ V for the 2^0 bit (LSB).

Figure 13–41(a) shows the first step in the conversion cycle with the MSB = 1. The output of the D/A is 8 V. Since this is greater than the analog input of 5 V, the output of the comparator is low, causing the MSB in the SAR to be reset to a 0.

Figure 13–41(b) shows the second step in the conversion cycle with the 2^2 bit equal to a 1. The output of the D/A is 4 V. Since this is less than the analog input of 5 V, the output of the comparator switches to its high level, causing this bit to be retained in the SAR.

Figure 13–41(c) shows the third step in the conversion cycle with the 2^1 bit equal to a 1. The output of the D/A is 6 V because there is a 1 on the 2^2 bit input and on the 2^1 bit input and 4 V + 2 V = 6 V. Since this is greater than the analog input of 5 V, the output of the comparator switches to its low level, causing this bit to be reset to a 0.

FIGURE 13–41
Successive-approximation conversion process.

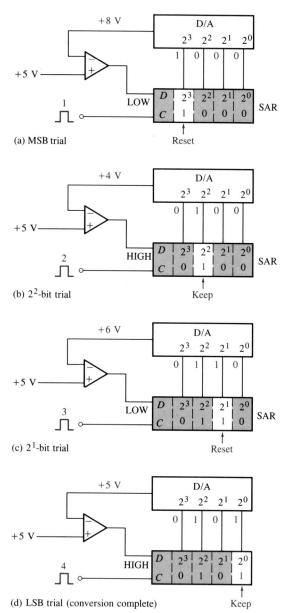

(a) MSB trial

(b) 2^2-bit trial

(c) 2^1-bit trial

(d) LSB trial (conversion complete)

Figure 13–41(d) shows the fourth and final step in the conversion cycle with the 2^0 bit equal to a 1. The output of the D/A is 5 V because there is a 1 on the 2^2 bit input and on the 2^0 bit input and 4 V + 1 V = 5 V.

The four bits have all been tried, thus completing the conversion cycle. At this point the binary code in the register is 0101, which is the binary value of the analog input of 5 V. Another conversion cycle now begins, and the basic process is repeated. The SAR is cleared at the beginning of each cycle.

A Specific A/D Converter

The ADC0801 is an example of a successive-approximation analog-to-digital converter. A block diagram is shown in Figure 13–42. This device operates from a +5 V supply and has a resolution of eight bits with a conversion time of 100 μs. Also, it has guaranteed monotonicity and an on-chip clock generator. The data outputs are tristate so that they can be interfaced with a microprocessor bus system.

FIGURE 13–42
The ADC0801 analog-to-digital converter.

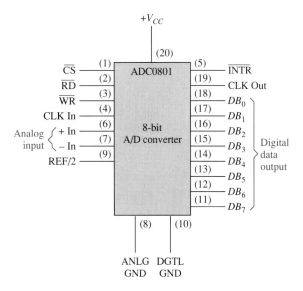

A detailed logic diagram of the ADC0801 is shown in Figure 13–43, and the basic operation of the device is as follows. The ADC0801 contains the equivalent of a 256-resistor D/A converter network. The successive-approximation logic sequences the network to match the analog differential input voltage ($+V_{IN} - -V_{IN}$) with an output from the resistive network. The MSB is tested first. After eight comparisons (sixty-four clock periods), an eight-bit binary code is transferred to an output latch, and the interrupt ($\overline{INTR}$) output goes low. The device can be operated in a free-running mode by connecting the $\overline{INTR}$ output to the write ($\overline{WR}$) input and holding the conversion start ($\overline{CS}$) low. To ensure start-up under all conditions, a low $\overline{WR}$ input is required during the power-up cycle. Taking $\overline{CS}$ low anytime after that will interrupt the conversion process.

When the $\overline{WR}$ input goes low, the internal successive-approximation register (SAR) and the eight-bit shift register are reset. As long as both $\overline{CS}$ and $\overline{WR}$ remain low, the analog-to-digital converter remains in a reset state. One to eight clock periods after $\overline{CS}$ or $\overline{WR}$ makes a low-to-high transition, conversion starts.

When the $\overline{CS}$ and $\overline{WR}$ inputs are low, the start flip-flop is set, and the interrupt flip-flop and eight-bit register are reset. The high is ANDed with the next clock pulse, which puts a high on the reset input of the start flip-flop. If either $\overline{CS}$ or $\overline{WR}$ has gone high, the set signal to the start flip-flop is removed, causing it to be reset. A high is placed on the D input of the eight-bit shift register, and the conversion process is started. If the $\overline{CS}$ and $\overline{WR}$ inputs are still low, the start flip-flop, the eight-bit shift register, and the SAR

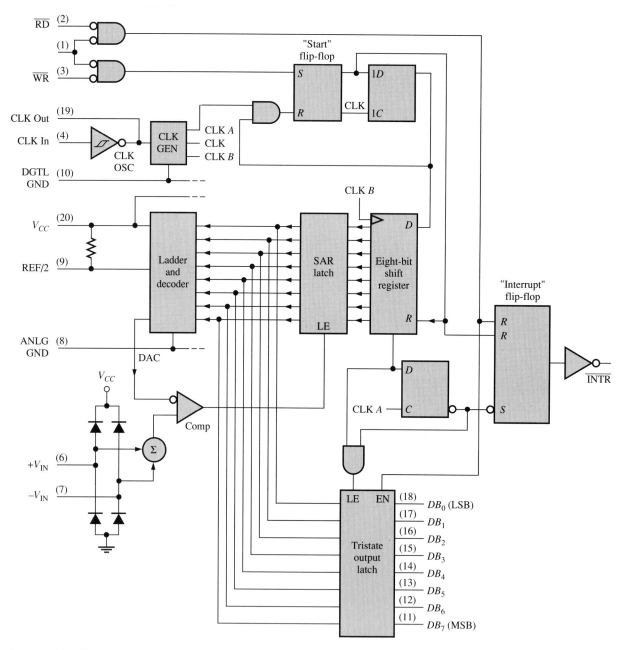

FIGURE 13-43
Logic diagram of the ADC0801 A/D converter.

remain reset. This action allows for wide $\overline{CS}$ and $\overline{WR}$ inputs, with conversion starting from one to eight clock periods after one of the inputs has gone high.

When the high input has been clocked through the eight-bit shift register, completing the SAR search, it is applied to an AND gate controlling the output latches and to the *D* input of a flip-flop. On the next clock pulse, the digital word is transferred to the tristate output latches, and the interrupt flip-flop is set. The output of the interrupt flip-flop is inverted to provide an $\overline{INTR}$ output that is high during conversion and low when conversion is complete.

When a low is at both the $\overline{CS}$ and $\overline{RD}$ inputs, the tristate output latch is enabled, the output code is applied to the DB_0 through DB_7 lines, and the interrupt flip-flop is reset. When either the $\overline{CS}$ or the $\overline{RD}$ input returns to a high, the DB_0 through DB_7 outputs are disabled. The interrupt flip-flop remains reset.

Several additional IC analog-to-digital converters are listed in Table 13–1.

TABLE 13–1
Several popular A/D converters

Device	Type	Description	Resolution	Conversion Time	Supply Voltages
MC14433P	A/D	Dual Slope	3½ Digits	40 ms	+5 V, +8 V
MC14443P	A/D	Single Slope	8 bits	300 ms	+5 V, +8 V
MC14447P	A/D	Single Slope	8 bits	300 ms	−5 V, +8 V
MC14559BCP	A/D	Successive Approximation	8 bits	—	±3 V to ±18 V
ADC0803	A/D	Successive Approximation	8 bits	100 μs	+5 V
ADC0808	A/D	Successive Approximation	8 bits	100 μs	+5 V
ADC0809	A/D	Successive Approximation	8 bits	100 μs	+5 V
ADC0817	A/D	Successive Approximation	8 bits	100 μs	+5 V
ADC0820	A/D	Flash Conversion	—	1.18 μs	+5 V

13–6 REVIEW QUESTIONS

1. What is the fastest method of analog-to-digital conversion?
2. Which A/D conversion method uses an up/down counter?
3. The successive-approximation converter has a fixed conversion time. (True or false)

13–7 VOLTAGE-TO-FREQUENCY (V/F) AND FREQUENCY-TO-VOLTAGE (F/V) CONVERTERS

Voltage-to-frequency converters convert an analog input voltage to a pulse stream or square wave in such a way that there is a linearly proportional relationship between the analog voltage and the frequency of the pulse stream. Frequency-to-voltage converters perform the inverse operation by converting a pulse stream to a voltage that is proportional to the pulse stream frequency. Actually, V/F and F/V converters can be used as A/D and D/A converters in certain applications. In other applications, V/F and F/V converters are used, for example, in high-noise immunity digital transmission and in digital voltmeters.

A Basic V/F Converter

The concept of voltage-to-frequency converters is shown in Figure 13–44. An analog voltage on the input is converted to a pulse signal with a frequency that is directly proportional to the amplitude of the input voltage. There are several ways to implement a V/F converter. For example, the VCO (voltage-controlled oscillator) with which you are already familiar can be used as one type of V/F converter. In this section, we will look at a relatively common implementation called the *charge-balance* V/F converter.

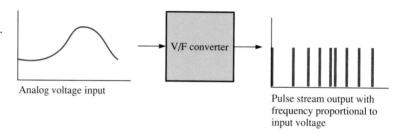

FIGURE 13–44
The basic V/F concept.

Figure 13–45 shows the diagram of a basic charge-balance V/F converter. It consists of an integrator, a comparator, a one-shot, a current source, and an electronic switch. The input resistor R_{in}, the integration capacitor C_{int}, and the one-shot timing capacitor C_{os} are components whose values are selected based on desired performance.

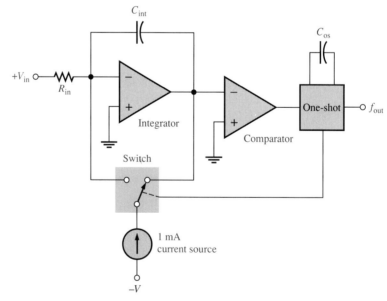

FIGURE 13–45
A basic voltage-to-frequency converter.

The basic operation of the V/F converter in Figure 13–46 is as follows: A positive input voltage produces an input current ($I_{in} = V_{in}/R_{in}$) which charges the capacitor C_{int}, as indicated in Figure 13–46(a). During this integrate mode, the integrator output voltage

DATA CONVERSION CIRCUITS

FIGURE 13–46
Basic operation of a V/F converter for a constant input voltage.

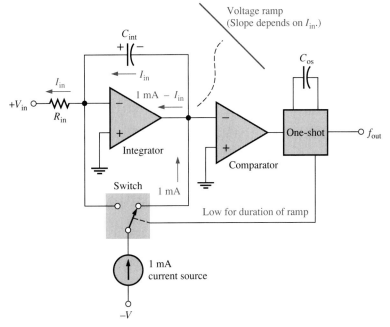

(a) V/F converter in the integrate mode

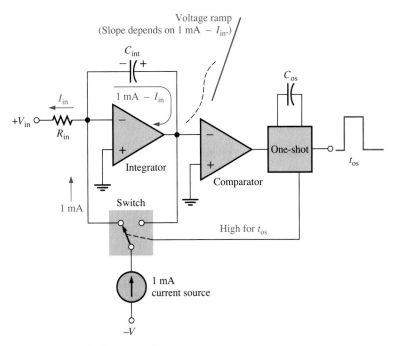

(b) V/F converter in the reset mode

is a downward ramp, as shown. When the integrator output voltage reaches zero, the comparator triggers the one-shot. The one-shot produces a pulse with a fixed width, t_{os}, that switches the 1 mA current source to the input of the integrator and initiates the reset mode.

During the reset mode, current through the capacitor is in the opposite direction as indicated in Figure 13–46(b). This produces an upward ramp on the integrator output as indicated. After the one-shot times out, the current source is switched back to the integrator output, initiating another integrate mode and the cycle repeats.

If the input voltage in held constant, the output waveform of the integrator is as shown in Figure 13–47(a), where the amplitude and the integrate time remain constant. The final output of the V/F converter is taken off the one-shot, as indicated in Figure 13–46. As long as the input voltage is constant, the output pulse stream has a constant frequency as indicated in Figure 13–47(b).

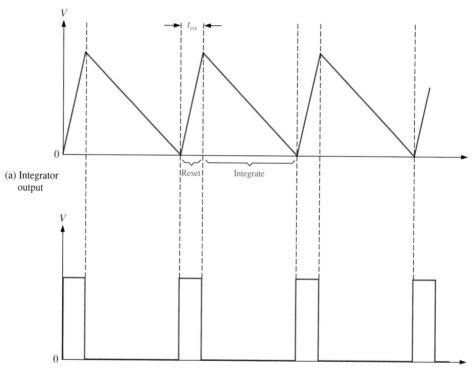

(a) Integrator output

(b) Final output (one-shot)

FIGURE 13–47
V/F converter waveforms for a constant input voltage.

WHEN THE INPUT VOLTAGE INCREASES An increase in the input voltage, V_{in}, causes the input current, I_{in}, to increase. In the basic relationship $I_C = (V_C/t)C$, the term V_C/t is the slope of the capacitor voltage. If the current increases, V_C/t also increases since C is constant. As applied to the V/F converter, this means that if the input current (I_{in}) increases, then

DATA CONVERSION CIRCUITS

the slope of the integrator output during the integrate mode will also increase and reduce the period of the final output voltage. Also, during the reset mode, the opposite current through the capacitor, 1 mA $- I_{in}$, is smaller, thus decreasing the slope of the upward ramp and reducing the amplitude of the integrator output voltage. This is illustrated in the waveform diagram of Figure 13–48 where the input voltage, and thus the input current, takes a step increase from one value to another. Notice that during reset, the positive-going slope of the integrator voltage is less, so it reaches a smaller amplitude during the time t_{os}. Remember, t_{os} does not change. Notice also that during integration, the negative-going slope of the integrator voltage is greater, so it reaches zero quicker. The net result of this increase in input voltage is that the output frequency increases an amount proportional to the increase in the input voltage. So, as the input voltage varies, the output frequency varies proportionally.

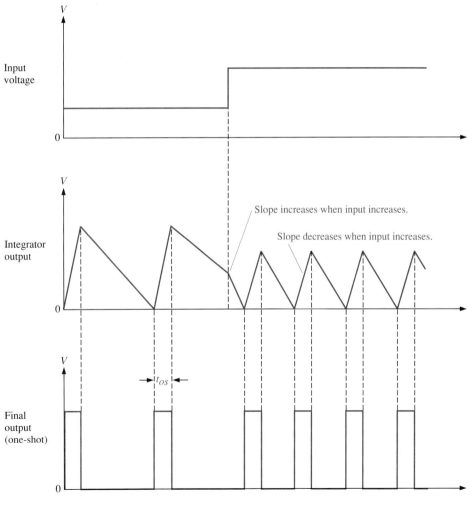

FIGURE 13–48
The output frequency increases when the input voltage increases.

The AD650 Integrated Circuit V/F Converter

The AD650 is a good example of a V/F converter very similar to the basic device we just discussed. The main differences in the AD650 are the output transistor and the comparator threshold voltage of -0.6 V instead of ground, as shown in Figure 13–49. The input resistor, integrating capacitor, one-shot capacitor, and the output pull-up resistor are external components, as indicated.

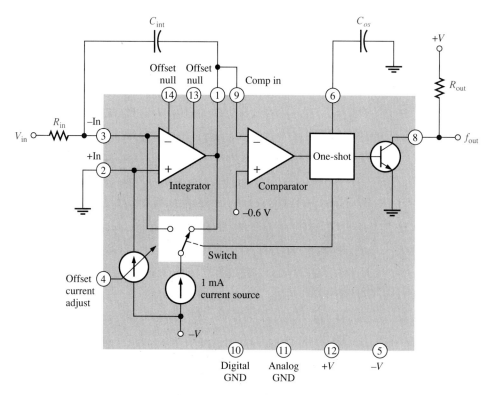

FIGURE 13–49
The AD650 V/F converter.

The values of the external components determine the operating characteristics of the device. The pulse width of the one-shot output is set by the following formula:

$$t_{os} = C_{os}(6.8 \times 10^3 \text{ s/F}) + 3 \times 10^{-7} \text{ s} \tag{13-1}$$

During the reset interval, the integrator output voltage increases by an amount expressed as

$$\Delta V = \frac{(1 \text{ mA} - I_{in})t_{os}}{C_{int}} \tag{13-2}$$

Data Conversion Circuits

The duration of the integrate interval when the integrator output is sloping downward is

$$t_{int} = \frac{\Delta V}{I_{in}/C_{int}}$$

$$= \frac{t_{os}(1 \text{ mA} - I_{in})/C_{int}}{I_{in}/C_{int}}$$

$$t_{int} = \left(\frac{1 \text{ mA}}{I_{in}} - 1\right) t_{os} \qquad (13\text{--}3)$$

The period of a full cycle consists of the reset interval plus the integrate interval.

$$T = t_{os} + t_{int}$$

$$= t_{os} + \left(\frac{1 \text{ mA}}{I_{in}} - 1\right) t_{os}$$

$$= \left(1 + \frac{1 \text{ mA}}{I_{in}} - 1\right) t_{os}$$

$$T = \left(\frac{1 \text{ mA}}{I_{in}}\right) t_{os}$$

Therefore, the output frequency can be expressed as

$$f_{out} = \frac{I_{in}}{(t_{os})1 \text{ mA}} \qquad (13\text{--}4)$$

As you can see in Equation (13–4), the output frequency is directly proportional to the input current; and since $I_{in} = V_{in}/R_{in}$, it is also directly proportional to the input voltage and inversely proportional to the input resistance. The output frequency is also inversely proportional to t_{os}, which depends on the value of C_{os}.

EXAMPLE 13–7 Determine the output frequency for the AD650 V/F converter in Figure 13–50 on page 662 when a constant input voltage of 5 V is applied.

SOLUTION

$$t_{os} = C_{os}(6.8 \times 10^3 \text{ s/F}) + 3 \times 10^{-7} \text{ s}$$
$$= 330 \text{ pF}(6.8 \times 10^3 \text{ s/F}) + 3 \times 10^{-7} \text{ s}$$
$$= 2.5 \text{ } \mu\text{s}$$

$$I_{in} = \frac{V_{in}}{R_{in}} = \frac{5 \text{ V}}{10 \text{ k}\Omega} = 500 \text{ } \mu\text{A}$$

$$f_{out} = \frac{I_{in}}{t_{os}(1 \text{ mA})} = \frac{500 \text{ } \mu\text{A}}{(2.5 \text{ } \mu\text{s})(1 \text{ mA})} = 200 \text{ kHz}$$

PRACTICE EXERCISE 13–7

What are the minimum and maximum output frequencies for the V/F converter in Figure 13–49 when a triangular wave with a minimum peak value of 1 V and a maximum peak value of 6 V is applied to the input?

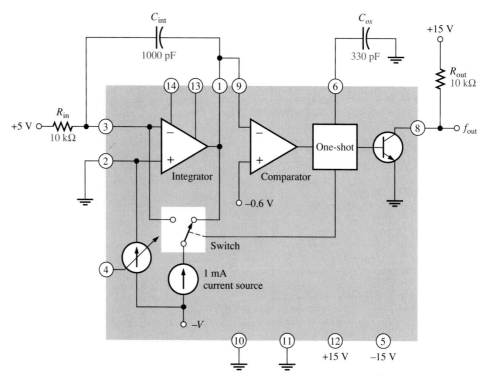

FIGURE 13–50

FIGURE 13–51
A basic frequency-to-voltage (F/V) converter.

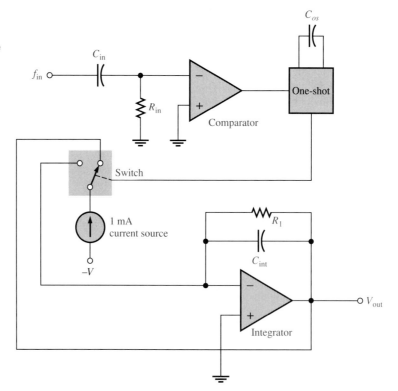

A Basic F/V Converter

Figure 13–51 shows a basic frequency-to-voltage converter. The elements are the same as those in the voltage-to-frequency converter of Figure 13–45, but they are connected differently.

When an input frequency is applied to the comparator input, it triggers the one-shot which produces a fixed pulse width (t_{os}) determined by C_{os}. This switches the 1 mA current source to the integrator input and C_{int} charges. Between one-shot pulses, C_{int} discharges through R_1. The higher the input frequency, the closer the one-shot pulses are together and the less C_{int} discharges. This causes the integrator output to increase as input frequency increases and to decrease as the input frequency decreases. The integrator output is the final voltage output of the F/V converter. F/V conversion action is illustrated by the waveforms in Figure 13–52. C_{int} and R_1 act as a filter and tend to smooth out the ripples on the integrator ouput as indicated by the dashed curve.

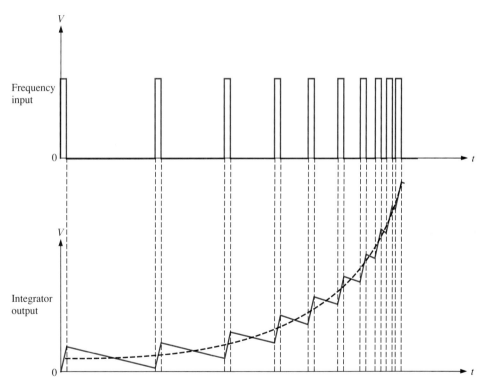

FIGURE 13–52
An example of frequency-to-voltage conversion.

Figure 13–53 shows the AD650 connected to function as a frequency-to-voltage converter. Compare this configuration with the voltage-to-frequency connection in Figure 13–49.

664   Chapter 13

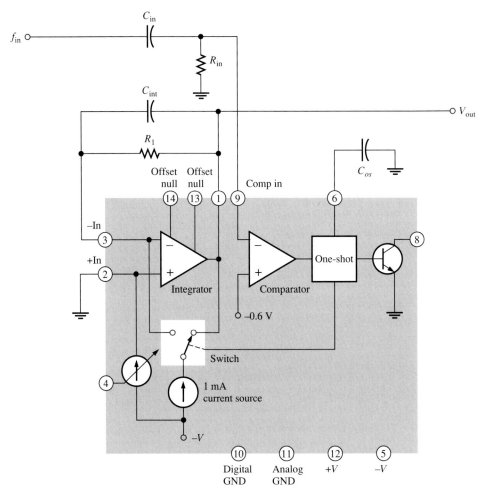

FIGURE 13–53
The AD650 connected as a frequency-to-voltage (F/V) converter.

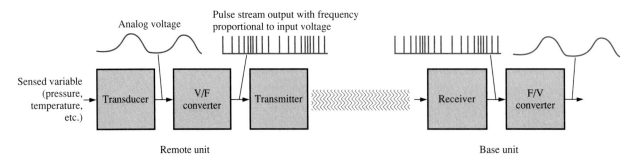

FIGURE 13–54
Basic application of V/F and F/V conversion.

Data Conversion Circuits

An Application

One application of V/F and F/V converters is in the remote sensing of a quantity (temperature, pressure, level) that is converted to an analog voltage by a transducer. The analog voltage is then converted to a pulse frequency by a V/F converter which is then transmitted by some method (radio link, fiber-optical link, telemetry) to a base unit receiver that includes an F/V converter. This basic application of V/F and F/V conversion is illustrated in Figure 13–54.

13–7 Review Questions

1. List the basic components in a typical V/F converter.
2. In a V/F converter, if the input voltage changes from 1 V to 6.5 V, what happens to the output?
3. Describe the basic differences between a V/F and a F/V converter in terms of inputs and outputs.

13–8 Troubleshooting

Basic testing of D/A and A/D converters includes checking their performance characteristics, such as monotonicity, offset, linearity, and gain, and checking for missing or incorrect codes. In this section, the fundamentals of testing these analog interfaces are introduced.

Testing D/A Converters

The concept of D/A converter testing is illustrated in Figure 13–55. In this basic method, a sequence of binary codes is applied to the inputs, and the resulting output is observed. The binary code sequence extends over the full range of values from 0 to $2^n - 1$ in ascending order, where n is the number of bits.

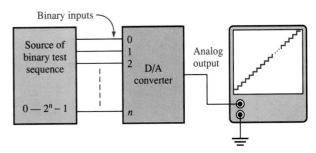

Figure 13–55
Basic test setup for a D/A converter.

The ideal output is a straight-line stairstep as indicated. As the number of bits in the binary code is increased, the resolution is improved. That is, the number of discrete steps increases, and the output approaches a straight-line linear ramp.

D/A Conversion Errors

Several D/A conversion errors to be checked for are shown in Figure 13–56, which uses a four-bit conversion for illustration purposes. A four-bit conversion produces fifteen discrete steps. Each graph in the figure includes an ideal stairstep ramp for comparison with the faulty outputs.

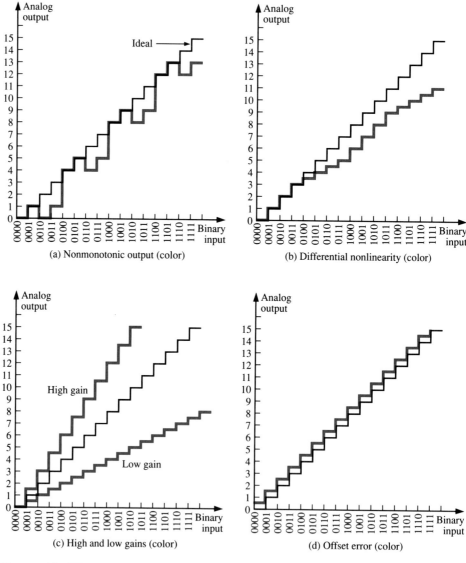

FIGURE 13–56
Illustrations of several D/A conversion errors.

NONMONOTONICITY The step reversals in Figure 13–56(a) indicate **nonmonotonic** performance, which is a form of nonlinearity. In this particular case, the error occurs because

DATA CONVERSION CIRCUITS

the 2^1 bit in the binary code is interpreted as a constant 0. That is, a short is causing the bit input line to be stuck in the low state.

DIFFERENTIAL NONLINEARITY Figure 13–56(b) illustrates differential nonlinearity in which the step amplitude is less than it should be for certain input codes. This particular output could be caused by the 2^2 bit's having an insufficient weight, perhaps because of a faulty input resistor. We could also see steps with amplitudes greater than normal if a particular binary weight were greater than it should be.

LOW OR HIGH GAIN Output errors caused by low or high gain are illustrated in Figure 13–56(c). In the case of low gain, all of the step amplitudes are less than ideal. In the case of high gain, all of the step amplitudes are greater than ideal. This situation may be caused by a faulty feedback resistor in the op-amp circuit.

OFFSET ERROR An offset error is illustrated in Figure 13–56(d). Notice that when the binary input is 0000, the output voltage is nonzero and that this amount of offset is the same for all steps in the conversion. A faulty op-amp may be the culprit in this situation.

EXAMPLE 13–8

The D/A converter output in Figure 13–57 is observed when a straight four-bit binary sequence is applied to the inputs. Identify the type of error, and suggest an approach to isolate the fault.

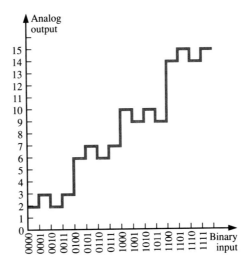

FIGURE 13–57

SOLUTION
The D/A converter in this case is nonmonotonic. Analysis of the output reveals that the device is converting the following sequence, rather than the actual binary sequence applied to the inputs.

0010, 0011, 0010, 0011, 0110, 0111, 0110, 0111, 1010, 1011, 1010, 1011, 1110, 1111, 1110, 1111

Apparently, the 2^1 bit (second from right) is stuck in the high state. To find the problem, first monitor the bit input pin of the device. If it is changing states, the fault is internal, most likely an open. If the external pin is not changing states and is always high, check for an external short to $+V$ that may be caused by a solder bridge somewhere on the circuit board. If no problem is found here, disconnect the source output from the D/A input pin, and see if the output signal is correct. If these checks produce no results, the fault is most likely internal to the D/A converter, perhaps a short to the supply voltage.

PRACTICE EXERCISE 13-8
Graph the D/A output if a straight binary sequence is applied and the most significant bit input of the D/A is stuck high.

TESTING A/D CONVERTERS

One method for testing A/D converter is shown in Figure 13-58. A D/A converter is used as part of the test setup to convert the A/D output back to analog form for comparison with the test input.

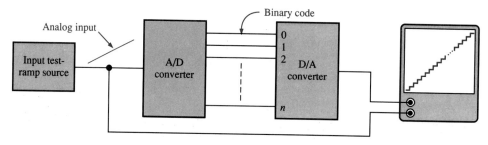

FIGURE 13-58
A method for testing A/D converters.

A test input in the form of a linear ramp is applied to the input of the A/D converter. The resulting binary output sequence is then applied to the D/A test unit and converted to a stairstep ramp. The input and output ramps are compared for any deviation.

A/D CONVERSION ERRORS

Again, a four-bit conversion is used to illustrate the principles. We assume that the test input is an ideal linear ramp.

MISSING CODE The stairstep output in Figure 13-59(a) indicates that the binary code 1001 does not appear on the output of the A/D converter. Notice that the 1000 value stays for two intervals and then the output jumps to the 1010 value.

In a flash A/D converter, for example, a failure of one of the comparators can cause a missing-code error.

Data Conversion Circuits

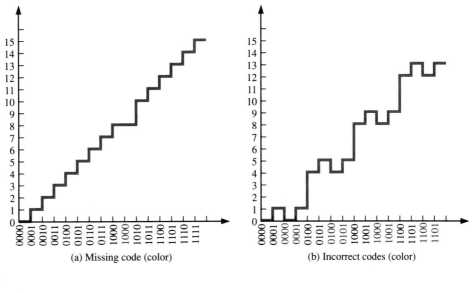

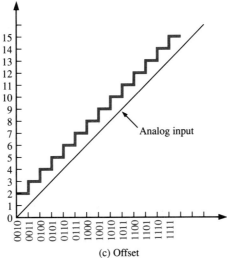

FIGURE 13–59
Illustrations of A/D conversion errors.

INCORRECT CODES The stairstep output in Figure 13–59(b) indicates that several of the binary code words coming out of the A/D converter are incorrect. Analysis indicates that the 2^1-bit line is stuck in the low state in this particular case.

OFFSET Offset conditions are shown in 13–59(c). In this situation, the A/D converter interprets the analog input voltage as greater than its actual value. This error is probably due to a faulty comparator circuit.

EXAMPLE 13-9

A four-bit flash A/D converter is shown in Figure 13–60(a). It is tested with a setup like the one in Figure 13–58. The resulting reconstructed analog output is shown in Figure 13–60(b). Identify the problem and the most probable fault.

FIGURE 13-60

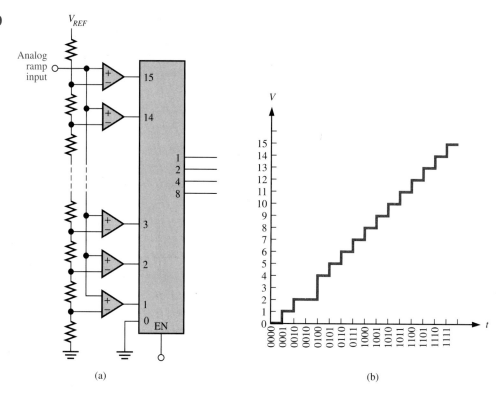

(a) (b)

SOLUTION
The binary code 0011 is missing from the A/D output, as indicated by the missing step. Most likely, the output of comparator 3 is stuck in its inactive state (low).

PRACTICE EXERCISE 13-9
If the output of comparator 15 is stuck in the high state, what will be the reconstructed analog output when the A/D converter is tested in a setup like the one in Figure 13–58?

13-8 REVIEW QUESTIONS

1. How do you detect nonmonotonic behavior in a D/A converter?
2. What effect does low gain have on a D/A output?
3. Name two types of output errors in an A/D converter.

13-9 A System Application

The solar panel control system introduced at the opening of this chapter includes analog multiplexers and A/D converters. The analog multiplexers accept position information from the potentiometers located on each solar unit and send it to the A/D converters where the analog position information is converted to digital form. The digital outputs of the A/D converters go to the digital controller for processing and control signals are sent back to the solar units to keep them properly positioned. In this section, you will

- *See how an analog multiplexer can be used to collect analog data.*
- *See how an A/D converter is used in a system application.*
- *Translate between a printed circuit board and a schematic.*
- *Analyze the analog board.*
- *Troubleshoot some common problems.*

A Brief Description of the System

The purpose of the solar panel control system is to maintain each solar panel at approximately a 90° angle with respect to the sun's rays. Two angular positions are required to properly align the panels. The azimuth position is along a curve from east to west parallel with the horizon. The elevation position is from the horizon to directly overhead. These angular movements for a solar unit are illustrated in Figure 13-61. There are two stepping motors that drive the solar panel to the proper position, one for azimuth and one for elevation. Also, there are two position transducers (potentiometers are used as the sensors in this case) that produce voltages proportional to the positions of the panel as determined by the angular positions of the motor shafts.

Figure 13-61

A solar panel with position controls and sensors for azimuth and elevation.

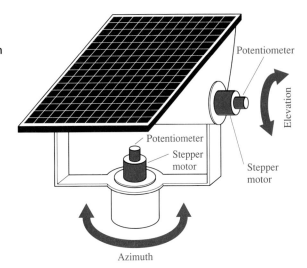

In this application, the movement of the solar panel is very slow. The azimuth angle begins at an easterly orientation at sunrise and turns through approximately 180° by sunset. The elevation angle tracks the arc of the sun each day and also adjusts for seasonal variation of the sun's relative position in the sky. On the first day of winter it tracks through the lowest arc and on the first day of summer it tracks through the highest arc.

A basic block diagram of the control electronics is shown in Figure 13–62. There are two 4-input analog multiplexers, one for the azimuth and one for the elevation. Each input has a dc voltage coming from the associated position potentiometer. Each voltage is proportional to the current angular position of the solar panel as measured by the potentiometer. Every five minutes the digital controller causes the multiplexer to quickly sequence through the four azimuth and the four elevation position voltages so that one voltage at a time is applied to the A/D converter. Each resulting digital code is processed by comparing the angle it represents to angular information about solar position based on time of day and date that is permanently stored in the digital controller's memory. Based on its computations, the digital controller issues the proper control signal to the appropriate stepping motor to update its position.

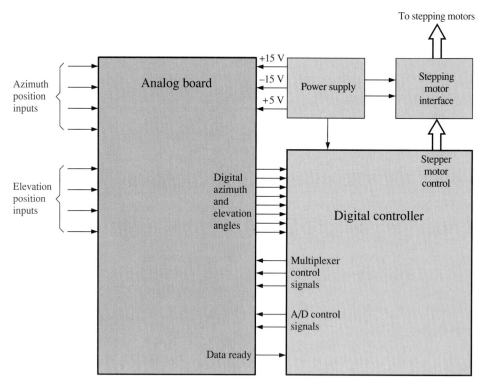

FIGURE 13–62
Basic block diagram of the solar panel control system.

THE POTENTIOMETER In this application, the potentiometer is used as a position transducer to convert the angular shaft position of the stepping motor to a proportional dc voltage.

DATA CONVERSION CIRCUITS

The potentiometer is mechanically linked to the associated motor so as the motor shaft turns, the wiper of the potentiometer slides along the resistive element. It is calibrated so that the smallest angle produces the smallest voltage. In Figure 13–63, a simplified diagram shows the basic construction and the schematic indicates minimum and maximum angles corresponding to the wiper position.

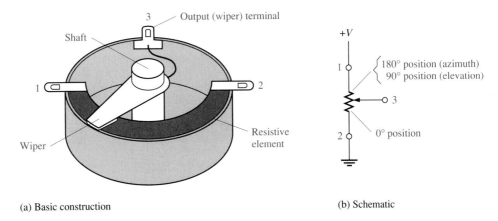

(a) Basic construction

(b) Schematic

FIGURE 13–63
The potentiometer as an angular transducer.

STEPPING MOTORS Although our focus is on the analog electronics in this system, a basic familiarization with stepping motors will help you have a better understanding of the overall system operation. A stepping motor is one in which the rotor shaft can be rotated in a series of incremental moves called steps or step angles. Stepping motors are available with step angles ranging from 0.9° to 30°. The motors used in this system are assumed to have a 2° step angle (180 steps/revolution). Rotation is achieved by applying a series of pulses to the windings. Generally, one pulse will rotate the shaft by one step angle (2° in our case). The sequence in which the pulses are applied to the windings determine the direction of rotation (CW or CCW). The speed of the motor is controlled by the rate at which the stepping pulses are applied. Obviously, in this system the motors move very slowly so speed is not a consideration.

BASIC OPERATION OF THE ANALOG CIRCUITS

A block diagram of the analog board is shown in Figure 13–64. DC analog voltages from the four azimuth potentiometers (remember there are four solar panels) are applied to the inputs of the upper AD9300, and dc voltages from the four elevation potentiometers are applied to the inputs of the lower AD9300.

The digital controller issues a high-level enable signal to the azimuth multiplexer. While the enable input is high, the digital controller sequentially switches each of the four inputs to the output by applying a 2-bit binary code to the channel select inputs. As each

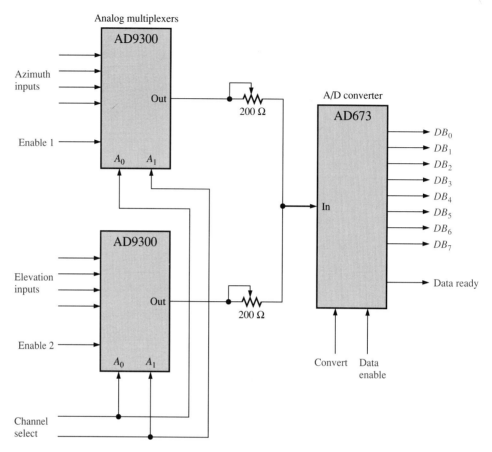

FIGURE 13–64
Simplified block diagram of the analog board.

input voltage appears on the multiplexer output, the digital controller issues a convert signal to the AD673 A/D converter to initiate a conversion. After the conversion is complete, the A/D converter then sends a data ready signal to the digital controller. The controller responds with a data enable signal, which places the 8-bit binary code on the outputs allowing the binary number to be processed by the controller.

After the first conversion, the digital controller advances to the next azimuth multiplexer input and repeats the operation. After all four azimuth inputs have been converted to digital and processed, the controller disables the azimuth multiplexer and enables the elevation multiplexer for conversion of its four inputs to digital. The 200 Ω variable resistors are used to adjust the input voltage to the A/D converter so that the highest output code (all 1s) corresponds to the maximum input voltage. Therefore, the full range of azimuth and elevation angles are converted into 256 discrete values.

The digital controller repeats the conversion sequence every eight minutes in this system. The basic timing diagram in Figure 13–65 graphically illustrates the operation.

DATA CONVERSION CIRCUITS

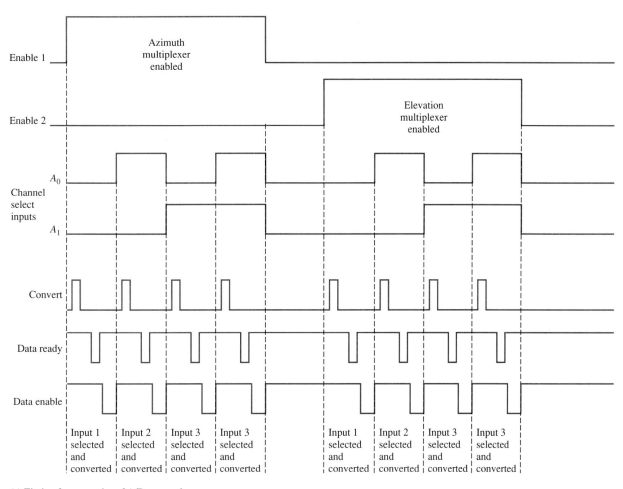

(a) Timing for one series of A/D conversions

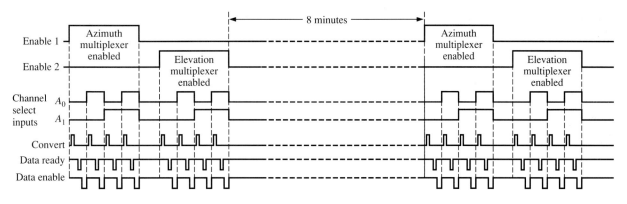

(b) A series of A/D conversions occurs every eight minutes.

FIGURE 13–65
Timing diagram for the A/D conversion sequence.

ON THE TEST BENCH

Now, so that you can take a closer look at the analog board, let's take it out of the system and put it on the test bench.

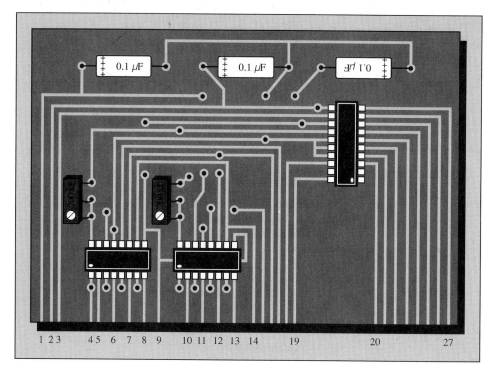

FIGURE 13–66

■ **ACTIVITY 1   RELATE THE PC BOARD TO THE SCHEMATIC**

Using the pin configuration diagrams on the data sheets for the AD9300 and the AD673 in Appendix A, complete the diagram in Figure 13–64 by adding pin numbers, ground connections, and supply voltage connections. Use the completed schematic diagram to locate and identify the components and the input and output functions on the analog board in Figure 13–66.

■ **ACTIVITY 2   ANALYZE THE SYSTEM**

For this application, assume that, at its limits, the azimuth potentiometer produces 0 V for a 0° orientation (due east) and 10 V for a 180° orientation (due west). At its limits, the elevation potentiometer produces 0 V for a 0° orientation (horizontal) and 10 V for a 90° orientation (vertical). The potentiometers rotate in 2° increments.

STEP 1   Calculate the voltage increment that will occur on the azimuth multiplexer input when the azimuth motor makes one step.

STEP 2   Calculate the voltage increment that will occur on the elevation multiplexer input when the elevation motor makes one step.

Data Conversion Circuits

Step 3 Determine the resolution in degrees to which the A/D converter can represent the input voltage range for both azimuth and elevation. Is this resolution adequate? Explain.

■ **Activity 3** **Write a Technical Report**

Describe the overall operation of the solar panel control system. List the functional requirements for each block in the system. Discuss the operation of the analog board and its role in the overall system. Use the results of Activity 2 if appropriate.

■ **Activity 4** **Troubleshoot the System for Each of the Following Problems By Stating the Probable Cause or Causes**

1. No voltage on one of the azimuth input lines to the analog multiplexer when there are 2.22 V on each of the other three inputs.
2. All four solar units are stuck at a given elevation position, but their azimuth position advances properly.
3. One solar unit is stuck at a given azimuth position, but its elevation position advances properly.

13–9 Review Questions

1. What is the purpose of the three capacitors on the analog board?
2. How many discrete values of the analog position voltage can be converted to digital?
3. Explain why this system repeats the conversion sequence every eight minutes.

Summary

- There are three basic types of analog switches: single pole–single throw (SPST), single pole–double throw (SPDT), and double pole–double throw (DPDT).
- An analog switch is typically a MOSFET that is opened or closed with a control input.
- The purpose of a sample-and-hold amplifier is to sample a voltage at a certain point in time and retain or hold that voltage for an interval of time.
- An analog quantity is one that has a continuous set of values over time.
- A digital quantity is one that has a set of discrete values over time.
- Two basic types of digital-to-analog converters are the binary-weighted-input converter and the $R/2R$ ladder converter.
- The $R/2R$ ladder D/A converter is easier to implement because only two resistor values are required compared to a different value for each input in the binary-weighted-input D/A converter.
- The number of bits in an analog-to-digital converter determines its resolution.
- The minimum sampling rate for A/D conversion is twice the maximum frequency component of the analog signal.
- The flash or simultaneous method of A/D conversion is the fastest.

- ☐ The successive-approximation method of A/D conversion is the most widely used.
- ☐ Other common methods of A/D conversion are single-slope, dual-slope, tracking, and stairstep ramp (counter method).
- ☐ In a voltage-to-frequency converter (V/F), the output frequency is directly proportional to the amplitude of the analog input voltage.
- ☐ In a frequency-to-voltage converter (F/V), the amplitude of the output voltage is directly proportional to the input frequency.
- ☐ Types of D/A conversion errors include nonmonotonicity, differential nonlinearity, low or high gain, and offset error.
- ☐ Types of A/D conversion errors include missing code, incorrect code, and offset.

Glossary

Acquisition time In an analog switch, the time required for the device to reach its final value when switched from hold to sample.

Analog Related to a continuous set of values.

Aperture jitter In an analog switch, the uncertainty in the aperture time.

Aperture time In an analog switch, the time to fully open after being switched from sample to hold.

Digital Related to a discrete set of values.

Droop In an analog switch, the change in the sampled value during the hold interval.

Feedthrough In an analog switch, the component of the output voltage which follows the input voltage after the switch opens.

Flash A method of A/D conversion.

Monotonicity In relation to D/A converters, the presence of all steps in the output when sequenced over the entire range of input bits.

Nonmonotonicity In relation to D/A converters, a step reversal or missing step in the output when sequenced over the entire range of input bits.

Nyquist rate In sampling theory, the minimum rate at which an analog voltage can be sampled for A/D conversion and equal to twice the maximum frequency component of the input signal.

Quantization The determination of a value for an analog quantity.

Quantization error The error resulting from the change in the analog voltage during the A/D conversion time.

Resolution In relation to D/A or A/D converters, the number of bits involved in the conversion. Also, for D/A converters, the reciprocal of the maximum number of discrete steps in the output.

Sample The process of taking the instantaneous value of a quantity at a specific point in time.

Successive approximation A method of A/D conversion.

FORMULAS

V/F CONVERTERS

(13–1) $\quad t_{os} = C_{os}(6.8 \times 10^3 \text{ s/F}) + 3 \times 10^{-7} \text{ s}$ $\qquad$ One-shot time

(13–2) $\quad \Delta V = \dfrac{(1 \text{ mA} - I_{in})t_{os}}{C_{int}}$ $\qquad$ Integrator output increase in reset interval

(13–3) $\quad t_{int} = \left(\dfrac{1 \text{ mA}}{I_{in}} - 1\right)t_{os}$ $\qquad$ Integrate interval

(13–4) $\quad f_{out} = \dfrac{I_{in}}{(t_{os})1 \text{ mA}}$ $\qquad$ Output frequency

SELF-TEST

1. An analog switch
   (a) changes an analog signal to digital
   (b) connects or disconnects an analog signal to the output
   (c) stores the value of an analog voltage at a certain point
   (d) combines two or more analog signals onto a single line
2. An analog multiplexer
   (a) produces the sum of several analog voltages on an output line
   (b) connects two or more analog signals to an output at the same time
   (c) connects two or more analog signals to an output one at a time in sequence
   (d) distributes two or more analog signals to different outputs in sequence
3. A basic sample-and-hold circuit contains
   (a) an analog switch and an amplifier
   (b) an analog switch, a capacitor, and an amplifier
   (c) an analog multiplexer and a capacitor
   (d) an analog switch, a capacitor, and input and output buffer amplifiers
4. In a sample/track-and-hold amplifier,
   (a) the voltage at the end of the sample interval is held
   (b) the voltage at the beginning of the sample interval is held
   (c) the average voltage during the sample interval is held
   (d) the output follows the input during the sample interval
   (e) a and d
5. In an analog switch, the aperture time is the time it takes for the switch to
   (a) fully open after the control switches from hold to sample
   (b) fully close after the control switches from sample to hold

(c) fully open after the control switches from sample to hold

(d) fully close after the control switches from hold to sample

6. In a binary-weighted-input D/A converter,

   (a) all of the input resistors are of equal value

   (b) there are only two input resistor values required

   (c) the number of different input resistor values equals the number of inputs

7. In a 4-bit binary-weighted-input D/A converter, if the lowest-valued input resistor is 1 kΩ, the highest-valued input resistor is

   (a) 2 kΩ (b) 4 kΩ (c) 8 kΩ (d) 16 kΩ

8. The advantage of an $R/2R$ ladder D/A converter is

   (a) it is more accurate   (b) it uses only two resistor values

   (c) it uses only one resistor value   (d) it can handle more inputs

9. In a D/A converter, monotonicity means that

   (a) the accuracy is within one-half of a least significant bit

   (b) there are no missing steps in the output

   (c) there is one bit missing from the input

   (d) there are no linear errors

10. An 8-bit A/D converter can represent

    (a) 144 discrete values of an analog input

    (b) 4096 discrete values of an analog input

    (c) a continuous set of values of an analog input

    (d) 256 discrete values of an analog input

11. An analog signal must be sampled at a minimum rate equal to

    (a) twice the maximum frequency

    (b) twice the minimum frequency

    (c) the maximum frequency

    (d) the minimum frequency

12. Quantization error in an A/D converter is due to

    (a) poor resolution

    (b) nonlinearity of the input

    (c) a missing bit in the output

    (d) a change in the input voltage during the conversion time

13. Quantization error can be avoided by

    (a) using a higher resolution A/D converter

    (b) using a sample-and-hold prior to the A/D converter

    (c) shortening the conversion time

    (d) using a flash A/D converter

Data Conversion Circuits

14. The type of A/D converter with the fastest conversion time is the
    - (a) dual-slope
    - (b) single-slope
    - (c) simultaneous
    - (d) successive-approximation

15. The output of a V/F converter
    - (a) has an amplitude proportional to the frequency of the input
    - (b) is a digital reproduction of the input voltage
    - (c) has a frequency that is inversely proportional to the amplitude of the input
    - (d) has a frequency that is directly proportional to the amplitude of the input

16. An element not found in the typical V/F converter is
    - (a) a linear amplifier
    - (b) a one-shot
    - (c) an integrator
    - (d) a comparator

Problems

Section 13–1 Analog Switches

1. Determine the output waveform for the analog switch in Figure 13–67(a) for each set of waveforms in parts (b), (c), and (d).

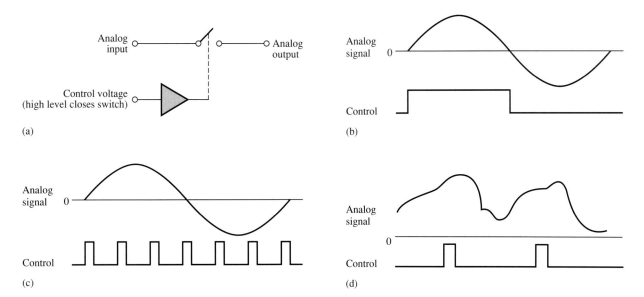

Figure 13–67

2. Determine the output of the 4-channel analog multiplexer in Figure 13–68 for the signal and control inputs shown.

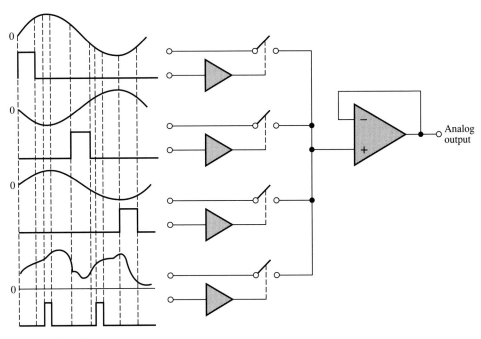

FIGURE 13–68

SECTION 13–2 SAMPLE-AND-HOLD AMPLIFIERS

3. Determine the output voltage waveform for the sample/track-and-hold amplifier in Figure 13–69 given the analog input and the control voltage waveforms shown. Sample is the high control level.

4. Repeat Problem 3 for the waveforms in Figure 13–70.

5. Determine the gain of each AD582 sample-and-hold amplifier in Figure 13–71 on page 684.

SECTION 13–3 INTERFACING THE ANALOG AND DIGITAL WORLDS

6. The analog signal in Figure 13–72 on page 685 is sampled at 2 ms intervals. Represent the signal by a series of 4-bit binary numbers.

7. Sketch the digital reproduction of the analog curve represented by the series of binary numbers developed in Problem 6.

8. Graph the analog signal represented by the following sequence of binary numbers: 1111, 1110, 1101, 1100, 1010, 1001, 1000, 0111, 0110, 0101, 0100, 0101, 0110, 0111, 1000, 1001, 1010, 1011, 1100, 1100, 1011, 1010, 1010.

DATA CONVERSION CIRCUITS

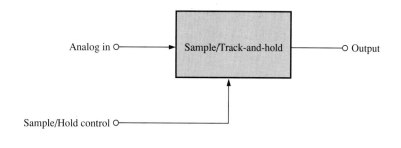

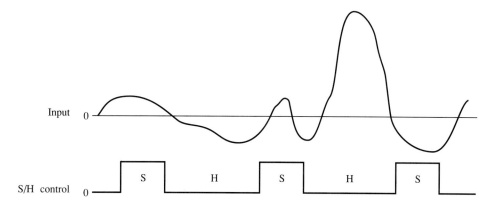

FIGURE 13–69

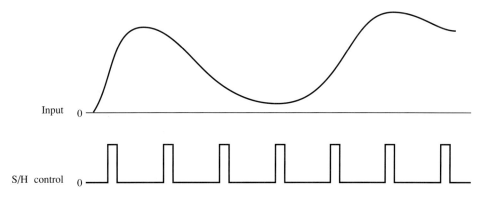

FIGURE 13–70

SECTION 13–4 DIGITAL-TO-ANALOG (D/A) CONVERSION

9. In a certain 4-bit D/A converter, the lowest-weighted resistor has a value of 10 kΩ. What should the values of the other input resistors be?
10. Determine the output of the D/A converter in Figure 13–73(a) on page 685 if the sequence of 4-bit numbers in part (b) is applied to the inputs.

684 CHAPTER 13

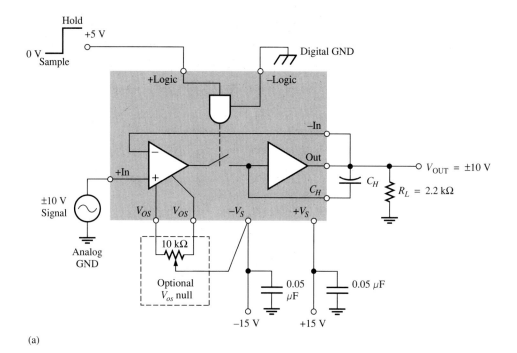

FIGURE 13-71

DATA CONVERSION CIRCUITS

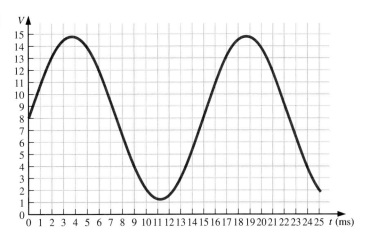

FIGURE 13–72

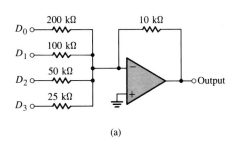

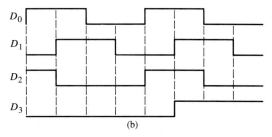

(a)

(b)

FIGURE 13–73

11. Repeat Problem 10 for the inputs in Figure 13–74.

FIGURE 13–74

12. Determine the resolution expressed as a percentage for each of the following D/A converters:

(a) 3-bit   (b) 10-bit   (c) 18-bit

SECTION 13–5 BASIC CONCEPTS OF ANALOG-TO-DIGITAL (A/D) CONVERSION

13. How many discrete values of an analog signal can each of the following A/D converters represent?

(a) 4-bit   (b) 5-bit   (c) 8-bit   (d) 16-bit

14. Determine the Nyquist rate for sinusoidal voltages with each of the following periods:
    (a) 10 s   (b) 1 ms   (c) 30 μs   (d) 1000 ns

15. What is the quantization error expressed in volts of an A/D converter with a sample-and-hold input for a sampled value of 3.2 V if the sample-and-hold has a droop of 100 mV/s? Assume that the conversion time of the A/D converter is 10 ms.

Section 13–6 Analog-to-Digital (A/D) Conversion Methods

16. Determine the binary output sequence of a 3-bit flash A/D converter for the analog input signal in Figure 13–75. The sampling rate is 100 kHz.

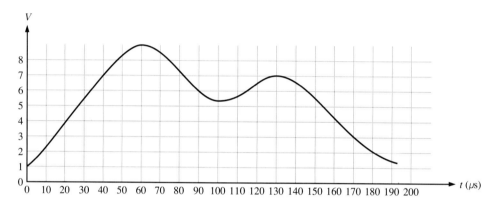

FIGURE 13–75

17. Repeat Problem 16 for the analog waveform in Figure 13–76.

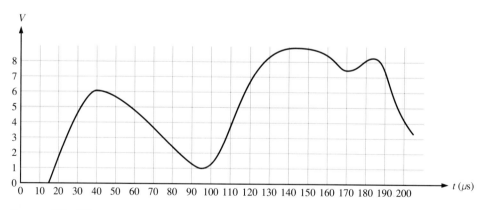

FIGURE 13–76

18. For a certain 4-bit successive-approximation A/D converter, the maximum ladder output is +8 V. If a constant +6 V is applied to the analog input, determine the sequence of binary states for the SAR.

Section 13–7 Voltage-to-Frequency (V/F) and Frequency-to-Voltage (F/V) Converters

19. The analog input to a V/F converter increases from 0.5 V to 3.5 V. Does the output frequency increase, decrease, or remain unchanged?

20. Assume that when the input to a certain V/F converter is 0 V, there is no output signal (0 Hz). Also, when a constant +2 V is applied to the input, the corresponding output frequency is 1 kHz. Now, if the input takes a step up to +4 V, what is the output frequency?

21. Calculate the value of the timing capacitor required to produce a 5 μs pulse width in an AD650 V/F converter.

22. Determine the increase in the integrator output voltage during the reset interval in the AD650 shown in Figure 13–77.

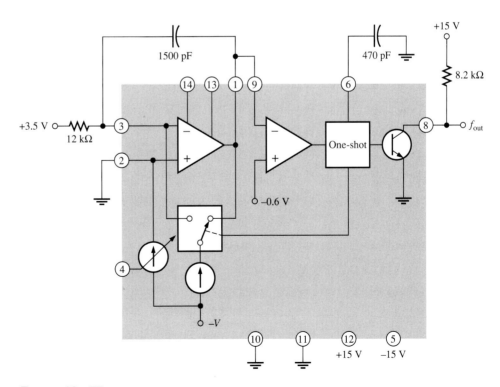

FIGURE 13–77

688   CHAPTER 13

23. Determine the minimum and maximum output frequencies of the AD6450 in Figure 13–77 for the portion of the input voltage shown within the shaded area in Figure 13–78.

FIGURE 13–78

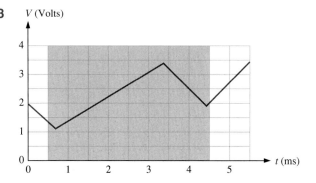

SECTION 13–8   TROUBLESHOOTING

24. A 4-bit D/A converter has failed in such a way that the MSB is stuck at 0. Draw the analog output when a straight binary sequence is applied to the inputs.

25. A straight binary sequence is applied to a 4-bit D/A converter and the output in Figure 13–79 is observed. What is the problem?

FIGURE 13–79

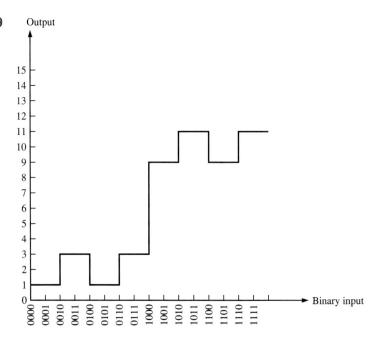

DATA CONVERSION CIRCUITS

26. An A/D converter produces the following sequence of binary numbers when a certain analog signal is applied to its input: 0000, 0001, 0010, 0011, 0100, 0101, 0110, 0111, 0110, 0101, 0100, 0011, 0010, 0001, 0000.

(a) Reconstruct the input from the digital codes as a D/A converter would.

(b) If the A/D converter failed so that the code 0111 were missing from the output, what would the reconstructed output look like?

Answers to Review Questions

Section 13–1
1. An analog switch connects an analog voltage on its input to its output when activated.
2. An analog multiplexer switches analog voltages from several lines onto a common output line in a time sequence.

Section 13–2
1. A sample-and-hold retains the value of an analog signal taken at a given point.
2. Droop is the decrease in the held voltage due to capacitor leakage.
3. Aperture time is the time required for an analog switch to fully open at the end of a sample pulse.
4. Acquisition time is the time required for the device to reach final value at the start of the sample pulse.

Section 13–3
1. Natural quantities are in analog form.
2. A/D conversion changes an analog quantity into digital form.
3. D/A conversion changes a digital quantity into analog form.

Section 13–4
1. Each input resistor must have a different value.
2. 6.67%

Section 13–5
1. A 12-bit A/D converter is more accurate than an 8-bit.
2. The time for a sampled analog value to be converted to digital is the conversion time.
3. Twice the maximum frequency, 200 Hz
4. The sample-and-hold keeps the sampled value constant during conversion.

Section 13–6
1. Flash is the fastest method of A/D conversion.
2. Tracking A/D conversion uses an up-down counter.
3. True

Section 13-7
1. V/F components: integrator, comparator, one-shot, current source, and switch
2. The output frequency increases proportionally.
3. The V/F converter has a voltage input and a frequency output. The F/V converter has a frequency input and a voltage output.

Section 13-8
1. Nonmonotonicity is indicated by a step reversal.
2. Step amplitudes are less than ideal.
3. Missing code and incorrect code.

Section 13-9
1. The capacitors are for power supply decoupling.
2. $2^8 = 256$ values can be converted to digital.
3. Eight minutes is the minimum interval between 2° steps required to track the sun through a 180° arc for 12 hours.

Answers to Practice Exercises

13-1 See Figure 13-80.

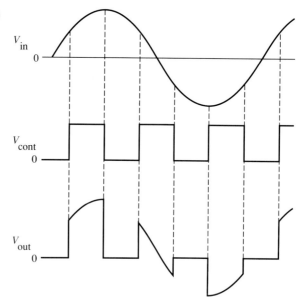

FIGURE 13-80

13-2 The same tones will be closer together.

13-3 See Figure 13-81.

DATA CONVERSION CIRCUITS

FIGURE 13–81

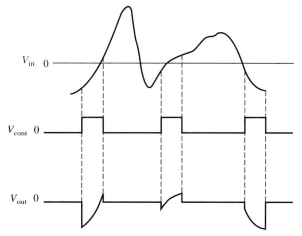

13–4 0.25 mV

13–5 0.00038%

13–6 010, 011, 011, 011, 010, 001, 000, 000, 001, 010, 011, 011

13–7 $f_{min} = 40$ kHz, $f_{max} = 240$ kHz

13–8 See Figure 13–82.

FIGURE 13–82

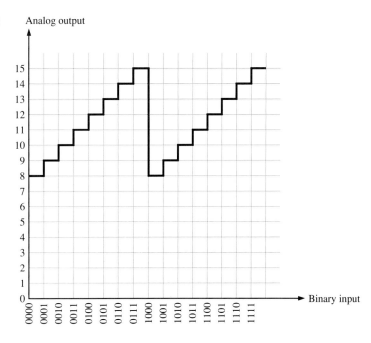

13–9 A constant 15 V output.

14

MEASUREMENT AND CONTROL CIRCUITS

14–1 RMS-TO-DC CONVERTERS
14–2 ANGLE-MEASURING CIRCUITS
14–3 TEMPERATURE-MEASURING CIRCUITS
14–4 STRAIN-MEASURING AND PRESSURE-MEASURING CIRCUITS
14–5 POWER-CONTROL CIRCUITS
14–6 A SYSTEM APPLICATION

After completing this chapter, you should be able to

☐ Describe the operation of a basic rms-to-dc converter.
☐ Name some applications for rms-to-dc converters.
☐ Explain what a synchro is and how it is used.
☐ Describe a typical resolver and how it functions.
☐ Explain how a resolver-to-digital converter (RDC) works.
☐ Show how angular positions are represented by a digital code.
☐ Discuss a simple RDC application.
☐ Explain what a thermocouple is and basically how it works.
☐ Discuss methods for measuring temperature with a thermocouple.
☐ Describe the basic operation of an integrated circuit thermocouple signal conditioner.
☐ Explain how a resistance temperature detector (RTD) works.
☐ Discuss methods for measuring temperature with an RTD.
☐ Explain the theory of a 3-wire bridge.
☐ Describe the basic operation of an integrated circuit RTD signal conditioner.
☐ Explain what a thermistor is and how it differs from an RTD.
☐ Explain how a strain gage works.
☐ Discuss basic strain gage circuits.
☐ Show how pressure measurements can be made using strain gages.
☐ Describe the basic operation of a zero-voltage switch and explain its purpose.

This chapter introduces several types of transducers and related circuits for measuring basic physical analog parameters such as angular position, temperature, strain, pressure, and flow rate. A **transducer** is an element that is designed to convert a physical parameter to an electrical parameter. For example, one type of temperature transducer converts temperature to

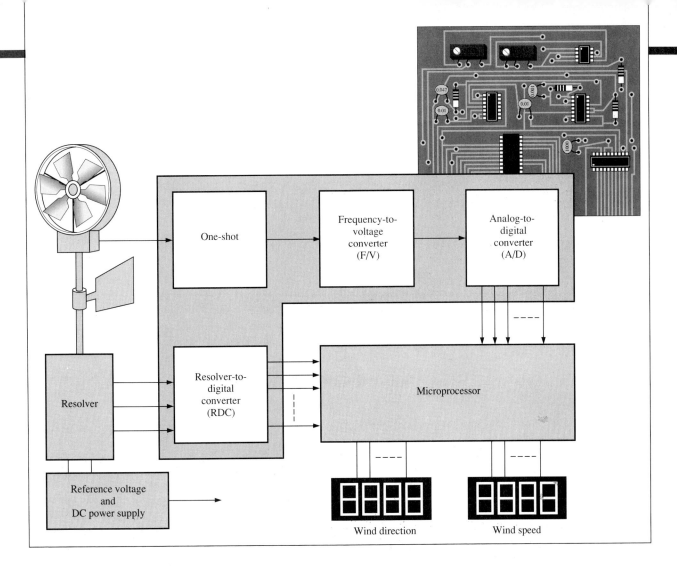

voltage, and another type converts temperature to resistance. Transducers and their associated circuits are important in many applications. The measurement of angular position is critical in robotics, radar, and industrial machine control. Temperature-measuring and pressure-measuring circuits are used in industry for monitoring temperatures and pressures of various fluids or gases in tanks and pipes, and they are used in automotive applications for measuring temperatures and pressures in various parts of the automobile. Strain measurement is important for testing the strength of materials under stress in such areas as aircraft design. Also, zero-voltage switching, an important technique in controlling various functions, is introduced in this chapter.

A System Application

The system application in this chapter focuses on the measurement of wind speed and direction. The input from the wind speed measurement part of the system (anemometer) is generated by a type of flow meter in the form of a propeller arrangement mounted on a wind vane. The wind causes the flow meter blades to rotate on a shaft at a rate proportional to the wind speed. A magnetic device senses each rotation and the circuitry produces a pulse. The frequency of the pulses indicates the speed of the wind. The input from the wind direction measurement part of the system is generated by the wind vane attached to the shaft of a resolver. The wind vane aligns itself with the direction of the wind, and the resolver produces electrical signals proportional to the angular position. The diagram on the preceding page shows a simplified diagram for the wind-measuring system that will be the focus of the system application in Section 14–6. Notice that some circuits from previous chapters are utilized in this application.

For the system application in Section 14–6, in addition to the other topics, be sure you understand

☐ Basic resolver and RDC operation.
☐ The 555 timer, frequency-to-voltage converters, and A/D converters.

14–1 RMS-TO-DC CONVERTERS

RMS-to dc converters are used in several basic areas. One important application is in noise measurement for determining thermal noise, transistor, and switch contact noise. Another application is in the measurement of signals from mechanical phenomena such as strain, vibration, and expansion or contraction. RMS-to-dc converters are also useful for accurate measurements of low-frequency, low duty-cycle pulse trains.

DEFINITION OF RMS

RMS stands for **root mean square** and is a basic measurement of the amplitude of an ac signal. In practical terms, the rms value of an ac signal voltage is equal to the value of dc voltage required to produce the same amount of heat in the same resistance as the ac voltage. For example, an ac voltage with an rms value of 1 V produces the same amount of heat in a given resistor as a 1 V dc voltage. Mathematically, the rms value is found by taking the square root of the average (**mean**) of the signal voltage squared, as expressed in the following formula:

$$V_{rms} = \sqrt{\text{avg}(V_{in}^2)} \qquad (14\text{–}1)$$

RMS-TO-DC CONVERSION

RMS-to-dc converters are electronic circuits that continuously compute the square of the input signal voltage, average it, and take the square root of the result. The output of an rms-to-dc converter is a dc voltage that is proportional to the rms value of the input signal. The block diagram in Figure 14–1 illustrates the basic conversion process.

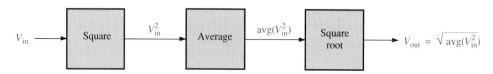

FIGURE 14–1
The rms-to-dc conversion process.

Measurement and Control Circuits

THE SQUARING CIRCUIT The squaring circuit is generally a linear multiplier with the signal applied to both inputs as shown in Figure 14–2. Linear multipliers were introduced in Chapter 12.

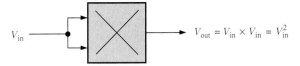

FIGURE 14–2
The squaring circuit is a linear multiplier.

V_{in} $V_{out} = V_{in} \times V_{in} = V_{in}^2$

THE AVERAGING CIRCUIT The simplest type of averaging circuit is a single-pole low-pass filter on the input of an op-amp voltage follower, as shown in Figure 14–3. The *RC* filter passes only the dc component (average value) of the squared input voltage. The overbar designates an average value.

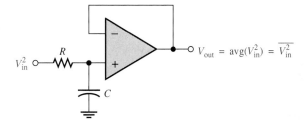

FIGURE 14–3
A basic averaging circuit.

$V_{out} = \text{avg}(V_{in}^2) = \overline{V_{in}^2}$

THE SQUARE ROOT CIRCUIT Recall from Chapter 12 that a square root circuit uses a linear multiplier in an op-amp feedback loop as shown in Figure 14–4.

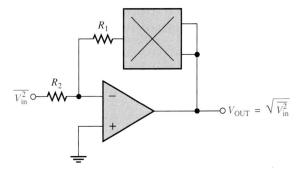

FIGURE 14–4
A square root circuit.

$V_{OUT} = \sqrt{\overline{V_{in}^2}}$

A Complete RMS-to-DC Converter

Figure 14–5 shows the three functional circuits combined to form an rms-to-dc converter. This combination is often referred to as an *explicit* rms-to-dc converter because of the straightforward method used in determining the rms value.

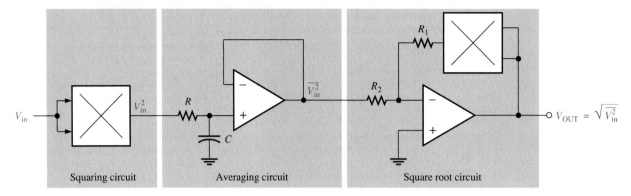

FIGURE 14–5
Explicit type of rms-to-dc converter.

Another method for achieving rms-to-dc conversion, sometimes called the *implicit* method, uses feedback to perform the square root operation. A basic circuit is shown in Figure 14–6. The first block squares the input voltage and divides by the output voltage. The averaging circuit produces the final dc output voltage, which is fed back to the squarer/divider circuit.

FIGURE 14–6
Implicit type of rms-to-dc converter.

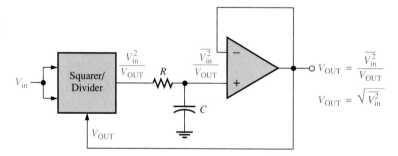

The operation of the circuit in Figure 14–6 can be understood better by going through the mathematical steps performed by the circuit as follows. The expression for the output of the squarer/divider is

$$\frac{V_{in}^2}{V_{OUT}}$$

The voltage at the noninverting (+) input to the voltage follower is

$$V_{in(NI)} = \frac{\overline{V_{in}^2}}{V_{OUT}}$$

where the overbar indicates average value. The final output voltage is

$$V_{OUT} = \frac{\overline{V_{in}^2}}{V_{OUT}}$$

$$V_{OUT}^2 = \overline{V_{in}^2}$$

$$V_{OUT} = \sqrt{\overline{V_{in}^2}} \tag{14-2}$$

THE AD637 IC RMS-TO-DC CONVERTER

As an example of a specific IC device, we will look at the AD637 rms-to-dc converter. This device is essentially an implicit type of converter except that it has an absolute value circuit at the input and it uses an inverting low-pass filter for averaging, as indicated in Figure 14–7. The averaging capacitor, C_{avg}, is an external component that can be selected to provide a minimum averaging error under various input conditions.

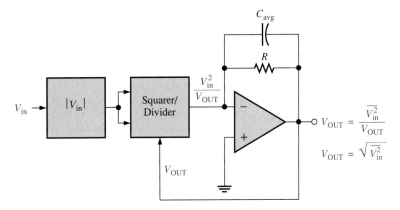

FIGURE 14–7
Basic diagram of the AD637 rms-to-dc converter.

The absolute value circuit in the first block is simply a full-wave rectifier that changes all the negative portions of an input voltage to positive. The squarer/divider circuit is actually implemented with log and antilog circuits as shown in Figure 14–8.

Notice that the second block in Figure 14–8 produces the log of the square of the input by taking the log of V_{in} and then multiplying it by two.

$$2 \log V_{in} = \log V_{in}^2$$

This equation is based on the fundamental rule of logarithms that states the log of a variable squared is equal to twice the log of the variable. The third block is a subtracter that subtracts the logarithm of the output voltage from the log of the input squared.

$$2 \log V_{in} - \log V_{OUT} = \log V_{in}^2 - \log V_{OUT} = \log \frac{V_{in}^2}{V_{OUT}}$$

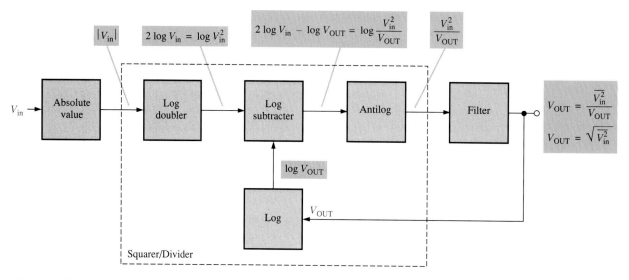

FIGURE 14–8
Internal function diagram of the AD637.

This equation is based on the fundamental rule of logarithms that states that the difference of two logarithmic terms equals the logarithm of the quotient of the two terms. The antilog circuit takes the antilog of $\log(V_{in}^2/V_{OUT})$ and produces an output equal to V_{in}^2/V_{OUT}, as indicated in the figure. The low-pass filter averages the output of the antilog circuit and produces the final output.

EXAMPLES OF RMS-TO-DC CONVERTER APPLICATIONS

In addition to the measurement applications mentioned in the section introduction, rms-to-dc converters are used in a variety of system applications. A couple of typical applications are in automatic gain control (AGC) circuits and rms voltmeters. We will look at each of these in a general way to give you an idea of how rms-to-dc converters are used.

AGC CIRCUITS Figure 14–9 shows a general diagram of an AGC circuit that incorporates an rms-to-dc converter. AGC circuits are used in audio systems to keep the output amplitude constant when the input signal level varies over a certain range and in signal generators to keep the output amplitude constant with variations in waveform, duty cycle, and frequency.

RMS VOLTMETERS Figure 14–10 basically illustrates the rms-to-dc converter in an rms voltmeter. The rms-to-dc converter produces a dc output that is the rms value of the input signal. This rms value is then converted to digital form by an A/D converter and displayed.

MEASUREMENT AND CONTROL CIRCUITS

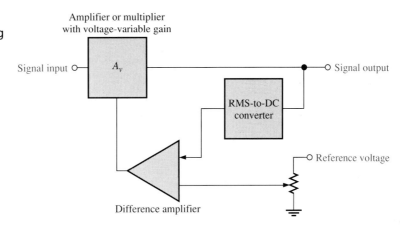

FIGURE 14–9
A simplified AGC circuit using an rms-to-dc converter.

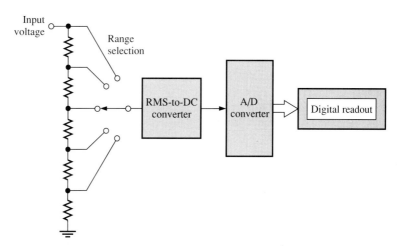

FIGURE 14–10
A simplified rms voltmeter.

14–1 REVIEW QUESTIONS

1. What is the basic purpose of an rms-to-dc converter?
2. What are the three internal functions performed by an rms-to-dc converter?

14–2 ANGLE-MEASURING CIRCUITS

In many applications, the angular position of a shaft or other mechanical mechanism must be measured and converted to an electrical signal for processing or display. Examples of this mechanical-to-electrical interfacing are found in radar and satellite antennas, wind vanes, solar systems, industrial machines including robots, and military fire control systems, to name a few. In this section, the circuits for interfacing angular position transducers, called synchros and resolvers, are intro-

duced. Transducers, in general, are devices that convert energy from one form to another. Before we get into the circuits used in angular measurements, a brief introduction to synchros will provide some background.

SYNCHROS

A **synchro** is an electromechanical transducer used for shaft angle measurement and positioning. There are several different types of synchros, but all can be thought of basically as rotating transformers. In physical appearance, a synchro resembles a small ac motor as shown in Figure 14–11(a) with a diameter ranging from about 0.5 in. to about 3.7 in.

FIGURE 14–11
A typical synchro and its basic winding structure.

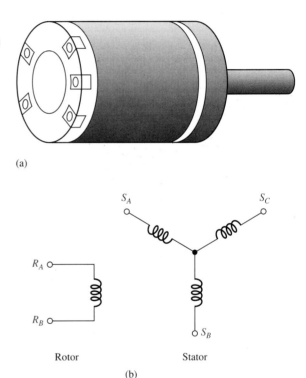

The basic synchro consists of a **rotor,** which can revolve within a fixed **stator** assembly. A shaft is connected to the rotor so that when the shaft rotates, the rotor also rotates. In most synchros, there is a rotor winding and three stator windings. The stator windings are connected as shown in Figure 14–11(b) and are separated by 120° around the stator. The windings are brought out to a terminal block at one end of the housing.

SYNCHRO VOLTAGES When a reference sine-wave voltage is applied across the rotor winding, the voltage induced across any one of the stator windings is proportional to the sine of the angle (θ) between the rotor winding and the stator winding. The angle θ is dependent on the shaft position.

Measurement and Control Circuits

The voltage induced across any two stator windings (between any two stator terminals) is the sum or difference of the two stator voltages. These three voltages, called *synchro format voltages*, are represented in Figure 14–12 and are derived using a basic trigonometric identity. The important thing is that each of the three synchro format voltages is a function of the shaft angle, θ, and can be used to determine the angular position at any time. As the shaft rotates, the format voltages change proportionally.

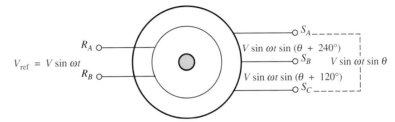

FIGURE 14–12
Synchro format voltages with a reference voltage applied to the rotor.

RESOLVERS

The **resolver** is a particular type of synchro that is often used in rotational systems to transduce the angular position. Resolvers differ from regular synchros in that the rotor and two stator windings are separated from each other by 90° rather than by 120°. The basic winding configuration of a simple resolver is shown in Figure 14–13.

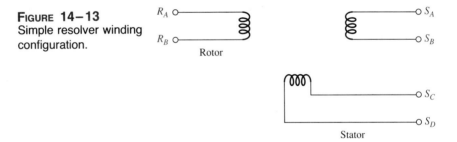

FIGURE 14–13
Simple resolver winding configuration.

RESOLVER VOLTAGES If a reference sine-wave voltage is applied to the rotor winding, the resulting voltages across the stator windings are as given in Figure 14–14. These voltages are a function of the shaft angle θ and are called *resolver format voltages*. One of the

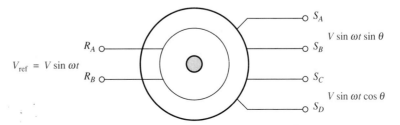

FIGURE 14–14
Resolver format voltages with a reference voltage applied to the rotor.

voltages is proportional to the sine of θ and the other voltage is proportioned to the cosine of θ. Notice that the resolver has a four-terminal output compared to the three-terminal output of the standard synchro.

BASIC OPERATION OF SYNCHRO-TO-DIGITAL AND RESOLVER-TO-DIGITAL CONVERTERS

Electronic circuits known as **synchro-to-digital converters (SDCs)** and **resolver-to-digital converters (RDCs)** are used to convert the format voltages from a synchro or resolver to a digital format. These devices may be considered a very specialized form of analog-to-digital converter.

All converters, both SDCs and RDCs, operate internally with resolver format voltages. Therefore, the output format voltages of a synchro must first be transformed into resolver format by a special type of transformer called the *Scott-T transformer*, as illustrated in Figure 14–15.

FIGURE 14–15
Inputs and outputs of a Scott-T transformer.

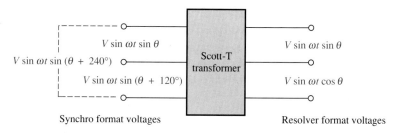

Some SDCs have internal Scott-T transformers, but others require a separate transformer. Other than the transformer, the basic operation and internal circuitry of SDCs and RDCs are the same, so we will focus on RDCs. A simplified block diagram of a tracking RDC is shown in Figure 14–16.

The two resolver format voltages, $V_1 = V \sin \omega t \sin \theta$ and $V_2 = V \sin \omega t \cos \theta$, are applied to the RDC inputs as indicated in Figure 14–16 (θ is the current shaft angle of the resolver). These resolver voltages go through buffers to special multiplier circuits. We will assume that the current state of the up/down counter represents some angle, ϕ. The digital code representing ϕ is applied to the multiplier circuits along with the resolver voltages. The cosine multiplier takes the cosine of ϕ and multiplies it times the resolver voltage V_1. The sine multiplier takes the sine of ϕ and multiplies it times the resolver voltage V_2. The resulting output of the cosine multiplier is

$$V_1 \cos \phi = V \sin \omega t \sin \theta \cos \phi$$

The resulting output of the sine multiplier is

$$V_2 \sin \phi = V \sin \omega t \cos \theta \sin \phi$$

These two voltages are subtracted by the error amplifier to produce the following error voltage:

$$V \sin \omega t \sin \theta \cos \phi - V \sin \omega t \sin \theta \sin \phi = V \sin \omega t (\sin \theta \cos \phi - \cos \theta \sin \phi)$$

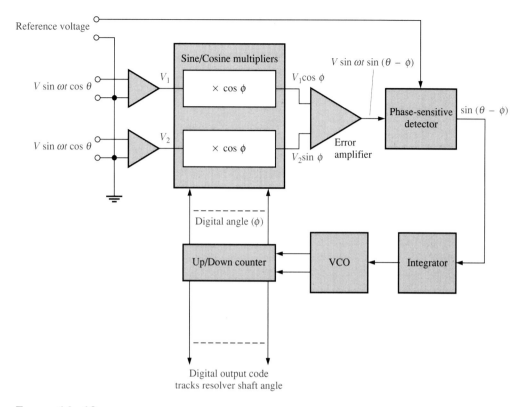

FIGURE 14–16
Simplified diagram of a resolver-to-digital converter (RDC).

A basic trigonometric identity reduces the error voltage expression to

$$V \sin \omega t \sin(\theta - \phi)$$

The phase sensitive detector produces a dc error voltage proportional to $\sin(\theta - \phi)$, which is applied to the integrator. The output of the integrator drives a voltage-controlled oscillator (VCO), which provides clock pulses to the up/down counter. When the counter reaches the value of the current shaft angle θ, then $\phi = \theta$ and we have

$$\sin(\theta - \phi) = 0$$

If the sine is zero, then the difference of the angles is 0°.

$$\theta - \phi = 0°$$

At this point, the angle stored in the counter equals the resolver shaft angle.

$$\phi = \theta$$

When the shaft angle changes, the counter will count up or down until its count equals the new shaft angle. Therefore, the RDC will continuously track the resolver shaft angle and produce an output digital code that equals the angle at all times.

REPRESENTATION OF ANGLES WITH A DIGITAL CODE

The most common method of representing an angular measurement with a digital code is given in Table 14–1 for word lengths up to 16 bits. A binary 1 in any bit position means that the corresponding angle is included, and a 0 means that the corresponding angle is not included.

TABLE 14–1
Bit weights for resolver-to-digital conversion

Bit Position	Angle (Degrees)
1 (MSB)	180.00000
2	90.00000
3	45.00000
4	22.50000
5	11.25000
6	5.62500
7	2.81250
8	1.40625
9	0.70313
10	0.35156
11	0.17578
12	0.08790
13	0.04395
14	0.02197
15	0.01099
16	0.00549

EXAMPLE 14–1

A certain RDC has an 8-bit digital output. What is the angle being measured if the output code is 01001101? The left-most bit is the MSB.

SOLUTION

Bit Position	Bit	Angle (Degrees)
1	0	0
2	1	90.00000
3	0	0
4	0	0
5	1	11.25000
6	1	5.62500
7	0	0
8	1	1.40625

To get the angle represented, add all the included angles (as indicated by the presence of a 1 in the output code).

$$90.00000° + 11.25000° + 5.62500° + 1.40625° = 108.28125°$$

Practice Exercise 14–1
What is the angular shaft position measured by a 12-bit RDC when it has a binary code of 100000100001 on its outputs?

A Specific RDC

To illustrate a typical integrated circuit device, we will look at Analog Devices 1S20, which is a 12-bit converter. The diagram for this device is shown in Figure 14–17 and, as you can see, it is basically the same as the general RDC in Figure 14–16 with some additions. Additional circuits include the latch and output buffers for controlling the data transfer to and interfacing with other digital systems. These additional circuits do not affect the conversion process.

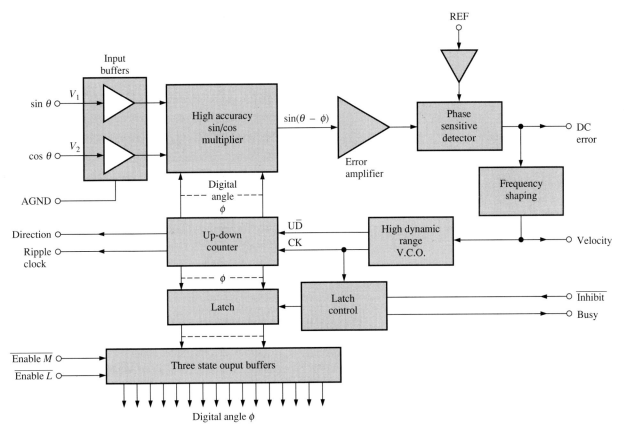

FIGURE 14–17
Diagram of the 1S20 resolver-to-digital converter.

Additional outputs include the Busy signal that indicates when the counter is in transition from one code to another as a result of the angle changing. The Direction output indicates the direction of rotation of the resolver shaft. The Ripple clock output indicates when the counter resets from all 1s to all 0s or vice versa, indicating that the shaft has completed a full revolution (360°). The Velocity output is proportional to the rate of change of the input angle.

Additional inputs include the $\overline{\text{Inhibit}}$ signal that is used to control the transfer of data from the counter to the latch. The two $\overline{\text{Enable}}$ inputs are used to enable the buffer outputs.

A SIMPLE RDC APPLICATION

The measurement of wind direction is one example of how an RDC can be used. As shown in Figure 14–18, a wind vane is fixed to the shaft of a resolver. As the wind vane moves to align with the direction of the wind, the resolver shaft rotates and its angle indicates the wind direction. The resolver output is applied to an RDC and the resulting digital output code, which represents the wind direction, drives a digital readout.

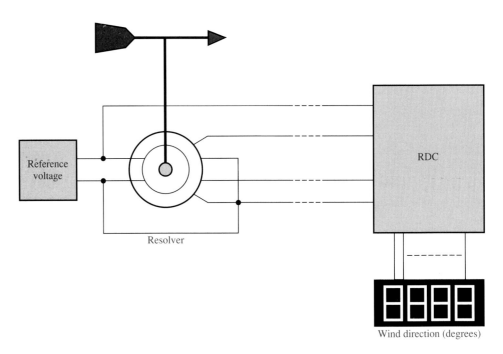

FIGURE 14–18
Measurement and display of wind direction with a resolver and an RDC.

14–2 REVIEW QUESTIONS

1. What is a transducer that converts a mechanical shaft position into electrical signals called?
2. What type of input does an RDC accept?
3. What type of output does an RDC produce?
4. What is the function of an RDC?

14-3 TEMPERATURE-MEASURING CIRCUITS

Temperature is perhaps the most common physical parameter that is measured and converted to electrical form. Several types of temperature sensors (transducers) respond to temperature and produce a corresponding indication by a change or alteration in a physical characteristic that can be detected by an electronic circuit. Common types of temperature sensors are thermocouples, resistance temperature detectors (RTDs), and thermistors. In this section, we will look at each of these sensors and at signal conditioning circuits that are required to interface the transducers to electronic equipment.

THE THERMOCOUPLE

The **thermocouple** is formed by joining two dissimilar metals. A small voltage, called the *Seebeck voltage,* is produced across the junction of the two metals when heated, as illustrated in Figure 14-19. The amount of voltage produced is dependent on the types of metals and is directly proportional to the temperature of the junction (positive temperature coefficient); however, this voltage is generally much less than 100 mV. The voltage versus temperature characteristic of thermocouples is somewhat nonlinear, but the amount of nonlinearity is predictable. Thermocouples are widely used in certain industries because they have a wide temperature range and can be used to measure very high temperatures.

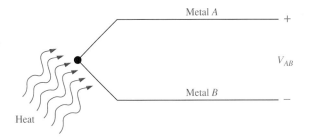

FIGURE 14-19
A voltage proportional to temperature is generated when a thermocouple is heated.

Some common metal combinations used in commercial thermocouples are copper/constantan (constantan is a copper-nickel alloy), iron/constantan, chromel/constantan (chromel is a nickel-chromium alloy), and chromel/alumel (alumel is a nickel-aluminum alloy). Each of these types of thermocouple has a different temperature range, coefficient, and voltage characteristic and is designated by the letter *T, J, E,* and *K,* respectively. The overall temperature range covered by thermocouples is from $-250°C$ to $2000°C$. Each type covers a different portion of this range.

THERMOCOUPLE-TO-ELECTRONICS INTERFACE When a thermocouple is connected to a signal-conditioning circuit, as illustrated in Figure 14-20, an unwanted thermocouple is effectively created at the point(s) where one or both of the thermocouple wires connect to the circuit terminals made of a dissimilar metal. The unwanted thermocouple junction is

FIGURE 14–20

Creation of an unwanted thermocouple in a thermocouple-to-electronics interface.

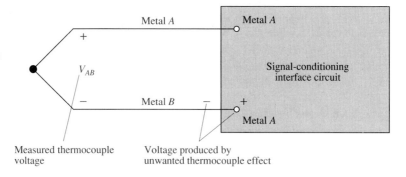

sometimes referred to as a **cold junction** in some references because it is normally at a significantly lower temperature than that being measured by the measuring thermocouple. These unwanted thermocouples can have an unpredictable effect on the overall voltage that is sensed by the circuit because the voltage produced by the unwanted thermocouple opposes the measured thermocouple voltage and its value depends on ambient temperature.

EXAMPLE OF A THERMOCOUPLE-TO-ELECTRONICS INTERFACE As shown in Figure 14–21, a copper/constantan thermocouple (known as type T) is used, in this case, to measure the temperature in an industrial temperature chamber. The copper thermocouple wire is connected to a copper terminal on the circuit board and the constantan wire is also connected to a copper terminal on the circuit board. The copper-to-copper connection is no problem because the metals are the same. The constantan-to-copper connection acts as an unwanted thermocouple that will produce a voltage in opposition to the thermocouple voltage because the metals are dissimilar.

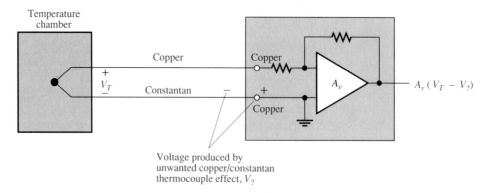

FIGURE 14–21

A simplified temperature-measuring circuit with an unwanted thermocouple at the junction of the constantan wire and the copper terminal.

MEASUREMENT AND CONTROL CIRCUITS

Since the unwanted thermocouple connection is not at a fixed temperature, its effects are unpredictable and it will introduce inaccuracy into the measured temperature. One method for eliminating an unwanted thermocouple effect is to add a reference thermocouple at a constant known temperature (usually 0°C). Figure 14–22 shows that by using a reference thermocouple that is held at a constant known temperature, the unwanted thermocouple at the circuit terminal is eliminated because both contacts to the circuit terminals are now copper-to-copper. The voltage produced by the reference thermocouple is a known constant value and can be compensated for in the circuitry.

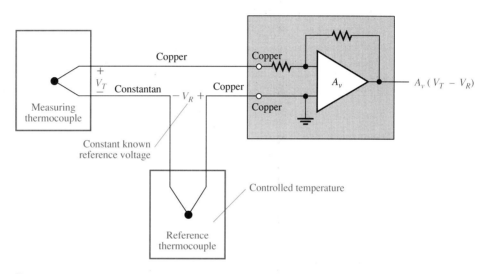

FIGURE 14–22
Using a reference thermocouple in a temperature-measuring circuit.

EXAMPLE 14–2

Suppose the thermocouple in Figure 14–21 is measuring 200°C in an industrial oven. The circuit board is in an area where the ambient temperature can vary from 15°C to 35°C. Using Table 14–2 for a type-T (copper/constantan) thermocouple, determine the voltage across the circuit input terminals at the ambient temperature extremes. What is the maximum percent error in the voltage at the circuit input terminals?

TABLE 14–2
Type-T thermocouple voltage

Temperature °C	Output mV
−200	−5.603
−100	−3.378
0	0.000
+100	4.277
+200	9.286
+300	14.860
+400	20.869

Solution

From Table 14–2, we know that the measuring thermocouple is producing 9.286 mV. To determine the voltage that the unwanted thermocouple is creating at 15°C, we must interpolate from the table. Since 15°C is 15% of 100°C, a linear interpolation gives the following voltage:

$$0.15(4.277 \text{ mV}) = 0.642 \text{ mV}$$

Since 35°C is 35% of 100°C, we get a voltage of

$$0.35(4.277 \text{ mV}) = 1.497 \text{ mV}$$

The voltage across the circuit input terminals at 15°C is

$$9.286 \text{ mV} - 0.642 \text{ mV} = 8.644 \text{ mV}$$

The voltage across the circuit input terminals at 35°C is

$$9.286 \text{ mV} - 1.497 \text{ mV} = 7.789 \text{ mV}$$

The maximum percent error in the voltage at the circuit input terminals is

$$\left(\frac{9.286 \text{ mV} - 7.789 \text{ mV}}{9.286 \text{ mV}}\right)100\% = 16.12\%$$

We can never be sure how much it is off because we have no control over the ambient temperature. Also, the linear interpolation may or may not be accurate depending on the linearity of the temperature characteristic of the unwanted thermocouple.

Practice Exercise 14–2
In the case of the circuit in Figure 14–21, if the temperature being measured goes up to 300°C, what is the maximum percent error in the voltage across the circuit input terminals?

■ **Example 14–3** Now, refer to the thermocouple circuit in Figure 14–22. Suppose the thermocouple is measuring 200°C. Again, the circuit board is in an area where the ambient temperature can vary from 15°C to 35°C. The reference thermocouple is held at exactly 0°C. Determine the voltage across the circuit input terminals at the ambient temperature extremes.

Solution
From Table 14–2 in Example 14–2, the thermocouple voltage is 0 V at 0°C. Since the reference thermocouple produces no voltage at 0°C and is completely independent of ambient temperature, there is no error in the measured voltage over the ambient temperature range. Therefore, the voltage across the circuit input terminals at both temperature extremes equals the measuring thermocouple voltage, which is 9.286 mV.

MEASUREMENT AND CONTROL CIRCUITS

PRACTICE EXERCISE 14–3
If the reference thermocouple were held at $-100°C$ instead of $0°C$, what would be the voltage across the circuit input terminals if the measuring thermocouple were at $400°C$?

COMPENSATION It is bulky and expensive to maintain a reference thermocouple at a fixed temperature (usually an ice bath is required). Another approach is to compensate for the unwanted thermocouple effect by adding a compensation circuit as shown in Figure 14–23. This is sometimes referred to as *cold-junction compensation*. The compensation circuit consists of a resistor and an integrated circuit temperature sensor with a temperature coefficient that matches that of the unwanted thermocouple.

The current source in the temperature sensor produces a current that creates a voltage drop, V_c, across the compensation resistor, R_c. The resistance is adjusted so that this voltage drop is equal and opposite the voltage produced by the unwanted thermocouple at a given temperature. When the ambient temperature changes, the current changes proportionally, so that the voltage across the compensation resistor is always approximately equal to the unwanted thermocouple voltage. Since the compensation voltage, V_c, is opposite in polarity to the unwanted thermocouple voltage, the unwanted voltage is effectively cancelled.

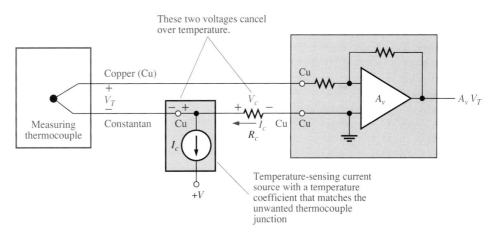

FIGURE 14–23
Simplified circuit for compensation of unwanted thermocouple effect.

A THERMOCOUPLE SIGNAL CONDITIONER

The functions shown in the circuit of Figure 14–23 plus others are available in single-package hybrid IC circuits, known as *thermocouple signal conditioners*. The 2B50 is a good example of this type of circuit. It is designed for interfacing a thermocouple with various types of electronic systems and provides gain, compensation, isolation, common-mode rejection, and other features in one package. A diagram showing the simplified internal structure package pin numbers is shown in Figure 14–24(a), and the physical package configuration is shown in Figure 14–24(b).

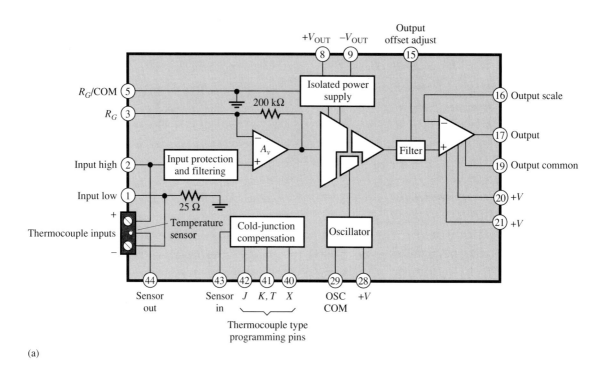

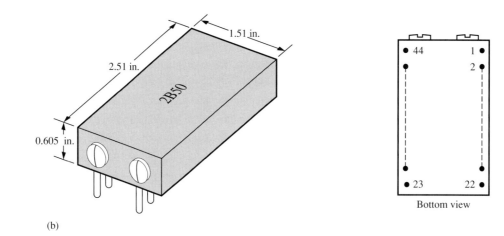

FIGURE 14–24
The 2B50 thermocouple signal conditioner.

A basic description of the 2B50 shown in Figure 14–25 follows. The thermocouple can be connected directly to the two screw terminals or to circuit board pins 1 and 2. An input filtering and protection circuit precedes the op-amp. The voltage gain can be set anywhere from 50 to 1000 by an external resistor, R_G, between pins 3 and 5 as shown. The internal isolated power supply (part of the isolation amplifier) provides convenient dc voltages for use with external input offset adjustment as shown in the figure. The

MEASUREMENT AND CONTROL CIRCUITS

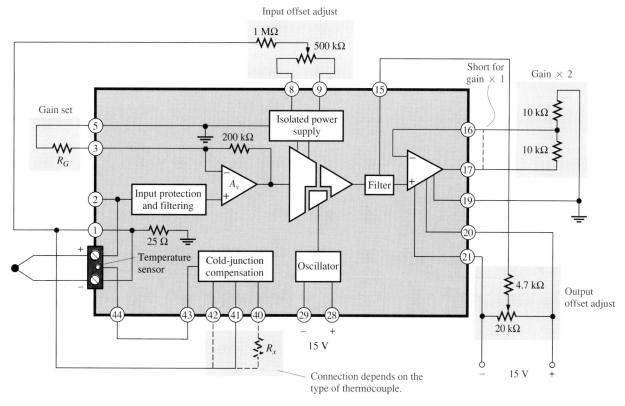

FIGURE 14–25
The 2B50 thermocouple signal conditioner externally configured for a typical application.

cold-junction compensation circuit provides for compensation of the thermocouple connection at the circuit input terminals. Compensating circuits for *J*-, *K*-, and *T*-type thermocouples are built into the device and are configured using connections to pins 41 or 42. (Refer to the pin designations that were shown in part (a) of Figure 14–24.) An external resistor connected to the *X* input (pin 40) configures the circuit for other types of thermocouples. The value of R_X is selected for the particular type of thermocouple based on the manufacturer's recommendation. The op-amp is followed by an isolation amplifier whose output is filtered and is applied to an output op-amp. If the output scale (pin 16) is connected to the output (pin 17), the output op-amp is a voltage follower and the overall gain of the 2B50 is equal to the gain set by R_G. In this configuration, the full-scale output range is ±5 V. If a greater output range is required, the overall gain can be doubled by connecting external resistors as shown in Figure 14–25. This gives the output op-amp a gain of two and an output range of ±10 V. The oscillator shown is associated with the isolation amplifier (refer to Chapter 11).

EXAMPLE 14–4 Determine the overall gain and the type of thermocouple in the signal conditioner of Figure 14–26.

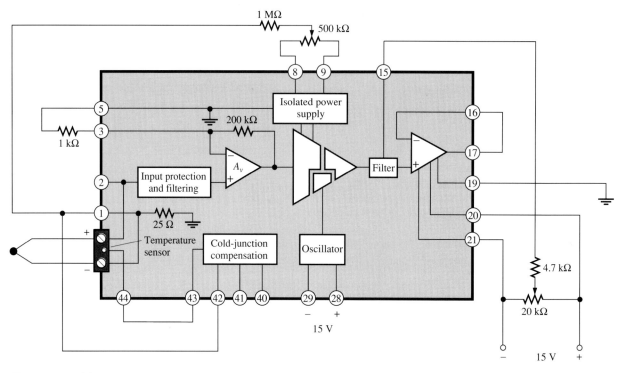

FIGURE 14–26

SOLUTION

The input op-amp is connected in a noninverting configuration. Recall from Chapter 5 that the voltage gain of this type of configuration is

$$A_v = 1 + \frac{R_f}{R_i}$$

In this case, R_f is the 200 kΩ internal resistor and R_i is the 1 kΩ external gain resistor, R_G. The gain of the input op-amp is, therefore,

$$A_v = 1 + \frac{200 \text{ k}\Omega}{1 \text{ k}\Omega} = 201$$

The output op-amp is connected as a voltage follower; its gain is 1. The overall gain of the device is

$$A_v \times 1 = 201$$

The external connection from pin 1 to pin 42 tells us that this device is set up for a type *J* thermocouple.

PRACTICE EXERCISE 14–4

What change is required to the circuit of Figure 14–26 to have an overall voltage gain of 21 and a *K*-type thermocouple?

Resistance Temperature Detectors (RTDs)

A second major type of temperature transducer is the **resistance temperature detector (RTD)**. The RTD is a resistive device in which the resistance changes directly with temperature (positive temperature coefficient). The RTD is more nearly linear than the thermocouple. RTDs are constructed in either a wire-wound configuration or by a metal film technique. The most common RTDs are made of either platinum, nickel, or nickel alloys.

Generally, RTDs are used to sense temperature in two basic ways. First, as shown in Figure 14–27(a), the RTD is driven by a current source and, since the current is constant, the change in voltage across it is proportional (by Ohm's law) to the change in its resistance with temperature. Second, as shown in Figure 14–27(b), the RTD is connected in a 3-wire bridge circuit; and the bridge output voltage is used to sense the change in the RTD resistance and, thus, the temperature.

FIGURE 14–27
Basic methods of employing an RTD in a temperature-sensing circuit.

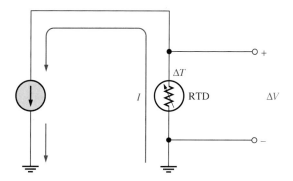

(a) A change in temperature, ΔT, produces a change in voltage, ΔV, across the RTD proportional to the change in RTD resistance when the current is constant.

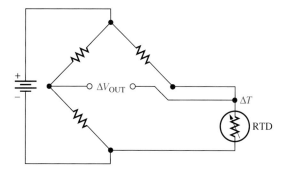

(b) A change in temperature, ΔT, produces a change in bridge output voltage, ΔV, proportional to the change in RTD resistance.

Theory of the 3-wire Bridge To avoid subjecting the three bridge resistors to the same temperature that the RTD is sensing, the RTD is usually remotely located to the point where temperature variations are to be measured and connected to the rest of the bridge by long wires. The resistance of the three bridge resistors must remain constant. The long extension wires to the RTD have resistance that can affect the accurate operation of the bridge.

Figure 14–28(a) shows the RTD connected in the bridge with a 2-wire configuration. Notice that the resistance of both of the long connecting wires appear in the same leg of the bridge as the RTD. Recall from your study of basic circuits that $V_{OUT} = 0$ V and the bridge is balanced when $R_{RTD} = R_3$ if $R_1 = R_2$. The wire resistances will throw the bridge off balance when $R_{RTD} = R_3$ and will cause an error in the output voltage for any value of the RTD resistance because they are in series with the RTD in the same leg of the bridge.

FIGURE 14–28

Comparison of 2-wire and 3-wire bridge connections in an RTD circuit.

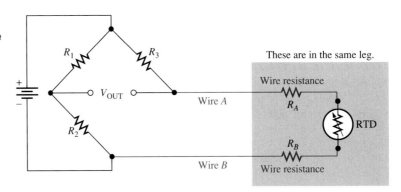

(a) Two-wire bridge connection

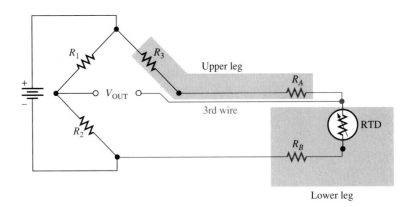

(b) Three-wire bridge connection

The 3-wire configuration in Figure 14–28(b) overcomes the wire resistance problem. By connecting a third wire to one end of the RTD as shown, the resistance of wire A is now placed in the same leg of the bridge as R_3 and the resistance of wire B is placed in the same leg of the bridge as the RTD. Because the wire resistances are now in opposite legs of the bridge, their effects will cancel if both wire resistances are the same (equal lengths of same type of wire). The resistance of the third wire has no effect; essentially no current goes through it because the output terminals of the bridge are open or are connected across a very high impedance. The balance condition is expressed as

$$R_{\text{RTD}} + R_B = R_3 + R_A$$

If $R_A = R_B$, then they cancel in the equation and the balance condition is completely independent of the wire resistances.

$$R_{\text{RTD}} = R_3$$

Basic RTD Temperature-Sensing Circuits

Two simplified RTD measurement circuits are shown in Figure 14–29. The circuit in part (a) is one implementation of an RTD driven by a constant current. The operation is as follows. From your study of basic op-amp circuits, recall that the input current and the current through the feedback path are essentially equal because the input impedance of the

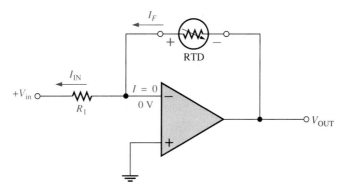
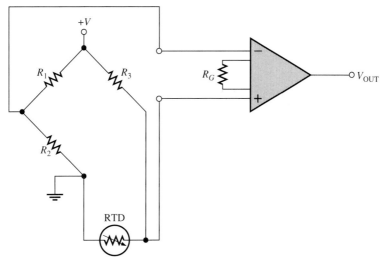

FIGURE 14–29
Basic RTD temperature-measuring circuits.

(a) Constant-current circuit

(b) Three-wire bridge circuit

op-amp is ideally infinite. Therefore, the constant current through the RTD is set by the constant input voltage, V_{IN}, and the input resistance, R_{IN}, because the inverting input is at virtual ground. The RTD is in the feedback path and, therefore, the output voltage of the op-amp is equal to the voltage across the RTD. As the resistance of the RTD changes with temperature, the voltage across the RTD also changes because the current is constant.

The circuit in Figure 14–29(b) shows a basic circuit in which an instrumentation amplifier is used to amplify the voltage across the 3-wire bridge circuit. The RTD forms one leg of the bridge; and as its resistance changes with temperature, the bridge output voltage also changes proportionally. The bridge is adjusted for balance ($V_{OUT} = 0$ V) at some reference temperature, say 0°C. This means that R_3, is selected to equal the resistance of the RTD at this reference temperature.

EXAMPLE 14–5

Determine the output voltage of the instrumentation amplifier in the RTD circuit in Figure 14–30 if the resistance of the RTD is 1320 Ω at the temperature being measured.

FIGURE 14–30

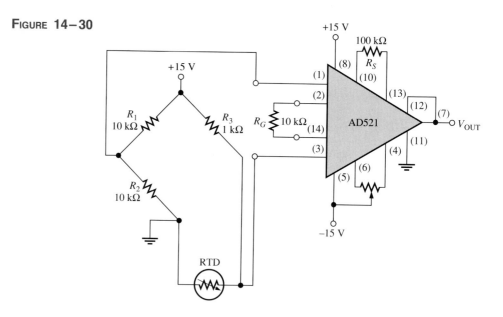

SOLUTION
The bridge output voltage is

$$V_{OUT} = \left(\frac{R_{RTD}}{R_3 + R_{RTD}}\right)15\text{ V} - \left(\frac{R_2}{R_1 + R_2}\right)15\text{ V}$$

$$= \left(\frac{1320\text{ }\Omega}{2320\text{ }\Omega}\right)15\text{ V} - \left(\frac{10\text{ k}\Omega}{20\text{ k}\Omega}\right)15\text{ V}$$

$$= 8.53\text{ V} - 7.5\text{ V}$$

$$= 1.03\text{ V}$$

MEASUREMENT AND CONTROL CIRCUITS

Using Equation (11–6), the voltage gain of the AD521 instrumentation amplifier is

$$A_v = \frac{R_S}{R_G} = \frac{100 \text{ k}\Omega}{10 \text{ k}\Omega} = 10$$

The output voltage from the amplifier is

$$V_{\text{OUT}} = (10)(1.03 \text{ V}) = 10.3 \text{ V}$$

PRACTICE EXERCISE 14–5

What must be the nominal resistance of the RTD in Figure 14–30 to balance the bridge at 25°C? What is the amplifier output voltage when the bridge is balanced?

AN RTD SIGNAL CONDITIONER

The 2B31 is a good example of a hybrid integrated circuit device that provides for either 3-wire bridge or current-source RTD measurements. Figure 14–31 is a simplified diagram of the 2B31. It includes an input instrumentation amplifier followed by a buffer amplifier that drives an active 2 kHz low-pass filter having a Bessel characteristic. The amplifier gain is set by external resistors, and the filter response can be externally adjusted up to 5 kHz. The filter eliminates 60 Hz or any noise picked up on the long RTD lines. The adjustable bridge excitation circuit is basically a precision dc power supply used to generate accurate bridge voltage or a constant current depending on the RTD sensing method.

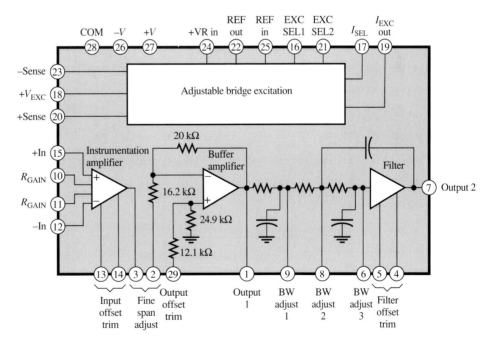

FIGURE 14–31
The 2B31 RTD signal conditioner.

Typical connections for the signal conditioner are shown in Figure 14–32. Part (a) is a constant-current configuration. The bridge excitation circuit is connected to operate as a constant-current source to supply current to the external RTD, and the resulting voltage across the RTD is applied to the inputs of the device. Voltage gain is set by R_G and R_F. Also, input and output offset adjustments can be implemented but are left off for simplicity.

Figure 14–32(b) shows a 3-wire RTD bridge connection. In this case, the bridge excitation circuit is connected to operate as a precision voltage supply. The bridge voltage can be externally adjusted by the value of the external resistor.

VOLTAGE GAIN OF THE 2B31 The voltage gain of the 2B31 is determined by the gain of the instrumentation amplifier and the gain of the buffer amplifier that follows it. The voltage gain for the instrumentation amplifier in this particular device is determined by a 94 kΩ internal resistance and the external resistor R_G.

$$A_{v(\text{inst})} = 1 + \frac{94 \text{ k}\Omega}{R_G}$$

The buffer amplifier is an inverting op-amp configuration. Its gain is determined by the 16.2 kΩ and the 20 kΩ internal resistors and the external resistor R_F.

The active filter circuit has a unity gain. Therefore, the overall voltage gain of the device is

$$A_v = \left(1 + \frac{94 \text{ k}\Omega}{R_G}\right)\left(\frac{20 \text{ k}\Omega}{R_F + 16.2 \text{ k}\Omega}\right)$$

■ **EXAMPLE 14–6**

Determine the voltage gain of a 2B31 for $R_G = 4.7$ kΩ and $R_F = 10$ kΩ.

SOLUTION

$$A_v = \left(1 + \frac{94 \text{ k}\Omega}{4.7 \text{ k}\Omega}\right)\left(\frac{20 \text{ k}\Omega}{10 \text{ k}\Omega + 16.2 \text{ k}\Omega}\right) = (21)(0.763) = 16.02$$

PRACTICE EXERCISE 14–6

If pins 2 and 3 of the 2B31 are shorted together, how is the gain affected for $R_G = 4.7$ kΩ?

THERMISTORS

A third major type of temperature transducer is the **thermistor,** which is a resistive device made from a semiconductive material such as nickel oxide or cobalt oxide. The resistance of a thermistor changes inversely with temperature (negative temperature coefficient.) The temperature characteristic for thermistors is more nonlinear than that for thermocouples or RTDs; in fact, a thermistor's temperature characteristic is essentially logarithmic. Also, like the RTD, the temperature range of a thermistor is more limited than that of a thermocouple. Thermistors have the advantage of a greater sensitivity than either thermocouples or RTDs and are generally less expensive. This means that their change in

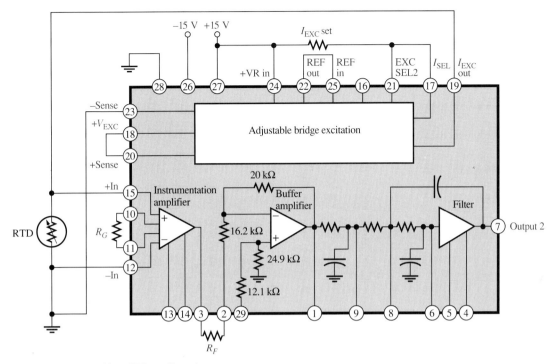

(a) Constant-current driven RTD configuration

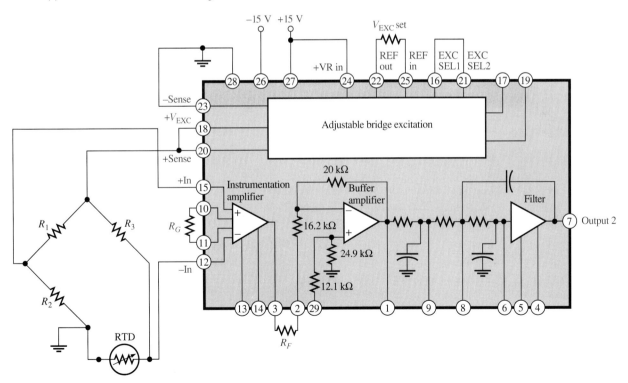

(b) Three-wire bridge RTD configuration

FIGURE 14–32
Two basic RTD temperature-measuring circuits using the 2B31 signal conditioner.

resistance per degree change in temperature is greater. Since they are both variable-resistance devices, the thermistor and the RTD can be used in similar circuits.

Like the RTD, thermistors can be used in constant-current-driven configurations or in bridges. In Figure 14–33, the general response of a thermistor in a constant-current op-amp circuit is compared to that of an RTD in a similar circuit. Both the RTD and the thermistor are exposed to the same temperature environment as indicated. It is assumed that at some reference temperature, the RTD and the thermistor have the same resistance and produce the same output voltage. In the RTD circuit, as the temperature increases from the reference value (becomes more negative), the op-amp's output voltage decreases from the reference value because the resistance of the RTD increases. In the thermistor circuit, as the temperature increases, the op-amp's output voltage increases from the reference value (becomes less negative) because the thermistor's resistance decreases due to its negative temperature coefficient. Also, for the same temperature change, the change in the output voltage of the thermistor circuit is greater than the corresponding change in the output voltage of the RTD circuit because of the greater sensitivity of the thermistor.

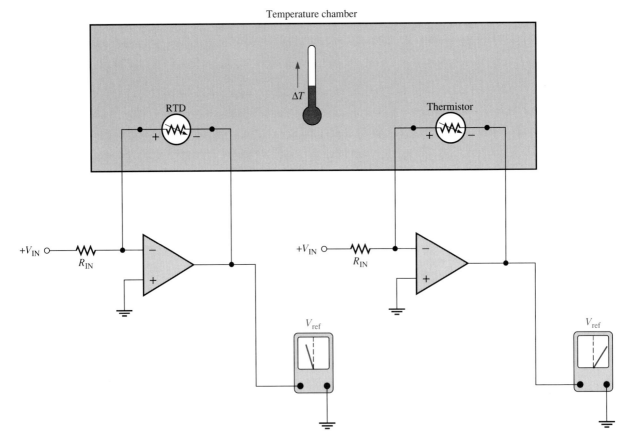

FIGURE 14–33
General comparison of the responses of a thermistor circuit to a similar RTD circuit.

14–3 REVIEW QUESTIONS

1. What is a thermocouple?
2. How can temperature be measured with a thermocouple?
3. What is an RTD and how does its operation differ from a thermocouple?
4. What is the primary operational difference between an RTD and a thermistor?
5. Of the three devices introduced in this section, which one would most likely be used to measure extremely high temperatures?

14–4 STRAIN-MEASURING AND PRESSURE-MEASURING CIRCUITS

In this section, methods of measuring two types of force-related parameters, strain and pressure, are examined. A variety of applications require the measurement of these two parameters. Also, other parameters, such as the flow rate of a fluid, can be measured indirectly by measuring strain or pressure.

THE STRAIN GAGE

Strain is the deformation, either expansion or compression, of a material due to a force acting on it. For example, a metal rod or bar will lengthen slightly when an appropriate force is applied as illustrated in Figure 14–34(a). Also, if a metal plate is bent, there is an expansion of the upper surface, called *tensile strain,* and a compression of the lower surface, called *compressive strain,* as shown in Figure 14–34(b).

FIGURE 14–34
Examples of strain.

(a) Strain occurs as length changes from L to $L + \Delta L$ when force is applied.

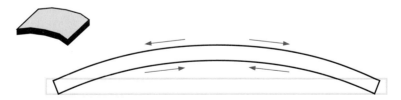

(b) Strain occurs when the flat plate is bent, causing the upper surface to expand and the lower surface to contract.

Most strain gages are based on the principle that the resistance of a material increases if its length increases and decreases if its length decreases. This is expressed by the following formula (which you should recall from your dc/ac circuits course).

$$R = \frac{\rho L}{A} \qquad (14\text{–}3)$$

This formula states that the resistance of a material, such as a length of wire, depends directly on the resistivity (ρ) and the length (L) and inversely on the cross-sectional area (A).

A **strain gage** is basically a long strip of resistive material that is bonded to the surface of an object on which strain is to be measured, such as a wing or tail section of an airplane under test. When a force acts on the object to cause a slight elongation, the strain gage also lengthens proportionally and its resistance increases. Most strain gages are formed in a pattern similar to that in Figure 14–35(a) to achieve enough length for a sufficient resistance value. It is then placed along the line of strain as indicated in Figure 14–35(b).

FIGURE 14–35
A simple strain gage and its placement.

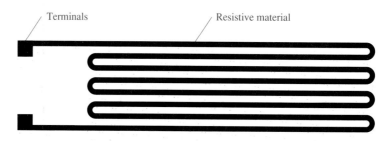

(a) Typical strain gage configuration.

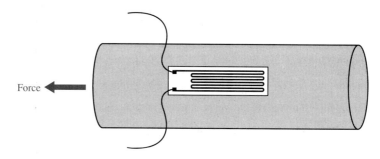

(b) The strain gage is bonded to the surface to be measured along the line of force. When the surface lengthens, the strain gage stretches.

THE GAGE FACTOR OF A STRAIN GAGE An important characteristic of strain gages is the **gage factor (GF)**, which is defined as the ratio of the fractional change in resistance to the fractional change in length along the axis of the gage. The concept of gage factor is illustrated in Figure 14–36 and expressed in Equation (14–4) where R is the nominal resistance and ΔR is the change in resistance due to strain. The fractional change in length ($\Delta L/L$) is usually in parts per million, called *microstrain* (designated ϵ).

$$GF = \frac{\Delta R/R}{\Delta L/L}$$

MEASUREMENT AND CONTROL CIRCUITS

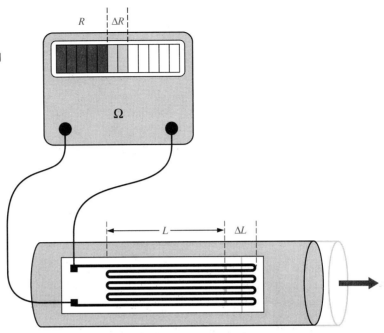

FIGURE 14–36
Illustration of gage factor. The ohmmeter symbol is not intended to represent a practical method for measuring ΔR.

■ **EXAMPLE 14–7**   A certain material being measured under stress undergoes a strain of 5 parts per million (5 $\mu\epsilon$). The strain gage has a nominal (unstrained) resistance of 320 Ω and a gage factor of 2. Determine the resistance change in the strain gage.

SOLUTION

$$GF = \frac{\Delta R/R}{\epsilon}$$

$$\Delta R = (GF)(R)(\epsilon)$$
$$\Delta R = 2(320 \text{ } \Omega)(5 \times 10^{-6}) = 3.2 \text{ m}\Omega$$

PRACTICE EXERCISE 14–7
If the strain in this example is 8 $\mu\epsilon$, how much does the resistance change?

BASIC STRAIN GAGE CIRCUITS

Because a strain gage exhibits a resistance change when the quantity it is sensing changes, it is typically used in circuits similar to those used for RTDs. The basic difference is that strain instead of temperature is being measured. Therefore, strain gages are usually applied in bridge circuits or in constant-current-driven circuits, as shown in Figure 14–37, just as RTDs and thermistors are. In fact, the 2B31 signal conditioner is designed for use with both strain gages and RTDs. Refer to Section 14–3 for the operation of the basic circuits.

FIGURE 14–37
Basic strain-measuring circuits.

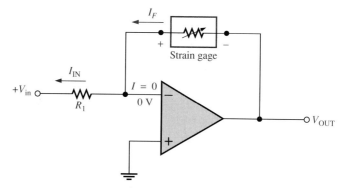

(a) Constant-current circuit

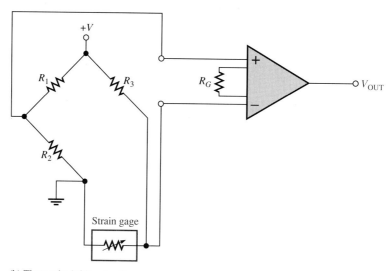

(b) Three-wire bridge circuit

PRESSURE TRANSDUCERS

Pressure transducers are devices that exhibit a change in resistance proportional to a change in pressure. Basically, pressure sensing is accomplished using a strain gage bonded to a flexible diaphragm as shown in Figure 14–38(a). Figure 14–38(b) shows the diaphragm with no net pressure exerted on it. When a net positive pressure exists on one side of the diaphragm, as shown in Figure 14–38(c), the diaphragm is pushed upward and its surface expands. This expansion causes the strain gage to lengthen and its resistance to increase.

Pressure transducers typically are manufactured using a foil strain gage bonded to a stainless steel diaphragm or by integrating semiconductor strain gages (resistors) in a silicon diaphragm. Either way, the basic principle remains the same.

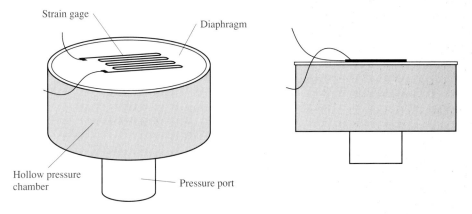

(a) Basic pressure gage construction

(b) With no net pressure on diaphragm, strain gage resistance is at its nominal value (side view).

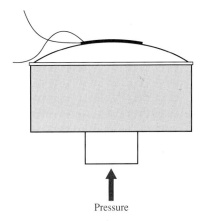

(c) Net pressure forces diaphragm to expand, causing elongation of the strain gage and thus an increase in its resistance.

FIGURE 14–38
A simplified pressure sensor constructed with a strain gage bonded to a flexible diaphragm.

Pressure transducers come in three basic configurations in terms of relative pressure measurement. The absolute pressure transducer measures applied pressure relative to a vacuum as illustrated in Figure 14–39(a). The gage pressure transducer measures applied pressure relative to the pressure of the surroundings (ambient pressure) as illustrated in Figure 14–39(b). The differential pressure transducer measures one applied pressure relative to another applied pressure as shown in Figure 14–39(c). Some transducer configurations include circuitry such as bridge completion circuits and op-amps within the same package as the sensor itself, as indicated.

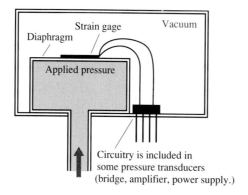
(a) Absolute pressure transducer

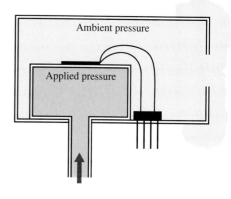
(b) Gage pressure transducer

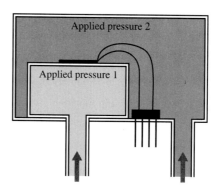
(c) Differential pressure transducer

FIGURE 14–39
Three basic types of pressure transducers.

Pressure-Measuring Circuits

Because pressure transducers are devices in which the resistance changes with the quantity being measured, they are usually in a bridge configuration as shown by the basic op-amp bridge circuit in Figure 14–40(a). In some cases, the complete circuitry is built into the transducer package, and in other cases the circuitry is external to the sensor. The symbols in parts (b) through (d) of Figure 14–40 are sometimes used to represent the complete pressure transducer with an amplified output. The symbol in part (b) represents the absolute pressure transducer, the symbol in part (c) represents the gage pressure transducer, and the symbol in part (d) represents the differential pressure transducer.

Flow Rate Measurement One common method of measuring the flow rate of a fluid through a pipe is the differential-pressure method. A flow restriction device such as a Venturi section (or other type of restriction such as an orifice) is placed in the flow stream. The Venturi section is formed by a narrowing of the pipe, as indicated in Figure 14–41. Although the velocity of the fluid increases as it flows through the narrow

MEASUREMENT AND CONTROL CIRCUITS

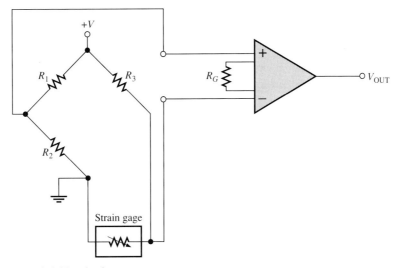

(a) Basic bridge circuit

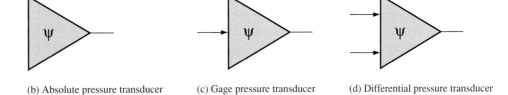

(b) Absolute pressure transducer   (c) Gage pressure transducer   (d) Differential pressure transducer

FIGURE 14–40
A basic pressure transducer circuit and symbols.

FIGURE 14–41
A basic method of flow rate measurement.

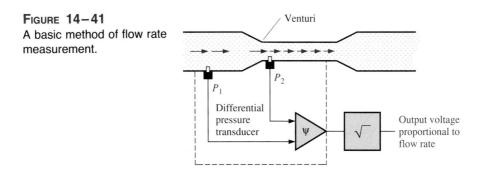

channel, the volume of fluid per minute (volumetric flow rate) is constant throughout the pipe.

Because the velocity of the fluid increases as it goes through the restricted area, the pressure also increases. If pressure is measured at a wide point and at a narrow point, the flow rate can be determined because flow rate is proportional to the square root of the differential pressure, as shown in Figure 14–41.

PRESSURE TRANSDUCER APPLICATIONS Pressure transducers are used anywhere there is a need to determine the pressure of a substance. In medical applications, pressure transducers are used for blood pressure measurement; in aircraft, pressure transducers are used for altitude pressure, cabin pressure, and hydraulic pressure; in automobiles, pressure transducers are used for fuel flow, oil pressure, brake line pressure, manifold pressure, and steering system pressure, to name a few applications.

14-4 REVIEW QUESTIONS

1. Describe a basic strain gage.
2. Describe a basic pressure gage.
3. List three types of pressure gages.

14-5 POWER-CONTROL CIRCUITS

Many types of integrated circuits are used for controlling power to a load. In this section, you will learn about the zero-voltage switch that is used for driving an SCR or triac to control the amount of power to a load. The main advantage of the zero-voltage switching method is the elimination of large switching transients that occur when an SCR or triac is turned on near the peak of the ac voltage. Zero-voltage switches are used in the control of heaters, lamps, valves, motors, to name a few applications.

THE ZERO-VOLTAGE SWITCH

A zero-voltage switch is used to control the amount of power to a load and to switch the power on only when the line voltage crosses the zero-voltage axis. This operation is illustrated in Figure 14-42 where the zero-voltage switch triggers an SCR in order to turn power on to a load. Recall from Chapter 3 that an SCR is turned on by a trigger pulse at the gate and turns off when the current falls to a low value. The load might be a resistive heating element, and the power is typically turned on for several cycles of the ac and then turned off for several cycles to maintain a certain temperature. The zero-voltage switch uses a sensing circuit to determine when to turn power on. In the case of a heat-control circuit, the sensing can be accomplished by a thermistor in a bridge network.

FIGURE 14-42
Basic operation of a zero-voltage switch.

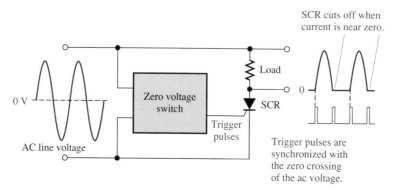

MEASUREMENT AND CONTROL CIRCUITS

THE REASON FOR SWITCHING AT ZERO VOLTAGE The key feature of a zero-voltage switch is that it generates trigger pulses for the control device only at the points where the ac line voltage crosses zero. The main reason for synchronizing the switching with the zero crossings is to prevent **radio frequency interference (RFI)**.

As demonstrated in Figure 14–43, if the SCR or triac were switched on somewhere near the peak of the ac cycle, for example, there would be a sudden inrush of current to the load. When there is a sudden transition of voltage or current, many high-frequency components are generated. (Recall that the rising and falling edges of a pulse contain high frequencies). By switching the SCR or triac on when the voltage across it is zero, the sudden increase in current is prevented because the current will increase sinusoidally with the ac voltage. Zero-voltage switching also prevents thermal shock to the load which, depending on the type of load, may shorten its life.

FIGURE 14–43
Comparison of zero-voltage switching to unsynchronized switching of power to a load.

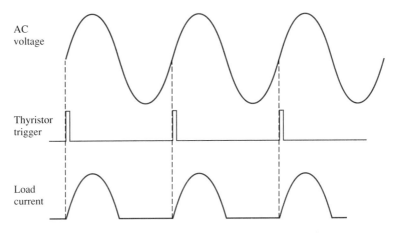

(a) Zero-voltage switching of load current

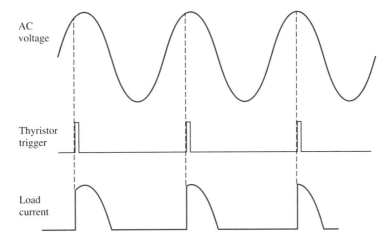

(b) Unsynchronized switching of load current produces current transients that cause RFI.

A Specific Zero-Voltage Switch

The CA3079 is an example of an integrated circuit zero-voltage switch. A block diagram of this device is shown in Figure 14–44. The limiter and the power supply allow the CA3079 to be operated directly from the ac power line with no external dc power required. The zero-crossing detector determines when the ac voltage crosses zero and produces synchronized pulses at each zero crossing. The on/off sensing amplifier, which is basically a comparator, accepts inputs from external sensing circuits (such as a temperature sensor) and provides an output to an AND gate that either enables or inhibits the zero-crossing pulses. The driver provides sufficient current to the gate of an external thyristor. The amount of drive current can be increased by connecting the current boost pin to the dc supply voltage pin.

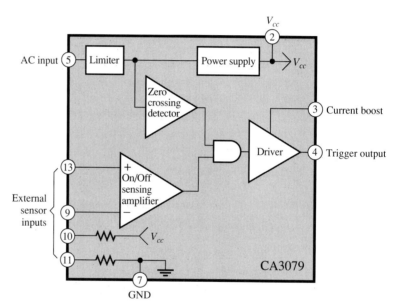

FIGURE 14–44
The CA3079 zero-voltage switch.

A CA3079 connected for a typical application is shown in Figure 14–45. The ac line voltage is applied across the triac and load and to the input of the CA3079A through a limiting resistor, R_{in}. The manufacturer recommends a 2 W, 10 kΩ resistor for R_{in} if the ac voltage is 120 V rms. A bridge circuit with a thermistor in one leg is the temperature sensor in this particular application. The variable resistor in the bridge, R_{set}, adjusts the temperature level. Notice that the other part of the bridge is formed by the internal resistors R_1 and R_2, which are connected using pins 9, 10, and 11. The output circuit in this case drives a triac which, when turned on, conducts current through a resistive heating element.

As long as the temperature of the heating element is below the set point, the triac is triggered on at each zero crossing of the ac voltage and turns off near the end of each half cycle when the load current drops to near zero. The load continues to get current on each half-cycle until the increasing temperature causes the thermistor resistance to decrease to a value where the bridge voltage at pin 13 is less than the voltage at pin 9. At this point,

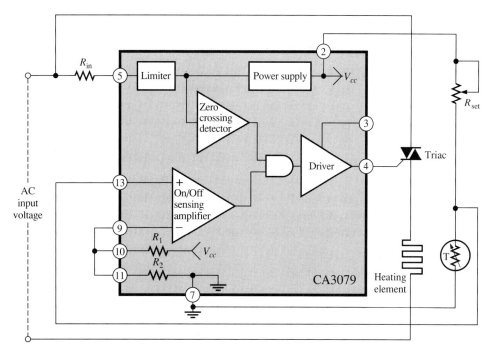

FIGURE 14–45
An example of a CA3079 application as a temperature controller.

the output of the on/off sensing amplifier goes low and inhibits the AND gate, thus preventing trigger pulses from getting to the triac. The triac remains off until the temperature of the load drops enough for the thermistor resistance to increase back to a point where the voltage at pin 13 is greater than the voltage at pin 9. At this point, the on/off sensing amplifier switches back to its high output state and enables the AND gate so that the trigger pulses again turn on the triac. Examine the circuit in Figure 14–45 carefully until you see how the input bridge circuit is formed.

14–5 Review Questions

1. Explain the basic purpose in zero-voltage switching.
2. How does an SCR differ from a triac in terms of delivering power to a load in a circuit such as in Figure 14–45?

14–6 A System Application

The wind speed and direction measurement system presented at the beginning of the chapter is a representation of a type of instrument that is typically found at a meteorological data gathering facility. This system is actually two systems in one because it measures two parameters, wind speed and wind direction, independently. In this system, you will find circuits that you learned about in this chapter and some that

734 CHAPTER 14

were studied in previous chapters. In fact, the measurement of wind direction was briefly mentioned in Section 14–2. In this system application, you will

☐ *See how resolvers and RDCs are used to measure wind direction.*
☐ *See how wind speed can be measured.*
☐ *Translate between the printed circuit board and a schematic.*
☐ *Analyze the measurement circuit board.*
☐ *Troubleshoot some common problems.*

A BRIEF DESCRIPTION OF THE SYSTEM

As you can see in the system block diagram in Figure 14–46, there are two transducers—the anemometer flow meter and the resolver. The flow meter used in this system is basically a propeller-type instrument in which the blades revolve as the wind blows across them. The faster the wind blows, the faster the blades revolve. A magnetic sensor detects each time one revolution is completed and produces a short duration pulse that triggers the 555 one-shot. The frequency of the pulse train produced by the one-shot increases as the wind speed increases. A frequency-to-voltage converter produces an

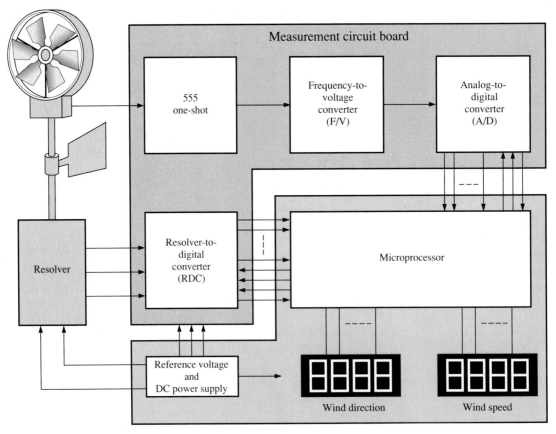

FIGURE 14–46
Block diagram of the wind-measuring system.

MEASUREMENT AND CONTROL CIRCUITS

output voltage that is proportional to the frequency and thus the wind speed. This voltage is converted to digital form and the resulting digital code goes to the microprocessor, which translates it to a binary number corresponding to the wind speed and produces a digital readout. A resolver and a resolver-to-digital converter (RDC) are used to measure the wind direction. The digital output of the RDC goes to the microprocessor where it is translated to an appropriate binary number and displayed.

As indicated in Figure 14–46, one circuit board in this system contains the measurement circuitry and another board contains the microprocessor, display circuitry, and power supply. Our focus in this section is on the measurement circuit board.

Now, so that you can take a closer look at the measurement circuit board, let's take it out of the system and put it on the test bench.

ON THE TEST BENCH

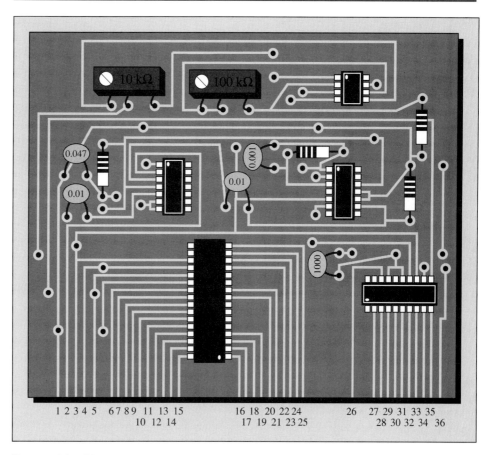

FIGURE 14–47

■ **ACTIVITY 1** **RELATE THE PC BOARD TO THE SCHEMATIC**

Locate and identify each component and each input/output pin on the pc board in Figure 14–47 after all of the inputs and outputs on the schematic in Figure 14–48

CHAPTER 14

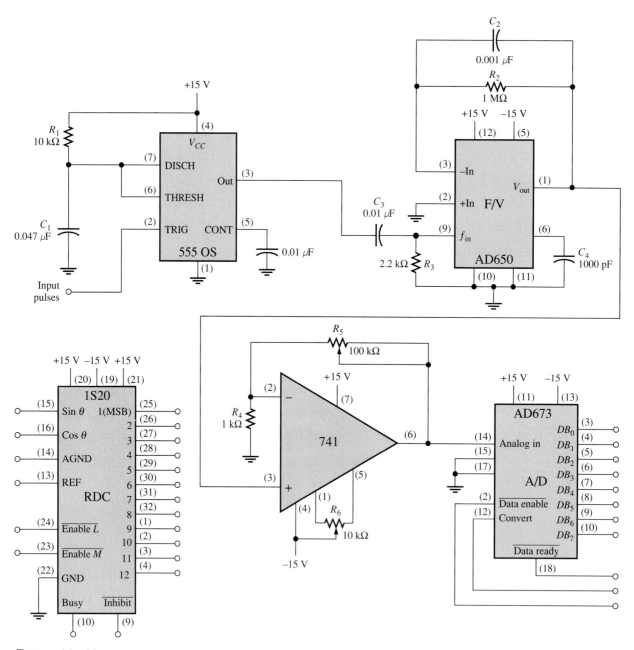

FIGURE 14–48

have been identified. Find and identify the pc connections on the back side of the board. Verify that the board and the schematic agree.

■ **ACTIVITY 2 WRITE A TECHNICAL REPORT**

Describe the overall operation of the measurement circuit board. Specify how each circuit works and its purpose.

Measurement and Control Circuits

 ■ **Activity 3** Troubleshoot the Circuit Board for Each of the Following Problems by Stating the Probable Cause or Causes

1. No pulses out of the one-shot.
2. There is a 100 mV level out of the F/V converter, but the output of the A/D indicates zero.
3. There are pulses out of the one-shot but no voltage out of the F/V converter.
4. For one complete revolution of the resolver shaft, the maximum angle represented by the RDC output code is 180.

14–6 Review Questions

1. Which components determine the pulse width of the one-shot?
2. What is the purpose of the 741 op-amp circuit?

Summary

- An rms-to-dc converter performs three basic functions: squaring, averaging, and taking the square root.
- Squaring is usually implemented with a linear multiplier.
- A simple averaging circuit is a low-pass filter that passes only the dc component of the input.
- A square root circuit utilizes a linear multiplier in the feedback loop of an op-amp.
- A synchro is a shaft angle transducer having three stator windings.
- A resolver is a type of synchro which, in its simplest form, has two stator windings.
- The output voltages of a synchro or resolver are called *format voltages* and are proportional to the shaft angle.
- A resolver-to-digital converter (RDC) converts resolver format voltages to a digital code that represents the angular position of the shaft.
- A thermocouple is a type of temperature transducer formed by the junction of two dissimilar metals.
- When the thermocouple junction is heated, a voltage is generated across the junction that is proportional to the temperature.
- Thermocouples can be used to measure very high temperatures.
- The resistance temperature detector (RTD) is a temperature transducer in which the resistance changes directly with temperature. It has a positive temperature coefficient.
- RTDs are typically employed in bridge circuits or in constant-current circuits to measure temperature. They have a more limited temperature range than thermocouples.
- The thermistor is a temperature transducer in which the resistance changes inversely with temperature. It has a negative temperature coefficient.
- Thermistors are more sensitive than RTDs or thermocouples, but their temperature range is limited.

- The strain gage is based on the fact that the resistance of a material increases when its length increases.
- The gage factor of a strain gage is the fractional change in resistance to the fractional change in length.
- Pressure transducers are constructed with strain gages bonded to a flexible diaphragm.
- An absolute pressure transducer measures pressure relative to a vacuum.
- A gage pressure transducer measures pressure relative to ambient pressure.
- A differential pressure transducer measures one pressure relative to another pressure.
- The flow rate of a liquid can be measured using a differential pressure gage.
- A zero-voltage switch generates pulses at the zero crossings of an ac voltage for triggering a thyristor used in power control.

Glossary

Cold junction A reference thermocouple held at a fixed temperature and used for compensation in thermocouple circuits.

Gage factor (*GF*) The ratio of the fractional change in resistance to the fractional change in length along the axis of the gage.

Mean value Average value.

Radio frequency interference (RFI) High frequencies produced when high values of current and voltage are rapidly switched on and off.

Resistance temperature detector (RTD) A type of temperature transducer in which resistance is directly proportional to temperature.

Resolver A type of synchro.

Resolver-to-digital converter (RDC) An electronic circuit that converts resolver voltages to a digital format which represents the angular position of the rotor shaft.

Root mean square (RMS) A measure of the amplitude of an ac signal that equals the value of a dc voltage required to produce the same amount of heat in the same resistance as the ac voltage.

Rotor The part of a synchro that is attached to the shaft and rotates. The rotor winding is located on the rotor.

Stator The part of a synchro that is fixed. The stator windings are located on the stator.

Strain The expansion or compression of a material caused by stress forces acting on it.

Strain gage A transducer formed by a resistive material in which a lengthening or shortening due to stress produces a proportional change in resistance.

Synchro An electromechanical transducer used for shaft angle measurement and control.

Synchro-to-digital converter (SDC) An electronic circuit that converts synchro voltages to a digital format which represents the angular position of the rotor shaft.

Thermistor A type of temperature transducer in which resistance is inversely proportional to temperature.

Thermocouple A type of temperature transducer formed by the junction of two dissimilar metals which produces a voltage proportional to temperature.

Transducer A device that converts a physical parameter into an electrical quantity.

FORMULAS

(14–1) $\quad V_{rms} = \sqrt{\text{avg}(V_{in}^2)} \quad$ Root-mean-square value

(14–2) $\quad V_{out} = \sqrt{V_{in}^2} \quad$ RMS-to-dc converter output

(14–3) $\quad R = \dfrac{\rho L}{A} \quad$ Resistance of a material

(14–4) $\quad GF = \dfrac{\Delta R/R}{\Delta L/L} \quad$ Gage factor of a strain gage

SELF-TEST

1. The rms value of an ac signal is equal to
   (a) the peak value
   (b) the dc value that produces the same heating effect
   (c) the square root of the average value
   (d) b and c

2. An explicit type of rms-to-dc converter contains
   (a) a squaring circuit   (b) an averaging circuit
   (c) a square root circuit   (d) a squarer/divider circuit
   (e) all of the above   (f) a, b, and c only

3. A synchro produces
   (a) three format voltages   (b) two format voltages
   (c) one format voltage   (d) one reference voltage

4. A resolver produces
   (a) three format voltages   (b) two format voltages
   (c) one format voltage   (d) none of these

5. A Scott-T transformer is used for
   (a) coupling the reference voltage to a synchro or resolver
   (b) changing resolver format voltages to synchro format voltages
   (c) changing synchro format voltages to resolver format voltages
   (d) isolating the rotor winding from the stator windings

6. The output of an RDC is a
   (a) sine wave with an amplitude proportional to the angular position of the resolver shaft

(b) digital code representing the angular position of the stator housing
(c) digital code representing the angular position of the resolver shaft
(d) sine wave with a frequency proportional to the angular position of the resolver shaft

7. A thermocouple
   (a) produces a change in resistance for a change in temperature
   (b) produces a change in voltage for a change in temperature
   (c) is made of two dissimilar metals
   (d) b and c

8. In a thermocouple circuit, where each of the thermocouple wires is connected to a copper circuit board terminal,
   (a) an unwanted thermocouple is produced
   (b) compensation is required
   (c) a reference thermocouple must be used
   (d) all of these
   (e) a and c

9. A typical thermocouple signal conditioner includes
   (a) an isolation amplifier
   (b) an instrumentation amplifier
   (c) cold-junction compensation
   (d) none of these
   (e) a and c

10. An RTD
    (a) produces a change in resistance for a change in temperature
    (b) as a negative temperature coefficient
    (c) has a wider temperature range than a thermocouple
    (d) all of these

11. The purpose of a 3-wire bridge is to eliminate
    (a) nonlinearity of an RTD
    (b) the effects of wire resistance in an RTD circuit
    (c) amplify the RTD resistance
    (d) none of these

12. A thermistor has
    (a) less sensitivity than an RTD
    (b) a greater temperature range than a thermocouple
    (c) a negative temperature coefficient
    (d) a positive temperature coefficient

13. Both RTDs and thermistors are used in
    (a) circuits that measure resistance
    (b) circuits that measure temperature
    (c) bridge circuits
    (d) constant-current-driven circuits
    (e) b, c, and d
    (f) b and c only

14. When the length of a strain gage increases,
    (a) it produces more voltage
    (b) its resistance increases
    (c) its resistance decreases
    (d) it produces an open circuit

15. A higher gage factor indicates that the strain gage is
    (a) less sensitive to a change in length
    (b) more sensitive to a change in length

(c) has more total resistance
(d) made of a physically larger conductor

16. Many types of pressure transducers are made with
    (a) thermistors   (b) RTDs
    (c) strain gages   (d) none of these

17. Gage pressure is measured relative to
    (a) ambient pressure   (b) a vacuum
    (c) a reference pressure

18. The flow rate of a liquid can be measured
    (a) with a string
    (b) with a temperature sensor
    (c) with an absolute pressure transducer
    (d) with a differential pressure transducer

19. Zero-voltage switching is commonly used in
    (a) determining thermocouple voltage   (b) SCR and triac power control circuits
    (c) in balanced bridge circuits   (d) RFI generation

20. A major disadvantage of unsynchronized switching of power to a load is
    (a) lack of efficiency   (b) possible damage to the thyristor
    (c) RFI generation

Problems

Section 14–1 RMS-to-DC Converters

1. A 5 V dc voltage is applied across a 1 kΩ resistor. To achieve the same power in the 1 kΩ resistor as produced by the dc voltage, what must be the rms value of a sine voltage?

2. Based on the fundamental definition of rms, determine the rms value for a symmetrical square wave with an amplitude of ± 1 V.

Section 14–2 Angle-Measuring Circuits

3. A certain RDC has an 8-bit digital output. What is the angle that is being measured if the output code is 10000111?

4. Repeat Problem 3 for an RDC output code of 00010101.

5. What is the purpose of the $\overline{\text{Enable } M}$ and $\overline{\text{Enable } L}$ inputs on a 1S20 RDC?

6. Explain the Direction and Velocity outputs on a 1S20 RDC.

Section 14–3 Temperature-Measuring Circuits

7. Three identical thermocouples are each exposed to a different temperature as follows: Thermocouple A is exposed to 450°C, thermocouple B is exposed to 420°C, and thermocouple C is exposed to 1200°C. Which thermocouple produces the most voltage?

8. You have two thermocouples. One is a K type and the other is a T type. In general, what do these letter designations tell you?

742 CHAPTER 14

9. Determine the output voltage of the op-amp in Figure 14–49 if the thermocouple is measuring a temperature of 400°C and the circuit itself is at 25°C. Refer to Table 14–2.

FIGURE 14–49

10. What should be the output voltage in Problem 9 if the circuit is properly compensated?
11. Determine the overall gain of the 2B50 thermocouple signal conditioner in Figure 14–50.
12. What type of thermocouple is the 2B50 in Figure 14–50 set up for?

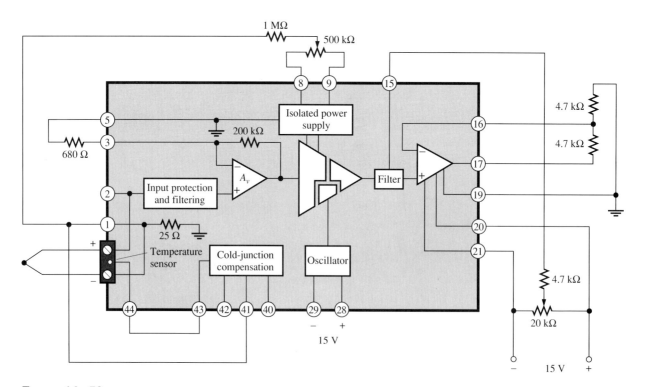

FIGURE 14–50

13. At what resistance value of the RTD will the bridge circuit in Figure 14–51 be balanced if the wires running to the RTD each have a resistance of 10 Ω?

FIGURE 14–51

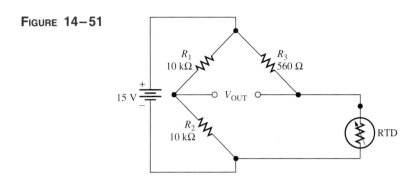

14. At what resistance value of the RTD will the bridge circuit in Figure 14–52 be balanced if the wires running to the RTD each have a resistance of 10 Ω?

FIGURE 14–52

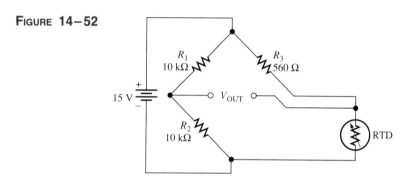

15. Explain the difference in the results of Problems 13 and 14.
16. Determine the output voltage of the instrumentation amplifier in Figure 14–53 if the resistance of the RTD is 697 Ω at the temperature being measured.

FIGURE 14–53

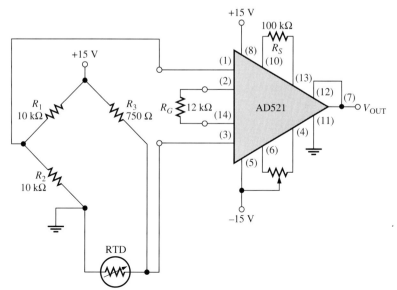

17. Determine the voltage gain of the 2B31 in Figure 14–54.

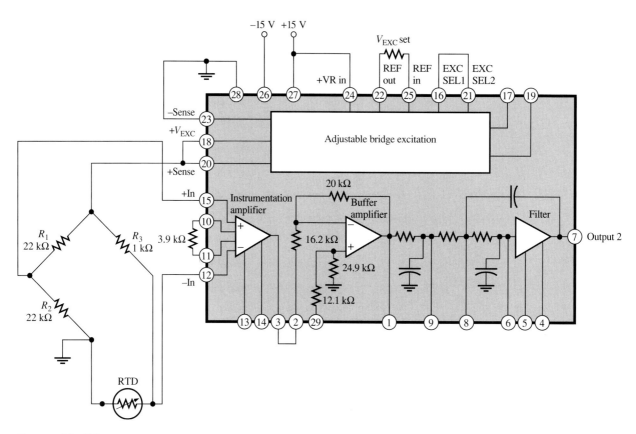

FIGURE 14–54

SECTION 14–4 STRAIN-MEASURING AND PRESSURE-MEASURING CIRCUITS

18. A certain material being measured undergoes a strain of 3 parts per million. The strain gage has a nominal resistance of 600 Ω and a gage factor of 2.5. Determine the resistance change in the strain gage.
19. Explain how a strain gage can be used to measure pressure.
20. Identify and compare the three symbols in Figure 14–55.

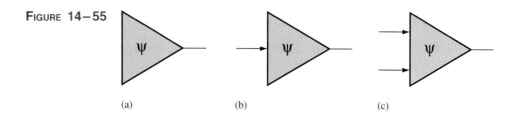

FIGURE 14–55

MEASUREMENT AND CONTROL CIRCUITS 745

Section 14–5 Power-Control Circuits

21. Show how to connect the CA3079 zero-voltage switch to a 3-wire RTD bridge in order to control a triac-driven heating element.

22. Assume the RTD in Problem 21 has a resistance of 200 Ω at 50°C, 400 Ω at 100°C, and 600 Ω at 150°C. To what value must you set the other external bridge resistor to maintain a temperature of 125°C?

Answers to Review Questions

Section 14–1
1. An rms-to-dc converter produces a dc output voltage that is equal to the rms value of the ac input voltage.
2. Internally, an rms-to-dc converter squares, averages, and takes the square root.

Section 14–2
1. Synchro
2. An RDC accepts resolver format voltages on its inputs.
3. An RDC produces a digital code representing the angular shaft position of the resolver.
4. An RDC converts the angular shaft position of a resolver into a digital code.

Section 14–3
1. A thermocouple is a temperature transducer formed by the junction of two dissimilar metals.
2. A thermocouple produces a voltage proportional to the temperature at the junction.
3. A RTD is a resistance temperature detector in which the resistance is proportional to the temperature, whereas the thermocouple produces a voltage.
4. An RTD has a positive temperature coefficient and a thermistor has a negative temperature coefficient.
5. The thermocouple has a greater temperature range than the RTD or thermistor.

Section 14–4
1. Basically, a strain gage is a resistive element whose dimensions can be altered by an applied force to produce a change in resistance.
2. Basically, a pressure gage is a strain gage bonded to a flexible diaphragm.
3. Absolute, gage, and differential

Section 14–5
1. Zero-voltage switching eliminates fast transitions in the current to a load, thus reducing RFI emissions and thermal shock to the load element.
2. An SCR is unidirectional and therefore allows current through the load only during half of the ac cycle. A triac is bidirectional and allows current during the complete cycle.

Section 14–6

1. R_1 and C_1 set the pulse width of the one-shot.
2. The noninverting op-amp provides gain for the output of the F/V converter and permits adjustment of the input to the A/D for calibration purposes.

Answers to Practice Exercises

14–1 182.90040°
14–2 10.07%
14–3 24.247 mV
14–4 Change R_G to 10 kΩ and move wire from pin 42 to pin 41.
14–5 1 kΩ; 0 V
14–6 The gain increases to 25.93.
14–7 5.12 mΩ

Appendix A
Data Sheets

Diodes

1N746–1N4372	Zener Diodes	748
1N1183A–1N1190A	Rectifier Diodes	750
1N4001–1N4007	Rectifier Diodes	751

BJTs and FETs

2N2222A	General-purpose BJT	753
2N3946/3947	General-purpose BJT	758
2N2843/2N2844	P-channel JFET	764

Thyristors

2N6394–2N6399	SCR	765

Linear Integrated Circuits

LM117–LM317	Adjustable Positive Voltage Regulator	767
MC1455	555-Equivalent Timer	769
MC1741	741 Op-Amp	771
MC7800	Fixed Positive Voltage Regulator	775
MC7900	Fixed Negative Voltage Regulator	778
AD295	Isolation Amplifier	781
AD521	Instrumentation Amplifier	783
AD650	Voltage-to-Frequency Converter	785
AD673	8-Bit A/D Converter	788
AD9300	Analog Multiplexer	791
1S20	Resolver-to-Digital Converter	794
2B31	Strain Gage/RTD Conditioner	796
2B50	Thermocouple Signal Conditioner	798

The data sheets in this appendix are either copyright of Motorola, Inc. or Analog Devices, Inc. Used by permission.

MOTOROLA SEMICONDUCTOR
TECHNICAL DATA

1.5KE6.8, A thru 1.5KE250, A
See Page 4-59

Designers Data Sheet

500-MILLIWATT HERMETICALLY SEALED GLASS SILICON ZENER DIODES

- Complete Voltage Range — 2.4 to 110 Volts
- DO-35 Package — Smaller than Conventional DO-7 Package
- Double Slug Type Construction
- Metallurgically Bonded Construction
- Oxide Passivated Die

Designer's Data for "Worst Case" Conditions

The Designer's Data sheets permit the design of most circuits entirely from the information presented. Limit curves — representing boundaries on device characteristics — are given to facilitate "worst case" design.

1N746 thru 1N759
1N957A thru 1N986A
1N4370 thru 1N4372

GLASS ZENER DIODES
500 MILLIWATTS
2.4–110 VOLTS

MAXIMUM RATINGS

Rating	Symbol	Value	Unit
DC Power Dissipation @ $T_L \leq 50°C$, Lead Length = 3/8"	P_D		
*JEDEC Registration		400	mW
*Derate above $T_L = 50°C$		3.2	mW/°C
Motorola Device Ratings		500	mW
Derate above $T_L = 50°C$		3.33	mW/°C
Operating and Storage Junction Temperature Range	T_J, T_{stg}		°C
*JEDEC Registration		−65 to +175	
Motorola Device Ratings		−65 to +200	

*Indicates JEDEC Registered Data.

MECHANICAL CHARACTERISTICS

MAXIMUM LEAD TEMPERATURE FOR SOLDERING PURPOSES: 230°C, 1/16" from case for 10 seconds

FINISH: All external surfaces are corrosion resistant with readily solderable leads.

POLARITY: Cathode indicated by color band. When operated in zener mode, cathode will be positive with respect to anode.

MOUNTING POSITION: Any

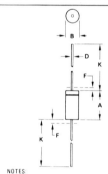

NOTES:
1. PACKAGE CONTOUR OPTIONAL WITHIN A AND B. HEAT SLUGS, IF ANY, SHALL BE INCLUDED WITHIN THIS CYLINDER, BUT NOT SUBJECT TO THE MINIMUM LIMIT OF B.
2. LEAD DIAMETER NOT CONTROLLED IN ZONE F TO ALLOW FOR FLASH, LEAD FINISH BUILDUP AND MINOR IRREGULARITIES OTHER THAN HEAT SLUGS.
3. POLARITY DENOTED BY CATHODE BAND.
4. DIMENSIONING AND TOLERANCING PER ANSI Y14.5, 1973.

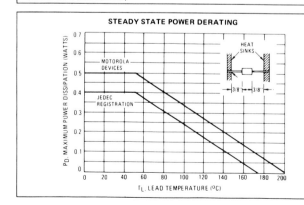

STEADY STATE POWER DERATING

DIM	MILLIMETERS		INCHES	
	MIN	MAX	MIN	MAX
A	3.05	5.08	0.120	0.200
B	1.52	2.29	0.060	0.090
D	0.46	0.56	0.018	0.022
F	—	1.27	—	0.050
K	25.40	38.10	1.000	1.500

All JEDEC dimensions and notes apply.

CASE 299-02
DO-204AH
GLASS

1N746 thru 1N759, 1N957A thru 1N986A, 1N4370 thru 1N4372

ELECTRICAL CHARACTERISTICS ($T_A = 25°C$, $V_F = 1.5$ V max at 200 mA for all types)

Type Number (Note 1)	Nominal Zener Voltage V_Z @ I_{ZT} (Note 2) Volts	Test Current I_{ZT} mA	Maximum Zener Impedance Z_{ZT} @ I_{ZT} (Note 3) Ohms	*Maximum DC Zener Current I_{ZM} (Note 4) mA		Maximum Reverse Leakage Current	
						$T_A = 25°C$ I_R @ $V_R = 1$ V μA	$T_A = 150°C$ I_R @ $V_R = 1$ V μA
1N4370	2.4	20	30	150	190	100	200
1N4371	2.7	20	30	135	165	75	150
1N4372	3.0	20	29	120	150	50	100
1N746	3.3	20	28	110	135	10	30
1N747	3.6	20	24	100	125	10	30
1N748	3.9	20	23	95	115	10	30
1N749	4.3	20	22	85	105	2	30
1N750	4.7	20	19	75	95	2	30
1N751	5.1	20	17	70	85	1	20
1N752	5.6	20	11	65	80	1	20
1N753	6.2	20	7	60	70	0.1	20
1N754	6.8	20	5	55	65	0.1	20
1N755	7.5	20	6	50	60	0.1	20
1N756	8.2	20	8	45	55	0.1	20
1N757	9.1	20	10	40	50	0.1	20
1N758	10	20	17	35	45	0.1	20
1N759	12	20	30	30	35	0.1	20

Type Number (Note 1)	Nominal Zener Voltage V_Z (Note 2) Volts	Test Current I_{ZT} mA	Maximum Zener Impedance (Note 3)			*Maximum DC Zener Current I_{ZM} (Note 4) mA		Maximum Reverse Current		
			Z_{ZT} @ I_{ZT} Ohms	Z_{ZK} @ I_{ZK} Ohms	I_{ZK} mA			I_R Maximum μA	Test Voltage Vdc 5% V_R	10%
1N957A	6.8	18.5	4.5	700	1.0	47	61	150	5.2	4.9
1N958A	7.5	16.5	5.5	700	0.5	42	55	75	5.7	5.4
1N959A	8.2	15	6.5	700	0.5	38	50	50	6.2	5.9
1N960A	9.1	14	7.5	700	0.5	35	45	25	6.9	6.6
1N961A	10	12.5	8.5	700	0.25	32	41	10	7.6	7.2
1N962A	11	11.5	9.5	700	0.25	28	37	5	8.4	8.0
1N963A	12	10.5	11.5	700	0.25	26	34	5	9.1	8.6
1N964A	13	9.5	13	700	0.25	24	32	5	9.9	9.4
1N965A	15	8.5	16	700	0.25	21	27	5	11.4	10.8
1N966A	16	7.8	17	700	0.25	19	37	5	12.2	11.5
1N967A	18	7.0	21	750	0.25	17	23	5	13.7	13.0
1N968A	20	6.2	25	750	0.25	15	20	5	15.2	14.4
1N969A	22	5.6	29	750	0.25	14	18	5	16.7	15.8
1N970A	24	5.2	33	750	0.25	13	17	5	18.2	17.3
1N971A	27	4.6	41	750	0.25	11	15	5	20.6	19.4
1N972A	30	4.2	49	1000	0.25	10	13	5	22.8	21.6
1N973A	33	3.8	58	1000	0.25	9.2	12	5	25.1	23.8
1N974A	36	3.4	70	1000	0.25	8.5	11	5	27.4	25.9
1N975A	39	3.2	80	1000	0.25	7.8	10	5	29.7	28.1
1N976A	43	3.0	93	1500	0.25	7.0	9.6	5	32.7	31.0
1N977A	47	2.7	105	1500	0.25	6.4	8.8	5	35.8	33.8
1N978A	51	2.5	125	1500	0.25	5.9	8.1	5	38.8	36.7
1N979A	56	2.2	150	2000	0.25	5.4	7.4	5	42.6	40.3
1N980A	62	2.0	185	2000	0.25	4.9	6.7	5	47.1	44.6
1N981A	68	1.8	230	2000	0.25	4.5	6.1	5	51.7	49.0
1N982A	75	1.7	270	2000	0.25	1.0	5.5	5	56.0	54.0
1N983A	82	1.5	330	3000	0.25	3.7	5.0	5	62.2	59.0
1N984A	91	1.4	400	3000	0.25	3.3	4.5	5	69.2	65.5
1N985A	100	1.3	500	3000	0.25	3.0	4.5	5	76	72
1N986A	110	1.1	750	4000	0.25	2.7	4.1	5	83.6	79.2

NOTE 1. TOLERANCE AND VOLTAGE DESIGNATION

Tolerance Designation

The type numbers shown have tolerance designations as follows:

    1N4370 series: ±10%, suffix A for ±5% units,
                       C for ±2%, D for ±1%.
    1N746 series: ±10%, suffix A for ±5% units,
                     C for ±2%, D for ±1%.
    1N957 series: ±10%, suffix A for ±10% units,
                     C for ±2%, D for ±1%,
              suffix B for ±5% units,
                     C for ±2%, D for ±1%.

MOTOROLA SEMICONDUCTOR TECHNICAL DATA

1N1183A thru 1N1190A

MEDIUM-CURRENT RECTIFIERS

...for applications requiring low forward voltage drop and rugged construction.

- High Surge Handling Ability
- Rugged Construction
- Reverse Polarity Available; Eliminates Need for Insulating Hardware in Many Cases
- Hermetically Sealed

20-AMP RECTIFIERS
SILICON
DIFFUSED-JUNCTION

*MAXIMUM RATINGS

Rating	Symbol	1N1183A	1N1184A	1N1186A	1N1188A	1N1190A	Unit
Peak Repetitive Reverse Voltage	V_{RRM}, V_{RWM}, V_R	50	100	200	400	600	Volts
Average Half-Wave Rectified Forward Current With Resistive Load @ T_A = 150°C	I_O	40	40	40	40	40	Amp
Peak One Cycle Surge Current (60 Hz and 150°C Case Temperature)	I_{FSM}	800	800	800	800	800	Amp
Operating Junction Temperature	T_J	−65 to +200					°C
Storage Temperature	T_{stg}	−65 to +200					°C

*ELECTRICAL CHARACTERISTICS (All Types) at 25°C Case Temperature

Characteristic	Symbol	Value	Unit
Maximum Forward Voltage at 100 Amp DC Forward Current	V_F	1.1	Volts
Maximum Reverse Current at Rated DC Reverse Voltage	I_R	5.0	mAdc

THERMAL CHARACTERISTICS

Characteristic	Symbol	Typical	Unit
Thermal Resistance, Junction to Case	$R_{\theta JC}$	1.0	°C/W

*Indicates JEDEC registered data.

MECHANICAL CHARACTERISTICS

CASE: Welded, hermetically sealed construction
FINISH: All external surfaces corrosion-resistant and the terminal lead is readily solderable
WEIGHT: 25 grams (approx.)
POLARITY: Cathode connected to case (reverse polarity available denoted by Suffix R, i.e.: 1N3212R)
MOUNTING POSITION: Any
MOUNTING TORQUE: 25 in-lb max

DIM	MILLIMETERS MIN	MILLIMETERS MAX	INCHES MIN	INCHES MAX
A	—	20.07	—	0.790
B	16.94	17.45	0.669	0.687
C	—	11.43	—	0.450
D	—	9.53	—	0.375
E	2.92	5.08	0.115	0.200
F	—	2.03	—	0.080
J	10.72	11.51	0.422	0.453
K	19.05	25.40	0.750	1.00
L	3.96	—	0.156	—
P	5.59	6.32	0.220	0.249
Q	3.56	4.45	0.140	0.175
R	—	16.94	—	0.667
S	—	2.26	—	0.089

CASE 42A-01
DO-203AB
METAL

1N4001 thru 1N4007

Designers Data Sheet

"SURMETIC"▲ RECTIFIERS

... subminiature size, axial lead mounted rectifiers for general-purpose low-power applications.

Designers Data for "Worst Case" Conditions

The Designers▲ Data Sheets permit the design of most circuits entirely from the information presented. Limit curves — representing boundaries on device characteristics — are given to facilitate "worst case" design.

LEAD MOUNTED SILICON RECTIFIERS

50-1000 VOLTS DIFFUSED JUNCTION

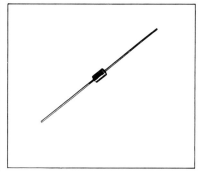

*MAXIMUM RATINGS

Rating	Symbol	1N4001	1N4002	1N4003	1N4004	1N4005	1N4006	1N4007	Unit
Peak Repetitive Reverse Voltage Working Peak Reverse Voltage DC Blocking Voltage	V_{RRM} V_{RWM} V_R	50	100	200	400	600	800	1000	Volts
Non-Repetitive Peak Reverse Voltage (halfwave, single phase, 60 Hz)	V_{RSM}	60	120	240	480	720	1000	1200	Volts
RMS Reverse Voltage	$V_{R(RMS)}$	35	70	140	280	420	560	700	Volts
Average Rectified Forward Current (single phase, resistive load, 60 Hz, see Figure 8, $T_A = 75°C$)	I_O	1.0							Amp
Non-Repetitive Peak Surge Current (surge applied at rated load conditions, see Figure 2)	I_{FSM}	30 (for 1 cycle)							Amp
Operating and Storage Junction Temperature Range	T_J, T_{stg}	−65 to +175							°C

*ELECTRICAL CHARACTERISTICS

Characteristic and Conditions	Symbol	Typ	Max	Unit
Maximum Instantaneous Forward Voltage Drop ($i_F = 1.0$ Amp, $T_J = 25°C$) Figure 1	v_F	0.93	1.1	Volts
Maximum Full-Cycle Average Forward Voltage Drop ($I_O = 1.0$ Amp, $T_L = 75°C$, 1 inch leads)	$V_{F(AV)}$	—	0.8	Volts
Maximum Reverse Current (rated dc voltage) $T_J = 25°C$ $T_J = 100°C$	I_R	0.05 1.0	10 50	µA
Maximum Full-Cycle Average Reverse Current ($I_O = 1.0$ Amp, $T_L = 75°C$, 1 inch leads)	$I_{R(AV)}$	—	30	µA

*Indicates JEDEC Registered Data.

MECHANICAL CHARACTERISTICS

CASE: Transfer Molded Plastic
MAXIMUM LEAD TEMPERATURE FOR SOLDERING PURPOSES: 350°C, 3/8" from case for 10 seconds at 5 lbs. tension
FINISH: All external surfaces are corrosion-resistant, leads are readily solderable
POLARITY: Cathode indicated by color band
WEIGHT: 0.40 Grams (approximately)

▲Trademark of Motorola Inc.

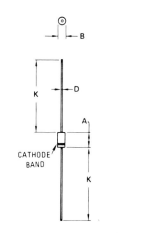

DIM	MILLIMETERS		INCHES	
	MIN	MAX	MIN	MAX
A	5.97	6.60	0.235	0.260
B	2.79	3.05	0.110	0.120
D	0.76	0.86	0.030	0.034
K	27.94	—	1.100	—

CASE 59-04
Does Not Conform to DO-41 Outline.

© MOTOROLA INC., 1982

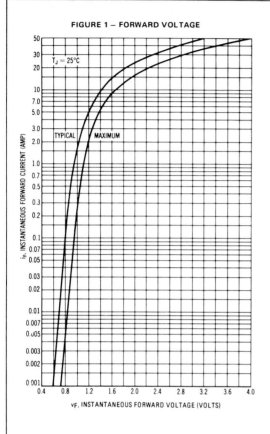

FIGURE 1 — FORWARD VOLTAGE

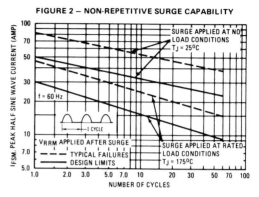

FIGURE 2 — NON-REPETITIVE SURGE CAPABILITY

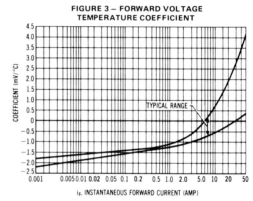

FIGURE 3 — FORWARD VOLTAGE TEMPERATURE COEFFICIENT

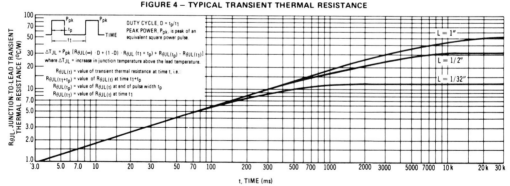

FIGURE 4 — TYPICAL TRANSIENT THERMAL RESISTANCE

The temperature of the lead should be measured using a thermocouple placed on the lead as close as possible to the tie point. The thermal mass connected to the tie point is normally large enough so that it will not significantly respond to heat surges generated in the diode as a result of pulsed operation once steady-state conditions are achieved. Using the measured value of T_L, the junction temperature may be determined by:

$$T_J = T_L + \Delta T_{JL}.$$

MOTOROLA Semiconductor Products Inc.

MAXIMUM RATINGS

Rating	Symbol	2N2219 2N2222	2N2218A 2N2219A 2N2221A 2N2222A	Unit
Collector-Emitter Voltage	V_{CEO}	30	40	Vdc
Collector-Base Voltage	V_{CBO}	60	75	Vdc
Emitter-Base Voltage	V_{EBO}	5.0	6.0	Vdc
Collector Current — Continuous	I_C	800	800	mAdc
		2N2218A 2N2219,A	2N2221A 2N2222,A	
Total Device Dissipation @ T_A = 25°C Derate above 25°C	P_D	0.8 4.57	0.4 2.28	Watt mW/°C
Total Device Dissipation @ T_C = 25°C Derate above 25°C	P_D	3.0 17.1	1.2 6.85	Watts mW/°C
Operating and Storage Junction Temperature Range	T_J, T_{stg}	−65 to +200		°C

2N2218A, 2N2219, A
2N2221A, 2N2222, A

JAN, JTX, JTXV AVAILABLE

2N2218, A/2N2219, A
CASE 79-04
TO-39 (TO-205AD)
STYLE 1

2N2221, A/2N2222, A
CASE 22-03
TO-18 (TO-206AA)
STYLE 1

3 Collector
2 Base
1 Emitter

GENERAL PURPOSE TRANSISTORS
NPN SILICON

THERMAL CHARACTERISTICS

Characteristic	Symbol	2N2218A 2N2219,A	2N2221A 2N2222,A	Unit
Thermal Resistance, Junction to Ambient	$R_{\theta JA}$	219	145.8	°C/W
Thermal Resistance, Junction to Case	$R_{\theta JC}$	58	437.5	°C/W

ELECTRICAL CHARACTERISTICS (T_A = 25°C unless otherwise noted.)

Characteristic		Symbol	Min	Max	Unit
OFF CHARACTERISTICS					
Collector-Emitter Breakdown Voltage (I_C = 10 mAdc, I_B = 0)	Non-A Suffix A-Suffix	$V_{(BR)CEO}$	30 40	— —	Vdc
Collector-Base Breakdown Voltage (I_C = 10 μAdc, I_E = 0)	Non-A Suffix A-Suffix	$V_{(BR)CBO}$	60 75	— —	Vdc
Emitter-Base Breakdown Voltage (I_E = 10 μAdc, I_C = 0)	Non-A Suffix A-Suffix	$V_{(BR)EBO}$	5.0 6.0	— —	Vdc
Collector Cutoff Current (V_{CE} = 60 Vdc, $V_{EB(off)}$ = 3.0 Vdc)	A-Suffix	I_{CEX}	—	10	nAdc
Collector Cutoff Current (V_{CB} = 50 Vdc, I_E = 0) (V_{CB} = 60 Vdc, I_E = 0) (V_{CB} = 50 Vdc, I_E = 0, T_A = 150°C) (V_{CB} = 60 Vdc, I_E = 0, T_A = 150°C)	Non-A Suffix A-Suffix Non-A Suffix A-Suffix	I_{CBO}	— — — —	0.01 0.01 10 10	μAdc
Emitter Cutoff Current (V_{EB} = 3.0 Vdc, I_C = 0)	A-Suffix	I_{EBO}	—	10	nAdc
Base Cutoff Current (V_{CE} = 60 Vdc, $V_{EB(off)}$ = 3.0 Vdc)	A-Suffix	I_{BL}	—	20	nAdc
ON CHARACTERISTICS					
DC Current Gain (I_C = 0.1 mAdc, V_{CE} = 10 Vdc)	2N2218A, 2N2221A(1) 2N2219,A, 2N2222,A(1)	h_{FE}	20 35	— —	—
(I_C = 1.0 mAdc, V_{CE} = 10 Vdc)	2N2218A, 2N2221A 2N2219,A, 2N2222,A		25 50	— —	
(I_C = 10 mAdc, V_{CE} = 10 Vdc)	2N2218A, 2N2221A(1) 2N2219,A, 2N2222,A(1)		35 75	— —	
(I_C = 10 mAdc, V_{CE} = 10 Vdc, T_A = −55°C)	2N2218A, 2N2221A 2N2219,A, 2N2222,A		15 35	— —	
(I_C = 150 mAdc, V_{CE} = 10 Vdc)(1)	2N2218A, 2N2221A 2N2219,A, 2N2222,A		40 100	120 300	

2N2218A/19/19A/21A/22/22A

ELECTRICAL CHARACTERISTICS (continued) (T_A = 25°C unless otherwise noted.)

Characteristic		Symbol	Min	Max	Unit
(I_C = 150 mAdc, V_{CE} = 1.0 Vdc)(1)	2N2218A, 2N2221A		20	—	
	2N2219,A, 2N2222,A		50	—	
(I_C = 500 mAdc, V_{CE} = 10 Vdc)(1)	2N2219, 2N2222		30	—	
	2N2218A, 2N2221A,		25	—	
	2N2219A, 2N2222A		40	—	
Collector-Emitter Saturation Voltage(1)		$V_{CE(sat)}$			Vdc
(I_C = 150 mAdc, I_B = 15 mAdc)	Non-A Suffix		—	0.4	
	A-Suffix		—	0.3	
(I_C = 500 mAdc, I_B = 50 mAdc)	Non-A Suffix		—	1.6	
	A-Suffix		—	1.0	
Base-Emitter Saturation Voltage(1)		$V_{BE(sat)}$			Vdc
(I_C = 150 mAdc, I_B = 15 mAdc)	Non-A Suffix		0.6	1.3	
	A-Suffix		0.6	1.2	
(I_C = 500 mAdc, I_B = 50 mAdc)	Non-A Suffix		—	2.6	
	A-Suffix		—	2.0	
SMALL-SIGNAL CHARACTERISTICS					
Current Gain — Bandwidth Product(2)		f_T			MHz
(I_C = 20 mAdc, V_{CE} = 20 Vdc, f = 100 MHz)	All Types, Except		250	—	
	2N2219A, 2N2222A		300	—	
Output Capacitance(3)		C_{obo}	—	8.0	pF
(V_{CB} = 10 Vdc, I_E = 0, f = 1.0 MHz)					
Input Capacitance(3)		C_{ibo}			pF
(V_{EB} = 0.5 Vdc, I_C = 0, f = 1.0 MHz)	Non-A Suffix		—	30	
	A-Suffix		—	25	
Input Impedance		h_{ie}			kohms
(I_C = 1.0 mAdc, V_{CE} = 10 Vdc, f = 1.0 kHz)	2N2218A, 2N2221A		1.0	3.5	
	2N2219A, 2N2222A		2.0	8.0	
(I_C = 10 mAdc, V_{CE} = 10 Vdc, f = 1.0 kHz)	2N2218A, 2N2221A		0.2	1.0	
	2N2219A, 2N2222A		0.25	1.25	
Voltage Feedback Ratio		h_{re}			X 10^{-4}
(I_C = 1.0 mAdc, V_{CE} = 10 Vdc, f = 1.0 kHz)	2N2218A, 2N2221A		—	5.0	
	2N2219A, 2N2222A		—	8.0	
(I_C = 10 mAdc, V_{CE} = 10 Vdc, f = 1.0 kHz)	2N2218A, 2N2221A		—	2.5	
	2N2219A, 2N2222A		—	4.0	
Small-Signal Current Gain		h_{fe}			—
(I_C = 1.0 mAdc, V_{CE} = 10 Vdc, f = 1.0 kHz)	2N2218A, 2N2221A		30	150	
	2N2219A, 2N2222A		50	300	
(I_C = 10 mAdc, V_{CE} = 10 Vdc, f = 1.0 kHz)	2N2218A, 2N2221A		50	300	
	2N2219A, 2N2222A		75	375	
Output Admittance		h_{oe}			µmhos
(I_C = 1.0 mAdc, V_{CE} = 10 Vdc, f = 1.0 kHz)	2N2218A, 2N2221A		3.0	15	
	2N2219A, 2N2222A		5.0	35	
(I_C = 10 mAdc, V_{CE} = 10 Vdc, f = 1.0 kHz)	2N2218A, 2N2221A		10	100	
	2N2219A, 2N2222A		15	200	
Collector Base Time Constant		$rb'C_c$			ps
(I_E = 20 mAdc, V_{CB} = 20Vdc, f = 31.8 MHz)	A-Suffix		—	150	
Noise Figure		NF			dB
(I_C = 100 µAdc, V_{CE} = 10 Vdc, R_S = 1.0 kohm, f = 1.0 kHz)	2N2222A		—	4.0	
Real Part of Common-Emitter High Frequency Input Impedance		$Re(h_{ie})$			Ohms
(I_C = 20 mAdc, V_{CE} = 20 Vdc, f = 300 MHz)	2N2218A, 2N2219A		—	60	
	2N2221A, 2N2222A				

(1) Pulse Test: Pulse Width ≤ 300 µs, Duty Cycle ≤ 2.0%.
(2) f_T is defined as the frequency at which $|h_{fe}|$ extrapolates to unity.
(3) 2N5581 and 2N5582 are Listed C_{cb} and C_{eb} for these conditions and values.

2N2218A/19/19A/21A/22/22A

ELECTRICAL CHARACTERISTICS (continued) ($T_A = 25°C$ unless otherwise noted.)

Characteristic		Symbol	Min	Max	Unit
SWITCHING CHARACTERISTICS					
Delay Time	($V_{CC} = 30$ Vdc, $V_{BE(off)} = 0.5$ Vdc, $I_C = 150$ mAdc, $I_{B1} = 15$ mAdc) (Figure 14)	t_d	—	10	ns
Rise Time		t_r	—	25	ns
Storage Time	($V_{CC} = 30$ Vdc, $I_C = 150$ mAdc, $I_{B1} = I_{B2} = 15$ mAdc) (Figure 15)	t_s	—	225	ns
Fall Time		t_f	—	60	ns
Active Region Time Constant ($I_C = 150$ mAdc, $V_{CE} = 30$ Vdc) (See Figure 12 for 2N2218A, 2N2219A, 2N2221A, 2N2222A)		T_A	—	2.5	ns

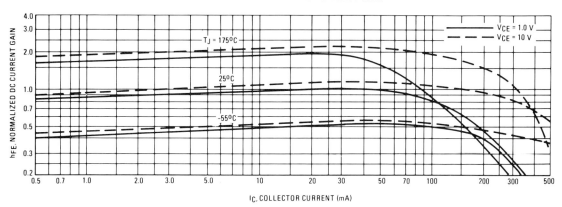

FIGURE 1 – NORMALIZED DC CURRENT GAIN

FIGURE 2 – COLLECTOR CHARACTERISTICS IN SATURATION REGION

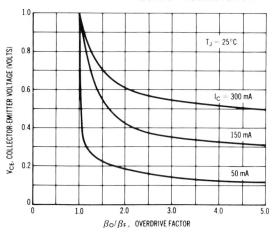

This graph shows the effect of base current on collector current. β_o (current gain at the edge of saturation) is the current gain of the transistor at 1 volt, and β_f (forced gain) is the ratio of I_c/I_{bf} in a circuit.

EXAMPLE: For type 2N2219, estimate a base current (I_{bf}) to insure saturation at a temperature of 25°C and a collector current of 150 mA.

Observe that at $I_c = 150$ mA an overdrive factor of at least 2.5 is required to drive the transistor well into the saturation region. From Figure 1, it is seen that h_{FE} @ 1 volt is approximately 0.62 of h_{FE} @ 10 volts. Using the guaranteed minimum gain of 100 @ 150 mA and 10 V, $\beta_o = 62$ and substituting values in the overdrive equation, we find:

$$\frac{\beta_o}{\beta_f} = \frac{h_{FE} @ 1.0 V}{I_c/I_{bf}} \qquad 2.5 = \frac{62}{150/I_{bf}} \qquad I_{bf} \approx 6.0 \text{ mA}$$

2N2218A/19/19A/21A/22/22A

FIGURE 3 — "ON" VOLTAGES

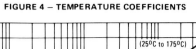

FIGURE 4 — TEMPERATURE COEFFICIENTS

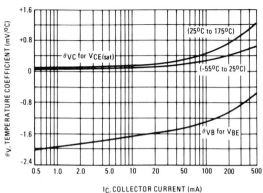

h PARAMETERS

V_{CE} = 10 Vdc, f = 1.0 kHz, T_A = 25°C

This group of graphs illustrates the relationship between h_{fe} and other "h" parameters for this series of transistors. To obtain these curves, a high-gain and a low-gain unit were selected and the same units were used to develop the correspondingly numbered curves on each graph.

FIGURE 5 — INPUT IMPEDANCE

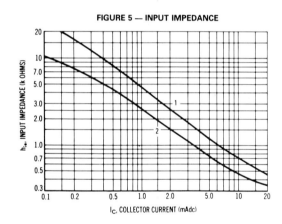

FIGURE 6 — VOLTAGE FEEDBACK RATIO

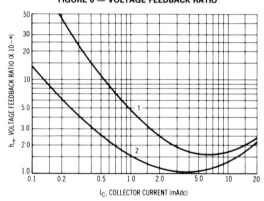

FIGURE 7 — CURRENT GAIN

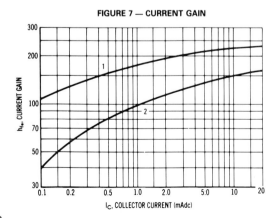

FIGURE 8 — OUTPUT ADMITTANCE

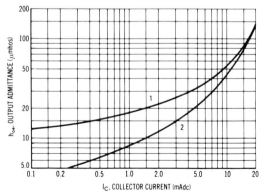

2N2218A/19/19A/21A/22/22A

SWITCHING TIME CHARACTERISTICS

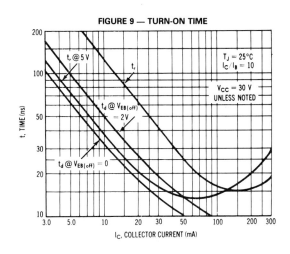

FIGURE 9 — TURN-ON TIME

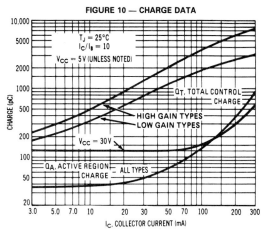

FIGURE 10 — CHARGE DATA

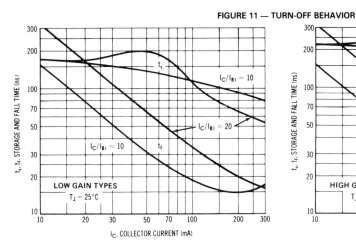

FIGURE 11 — TURN-OFF BEHAVIOR

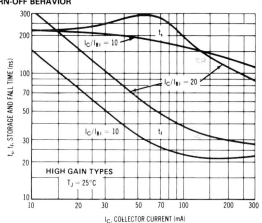

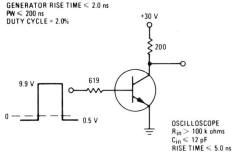

FIGURE 12 — DELAY AND RISE TIME EQUIVALENT TEST CIRCUIT

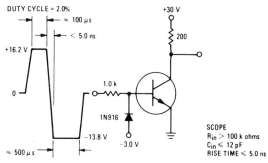

FIGURE 13 — STORAGE TIME AND FALL TIME EQUIVALENT TEST CIRCUIT

2N3946
2N3947

CASE 22-03, STYLE 1
TO-18 (TO-206AA)

GENERAL PURPOSE TRANSISTORS

NPN SILICON

MAXIMUM RATINGS

Rating	Symbol	Value	Unit
Collector-Emitter Voltage	V_{CEO}	40	Vdc
Collector-Base Voltage	V_{CBO}	60	Vdc
Emitter-Base Voltage	V_{EBO}	6.0	Vdc
Collector Current — Continuous	I_C	200	mAdc
Total Device Dissipation @ T_A = 25°C Derate above 25°C	P_D	0.36 2.06	Watt mW/°C
Total Device Dissipation @ T_C = 25°C Derate above 25°C	P_D	1.2 6.9	Watts mW/°C
Operating and Storage Junction Temperature Range	T_J, T_{stg}	−65 to +200	°C

THERMAL CHARACTERISTICS

Characteristic	Symbol	Max	Unit
Thermal Resistance, Junction to Case	$R_{\theta JC}$	0.15	°C/mW
Thermal Resistance, Junction to Ambient	$R_{\theta JA}$	0.49	°C/mW

ELECTRICAL CHARACTERISTICS (T_A = 25°C unless otherwise noted.)

Characteristic		Symbol	Min	Max	Unit
OFF CHARACTERISTICS					
Collector-Emitter Breakdown Voltage(1) (I_C = 10 mAdc)		$V_{(BR)CEO}$	40	—	Vdc
Collector-Base Breakdown Voltage (I_C = 10 μAdc, I_E = 0)		$V_{(BR)CBO}$	60	—	Vdc
Emitter-Base Breakdown Voltage (I_E = 10 μAdc, I_C = 0)		$V_{(BR)EBO}$	6.0	—	Vdc
Collector Cutoff Current (V_{CE} = 40 Vdc, V_{EB} = 3.0 Vdc) (V_{CE} = 40 Vdc, V_{EB} = 3.0 Vdc, T_A = 150°C)		I_{CEX}	— —	0.010 15	μAdc
Base Cutoff Current (V_{CE} = 40 Vdc, V_{EB} = 3.0 Vdc)		I_{BL}	—	.025	μAdc
ON CHARACTERISTICS					
DC Current Gain(1) (I_C = 0.1 mAdc, V_{CE} = 1.0 Vdc)	2N3946 2N3947	h_{FE}	30 60	—	—
(I_C = 1.0 mAdc, V_{CE} = 1.0 Vdc)	2N3946 2N3947		45 90	— —	
(I_C = 10 mAdc, V_{CE} = 1.0 Vdc)	2N3946 2N3947		50 100	150 300	
(I_C = 50 mAdc, V_{CE} = 1.0 Vdc)	2N3946 2N3947		20 40	— —	
Collector-Emitter Saturation Voltage(1) (I_C = 10 mAdc, I_B = 1.0 mAdc) (I_C = 50 mAdc, I_B = 5.0 mAdc)		$V_{CE(sat)}$	— —	0.2 0.3	Vdc
Base-Emitter Saturation Voltage(1) (I_C = 10 mAdc, I_B = 1.0 mAdc) (I_C = 50 mAdc, I_B = 5.0 mAdc)		$V_{BE(sat)}$	0.6 —	0.9 1.0	Vdc
SMALL-SIGNAL CHARACTERISTICS					
Current-Gain — Bandwidth Product (I_C = 10 mAdc, V_{CE} = 20 Vdc, f = 100 MHz)	2N3946 2N3947	f_T	250 300	— —	MHz
Output Capacitance (V_{CB} = 10 Vdc, I_E = 0, f = 1.0 MHz)		C_{obo}	—	4.0	pF

2N3946, 2N3947

ELECTRICAL CHARACTERISTICS (continued) ($T_A = 25°C$ unless otherwise noted.)

Characteristic		Symbol	Min	Max	Unit
Input Capacitance ($V_{EB} = 1.0$ Vdc, $I_C = 0$, f = 1.0 MHz)		C_{ibo}	—	8.0	pF
Input Impedance ($I_C = 1.0$ mA, $V_{CE} = 10$ V, f = 1.0 kHz) 2N3946		h_{ie}	0.5	6.0	kohms
2N3947			2.0	12	
Voltage Feedback Ratio ($I_C = 1.0$ mA, $V_{CE} = 10$ V, f = 1.0 kHz) 2N3946		h_{re}	—	10	$\times 10^{-4}$
2N3947			—	20	
Small Signal Current Gain ($I_C = 1.0$ mA, $V_{CE} = 10$ V, f = 1.0 kHz) 2N3946		h_{fe}	50	250	—
2N3947			100	700	
Output Admittance ($I_C = 1.0$ mA, $V_{CE} = 10$ V, f = 1.0 kHz) 2N3946		h_{oe}	1.0	30	µmhos
2N3947			5.0	50	
Collector Base Time Constant ($I_C = 10$ mA, $V_{CE} = 20$ V, f = 31.8 MHz)		$r_b'C_c$	—	200	ps
Noise Figure ($I_C = 100$ µA, $V_{CE} = 5.0$ V, $R_g = 1.0$ kΩ, f = 1.0 kHz)		NF	—	5.0	dB
SWITCHING CHARACTERISTICS					
Delay Time	$V_{CC} = 3.0$ Vdc, $V_{OB} = 0.5$ Vdc, $I_C = 10$ mAdc, $I_{B1} = 1.0$ mA	t_d	—	35	ns
Rise Time		t_r	—	35	ns
Storage Time	$V_{CC} = 3.0$ V, $I_C = 10$ mA, 2N3946	t_s	—	300	ns
	2N3947		—	375	
Fall Time	$I_{B1} = I_{B2} = 1.0$ mAdc	t_f	—	75	ns

(1) Pulse Test: PW ≤ 300 µs, Duty Cycle ≤ 2%.

TYPICAL SWITCHING CHARACTERISTICS
($T_A = 25°C$ unless otherwise noted)

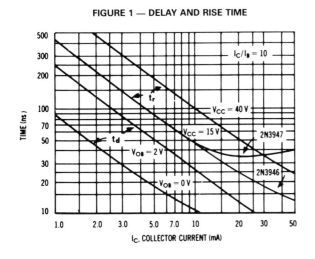

FIGURE 1 — DELAY AND RISE TIME

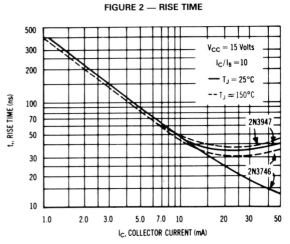

FIGURE 2 — RISE TIME

2N3946, 2N3947

FIGURE 3 — STORAGE AND FALL TIMES

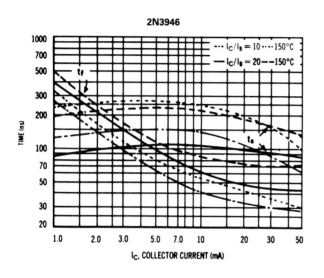
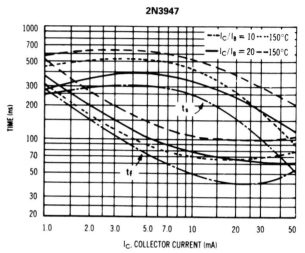

FIGURE 4 — TURN-ON TIME EQUIVALENT TEST CIRCUIT

FIGURE 5 — TURN-OFF TIME EQUIVALENT TEST CIRCUIT

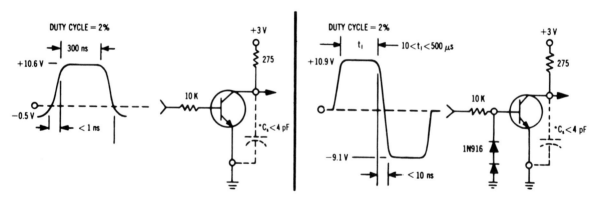

*TOTAL SHUNT CAPACITANCE OF TEST JIG AND CONNECTORS

2N3946, 2N3947

AUDIO SMALL-SIGNAL CHARACTERISTICS

FIGURE 6 — NOISE FIGURE VARIATIONS
V_{CE} = 5.0 V, T_A = 25°C

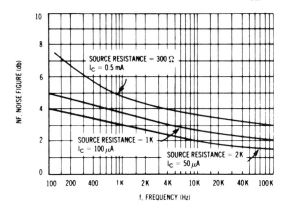
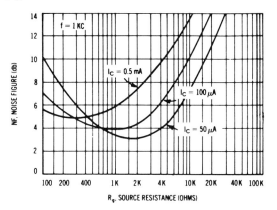

h PARAMETERS
V_{CE} = 10 V, T_A = 25°C, f = 1.0 kc

FIGURE 7 — CURRENT GAIN

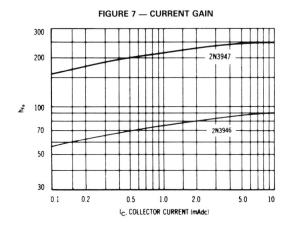

FIGURE 8 — OUTPUT CAPACITANCE

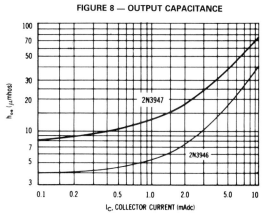

FIGURE 9 — INPUT IMPEDANCE

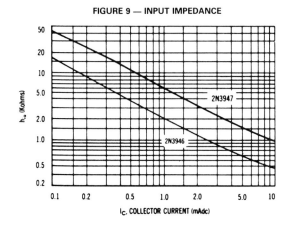

FIGURE 10 — VOLTAGE FEEDBACK RATIO

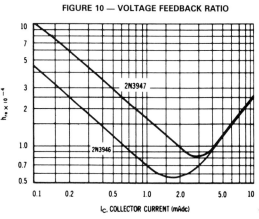

2N3946, 2N3947

FIGURE 11 — CURRENT GAIN CHARACTERISTICS

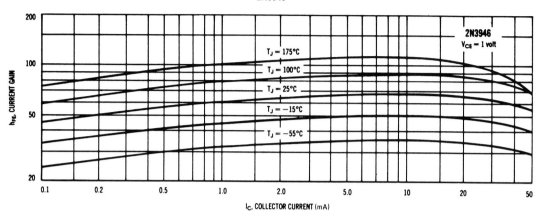

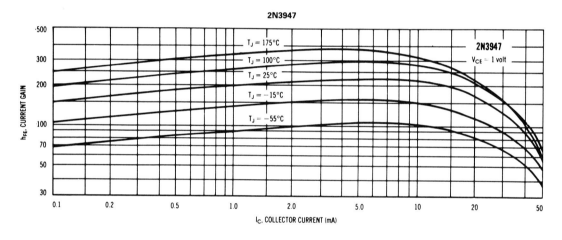

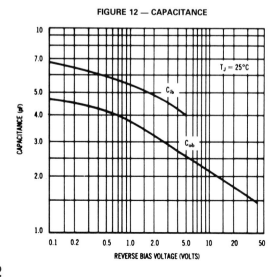

FIGURE 12 — CAPACITANCE

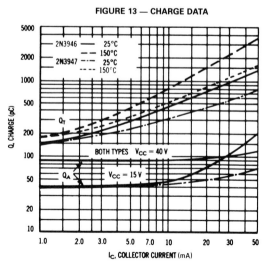

FIGURE 13 — CHARGE DATA

2N3946, 2N3947

FIGURE 14 — COLLECTOR SATURATION REGION
2N3946

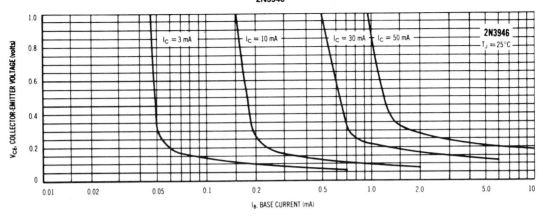

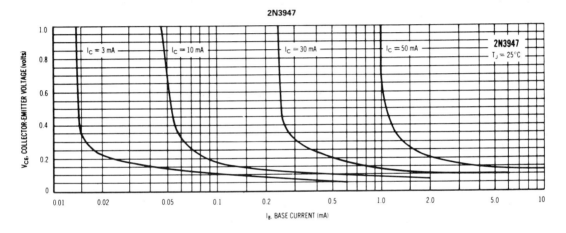

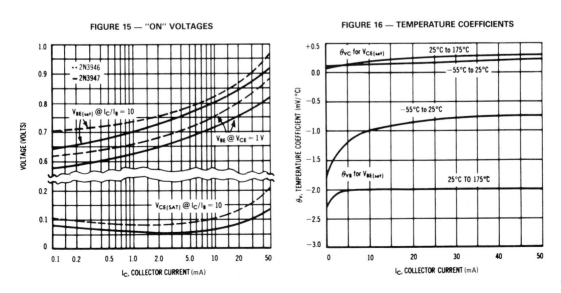

FIGURE 15 — "ON" VOLTAGES

FIGURE 16 — TEMPERATURE COEFFICIENTS

These devices may no longer be available. Please contact nearest Motorola Semiconductor sales office for the current status.

2N2843
2N2844

CASE 22-03, STYLE 12
TO-18 (TO-206AA)

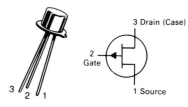

JFETs
GENERAL PURPOSE

P-CHANNEL — DEPLETION

MAXIMUM RATINGS

Rating	Symbol	Value	Unit
Drain-Source Voltage	V_{DS}	30	Vdc
Drain-Gate Voltage	V_{DG}	30	Vdc
Gate-Source Voltage	V_{GS}	30	Vdc
Drain Current	I_D	50	mA
Total Device Dissipation @ T_A = 25°C Derate above 25°C	P_D	300 1.7	mW mW/°C
Storage Temperature Range	T_{stg}	−60 to +200	°C

ELECTRICAL CHARACTERISTICS (T_A = 25°C unless otherwise noted.)

Characteristic		Symbol	Min	Max	Unit
OFF CHARACTERISTICS					
Gate-Source Breakdown Voltage (I_G = 1.0 μA)		$V_{(BR)GSS}$	30	—	Vdc
Gate Reverse Current (V_{GS} = 5.0 V)		I_{GSS}	—	10	nA
Gate Source Cutoff Voltage (V_{DS} = −5.0 V, I_D = −1.0 μA)		$V_{GS(off)}$	—	1.7	Vdc
ON CHARACTERISTICS					
Zero-Gate-Voltage Drain Current (V_{DS} = −5.0 V)	2N2843 2N2844	I_{DSS}*	−200 −440	−1000 −2200	μA
SMALL-SIGNAL CHARACTERISTICS					
Forward Transfer Admittance (V_{DS} = −5.0 V, f = 1.0 kHz)	2N2843 2N2844	$\|Y_{fs}\|$*	540 1400	— —	μmhos
Input Capacitance (V_{DS} = −5.0 V, V_{GS} = 1.0 V, f = 140 kHz)	2N2843 2N2844	C_{iss}	— —	17 30	pF
FUNCTIONAL CHARACTERISTICS					
Noise Figure (V_{DS} = −5.0 V, f = 1.0 kHz, R_G = 1.0 meg)		NF	—	3.0	dB

*Pulse Width ≤ 630 ms, Duty Cycle = 10%.

2N6394 MCR220-5
thru MCR220-7
2N6399 MCR220-9

THYRISTORS
12 AMPERES RMS
50-800 VOLTS

SILICON CONTROLLED RECTIFIERS

... designed primarily for half-wave ac control applications, such as motor controls, heating controls and power supplies; or wherever half-wave silicon gate-controlled, solid-state devices are needed.

- Glass Passivated Junctions and Center Gate Fire for Greater Parameter Uniformity and Stability
- Small, Rugged, Thermowatt▲ Construction for Low Thermal Resistance, High Heat Dissipation and Durability
- Blocking Voltage to 800 Volts

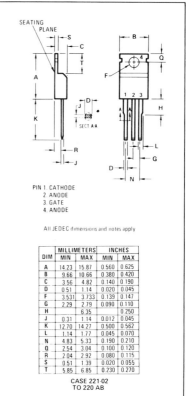

PIN 1. CATHODE
2. ANODE
3. GATE
4. ANODE

All JEDEC dimensions and notes apply

DIM	MILLIMETERS MIN	MILLIMETERS MAX	INCHES MIN	INCHES MAX
A	14.23	15.87	0.560	0.625
B	9.66	10.66	0.380	0.420
C	3.56	4.82	0.140	0.190
D	0.51	1.14	0.020	0.045
F	3.531	3.733	0.139	0.147
G	2.29	2.79	0.090	0.110
H		6.35		0.250
J	0.31	1.14	0.012	0.045
K	12.70	14.27	0.500	0.562
L	1.14	1.77	0.045	0.070
N	4.83	5.33	0.190	0.210
Q	2.54	3.04	0.100	0.120
R	2.04	2.92	0.080	0.115
S	0.51	1.39	0.020	0.055
T	5.85	6.85	0.230	0.270

CASE 221-02
TO 220 AB

*MAXIMUM RATINGS

Rating	Symbol	Value	Unit
Peak Reverse Voltage (1)	V_{RRM}		Volts
2N6394		50	
2N6395		100	
2N6396		200	
MCR220-5		300	
2N6397		400	
MCR220-7		500	
2N6398		600	
MCR220-9		700	
2N6399		800	
Forward Current RMS $T_J = 125°C$ (All Conduction Angles)	$I_{T(RMS)}$	12	Amps
Peak Forward Surge Current (1/2 cycle, Sine Wave, 60 Hz, $T_J = 125°C$)	I_{TSM}	100	Amps
Circuit Fusing Considerations ($T_J = -40$ to $+125°C$, t = 1.0 to 8.3 ms)	I^2t	40	A^2s
Forward Peak Gate Power	P_{GM}	20	Watts
Forward Average Gate Power	$P_{G(AV)}$	0.5	Watt
Forward Peak Gate Current	I_{GM}	2.0	Amps
Operating Junction Temperature Range	T_J	-40 to $+125$	°C
Storage Temperature Range	T_{stg}	-40 to $+150$	°C

THERMAL CHARACTERISTICS

Characteristic	Symbol	Max	Unit
Thermal Resistance, Junction to Case	$R_{\theta JC}$	2.0	°C/W

(1) V_{RRM} for all types can be applied on a continuous dc basis without incurring damage. Ratings apply for zero or negative gate voltage. Devices should not be tested for blocking capability in a manner such that the voltage supplied exceeds the rated blocking voltage.

*Indicates JEDEC Registered Data.
▲Trademark of Motorola Inc.

©MOTOROLA INC., 1975   DS 6565 R1

2N6394 thru 2N6399 • MCR220-5 • MCR220-7 • MCR220-9

ELECTRICAL CHARACTERISTICS (T_C = 25°C unless otherwise noted.)

Characteristic	Symbol	Min	Typ	Max	Unit
*Peak Forward Blocking Voltage (T_J = 125°C)	V_{DRM}				Volts
2N6394		50	—	—	
2N6395		100	—	—	
2N6396		200	—	—	
MCR220-5		300	—	—	
2N6397		400	—	—	
MCR220-7		500	—	—	
2N6398		600	—	—	
MCR220-9		700	—	—	
2N6399		800	—	—	
* Peak Forward Blocking Current (Rated V_{DRM} @ T_J = 125°C)	I_{DRM}	—	—	2.0	mA
* Peak Reverse Blocking Current (Rated V_{RRM} @ T_J = 125°C)	I_{RRM}	—	—	2.0	mA
* Forward "On" Voltage (I_{TM} = 24 A Peak)	V_{TM}	—	1.7	2.2	Volts
* Gate Trigger Current (Continuous dc) (Anode Voltage = 12 Vdc, R_L = 100 Ohms)	I_{GT}	—	5.0	30	mA
* Gate Trigger Voltage (Continuous dc) (Anode Voltage = 12 Vdc, R_L = 100 Ohms)	V_{GT}	—	0.7	1.5	Volts
* Gate Non-Trigger Voltage (Anode Voltage = Rated V_{DRM}, R_L = 100 Ohms, T_J = 125°C)	V_{GD}	0.2	—	—	Volts
* Holding Current (Anode Voltage = 12 Vdc)	I_H	—	6.0	40	mA
Turn-On Time (I_{TM} = 12 A, I_{GT} = 40 mAdc)	t_{gt}	—	1.0	2.0	μs
Turn-Off Time (V_{DRM} = rated voltage) (I_{TM} = 12 A, I_R = 12 A) (I_{TM} = 12 A, I_R = 12 A, T_J = 125°C)	t_q	— —	15 35	— —	μs
Forward Voltage Application Rate (T_J = 125°C)	dv/dt	—	50	—	V/μs

*Indicates JEDEC Registered Data.

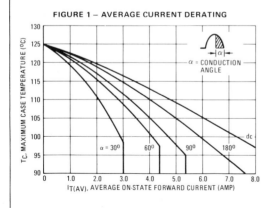

FIGURE 1 — AVERAGE CURRENT DERATING

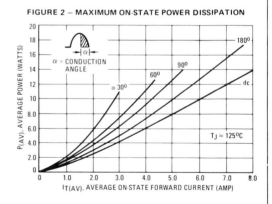

FIGURE 2 — MAXIMUM ON-STATE POWER DISSIPATION

MOTOROLA *Semiconductor Products Inc.*

MOTOROLA SEMICONDUCTOR
TECHNICAL DATA

**LM117
LM217
LM317**

THREE-TERMINAL ADJUSTABLE POSITIVE VOLTAGE REGULATORS

SILICON MONOLITHIC
INTEGRATED CIRCUIT

THREE-TERMINAL ADJUSTABLE OUTPUT POSITIVE VOLTAGE REGULATORS

The LM117/217/317 are adjustable 3-terminal positive voltage regulators capable of supplying in excess of 1.5 A over an output voltage range of 1.2 V to 37 V. These voltage regulators are exceptionally easy to use and require only two external resistors to set the output voltage. Further, they employ internal current limiting, thermal shutdown and safe area compensation, making them essentially blow-out proof.

The LM117 series serve a wide variety of applications including local, on card regulation. This device can also be used to make a programmable output regulator, or by connecting a fixed resistor between the adjustment and output, the LM117 series can be used as a precision current regulator.

- Output Current in Excess of 1.5 Ampere in K and T Suffix Packages
- Output Current in Excess of 0.5 Ampere in H Suffix Package
- Output Adjustable between 1.2 V and 37 V
- Internal Thermal Overload Protectiion
- Internal Short-Circuit Current Limiting Constant with Temperature
- Output Transistor Safe-Area Compensation
- Floating Operation for High Voltage Applications
- Standard 3-lead Transistor Packages
- Eliminates Stocking Many Fixed Voltages

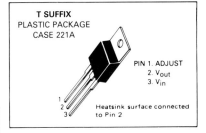

K SUFFIX
METAL PACKAGE
CASE 1

(Bottom View)
CASE IS OUTPUT

Pins 1 and 2 electrically isolated from case.
Case is third electrical connection.

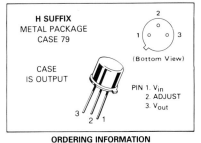

T SUFFIX
PLASTIC PACKAGE
CASE 221A

PIN 1. ADJUST
2. V_{out}
3. V_{in}

Heatsink surface connected to Pin 2

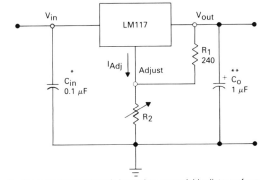

STANDARD APPLICATION

H SUFFIX
METAL PACKAGE
CASE 79

(Bottom View)
CASE IS OUTPUT

PIN 1. V_{in}
2. ADJUST
3. V_{out}

* = C_{in} is required if regulator is located an appreciable distance from power supply filter.

** = C_O is not needed for stability, however it does improve transient response.

$$V_{out} = 1.25 \text{ V} (1 + \frac{R_2}{R_1}) + I_{Adj} R_2$$

Since I_{Adj} is controlled to less than 100 µA, the error associated with this term is negligible in most applications.

ORDERING INFORMATION

Device	Tested Operating Temperature Range	Package
LM117H LM117K	$T_J = -55°C$ to $+150°C$	Metal Can Metal Power
LM217H LM217K	$T_J = -25°C$ to $+150°C$	Metal Can Metal Power
LM317H LM317K LM317T	$T_J = 0°C$ to $+125°C$	Metal Can Metal Power Plastic Power
LM317BT#	$T_J = -40°C$ to $+125°C$	Plastic Power

#Automotive temperature range selections are available with special test conditions and additional tests.
Contact your local Motorola sales office for information.

LM117, LM217, LM317

MAXIMUM RATINGS

Rating	Symbol	Value	Unit
Input-Output Voltage Differential	$V_I - V_O$	40	Vdc
Power Dissipation	P_D	Internally Limited	
Operating Junction Temperature Range LM117 LM217 LM317	T_J	 −55 to +150 −25 to +150 0 to +150	°C
Storage Temperature Range	T_{stg}	−65 to +150	°C

ELECTRICAL CHARACTERISTICS ($V_I - V_O$ = 5.0 V; I_O = 0.5 A for K and T packages; I_O = 0.1 A for H package; $T_J = T_{low}$ to T_{high} [see Note 1]; I_{max} and P_{max} per Note 2; unless otherwise specified.)

Characteristic	Figure	Symbol	LM117/217 Min	LM117/217 Typ	LM117/217 Max	LM317 Min	LM317 Typ	LM317 Max	Unit
Line Regulation (Note 3) T_A = 25°C, 3.0 V ≤ $V_I - V_O$ ≤ 40 V	1	Reg$_{line}$	—	0.01	0.02	—	0.01	0.04	%/V
Load Regulation (Note 3) T_A = 25°C, 10 mA ≤ I_O ≤ I_{max} V_O ≤ 5.0 V V_O ≥ 5.0 V	2	Reg$_{load}$	 — —	 5.0 0.1	 15 0.3	 — —	 5.0 0.1	 25 0.5	 mV %/V_O
Thermal Regulation (T_A = +25°C) 20 ms Pulse	—	—	—	0.02	0.07	—	0.03	0.07	%/W
Adjustment Pin Current	3	I_{Adj}	—	50	100	—	50	100	μA
Adjustment Pin Current Change 2.5 V ≤ $V_I - V_O$ ≤ 40 V 10 mA ≤ I_L ≤ I_{max}, P_D ≤ P_{max}	1,2	ΔI_{Adj}	—	0.2	5.0	—	0.2	5.0	μA
Reference Voltage (Note 4) 3.0 V ≤ $V_I - V_O$ ≤ 40 V 10 mA ≤ I_O ≤ I_{max}, P_D ≤ P_{max}	3	V_{ref}	1.2	1.25	1.3	1.2	1.25	1.3	V
Line Regulation (Note 3) 3.0 V ≤ $V_I - V_O$ ≤ 40 V	1	Reg$_{line}$	—	0.02	0.05	—	0.02	0.07	%/V
Load Regulation (Note 3) 10 mA ≤ I_O ≤ I_{max} V_O ≤ 5.0 V V_O ≥ 5.0 V	2	Reg$_{load}$	 — —	 20 0.3	 50 1.0	 — —	 20 0.3	 70 1.5	 mV %/V_O
Temperature Stability (T_{low} ≤ T_J ≤ T_{high})	3	T_S	—	0.7	—	—	0.7	—	%/V_O
Minimum Load Current to Maintain Regulation ($V_I - V_O$ = 40 V)	3	I_{Lmin}	—	3.5	5.0	—	3.5	10	mA
Maximum Output Current $V_I - V_O$ ≤ 15 V, P_D ≤ P_{max} K and T Packages H Package $V_I - V_O$ = 40 V, P_D ≤ P_{max}, T_A = 25°C K and T Packages H Package	3	I_{max}	 1.5 0.5 0.25 —	 2.2 0.8 0.4 0.07	 — — — —	 1.5 0.5 0.15 —	 2.2 0.8 0.4 0.07	 — — — —	A
RMS Noise, % of V_O T_A = 25°C, 10 Hz ≤ f ≤ 10 kHz	—	N	—	0.003	—	—	0.003	—	%/V_O
Ripple Rejection, V_O = 10 V, f = 120 Hz (Note 5) Without C_{Adj} C_{Adj} = 10 μF	4	RR	 — 66	 65 80	 — —	 — 66	 65 80	 — —	dB
Long-Term Stability, T_J = T_{high} (Note 6) T_A = 25°C for Endpoint Measurements	3	S	—	0.3	1.0	—	0.3	1.0	%/1.0 k Hrs.
Thermal Resistance Junction to Case H Package K Package T Package	—	$R_{\theta JC}$	 — — —	 12 2.3 —	 15 3.0 5.0	 — — —	 12 2.3 —	 15 3.0 —	°C/W

NOTES: (1) T_{low} = −55°C for LM117 T_{high} = +150°C for LM117
      = −25°C for LM217   = +150°C for LM217
      = 0°C for LM317   = +125°C for LM317
(2) I_{max} = 1.5 A for K and T Packages
       = 0.5 A for H Package
    P_{max} = 20 W for K Package
       = 20 W for T Package
       = 2.0 W for H Package
(3) Load and line regulation are specified at constant junction temperature. Changes in V_O due to heating effects must be taken into account separately. Pulse testing with low duty cycle is used.
(4) Selected devices with tightened tolerance reference voltage available.
(5) C_{ADJ}, when used, is connected between the adjustment pin and ground.
(6) Since Long-Term Stability cannot be measured on each device before shipment, this specification is an engineering estimate of average stability from lot to lot.

MOTOROLA SEMICONDUCTOR
TECHNICAL DATA

MC1455

TIMING CIRCUIT

SILICON MONOLITHIC
INTEGRATED CIRCUIT

TIMING CIRCUIT

The MC1455 monolithic timing circuit is a highly stable controller capable of producing accurate time delays, or oscillation. Additional terminals are provided for triggering or resetting if desired. In the time delay mode of operation, the time is precisely controlled by one external resistor and capacitor. For astable operation as an oscillator, the free running frequency and the duty cycle are both accurately controlled with two external resistors and one capacitor. The circuit may be triggered and reset on falling waveforms, and the output structure can source or sink up to 200 mA or drive MTTL circuits.

- Direct Replacement for NE555 Timers
- Timing From Microseconds Through Hours
- Operates in Both Astable and Monostable Modes
- Adjustable Duty Cycle
- High Current Output Can Source or Sink 200 mA
- Output Can Drive MTTL
- Temperature Stability of 0.005% per °C
- Normally "On" or Normally "Off" Output

G SUFFIX
METAL PACKAGE
CASE 601

1. Ground   5. Control Voltage
2. Trigger   6. Threshold
3. Output   7. Discharge
4. Reset   8. V_{CC}

P1 SUFFIX
PLASTIC PACKAGE
CASE 626

U SUFFIX
CERAMIC PACKAGE
CASE 693

D SUFFIX
PLASTIC PACKAGE
CASE 751
(SO-8)

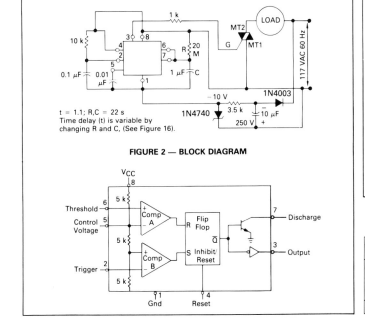

FIGURE 1 — 22-SECOND SOLID-STATE TIME DELAY RELAY CIRCUIT

t = 1.1; R,C = 22 s
Time delay (t) is variable by changing R and C, (See Figure 16).

FIGURE 2 — BLOCK DIAGRAM

ORDERING INFORMATION

Device	Alternate	Temperature Range	Package
MC1455G	—	0°C to +70°C	Metal Can
MC1455P1	NE555V	0°C to +70°C	Plastic DIP
MC1455D	—	0°C to +70°C	SO-8
MC1455U	—	0°C to +70°C	Ceramic DIP
MC1455BP1	—	−40°C to +85°C	Plastic DIP

MC1455

MAXIMUM RATINGS (T_A = +25°C unless otherwise noted.)

Rating	Symbol	Value	Unit
Power Supply Voltage	V_{CC}	+18	Vdc
Discharge Current (Pin 7)	I_7	200	mA
Power Dissipation (Package Limitation)	P_D		
Metal Can		680	mW
Derate above T_A = +25°C		4.6	mW/°C
Plastic Dual In-Line Package		625	mW
Derate above T_A = +25°C		5.0	mW/°C
Operating Temperature Range (Ambient)	T_A		°C
MC1455B		−40 to +85	
MC1455		0 to +70	
Storage Temperature Range	T_{stg}	−65 to +150	°C

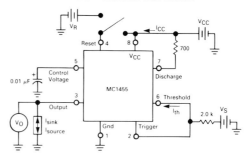

FIGURE 3 — GENERAL TEST CIRCUIT

Test Circuit for Measuring dc Parameters: (to set output and measure parameters)
a) When V_S = 2.3 V_{CC}, V_O is low.
b) When V_S = 1.3 V_{CC}, V_O is high.
c) When V_O is low, pin 7 sinks current. To test for Reset, set V_O high, apply Reset voltage, and test for current flowing into pin 7. When Reset is not in use, it should be tied to V_{CC}.

ELECTRICAL CHARACTERISTICS (T_A = +25°C, V_{CC} = +5.0 V to +15 V unless otherwise noted.)

Characteristics	Symbol	Min	Typ	Max	Unit
Operating Supply Voltage Range	V_{CC}	4.5	—	16	V
Supply Current	I_{CC}				mA
V_{CC} = 5.0 V, R_L = ∞		—	3.0	6.0	
V_{CC} = 15 V, R_L = ∞		—	10	15	
Low State, (Note 1)					
Timing Error (Note 2)					
R = 1.0 kΩ to 100 kΩ					
Initial Accuracy C = 0.1 μF		—	1.0	—	%
Drift with Temperature		—	50	—	PPM/°C
Drift with Supply Voltage		—	0.1	—	%/Volt
Threshold Voltage	V_{th}	—	2/3	—	xV_{CC}
Trigger Voltage	V_T				V
V_{CC} = 15 V		—	5.0	—	
V_{CC} = 5.0 V		—	1.67	—	
Trigger Current	I_T	—	0.5	—	μA
Reset Voltage	V_R	0.4	0.7	1.0	V
Reset Current	I_R	—	0.1	—	mA
Threshold Current (Note 3)	I_{th}	—	0.1	0.25	μA
Discharge Leakage Current (Pin 7)	I_{dis}	—	—	100	nA
Control Voltage Level	V_{CL}				V
V_{CC} = 15 V		9.0	10	11	
V_{CC} = 5.0 V		2.6	3.33	4.0	
Output Voltage Low	V_{OL}				V
(V_{CC} = 15 V)					
I_{sink} = 10 mA		—	0.1	0.25	
I_{sink} = 50 mA		—	0.4	0.75	
I_{sink} = 100 mA		—	2.0	2.5	
I_{sink} = 200 mA		—	2.5	—	
(V_{CC} = 5.0 V)					
I_{sink} = 8.0 mA		—	—	—	
I_{sink} = 5.0 mA		—	0.25	0.35	
Output Voltage High	V_{OH}				V
(I_{source} = 200 mA)					
V_{CC} = 15 V		—	12.5	—	
(I_{source} = 100 mA)					
V_{CC} = 15 V		12.75	13.3	—	
V_{CC} = 5.0 V		2.75	3.3	—	
Rise Time of Output	t_{OLH}	—	100	—	ns
Fall Time of Output	t_{OHL}	—	100	—	ns

NOTES:
1. Supply current when output is high is typically 1.0 mA less.
2. Tested at V_{CC} = 5.0 V and V_{CC} = 15 V. Monostable mode
3. This will determine the maximum value of R_A + R_B for 15 V operation. The maximum total R = 20 megohms.

MOTOROLA SEMICONDUCTOR TECHNICAL DATA

MC1741
MC1741C

OPERATIONAL AMPLIFIER

SILICON MONOLITHIC INTEGRATED CIRCUIT

INTERNALLY COMPENSATED, HIGH PERFORMANCE OPERATIONAL AMPLIFIERS

...designed for use as a summing amplifier, integrator, or amplifier with operating characteristics as a function of the external feedback components.

- No Frequency Compensation Required
- Short-Circuit Protection
- Offset Voltage Null Capability
- Wide Common-Mode and Differential Voltage Ranges
- Low-Power Consumption
- No Latch Up

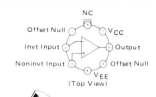

(Top View)

G SUFFIX
METAL PACKAGE
CASE 601

MAXIMUM RATINGS ($T_A = +25°C$ unless otherwise noted)

Rating	Symbol	MC1741C	MC1741	Unit
Power Supply Voltage	V_{CC}	+18	+22	Vdc
	V_{EE}	−18	−22	Vdc
Input Differential Voltage	V_{ID}	±30		Volts
Input Common Mode Voltage (Note 1)	V_{ICM}	±15		Volts
Output Short Circuit Duration (Note 2)	t_S	Continuous		
Operating Ambient Temperature Range	T_A	0 to +70	−55 to +125	°C
Storage Temperature Range Metal and Ceramic Packages Plastic Packages	T_{stg}	 −65 to +150 −55 to +125		°C

NOTES:
1. For supply voltages less than +15 V, the absolute maximum input voltage is equal to the supply voltage.
2. Supply voltage equal to or less than 15 V.

P1 SUFFIX
PLASTIC PACKAGE
CASE 626

U SUFFIX
CERAMIC PACKAGE
CASE 693

D SUFFIX
PLASTIC PACKAGE
CASE 751
(SO-8)

PIN CONNECTIONS

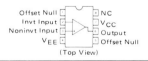

(Top View)

EQUIVALENT CIRCUIT SCHEMATIC

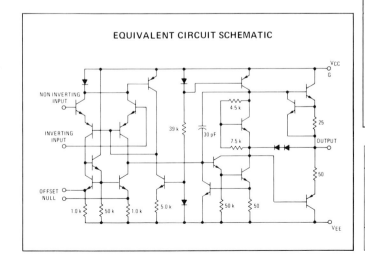

ORDERING INFORMATION

Device	Alternate	Temperature Range	Package
MC1741CD	—	0°C to +70°C	SO-8
MC1741CG	LM741CH, μA741HC		Metal Can
MC1741CP1	LM741CN, μA741TC		Plastic DIP
MC1741CU	—		Ceramic DIP
MC1741G	—	−55°C to +125°C	Metal Can
MC1741U	—		Ceramic DIP

MC1741, MC1741C

ELECTRICAL CHARACTERISTICS (V_{CC} = +15 V, V_{EE} = −15 V, T_A = 25°C unless otherwise noted).

Characteristic	Symbol	MC1741 Min	MC1741 Typ	MC1741 Max	MC1741C Min	MC1741C Typ	MC1741C Max	Unit
Input Offset Voltage ($R_S \leq$ 10 k)	V_{IO}	—	1.0	5.0	—	2.0	6.0	mV
Input Offset Current	I_{IO}	—	20	200	—	20	200	nA
Input Bias Current	I_{IB}	—	80	500	—	80	500	nA
Input Resistance	r_i	0.3	2.0	—	0.3	2.0	—	MΩ
Input Capacitance	C_i	—	1.4	—	—	1.4	—	pF
Offset Voltage Adjustment Range	V_{IOR}	—	±15	—	—	±15	—	mV
Common Mode Input Voltage Range	V_{ICR}	±12	±13	—	±12	±13	—	V
Large Signal Voltage Gain (V_O = ±10 V, $R_L \geq$ 2.0 k)	A_v	50	200	—	20	200	—	V/mV
Output Resistance	r_o	—	75	—	—	75	—	Ω
Common Mode Rejection Ratio ($R_S \leq$ 10 k)	CMRR	70	90	—	70	90	—	dB
Supply Voltage Rejection Ratio ($R_S \leq$ 10 k)	PSRR	—	30	150	—	30	150	μV/V
Output Voltage Swing ($R_L \geq$ 10 k) ($R_L \geq$ 2 k)	V_O	±12 ±10	±14 ±13	— —	±12 ±10	±14 ±13	— —	V
Output Short-Circuit Current	I_{os}	—	20	—	—	20	—	mA
Supply Current	I_D	—	1.7	2.8	—	1.7	2.8	mA
Power Consumption	P_C	—	50	85	—	50	85	mW
Transient Response (Unity Gain − Non-Inverting) (V_I = 20 mV, $R_L \geq$ 2 k, $C_L \leq$ 100 pF) Rise Time	t_{TLH}	—	0.3	—	—	0.3	—	μs
(V_I = 20 mV, $R_L \geq$ 2 k, $C_L \leq$ 100 pF) Overshoot	os	—	15	—	—	15	—	%
(V_I = 10 V, $R_L \geq$ 2 k, $C_L \leq$ 100 pF) Slew Rate	SR	—	0.5	—	—	0.5	—	V/μs

ELECTRICAL CHARACTERISTICS (V_{CC} = +15 V, V_{EE} = −15 V, T_A = T_{low} to T_{high} unless otherwise noted).

Characteristic	Symbol	MC1741 Min	MC1741 Typ	MC1741 Max	MC1741C Min	MC1741C Typ	MC1741C Max	Unit
Input Offset Voltage ($R_S \leq$ 10 kΩ)	V_{IO}	—	1.0	6.0	—	—	7.5	mV
Input Offset Current (T_A = 125°C) (T_A = −55°C) (T_A = 0°C to +70°C)	I_{IO}	— — —	7.0 85 —	200 500 —	— — —	— — —	— — 300	nA
Input Bias Current (T_A = 125°C) (T_A = −55°C) (T_A = 0°C to +70°C)	I_{IB}	— — —	30 300 —	500 1500 —	— — —	— — —	— — 800	nA
Common Mode Input Voltage Range	V_{ICR}	±12	±13	—	—	—	—	V
Common Mode Rejection Ratio ($R_S \leq$ 10 k)	CMRR	70	90	—	—	—	—	dB
Supply Voltage Rejection Ratio ($R_S \leq$ 10 k)	PSRR	—	30	150	—	—	—	μV/V
Output Voltage Swing ($R_L \geq$ 10 k) ($R_L \geq$ 2 k)	V_O	±12 ±10	±14 ±13	— —	— ±10	— ±13	— —	V
Large Signal Voltage Gain ($R_L \geq$ 2 k, V_{out} = ±10 V)	A_v	25	—	—	15	—	—	V/mV
Supply Currents (T_A = 125°C) (T_A = −55°C)	I_D	— —	1.5 2.0	2.5 3.3	— —	— —	— —	mA
Power Consumption (T_A = +125°C) (T_A = −55°C)	P_C	— —	45 60	75 100	— —	— —	— —	mW

*T_{high} = 125°C for MC1741 and 70°C for MC1741C
T_{low} = −55°C for MC1741 and 0°C for MC1741C

MC1741, MC1741C

FIGURE 1 — BURST NOISE versus SOURCE RESISTANCE

FIGURE 2 — RMS NOISE versus SOURCE RESISTANCE

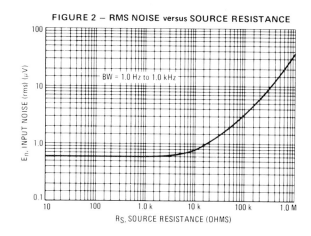

FIGURE 3 — OUTPUT NOISE versus SOURCE RESISTANCE

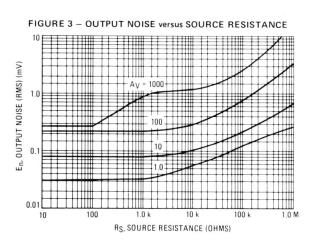

FIGURE 4 — SPECTRAL NOISE DENSITY

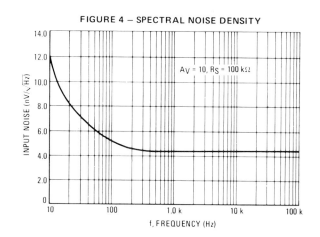

FIGURE 5 — BURST NOISE TEST CIRCUIT

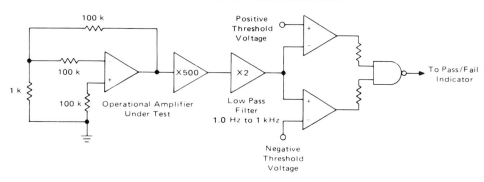

Unlike conventional peak reading or RMS meters, this system was especially designed to provide the quick response time essential to burst (popcorn) noise testing.

The test time employed is 10 seconds and the 20 µV peak limit refers to the operational amplifier input thus eliminating errors in the closed-loop gain factor of the operational amplifier under test.

MC1741, MC1741C

TYPICAL CHARACTERISTICS
(V_{CC} = +15 Vdc, V_{EE} = -15 Vdc, T_A = +25°C unless otherwise noted)

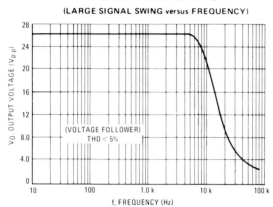

FIGURE 6 – POWER BANDWIDTH
(LARGE SIGNAL SWING versus FREQUENCY)

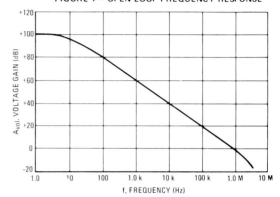

FIGURE 7 – OPEN LOOP FREQUENCY RESPONSE

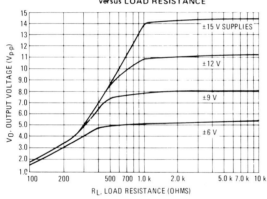

FIGURE 8 – POSITIVE OUTPUT VOLTAGE SWING
versus LOAD RESISTANCE

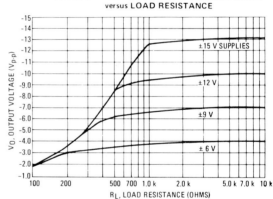

FIGURE 9 – NEGATIVE OUTPUT VOLTAGE SWING
versus LOAD RESISTANCE

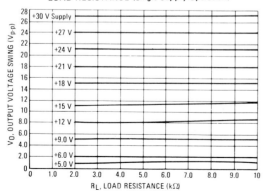

FIGURE 10 – OUTPUT VOLTAGE SWING versus
LOAD RESISTANCE (Single Supply Operation)

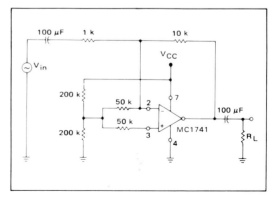

FIGURE 11 – SINGLE SUPPLY INVERTING AMPLIFIER

MOTOROLA SEMICONDUCTOR
TECHNICAL DATA

MC7800 Series

THREE-TERMINAL POSITIVE VOLTAGE REGULATORS

These voltage regulators are monolithic integrated circuits designed as fixed-voltage regulators for a wide variety of applications including local, on-card regulation. These regulators employ internal current limiting, thermal shutdown, and safe-area compensation. With adequate heatsinking they can deliver output currents in excess of 1.0 ampere. Although designed primarily as a fixed voltage regulator, these devices can be used with external components to obtain adjustable voltages and currents.

- Output Current in Excess of 1.0 Ampere
- No External Components Required
- Internal Thermal Overload Protection
- Internal Short-Circuit Current Limiting
- Output Transistor Safe-Area Compensation
- Output Voltage Offered in 2% and 4% Tolerance

THREE-TERMINAL POSITIVE FIXED VOLTAGE REGULATORS

SILICON MONOLITHIC INTEGRATED CIRCUITS

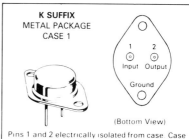

K SUFFIX
METAL PACKAGE
CASE 1

(Bottom View)

Pins 1 and 2 electrically isolated from case. Case is third electrical connection.

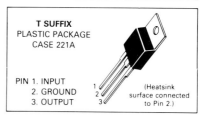

T SUFFIX
PLASTIC PACKAGE
CASE 221A

PIN 1. INPUT
2. GROUND
3. OUTPUT

(Heatsink surface connected to Pin 2.)

REPRESENTATIVE SCHEMATIC DIAGRAM

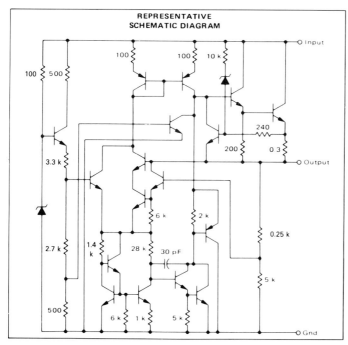

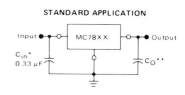

STANDARD APPLICATION

A common ground is required between the input and the output voltages. The input voltage must remain typically 2.0 V above the output voltage even during the low point on the input ripple voltage.

XX = these two digits of the type number indicate voltage.

* = C_{in} is required if regulator is located an appreciable distance from power supply filter.

** = C_O is not needed for stability; however, it does improve transient response.

XX indicates nominal voltage

ORDERING INFORMATION

Device	Output Voltage Tolerance	Tested Operating Junction Temp. Range	Package
MC78XXK	4%	−55 to +150°C	Metal Power
MC78XXAK*	2%		
MC78XXCK	4%	0 to +125°C	
MC78XXACK*	2%		
MC78XXCT	4%		Plastic Power
MC78XXACT	2%		
MC78XXBT	4%	−40 to +125°C	

*2% regulators in Metal Power packages are available in 5, 12 and 15 volt devices.

TYPE NO./VOLTAGE

MC7805	5.0 Volts	MC7812	12 Volts
MC7806	6.0 Volts	MC7815	15 Volts
MC7808	8.0 Volts	MC7818	18 Volts
MC7809	9.0 Volts	MC7824	24 Volts

MC7800 Series

MAXIMUM RATINGS ($T_A = +25°C$ unless otherwise noted.)

Rating	Symbol	Value	Unit
Input Voltage (5.0 V – 18 V) (24 V)	V_{in}	35 40	Vdc
Power Dissipation and Thermal Characteristics Plastic Package $\quad T_A = +25°C$ $\quad$ Derate above $T_A = +25°C$ $\quad$ Thermal Resistance, Junction to Air $\quad T_C = +25°C$ $\quad$ Derate above $T_C = +75°C$ (See Figure 1) $\quad$ Thermal Resistance, Junction to Case Metal Package $\quad T_A = +25°C$ $\quad$ Derate above $T_A = +25°C$ $\quad$ Thermal Resistance, Junction to Air $\quad T_C = +25°C$ $\quad$ Derate above $T_C = +65°C$ (See Figure 2) $\quad$ Thermal Resistance, Junction to Case	P_D $1/\theta_{JA}$ θ_{JA} P_D $1/\theta_{JC}$ θ_{JC} P_D $1/\theta_{JA}$ θ_{JA} P_D $1/\theta_{JC}$ θ_{JC}	Internally Limited 15.4 65 Internally Limited 200 5.0 Internally Limited 22.5 45 Internally Limited 182 5.5	Watts mW/°C °C/W Watts mW/°C °C/W Watts mW/°C °C/W Watts mW/°C °C/W
Storage Junction Temperature Range	T_{stg}	–65 to +150	°C
Operating Junction Temperature Range $\quad$ MC7800,A $\quad$ MC7800C,AC $\quad$ MC7800B	T_J	 –55 to +150 0 to +150 –40 to +150	°C

DEFINITIONS

Line Regulation — The change in output voltage for a change in the input voltage. The measurement is made under conditions of low dissipation or by using pulse techniques such that the average chip temperature is not significantly affected.

Load Regulation — The change in output voltage for a change in load current at constant chip temperature.

Maximum Power Dissipation — The maximum total device dissipation for which the regulator will operate within specifications.

Quiescent Current — That part of the input current that is not delivered to the load.

Output Noise Voltage — The rms ac voltage at the output, with constant load and no input ripple, measured over a specified frequency range.

Long Term Stability — Output voltage stability under accelerated life test conditions with the maximum rated voltage listed in the devices' electrical characteristics and maximum power dissipation.

MC7800 Series

MC7805, B, C
ELECTRICAL CHARACTERISTICS (V_{in} = 10 V, I_O = 500 mA, T_J = T_{low} to T_{high} [Note 1] unless otherwise noted).

Characteristic	Symbol	MC7805 Min	MC7805 Typ	MC7805 Max	MC7805B Min	MC7805B Typ	MC7805B Max	MC7805C Min	MC7805C Typ	MC7805C Max	Unit
Output Voltage (T_J = +25°C)	V_O	4.8	5.0	5.2	4.8	5.0	5.2	4.8	5.0	5.2	Vdc
Output Voltage (5.0 mA ≤ I_O ≤ 1.0 A, P_O ≤ 15 W)	V_O										Vdc
7.0 Vdc ≤ V_{in} ≤ 20 Vdc		—	—	—	—	—	—	4.75	5.0	5.25	
8.0 Vdc ≤ V_{in} ≤ 20 Vdc		4.65	5.0	5.35	4.75	5.0	5.25	—	—	—	
Line Regulation (T_J = +25°C, Note 2)	Reg$_{line}$										mV
7.0 Vdc ≤ V_{in} ≤ 25 Vdc		—	2.0	50	—	7.0	100	—	7.0	100	
8.0 Vdc ≤ V_{in} ≤ 12 Vdc		—	1.0	25	—	2.0	50	—	2.0	50	
Load Regulation (T_J = +25°C, Note 2)	Reg$_{load}$										mV
5.0 mA ≤ I_O ≤ 1.5 A		—	25	100	—	40	100	—	40	100	
250 mA ≤ I_O ≤ 750 mA		—	8.0	25	—	15	50	—	15	50	
Quiescent Current (T_J = +25°C)	I_B	—	3.2	6.0	—	4.3	8.0	—	4.3	8.0	mA
Quiescent Current Change	ΔI_B										mA
7.0 Vdc ≤ V_{in} ≤ 25 Vdc		—	—	—	—	—	—	—	—	1.3	
8.0 Vdc ≤ V_{in} ≤ 25 Vdc		—	0.3	0.8	—	—	1.3	—	—	—	
5.0 mA ≤ I_O ≤ 1.0 A		—	0.04	0.5	—	—	0.5	—	—	0.5	
Ripple Rejection 8.0 Vdc ≤ V_{in} ≤ 18 Vdc, f = 120 Hz	RR	68	75	—	68	—	—	68	—	—	dB
Dropout Voltage (I_O = 1.0 A, T_J = +25°C)	$V_{in} - V_O$	—	2.0	2.5	—	2.0	—	—	2.0	—	Vdc
Output Noise Voltage (T_A = +25°C) 10 Hz ≤ f ≤ 100 kHz	V_n	—	10	40	—	10	—	—	10	—	μV/V_O
Output Resistance f = 1.0 kHz	r_O	—	17	—	—	17	—	—	17	—	mΩ
Short-Circuit Current Limit (T_A = +25°C) V_{in} = 35 Vdc	I_{sc}	—	0.2	1.2	—	0.2	—	—	0.2	—	A
Peak Output Current (T_J = +25°C)	I_{max}	1.3	2.5	3.3	—	2.2	—	—	2.2	—	A
Average Temperature Coefficient of Output Voltage	TCV$_O$	—	±0.6	—	—	-1.1	—	—	-1.1	—	mV/°C

MC7805A, AC
ELECTRICAL CHARACTERISTICS (V_{in} = 10 V, I_O = 1.0 A, T_J = T_{low} to T_{high} [Note 1] unless otherwise noted).

Characteristics	Symbol	MC7805A Min	MC7805A Typ	MC7805A Max	MC7805AC Min	MC7805AC Typ	MC7805AC Max	Unit
Output Voltage (T_J = +25°C)	V_O	4.9	5.0	5.1	4.9	5.0	5.1	Vdc
Output Voltage (5.0 mA ≤ I_O ≤ 1.0 A, P_O ≤ 15 W) 7.5 Vdc ≤ V_{in} ≤ 20 Vdc	V_O	4.8	5.0	5.2	4.8	5.0	5.2	Vdc
Line Regulation (Note 2)	Reg$_{line}$							mV
7.5 Vdc ≤ V_{in} ≤ 25 Vdc, I_O = 500 mA		—	2.0	10	—	7.0	50	
8.0 Vdc ≤ V_{in} ≤ 12 Vdc		—	3.0	10	—	10	50	
8.0 Vdc ≤ V_{in} ≤ 12 Vdc, T_J = +25°C		—	1.0	4.0	—	2.0	25	
7.3 Vdc ≤ V_{in} ≤ 20 Vdc, T_J = +25°C		—	2.0	10	—	7.0	50	
Load Regulation (Note 2)	Reg$_{load}$							mV
5.0 mA ≤ I_O ≤ 1.5 A, T_J = +25°C		—	2.0	25	—	25	100	
5.0 mA ≤ I_O ≤ 1.0 A		—	2.0	25	—	25	100	
250 mA ≤ I_O ≤ 750 mA, T_J = +25°C		—	1.0	15	—	—	—	
250 mA ≤ I_O ≤ 750 mA		—	1.0	25	—	8.0	50	
Quiescent Current	I_B							mA
T_J = +25°C		—	—	5.0	—	—	6.0	
		—	3.2	4.0	—	4.3	6.0	
Quiescent Current Change	ΔI_B							mA
8.0 Vdc ≤ V_{in} ≤ 25 Vdc, I_O = 500 mA		—	0.3	0.5	—	—	0.8	
7.5 Vdc ≤ V_{in} ≤ 20 Vdc, T_J = +25°C		—	0.2	0.5	—	—	0.8	
5.0 mA ≤ I_O ≤ 1.0 A		—	0.04	0.2	—	—	0.5	
Ripple Rejection	RR							dB
8.0 Vdc ≤ V_{in} ≤ 18 Vdc, f = 120 Hz, T_J = +25°C		68	75	—	—	—	—	
8.0 Vdc ≤ V_{in} ≤ 18 Vdc, f = 120 Hz, I_O = 500 mA		68	75	—	—	68	—	
Dropout Voltage (I_O = 1.0 A, T_J = +25°C)	$V_{in} - V_O$	—	2.0	2.5	—	2.0	—	Vdc
Output Noise Voltage (T_A = +25°C) 10 Hz ≤ f ≤ 100 kHz	V_n	—	10	40	—	10	—	μV/V_O
Output Resistance (f = 1.0 kHz)	r_O	—	2.0	—	—	17	—	mΩ
Short-Circuit Current Limit (T_A = +25°C) V_{in} = 35 Vdc	I_{sc}	—	0.2	1.2	—	0.2	—	A
Peak Output Current (T_J = +25°C)	I_{max}	1.3	2.5	3.3	—	2.2	—	A
Average Temperature Coefficient of Output Voltage	TCV$_O$	—	±0.6	—	—	-1.1	—	mV/°C

NOTES: 1. T_{low} = -55°C for MC78XX, A; T_{high} = +150°C for MC78XX, A
   = 0° for MC78XXC, AC; = +125°C for MC78XXC, AC, B
   = -40°C for MC78XXB
2. Load and line regulation are specified at constant junction temperature. Changes in V_O due to heating effects must be taken into account separately. Pulse testing with low duty cycle is used.

MOTOROLA SEMICONDUCTOR TECHNICAL DATA

MC7900 Series

THREE-TERMINAL NEGATIVE VOLTAGE REGULATORS

The MC7900 Series of fixed output negative voltage regulators are intended as complements to the popular MC7800 Series devices. These negative regulators are available in the same seven-voltage options as the MC7800 devices. In addition, one extra voltage option commonly employed in MECL systems is also available in the negative MC7900 Series.

Available in fixed output voltage options from -5.0 to -24 volts, these regulators employ current limiting, thermal shutdown, and safe-area compensation — making them remarkably rugged under most operating conditions. With adequate heat-sinking they can deliver output currents in excess of 1.0 ampere.

- No External Components Required
- Internal Thermal Overload Protection
- Internal Short-Circuit Current Limiting
- Output Transistor Safe-Area Compensation
- Available in 2% Voltage Tolerance (See Ordering Information)

THREE-TERMINAL NEGATIVE FIXED VOLTAGE REGULATORS

K SUFFIX
METAL PACKAGE
CASE 1

1 — Gnd
2 — Output
Case Input
(Bottom View)

T SUFFIX
PLASTIC PACKAGE
CASE 221A

PIN 1. GROUND
2. INPUT
3. OUTPUT

(Heatsink surface connected to Pin 2)

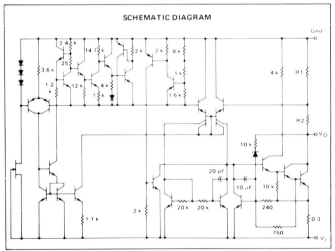

SCHEMATIC DIAGRAM

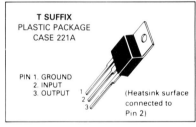

STANDARD APPLICATION

A common ground is required between the input and the output voltages. The input voltage must remain typically 2.0 V more negative even during the high point on the input ripple voltage.

XX = these two digits of the type number indicate voltage.

* = C_{in} is required if regulator is located an appreciable distance from power supply filter.

** = C_O improves stability and transient response.

ORDERING INFORMATION

Device	Output Voltage Tolerance	Tested Operating Junction Temp. Range	Package
MC79XXCK	4%	$T_J = 0°C$ to $+125°C$	Metal Power**
MC79XXACK*	2%		
MC79XXCT	4%		Plastic Power
MC79XXACT*	2%		
MC79XXBT#	4%	$T_J = -40°C$ to $+125°C$	

XX indicates nominal voltage.

*2% output voltage tolerance available in 5, 12 and 15 volt devices.
**Metal power package available in 5, 12 and 15 volt devices.
#Automotive temperature range selections are available with special test conditions and additional tests in 5, 12 and 15 volt devices. Contact your local Motorola sales office for information.

DEVICE TYPE / NOMINAL OUTPUT VOLTAGE

MC7905	5.0 Volts	MC7912	12 Volts
MC7905.2	5.2 Volts	MC7915	15 Volts
MC7906	6.0 Volts	MC7918	18 Volts
MC7908	8.0 Volts	MC7924	24 Volts

MC7900 Series

MAXIMUM RATINGS (T_A = +25°C unless otherwise noted.)

Rating	Symbol	Value	Unit
Input Voltage (–5.0 V ≥ V_O ≥ –18 V) (24 V)	V_I	–35 –40	Vdc

Power Dissipation
 Plastic Package
   T_A = +25°C
   Derate above T_A = +25°C
   T_C = +25°C
   Derate above T_C = +95°C (See Figure 1)
 Metal Package
   T_A = +25°C
   Derate above T_A = +25°C
   T_C = +25°C
   Derate above T_C = +65°C
	
	 P_D
	 $1/R_{\theta JA}$
	 P_D
	 $1/R_{\theta JC}$
	
	 P_D
	 $1/R_{\theta JA}$
	 P_D
	 $1/R_{\theta JC}$
		
		 Internally Limited
		 15.4
		 Internally Limited
		 200
		
		 Internally Limited
		 22.2
		 Internally Limited
		 182
			
			 Watts
			 mW/°C
			 Watts
			 mW/°C
			
			 Watts
			 mW/°C
			 Watts
			 mW/°C

Storage Junction Temperature Range	T_{stg}	–65 to +150	°C
Junction Temperature Range	T_J	0 to +150	°C

THERMAL CHARACTERISTICS

Characteristic	Symbol	Max	Unit
Thermal Resistance, Junction to Ambient — Plastic Package — Metal Package	$R_{\theta JA}$	65 45	°C/W
Thermal Resistance, Junction to Case — Plastic Package — Metal Package	$R_{\theta JC}$	5.0 5.5	°C/W

MC7905C ELECTRICAL CHARACTERISTICS (V_I = –10 V, I_O = 500 mA, 0°C < T_J < +125°C unless otherwise noted.)

Characteristic	Symbol	Min	Typ	Max	Unit
Output Voltage (T_J = +25°C)	V_O	–4.8	–5.0	–5.2	Vdc

Line Regulation (Note 1)
 (T_J = +25°C, I_O = 100 mA)
   –7.0 Vdc ≥ V_I ≥ –25 Vdc
   –8.0 Vdc ≥ V_I ≥ –12 Vdc
 (T_J = +25°C, I_O = 500 mA)
   –7.0 Vdc ≥ V_I ≥ –25 Vdc
   –8.0 Vdc ≥ V_I ≥ –12 Vdc
	Reg$_{line}$	
		
		 —
		 —
		
		 —
		 —
			
			 70
			 20
			
			 35
			 8.0
				
				 50
				 25
				
				 100
				 50
					mV

Load Regulation (T_J = +25°C) (Note 1)
   5.0 mA ≤ I_O ≤ 1.5 A
   250 mA ≤ I_O ≤ 750 mA
	Reg$_{load}$	
		 —
		 —
			
			 11
			 4.0
				
				 100
				 50
					mV

Output Voltage –7.0 Vdc ≥ V_I ≥ –20 Vdc, 5.0 mA ≤ I_O ≤ 1.0 A, P ≤ 15 W	V_O	–4.75	—	–5.25	Vdc
Input Bias Current (T_J = +25°C)	I_{IB}	—	4.3	8.0	mA

Input Bias Current Change
   –7.0 Vdc ≥ V_I ≥ –25 Vdc
   5.0 mA ≤ I_O ≤ 1.5 A
	ΔI_{IB}	
		 —
		 —
			
			 —
			 —
				
				 1.3
				 0.5
					mA

Output Noise Voltage (T_A = +25°C, 10 Hz ≤ f ≤ 100 kHz)	e_{on}	—	40	—	μV
Ripple Rejection (I_O = 20 mA, f = 120 Hz)	RR	—	70	—	dB
Dropout Voltage I_O = 1.0 A, T_J = +25°C	$V_I - V_O$	—	2.0	—	Vdc
Average Temperature Coefficient of Output Voltage I_O = 5.0 mA, 0°C ≤ T_J ≤ +125°C	$\Delta V_O / \Delta T$	—	–1.0	—	mV/°C

Note
1. Load and line regulation are specified at constant junction temperature. Changes in V_O due to heating effects must be taken into account separately. Pulse testing with low duty cycle is used.

MC7900 Series

MC7912C ELECTRICAL CHARACTERISTICS ($V_I = -19$ V, $I_O = 500$ mA, $0°C < T_J < +125°C$ unless otherwise noted.)

Characteristic	Symbol	Min	Typ	Max	Unit
Output Voltage ($T_J = +25°C$)	V_O	−11.5	−12	−12.5	Vdc
Line Regulation (Note 1)	Reg$_{line}$				mV
($T_J = +25°C$, $I_O = 100$ mA)					
−14.5 Vdc ≥ V_I ≥ −30 Vdc		—	13	120	
−16 Vdc ≥ V_I ≥ −22 Vdc		—	6.0	60	
($T_J = +25°C$, $I_O = 500$ mA)					
−14.5 Vdc ≥ V_I ≥ −30 Vdc		—	55	240	
−16 Vdc ≥ V_I ≥ −22 Vdc		—	24	120	
Load Regulation ($T_J = +25°C$) (Note 1)	Reg$_{load}$				mV
5.0 mA ≤ I_O ≤ 1.5 A		—	46	240	
250 mA ≤ I_O ≤ 750 mA		—	17	120	
Output Voltage	V_O	−11.4	—	−12.6	Vdc
−14.5 Vdc ≥ V_I ≥ −27 Vdc, 5.0 mA ≤ I_O ≤ 1.0 A, P ≤ 15 W					
Input Bias Current ($T_J = +25°C$)	I_{IB}	—	4.4	8.0	mA
Input Bias Current Change	ΔI_{IB}				mA
−14.5 Vdc ≥ V_I ≥ −30 Vdc		—	—	1.0	
5.0 mA ≤ I_O ≤ 1.5 A		—	—	0.5	
Output Noise Voltage ($T_A = +25°C$, 10 Hz ≤ f ≤ 100 kHz)	e_{on}	—	75	—	μV
Ripple Rejection ($I_O = 20$ mA, f = 120 Hz)	RR	—	61	—	dB
Dropout Voltage	$V_I - V_O$		2.0		Vdc
$I_O = 1.0$ A, $T_J = +25°C$					
Average Temperature Coefficient of Output Voltage	$\Delta V_O/\Delta T$	—	−1.0	—	mV/°C
$I_O = 5.0$ mA, $0°C ≤ T_J ≤ +125°C$					

MC7912AC ELECTRICAL CHARACTERISTICS ($V_I = -19$ V, $I_O = 500$ mA, $0°C < T_J < +125°C$ unless otherwise noted.)

Characteristic	Symbol	Min	Typ	Max	Unit
Output Voltage ($T_J = +25°C$)	V_O	−11.75	−12	−12.25	Vdc
Line Regulation (Note 1)	Reg$_{line}$				mV
−16 Vdc ≥ V_I ≥ −22 Vdc; $I_O = 1.0$ A, $T_J = 25°C$		—	6.0	60	
−16 Vdc ≥ V_I ≥ −22 Vdc; $I_O = 1.0$ A,		—	24	120	
−14.8 Vdc ≥ V_I ≥ −30 Vdc, $I_O = 500$ mA		—	24	120	
−14.5 Vdc ≥ V_I ≥ −27 Vdc, $I_O = 1.0$ A, $T_J = 25°C$		—	13	120	
Load Regulation (Note 1)	Reg$_{load}$				mV
5.0 mA ≤ I_O ≤ 1.5 A, $T_J = 25°C$		—	46	150	
250 mA ≤ I_O ≤ 750 mA		—	17	75	
5.0 mA ≤ I_O ≤ 1.0 A		—	35	150	
Output Voltage	V_O	−11.5	—	−12.5	Vdc
−14.8 Vdc ≥ V_I ≥ −27 Vdc, 5.0 mA ≤ I_O ≤ 1.0 A, P ≤ 15 W					
Input Bias Current	I_{IB}	—	4.4	8.0	mA
Input Bias Current Change	ΔI_{IB}				mA
−15 Vdc ≥ V_I ≥ −30 Vdc		—	—	0.8	
5.0 mA ≤ I_O ≤ 1.0 A		—	—	0.5	
5.0 mA ≤ I_O ≤ 1.5 A, $T_J = 25°C$		—	—	0.5	
Output Noise Voltage ($T_A = +25°C$, 10 Hz ≤ f ≤ 100 kHz)	e_{on}	—	75	—	μV
Ripple Rejection ($I_O = 20$ mA, f = 120 Hz)	RR	—	61	—	dB
Dropout Voltage	$V_I - V_O$		2.0		Vdc
$I_O = 1.0$ A, $T_J = +25°C$					
Average Temperature Coefficient of Output Voltage	$\Delta V_O/\Delta T$	—	−1.0	—	mV/°C
$I_O = 5.0$ mA, $0°C ≤ T_J ≤ +125°C$					

Note

1. Load and line regulation are specified at constant junction temperature. Changes in V_O due to heating effects must be taken into account separately. Pulse testing with low duty cycle is used.

Precision, Hybrid Isolation Amplifier
AD295

FEATURES
Low Nonlinearity: ±0.012% max (AD295C)
Low Gain Drift: ±60ppm/°C max
Floating Input and Output Power: ±15V dc @ 5mA
3-Port Isolation: ±2500V CMV (Input to Output)
Complies with NEMA ICS1-111
Gain Adjustable: 1V/V to 1000V/V
User Configurable Input Amplifier

APPLICATIONS
Motor Controls
Process Signal Isolator
High Voltage Instrumentation Amplifier
Multichannel Data Acquisition Systems
Off Ground Signal Measurements

AD295 FUNCTIONAL BLOCK DIAGRAM

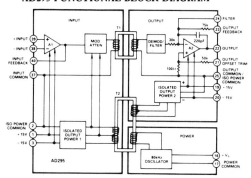

GENERAL DESCRIPTION
The AD295 is a high accuracy, high reliability hybrid isolation amplifier designed for industrial, instrumentation and medical applications. Three performance versions are available offering guaranteed nonlinearity error at 10V p-p output: ±0.05% max (AD295A), ±0.025% max (AD295B), ±0.012% max (AD295C). Using a pulse width modulation technique the AD295 provides 3-port isolation between input, output and power supply ports. Using this technique, the AD295 interrupts ground loops and leakage paths and minimizes the effect of high voltage transients. Additionally, floating (isolated) power ±15V dc @ 5mA is available at both the input and output. The AD295's gain can be programmed at the input, output or both sections allowing for user flexibility. An uncommitted input amplifier allows configuration as a buffer, inverter, subtractor or differential amplifier.

The AD295 is provided in an epoxy sealed ceramic 40-pin package that insures quality performance, high stability and accuracy. Input/output pin spacing complies with NEMA (ICS1-111) separation specifications required for many industrial applications.

WHERE TO USE THE MODEL AD295
Industrial: The AD295 is designed for measuring signals in harsh industrial environments. The AD295 provides high accuracy with complete galvanic isolation and protection from transients or where ground fault currents or high common-mode voltages are present. The AD295 can be applied in process controllers, current loop receivers, motor controls and weighing systems.

Instrumentation: In data acquisition systems the AD295 provides common-mode rejection for conditioning thermocouples, strain gauges or other low-level signals where high performance and system protection is required.

Medical: In biomedical and patient monitoring equipment like diagnostic systems and blood pressure monitors, the AD295 provides protection from lethal ground fault currents. Low level signal recording and monitoring is achieved with the AD295's low input noise (2µV p-p @ G=1000V/V) and high CMR (106dB @ 60Hz).

DESIGN FEATURES AND USER BENEFITS
Isolated Power: Isolated power supply sections at the input and output provide ±15V dc @ 5mA. Isolated power is load regulated to 4%. This feature permits the AD295 to excite floating signal conditioners, front-end buffer amplifiers and remote transducers at the input and external circuitry at the output. This eliminates the need for a separate dc/dc converter.

Input Amplifier: The uncommitted input amplifier allows the user to configure the input as a buffer, inverter, subtractor or differential amplifier to meet the application need.

Adjustable Gain: Gain can be selected at the input, output or both. Thus, circuit response can be tailored to the user's application. The AD295 provides the user with flexibility for circuit optimization without requiring external active components.

Three-Port Isolation: Provides true galvanic isolation between input, output and power supply ports. Eliminates the need for power supply and output ports being returned through a common ground.

Wide Operating Temperature: The AD295 is designed to operate over the −40°C to +100°C temperature range with rated performance over −25°C to +85°C.

Leakage: The low coupling capacitance between input and output yields a ground leakage current of less than 2µA rms at 115V ac, 60Hz. The AD295 meets standards established by UL STD 544.

781

SPECIFICATIONS (typical @ +25°C, & $V_S = +15V$ unless otherwise noted)

Model	AD295A	AD295B	AD295C
GAIN			
Range	1V/V to 1000V/V	*	*
Open Loop	100dB		
Accuracy G = 1V/V	± 1.5%	*	*
vs. Temperature (−25°C to +85°C)			
G = 1V/V to 100V/V	± 60ppm/°C max	*	*
Nonlinearity (± 5V Swing) G = 1V–100V/V	± 0.05% max	± 0.025% max	± 0.012% max
INPUT VOLTAGE RATINGS			
Linear Differential Range	± 10V min	*	*
Max Safe Differential Input	± 15V	*	*
Max CMV (Input to Output)			
Continuous ac or dc	± 2500V peak	*	*
ac, 60Hz, 1 Minute Duration	2500V rms	*	*
Max CMV (Input to Power Common/Output to Power Common)			
Continuous ac or dc	± 2000V peak	*	*
ac, 60Hz, 1 Minute Duration	2000V rms	*	*
CMR, Input to Output 60Hz, G = 1V/V			
$R_S \leq 1k\Omega$ Balanced Source Impedance	106dB	*	*
$R_S \leq 1k$ Source Impedance Imbalance	103dB min	*	*
Max Leakage Current, Input to Output			
@ 115V ac, 60Hz	2μA rms max	*	*
INPUT IMPEDANCE			
Differential	$5 \times 10^{10}\Omega$/33pF	*	*
Common Mode	$10^9\Omega$/20pF	*	*
INPUT BIAS CURRENT			
Initial, @ +25°C	5nA max	*	*
vs. Temperature (−25°C to +85°C)	−25pA/°C max	*	*
INPUT DIFFERENCE CURRENT			
Initial, @ +25°C	± 2nA max	*	*
vs. Temperature (−25°C to +85°C)	± 5pA/°C max	*	*
INPUT NOISE (Gain = 1000V/V)			
Voltage			
0.01Hz to 10Hz	2μV p-p	*	*
10Hz to 1kHz	1μV rms	*	*
Current			
0.01Hz to 10Hz	10pA p-p	*	*
FREQUENCY RESPONSE			
Small Signal (−3dB)			
G = 1V/V to 100V/V	4.5kHz	*	*
G = 1000V/V	600Hz	*	*
Full Power, 20V p-p Output			
G = 1V/V to 100V/V	1.4kHz	*	*
G = 1000V/V	200Hz	*	*
Slew Rate G = 1V/V to 100V/V	0.1V/μs	*	*
Settling Time G = 1V/V			
(to ± 0.1% for 10V Step)	550μs	*	*
(to ± 0.1% for 20V Step)	700μs	*	*
OFFSET VOLTAGE, REFERRED TO INPUT			
Initial @ +25°C (Adjustable to Zero)	$\pm \left(3 + \frac{15}{G_{IN}}\right)$ mV max	*	*
vs. Temperature (−25°C to +85°C)	$\pm \left(10 + \frac{450}{G_{IN}}\right) \mu V/°C$ max	$\pm \left(3 + \frac{300}{G_{IN}}\right) \mu V/°C$ max	$\pm \left(1.5 + \frac{150}{G_{IN}}\right) \mu V/°C$ max
vs. Supply	$\pm \left(1 + \frac{200}{G_{IN}}\right) \mu V/\%$	*	*
RATED OUTPUT			
Voltage, 2kΩ Load	± 10V min	*	*
Output Impedance	2Ω (dc to 100Hz)	*	*
Output Ripple (10Hz to 10kHz)	6mV p-p	*	*
(10Hz to 100kHz)	40mV p-p	*	*
ISOLATED POWER SUPPLIES[1] (V_{ISO1} & V_{ISO2})			
Voltage	± 15V dc	*	*
Accuracy	± 5%	*	*
Current[2]	± 5mA max	*	*
Load Regulation (No Load to Full Load)	−4%	*	*
Ripple, 100kHz BW	12mV p-p	*	*
POWER SUPPLY (+ V_S)			
Voltage, Rated Performance	+15V dc ± 3%	*	*
Voltage, Operating	+12V dc to +16V dc	*	*
Current, Quiescent ($V_S = +15V$)	40mA	*	*
With V_{ISO} Loaded	45mA	*	*
TEMPERATURE RANGE			
Rated Performance	−25°C to +85°C	*	*
Operating	−40°C to +100°C	*	*
Storage	−40°C to +100°C	*	*
CASE DIMENSIONS	2.7″ × 0.88″ × 0.375″	*	*

NOTES
[1] V_{ISO} accuracy and regulation 10%.
[2] ± 10mA can be supplied by V_{ISO1}, if V_{ISO2} is not used.

*Specifications same as AD295A
Specifications subject to change without notice.

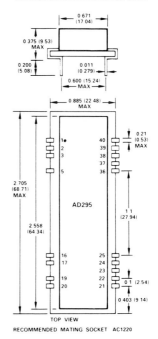

OUTLINE DIMENSIONS
Dimensions shown in inches and (mm).

RECOMMENDED MATING SOCKET AC1220

PIN DESIGNATIONS

PIN	FUNCTION	PIN	FUNCTION
1	+15V (+V_{ISO1})	40	INPUT FEEDBACK
2	V_{ISO1} COM	39	+ INPUT
3	−15V (−V_{ISO1})	38	− INPUT
		37	INPUT COM
5	NO CONNECTION	36	NO CONNECTION
16	+V_S	25	OUTPUT COM, V_{ISO2} COM
17	POWER COMMON	24	FILTER
		23	OUTPUT FEEDBACK
19	+15V (+V_{ISO2})	22	OUTPUT
20	−15V (−V_{ISO2})	21	OUTPUT OFFSET TRIM

Integrated Circuit
Precision Instrumentation Amplifier
AD521

FEATURES
Programmable Gains from 0.1 to 1000
Differential Inputs
High CMRR: 110dB min
Low Drift: 2µV/°C max (L)
Complete Input Protection, Power ON and Power OFF
Functionally Complete with the Addition of Two Resistors
Internally Compensated
Gain Bandwidth Product: 40MHz
Output Current Limited: 25mA
Very Low Noise: 0.5µV p-p, 0.1Hz to 10Hz, RTI @ G = 1000
Chips are Available

AD521 PIN CONFIGURATION

Pin	Signal	Pin	Signal
1	+ INPUT	14	R GAIN
2	R GAIN	13	R SCALE
3	− INPUT	12	SENSE
4	OFFSET TRIM	11	REF
5	V−	10	R SCALE
6	OFFSET TRIM	9	COMP.
7	OUTPUT	8	V+

PRODUCT DESCRIPTION
The AD521 is a second generation, low cost, monolithic IC instrumentation amplifier developed by Analog Devices. As a true instrumentation amplifier, the AD521 is a gain block with differential inputs and an accurately programmable input/output gain relationship.

The AD521 IC instrumentation amplifier should not be confused with an operational amplifier, although several manufacturers (including Analog Devices) offer op amps which can be used as building blocks in variable gain instrumentation amplifier circuits. Op amps are general-purpose components which, when used with precision-matched external resistors, can perform the instrumentation amplifier function.

An instrumentation amplifier is a precision differential voltage gain device optimized for operation in a real world environment, and is intended to be used wherever acquisition of a useful signal is difficult. It is characterized by high input impedance, balanced differential inputs, low bias currents and high CMR.

As a complete instrumentation amplifier, the AD521 requires only two resistors to set its gain to any value between 0.1 and 1000. The ratio matching of these resistors does not affect the high CMRR (up to 120dB) or the high input impedance (3 × $10^9 \Omega$) of the AD521. Furthermore, unlike most operational amplifier-based instrumentation amplifiers, the inputs are protected against overvoltages up to ±15 volts beyond the supplies.

The AD521 IC instrumentation amplifier is available in four different versions of accuracy and operating temperature range. The economical "J" grade, the low drift "K" grade, and the lower drift, higher linearity "L" grade are specified from 0 to +70°C. The "S" grade guarantees performance to specification over the extended temperature range: −55°C to +125°C.

PRODUCT HIGHLIGHTS
1. The AD521 is a true instrumentation amplifier in integrated circuit form, offering the user performance comparable to many modular instrumentation amplifiers at a fraction of the cost.
2. The AD521 has low guaranteed input offset voltage drift (2µV/°C for L grade) and low noise for precision, high gain applications.
3. The AD521 is functionally complete with the addition of two resistors. Gain can be preset from 0.1 to more than 1000.
4. The AD521 is fully protected for input levels up to 15V beyond the supply voltages and 30V differential at the inputs.
5. Internally compensated for all gains, the AD521 also offers the user the provision for limiting bandwidth.
6. Offset nulling can be achieved with an optional trim pot.
7. The AD521 offers superior dynamic performance with a gain-bandwidth product of 40MHz, full peak response of 100kHz (independent of gain) and a settling time of 5µs to 0.1% of a 10V step.

783

SPECIFICATIONS (typical @ $V_S = \pm15V$, $R_L = 2k\Omega$ and $T_A = 25°C$ unless otherwise specified)

MODEL	AD521JD	AD521KD	AD521LD	AD521SD (AD521SD/883B)
GAIN				
Range (For Specified Operation, Note 1)	1 to 1000	*	*	*
Equation	$G = R_S/R_G$ V/V	*	*	*
Error from Equation	$(\pm0.25 - 0.004G)\%$	*	*	*
Nonlinearity (Note 2)				
$1 \leq G \leq 1000$	0.2% max	*	0.1% max	*
Gain Temperature Coefficient	$\pm(3 \pm 0.05G)$ ppm/°C	*	*	$\pm(15 \pm 0.4G)$ ppm/°C
OUTPUT CHARACTERISTICS				
Rated Output	$\pm10V$, $\pm10mA$ min	*	*	*
Output at Maximum Operating Temperature	$\pm10V$ @ 5mA min	*	*	*
Impedance	0.1Ω	*	*	*
DYNAMIC RESPONSE				
Small Signal Bandwidth ($\pm3dB$)				
G = 1	>2MHz	*	*	*
G = 10	300kHz	*	*	*
G = 100	200kHz	*	*	*
G = 1000	40kHz	*	*	*
Small Signal, $\pm1.0\%$ Flatness				
G = 1	75kHz	*	*	*
G = 10	26kHz	*	*	*
G = 100	24kHz	*	*	*
G = 1000	6kHz	*	*	*
Full Peak Response (Note 3)	100kHz	*	*	*
Slew Rate, $1 \leq G \leq 1000$	10V/μs	*	*	*
Settling Time (any 10V step to within 10mV of Final Value)				
G = 1	7μs	*	*	*
G = 10	5μs	*	*	*
G = 100	10μs	*	*	*
G = 1000	35μs	*	*	*
Differential Overload Recovery ($\pm30V$ Input to within 10mV of Final Value) (Note 4)				
G = 1000	50μs	*	*	*
Common Mode Step Recovery (30V Input to within 10mV of Final Value) (Note 5)				
G = 1000	10μs	*	*	*
VOLTAGE OFFSET (may be nulled)				
Input Offset Voltage (V_{OS_I})	3mV max (2mV typ)	1.5mV max (0.5mV typ)	1.0mV max (0.5mV typ)	**
vs. Temperature	15μV/°C max (7μV/°C typ)	5μV/°C max (1.5μV/°C typ)	2μV/°C max	**
vs. Supply	3μV/%	*	*	**
Output Offset Voltage (V_{OS_O})	400mV max (200mV typ)	200mV max (30mV typ)	100mV max	**
vs. Temperature	400μV/°C max (150μV/°C typ)	150μV/°C max (50μV/°C typ)	75μV/°C max	**
vs. Supply (Note 6)	$0.005 V_{OS_O}/\%$	*	*	**
INPUT CURRENTS				
Input Bias Current (either input)	80nA max	40nA max	**	**
vs. Temperature	1nA/°C max	500pA/°C max	**	**
vs. Supply	2%/V	*	*	*
Input Offset Current	20nA max	10nA max	**	**
vs. Temperature	250pA/°C max	125pA/°C max	**	**
INPUT				
Differential Input Impedance (Note 7)	$3 \times 10^9 \Omega \| 1.8pF$	*	*	*
Common Mode Input Impedance (Note 8)	$6 \times 10^{10} \Omega \| 3.0pF$	*	*	*
Input Voltage Range for Specified Performance (with respect to ground)	$\pm10V$	*	*	*
Maximum Voltage without Damage to Unit, Power ON or OFF Differential Mode (Note 9)	30V	*	*	*
Voltage at either input (Note 9)	$V_S \pm 15V$	*	*	*
Common Mode Rejection Ratio, DC to 60Hz with $1k\Omega$ source unbalance				
G = 1	70dB min (74dB typ)	74dB min (80dB typ)	**	**
G = 10	90dB min (94dB typ)	94dB min (100dB typ)	**	**
G = 100	100dB min (104dB typ)	104dB min (114dB typ)	**	**
G = 1000	100dB min (110dB typ)	110dB min (120dB typ)	**	**
NOISE				
Voltage RTO (p-p) @ 0.1Hz to 10Hz (Note 10)	$\sqrt{(0.5G)^2 + (225)^2}$ μV	*	*	*
RMS RTO, 10Hz to 10kHz	$\sqrt{(1.2G)^2 + (50)^2}$ μV	*	*	*
Input Current, rms, 10Hz to 10kHz	15pA (rms)	*	*	*
REFERENCE TERMINAL				
Bias Current	3μA	*	*	*
Input Resistance	$10M\Omega$	*	*	*
Voltage Range	$\pm10V$	*	*	*
Gain to Output	1	*	*	*
POWER SUPPLY				
Operating Voltage Range	$\pm5V$ to $\pm18V$	*	*	*
Quiescent Supply Current	5mA max	*	*	*
TEMPERATURE RANGE				
Specified Performance	0 to +70°C	*	*	-55°C to +125°C
Operating	-25°C to +85°C	*	*	
Storage	-65°C to +150°C	*	*	-55°C to +125°C
PACKAGE OPTION[1]				
Ceramic (D-14)	AD521JD	AD521KD	AD521LD	AD521SD

NOTES
[1] See Section 20 for package outline information.
*Specifications same as AD521JD.
**Specifications same as AD521KD.
Specifications subject to change without notice.

Voltage-to-Frequency and Frequency-to-Voltage Converter
AD650

FEATURES
V/F Conversion to 1MHz
Reliable Monolithic Construction
Very Low Nonlinearity
  0.002% typ at 10kHz
  0.005% typ at 100kHz
  0.07% typ at 1MHz
Input Offset Trimmable to Zero
CMOS or TTL Compatible
Unipolar, Bipolar, or Differential V/F
V/F or F/V Conversion
Available in Surface Mount
MIL-STD-883-Compliant Versions Available

PIN CONFIGURATION

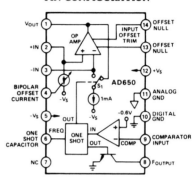

PRODUCT DESCRIPTION
The AD650 V/F/V (voltage-to-frequency or frequency-to-voltage converter) provides a combination of high frequency operation and low nonlinearity previously unavailable in monolithic form. The inherent monotonicity of the V/F transfer function makes the AD650 useful as a high-resolution analog-to-digital converter. A flexible input configuration allows a wide variety of input voltage and current formats to be used, and an open-collector output with separate digital ground allows simple interfacing to either standard logic families or opto-couplers.

The linearity error of the AD650 is typically 20ppm (0.002% of full scale) and 50ppm (0.005%) maximum at 10kHz full scale. This corresponds to approximately 14-bit linearity in an analog-to-digital converter circuit. Higher full-scale frequencies or longer count intervals can be used for higher resolution conversions. The AD650 has a useful dynamic range of six decades allowing extremely high resolution measurements. Even at 1MHz full scale, linearity is guaranteed less than 1000ppm (0.1%) on the AD650KN, KP, BD and SD grades.

In addition to analog-to-digital conversion, the AD650 can be used in isolated analog signal transmission applications, phased-locked-loop circuits, and precision stepper motor speed controllers. In the F/V mode, the AD650 can be used in precision tachometer and FM demodulator circuits.

The input signal range and full-scale output frequency are user-programmable with two external capacitors and one resistor. Input offset voltage can be trimmed to zero with an external potentiometer.

The AD650JN and AD650KN are offered in a plastic 14-pin DIP package. The AD650JP and AD650KP are available in a 20-pin plastic leaded chip carrier (PLCC). Both plastic packaged versions of the AD650 are specified for the commerical (0 to +70°C) temperature range. For industrial temperature range (−25°C to +85°C) applications, the AD650AD and AD650BD are offered in a ceramic package. The AD650SD is specified for the full −55°C to +125°C extended temperature range.

PRODUCT HIGHLIGHTS
1. In addition to very high linearity, the AD650 can operate at full scale output frequency up to 1MHz. The combination of these two features makes the AD650 an inexpensive solution for applications requiring high resolution monotonic A/D conversion.

2. The AD650 has a very versatile architecture that can be configured to accommodate bipolar, unipolar, or differential input voltages, or unipolar input currents.

3. TTL or CMOS compatibility is achieved using an open collector frequency output. The pullup resistor can be connected to voltages up to +30V, or +15V or +5V for conventional CMOS or TTL logic levels.

4. The same components used for V/F conversion can also be used for F/V conversion by adding a simple logic biasing network and reconfiguring the AD650.

5. The AD650 provides separate analog and digital grounds. This feature allows prevention of ground loops in real-world applications.

6. The AD650 is available in versions compliant with MIL-STD-883. Refer to the Analog Devices Military Products Databook or current AD650/883B data sheet for detailed specifications.

AD650—SPECIFICATIONS (@ +25°C with $V_S = \pm 15V$ unless otherwise noted)

Model	AD650J/AD650A Min	Typ	Max	AD650K/AD650B Min	Typ	Max	AD650S Min	Typ	Max	Units
DYNAMIC PERFORMANCE										
Full Scale Frequency Range			1			1			1	MHz
Nonlinearity[1] f_{max} = 10kHz		0.002	0.005		0.002	0.005		0.002	0.005	%
100kHz		0.005	**0.02**		0.005	**0.02**		0.005	**0.02**	%
500kHz		0.02	0.05		0.02	0.05		0.02	0.05	%
1MHz		0.1			0.05	**0.1**		0.05	**0.1**	%
Full Scale Calibration Error[2], 100kHz		±5			±5			±5		%
1MHz		±10			±10			±5		%
vs. Supply[3]	−0.015		+0.015	−0.015		+0.015	−0.015		+0.015	% of FSR/V
vs. Temperature										
A, B, and S Grades										
at 10kHz			±75			±75			±75	ppm/°C
at 100kHz			±150			±150			±150	ppm/°C
J and K Grades										
at 10kHz		±75			±75					ppm/°C
at 100kHz		±150			±150					ppm/°C
BIPOLAR OFFSET CURRENT										
Activated by 1.24kΩ between pins 4 and 5	0.45	0.5	0.55	0.45	0.5	0.55	0.45	0.5	0.55	mA
DYNAMIC RESPONSE										
Maximum Settling Time for Full Scale										
Step Input		1 Pulse of New Frequency Plus 1μs			1 Pulse of New Frequency Plus 1μs			1 Pulse of New Frequency Plus 1μs		
Overload Recovery Time										
Step Input		1 Pulse of New Frequency Plus 1μs			1 Pulse of New Frequency Plus 1μs			1 Pulse of New Frequency Plus 1μs		
ANALOG INPUT AMPLIFIER (V/F Conversion)										
Current Input Range (Figure 1)	0		+0.6	0		+0.6	0		+0.6	mA
Voltage Input Range (Figure 5)	−10		0	−10		0	−10		0	V
Differential Impedance		2MΩ‖10pF			2MΩ‖10pF			2MΩ‖10pF		
Common Mode Impedance		1000MΩ‖10pF			1000MΩ‖10pF			1000MΩ‖10pF		
Input Bias Current										
Noninverting Input		40	100		40	100		40	100	nA
Inverting Input		±8	±20		±8	±20		±8	±20	nA
Input Offset Voltage										
(Trimmable to Zero)			±4			±4			±4	mV
vs. Temperature (T_{min} to T_{max})		±30			±30			±30		μV/°C
Safe Input Voltage		±V_S			±V_S			±V_S		V
COMPARATOR (F/V Conversion)										
Logic "0" Level	−V_S		1	−V_S		1	−V_S		+1	V
Logic "1" Level	0		+V_S	0		+V_S	0		+V_S	V
Pulse Width Range[4]	0.1		$(0.3 \cdot t_{OS})$	0.1		$(0.3 \cdot t_{OS})$	0.1		$(0.3 \cdot t_{OS})$	μs
Input Impedance		250			250			250		kΩ
OPEN COLLECTOR OUTPUT (V/F Conversion)										
Output Voltage in Logic "0"										
I_{SINK} = 8mA, T_{min} to T_{max}			0.4			0.4			0.4	V
Output Leakage Current in Logic "1"			100			100			100	nA
Voltage Range[5]	0		+36	0		+36	0		+36	V
AMPLIFIER OUTPUT (F/V Conversion)										
Voltage Range (150Ω min load resistance)	0		+10	0		+10	0		+10	V
Source Current (750Ω max load resistance)	10			10			10			mA
Capacitive Load (Without Oscillation)			100			100			100	pF
POWER SUPPLY										
Voltage, Rated Performance	±9		±18	±9		±18	±9		±18	V
Quiescent Current		8			8			8		mA
TEMPERATURE RANGE										
Rated Performance – N Package	0		+70	0		+70				°C
D Package	−25		+85	−25		+85	−55		+125	°C
Storage – N Package	−25		+85	−25		+85				°C
D Package	−65		+150	−65		+150	−65		+150	°C
PACKAGE OPTIONS[6]										
PLCC (P-20A)		AD650JP			AD650KP					
Plastic DIP (N-14)		AD650JN			AD650KN					
Ceramic DIP (D-14)		AD650AD			AD650BD			AD650SD		

NOTES
[1] Nonlinearity is defined as deviation from a straight line from zero to full scale, expressed as a fraction of full scale.
[2] Full scale calibration error adjustable to zero.
[3] Measured at full scale output frequency of 100kHz.
[4] Refer to F/V conversion section of the text.
[5] Referred to digital ground.
[6] D = Ceramic DIP; N = Plastic DIP; P = Plastic Leaded Chip Carrier.
For outline information see Package Information section.

Specifications shown in boldface are tested on all production units at final electrical test. Results from those tests are used to calculate outgoing quality levels. All min and max specifications are guaranteed, although only those shown in boldface are tested on all production units.

Specifications subject to change without notice.

Unipolar Operation—AD650

ABSOLUTE MAXIMUM RATINGS

Total Supply Voltage $+V_S$ to $-V_S$ 36V
Storage Temperature Ceramic . . . . $-55°C$ to $+165°C$
                       Plastic . . . . $-25°C$ to $+125°C$
Differential Input Voltage (Pins 2 & 3) . . . . $\pm 10V$
Maximum Input Voltage . . . . $\pm V_S$
Open Collector Output   Voltage Above Digital GND . . 36V
                    Current 50mA
Amplifier Short Ckt to Ground . . . . Indefinite
Comparator Input Voltage (Pin 9) . . . . $\pm V_S$

ORDERING GUIDE

Part[1] Number	Gain Tempco ppm/°C 100kHz	1MHz Linearity	Specified Temperature Range °C	Package
AD650JN	150 typ	0.1% typ	0 to +70	Plastic DIP
AD650KN	150 typ	0.1% max	0 to +70	Plastic DIP
AD650JP	150 typ	0.1% typ	0 to +70	PLCC
AD650KP	150 typ	0.1% max	0 to +70	PLCC
AD650AD	150 max	0.1% typ	−25 to +85	Ceramic
AD650BD	150 max	0.1% max	−25 to +85	Ceramic
AD650SD	150 max	0.1% max	−55 to +125	Ceramic

NOTE
[1]For details on grade and package offerings screened in accordance with MIL-STD-883, refer to the Analog Devices Military Products Databook or current AD650/883B data sheet.

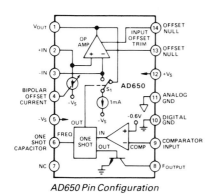

AD650 Pin Configuration

8-Bit A/D Converter
AD673*

FEATURES
Complete 8-Bit A/D Converter with Reference, Clock and Comparator
30µs Maximum Conversion Time
Full 8- or 16-Bit Microprocessor Bus Interface
Unipolar and Bipolar Inputs
No Missing Codes Over Temperature
Operates on +5V and −12V to −15V Supplies
MIL-STD-883 Compliant Version Available

FUNCTIONAL BLOCK DIAGRAM

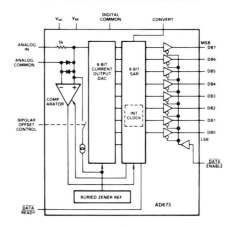

PRODUCT DESCRIPTION

The AD673 is a complete 8-bit successive approximation analog-to-digital converter consisting of a DAC, voltage reference, clock, comparator, successive approximation register (SAR) and 3 state output buffers–all fabricated on a single chip. No external components are required to perform a full accuracy 8-bit conversion in 20µs.

The AD673 incorporates advanced integrated circuit design and processing technologies. The successive approximation function is implemented with I^2L (integrated injection logic). Laser trimming of the high stability SiCr thin film resistor ladder network insures high accuracy, which is maintained with a temperature compensated sub-surface Zener reference.

Operating on supplies of +5V and −12V to −15V, the AD673 will accept analog inputs of 0 to +10V or −5V to +5V. The trailing edge of a positive pulse on the CONVERT line initiates the 20µs conversion cycle. DATA READY indicates completion of the conversion.

The AD673 is available in two versions. The AD673J as specified over the 0 to +70°C temperature range and the AD673S guarantees ±½LSB relative accuracy and no missing codes from −55°C to +125°C.

Two package configurations are offered. All versions are also offered in a 20-pin hermetically sealed ceramic DIP. The AD673J is also available in a 20-pin plastic DIP.

PRODUCT HIGHLIGHTS

1. The AD673 is a complete 8-bit A/D converter. No external components are required to perform a conversion.
2. The AD673 interfaces to many popular microprocessors without external buffers or peripheral interface adapters.
3. The device offers true 8-bit accuracy and exhibits no missing codes over its entire operating temperature range.
4. The AD673 adapts to either unipolar (0 to +10V) or bipolar (−5V to +5V) analog inputs by simply grounding or opening a single pin.
5. Performance is guaranteed with +5V and −12V or −15V supplies.
6. The AD673 is available in a version compliant with MIL-STD-883. Refer to the Analog Devices Military Products Databook or current /883B data sheet for detailed specifications.

*Protected by U.S. Patent Nos. 3,940,760; 4,213,806; 4,136,349; 4,400,689; and 4,400,690

AD673—SPECIFICATIONS

($T_A = 25°C$, $V+ = +5V$, $V- = -12V$ or $-15V$, all voltages measured with respect to digital common, unless otherwise indicated.)

Model	AD673J Min	AD673J Typ	AD673J Max	AD673S Min	AD673S Typ	AD673S Max	Units
RESOLUTION		8			8		Bits
RELATIVE ACCURACY,[1]			±1/2			±1/2	LSB
$T_A = T_{min}$ to T_{max}			±1/2			±1/2	LSB
FULL SCALE CALIBRATION[2]		±2			±2		LSB
UNIPOLAR OFFSET			±1/2			±1/2	LSB
BIPOLAR OFFSET			±1/2			±1/2	LSB
DIFFERENTIAL NONLINEARITY,[3]	8			8			Bits
$T_A = T_{min}$ to T_{max}	8			8			Bits
TEMPERATURE RANGE	0		+70	-55		+125	°C
TEMPERATURE COEFFICIENTS							
Unipolar Offset			±1			±1	LSB
Bipolar Offset			±1			±1	LSB
Full Scale Calibration[2]			±2			±2	LSB
POWER SUPPLY REJECTION							
Positive Supply							
$+4.5 \leq V+ \leq +5.5V$			±2			±2	LSB
Negative Supply							
$-15.75V \leq V- \leq -14.25V$			±2			±2	LSB
$-12.6V \leq V- \leq -11.4V$			±2			±2	LSB
ANALOG INPUT IMPEDANCE	3.0	5.0	7.0	3.0	5.0	7.0	kΩ
ANALOG INPUT RANGES							
Unipolar	0		+10	0		+10	V
Bipolar	-5		+5	-5		+5	V
OUTPUT CODING							
Unipolar		Positive True Binary			Positive True Binary		
Bipolar		Positive True Offset Binary			Positive True Offset Binary		
LOGIC OUTPUT							
Output Sink Current							
($V_{OUT} = 0.4V$ max, T_{min} to T_{max})	3.2			3.2			mA
Output Source Current[4]							
($V_{OUT} = 2.4V$ min, T_{min} to T_{max})	0.5			0.5			mA
Output Leakage			±40			±40	μA
LOGIC INPUTS							
Input Current			±100			±100	μA
Logic "1"	2.0			2.0			V
Logic "0"			0.8			0.8	V
CONVERSION TIME, T_A and							
T_{min} to T_{max}	10	20	30	10	20	30	μs
POWER SUPPLY							
V+	+4.5	+5.0	+7.0	+4.5	+5.0	+7.0	V
V-	-11.4	-15	-16.5	-11.4	-15	-16.5	V
OPERATING CURRENT							
V+		15	20		15	20	mA
V-		9	15		9	15	mA

NOTES

[1]Relative accuracy is defined as the deviation of the code transition points from the ideal transfer point on a straight line from the zero to the full scale of the device.

[2]Full scale calibration is guaranteed trimmable to zero with an external 200Ω potentiometer in place of the 15Ω fixed resistor.
Full scale is defined as 10 volts minus 1LSB, or 9.961 volts.

[3]Defined as the resolution for which no missing codes will occur.

[4]The data output lines have active pull-ups to source 0.5mA. The DATA READY line is open collector with a nominal 6kΩ internal pull-up resistor.

Specifications subject to change without notice.

Specifications shown in boldface are tested on all production units at final electrical test. Results from those tests are used to calculate outgoing quality levels. All min and max specifications are guaranteed, although only those shown in boldface are tested on all production units.

AD673

ABSOLUTE MAXIMUM RATINGS
V+ to Digital Common . . . . 0 to +7V
V− to Digital Common . . . . 0 to −16.5V
Analog Common to Digital Common . . . . ±1V
Analog Input to Analog Common . . . . ±15V
Control Inputs . . . . 0 to V+
Digital Outputs (High Impedance State) . . . . . 0 to V+
Power Dissipation 800mW

ORDERING GUIDE

Model	Temperature Range	Relative Accuracy	Package Options[1]
AD673JN	0 to +70°C	±1/2LSB max	Plastic DIP (N-20)
AD673JD	0 to +70°C	±1/2LSB max	Ceramic DIP (D-20)
AD673SD[2]	−55°C to +125°C	±1/2LSB max	Ceramic DIP (D-20)
AD673JP	0 to +70°C	±1/2LSB max	PLCC (P-20A)

NOTES
[1]D = Ceramic DIP; N = Plastic DIP; P = Plastic Leaded Chip Carrier. For outline information see Package Information section.
[2]For details on grade and package offering screened in accordance with MIL-STD-883, refer to Analog Devices Military Products Databook.

FUNCTIONAL DESCRIPTION
A block diagram of the AD673 is shown in Figure 1. The positive CONVERT pulse must be at least 500ns wide. $\overline{DR}$ goes high within 1.5µs after the leading edge of the convert pulse indicating that the internal logic has been reset. The negative edge of the CONVERT pulse initiates the conversion. The internal 8-bit current output DAC is sequenced by the integrated injection logic (I^2L) successive approximation register (SAR) from its most significant bit to least significant bit to provide an output current which accurately balances the input signal current through the 5kΩ resistor. The comparator determines whether the addition of each successively weighted bit current causes the DAC current sum to be greater or less than the input current; if the sum is more, the bit is turned off. After testing all bits, the SAR contains a 8-bit binary code which accurately represents the input signal to within (0.05% of full scale).

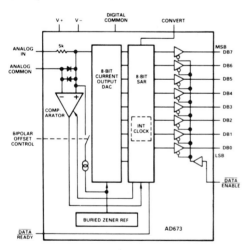

Figure 1. AD673 Functional Block Diagram

The SAR drives $\overline{DR}$ low to indicate that the conversion is complete and that the data is available to the output buffers. $\overline{DATA\ ENABLE}$ can then be activated to enable the 8-bits of data desired. $\overline{DATA\ ENABLE}$ should be brought high prior to the next conversion to place the output buffers in the high impedance state.

The temperature compensated buried Zener reference provides the primary voltage reference to the DAC and ensures excellent stability with both time and temperature. The bipolar offset input controls a switch which allows the positive bipolar offset current (exactly equal to the value of the MSB less ½LSB) to be injected into the summing (+) node of the comparator to offset the DAC output. Thus the nominal 0 to +10V unipolar input range becomes a −5V to +5V range. The 5kΩ thin film input resistor is trimmed so that with a full scale input signal, an input current will be generated which exactly matches the DAC output with all bits on.

UNIPOLAR CONNECTION
The AD673 contains all the active components required to perform a complete A/D conversion. Thus, for many applications, all that is necessary is connection of the power supplies (+5V and −12V to −15V), the analog input and the convert pulse. However, there are some features and special connections which should be considered for achieving optimum performance. The functional pin-out is shown in Figure 2.

The standard unipolar 0 to +10V range is obtained by shorting the bipolar offset control pin (pin 16) to digital common (pin 17).

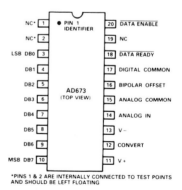

*PINS 1 & 2 ARE INTERNALLY CONNECTED TO TEST POINTS AND SHOULD BE LEFT FLOATING

Figure 2. AD673 Pin Connections

4 × 1 Wideband Video Multiplexer

AD9300

FEATURES
34MHz Full Power Bandwidth
±0.1dB Gain Flatness to 8MHz
72dB Crosstalk Rejection @ 10MHz
0.03°/0.01% Differential Phase/Gain
Cascadable for Switch Matrices
MIL-STD-883 Compliant Versions Available

APPLICATIONS
Video Routing
Medical Imaging
Electro-Optics
ECM Systems
Radar Systems
Data Acquisition

FUNCTIONAL BLOCK DIAGRAM
(Based on Cerdip)

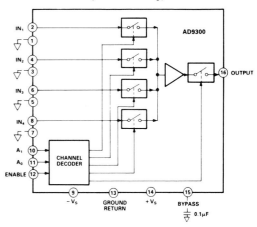

GENERAL DESCRIPTION

The AD9300 is a monolithic high-speed video signal multiplexer useable in a wide variety of applications.

Its four channels of video input signals can be randomly switched at megahertz rates to the single output. In addition, multiple devices can be configured in either parallel or cascade arrangements to form switch matrices. This flexibility in using the AD9300 is possible because the output of the device is in a high-impedance state when the chip is not enabled; when the chip is enabled, the unit acts as a buffer with a high input impedance and low output impedance.

An advanced bipolar process provides fast, wideband switching capabilities while maintaining crosstalk rejection of 72dB at 10MHz. Full power bandwidth is a minimum 27MHz. The device can be operated from ±10V to ±15V power supplies.

The AD9300K is available in a 16-pin ceramic DIP and a 20-pin PLCC and is designed to operate over the commercial temperature range of 0 to +70°C. The AD9300TQ is a hermetic 16-pin ceramic DIP for military temperature range (−55°C to +125°C) applications. This part is also available processed to MIL-STD-883. The AD9300 is available in a 20-pin LCC as the model AD9300TE, which operates over a temperature range of −55°C to +125°C.

The AD9300 Video Multiplexer is available in versions compliant with MIL-STD-883. Refer to the Analog Devices *Military Products Databook* or current AD9300/883B data sheet for detailed specifications.

PIN DESIGNATIONS

DIP

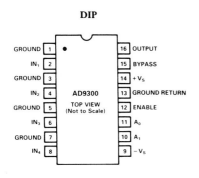

LCC and PLCC

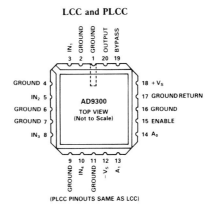

(PLCC PINOUTS SAME AS LCC)

AD9300–SPECIFICATIONS

ELECTRICAL CHARACTERISTICS ($\pm V_S = \pm 12V \pm 5\%$; $C_L = 10pF$; $R_L = 2k\Omega$, unless otherwise noted)

Parameter (Conditions)	Temp	Test Level	COMMERCIAL 0°C to +70°C AD9300KQ/KP Min	Typ	Max	Units
INPUT CHARACTERISTICS						
Input Offset Voltage	+25°C	I		3	10	mV
Input Offset Voltage	Full	VI			14	mV
Input Offset Voltage Drift[2]	Full	V		75		µV/°C
Input Bias Current	+25°C	I		15	37	µA
Input Bias Current	Full	VI			55	µA
Input Resistance	+25°C	V		3.0		MΩ
Input Capacitance	+25°C	V		2		pF
Input Noise Voltage (dc to 8MHz)	+25°C	V		16		µV rms
TRANSFER CHARACTERISTICS						
Voltage Gain[3]	+25°C	I	0.990	0.994		V/V
Voltage Gain[3]	Full	VI	0.985			V/V
DC Linearity[4]	+25°C	V		0.01		%
Gain Tolerance ($V_{IN} = \pm 1V$)						
dc to 5MHz	+25°C	I		0.05	0.1	dB
5MHz to 8MHz	+25°C	I		0.1	0.3	dB
Small-Signal Bandwidth	+25°C	V		350		MHz
($V_{IN} = 100mV$ p-p)						
Full Power Bandwidth[5]	+25°C	I	27	34		MHz
($V_{IN} = 2V$ p-p)						
Output Swing	Full	VI	±2			V
Output Current (Sinking @ = 25°C)	+25°C	V		5		mA
Output Resistance	+25°C	IV, V		9	15	Ω
DYNAMIC CHARACTERISTICS						
Slew Rate[6]	+25°C	I	170	215		V/µs
Settling Time						
(to 0.1% on ±2V Output)	+25°C	IV		70	100	ns
Overshoot						
To T-Step[7]	+25°C	V		<0.1		%
To Pulse[8]	+25°C	V		<10		%
Differential Phase[9]	+25°C	IV		0.03	0.1	°
Differential Gain[9]	+25°C	IV		0.01	0.1	%
Crosstalk Rejection						
Three Channels[10]	+25°C	IV	68	72		dB
One Channel[11]	+25°C	IV	70	76		dB
SWITCHING CHARACTERISTICS[12]						
A_X Input to Channel HIGH Time[13]	+25°C	I		40	50	ns
(t_{HIGH})						
A_X Input to Channel LOW Time[15]	+25°C	I		35	45	ns
(t_{LOW})						
Enable to Channel ON Time[15]	+25°C	I		35	45	ns
(t_{ON})						
Enable to Channel OFF Time[16]	+25°C	I		35	45	ns
(t_{OFF})						
Switching Transient[17]	+25°C	V		60		mV

EXPLANATION OF TEST LEVELS

Test Level I – 100% production tested.
Test Level II – 100% production tested at +25°C, and sample tested at specified temperatures.
Test Level III – Sample tested only.
Test Level IV – Parameter is guaranteed by design and characterization testing.
Test Level V – Parameter is a typical value only.
Test Level VI – All devices are 100% production tested at +25°C. 100% production tested at temperature extremes for military temperature devices; sample tested at temperature extremes for commercial/industrial devices.

AD9300

Parameter (Conditions)	Temp	Test Level	COMMERCIAL 0°C to +70°C AD9300KQ/KP			Units
			Min	Typ	Max	
DIGITAL INPUTS						
Logic "1" Voltage	Full	VI	2			V
Logic "0" Voltage	Full	VI			0.8	V
Logic "1" Current	Full	VI			5	µA
Logic "0" Current	Full	VI			1	µA
POWER SUPPLY						
Positive Supply Current (+12V)	+25°C	I		13	16	mA
Positive Supply Current (+12V)	Full	VI		13	16	mA
Negative Supply Current (−12V)	+25°C	I		12.5	15	mA
Negative Supply Current (−12V)	Full	VI		12.5	16	mA
Power Supply Rejection Ratio ($\pm V_S = \pm 12V \pm 5\%$)	Full	VI	67	75		dB
Power Dissipation ($\pm 12V$)[19]	+25°C	V		306		mW

NOTES
[1] Permanent damage may occur if any one absolute maximum rating is exceeded. Functional operation is not implied, and device reliability may be impaired by exposure to higher-than-recommended voltages for extended periods of time.
[2] Measured at extremes of temperature range.
[3] Measured as slope of V_{OUT} versus V_{IN} with $V_{IN} = \pm 1V$.
[4] Measured as worst deviation from end-point fit with $V_{IN} = \pm 1V$.
[5] Full Power Bandwith (FPBW) based on Slew Rate (SR). FPBW = $SR \cdot 2\pi V_{PEAK}$
[6] Measured between 20% and 80% transition points of $\pm 1V$ output.
[7] T-Step = $Sin^2 X$ Step, when Step between 0V and +700mV points has 10%-to-90% risetime = 125ns.
[8] Measured with a pulse input having slew rate >250V/µs.
[9] Measured at output between 0.28Vdc and 1.0Vdc with V_{IN} = 284mV p-p at 3.58MHz and 4.43MHz.
[10] This specification is critically dependent on circuit layout. Value shown is measured with selected channel grounded and 10MHz 2V p-p signal applied to remaining three channels. If selected channel is grounded through 75Ω, value is approximately 6dB higher.
[11] This specification is critically dependent on circuit layout. Value shown is measured with selected channel grounded and 10MHz 2V p-p signal applied to one other channel. If selected channel is grounded through 75Ω, value is approximately 6dB higher. Minimum specification in ( ) applies to DIPs.
[12] Consult system timing diagram.
[13] Measured from address change to 90% point of −2V to +2V output LOW-to-HIGH transition.
[14] Measured from address change to 90% point of +2V to −2V output HIGH-to-LOW transition.
[15] Measured from 50% transition point of ENABLE input 90% transition of 0V to −2V and 0V to +2V output.
[16] Measured from 50% transition point of ENABLE input to 10% transition of +2V to 0V and −2V to 0V output.
[17] Measured while switching between two grounded channels.
[18] Maximum power dissipation is a package-dependent parameter related to the following typical thermal impedances:
 16-Pin Ceramic θ_{JA} = 87°C/W; θ_{JC} = 25°C/W
 20-Pin LCC θ_{JA} = 74°C/W; θ_{JC} = 10°C/W
 20-Pin PLCC θ_{JA} = 71°C/W; θ_{JC} = 26°C/W

Specifications subject to change without notice.

ABSOLUTE MAXIMUM RATINGS[1]

Supply Voltages ($\pm V_S$) .... ±16V
Analog Input Voltage Each Input
 (IN_1 thru IN_4) .... ±3.5V
Differential Voltage Between Any Two
 Inputs (IN_1 thru IN_4) 5V
Digital Input Voltages (A_0, A_1, ENABLE) . −0.5V to +5.5V

Output Current
 Sinking 6.0mA
 Sourcing 6.0mA
Operating Temperature Range
 AD9300KQ/KP .... 0°C to +70°C
Storage Temperature Range .... −65°C to +150°C
Junction Temperature .... +175°C
Lead Soldering (10sec) .... +300°C

ORDERING INFORMATION

Device	Temperature Range	Description	Package Option[1]
AD9300KQ	0 to +70°C	16-Pin Cerdip, Commercial	Q-16
AD9300TE/883B[2]	−55°C to +125°C	20-Pin LCC, Military Temperature	E-20A
AD9300TQ/883B[2]	−55°C to +125°C	16-Pin Cerdip, Military Temperature	Q-16
AD9300KP	0 to +70°C	20-Pin PLCC, Commercial	P-20A

NOTES
[1] E = Ceramic Leadless Chip Carrier; P = Plastic Leaded Chip Carrier; Q = Cerdip. For outline information see Package Information section.
[2] For specifications, refer to Analog Devices *Military Products Databook*.

Hybrid, Tracking Resolver-to-Digital Converters

1S20/1S40/1S60/1S61

FEATURES
- Low Cost
- 32-Pin Hybrid
- High Tracking Rate 170rps at 12 Bits
- Velocity Output
- DC Error Output
- Logic Outputs for Extension Pitch Counter

APPLICATIONS
- Numerical Control of Machine Tools
- Robotics

1S20/1S40/1S60/1S61 FUNCTIONAL BLOCK DIAGRAM

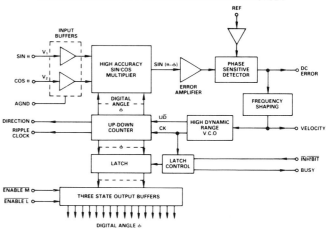

GENERAL DESCRIPTION

The 1S20/40/60/61 are a series of low cost hybrid converters with a high tracking rate and all essential features for numerically controlled machine applications. These converters are housed in a 32-pin triple DIP ceramic package measuring $1.1'' \times 1.7'' \times 0.205''$ ($28 \times 43.2 \times 5.2$mm).

The 1S20/40/60/61 convert resolver format input signals into a parallel natural binary digital word. Typically, these signals would be obtained from a brushless resolver and the resolver/converter combination gives a parallel absolute angular output word similar to that provided by an absolute encoder. The ratiometric conversion principle of the 1S20/40/60/61 series ensures high noise immunity and tolerance of lead length when the converter is at a distance from the resolver.

The output word is in three-state digital logic form with a high and low byte enable input so that the converter can communicate with an 8- or 16-bit digital highway. In this series there are 12-, 14- and two 16-bit resolution (± 4 arc mins and ± 10 arc mins accuracy) models available.

Repeatability is 1LSB for all models under constant temperature conditions.

The 1S20/40/60/61 are available with three frequency options covering the range 400Hz to 10kHz.

Models Available

Four models are available in this range and three frequency options for each model.

1S20 is a 12-bit up to 170 revolutions per second
1S40 is a 14-bit up to 42.5 revolutions per second
1S60 is a 16-bit up to 10.5 revolutions per second
1S61 is a 16-bit up to 10.5 revolutions per second

APPLICATIONS/USER BENEFITS

The 1S20/40/60/61 has been specifically designed for the numerically controlled machine and robot industry. Using the type 2 servo loop tracking principle ideally suits these converters to the electrically noisy environment found in these industrial applications.

By using hybrid construction techniques, small size, low power and high reliability are further benefits offered by these converters. This small size with the three-state digital outputs makes these converters ideal for multichannel operation.

The layout of the connections simplifies the parallel connection to a digital highway.

The provision of the digital outputs of DIRECTION and RIPPLE CLOCK allow simple extension counters for multi-pitch operation to be implemented.

Analog outputs of velocity and dc error for control loop stabilization and bite (built in test) provide two more features required in these applications.

SPECIFICATIONS (typical @ +25°C, unless otherwise specified)

Models	1S20	1S40	1S60	1S61	Units
RESOLUTION	12	14	16	16	Bits
ACCURACY[1]	±8.5	±5.3	±4.0	±10	arc-mins
REPEATABILITY[2]	1	*	*	*	LSB
SIGNAL AND REFERENCE FREQUENCY[3]	400-10k	*	*	*	Hz
DIGITAL OUTPUT	Parallel natural binary				
Max Load	20	*	*	*	LSTTL
TRACKING RATE (min)					
400Hz – 2.6kHz	50	12.5	3.0	3.0	rps
2.6kHz – 5kHz	90	22.5	5.5	5.5	rps
5kHz – 10kHz	170	42.5	10.5	10.5	rps
SETTLING TIME					
400Hz – 2.6kHz	150	180	350	350	ms
2.6kHz – 5kHz	40	50	130	130	ms
5kHz – 10kHz	20	25	60	60	ms
ACCELERATION CONSTANT (K_a)					
400Hz – 2.6kHz	9,500	*	*	*	sec^{-2}
2.6kHz – 5kHz	144,000	*	*	*	sec^{-2}
5kHz – 10kHz	713,000	*	*	*	sec^{-2}
SIGNAL VOLTAGE	2.0	*	*	*	V rms
SIGNAL INPUT IMPEDANCE	>10	*	*	*	MΩ
REFERENCE VOLTAGE	2.0	*	*	*	V rms
REFERENCE INPUT IMPEDANCE	125	*	*	*	kΩ
ALLOWABLE PHASE SHIFT[4] (Signal to Reference)	±10	*	*	*	Degrees
BUSY OUTPUT[5]	Logic "Hi" when Busy				
Max Load	20	*	*	*	LSTTL
BUSY WIDTH	430	*	*	*	ns
ENABLE INPUTS	Logic "Lo" to ENABLE				
Load	1	*	*	*	LSTTL
ENABLE AND DISABLE TIMES	120 (typ)	*	*	*	ns
	220 (max)	*	*	*	ns
INHIBIT INPUT	Logic "Lo" to INHIBIT				
Load	1	*	*	*	LSTTL
DIRECTION OUTPUT (DIR)[5]	Logic "Hi" when counting up				
	Logic "Lo" when counting down				
Max Load	20	*	*	*	LSTTL
RIPPLE CLOCK[5]	Negative pulse indicating when internal counters change from all "1's" to all "0's" or vice versa.				
Max Load	20	*	*	*	LSTTL
VELOCITY OUTPUT[6] (at specified min tracking rate).					
Polarity	positive for increasing angle	*	*	*	–
Output Voltage[7]	±10	*	*	*	V dc
Accuracy	±10	*	*	*	% FSD
Zero Offset	±8	*	*	*	mV
DC ERROR OUTPUT VOLTAGE[6]	40	10	2.5	2.5	mV/LSB
POWER SUPPLIES					
$+V_s$	+11.5 to +16	*	*	*	V
$-V_s$	−11.5 to −16	*	*	*	V
+5V	+4.75 to +5.25	*	*	*	V
POWER SUPPLY CONSUMPTION[7]					
$+V_s$	20, 30 (max)	*	*	*	mA
$-V_s$	20, 30 (max)	*	*	*	mA
+5V	105, 125 (max)	*	*	*	mA
POWER DISSIPATION[7]	1.1, 1.5 (max)	*	*	*	W
TEMPERATURE RANGE					
Operating	0 to +70	*	*	*	°C
Storage	−55 to +125	*	*	*	°C
PACKAGE OPTION[8]	DH-32E	*	*	*	
WEIGHT	1 (28)	*	*	*	oz. (grms)

NOTES

[1] Specified over the operating temperature range and for:
  a). ±10% signal and reference amplitude variation.
  b). 10% signal and reference harmonic distortion.
  c). ±10% on frequency range of option.
[2] Specified at constant temperature. Over the operating temperature range, worst case repeatability could be up to 1.5 arc mins for all models.
[3] See frequency range options.
[4] For no additional error with a static input, see "Dynamic Accuracy vs. Resolver Phase Shift"
[5] See timing diagram.
[6] These outputs should be connected via buffers or comparator inputs (max load 100pF).
[7] $±V_s = ±15$ volts.
[8] See Section 14 for package outline information.
*Specifications same as 1S20.
Specifications subject to change without notice.

High Performance, Economy Strain Gage/RTD Conditioners
2B30/2B31

FEATURES
Low Cost
Complete Signal Conditioning Function
Low Drift: $0.5\mu V/°C$ max ("L"); **Low Noise:** $1\mu V$ p-p max
Wide Gain Range: 1 to 2000V/V
Low Nonlinearity: 0.0025% max ("L")
High CMR: 140dB min (60Hz, G = 1000V/V)
Input Protected to 130V rms
Adjustable Low Pass Filter: 60dB/Decade Roll-Off (from 2Hz)
Programmable Transducer Excitation: Voltage (4V to 15V @ 100mA) or Current ($100\mu A$ to 10mA)

APPLICATIONS
Measurement and Control of:
  Pressure, Temperature, Strain/Stress, Force, Torque
Instrumentation: Indicators, Recorders, Controllers
Data Acquisition Systems
Microcomputer Analog I/O

2B31 FUNCTIONAL BLOCK DIAGRAM

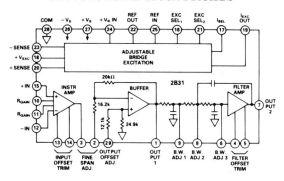

GENERAL DESCRIPTION
Models 2B30 and 2B31 are high performance, low cost, compact signal conditioning modules designed specifically for high accuracy interface to strain gage-type transducers and RTD's (resistance temperature detectors). The 2B31 consists of three basic sections: a high quality instrumentation amplifier; a three-pole low pass filter, and an adjustable transducer excitation. The 2B30 has the same amplifier and filter as the 2B31, but no excitation capability.

Available with low offset drift of $0.5\mu V/°C$ max (RTI, G = 1000V/V) and excellent linearity of 0.0025% max, both models feature guaranteed low noise performance ($1\mu V$ p-p max) and outstanding 140dB common mode rejection (60Hz, CMV = $\pm 10V$, G = 1000V/V) enabling the 2B30/2B31 to maintain total amplifier errors below 0.1% over a 20°C temperature range. The low pass filter offers 60dB/decade roll-off from 2Hz to reduce normal-mode noise bandwidth and improve system signal-to-noise ratio. The 2B31's regulated transducer excitation stage features a low output drift (0.015%/°C max) and a capability of either constant voltage or constant current operation.

Gain, filter cutoff frequency, output offset level and bridge excitation (2B31) are all adjustable, making the 2B30/2B31 the industry's most versatile high-accuracy transducer-interface modules. Both models are offered in three accuracy selections, J/K/L, differing only in maximum nonlinearity and offset drift specifications.

APPLICATIONS
The 2B30/2B31 may be easily and directly interfaced to a wide variety of transducers for precise measurement and control of pressure, temperature, stress, force and torque. For applications in harsh industrial environments, such characteristics as high CMR, input protection, low noise, and excellent temperature stability make 2B30/2B31 ideally suited for use in indicators, recorders, and controllers.

The combination of low cost, small size and high performance of the 2B30/2B31 offers also exceptional quality and value to the data acquisition system designer, allowing him to assign a conditioner to each transducer channel. The advantages of this approach over low level multiplexers include significant improvements in system noise and resolution, and elimination of crosstalk and aliasing errors.

DESIGN FEATURES AND USER BENEFITS
High Noise Rejection: The true differential input circuitry with high CMR (140dB) eliminating common-mode noise pickup errors, input filtering minimizing RFI/EMI effects, output low pass filtering ($f_c=2Hz$) rejecting 50/60Hz line frequency pickup and series-mode noise.

Input and Output Protection: Input protected for shorts to power lines (130V rms), output protected for shorts to ground and either supply.

Ease of Use: Direct transducer interface with minimum external parts required, convenient offset and span adjustment capability.

Programmable Transducer Excitation: User-programmable adjustable excitation source-constant voltage (4V to 15V @ 100mA) or constant current ($100\mu A$ to 10mA) to optimize transducer performance.

Adjustable Low Pass Filter: The three-pole active filter ($f_c=2Hz$) reducing noise bandwidth and aliasing errors with provisions for external adjustment of cutoff frequency.

SPECIFICATIONS
(typical @ +25°C and V_S = ±15V unless otherwise noted)

MODEL	2B30J / 2B31J	2B30K / 2B31K	2B30L / 2B31L
GAIN[1]			
Gain Range	1 to 2000V/V	*	*
Gain Equation	$G = (1 + 94k\Omega/R_G) [20k\Omega/(R_F + 16.2k\Omega)]$	*	*
Gain Equation Accuracy	±2%	*	*
Fine Gain (Span) Adjust Range	±20%	*	*
Gain Temperature Coefficient	±25ppm/°C max (±10ppm/°C typ)	*	*
Gain Nonlinearity	±0.01% max	±0.005% max	±0.0025% max
OFFSET VOLTAGES[1]			
Total Offset Voltage, Referred to Input			
Initial @ +25°C	Adjustable to Zero (±0.5mV typ)	*	*
Warm-Up Drift, 10 Min., G = 1000	Within ±5µV (RTI) of Final Value	*	*
vs. Temperature			
G = 1V/V	±150µV/°C max	±75µV/°C max	±50µV/°C max
G = 1000V/V	±3µV/°C max	±1µV/°C max	±0.5µV/°C max
At Other Gains	±(3 + 150/G)µV/°C max	±(1 + 75/G)µV/°C max	±(0.5 + 50/G)µV/°C max
vs. Supply, G = 1000V/V[5]	±25µV/V		
vs. Time, G = 1000V/V	±5µV/month	*	*
Output Offset Adjust Range	±10V	*	*
INPUT BIAS CURRENT			
Initial @ +25°C	+200nA max (100nA typ)	*	*
vs. Temperature (0 to +70°C)	−0.6nA/°C	*	*
INPUT DIFFERENCE CURRENT			
Initial @ +25°C	±5nA	*	*
vs. Temperature (0 to +70°C)	±40pA/°C	*	*
INPUT IMPEDANCE			
Differential	100MΩ‖47pF	*	*
Common Mode	100MΩ‖47pF	*	*
INPUT VOLTAGE RANGE			
Linear Differential Input	±10V	*	*
Maximum Differential or CMV Input Without Damage	130V rms	*	*
Common Mode Voltage	±10V	*	*
CMR, 1kΩ Source Imbalance			
G = 1V/V, dc to 60Hz[1]	90dB	*	*
G = 100V/V to 2000V/V, 60Hz[1]	140dB min	*	*
dc[2]	90dB min (112 typ)	*	*
INPUT NOISE			
Voltage, G = 1000V/V			
0.01Hz to 2Hz	1µV p-p max	*	*
10Hz to 100Hz[2]	1µV p-p	*	*
Current, G = 1000			
0.01Hz to 2Hz	70pA p-p	*	*
10Hz to 100Hz[2]	30pA rms	*	*
RATED OUTPUT[1]			
Voltage, 2kΩ Load[3]	±10V min	*	*
Current	±5mA min	*	*
Impedance, dc to 2Hz, G = 100V/V	0.1Ω	*	*
Load Capacitance	0.01µF max	*	*
DYNAMIC RESPONSE (Unfiltered)[2]			
Small Signal Bandwidth			
−3dB Gain Accuracy, G = 100V/V	30kHz	*	*
G = 1000V/V	5kHz	*	*
Slew Rate	1V/µs	*	*
Full Power	15kHz	*	*
Settling Time, G = 100, ±10V Output Step to ±0.1%	30µs	*	*
LOW PASS FILTER (Bessel)			
Number of Poles	3	*	*
Gain (Pass Band)	+1	*	*
Cutoff Frequency (−3dB Point)	2Hz	*	*
Roll-Off	60dB/decade	*	*
Offset (at 25°C)	±5mV	*	*
Settling Time, G = 100V/V, ±10V Output Step to ±0.1%	600ms	*	*
BRIDGE EXCITATION (See Table 1)			
POWER SUPPLY[4]			
Voltage, Rated Performance	±15V dc	*	*
Voltage, Operating	±(12 to 18)V dc	*	*
Current, Quiescent[6]	±15mA	*	*
TEMPERATURE RANGE			
Rated Performance	0 to +70°C	*	*
Operating	−25°C to +85°C	*	*
Storage	−55°C to +125°C	*	*
CASE SIZE	2" x 2" x 0.4" (51 x 51 x 10.2mm)	*	*

NOTES
*Specifications same as 2B30J/2B31J.
[1] Specifications referred to output at pin 7 with 3.75k, 1%, 25ppm/°C fine span resistor installed and internally set 2Hz filter cutoff frequency.
[2] Specifications referred to the unfiltered output at pin 1.
[3] Protected for shorts to ground and/or either supply voltage.
[4] Recommended power supply ADI model 902-2 or model 2B35.
[5] Tracking power supplies.
[6] Does not include bridge excitation and load currents.
Specifications subject to change without notice.

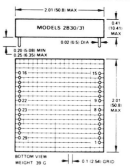

OUTLINE DIMENSIONS
Dimensions shown in inches and (mm).

PIN DESIGNATIONS

PIN	FUNCTION	PIN	FUNCTION
1	OUTPUT 1 (UNFILTERED)	16	EXC SEL 1
2	FINE GAIN (SPAN) ADJ	17	I SEL
3	FINE GAIN (SPAN) ADJ	18	V_{EXC} OUT
4	FILTER OFFSET TRIM	19	I_{EXC} OUT
5	FILTER OFFSET TRIM	20	SENSE HIGH (+)
6	BANDWIDTH ADJ 3	21	EXC SEL 2
7	OUTPUT 2 (FILTERED)	22	REF OUT
8	BANDWIDTH ADJ 2	23	SENSE LOW (−)
9	BANDWIDTH ADJ 1	24	REGULATOR +V_R IN
10	R_{GAIN}	25	REF IN
11	R_{GAIN}	26	−V_S
12	−INPUT	27	+V_S
13	INPUT OFFSET TRIM	28	COMMON
14	INPUT OFFSET TRIM	29	OUTPUT OFFSET TRIM
15	+INPUT		

Note: Pins 16 thru 25 are not connected in Model 2B30.

AC1211/AC1213 MOUNTING CARDS

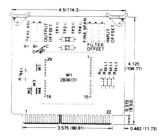

AC1211/AC1213 CONNECTOR DESIGNATIONS

PIN	FUNCTION	PIN	FUNCTION
A	REGULATOR +V_R IN	1	EXC SEL 1
B	SENSE LOW (−)	2	I SEL
C	REF OUT	3	V_{EXC} OUT
D	REF IN	4	I_{EXC} OUT
E		5	SENSE HIGH (+)
F		6	EXC SEL 2
H		7	OUTPUT OFFSET TRIM
J		8	
K	−V_S	9	−V_S
L	+V_S	10	+V_S
M		11	
N	COMMON	12	COMMON
P		13	
R	FINE GAIN ADJ	14	FINE GAIN ADJ
S	FILTER OFFSET TRIM	15	
T	FILTER OFFSET TRIM	16	R_{GAIN}
U	OUTPUT 2 (FILTERED)	17	R_{GAIN}
V	−INPUT	18	
W	INPUT OFFSET TRIM	19	OUTPUT 1 (UNFILTERED)
X	INPUT OFFSET TRIM	20	BANDWIDTH ADJ 1
Y		21	BANDWIDTH ADJ 3
Z	+INPUT	22	BANDWIDTH ADJ 2

The AC1211/AC1213 mounting card is available for the 2B30/2B31. The AC1211/AC1213 is an edge connector card with pin receptacles for plugging in the 2B30/2B31. In addition, it has provisions for installing the gain resistors and the bridge excitation, offset adjustment and filter cutoff programming components. The AC1211/AC1213 is provided with a Cinch 251-22-30-160 (or equivalent) edge connector. The AC1213 includes the adjustment pots; no pots are provided with the AC1211.

Isolated, Thermocouple Signal Conditioner
2B50

FEATURES
Accepts J, K, T, E, R, S or B Thermocouple Types
Internally Provided Cold Junction Compensation
High CMV Isolation: ±1500V pk
High CMR: 160dB min @ 60Hz
Low Drift: ±1µV/°C max (2B50B)
High Linearity: ±0.01% max (2B50B)
Input Protection and Filtering
Screw Terminal Input Connections

APPLICATIONS
Precision Thermocouple Signal Conditioning for:
　Process Control and Monitoring
　Industrial Automation
　Energy Management
　Data Acquisition Systems

2B50 FUNCTIONAL BLOCK DIAGRAM

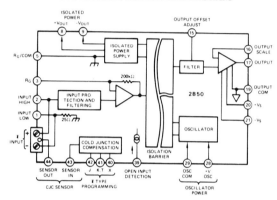

GENERAL DESCRIPTION
The model 2B50 is a high performance thermocouple signal conditioner providing input protection, isolation and common mode rejection, amplification, filtering and integral cold junction compensation in a single, compact package.

The 2B50 has been designed to condition low level analog signals, such as those produced by thermocouples, in the presence of high common mode voltages. Featuring direct thermocouple connection via screw terminals and internally provided reference junction temperature sensor, the 2B50 may be jumper programmed to provide cold junction compensation for thermocouple types J, K, T, and B, or resistor programmed for types E, R, and S.

The high performance of the 2B50 is accomplished by the use of reliable transformer isolation techniques. This assures complete input to output galvanic isolation (±1500V pk) and excellent common mode rejection (160dB @ 60Hz).

Other key features include: input protection (220V rms), filtering (NMR of 70dB @ 60Hz), low drift amplification (±1µV/°C max - 2B50B), and high linearity (±0.01% max - 2B50B).

APPLICATIONS
The 2B50 has been designed to provide thermocouple signal conditioning in data acquisition systems, computer interface systems, and temperature measurement and control instrumentation.

In thermocouple temperature measurement applications, outstanding features such as low drift, high noise rejection, and 1500V isolation make the 2B50 an ideal choice for systems used in harsh industrial environments.

DESIGN FEATURES AND USER BENEFITS
High Reliability: To assure high reliability and provide isolation protection to electronic instrumentation, the 2B50 has been conservatively designed to meet the IEEE Standard for transient voltage protection (472-1974: SWC) and provide 220V rms differential input protection.

High Noise Rejection: The 2B50 features internal filtering circuitry for elimination of errors caused by RFI/EMI, series mode noise, and 50Hz/60Hz pickup.

Ease of Use: Internal compensation enables the 2B50 to be used with seven different thermocouple types. Unique circuitry offers a choice of internal or remote reference junction temperature sensing. Thermocouple connections may be made either by screw terminals or, in applications requiring PC Board connections, by terminal pins.

Small Package: 1.5" × 2.5" × 0.6" size conserves board space.

SPECIFICATIONS (typical @ +25°C and V_S = ±15V unless otherwise noted)

MODEL	2B50A	2B50B
INPUT SPECIFICATIONS		
Thermocouple Types		
Jumper Configurable Compensation	J, K, T, or B	*
Resistor Configurable Compensation	R, S, or E	*
Input Span Range	±5mV to ±100mV	*
Gain Range	50V/V to 1000V/V	*
Gain Equation	$1 + (200k\Omega/R_G)$	*
Gain Error	±0.25%	*
Gain Temperature Coefficient	±35ppm/°C max	±25ppm/°C max
Gain Nonlinearity[1]	±0.025% max	±0.01% max
Offset Voltage		
Input Offset (Adjustable to Zero)	±50μV	*
vs. Temperature	±2.5μV/°C max	±1μV/°C max
vs. Time	±1.5μV/month	*
Output Offset (Adjustable to Zero)	±10mV	*
vs. Temperature	±30μV/°C	*
Total Offset Drift	$\pm \left(2.5 + \frac{30}{G}\right) \mu V/°C$	$\pm \left(1 + \frac{30}{G}\right) \mu V/°C$
Input Noise Voltage		
0.01Hz to 100Hz, R_S = 1kΩ	1μV p-p	*
Maximum Safe Differential Input Voltage	220V rms, Continuous	*
CMV, Input to Output		
Continuous, ac or dc	±1500V pk max	*
Common Mode Rejection		
@ 60Hz, 1kΩ Source Unbalance	160dB min	*
Normal Mode Rejection @ 60Hz	70dB min	*
Bandwidth	dc to 2.5Hz (-3dB)	*
Input Impedance	100MΩ	*
Input Bias Current[2]	±5nA	*
Open Input Detection	Downscale	*
Response Time[3], G = 250	1.4sec	*
Cold Junction Compensation		
Initial Accuracy[4]	±0.5°C	*
vs. Temperature[5] (+5°C to +45°C)	±0.01°C/°C	*
OUTPUT SPECIFICATIONS		
Output Voltage Range[6]	±5V @ ±2mA	*
Output Resistance	0.1Ω	*
Output Protection	Continuous Short to Ground	*
POWER SUPPLY		
Voltage		
Output ±V_S (Rated Performance)	±15V dc ±10% @ ±0.5mA	*
(Operating)	±12V to ±18V dc max	*
Oscillator +V_{OSC} (Rated Performance)	+13V to +18V @ 15mA	*
ENVIRONMENTAL		
Temperature Range, Rated Performance	0 to +70°C	*
Operating	-25°C to +85°C	*
Storage Temperature Range	-55°C to +85°C	*
RFI Effect (5W @ 470MHz @ 3ft)		
Error	±0.5% of Span	*
PHYSICAL		
Case Size	1.5" × 2.5" × 0.6"	*

OUTLINE DIMENSIONS
Dimensions shown in inches and (mm).

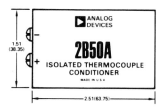

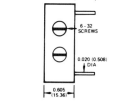

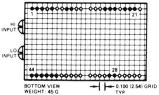

BOTTOM VIEW WEIGHT: 45 G — 0.100 (2.54) GRID TYP
NOTE: TERMINAL PINS INSTALLED ONLY IN SHADED HOLE LOCATIONS.

PIN DESIGNATIONS

PIN	FUNCTION	PIN	FUNCTION
1	INPUT LO	23	
2	INPUT HI	24	
3	R_G	25	
4		26	
5	R_G/COM	27	+V OSC
6		28	
7		29	OSC COM
8	+V ISO OUT	30	
9	-V ISO OUT	31	
10		32	
11		33	
12		34	
13		35	
14		36	
15	OUTPUT OFFSET ADJUST	37	
16	OUTPUT SCALE	38	
17	OUTPUT	39	OPEN INPUT DET
18		40	X T TYPE
19	OUTPUT COM	41	K,T PROGRAMMING
20	+V_S	42	J
21	-V_S	43	CJC SENSOR IN
22		44	CJC SENSOR OUT

MATING SOCKET: AC1218

NOTES
*Specifications same as 2B50A.
[1] Gain nonlinearity is specified as a percentage of output signal span representing peak deviation from the best straight line; e.g., nonlinearity at an output span of 10V pk-pk (±5V) is ±0.01% or ±1mV.
[2] Does not include open circuit detection current of 20nA (optional by jumper connection).
[3] Open input response time is dependent upon gain.
[4] When used with internally provided CJC sensor.
[5] Compensation error contributed by ambient temperature changes at the module.
[6] Output swing of ±10V may be obtained through output scaling (Figure 5).

Specifications subject to change without notice.

Appendix B
Derivations of Selected Equations

Equation (2–1)

The average value of a half-wave rectified sine wave is the area under the curve divided by the period (2π). The equation for a sine wave is

$$v = V_p \sin \theta$$

$$V_{avg} = \frac{\text{area}}{2\pi}$$

$$= \frac{1}{2\pi} \int_0^{\pi} V_p \sin \theta \, d\theta$$

$$= \frac{V_p}{2\pi}(-\cos \theta)\Big|_0^{\pi}$$

$$= \frac{V_p}{2\pi}[\cos \pi - (-\cos 0)]$$

$$= \frac{V_p}{2\pi}[-(-1) - (-1)]$$

$$= \frac{V_p}{2\pi}(2)$$

$$V_{avg} = \frac{V_p}{\pi}$$

Equation (4–3)

The Shockley equation for the base-emitter pn junction is

$$I_E = I_R(e^{VQ/kT} - 1)$$

where I_E = the total forward current across the base-emitter junction.
I_R = the reverse saturation current.
V = the voltage across the depletion layer.
Q = the charge on an electron.
k = a number known as Boltzmann's constant.
T = the absolute temperature.

Derivations of Selected Equations

At ambient temperature, $Q/kT \cong 40$, so

$$I_E = I_R(e^{40V} - 1)$$

Differentiating, we get

$$\frac{dI_E}{dV} = 40 I_R e^{40V}$$

Since $I_R e^{40V} = I_E + I_R$,

$$\frac{dI_E}{dV} = 40(I_E + I_R)$$

Assuming $I_R \ll I_E$,

$$\frac{dI_E}{dV} \cong 40 I_E$$

The ac resistance r_e of the base-emitter junction can be expressed as dV/dI_E.

$$r_e = \frac{dV}{dI_E} \cong \frac{1}{40 I_E} \cong \frac{25 \text{ mV}}{I_E}$$

Equation (6–11)

The formula for open-loop gain in Equation (6–7) can be expressed in complex notation as

$$A_{ol} = \frac{A_{ol(mid)}}{1 + jf/f_{c(ol)}}$$

Substituting the above expression into the equation $A_{cl} = A_{ol}/(1 + BA_{ol})$, we get a formula for the total closed-loop gain.

$$A_{cl} = \frac{A_{ol(mid)}/(1 + jf/f_{c(ol)})}{1 + BA_{ol(mid)}/(1 + jf/f_{c(ol)})}$$

Multiplying the numerator and denominator by $1 + jf/f_{c(ol)}$ yields

$$A_{cl} = \frac{A_{ol(mid)}}{1 + BA_{ol(mid)} + jf/f_{c(ol)}}$$

Dividing the numerator and denominator by $1 + BA_{ol(mid)}$ gives

$$A_{cl} = \frac{A_{ol(mid)}/(1 + BA_{ol(mid)})}{1 + j[f/(f_{c(ol)}(1 + BA_{ol(mid)}))]}$$

The above expression is of the form of the first equation

$$A_{cl} = \frac{A_{cl(mid)}}{1 + jf/f_{c(cl)}}$$

where $f_{c(cl)}$ is the closed-loop critical frequency. Thus,

$$f_{c(cl)} = f_{c(ol)}(1 + BA_{ol(mid)})$$

Equation (9–1)

$$\frac{V_{out}}{V_{in}} = \frac{R(-jX)/(R - jX)}{(R - jX) + R(-jX)/(R - jX)}$$

$$= \frac{R(-jX)}{(R - jX)^2 - jRX}$$

Multiplying the numerator and denominator by j,

$$\frac{V_{out}}{V_{in}} = \frac{RX}{j(R - jX)^2 + RX}$$

$$= \frac{RX}{RX + j(R^2 - j2RX - X^2)}$$

$$= \frac{RX}{RX + jR^2 + 2RX - jX^2}$$

$$\frac{V_{out}}{V_{in}} = \frac{RX}{3RX + j(R^2 - X^2)}$$

For a 0° phase angle there can be no j term. Recall from complex numbers in ac theory that a *nonzero* angle is associated with a complex number having a j term. Therefore, at f_r the j term is 0.

$$R^2 - X^2 = 0$$

Thus,

$$\frac{V_{out}}{V_{in}} = \frac{RX}{3RX}$$

Cancelling, we get

$$\frac{V_{out}}{V_{in}} = \frac{1}{3}$$

Equation (9–2)

From the derivation of Equation (9–1),

$$R^2 - X^2 = 0$$
$$R^2 = X^2$$
$$R = X$$

Since $X = \dfrac{1}{2\pi f_r C}$,

$$R = \frac{1}{2\pi f_r C}$$

$$f_r = \frac{1}{2\pi RC}$$

EQUATIONS (9–3) AND (9–4)

The feedback network in the phase-shift oscillator consists of three RC stages, as shown in Figure B–1. An expression for the attenuation is derived using the mesh analysis method for the loop assignment shown. All Rs are equal in value, and all Cs are equal in value.

$$(R - j1/2\pi fC)I_1 - RI_2 + 0I_3 = V_{in}$$
$$-RI_1 + (2R - j1/2\pi fC)I_2 - RI_3 = 0$$
$$0I_1 - RI_2 + (2R - j1/2\pi fC)I_3 = 0$$

FIGURE B–1

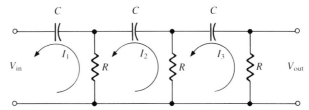

In order to get V_{out}, we must solve for I_3 using determinants:

$$I_3 = \frac{\begin{vmatrix} (R - j1/2\pi fC) & -R & V_{in} \\ -R & (2R - j1/2\pi fC) & 0 \\ 0 & -R & 0 \end{vmatrix}}{\begin{vmatrix} (R - j1/2\pi fC) & -R & 0 \\ -R & (2R - j1/2\pi fC) & -R \\ 0 & -R & (2R - j1/2\pi fC) \end{vmatrix}}$$

$$I_3 = \frac{R^2 V_{in}}{(R - j1/2\pi fC)(2R - j1/2\pi fC)^2 - R^2(2R - j1/2\pi fC) - R^2(R - 1/2\pi fC)}$$

$$\frac{V_{out}}{V_{in}} = \frac{RI_3}{V_{in}}$$

$$= \frac{R^3}{(R - j1/2\pi fC)(2R - j1/2\pi fC)^2 - R^3(2 - j1/2\pi fRC) - R^3(1 - 1/2\pi fRC)}$$

$$= \frac{R^3}{R^3((1 - j1/2\pi fRC)(2 - j1/2\pi fRC)^2 - R^3[(2 - j1/2\pi fRC) - (1 - j1/2\pi RC)]}$$

$$= \frac{R^3}{R^3(1 - j1/2\pi fRC)(2 - j1/2\pi fRC)^2 - R^3(3 - j1/2\pi fRC)}$$

$$\frac{V_{out}}{V_{in}} = \frac{1}{(1 - j1/2\pi fRC)(2 - j1/2\pi fRC)^2 - (3 - j1/2\pi fRC)}$$

APPENDIX B

Expanding and combining the real terms and the j terms separately,

$$\frac{V_{out}}{V_{in}} = \frac{1}{\left(1 - \dfrac{5}{4\pi^2 f^2 R^2 C^2}\right) - j\left(\dfrac{6}{2\pi f RC} - \dfrac{1}{(2\pi f)^3 R^3 C^3}\right)}$$

For oscillation in the phase-shift amplifier, the phase shift through the RC network must equal 180°. For this condition to exist, the j term must be 0 at the frequency of oscillation f_0.

$$\frac{6}{2\pi f_0 RC} - \frac{1}{(2\pi f_0)^3 R^3 C^3} = 0$$

$$\frac{6(2\pi)^2 f_0^2 R^2 C^2 - 1}{(2\pi)^3 f_0^3 R^3 C^3} = 0$$

$$6(2\pi)^2 f_0^2 R^2 C^2 - 1 = 0$$

$$f_0^2 = \frac{1}{6(2\pi)^2 R^2 C^2}$$

$$f_0 = \frac{1}{2\pi \sqrt{6} RC}$$

Since the j term is 0,

$$\frac{V_{out}}{V_{in}} = \frac{1}{1 - \dfrac{5}{4\pi^2 f_0^2 R^2 C^2}} = \frac{1}{1 - \dfrac{5}{\left(\dfrac{1}{\sqrt{6}RC}\right)^2 R^2 C^2}}$$

$$= \frac{1}{1 - 30} = -\frac{1}{29}$$

The negative sign results from the 180° inversion. Thus, the value of attenuation for the feedback network is

$$B = \frac{1}{29}$$

Answers to Self-Tests

Chapter 1
1. (b) 2. (c) 3. (d) 4. (c) 5. (b)
6. (a) 7. (b) 8. (d) 9. (c) 10. (a)
11. (d) 12. (c) 13. (a) 14. (d) 15. (f)
16. (a) 17. (d) 18. (b)

Chapter 2
1. (c) 2. (b) 3. (a) 4. (a) 5. (c)
6. (d) 7. (b) 8. (d) 9. (a) 10. (d)
11. (b) 12. (a) 13. (c) 14. (a) 15. (c)
16. (d) 17. (b) 18. (a)

Chapter 3
1. (b) 2. (a) 3. (c) 4. (d) 5. (b)
6. (a) 7. (c) 8. (d) 9. (b) 10. (a)
11. (d) 12. (c) 13. (a) 14. (c)

Chapter 4
1. (c) 2. (b) 3. (a) 4. (d) 5. (b)
6. (a) 7. (c) 8. (d) 9. (a) 10. (c)
11. (c) 12. (b) 13. (d) 14. (a) 15. (c)
16. (b)

Chapter 5
1. (c) 2. (b) 3. (d) 4. (b) 5. (a)
6. (c) 7. (b) 8. (a) 9. (d) 10. (c)
11. (d) 12. (a) 13. (b) 14. (c) 15. (c)
16. (d) 17. (b) 18. (c) 19. (a) 20. (c)
21. (d)

Chapter 6
1. (c) 2. (b) 3. (a) 4. (b) 5. (d)
6. (a) 7. (d) 8. (c) 9. (b) 10. (a)
11. (d) 12. (d) 13. (b) 14. (c) 15. (b)

Chapter 7
1. (c) 2. (a) 3. (c) 4. (e) 5. (b)
6. (d) 7. (c) 8. (a) 9. (c) 10. (a)

11. (b) 12. (c) 13. (b) 14. (d) 15. (d)
16. (a) 17. (d) 18. (c)

Chapter 8
1. (c) 2. (d) 3. (a) 4. (b) 5. (c)
6. (c) 7. (b) 8. (a) 9. (d) 10. (b)
11. (a) 12. (c) 13. (b) 14. (d)

Chapter 9
1. (b) 2. (a) 3. (c) 4. (b) 5. (d)
6. (c) 7. (b) 8. (d) 9. (a) 10. (b)
11. (c) 12. (e) 13. (c) 14. (d) 15. (d)

Chapter 10
1. (c) 2. (d) 3. (c) 4. (b) 5. (d)
6. (a) 7. (c) 8. (a) 9. (g) 10. (c)

Chapter 11
1. (d) 2. (b) 3. (a) 4. (e) 5. (c)
6. (b) 7. (a) 8. (c) 9. (d) 10. (c)
11. (a) 12. (b) 13. (f) 14. (b) 15. (b)
16. (c)

Chapter 12
1. (b) 2. (c) 3. (d) 4. (d) 5. (c)
6. (a) 7. (b) 8. (c) 9. (b) 10. (e)
11. (c) 12. (b) 13. (c) 14. (c) 15. (a)

Chapter 13
1. (b) 2. (c) 3. (d) 4. (e) 5. (c)
6. (c) 7. (c) 8. (b) 9. (b) 10. (d)
11. (a) 12. (d) 13. (b) 14. (c) 15. (d)
16. (a)

Chapter 14
1. (b) 2. (f) 3. (a) 4. (b) 5. (c)
6. (c) 7. (d) 8. (d) 9. (e) 10. (a)
11. (b) 12. (c) 13. (e) 14. (b) 15. (b)
16. (c) 17. (a) 18. (d) 19. (b) 20. (c)

Answers to Selected Odd-Numbered Problems

Chapter 1
No problem set

Chapter 2
1. 63.66 V
3. Yes
5. 47.75 V
7. 346.3 V
9. 78.54 V
11. See Figure ANS–1.
13. (a) A sine wave with a positive peak at +0.7 V, a negative peak at −7.3 V, and a dc value of −3.3 V
    (b) A sine wave with a positive peak at +29.3 V, a negative peak at −0.7 V, and a dc value of +14.3 V
15. 15.02 V
17. $I_{L(min)} = 0$ A, $I_{L(max)} = 32.08$ mA
19. 10.7%, 6.33%
21. A decrease of 10 pF

Figure ANS–1

V_{in} 50.9 V, 0

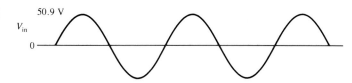

V_{AB} 0.7 V, 0, −50.2 V

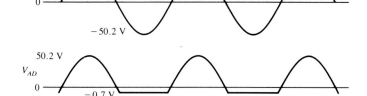

V_{AD} 50.2 V, 0, −0.7 V

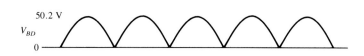

V_{BD} 50.2 V, 0

≈ 31.9 V

V_{CD} 0

23. 25.34 pF each
25. Increase
27. $V_{RRM} = 50$ V
29. 62.5 mΩ
31. (a) Functioning properly
    (b) Open diode
    (c) Functioning properly
    (d) Open diode
33. (a) Blown (open) fuse. Replace fuse.
    (b) Open transformer winding or connection. Check connections to transformer and, if they are OK, replace transformer.
    (c) Open transformer winding or connection. Check connections to transformer and, if they are OK, replace transformer.
    (d) Some primary windings in the transformer are shorted. Replace transformer.
    (e) Some secondary windings in the transformer are shorted. Replace transformer.
    (f) Capacitor C_1 open. Replace C_1.
    (g) Capacitor C_1 leaky. Replace C_1.
    (h) One of the diodes is open. Isolate and replace open diode.
    (i) C_2 is shorted, IC regulator is bad, fuse is blown, C_1 is shorted, transformer winding is open, or diode bridge is completely open (at least two diodes open).
35. See Figure ANS-2.
37. 0.32 or 3.125:1

CHAPTER 3

1. 4.865 mA
3. 125
5. $I_B = 26$ μA, $I_E = 1.3$ mA, $IC = 1.275$ mA
7. $I_B = 13.64$ μA, $I_C = 682$ μA, $V_C = 9.32$ V
9. $V_{CE} = 3.45$ V; Q point: $I_C = 15$ mA, $V_{CE} = 3.45$ V
11. 33.33
13. 0.5 mA, 3.33 μA, 4.03 V
15. 0.0031 mA
17. (a) Widen (b) Increase
19. -5 V
21. 10 mA
23. 4 V
25. The gate is insulated from the channel by an SiO$_2$ layer
27. 3 V
29. (a) Depletion (b) Enhancement
    (c) Zero bias (d) Depletion

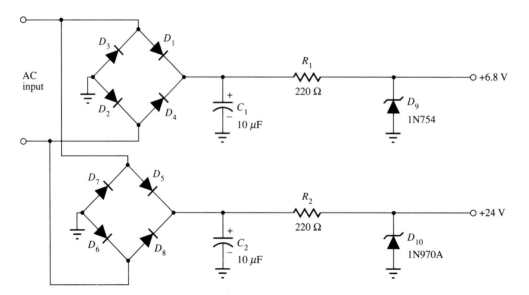

FIGURE ANS-2

31. 0.6

33. 1 ms

35. 2.4 kΩ

37. Positive probe on emitter, negative probe on base: very high resistance; positive probe on base, negative probe on emitter: very low resistance

39. (a) 36 (b) 124

41. Q_2 shorted collector-to-emitter, Q_4 shorted collector-to-emitter, R_1 open, R_6 open, Q_1 open, or Q_3 open

Chapter 4

1. 198.8

3. (a) 3.25 V (b) 2.55 V
     (c) 2.55 mA (d) $\cong$ 2.55 mA
     (e) 9.59 V (f) 7.04 V

5. $A_{v(\text{max})} = 92.3$, $A_{v(\text{min})} = 2.91$

7. 0.978

9. A_v is reduced slightly.

11. $R_{\text{in}} = 2.06\ \Omega$, $A_v = 583$, $A_i \cong 1$, $A_p = 583$

13. (a) 0.934 (b) 0.301

15. 30 dB, 31.62

17. (a) 7.71 mW (b) 53.56 mW

19. 10 V, 625 mA

21. 50.33 kHz

23. (a) 527.9 kHz (b) 758.7 kHz

25. Cutoff, 10 V

27.

DC Values								
V_{B1}	V_{B2}	V_{E1}	V_{E2}	V_{C1}	V_{C2}	V_{IN}	V_{OUT}
2.88 V	2.88 V	2.18 V	2.18 V	7.85 V	7.85 V	0 V	0 V

AC Values (rms)							
V_{b1}	V_{b2}	V_{e1}	V_{e2}	V_{c1}	V_{c2}	V_{in}	V_{out}
20.9 μV	1.96 mV	0 V	0 V	1.96 mV	0.592 V	20.9 μV	0.592 V

Chapter 5

1. *Practical op-amp:* High open-loop gain, high input impedance, low output impedance, large bandwidth, high CMRR.
Ideal op-amp: Infinite open-loop gain, infinite input impedance, zero output impedance, infinite bandwidth, infinite CMRR.

3. (a) Single-ended input; differential output
   (b) Single-ended input; single-ended output
   (c) Differential input; signal-ended output
   (d) Differential input; differential output

5. V_1: differential output voltage
     V_2: noninverting input voltage
     V_3: single-ended output voltage
     V_4: differential input voltage
     I_1: bias current

7. 8.1 μA

9. 107.96 dB

11. 0.3

13. 40 μs

15. $B = 0.0099$, $V_f = 49.5$ mV

17. (a) 11 (b) 101 (c) 47.81 (d) 23

19. (a) 1 (b) -1 (c) 22 (d) -10

21. (a) 0.45 mA (b) 0.45 mA
     (c) -9.9 V (d) -10

23. (a) $Z_{\text{in(VF)}} = 1.32 \times 10^{12}\ \Omega$
     $Z_{\text{out(VF)}} = 0.455$ mΩ
     (b) $Z_{\text{in(VF)}} = 5 \times 10^{11}\ \Omega$
     $Z_{\text{out(VF)}} = 0.6$ mΩ

(c) $Z_{in(VF)} = 40,000$ MΩ
$Z_{out(VF)} = 1.5$ mΩ

25. (a) 75 Ω placed in feedback path
    (b) 150 μV

27. 200 μV

29. (a) R_1 open or op-amp faulty
    (b) R_2 open
    (c) Nonzero output offset voltage; R_4 faulty or in need of adjustment

Chapter 6

1. 70 dB
3. 1.67 kΩ
5. (a) 79,603 (b) 56,569
   (c) 7960   (d) 80
7. (a) $-0.67°$ (b) $-2.69°$
   (c) $-5.71°$ (d) $-45°$
   (e) $-71.22°$ (f) $-84.29°$
9. (a) 0 dB/decade   (b) -20 dB/decade
   (c) -40 dB/decade (d) -60 dB/decade
11. 4.05 MHz
13. 21.14 MHz
15. Circuit (b) has smaller BW 97.5 kHz).
17. (a) 150° (b) 120° (c) 60°
    (d) 0°   (e) $-30°$
19. (a) Unstable (b) Stable
    (c) Marginally stable
21. 25 Hz

Chapter 7

1. 24 V, with distortion
3. $V_{UTP} = +2.77$ V, $V_{LTP} = -2.77$ V

5. See Figure ANS–3.

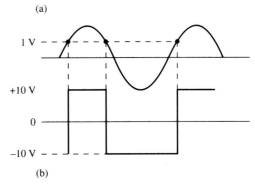

Figure ANS–3

7. $+8.57$ V and -0.968 V
9. (a) -2.5 V
   (b) -3.52 V
11. 110 kΩ
13. $V_{OUT} = -3.57$ V, $I_f = 0.357$ mA
15. -4.46 mV/μs
17. 1 mA
19. See Figure ANS–4.
21. See Figure ANS–5.
23. The output is not correct because the output should also be high when the input goes below

Figure ANS–4

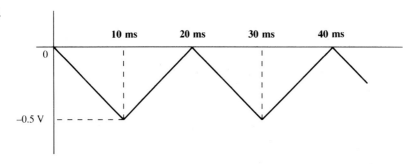

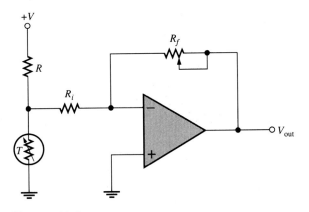

FIGURE ANS–5

+2 V. Possible faults: Op-amp2 bad, diode D_2 open, noninverting input (+) of op-amp2 not properly set at +2 V, or V_{in} is not reaching inverting input.

25. Output is not correct. R_2 is open.

CHAPTER 8

1. (a) Band pass (b) High pass
   (c) Low pass   (d) Band stop
3. 48.23 kHz, No
5. 700 Hz, 5.04
7. (a) 1, not Butterworth
   (b) 1.44, approximate Butterworth
   (c) 1st stage: 1.67
       2nd stage: 1.67
       Not Butterworth
9. (a) Chebyshev (b) Butterworth
   (c) Bessel   (d) Butterworth
11. 189.8 Hz
13. Add another identical stage and change R_1/R_2 ratio to 0.068 for first stage, 0.586 for second stage, and 1.482 for third stage.
15. Exchange positions of resistors and capacitors.
17. (a) Decrease R_1 and R_2 or C_1 and C_2.
    (b) Increase R_3 or decrease R_4.
19. (a) f_0 = 4949 Hz, BW = 3848 Hz
    (b) f_0 = 448.6 Hz, BW = 96.5 Hz
    (c) f_0 = 15.92 kHz, BW = 837.9 Hz
21. Sum the low-pass and high-pass outputs with a two-input adder.

CHAPTER 9

1. An oscillator requires no input (other than dc power).
3. 1/15
5. 733.33 mV
7. 50 kΩ
9. 7.5 V, 3.94
11. 136 kΩ, 1693 Hz
13. Change R_1 to 3.54 kΩ
15. R_4 = 65.8 kΩ
17. 3.33 V, 6.67 V
19. 0.0076 μF
21. 13.6 mS
23. 0.01 μF, 9.1 kΩ

CHAPTER 10

1. 0.00417%
3. 1.01%
5. See Figure ANS–6.

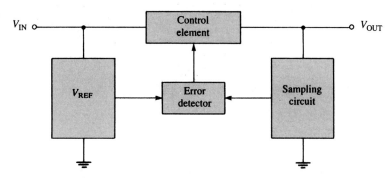

FIGURE ANS–6

7. 8.5 V
9. 9.57 V
11. 500 mA
13. 10 mA
15. $I_{L(max)} = 250$ mA, $P_{R1} = 6.25$ W
17. 40%
19. Increases
21. 14.25 V
23. 1.3 mA
25. 2.8 Ω
27. $R_{lim} = 0.35$ Ω
29. See Figure ANS-7.

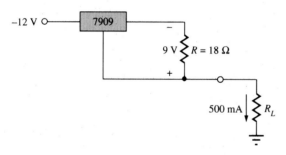

FIGURE ANS-7

CHAPTER 11
1. $A_{v(1)} = A_{v(2)} = 101$
3. 1.005 V
5. 9.1
7. Change R_G to 1.8 kΩ.
9. 225
11. Change the 18 kΩ resistor to 270 kΩ.
13. Connect output (pin 22) directly to pin 23, and connect pin 38 directly to pin 40 to make $R_F = 0$.
15. 500 μA, 5 V
17. $A_v \cong 17$
19. See Figure ANS-8.

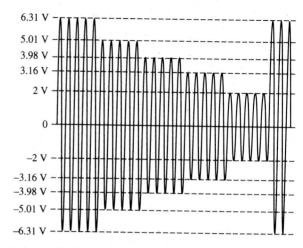

FIGURE ANS-8

21. See Figure ANS-9.

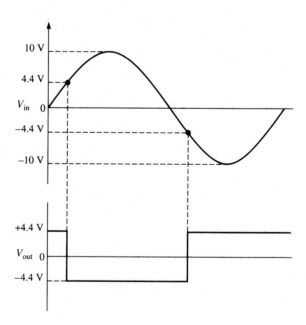

FIGURE ANS-9

23. (a) −0.301
(b) 0.301
(c) 1.699
(d) 2.114

25. The output of a log amplifier is limited to 0.7 V because of the transistor's pn junction.

27. −157 mV

29. $V_{in(max)} = -147$ mV, $V_{in(min)} = -89.2$ mV; the 1 V input peak is reduced 85% whereas the 100 mV input peak is reduced only 10%.

31. $V_{out(1)} = -67.2$ mV, $V_{out(2)} = +17.3$ mV, $V_{out(3)} = -49.9$ mV, $V_{out(4)} = +49.9$ mV, $V_{out(5)} = -6.8$ mV, $V_{out(6)} = 2$ V; Yes

33. Probe 1: ≈ 0 V
Probe 2: ≈ 0 V
Probe 3: 20 mV @ 1 kHz

Chapter 12

1. See Figure ANS–10.

3. 1135 kHz

5. RF: 91.2 MHz, IF: 10.7 MHz

7. 739 μA

9. −8.12 V

11. (a) 0.28 V
(b) 1.024 V
(c) 2.07 V
(d) 2.49 V

13. $f_{diff} = 8$ kHz, $f_{sum} = 10$ kHz

15. $f_{diff} = 800$ kHz, $f_{sum} = 2.8$ MHz, $f_1 = 1.8$ MHz

17. $f_c = 850$ kHz, $f_m = 3$ kHz

19. $V_{out} = 15$ mV cos [(1100 kHz)$2\pi t$] − 15 mV cos [(5500 kHz)$2\pi t$]

21. See Figure ANS–11.

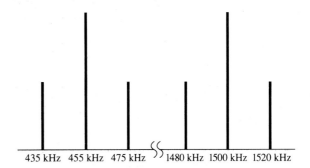

Figure ANS–11

23. See Figure ANS–12.

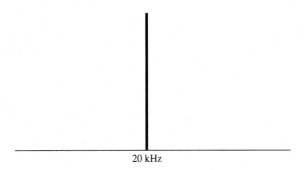

Figure ANS–12

25. The IF amplifier has a 450 kHz–460 kHz bandpass. The audio/power amplifiers have a 10 Hz–5 kHz bandpass.

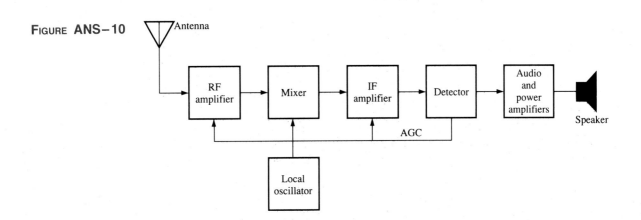

Figure ANS–10

27. The modulating input signal is applied to the control terminal of the VCO. As the input signal amplitude varies, the output frequency of the VCO will vary proportionally.

29. Varactor

31. (a) 10 MHz (b) 48.3 mV

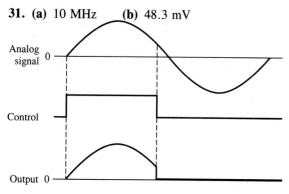

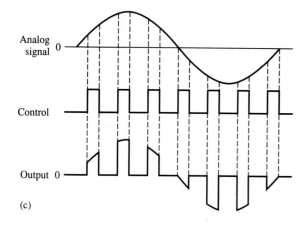

FIGURE ANS–13

33. 1005 Hz

35. $f_o = 233$ kHz, $f_{lock} = \pm 103.6$ kHz, $f_{cap} \cong \pm 4.56$ kHz

CHAPTER 13

1. See Figure ANS–13.

3. See Figure ANS–14.

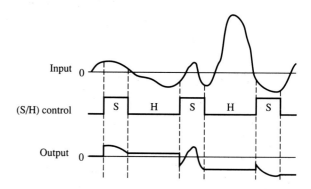

FIGURE ANS–14

5. (a) 1 (b) 11

7. See Figure ANS–15.

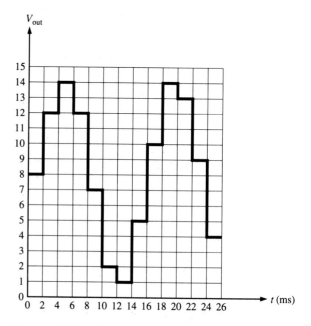

FIGURE ANS–15

9. 5 kΩ, 2.5 kΩ, 1.25 kΩ

11.

D_3	D_2	D_1	D_0	V_{out}
0	0	0	0	0 V
0	0	1	1	-0.50 V $+ (-0.25$ V$) = -0.75$ V
1	0	0	0	-2.00 V
1	1	1	1	-2.00 V $+ (-1.00$ V$) + (-0.50$ V$) + (-0.25$ V$) = -3.75$ V
1	1	1	0	-2.00 V $+ (-1.00$ V$) + (-0.50$ V$) = -3.50$ V
0	1	0	0	-1.00 V
0	0	0	0	0 V
0	0	0	1	-0.25 V
1	0	1	1	-2.00 V $+ (-0.50$ V$) + (-0.25$ V$) = -2.75$ V
1	1	1	0	-2.00 V $+ (-1.00$ V$) + (-0.50$ V$) = -3.50$ V
1	1	0	1	-2.00 V $+ (-1.00$ V$) + (-0.25$ V$) = -3.25$ V
0	1	0	0	-1.00 V
1	0	1	1	-2.00 V $+ (-0.50$ V$) + (-0.25$ V$) = -2.75$ V
0	0	0	1	-0.25 V
0	0	1	1	-0.50 V $+ (-0.25$ V$) = -0.75$ V

13. (a) 16 (b) 32 (c) 256 (d) 65,536

15. 1 mV

17.

Sampling Time (μS)	Binary Output
0	000
10	000
20	001
30	100
40	101
50	101
60	100
70	011
80	010
90	001
100	001
110	011
120	110
130	111
140	111
150	111
160	111
170	111
180	111
190	110
200	100

19. f_{out} increases.

21. 691 pF (Use standard 680 pF).

23. $f_{out(min)} = 2.84$ kHz, $f_{out(max)} = 8.44$ kHz

25. The D_0 (LSB) is stuck high and the D_2 is stuck low.

Chapter 14

1. 5 V

3. 189.84°

5. $\overline{\text{Enable } M}$ and $\overline{\text{Enable } L}$ enable the tri-state output buffers.

7. Thermocouple C

9. -4.36 V

11. 590

13. 540 Ω

15. The effects of the wire resistances are cancelled in the 3-wire bridge.

17. 31

19. Pressure is measured by a strain gage bonded to a flexible diaphragm.

21. See Figure ANS–16.

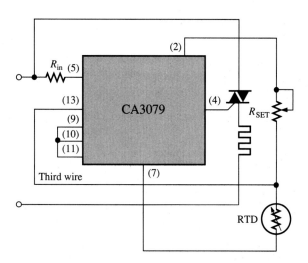

FIGURE ANS–16

Glossary

Acquisition time In an analog switch, the time required for the device to reach its final value when switched from hold to sample.

Active filter A frequency-selective circuit consisting of active devices such as transistors or op-amps coupled with reactive components.

A/D conversion A process whereby information in analog form is converted into digital form.

AGC Automatic gain control.

Alpha (α) The ratio of collector current to emitter current in a bipolar junction transistor.

Amplification The process of increasing the power, voltage, or current by electronic means.

Amplifier An electronic circuit having the capability to amplify power, voltage, or current.

Amplitude modulation (AM) A communication method in which a lower frequency signal modulates (varies) the amplitude of a higher frequency signal (carrier).

Analog Characterized by a linear process in which a variable takes on a continuous set of values.

Anode The more positive terminal of a diode and certain other electronic devices.

Aperture jitter In an analog switch, the uncertainty in the aperture time.

Aperture time In an analog switch, the time to fully open after being switched from sample to hold.

Astable Characterized by having no stable states; a type of oscillator.

Atom The smallest particle of an element that possesses the unique characteristics of that element.

Atomic number The number of electrons in a neutral atom.

Atomic weight The number of protons and neutrons in the nucleus of an atom.

Attenuation The reduction in the level of power, current, or voltage.

Audio Related to the frequency range of sound waves that can be heard by the human ear, generally considered to be in the 200 Hz to 20 kHz range.

Balanced modulation A form of amplitude modulation in which the carrier is suppressed; sometimes known as suppressed-carrier modulation.

Band-pass filter A type of filter that passes a range of frequencies lying between a certain lower frequency and a certain higher frequency.

Band-stop filter A type of filter that blocks or rejects a range of frequencies lying between a certain lower frequency and a certain higher frequency.

Bandwidth The characteristic of certain electronic circuits that specifies the usable range of frequencies that pass from input to output.

Barrier potential The inherent voltage across the depletion layer of a pn junction.

Base One of the semiconductor regions in a bipolar junction transistor. The base is very narrow compared to the other regions.

BASIC A computer programming language: Beginner's All-purpose Symbolic Instruction Code.

Bessel A type of filter response having a linear phase characteristic and less than 20 dB/decade/pole roll-off.

Beta (β) The ratio of collector current to base current in a bipolar junction transistor; current gain.

Bias The application of a dc voltage to a transistor or other device to produce a desired mode of operation.

Bipolar Characterized by both free electrons and holes as current carriers.

Bipolar junction transistor (BJT) A transistor constructed with three doped semiconductor regions separated by two pn junctions.

Bode plot An idealized graph of the gain in dB versus frequency used to graphically illustrate the response of an amplifier or filter.

Bounding The process of limiting the output range of an amplifier or other circuit.

Bridge rectifier A type of full-wave rectifier using four diodes arranged in a bridge configuration.

Butterworth A type of filter response characterized by flatness in the pass band and a 20 dB/decade/pole roll-off.

Carrier The high frequency (RF) signal that carries modulated information in AM, FM, or other systems.

Cascade An arrangement of circuits in which the output of one circuit becomes the input to the next.

Cathode The more negative terminal of a diode and certain other electronic devices.

Center tap A connection at the midpoint of the secondary of a transformer.

Center-tapped rectifier A type of full-wave rectifier using a center-tapped transformer and two diodes.

Channel The conductive path between the drain and source in an FET.

Characteristic curve A graph showing current vs. voltage in a diode or transistor.

Chebyshev A type of filter response characterized by ripples in the pass band and a greater than 20 dB/decade/pole roll-off.

Clamper A circuit using a diode and a capacitor which adds a dc level to an ac voltage; a dc restorer.

Class A A category of amplifier circuit that conducts for the entire input cycle and produces an output signal that is a replica of the input signal in terms of its waveshape.

Class B A category of amplifier circuit that conducts for half of the input cycle.

Class C A category of amplifier circuit that conducts for a very small portion of the input cycle.

Clipper See Limiter.

Closed-loop An op-amp connection in which the output is connected back to the input through a feedback circuit.

Closed-loop gain The overall voltage gain with external feedback.

Coherent light Light having only one wavelength.

Cold junction A reference thermocouple held at a fixed temperature and used for compensation in thermocouple circuits.

Collector One of the three semiconductor regions of a BJT.

Common-base A type of BJT amplifier configuration in which the base is the common (grounded) terminal.

Common-collector A type of BJT amplifier configuration in which the collector is the common (grounded) terminal.

Common-drain An FET amplifier configuration in which the drain is the grounded terminal.

Common-emitter A type of BJT amplifier configuration in which the emitter is the common (grounded) terminal.

Common-gate An FET amplifier configuration in which the gate is the grounded terminal.

Common mode A condition characterized by the presence of the same signal on both op-amp inputs.

Common-mode rejection ratio (CMRR) The ratio of open-loop gain to common-mode gain; a measure of an op-amp's ability to reject common-mode signals.

Common-source An FET amplifier configuration in which the source is the grounded terminal.

Comparator A circuit which compares two input voltages and produces an output in either of two states indicating the greater or less than relationship of the inputs.

Compensation The process of modifying the roll-off rate of an amplifier to ensure stability.

Complementary pair Two transistors, one npn and one pnp, having matched characteristics.

Conduction electron A free electron.

Conductor A material that conducts electrical current very well.

Covalent Related to the bonding of two or more atoms by the interaction of their valence electrons.

Critical frequency The frequency at which the response of an amplifier or filter is 3 dB less than at midrange.

Crossover distortion Distortion in the output of a class B push-pull amplifier at the point where each transistor changes from the cutoff state to the *on* state.

GLOSSARY

Crystal The pattern or arrangement of atoms forming a solid material; a quartz device that operates on the piezoelectric effect and exhibits very stable resonant properties.

Current The rate of flow of electrons.

Current gain The ratio of output current to input current.

Cutoff The nonconducting state of a transistor.

Cutoff frequency Another term for critical frequency.

Damping factor A filter characteristic that determines the type of response.

Dark current The amount of thermally generated reverse current in a photodiode in the absence of light.

Darlington pair A configuration of two transistors in which the collectors are connected and the emitter of the first drives the base of the second to achieve beta multiplication.

Decade A ten times increase or decrease in the value of a quantity such as frequency.

Decibel (dB) The unit of the logarithmic expression of a ratio, such as power or voltage.

Demodulation The process in which the information signal is recovered from the IF carrier signal; the reverse of modulation.

Depletion In a MOSFET, the process of removing or depleting the channel of charge carriers and thus decreasing the channel conductivity.

Depletion layer The area near a pn junction that has no majority carriers.

Derivative The instantaneous rate of change of a function, determined mathematically.

Diac A two-terminal four-layer semiconductor device (thyristor) that can conduct current in either direction when properly activated.

Differential amplifier (diff-amp) An amplifier that produces an output voltage proportional to the difference of the two input voltages.

Differentiator A circuit that produces an output which approximates the instantaneous rate of change of the input function.

Digital Characterized by a process in which a variable takes on either of two values; related to a discrete set of values.

Diode A two-terminal electronic device that permits current in only one direction.

Diode drop The voltage across the diode when it is forward-biased. Approximately the same as the barrier potential.

Discrete device An individual electrical or electronic component that must be used in combination with other components to form a complete functional circuit.

Discriminator A type of FM modulator.

Doping The process of imparting impurities to an intrinsic semiconductor material in order to control its conduction characteristics.

Drain One of the three terminals of an FET.

Droop In an analog switch, the change in the sampled value during the hold interval.

Electrical isolation The state in which two or more electrical or electronic circuits are not directly connected together by a current path and in which energy is transferred by magnetic or optical methods.

Electroluminescence The process of releasing light energy by the recombination of electrons in a semiconductor.

Electron The basic particle of negative electrical charge.

Electron-hole pair The conduction electron and the hole created when the electron leaves the valence band.

Emitter One of the three semiconductor regions of a BJT.

Emitter-follower A popular term for a common-emitter amplifier.

Energy The ability to do work.

Enhancement In a MOSFET, the process of creating a channel or increasing the conductivity of the channel by the addition of charge carriers.

Feedback The process of returning a portion of a circuit's output back to the input in such a way as to oppose a change in the output.

Feedforward A method of frequency compensation in op-amp circuits.

Feedthrough In an analog switch, the component of the output voltage which follows the input voltage after the switch opens.

Field-Effect Transistor (FET) A voltage-controlled device in which the voltage at the gate terminal controls the amount of current through the device.

Filter A type of circuit that passes or blocks certain frequencies to the exclusion of all others.

Flash A method of A/D conversion.

Fold-back current limiting A method of current limiting in voltage regulators.

Forced-commutation A method of turning off an SCR.

Forward bias The condition in which a pn junction conducts current.

Free electron An electron that has acquired enough energy to break away from the valence band of the parent atom; also called a conduction electron.

Frequency modulation (FM) A communication method in which a lower frequency intelligence carrying signal modulates (varies) the frequency of a higher frequency signal.

Full-wave rectifier A circuit that converts an ac sine wave input voltage into a pulsating dc voltage with two pulses occurring for each input cycle.

Fuse A protective device that opens when the current exceeds a rated limit.

Gage factor (GF) The ratio of the fractional change in resistance to the fractional change in length along the axis of the gage.

Gain The amount by which an electrical signal is increased or amplified.

Gain-bandwidth product A characteristic of amplifiers whereby the product of the gain and the bandwidth is always constant.

Gate One of the three terminals of an FET.

Germanium A semiconductor material.

Half-wave rectifier A circuit that converts an ac sine wave input voltage into a pulsating dc voltage with one pulse occurring for each input cycle.

Harmonics The frequencies contained in a composite waveform which are integer multiples of the repetition frequency (fundamental).

High-pass filter A type of filter that passes frequencies above a certain frequency while rejecting lower frequencies.

Hole The absence of an electron in the valence band of an atom.

Hysteresis Characteristic of a circuit in which two different trigger levels create an offset or lag in the switching action.

Index of refraction A property of light-conducting materials that specifies how much a light ray will bend when passing from one material to another.

Infrared Light that has a range of wavelengths greater than visible light.

Input The terminal of a circuit to which an electrical signal is first applied.

Instrumentation Related to an arrangement of instruments for the purpose of monitoring or measuring certain quantities.

Insulator A material that does not conduct current.

Integral The area under the curve of a function, determined mathematically.

Integrated circuit (IC) A type of circuit in which all the components are constructed on a single tiny chip of silicon.

Integrator A circuit that produces an output which approximates the area under the curve of the input function.

Interfacing The process of making the output of one type of circuit compatible with the input of another so that they can operate properly together.

Intrinsic semiconductor A pure or natural material with relatively few electrons.

Inversion The conversion of a quantity to its opposite value.

Inverting amplifier An op-amp closed-loop configuration in which the input signal is applied to the inverting input.

Ion An atom that has gained or lost a valence electron resulting in a net positive or negative charge.

Ionization The removal or addition of an electron from or to a neutral atom so that the resulting atom (called an ion) has a net positive or negative charge.

Junction field-effect transistor (JFET) One of two major types of field-effect transistor that operates with a reversed-biased junction to control current in a channel.

Large-signal A signal that operates an amplifier over a significant portion of its load line.

Laser *L*ight *A*mplification by *S*timulated *E*mission of *R*adiation.

Light-Activated Silicon-Controlled Rectifier (LASCR) A four-layer semiconductor device (thyristor) that conducts current in one direction when activated by a suf-

sufficient amount of light and continues to conduct until the current falls below a specified value.

Light-emitting diode (LED) A type of diode that emits light when there is forward current.

Limiter A diode circuit that clips off or removes part of a waveform above and/or below a specified level.

Linear Characterized by a straight-line relationship.

Line regulation The percentage change in output voltage for a given change in line (input) voltage.

Loading The amount of current drawn from the output of a circuit through a load impedance.

Load line A straight line on the characteristic curve of an amplifier that represents the operating range of the amplifier's voltages and currents.

Load regulation The percentage change in output voltage for a given change in load current.

Logarithm An exponent; the logarithm of a quantity is the exponent or power to which a given number called the base must be raised in order to equal the quantity.

Low-pass filter A type of filter that passes frequencies below a certain frequency while rejecting higher frequencies.

Lumen Unit of light measurement.

Majority carrier The most numerous charge carrier in a semiconductor material (either free electrons or holes).

Mean value Average value.

Midrange The frequency range of an amplifier lying between the lower and upper critical frequencies.

Minority carrier The least numerous charge carrier in a semiconductor material (either free electrons or holes).

Mixer A device for down-converting frequencies in a receiver system.

Modulation The process in which a signal containing information is used to modify the amplitude, frequency, or phase of a much higher-frequency signal called the carrier.

Monochromatic Related to light of a single frequency; one color.

Monostable Characterized by having one stable state.

Monotonicity In relation to D/A converters, the presence of all steps in the output when sequenced over the entire range of input bits.

MOSFET Metal oxide semiconductor field-effect transistor; one of two major types of FET.

Multiplier A linear device that produces an output voltage proportional to the product of two input voltages.

Multistage Characterized by having more than one stage; a cascaded arrangement of two or more amplifiers.

Multivibrator A type of circuit that can operate as an oscillator or as a one-shot.

Natural logarithm The exponent to which the base e ($e = 2.71828$) must be raised in order to equal a given quantity.

Negative feedback The process of returning a portion of the output signal to the input of an amplifier such that it is out-of-phase with the input signal.

Neutron An uncharged particle found in the nucleus of an atom.

Noise An unwanted signal.

Noninverting amplifier An op-amp closed-loop configuration in which the input signal is applied to the noninverting input.

Nonmonotonicity In relation to D/A converters, a step reversal or missing step in the output when sequenced over the entire range of input bits.

Nucleus The central part of an atom containing protons and neutrons.

Nyquist rate In sampling theory, the minimum rate at which an analog voltage can be sampled for A/D conversion and equal to twice the maximum frequency component of the input signal.

Octave A two times increase or decrease in the value of a quantity such as frequency.

One-shot A monostable multivibrator.

Open-loop A condition in which an op-amp has no feedback.

Open-loop gain The gain of an op-amp without feedback.

Operational amplifier (op-amp) A type of amplifier that has very high voltage gain, very high input impedance, very low output impedance, and good rejection of common-mode signals.

Orbit The path an electron takes as it circles around the nucleus of an atom.

Oscillator An electronic circuit based on positive feedback that produces a time-varying output signal without an external input signal.

Output The terminal of a circuit from which the final voltage is obtained.

Pentavalent atom An atom with five valence electrons.

Phase The relative angular displacement of a time-varying function relative to a reference.

Phase-locked loop (PLL) A device for locking onto and tracking the frequency of an incoming signal.

Phase margin The difference between the total phase shift through an amplifier and 180 degrees. The additional amount of phase shift that can be allowed before instability occurs.

Photodiode A diode in which the reverse current varies directly with the amount of light.

Photon A particle of light energy.

Phototransistor A transistor in which base current is produced when light strikes the photosensitive semiconductor base region.

Pinch-off voltage The value of the drain-to-source voltage of an FET at which the drain current becomes constant when the gate-to-source voltage is zero.

PN junction The boundary between two different types of semiconductor materials.

Pole A network containing one resistor and one capacitor that contributes 20 dB/decade to a filter's roll-off.

Positive feedback The return of a portion of the output signal to the input such that it sustains the output.

Power gain The ratio of output power to input power; the product of voltage gain and current gain.

Power supply The circuit that supplies the proper dc voltage and current to operate a system.

Programmable Unijunction Transistor (PUT) A type of three-terminal thyristor more like an SCR than a UJT that is triggered into conduction when the voltage at the anode exceeds the voltage at the gate.

Proton The basic particle of positive charge.

Push-Pull A type of class B amplifier with two transistors in which one transistor conducts for one half-cycle and the other conducts for the other half-cycle.

Q point The dc operating (bias) point of an amplifier specified by voltage and current values.

Quality factor The ratio of a band-pass filter's center frequency to its bandwidth.

Quantization The determination of a value for an analog quantity.

Quantization error The error resulting from the change in the analog voltage during the A/D conversion time.

Radiation The process of emitting electromagnetic or light energy.

Radio frequency interference (RFI) High frequencies produced when high values of current and voltage are rapidly switched on and off.

Recombination The process of a free (conduction band) electron falling into a hole in the valence band of an atom.

Rectifier An electronic circuit that converts ac into pulsating dc; one part of a power supply.

Regulator An electronic device or circuit that maintains an essentially constant output voltage for a range of input voltage or load values; one part of a power supply.

Resistance temperature detector (RTD) A type of temperature transducer in which resistance is directly proportional to temperature.

Resolution In relation to D/A or A/D converters, the number of bits involved in the conversion; also, for D/A converters, the reciprocal of the maximum number of discrete steps in the output.

Resolver A type of synchro.

Resolver-to-digital converter (RDC) An electronic circuit that converts resolver voltages to a digital format which represents the angular position of the rotor shaft.

Reverse bias The condition in which a pn junction blocks current.

RF Radio frequency.

Ripple factor A measure of effectiveness of a power supply filter in reducing the ripple voltage.

Ripple voltage The small variation in the dc output voltage of a filtered rectifier caused by the charging and discharging of the filter capacitor.

Roll-off The decrease in the gain of an amplifier above or below the critical frequencies.

Root mean square (RMS) A measure of the amplitude of an ac signal which equals the value of a dc voltage required to produce the same amount of heat in the same resistance as the ac voltage.

Glossary

Rotor The part of a synchro that is attached to the shaft and rotates. The rotor winding is located on the rotor.

Sample The process of taking the value of a quantity at a specific point in time.

Saturation The state of a BJT in which the collector current has reached a maximum and is independent of the base current.

Schematic A symbolized diagram representing an electrical or electronic circuit.

Schmitt trigger A comparator with hysteresis; an electronic device that changes its output state at two different input-voltage levels.

Schottky diode A diode using only majority carriers and intended for high-frequency operation.

Semiconductor A material that lies between conductors and insulators in its conductive properties.

Shockley diode The type of two-terminal thyristor that conducts current when the anode-to-cathode voltage reaches a specified "breakdown" value.

Signal compression The process of scaling down the amplitude of a signal voltage.

Silicon A semiconductor material used in diodes and transistors.

Silicon-controlled rectifier (SCR) A type of three-terminal thyristor that conducts current when triggered on by a voltage at the single gate terminal and remains on until the anode current falls below a specified value.

Silicon-controlled switch (SCS) A type of four-terminal thyristor that has two gate terminals that are used to trigger the device on and off.

Slew rate The rate of change of the output voltage of an op-amp in response to a step input.

Source One of the three terminals of an FET.

Source-follower The common-drain amplifier.

Spectral Pertaining to a range of frequencies.

Stability A measure of how well an amplifier maintains its design values (Q-point, gain, etc.) over changes in beta and temperature; a condition in which an amplifier circuit does not oscillate.

Stage One of the amplifier circuits in a multistage configuration.

Standoff ratio The characteristic of a UJT that determines its turn-on point.

Stator The part of a synchro that is fixed. The stator windings are located on the stator.

Step A fast voltage transition from one level to another.

Strain The expansion or compression of a material caused by stress forces acting on it.

Strain gage A transducer formed by a resistive material in which a lengthening or shortening due to stress produces a proportional change in resistance.

Successive approximation A method of A/D conversion.

Switch An electrical or electronic device for opening and closing a current path.

Synchro An electromechanical transducer used for shaft angle measurement and control.

Synchro-to-digital converter (SDC) An electronic circuit that converts synchro voltages to a digital format which represents the angular position of the rotor shaft.

Terminal The external contact point on an electrical or electronic device.

Thermal overload A condition in a rectifier where the internal power dissipation of the circuit exceeds a certain maximum due to excessive current.

Thermistor A temperature-sensitive resistor with a negative temperature coefficient; a type of temperature transducer in which resistance is inversely proportional to temperature.

Thermocouple A type of temperature transducer formed by the junction of two dissimilar metals which produces a voltage proportional to temperature.

Thyristor A class of four-layer (pnpn) semiconductor devices.

Transconductance The ratio of a change in drain current for a change in gate-to-source voltage in an FET.

Transducer A device that converts a physical parameter into an electrical quantity.

Transistor A semiconductor device used for amplification and switching applications.

Triac A three-terminal thyristor that can conduct current in either direction when properly activated.

Trigger The activating input of some electronic devices and circuits.

Trim To precisely adjust or fine tune a value.

Trivalent atom An atom with three valance electrons.

Troubleshooting The process and technique of identifying and locating faults in an electronic circuit or system.

Tunnel diode A diode exhibiting a negative resistance characteristic.

Unijunction Transistor (UJT) A three-terminal single pn junction device that exhibits a negative resistance characteristic.

Valance Related to the outer shell or orbit of an atom.

Varactor A variable capacitance diode.

Voltage-controlled oscillator (VCO) An oscillator for which the output frequency is dependent on a controlling input voltage.

Voltage follower A closed-loop, noninverting op-amp with a voltage gain of one.

Voltage gain The ratio of output voltage to input voltage.

Voltage regulation The process of maintaining an essentially constant output voltage over variations in input voltage or load.

Wavelength The distance in space occupied by one cycle of an electromagnetic or light wave.

Zener diode A diode designed for limiting the voltage across its terminals in reverse bias.

Index

Adder, 669
AGC, 556
Alpha, 101
Amplification, 109
Amplifier
  antilog, 526
  audio, 556, 558, 580
  averaging, 335
  capacitively-coupled, 185
  Class A, 187–90
  Class B, 191–96
  Class C, 197–200
  common-base, 174
  common-collector, 170–73
  common-drain, 180–82
  common-emitter, 164–70
  common-gate, 182
  common-source, 177–80
  differential, 226
  direct-coupled, 234
  intermediate frequency (IF), 556, 558, 577
  instrumentation, 502
  isolation, 508
  large-signal, 187
  log, 522
  multistage, 184, 289
  operational transconductance, 514
  power, 191
  push-pull, 191
  radio-frequency (RF), 197, 555, 558
  sample-and-hold, 623
  small-signal, 187
Amplitude modulation (AM), 163, 207, 519, 554, 566, 570
Analog, 627
Analog divider, 533
Analog multiplier, 527
Analog switch, 618
Analog-to-digital converter, 331, 352, 642
Anode, 16

Antenna, 555
Astable multivibrator, 432
Atom, 2
Atomic
  bond, 6
  number, 3
  weight, 3
Attenuation, 187, 414
Avalanche effect, 21, 54
Average value, 35, 37

Balanced modulation, 568
Bandwidth, 224, 243, 283, 290
Barrier potential, 14, 24, 35
Base, 98
Beta, 101, 114
Bias
  BJT, 99
  drain-feedback, 130
  forward, 16
  forward-reverse, 99
  midpoint, 187
  reverse, 19
  self, 127
  voltage-divider, 103, 130
  zero, 129
BIFET, 257
Bipolar, 98
Bipolar junction transistor (BJT), 98–117
Bode plot, 283
Bounded comparator, 326
Breakdown
  avalanche, 21, 54
  reverse, 22
  zener, 54
Breakover potential, 135
Bridge rectifier, 41
Bulk resistance, 26
Bypass capacitor, 164

Capacitance
  depletion layer, 20
  input, 123
  transistor, 241
Capacitive-coupling, 185
Capacitor
  bypass, 164
  compensating, 304
  coupling, 185
Carrier, 554, 557, 568, 569, 570
Cathode, 16
Center-tap, 38
Channel, 117
Clamper, 52
Clapp oscillator, 202
Clipper, 48
Closed-loop gain, 248, 282, 290
CMRR, 230, 239
Collector, 98
Collector characteristic curve, 106
Colpitts oscillator, 201
Common mode, 229, 231, 239
Common-mode rejection ratio (CMRR), 230, 232, 239
Comparator, 320
Compensation, 254, 298, 302
Conduction, 7
Conduction band, 7
Conduction electron, 7
Conductor, 10
Constant current source, 344
Covalent bond, 6
Critical frequency, 378
Crossover distortion, 192
Crystal, 203
Crystal oscillator, 203
Current, 9, 17, 20, 101
Current gain, 109, 171
Current limiting, 484
Current regulator, 484
Current-to-voltage converter, 345
Curve tracer, 142
Cutoff, 107, 113, 121
Cutoff frequency, 283

Damping factor, 376
Dark current, 67
Darlington, 173
Data sheet, 68, 235
dB, 184
DC restorer, 53
Decade, 283

Decibel (dB), 184
Deemphasis network, 558
Demodulation, 576, 584, 594
D-MOSFET, 177
Depletion, 14, 119, 124
Depletion-layer capacitance, 20
Detector, 556
Diac, 135
Differentiator, 341
Differential amplifier, 226
Digital-to-analog converter, 337, 632
Diode, 21–26
  light-emitting, 65
  photo, 66
  rectifier, 34
  varactor, 62–65
  zener, 54–62
Discriminator, 558
Divide circuit, 564
Doping, 12, 98
Drain, 117

Efficiency, 190, 195, 199
Electrocardiograph, 501, 525
Electroluminescence, 65
Electron, 2
Electron current, 8
Electron-hole pair, 8
Electron shell, 4
Electrostatic discharge (ESD), 127
Emitter, 98
Emitter-follower, 171
E-MOSFET, 178
Energy band, 7
Enhancement, 125, 126
Equivalent circuit, 56

Feedback, 242, 258, 388, 393, 415, 419
Feedforward compensation, 305
Field-effect transistor (FET), 176
Filter
  active, 370
  band-pass, 374, 386
  band-stop, 375, 393
  Bessel, 376
  Butterworth, 376
  capacitor-input, 43
  Chebyshev, 376
  high-pass, 373, 383
  LC, 46
  low-pass, 372, 379

INDEX

Filter, *continued*
   multiple-feedback, 388, 393
   π-type, 47
   rectifier, 43, 72
   Sallen-Key, 380, 385
   state variable, 390, 393
   T-type, 47
Floating, 137, 478
Fold-back current limiting, 463
Forward bias, 16
Free electron, 5
Frequency
   center, 423
   critical, 240, 282, 378
   difference, 567
   lower-side, 568
   resonant, 418
   sum, 567
   unity gain, 283
   upper-side, 568
Frequency measurement, 395
Frequency modulation (FM), 557, 581
Frequency spectra, 568, 576
Frequency-to-voltage converter, 663
Full-wave rectifier, 37–43
Function generator, 413, 426, 443

Gain, 109, 142, 164, 169, 171, 178, 180, 182, 184, 186, 201, 224, 231, 243, 245, 282, 511, 514
Gain-bandwidth product, 291
Gate, 117
Germanium, 5, 8

Half-wave rectifier, 34–37
Hartley oscillator, 202
Helium, 3
Heterodyne, 207
Holding current, 133
Hole, 8
Hole current, 9
Hydrogen, 3
Hysteresis, 324

Impedance
   input, 237, 249, 253
   Miller, 253
   output, 250
Input bias current, 237, 255, 256
Input offset current, 238
Input offset voltage, 235, 257
Input regulation, 58

Input resistance, 103, 123, 167, 171, 179, 181, 182
Insulator, 10
Integrator, 338
Intrinsic, 6, 10
Ion, 5
Ionization, 5

JFET, 117–24, 119, 127
Junction, 14

Large signal, 187
LC filter, 46
Leakage current, 142
LED, 65
Limiter, 558
Line regulation, 58, 456
Loading, 186
Load line, 108, 111, 188
Load regulation, 60, 457
Loop gain, 201, 293

Majority carrier, 12
Miller's theorem, 253
Minority carrier, 12
Mixer, 556, 558, 573
Monotonicity, 637
MOSFET, 124–27, 129
Multiplier, 527, 559, 566, 573
Multistage amplifier, 184–87
Multivibrator, 432

Negative feedback, 242, 290
Neutron, 2
Noise, 323
n-type semiconductor, 12
Nucleus, 2
Nyquist rate, 640

Octave, 283
One-shot, 438
Open-loop gain, 239, 282, 287
Operational amplifier, 224–64, 282–305, 320–56
Orbit, 4
Oscillator, 200–5, 414
   Clapp, 202
   Colpitts, 201
   crystal, 203
   Hartley, 202
   LC, 201
   local, 556, 558
   phase shift, 422

Oscillator, *continued*
  RC, 201
  relaxation, 228
  sawtooth, 427
  square-wave, 428
  triangular-wave, 425
  twin-T, 423
  voltage-controlled, 427, 436, 588
  Wein-bridge, 417
OTA, 514
Output impedance, 238, 250, 253
Output power, 189, 195, 199
Overload, 462, 477

Peak detector, 346
Peak inverse voltage, 36, 39, 41
Pentavalent atom, 12
Phase detector, 587
Phase error, 587
Phase inversion, 167
Phase lag compensation, 298
Phase-locked loop, 586
Phase margin, 293
Phase shift, 285
Phase shift oscillator, 422
Piezoelectric effect, 203
Pinch-off, 120
PIV, 36, 39, 41
pn junction, 14
Point-of-measurement, 139
Pole, 373, 377
Positive feedback, 200, 292, 415, 419
Power, 189, 195, 198, 199
Power gain, 169, 172, 175
Power supply, 33, 76, 489
Pressure transducer, 726
Priority encoder, 332
Programmable UJT (PUT), 427
Proton, 2
p-type semiconductor, 12
Push-pull amplifier, 191

Q (quality factor), 374
Q point, 108, 187
Quadrant, 559
Quartz, 203

Radio frequency (RF), 197
Receiver, 163, 207, 555, 557
Recombination, 8
Rectifier, 34–43

Rectifier diode, 34
Regulation, 58, 60, 61
Relaxation oscillator, 228
Resistance temperature detector (RTD), 715
Resolver, 701
Resolver-to-digital converter (RDC), 702
Resonant frequency, 418
Reverse bias, 19
Reverse breakdown, 21
Reverse current, 20
Ripple factor, 44
Ripple voltage, 44
RMS-to-dc conversion, 694
Roll-off, 287

Sample-and-hold, 623
Sampling theory, 638
Saturation, 108, 113
Sawtooth, 427
Scaling adder, 336
Schmitt trigger, 325, 521
Scott-T transformer, 702
SCR, 133
Self bias, 127
Semiconductor, 8, 10, 11
Series regulator, 458
Shunt regulator, 464
Signal compression, 528
Signal tracing, 205
Silicon, 5, 8
Silicon-controlled rectifier (SCR), 133
Single-ended, 228
Slew rate, 240
Small-signal operation, 187
Source, 117
Source-follower, 180
Square root circuit, 565
Squaring circuit, 564
Stability, 292, 294
Stage, 378
Stand-off ratio, 132
Step response, 240
Stereo, 281, 306, 455
Strain gage, 723
Successive approximation, 649
Superheterodyne, 163, 555, 557
Suppressed carrier, 569
Surge current, 45
Switch, transistor, 112
Switching regulator, 468
Synchro, 700

INDEX

Tank circuit, 64
Thermal overload, 477
Thermocouple, 707
Thermistor, 720
Three-terminal regulator, 476
Thyristor, 133
Time delay, 440
Timer, 431, 438
Transconductance, 176, 514
Transformer, 38, 508
Transient current, 19
Transistor,
    bipolar junction, 98–117
    external pass, 481
    field-effect, 117, 119, 124
    unijunction, 131
Transistor capacitance, 241
Transistor currents, 101
Transistor packaging, 135
Transistor ratings, 114
Transmitter, 583
Triac, 134
Triangular wave, 425
Trivalent atom, 12
Troubleshooting, 72, 137, 205, 258, 347, 665
Tuned circuit, 198
Turns ratio, 39

UJT, 131
Unijunction transistor (UJT), 131
Unity gain, 283
Unity-gain bandwidth, 283

Valence electron, 4
Varactor diode, 62
VCO, 427, 436, 588
Virtual ground, 247
Voltage
    divider, 103
    follower, 246, 252, 260
    gain, 109, 164, 171, 174, 178, 180, 182, 184, 186, 243, 245, 248, 282, 334, 511
    inverter, 473
    regulator, 58, 456
Voltage-controlled oscillator (VCO), 427, 436, 588
Voltage-to-current converter, 346
Voltage-to-frequency converter, 656

Wein-bridge oscillator, 417
Window comparator, 329

Zener diode, 54–62
Zero-level detection, 320
Zero-voltage switch, 730